THE CALCULUS LIFESAVER

The CALCULUS LIFESAVER

All the tools you need to excel at calculus

ADRIAN BANNER

PRINCETON UNIVERSITY PRESS
Princeton and Oxford

To Yarry

CONTENTS

WELCOME!

This book is designed to help you learn the major concepts of single-variable calculus, while also concentrating on problem-solving techniques. Whether this is your first exposure to calculus, or you are studying for a test, or you've already taken calculus and want to refresh your memory, I hope that this book will be a useful resource.

The inspiration for this book came from my students at Princeton University. Over the past few years, they have found early drafts to be helpful as a study guide in conjunction with lectures, review sessions and their textbook. Here are some of the questions that they've asked along the way, which you might also be inclined to ask:

- **Why is this book so long?** I assume that you, the reader, are motivated to the extent that you'd like to master the subject. Not wanting to get by with the bare minimum, you're prepared to put in some time and effort reading—and understanding—these detailed explanations.

- **What do I need to know before I start reading?** You need to know some basic algebra and how to solve simple equations. Most of the precalculus you need is covered in the first two chapters.

- **Help! The final is in one week, and I don't know anything! Where do I start?** The next three pages describe how to use this book to study for an exam.

- **Where are all the worked solutions to examples? All I see is lots of words with a few equations.** Looking at a worked solution doesn't tell you how to think of it in the first place. So, I usually try to give a sort of "inner monologue"—what should be going through your head as you try to solve the problem. You end up with all the pieces of the solution, but you still need to write it up properly. My advice is to read the solution, then come back later and try to work it out again by yourself.

- **Where are the proofs of the theorems?** Most of the theorems in this book are justified in some way. More formal proofs can be found in Appendix A.

- **The topics are out of order! What do I do?** There's no standard order for learning calculus. The order I have chosen works, but you might have to search the table of contents to find the topics you need and ignore the rest for now. I may also have missed out some topics too—why not try emailing me at adrian@calclifesaver.com and you never know, I just might write an extra section or chapter for you (and for the next edition, if there is one!).

- **Some of the methods you use are different from the methods I learned. Who is right—my instructor or you?** Hopefully we're both right! If in doubt, ask your instructor what's acceptable.

- **Where's all the calculus history and fun facts in the margins?** Look, there's a little bit of history in this book, but let's not get too distracted here. After you get this stuff down, read a book on the history of calculus. It's interesting stuff, and deserves more attention than a couple of sentences here and there.

- **Could my school use this book as a textbook?** Paired with a good collection of exercises, this book could function as a textbook, as well as being a study guide. Your instructor might also find the book useful to help prepare lectures, particularly in regard to problem-solving techniques.

- **What's with these videos?** You can find videos of a year's supply of my review sessions, which reference a lot (but not all!) of the sections and examples from this book, at this website:

www.calclifesaver.com

How to Use This Book to Study for an Exam

There's a good chance you have a test or exam coming up soon. I am sympathetic to your plight: you don't have time to read the whole book! There's a table on the next page that identifies the main sections that will help you to review for the exam. Also, throughout the book, the following icons appear in the margin to allow you quickly to identify what's relevant:

- A worked-out example begins on this line.

- Here's something really important.

- You should try this yourself.

- Beware: this part of the text is mostly for interest. If time is limited, skip to the next section.

Also, some important formulas or theorems have boxes around them: learn these well.

Two all-purpose study tips

- Write out your own summary of all the important points and formulas to memorize. Math isn't about memorization, but there are some key formulas and methods that you should have at your fingertips. The act of making the summary is often enough to solidify your understanding. This is the main reason why I don't summarize the important points at the end of a chapter: it's much more valuable if you do it yourself.

- Try to get your hands on similar exams—maybe your school makes previous years' finals available, for example—and take these exams under proper conditions. That means no breaks, no food, no books, no phone calls, no emails, no messaging, and so on. Then see if you can get a solution key and grade it, or ask someone (nicely!) to grade it for you.

You'll be on your way to that A if you do both of these things.

Key sections for exam review (by topic)

Topic	Subtopic	Section(s)
Precalculus	Lines	1.5
	Other common graphs	1.6
	Trig basics	2.1
	Trig with angles outside $[0, \pi/2]$	2.2
	Trig graphs	2.3
	Trig identities	2.4
	Exponentials and logs	9.1
Limits	Sandwich principle	3.6
	Polynomial limits	all of Chapter 4
	Derivatives in disguise	6.5
	Trig limits	7.1 (skip 7.1.5)
	Exponential and log limits	9.4
	L'Hôpital's Rule	14.1
	Overview of limit problems	14.2
Continuity	Definition	5.1
	Intermediate Value Theorem	5.1.4
Differentiation	Definition	6.1
	Rules (e.g., product/quotient/chain rule)	6.2
	Finding tangent lines	6.3
	Derivatives of piecewise-defined functions	6.6
	Sketching the derivative	6.7
	Trig functions	7.2, 7.2.1
	Implicit differentiation	8.1
	Exponentials and logs	9.3
	Logarithmic differentiation	9.5
	Hyperbolic functions	9.7
	Inverse functions in general	10.1
	Inverse trig functions	10.2
	Inverse hyperbolic functions	10.3
	Differentiating definite integrals	17.5

Topic	Subtopic	Section(s)
Applications of differentiation	Related rates	8.2
	Exponential growth and decay	9.6
	Finding global maxima and minima	11.1.3
	Rolle's Theorem/Mean Value Theorem	11.2, 11.3
	Classifying critical points	11.5, 12.1.1
	Finding inflection points	11.4, 12.1.2
	Sketching graphs	12.2, 12.3
	Optimization	13.1
	Linearization/differentials	13.2
	Newton's method	13.3
Integration	Definition	16.2 (skip 16.2.1)
	Basic properties	16.3
	Finding areas	16.4
	Estimating definite integrals	16.5, Appendix B
	Average values/Mean Value Theorem	16.6
	Basic examples	17.4, 17.6
	Substitution	18.1
	Integration by parts	18.2
	Partial fractions	18.3
	Trig integrals	19.1, 19.2
	Trig substitutions	19.3 (skip 19.3.6)
	Overview of integration techniques	19.4
Motion	Velocity and acceleration	6.4
	Constant acceleration	6.4.1
	Simple harmonic motion	7.2.2
	Finding displacements	16.1.1
Improper integrals	Basics	20.1, 20.2
	Problem-solving techniques	all of Chapter 21
Infinite series	Basics	22.1.2, 22.2
	Problem-solving techniques	all of Chapter 23
Taylor series and power series	Estimation and error estimates	all of Chapter 25
	Power/Taylor series problems	all of Chapter 26
Differential equations	Separable first-order	30.2
	First-order linear	30.3
	Constant coefficients	30.4
	Modeling	30.5
Miscellaneous topics	Parametric equations	27.1
	Polar coordinates	27.2
	Complex numbers	28.1–28.5
	Volumes	29.1, 29.2
	Arc lengths	29.3
	Surface areas	29.4

Unless specified otherwise, the Section(s) column includes all subsections; for example, 6.2 includes 6.2.1 through 6.2.7.

ACKNOWLEDGMENTS ⎯⎯⎯⎯⎯

There are many people I'd like to thank for supporting and helping me during the writing of this book. My students have been a source of education, entertainment, and delight; I have benefited greatly from their suggestions. I'd particularly like to thank my editor Vickie Kearn, my production editor Linny Schenck, and my designer Lorraine Doneker for all their help and support, and also Gerald Folland for his numerous excellent suggestions which have greatly improved this book. Ed Nelson, Maria Klawe, Christine Miranda, Lior Braunstein, Emily Sands, Jamaal Clue, Alison Ralph, Marcher Thompson, Ioannis Avramides, Kristen Molloy, Dave Uppal, Nwanneka Onvekwusi, Ellen Zuckerman, Charles MacCluer, and Gary Slezak brought errors and omissions to my attention.

The following faculty and staff members of the Princeton University Mathematics Department have been very supportive: Eli Stein, Simon Kochen, Matthew Ferszt, and Scott Kenney. Thank you also to all of my colleagues at INTECH for their support, in particular Bob Fernholz, Camm Maguire, Marie D'Albero, and Vassilios Papathanakos, who made some excellent last-minute suggestions. I'd also like to pay tribute to my 11th- and 12th-grade math teacher, William Pender, who is surely the best calculus teacher in the world. Many of the methods in this book were inspired by his teaching. I hope he forgives me for not putting arrows on my curves, not labeling all my axes, and neglecting to write "for some constant C" after every $+C$.

My friends and family have been fantastic in their support, especially my parents Freda and Michael, sister Carly, grandmother Rena, and in-laws Marianna and Michael. Finally, a very special thank you to my wife Amy for putting up with me while I wrote this book and always being there for me (and also for drawing the mountain-climber!).

CHAPTER 1 _____

Functions, Graphs, and Lines

Trying to do calculus without using functions would be one of the most point-less things you could do. If calculus had an ingredients list, functions would be first on it, and by some margin too. So, the first two chapters of this book are designed to jog your memory about the main features of functions. This chapter contains a review of the following topics:

- functions: their domain, codomain, and range, and the vertical line test;
- inverse functions and the horizontal line test;
- composition of functions;
- odd and even functions;
- graphs of linear functions and polynomials in general, as well as a brief survey of graphs of rational functions, exponentials, and logarithms; and
- how to deal with absolute values.

Trigonometric functions, or trig functions for short, are dealt with in the next chapter. So, let's kick off with a review of what a function actually is.

1.1 Functions

A *function* is a rule for transforming an object into another object. The object you start with is called the *input*, and comes from some set called the *domain*. What you get back is called the *output*; it comes from some set called the *codomain*.

Here are some examples of functions:

- Suppose you write $f(x) = x^2$. You have just defined a function f which transforms any number into its square. Since you didn't say what the domain or codomain are, it's assumed that they are both $\mathbb{R}$, the set of all real numbers. So you can square any real number, and get a real number back. For example, f transforms 2 into 4; it transforms $-1/2$ into $1/4$; and it transforms 1 into 1. This last one isn't much of a change at all, but that's no problem: the transformed object doesn't have to be different from the original one. When you write $f(2) = 4$, what you really mean is that f transforms 2 into 4. By the way, f is the transformation rule, while $f(x)$ is the result of applying the transformation rule to the variable x. So it's technically not correct to say "$f(x)$ is a function"; it should be "f is a function."

- Now, let $g(x) = x^2$ with domain consisting only of numbers greater than or equal to 0. (Such numbers are called *nonnegative*.) This seems like the same function as f, but it's not: the domains are different. For example, $f(-1/2) = 1/4$, but $g(-1/2)$ isn't defined. The function g just chokes on anything not in the domain, refusing even to touch it. Since g and f have the same rule, but the domain of g is smaller than the domain of f, we say that g is formed by *restricting the domain* of f.

- Still letting $f(x) = x^2$, what do you make of $f(\text{horse})$? Obviously this is undefined, since you can't square a horse. On the other hand, let's set

$$h(x) = \text{number of legs } x \text{ has,}$$

where the domain of h is the set of all animals. So $h(\text{horse}) = 4$, while $h(\text{ant}) = 6$ and $h(\text{salmon}) = 0$. The codomain could be the set of all nonnegative integers, since animals don't have negative or fractional numbers of legs. By the way, what is $h(2)$? This isn't defined, of course, since 2 isn't in the domain. How many legs does a "2" have, after all? The question doesn't really make sense. You might also think that $h(\text{chair}) = 4$, since most chairs have four legs, but that doesn't work either, since a chair isn't an animal, and so "chair" is not in the domain of h. That is, $h(\text{chair})$ is undefined.

- Suppose you have a dog called Junkster. Unfortunately, poor Junkster has indigestion. He eats something, then chews on it for a while and tries to digest it, fails, and hurls. Junkster has transformed the food into ... something else altogether. We could let

$$j(x) = \text{color of barf when Junkster eats } x,$$

where the domain of j is the set of foods that Junkster will eat. The codomain is the set of all colors. For this to work, we have to be confident that whenever Junkster eats a taco, his barf is always the same color (say, red). If it's sometimes red and sometimes green, that's no good: **a function must assign a unique output for each valid input**.

Now we have to look at the concept of the *range* of a function. The range is the set of all outputs that could possibly occur. You can think of the function working on transforming everything in the domain, one object at a time; the collection of transformed objects is the range. You might get duplicates, but that's OK.

So why isn't the range the same thing as the codomain? Well, the range is actually a subset of the codomain. The codomain is a set of **possible** outputs, while the range is the set of **actual** outputs. Here are the ranges of the functions we looked at above:

- If $f(x) = x^2$ with domain $\mathbb{R}$ and codomain $\mathbb{R}$, the range is the set of nonnegative numbers. After all, when you square a number, the result cannot be negative. How do you know the range is **all** the nonnegative numbers? Well, if you square every number, you definitely cover all nonnegative numbers. For example, you get 2 by squaring $\sqrt{2}$ (or $-\sqrt{2}$).

- If $g(x) = x^2$, where the domain of g is only the nonnegative numbers but the codomain is still all of $\mathbb{R}$, the range will again be the set of nonnegative numbers. When you square every nonnegative number, you still cover all the nonnegative numbers.

- If $h(x)$ is the number of legs the animal x has, then the range is all the possible numbers of legs that **any** animal can have. I can think of animals that have 0, 2, 4, 6, and 8 legs, as well as some creepy-crawlies with more legs. If you include individual animals which have lost one or more legs, you can also include 1, 3, 5, and 7 in the mix, as well as other possibilities. In any case, the range of this function isn't so clear-cut; you probably have to be a biologist to know the real answer.

- Finally, if $j(x)$ is the color of Junkster's barf when he eats x, then the range consists of all possible barf-colors. I dread to think what these are, but probably bright blue isn't among them.

1.1.1 Interval notation

In the rest of this book, our functions will always have codomain $\mathbb{R}$, and the domain will always be as much of $\mathbb{R}$ as possible (unless stated otherwise). So we'll often be dealing with subsets of the real line, especially connected intervals such as $\{x : 2 \leq x < 5\}$. It's a bit of a pain to write out the full set notation like this, but it sure beats having to say "all the numbers between 2 and 5, including 2 but not 5." We can do even better using interval notation.

We'll write $[a, b]$ to mean the set of all numbers between a and b, including a and b themselves. So $[a, b]$ means the set of all x such that $a \leq x \leq b$. For example, $[2, 5]$ is the set of all real numbers between 2 and 5, including 2 and 5. (It's not just the set consisting of 2, 3, 4, and 5: don't forget that there are loads of fractions and irrational numbers between 2 and 5, such as $5/2$, $\sqrt{7}$, and π.) An interval such as $[a, b]$ is called *closed*.

If you don't want the endpoints, change the square brackets to parentheses. In particular, (a, b) is the set of all numbers between a and b, not including a or b. So if x is in the interval (a, b), we know that $a < x < b$. The set $(2, 5)$ includes all real numbers between 2 and 5, but not 2 or 5. An interval of the form (a, b) is called *open*.

You can mix and match: $[a, b)$ consists of all numbers between a and b, including a but not b. And $(a, b]$ includes b but not a. These intervals are closed at one end and open at the other. Sometimes such intervals are called *half-open*. An example is the set $\{x : 2 \leq x < 5\}$ from above, which can also be written as $[2, 5)$.

There's also the useful notation (a, ∞) for all the numbers greater than a not including a; $[a, \infty)$ is the same thing but with a included. There are three other possibilities which involve $-\infty$; all in all, the situation looks like this:

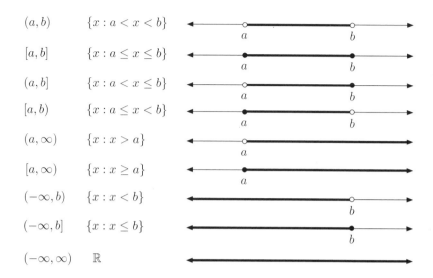

(a, b)	$\{x : a < x < b\}$
$[a, b]$	$\{x : a \le x \le b\}$
$(a, b]$	$\{x : a < x \le b\}$
$[a, b)$	$\{x : a \le x < b\}$
(a, ∞)	$\{x : x > a\}$
$[a, \infty)$	$\{x : x \ge a\}$
$(-\infty, b)$	$\{x : x < b\}$
$(-\infty, b]$	$\{x : x \le b\}$
$(-\infty, \infty)$	$\mathbb{R}$

1.1.2 Finding the domain

Sometimes the definition of a function will include the domain. (This was the case, for example, with our function g from Section 1.1 above.) Most of the time, however, the domain is not provided. The basic convention is that the domain consists of as much of the set of real numbers as possible. For example, if $k(x) = \sqrt{x}$, the domain can't be all of $\mathbb{R}$, since you can't take the square root of a negative number. The domain must be $[0, \infty)$, which is just the set of all numbers greater than or equal to 0.

OK, so square roots of negative numbers are bad. What else can cause a screw-up? Here's a list of the three most common possibilities:

1. The denominator of a fraction can't be zero.
2. You can't take the square root (or fourth root, sixth root, and so on) of a negative number.
3. You can't take the logarithm of a negative number or of 0. (Remember logs? If not, see Chapter 9!)

You might recall that $\tan(90°)$ is also a problem, but this is really a special case of the first item above. You see,

$$\tan(90°) = \frac{\sin(90°)}{\cos(90°)} = \frac{1}{0},$$

so the reason $\tan(90°)$ is undefined is really that a hidden denominator is zero. Here's another example: if we try to define

$$f(x) = \frac{\log_{10}(x + 8)\sqrt{26 - 2x}}{(x - 2)(x + 19)},$$

then what is the domain of f? Well, for $f(x)$ to make sense, here's what needs to happen:

- We need to take the square root of $(26 - 2x)$, so this quantity had better be nonnegative. That is, $26 - 2x \ge 0$. This can be rewritten as $x \le 13$.

- We also need to take the logarithm of $(x+8)$, so this quantity needs to be positive. (Notice the difference between logs and square roots: you can take the square root of 0, but you can't take the log of 0.) Anyway, we need $x + 8 > 0$, so $x > -8$. So far, we know that $-8 < x \le 13$, so the domain is at most $(-8, 13]$.
- The denominator can't be 0; this means that $(x-2) \ne 0$ and $(x+19) \ne 0$. In other words, $x \ne 2$ and $x \ne -19$. This last one isn't a problem, since we already know that x lies in $(-8, 13]$, so x can't possibly be -19. We do have to exclude 2, though.

So we have found that the domain is the set $(-8, 13]$ except for the number 2. This set could be written as $(-8, 13] \backslash \{2\}$. Here the backslash means "not including."

1.1.3 Finding the range using the graph

Let's define a new function F by specifying that its domain is $[-2, 1]$ and that $F(x) = x^2$ on this domain. (Remember, the codomain of any function we look at will always be the set of all real numbers.) Is F the same function as f, where $f(x) = x^2$ for all real numbers x? The answer is no, since the two functions have different domains (even though they have the same rule). As in the case of the function g from Section 1.1 above, the function F is formed by restricting the domain of f.

Now, what is the range of F? Well, what happens if you square every number between -2 and 1 inclusive? You should be able to work this out directly, but this is a good opportunity to see how to use a graph to find the range of a function. The idea is to sketch the graph of the function, then imagine two rows of lights shining from the far left and far right of the graph horizontally toward the y-axis. The curve will cast two shadows, one on the left side and one on the right side of the y-axis. The range is the union of both shadows: that is, if any point on the y-axis lies in either the left-hand or the right-hand shadow, it is in the range of the function. Let's see how this works with our function F:

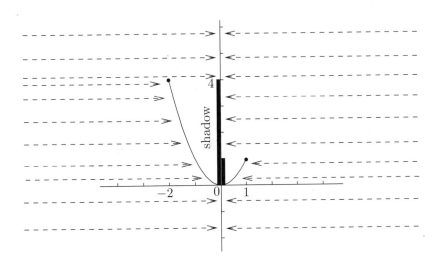

The left-hand shadow covers all the points on the y-axis between 0 and 4 inclusive, which is $[0, 4]$; on the other hand, the right-hand shadow covers the points between 0 and 1 inclusive, which is $[0, 1]$. The right-hand shadow doesn't contribute anything extra: the total coverage is still $[0, 4]$. This is the range of F.

1.1.4 The vertical line test

In the last section, we used the graph of a function to find its range. The graph of a function is very important: it really shows you what the function "looks like." We'll be looking at techniques for sketching graphs in Chapter 12, but for now I'd like to remind you about the vertical line test.

    You can draw any figure you like on a coordinate plane, but the result may not be the graph of a function. So what's special about the graph of a function? What is the graph of a function f, anyway? Well, it's the collection of all points with coordinates $(x, f(x))$, where x is in the domain of f. Here's another way of looking at this: start with some number x. If x is in the domain, you plot the point $(x, f(x))$, which of course is at a height of $f(x)$ units above the point x on the x-axis. If x isn't in the domain, you don't plot anything. Now repeat for every real number x to build up the graph.

    Here's the key idea: you can't have two points with the same x-coordinate. In other words, no two points on the graph can lie on the same vertical line. Otherwise, how would you know which of the two or more heights above the point x on the x-axis corresponds to the value of $f(x)$? So, this leads us to the *vertical line test*: if you have some graph and you want to know whether it's the graph of a function, see whether any vertical line intersects the graph more than once. If so, it's not the graph of a function; but if no vertical line intersects the graph more than once, you are indeed dealing with the graph of a function. For example, the circle of radius 3 units centered at the origin has a graph like this:

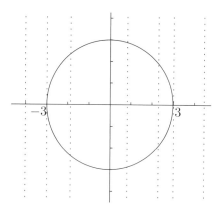

Such a commonplace object should be a function, right? No, check the vertical lines that are shown in the diagram. Sure, to the left of -3 or to the right of 3, there's no problem—the vertical lines don't even hit the graph, which is fine. Even at -3 or 3, the vertical lines only intersect the curve in one point each, which is also fine. The problem is when x is in the interval $(-3, 3)$. For any of these values of x, the vertical line through $(x,0)$ intersects the circle twice, which screws up the circle's potential function-status. You just don't know whether $f(x)$ is the top point or the bottom point.

The best way to salvage the situation is to chop the circle in half horizontally and choose only the top or the bottom half. The equation for the whole circle is $x^2 + y^2 = 9$, whereas the equation for the top semicircle is $y = \sqrt{9 - x^2}$. The bottom semicircle has equation $y = -\sqrt{9 - x^2}$. These last two are functions, both with domain $[-3, 3]$. If you felt like chopping in a different way, you wouldn't actually have to take semicircles—you could chop and change between the upper and lower semicircles, as long as you don't violate the vertical line test. For example, here's the graph of a function which also has domain $[-3, 3]$:

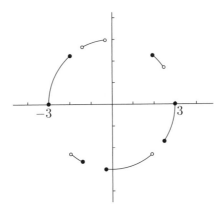

The vertical line test checks out, so this is indeed the graph of a function.

1.2 Inverse Functions

Let's say you have a function f. You present it with an input x; provided that x is in the domain of f, you get back an output, which we call $f(x)$. Now we try to do things all backward and ask this question: if you pick a number y, what input can you give to f in order to get back y as your output?

Here's how to state the problem in math-speak: given a number y, what x in the domain of f satisfies $f(x) = y$? The first thing to notice is that y has to be in the range of f. Otherwise, by definition there are no values of x such that $f(x) = y$. There would be nothing in the domain that f would transform into y, since the range is **all** the possible outputs.

On the other hand, if y is in the range, there might be many values that work. For example, if $f(x) = x^2$ (with domain $\mathbb{R}$), and we ask what value of x transforms into 64, there are obviously two values of x: 8 and -8. On the other hand, if $g(x) = x^3$, and we ask the same question, there's only one value of x, which is 4. The same would be true for any number we give to g to transform, because any number has only one (real) cube root.

So, here's the situation: we're given a function f, and we pick y in the range of f. Ideally, there will be exactly one value of x which satisfies $f(x) = y$. If this is true for every value of y in the range, then we can define a new function which reverses the transformation. Starting with the output y, the new function finds the one and only input x which leads to the output. The new function is called the *inverse function of f*, and is written as f^{-1}. Here's a summary of the situation in mathematical language:

1. Start with a function f such that for any y in the range of f, there is exactly one number x such that $f(x) = y$. That is, different inputs give different outputs. Now we will define the inverse function f^{-1}.
2. The domain of f^{-1} is the same as the range of f.
3. The range of f^{-1} is the same as the domain of f.
4. The value of $f^{-1}(y)$ is the number x such that $f(x) = y$. So,

$$\text{if} \quad f(x) = y, \qquad \text{then} \quad f^{-1}(y) = x.$$

The transformation f^{-1} acts like an undo button for f: if you start with x and transform it into y using the function f, then you can undo the effect of the transformation by using the inverse function f^{-1} on y to get x back.

This raises some questions: how do you see if there's only one value of x that satisfies the equation $f(x) = y$? If so, how do you find the inverse, and what does its graph look like? If not, how do you salvage the situation? We'll answer these questions in the next three sections.

1.2.1 The horizontal line test

For the first question—how to see that there's only one value of x that works for any y in the range—perhaps the best way is to look at the graph of your function. We want to pick y in the range of f and hopefully only have one value of x such that $f(x) = y$. What this means is that the horizontal line through the point $(0, y)$ should intersect the graph exactly once, at some point (x, y). That x is the one we want. If the horizontal line intersects the curve more than once, there would be multiple potential inverses x, which is bad. In that case, the only way to get an inverse function is to restrict the domain; we'll come back to this very shortly. What if the horizontal line doesn't intersect the curve at all? Then y isn't in the range after all, which is OK.

So, we have just described the *horizontal line test*: if every horizontal line intersects the graph of a function at most once, the function has an inverse. If even one horizontal line intersects the graph more than once, there isn't an inverse function. For example, look at the graphs of $f(x) = x^3$ and $g(x) = x^2$:

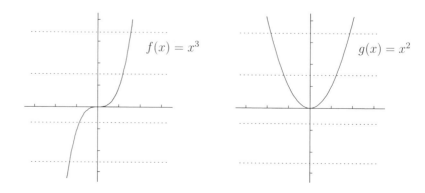

No horizontal line hits $y = f(x)$ more than once, so f has an inverse. On the other hand, some of the horizontal lines hit the curve $y = g(x)$ twice, so g has no inverse. Here's the problem: if you want to solve $y = x^2$ for x, where y is positive, then there are two solutions, $x = \sqrt{y}$ and $x = -\sqrt{y}$. You don't know which one to take!

1.2.2 Finding the inverse

Now let's move on to the second of our questions: how do you find the inverse of a function f? Well, you write down $y = f(x)$ and try to solve for x. In our example of $f(x) = x^3$, we have $y = x^3$, so $x = \sqrt[3]{y}$. This means that $f^{-1}(y) = \sqrt[3]{y}$. If the variable y here offends you, by all means switch it to x: you can write $f^{-1}(x) = \sqrt[3]{x}$ if you prefer. Of course, solving for x is not always easy and in fact is often impossible. On the other hand, if you know what the graph of your function looks like, the graph of the inverse function is easy to find. The idea is to draw the line $y = x$ on the graph, then pretend that this line is a two-sided mirror. The inverse function is the reflection of the original function in this mirror. When $f(x) = x^3$, here's what f^{-1} looks like:

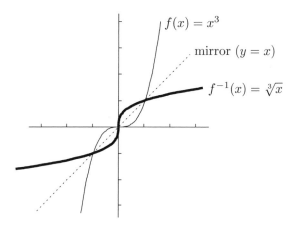

The original function f is reflected in the mirror $y = x$ to get the inverse function. Note that the domain and range of both f and f^{-1} are the whole real line.

1.2.3 Restricting the domain

Finally, we'll address our third question: if the horizontal line test fails and there's no inverse, what can be done? Our problem is that there are multiple values of x that give the same y. The only way to get around the problem is to throw away all but one of these values of x. That is, we have to decide which one of our values of x we want to keep, and throw the rest away. As we saw in Section 1.1 above, this is called *restricting the domain* of our function. Effectively, we ghost out part of the curve so that what's left no longer fails the horizontal line test. For example, if $g(x) = x^2$, we can ghost out the left half of the graph like this:

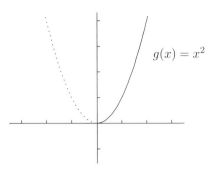

The new (unghosted) curve has the reduced domain $[0, \infty)$ and satisfies the horizontal line test, so there is an inverse function. More precisely, the function h, which has domain $[0, \infty)$ and is defined by $h(x) = x^2$ on this domain, has an inverse. Let's play the reflection game to see what it looks like:

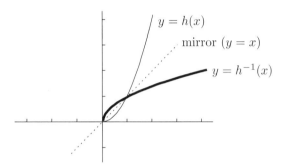

To find the equation of the inverse, we have to solve for x in the equation $y = x^2$. Clearly the solution is $x = \sqrt{y}$ or $x = -\sqrt{y}$, but which one do we need? We know that the range of the inverse function is the same as the domain of the original function, which we have restricted to be $[0, \infty)$. So we need a nonnegative number as our answer, and that has to be $x = \sqrt{y}$. That is, $h^{-1}(y) = \sqrt{y}$. Of course, we could have ghosted out the right half of the original graph to restrict the domain to $(-\infty, 0]$. In that case, we'd get a function j which has domain $(-\infty, 0]$ and again satisfies $j(x) = x^2$, but only on this domain. This function also has an inverse, but the inverse is now the negative square root: $j^{-1}(y) = -\sqrt{y}$.

By the way, if you take the original function g given by $g(x) = x^2$ with domain $(-\infty, \infty)$, which fails the horizontal line test, and try to reflect it in the mirror $y = x$, you get the following picture:

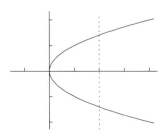

Notice that the graph fails the vertical line test, so it's not the graph of a function. This illustrates the connection between the vertical and horizontal line tests—when horizontal lines are reflected in the mirror $y = x$, they become vertical lines.

1.2.4 Inverses of inverse functions

One more thing about inverse functions: if f has an inverse, it's true that $f^{-1}(f(x)) = x$ for all x in the domain of f, and also that $f(f^{-1}(y)) = y$ for all y in the range of f. (Remember, the range of f is the same as the domain of f^{-1}, so you can indeed take $f^{-1}(y)$ for y in the range of f without causing any screwups.)

For example, if $f(x) = x^3$, then f has an inverse given by $f^{-1}(x) = \sqrt[3]{x}$, and so $f^{-1}(f(x)) = \sqrt[3]{x^3} = x$ for any x. Remember, the inverse function is like an undo button. We use x as an input to f, and then give the output to f^{-1}; this undoes the transformation and gives us back x, the original number. Similarly, $f(f^{-1}(y)) = (\sqrt[3]{y})^3 = y$. So f^{-1} is the inverse function of f, and f is the inverse function of f^{-1}. In other words, the inverse of the inverse is the original function.

Now, you have to be careful in the case where you restrict the domain. Let $g(x) = x^2$; we've seen that you need to restrict the domain to get an inverse. Let's say we restrict the domain to $[0, \infty)$ and carelessly continue to refer to the function as g instead of h, as in the previous section. We would then say that $g^{-1}(x) = \sqrt{x}$. If you calculate $g(g^{-1}(x))$, you find that this is $(\sqrt{x})^2$, which equals x, provided that $x \geq 0$. (Otherwise you can't take the square root in the first place.)

On the other hand, if you work out $g^{-1}(g(x))$, you get $\sqrt{x^2}$, which is not always the same thing as x. For example, if $x = -2$, then $x^2 = 4$ and so $\sqrt{x^2} = \sqrt{4} = 2$. So it's not true in general that $g^{-1}(g(x)) = x$. The problem is that -2 isn't in the restricted-domain version of g. Technically, you can't even compute $g(-2)$, since -2 is no longer in the domain of g. We really should be working with h, not g, so that we remember to be more careful. Nevertheless, in practice, mathematicians will often restrict the domain without changing letters! So it will be useful to summarize the situation as follows:

If the domain of a function f can be restricted so that f has an inverse f^{-1}, then

- $f(f^{-1}(y)) = y$ for all y in the range of f; but
- $f^{-1}(f(x))$ may not equal x; in fact, $f^{-1}(f(x)) = x$ only when x is in the restricted domain.

We'll be revisiting these important points in the context of inverse trig functions in Section 10.2.6 of Chapter 10.

1.3 Composition of Functions

Let's say we have a function g given by $g(x) = x^2$. You can replace x by anything you like, as long as it makes sense. For example, you can write

$g(y) = y^2$, or $g(x + 5) = (x + 5)^2$. This last example shows that you need to be very careful with parentheses. It would be wrong to write $g(x+5) = x+5^2$, since this is just $x + 25$, which is not the same thing as $(x + 5)^2$. If in doubt, use parentheses. That is, if you need to write out $f(\text{something})$, replace every instance of x by (something), making sure to include the parentheses. Just about the only time you don't need to use parentheses is when the function is an exponential function—for example, if $h(x) = 3^x$, then you can just write $h(x^2 + 6) = 3^{x^2+6}$. You don't need parentheses since you're already writing the $x^2 + 6$ as a superscript.

Now consider the function f defined by $f(x) = \cos(x^2)$. If I give you a number x, how do you compute $f(x)$? Well, first you square it, then you take the cosine of the result. Since we can decompose the action of $f(x)$ into these two separate actions which are performed one after the other, we might as well describe those actions as functions themselves. So, let $g(x) = x^2$ and $h(x) = \cos(x)$. To simulate what f does when you use x as an input, you could first give x to g to square it, and then instead of taking the result back you could ask g to give its result to h instead. Then h spits out a number, which is the final answer. The answer will, of course, be the cosine of what came out of g, which was the square of the original x. This behavior exactly mimics f, so we can write $f(x) = h(g(x))$. Another way of expressing this is to write $f = h \circ g$; here the circle means "composed with." That is, f is h composed with g, or in other words, f is the *composition* of h and g. What's tricky is that you write h before g (reading from left to right as usual!) but you apply g first. I agree that it's confusing, but what can I say—you just have to deal with it.

It's useful to practice composing two or more functions together. For example, if $g(x) = 2^x$, $h(x) = 5x^4$, and $j(x) = 2x - 1$, what is a formula for the function $f = g \circ h \circ j$? Well, just replace one thing at a time, starting with j, then h, then g. So:

$$f(x) = g(h(j(x))) = g(h(2x - 1)) = g(5(2x - 1)^4) = 2^{5(2x-1)^4}.$$

You should also practice reversing the process. For example, suppose you start off with

$$f(x) = \frac{1}{\tan(5 \log_2(x + 3))}.$$

How would you decompose f into simpler functions? Zoom in to where you see the quantity x. The first thing you do is add 3, so let $g(x) = x + 3$. Then you have to take the base 2 logarithm of the resulting quantity, so set $h(x) = \log_2(x)$. Next, multiply by 5, so set $j(x) = 5x$. Then take the tangent, so put $k(x) = \tan(x)$. Finally, take reciprocals, so let $m(x) = 1/x$. With all these definitions, you should check that

$$f(x) = m(k(j(h(g(x))))).$$

Using the composition notation, you can write

$$f = m \circ k \circ j \circ h \circ g.$$

This isn't the only way to break down f. For example, we could have combined h and j into another function n, where $n(x) = 5\log_2(x)$. Then you should check that $n = j \circ h$, and

$$f = m \circ k \circ n \circ g.$$

Perhaps the original decomposition (involving j and h) is better because it breaks down f into more elementary steps, but the second one (involving n) isn't wrong. After all, $n(x) = 5\log_2(x)$ is still a pretty simple function of x.

Beware: composition of functions isn't the same thing as multiplying them together. For example, if $f(x) = x^2\sin(x)$, then f is not the composition of two functions. To calculate $f(x)$ for any given x, you actually have to find both x^2 and $\sin(x)$ (it doesn't matter which one you find first, unlike with composition) and then multiply these two things together. If $g(x) = x^2$ and $h(x) = \sin(x)$, then we'd write $f(x) = g(x)h(x)$, or $f = gh$. Compare this to the composition of the two functions, $j = g \circ h$, which is given by

$$j(x) = g(h(x)) = g(\sin(x)) = (\sin(x))^2$$

or simply $j(x) = \sin^2(x)$. The function j is a completely different function from the product $x^2\sin(x)$. It's also different from the function $k = h \circ g$, which is also a composition of g and h but in the other order:

$$k(x) = h(g(x)) = h(x^2) = \sin(x^2).$$

This is yet another completely different function. The moral of the story is that products and compositions are not the same thing, and furthermore, the order of the functions matters when you compose them, but not when you multiply them together.

One simple but important example of composition of functions occurs when you compose some function f with $g(x) = x - a$, where a is some constant number. You end up with a new function h given by $h(x) = f(x-a)$. A useful point to note is that the graph of $y = h(x)$ is the same as the graph of $y = f(x)$, except that it's shifted over a units to the right. If a is negative, then the shift is to the left. (The way to think of this, for example, is that a shift of -3 units to the right is the same as a shift of 3 units to the left.) So, how would you sketch the graph of $y = (x - 1)^2$? This is the same as $y = x^2$, but with x replaced by $x - 1$. So the graph of $y = x^2$ needs to be shifted to the right by 1 unit, and looks like this:

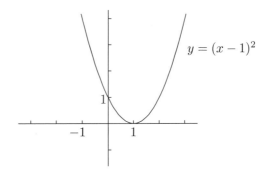

$y = (x - 1)^2$

Similarly, the graph of $y = (x + 2)^2$ is the graph of $y = x^2$ shifted to the left by 2 units, since you can interpret $(x + 2)$ as $(x - (-2))$.

1.4 Odd and Even Functions

Some functions have some symmetry properties that make them easier to deal with. Consider the function f given by $f(x) = x^2$. Pick any positive number you like (I'll choose 3) and hit it with f (I get 9). Now take the negative of that number, -3 in my case, and hit that with f (I get 9 again). You should get the same answer both times, as I did, regardless of which number you chose. You can express this phenomenon by writing $f(-x) = f(x)$ for all x. That is, if you give x to f as an input, you get back the same answer as if you used the input $-x$ instead. Notice that $g(x) = x^4$ and $h(x) = x^6$ also have this property—in fact, $j(x) = x^n$, where n is any even number (n could in fact be negative), has the same property. Inspired by this, we say that a function f is *even* if $f(-x) = f(x)$ for all x in the domain of f. It's not good enough for this equation to be true for some values of x; it has to be true for **all** x in the domain of f.

Now, let's say we play the same game with $f(x) = x^3$. Take your favorite positive number (I'll stick with 3) and hit that with f (I get 27). Now try again with the negative of your number, -3 in my case; I get -27, and you should also get the negative of what you got before. You can express this mathematically as $f(-x) = -f(x)$. Once again, the same property holds for $j(x) = x^n$ when n is any odd number (and once again, n could be negative). So, we say that a function f is *odd* if $f(-x) = -f(x)$ for all x in the domain of f.

In general, a function might be odd, it might be even, or it might be neither odd nor even. Don't forget this last point! Most functions are neither odd nor even. On the other hand, there's only one function that's both odd and even, which is the rather boring function given by $f(x) = 0$ for all x (we'll call this the "zero function"). Why is this the only odd and even function? Let's convince ourselves. If the function f is even, then $f(-x) = f(x)$ for all x. But if it's also odd, then $f(-x) = -f(x)$ for all x. Take the first of these equations and subtract the second from it. You should get $0 = 2f(x)$, which means that $f(x) = 0$. This is true for all x, so the function f must just be the zero function. One other nice observation is that if a function f is odd, and the number 0 is in its domain, then $f(0) = 0$. Why is it so? Because $f(-x) = -f(x)$ is true for all x in the domain of f, so let's try it for $x = 0$. You get $f(-0) = -f(0)$. But -0 is the same thing as 0, so we have $f(0) = -f(0)$. This simplifies to $2f(0) = 0$, or $f(0) = 0$ as claimed.

Anyway, starting with a function f, how can you tell if it is odd, even, or neither? And so what if it is odd or even anyway? Let's look at this second question before coming back to the first one. One nice thing about knowing that a function is odd or even is that it's easier to graph the function. In fact, if you can graph the right-hand half of the function, the left-hand half is a piece of cake! Let's say that f is an even function. Then since $f(x) = f(-x)$, the graph of $y = f(x)$ is at the same height above the x-coordinates x and $-x$. This is true for all x, so the situation looks something like this:

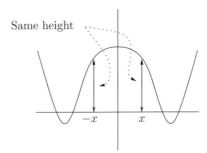

We can conclude that **the graph of an even function has mirror symmetry about the y-axis**. So, if you graph the right half of a function which you know is even, you can get the left half by reflecting the right half about the y-axis. Check the graph of $y = x^2$ to make sure that it has this mirror symmetry.

On the other hand, let's say that f is an odd function. Since we have $f(-x) = -f(x)$, the graph of $y = f(x)$ is at the same height **above** the x-coordinate x as it is **below** the x-coordinate $-x$. (Of course, if $f(x)$ is negative, then you have to switch the words "above" and "below.") In any case, the picture looks like this:

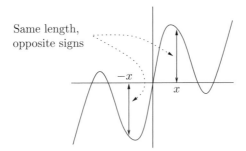

The symmetry is now a point symmetry about the origin. That is, **the graph of an odd function has 180° point symmetry about the origin**. This means that if you only have the right half of a function which you know is odd, you can get the left half as follows. Pretend that the curve is sitting on top of the paper, so you can pick it up if you like but you can't change its shape. Instead of picking it up, put a pin through the curve at the origin (remember, odd functions must pass through the origin if they are defined at 0) and then spin the whole curve around half a revolution. This is what the left-hand half of the graph looks like. (This doesn't work so well if the curve isn't continuous, that is, if the curve isn't all in one piece!) Check to see that the above graph and also the graph of $y = x^3$ have this symmetry.

Now, suppose f is defined by the equation $f(x) = \log_5(2x^6 - 6x^2 + 3)$. How do you tell if f is odd, even, or neither? The technique is to calculate $f(-x)$ by replacing every instance of x with $(-x)$, making sure not to forget the parentheses around $-x$, and then simplifying the result. If you end up with the original expression $f(x)$, then f is even; if you end up with the negative of the original expression $f(-x)$, then f is odd; if you end up with a mess that isn't either $f(x)$ or $-f(x)$, then f is neither (or you didn't simplify enough!).

In the example above, you'd write

$$f(-x) = \log_5(2(-x)^6 - 6(-x)^2 + 3) = \log_5(2x^6 - 6x^2 + 3),$$

which is actually equal to the original $f(x)$. So the function f is even. How about

$$g(x) = \frac{2x^3 + x}{3x^2 + 5} \qquad \text{and} \qquad h(x) = \frac{2x^3 + x - 1}{3x^2 + 5}?$$

Well, for g, we have

$$g(-x) = \frac{2(-x)^3 + (-x)}{3(-x)^2 + 5} = \frac{-2x^3 - x}{3x^2 + 5}.$$

Now you have to observe that you can take the minus sign out front and write

$$g(-x) = -\frac{2x^3 + x}{3x^2 + 5},$$

which, you notice, is equal to $-g(x)$. That is, apart from the minus sign, we get the original function back. So, g is an odd function. How about h? We have

$$h(-x) = \frac{2(-x)^3 + (-x) - 1}{3(-x)^2 + 5} = \frac{-2x^3 - x - 1}{3x^2 + 5}.$$

Once again, we take out the minus sign to get

$$h(-x) = -\frac{2x^3 + x + 1}{3x^2 + 5}.$$

Hmm, this doesn't appear to be the negative of the original function, because of the $+1$ term in the numerator. It's not the original function either, so the function h is neither odd nor even.

Let's look at one more example. Suppose you want to prove that the product of two odd functions is always an even function. How would you go about doing this? Well, it helps to have names for things, so let's say we have two odd functions f and g. We need to look at the product of these functions, so let's call the product h. That is, we define $h(x) = f(x)g(x)$. So, our task is to show that h is even. We'll do this by showing that $h(-x) = h(x)$, as usual. It will be helpful to note that $f(-x) = -f(x)$ and $g(-x) = -g(x)$, since f and g are odd. Let's start with $h(-x)$. Since h is the product of f and g, we have $h(-x) = f(-x)g(-x)$. Now we use the oddness of f and g to express this last term as $(-f(x))(-g(x))$. The minus signs now come out front and cancel out, so this is the same thing as $f(x)g(x)$ which of course equals $h(x)$. We could (and should) express all this text mathematically like this:

$$h(-x) = f(-x)g(-x) = (-f(x))(-g(x)) = f(x)g(x) = h(x).$$

Anyway, since $h(-x) = h(x)$, the function h is even. Now you should try to prove that the product of two even functions is always even, and also that the product of an odd and an even function must be odd. Go on, do it now!

1.5   Graphs of Linear Functions

Functions of the form $f(x) = mx + b$ are called *linear*. There's a good reason for this: the graphs of these functions are lines. (As far as we're concerned, the word "line" always means "straight line.") The slope of the line is given by m. Imagine for a moment that you are in the page, climbing the line as if it were a mountain. You start at the left side of the page and head to the right, like this:

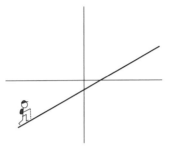

If the slope m is positive, as it is in the above picture, then you are heading uphill. The bigger m is, the steeper the climb. On the other hand, if the slope is negative, then you are heading downhill. The more negative the slope, the steeper the downhill grade. If the slope is zero, then the line is flat, or horizontal—you're going neither uphill nor downhill, just trudging along a flat line.

To sketch the graph of a linear function, you only need to identify two points on the graph. This is because there's only one line that goes through two different points. You just put your ruler on the points and draw the line. One point is easy to find, namely, the y-intercept. Set $x = 0$ in the equation $y = mx + b$, and you see that $y = m \times 0 + b = b$. That is, the y-intercept is equal to b, so the line goes through $(0, b)$. To find another point, you could find the x-intercept by setting $y = 0$ and finding what x is. This works pretty well except in two cases. The first case is when $b = 0$, in which case we are just dealing with $y = mx$. This goes through the origin, so the x-intercept and the y-intercept are both zero. To get another point, you'll just have to substitute in $x = 1$ and see that $y = m$. So, the line $y = mx$ goes through the origin and $(1, m)$. For example, the line $y = -2x$ goes through the origin and also through $(1, -2)$, so it looks like this:

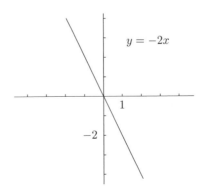

The other bad case is when $m = 0$. But then we just have $y = b$, which is a horizontal line through $(0, b)$.

For a more interesting example, consider $y = \frac{1}{2}x - 1$. The y-intercept is -1, and the slope is $\frac{1}{2}$. To sketch the line, find the x-intercept by setting $y = 0$. We get $0 = \frac{1}{2}x - 1$, which simplifies to $x = 2$. So, the line looks like this:

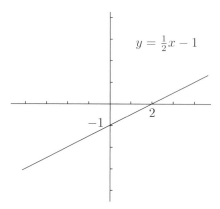

Now, let's suppose you know that you have a line in the plane, but you don't know its equation. If you know it goes through a certain point, and you know what its slope is, then you can find the equation of the line. You really, really, really need to know how to do this, since it comes up a lot. This formula, called the *point-slope* form of a linear function, is what you need to know:

> If a line goes through (x_0, y_0) and has slope m, then its equation is $\quad y - y_0 = m(x - x_0)$.

For example, what is the equation of the line through $(-2, 5)$ which has slope -3? It is $y - 5 = -3(x - (-2))$, which you can expand and simplify down to $y = -3x - 1$.

Sometimes you don't know the slope of the line, but you do know two points that it goes through. How do you find the equation? The technique is to find the slope, then use the previous idea with one of the points (your choice) to find the equation. First, you need to know this:

> If a line goes through (x_1, y_1) and (x_2, y_2), its slope is equal to $\dfrac{y_2 - y_1}{x_2 - x_1}$.

So, what is the equation of the line through $(-3, 4)$ and $(2, -6)$? Let's find the slope first:
$$\text{slope} = \frac{-6 - 4}{2 - (-3)} = \frac{-10}{5} = -2.$$

We now know that the line goes through $(-3, 4)$ and has slope -2, so its equation is $y - 4 = -2(x - (-3))$, or after simplifying, $y = -2x - 2$. Alternatively, we could have used the other point $(2, -6)$ with slope -2 to see that the equation of the line is $y - (-6) = -2(x - 2)$, which simplifies to $y = -2x - 2$. Thankfully this is the same equation as before—it doesn't matter which point you pick, as long as you have used both points to find the slope.

1.6 Common Functions and Graphs

Here are the most important functions you should know about.

1. Polynomials: these are functions built out of nonnegative integer powers of x. You start with the building blocks 1, x, x^2, x^3, and so on, and you are allowed to multiply these basic functions by numbers and add a finite number of them together. For example, the polynomial $f(x) = 5x^4 - 4x^3 + 10$ is formed by taking 5 times the building block x^4, and -4 times the building block x^3, and 10 times the building block 1, and adding them together. You might also want to include the intermediate building blocks x^2 and x, but since they don't appear, you need to take 0 times of each. The amount that you multiply the building block x^n by is called the *coefficient* of x^n. For example, in the polynomial f above, the coefficient of x^4 is 5, the coefficient of x^3 is -4, the coefficients of x^2 and x are both 0, and the coefficient of 1 is 10. (Why allow x and 1, by the way? They seem different from the other blocks, but they're not really: $x = x^1$ and $1 = x^0$.) The highest number n such that x^n has a nonzero coefficient is called the *degree* of the polynomial. For example, the degree of the above polynomial f is 4, since no power of x greater than 4 is present. The mathematical way to write a general polynomial of degree n is

$$p(x) = a_n x^n + a_{n-1} x^{n-1} + \cdots + a_2 x^2 + a_1 x + a_0,$$

where a_n is the coefficient of x^n, a_{n-1} is the coefficient of x^{n-1}, and so on down to a_0, which is the coefficient of 1.

Since the functions x^n are the building blocks of all polynomials, you should know what their graphs look like. The even powers mostly look similar to each other, and the same can be said for the odd powers. Here's what the graphs look like, from x^0 up to x^7:

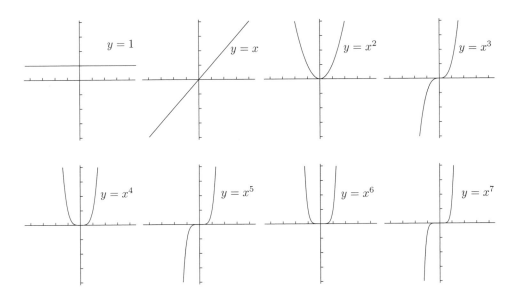

Sketching the graphs of more general polynomials is more difficult. Even finding the x-intercepts is often impossible unless the polynomial is very simple. There is one aspect of the graph that is fairly straightforward, which is what happens at the far left and right sides of the graph. This is determined by the so-called *leading coefficient*, which is the coefficient of the highest-degree term. This is basically the number a_n defined above. For example, in our polynomial $f(x) = 5x^4 - 4x^3 + 10$ from above, the leading coefficient is 5. In fact, it only matters whether the leading coefficient is positive or negative. It also matters whether the degree of the polynomial is odd or even; so there are four possibilities for what the edges of the graph can look like:

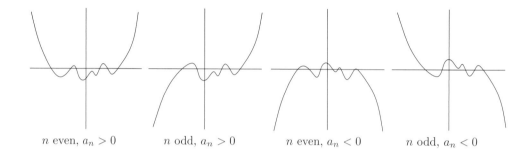

n even, $a_n > 0$ n odd, $a_n > 0$ n even, $a_n < 0$ n odd, $a_n < 0$

The wiggles in the center of these diagrams aren't relevant—they depend on the other terms of the polynomial. The diagram is just supposed to show what the graphs look like near the left and right edges. In this sense, the graph of our polynomial $f(x) = 5x^4 - 4x^3 + 10$ looks like the leftmost picture above, since $n = 4$ is even and $a_n = 5$ is positive.

Let's spend a little time on degree 2 polynomials, which are called *quadratics*. Instead of writing $p(x) = a_2 x^2 + a_1 x + a_0$, it's easier to write the coefficients as a, b, and c, so we have $p(x) = ax^2 + bx + c$. Quadratics have two, one, or zero (real) roots, depending on the sign of the *discriminant*. The discriminant, which is often written as Δ, is given by $\Delta = b^2 - 4ac$. There are three possibilities. If $\Delta > 0$, then there are two roots; if $\Delta = 0$, there is one root, which is called a *double root*; and if $\Delta < 0$, then there are no roots. In the first two cases, the roots are given by

$$\frac{-b \pm \sqrt{b^2 - 4ac}}{2a}.$$

Notice that the expression in the square root is just the discriminant. An important technique for dealing with quadratics is *completing the square*. Here's how it works. We'll use the example of the quadratic $2x^2 - 3x + 10$. The first step is to take out the leading coefficient as a factor. So our quadratic becomes $2(x^2 - \frac{3}{2}x + 5)$. This reduces the situation to dealing with a *monic* quadratic, which is a quadratic with leading coefficient equal to 1. So, let's worry about $x^2 - \frac{3}{2}x + 5$. The main technique now is to take the coefficient of x, which in our example is $-\frac{3}{2}$, divide it by 2 to get $-\frac{3}{4}$, and square it. We get $\frac{9}{16}$. We wish that the constant term were $\frac{9}{16}$ instead of 5, so let's do some mental gymnastics:

$$x^2 - \frac{3}{2}x + 5 = x^2 - \frac{3}{2}x + \frac{9}{16} + 5 - \frac{9}{16}.$$

Why on earth would we want to add and subtract $\frac{9}{16}$? Because the first three terms combine to form $(x - \frac{3}{4})^2$. So, we have

$$x^2 - \frac{3}{2}x + 5 = \left(x^2 - \frac{3}{2}x + \frac{9}{16} \right) + 5 - \frac{9}{16} = \left(x - \frac{3}{4} \right)^2 + 5 - \frac{9}{16}.$$

Now we just have to work out the last little bit, which is just arithmetic: $5 - \frac{9}{16} = \frac{71}{16}$. Putting it all together, and restoring the factor of 2, we have

$$2x^2 - 3x + 10 = 2 \left(x^2 - \frac{3}{2}x + 5 \right) = 2 \left(\left(x - \frac{3}{4} \right)^2 + \frac{71}{16} \right)$$

$$= 2 \left(x - \frac{3}{4} \right)^2 + \frac{71}{8}.$$

It turns out that this is a much nicer form to deal with in a number of situations. Make sure you know how to complete the square, since we'll be using this technique a lot in Chapters 18 and 19.

2. Rational functions: these are functions of the form

$$\frac{p(x)}{q(x)},$$

where p and q are polynomials. Rational functions will pop up in many different contexts, and the graphs can look very different depending on the polynomials p and q. The simplest examples of rational functions are polynomials themselves, which arise when $q(x)$ is the constant polynomial 1. The next simplest examples are the functions $1/x^n$, where n is a positive integer. Let's look at some of the graphs of these functions:

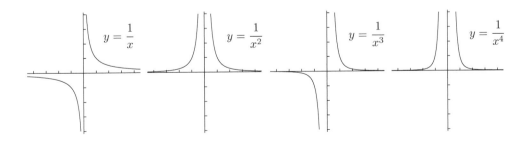

The odd powers look similar to each other, and the even powers look similar to each other. It's worth knowing what these graphs look like.

3. Exponentials and logarithms: you need to know what graphs of exponentials look like. For example, here is $y = 2^x$:

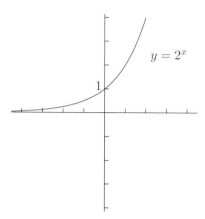

The graph of $y = b^x$ for any other base $b > 1$ looks similar to this. Things to notice are that the domain is the whole real line, the y-intercept is 1, the range is $(0, \infty)$, and there is a horizontal asymptote on the left at $y = 0$. In particular, the curve $y = b^x$ does **not**, I repeat, not touch the x-axis, no matter what it looks like on your graphing calculator! (We'll be looking at asymptotes again in Chapter 3.) The graph of $y = 2^{-x}$ is just the reflection of $y = 2^x$ in the y-axis:

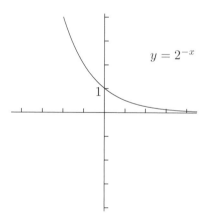

How about when the base is less than 1? For example, consider the graph of $y = \left(\frac{1}{2}\right)^x$. Notice that $\left(\frac{1}{2}\right)^x = 1/2^x = 2^{-x}$, so the above graph of $y = 2^{-x}$ is also the graph of $y = \left(\frac{1}{2}\right)^x$, since 2^{-x} and $\left(\frac{1}{2}\right)^x$ are equal for any x. The same sort of thing happens for $y = b^x$ for any $0 < b < 1$, not just $b = \frac{1}{2}$.

Now, notice that the graph of $y = 2^x$ satisfies the horizontal line test, so there is an inverse function. This is in fact the base 2 logarithm, which is written $y = \log_2(x)$. Using the line $y = x$ as a mirror, the graph of $y = \log_2(x)$ looks like this:

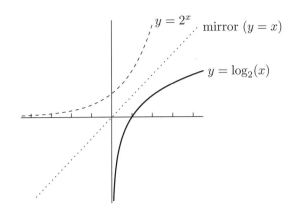

The domain is $(0, \infty)$; note that this backs up what I said earlier about not being able to take logarithms of a negative number or of 0. The range is all of $(-\infty, \infty)$, and there's a vertical asymptote at $x = 0$. The graphs of $\log_{10}(x)$, and indeed $\log_b(x)$ for any $b > 1$, are very similar to this one. The log function is very important in calculus, so you should really know how to draw the above graph. We'll look at other properties of logarithms in Chapter 9.

4. Trig functions: these are so important that the entire next chapter is devoted to them.

5. Functions involving absolute values: let's take a close look at the *absolute value function* f given by $f(x) = |x|$. Here's the definition of $|x|$:

$$|x| = \begin{cases} x & \text{if } x \geq 0, \\ -x & \text{if } x < 0. \end{cases}$$

Another way of looking at $|x|$ is that it is the distance between x and 0 on the number line. More generally, you should learn this nice fact:

$|x - y|$ is the distance between x and y on the number line.

For example, suppose that you need to identify the region $|x - 1| \leq 3$ on the number line. You can interpret the inequality as "the distance between x and 1 is less than or equal to 3." That is, we are looking for all the points that are no more than 3 units away from the number 1. So, let's take a number line and mark in the number 1 as follows:

$$1$$

The points which are no more than 3 units away extend to -2 on the left and 4 on the right, so the region we want looks like this:

$$\overset{\text{3 units}}{\longleftarrow} \quad \overset{\text{3 units}}{\longrightarrow}$$
$$-2 \qquad 1 \qquad 4$$

So, the region $|x - 1| \leq 3$ can also be described as $[-2, 4]$.

It's also true that $|x| = \sqrt{x^2}$. To check this, suppose that $x \geq 0$; then $\sqrt{x^2} = x$, no problem. If instead $x < 0$, then it can't be true that $\sqrt{x^2} = x$, since the left-hand side is positive but the right-hand side is negative. The correct equation is $\sqrt{x^2} = -x$; now the right-hand side is positive, since it's minus a negative number. If you look back at the definition of $|x|$, you'll see that we have just proved that $|x| = \sqrt{x^2}$. Even so, to deal with $|x|$, it's much better to use the piecewise definition than to write it as $\sqrt{x^2}$.

Finally, let's take a look at some graphs. If you know what the graph of a function looks like, you can get the graph of the absolute value of that function by reflecting everything below the x-axis up to above the x-axis, using the x-axis as your mirror. For example, here's the graph of $y = |x|$, which comes from reflecting the bottom portion of $y = x$ in the x-axis:

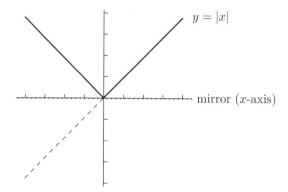

How about the graph of $y = |\log_2(x)|$? Using the reflection of the graph of $y = \log_2(x)$ above, this is what the absolute value version looks like:

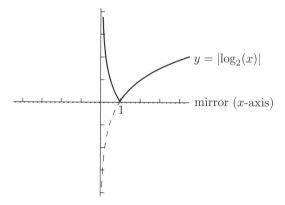

Anyway, that's all I have to say about functions, apart from trig functions which are the subject of the next chapter. Hopefully you've seen a lot of the stuff in this chapter before. Most of the material in this chapter is used over and over again in calculus, so make sure you really get on top of it all as soon as you can!

CHAPTER 2 ───────────────────

Review of Trigonometry

To do calculus, you really need to know trigonometry. Truth be told, we won't see much trig at first, but when it comes, it doesn't let up. So we might as well do a thorough review of the most important aspects of trig:

- angles in radians and the basics of the trig functions;
- trig functions on the real line (not just angles between $0°$ and $90°$);
- graphs of trig functions; and
- trig identities.

Time to refresh your memory....

2.1 The Basics

The first thing I want to remind you about is the notion of radians. Instead of saying that there are 360 degrees in a full revolution, we'll say that there are 2π *radians*. This may seem a bit wacky, but there is a reason: the circumference of a circle of radius 1 unit is 2π units. In fact, the arc length of a wedge of this circle is exactly the angle of the wedge: arc length = wedge angle θ^R

θ units (length)

$C = \pi d = 2\pi r$

arc of a circle $S = r\theta^R$

$180° = \pi$ radians

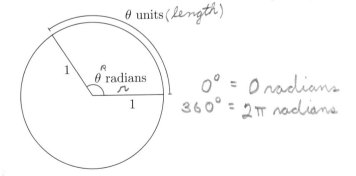

1
θ radians R
r
1

$0° = 0$ radians
$360° = 2\pi$ radians

This picture is pretty and all, but the main thing is to be comfortable with the most common angles in both degree and radian form. First, you should become absolutely comfortable with the idea that $90°$ is the same as $\pi/2$

Changing degrees into radians

$$\frac{\theta^{\circ}}{180} = \frac{\theta^{R}}{\pi}$$

radians, and similarly that $180°$ is the same as π radians and $270°$ is the same as $3\pi/2$ radians. Once you have that in mind, try to be comfortable converting all the angles in the following picture back and forth between degrees and radians:

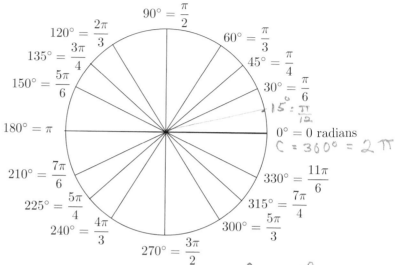

$15° : \frac{\pi}{12}$

$C = 360° = 2\pi$

More generally, you can also use the formula

$$\frac{\theta^{\circ}}{180} = \frac{\theta^{R}}{\pi}$$ *change degrees to radians*

$$\text{angle in radians} = \frac{\pi}{180} \times \text{angle in degrees}$$

if you need to. For example, to see what $5\pi/12$ radians is in degrees, solve

$$\frac{5\pi}{12} = \frac{\pi}{180} \times \text{angle in degrees}$$

to see that $5\pi/12$ radians is the same as $(180/\pi) \times (5\pi/12) = 75°$. In fact, you can think of this conversion from radians to degrees as a sort of change of units, like changing from miles to kilometers. The conversion factor is that π radians is the same as 180 degrees.

We have only looked at angles so far, so let's move on to trig functions. Obviously you have to know how the trig functions are defined in terms of triangles. Suppose you have a right-angled triangle and one of the angles, other than the right angle, is labeled θ, like this:

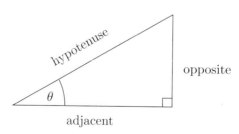

Then the basic formulas are

$$\sin(\theta) = \frac{\text{opposite}}{\text{hypotenuse}}, \quad \cos(\theta) = \frac{\text{adjacent}}{\text{hypotenuse}}, \quad \text{and} \quad \tan(\theta) = \frac{\text{opposite}}{\text{adjacent}}.$$

Of course, if the angle θ is moved, then the opposite and adjacent have to be moved as well:

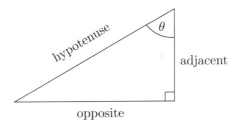

The opposite is, unsurprisingly, opposite the angle θ and the adjacent is next to it. The hypotenuse never changes, though: it is the longest side and is always across from the right angle.

We'll also be using the reciprocal functions csc, sec, and cot, which are defined as follows:

$$\csc(x) = \frac{1}{\sin(x)}, \qquad \sec(x) = \frac{1}{\cos(x)}, \qquad \text{and} \qquad \cot(x) = \frac{1}{\tan(x)}.$$

Now, a piece of advice if you ever plan to take a calculus exam (or even if you don't!): learn the values of the trig functions at the common angles 0, $\pi/6$, $\pi/4$, $\pi/3$, and $\pi/2$. For example, without thinking, can you simplify $\sin(\pi/3)$? How about $\tan(\pi/4)$? If you can't, then at best you're wasting time trying to use a triangle to find the answer, and at worst you're throwing away easy points by not simplifying your answer all the way. The solution is to memorize the following table:

	0	$\dfrac{\pi}{6}$	$\dfrac{\pi}{4}$	$\dfrac{\pi}{3}$	$\dfrac{\pi}{2}$
sin	0	$\dfrac{1}{2}$	$\dfrac{1}{\sqrt{2}}$	$\dfrac{\sqrt{3}}{2}$	1
cos	1	$\dfrac{\sqrt{3}}{2}$	$\dfrac{1}{\sqrt{2}}$	$\dfrac{1}{2}$	0
tan	0	$\dfrac{1}{\sqrt{3}}$	1	$\sqrt{3}$	$\star$

The star means that $\tan(\pi/2)$ is undefined. In fact, the tan function has a vertical asymptote at $\pi/2$ (this will be clear from the graph, which we'll look at in Section 2.3 below). Anyway, you need to be able to quote any of the entries in this table, both forward and backward! What this means is that you have to be able to answer two types of questions. Here are examples of each of these types:

1. What is $\sin(\pi/3)$? (Using the table, the answer is $\sqrt{3}/2$.)
2. What angle between 0 and $\pi/2$ has a sine equal to $\sqrt{3}/2$? (The answer is obviously $\pi/3$.)

Of course, you have to be able to answer these two types of questions for each entry in the table. Please, please, I beg of you, learn this table! Math isn't about memorization, but there are a few things that are worth memorizing and this table is definitely on the list. So make flash cards, get your friends to quiz you, spend one minute a day, whatever works for you, but learn the table.

2.2   Extending the Domain of Trig Functions

The above table (did you learn it yet?) only covers some angles ranging from 0 to $\pi/2$. It's possible to take sin or cos of any angle at all, even a negative one. For tan, we have to be a little more careful—for example, we saw above that $\tan(\pi/2)$ is undefined. Still, we'll be able to take tan of just about every angle, even most negative ones.

Let's first look at angles between 0 and 2π (remember that 2π is the same as 360°). Suppose you want to calculate $\sin(\theta)$ (or $\cos(\theta)$, or $\tan(\theta)$), where θ is between 0 and 2π. To see what this even means, start by drawing a coordinate plane with some slightly weird labels:

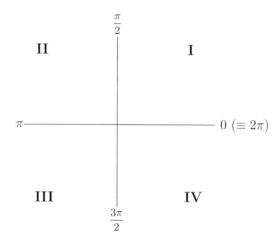

Notice that the axes divide the plane into four quadrants, which are creatively labeled from 1 to 4 (in Roman numerals), and that the labeling goes counterclockwise. These quadrants are called the first, second, third, and fourth quadrants, respectively. The next step is to draw a ray (that's half a line) starting at the origin. Which ray? It depends on θ. Just imagine yourself standing at the origin, looking to the right along the positive x-axis. Now turn counterclockwise an angle of θ, then march forward in a straight line. Your trail is the ray you're looking for.

Now the other labels on the above picture (and the one on page 26) make a lot of sense. Indeed, if you turn an angle of $\pi/2$, you are facing up the page and you trace out the positive y-axis as you walk along. If you had instead turned an angle of π, you'd get the negative x-axis; and if you had turned $3\pi/2$, you'd get the negative y-axis. Finally, if you had turned 2π, that would put you back to where you started, facing along the positive x-axis. It's the same as if you hadn't turned at all! That's why the picture says $0 \equiv 2\pi$. As far as angles are concerned, 0 and 2π are equivalent.

OK, let's take some angle θ and draw in the appropriate ray. Perhaps it might be somewhere in the third quadrant, like this:

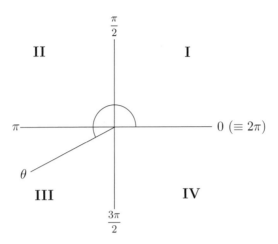

Notice that we label the **ray** as θ, not the angle itself. Anyway, now we pick some point on the ray and drop a perpendicular from that point to the x-axis:

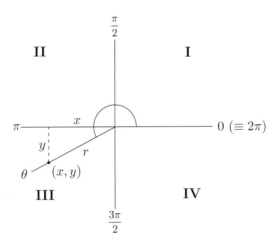

We're interested in three quantities: the x- and y-coordinates of the point (which are called x and y, of course!) and also the distance from the point to the origin, which is called r. Note that x and y could both potentially be negative—in fact, they both are negative in the above picture—but r is always positive, since it's a distance. In fact, by Pythagoras' Theorem, we have $r = \sqrt{x^2 + y^2}$, regardless of the signs of x and y. (The squares kill off any minus signs around.)

Armed with these three quantities, we can define the three trig functions as follows:

$$\sin(\theta) = \frac{y}{r}, \qquad \cos(\theta) = \frac{x}{r}, \qquad \text{and} \qquad \tan(\theta) = \frac{y}{x}.$$

These are just the regular formulas from Section 2.1 above, with the quantities x, y, and r interpreted as the adjacent, opposite, and hypotenuse, respectively. But wait, you say—what happens if you choose a different point on the ray? It doesn't matter, because your new triangle will be similar to the old one and the above ratios are unaffected. In fact, it is often convenient to assume that $r = 1$, so that the point (x, y) lies on the so-called *unit circle* (that's the circle of radius 1 centered at the origin).

Now let's look at an example. Suppose we want to find $\sin(7\pi/6)$. Which quadrant is $7\pi/6$ in? We need to decide where $7\pi/6$ fits in the list 0, $\pi/2$, π, $3\pi/2$, 2π. In fact, $7/6$ is greater than 1 but less than $3/2$, so $7\pi/6$ fits between π and $3\pi/2$. In fact, the picture looks pretty much like the above example:

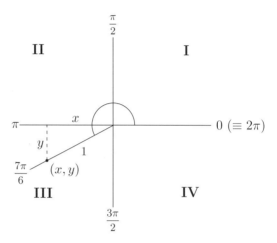

So the angle $7\pi/6$ is in the third quadrant. We've chosen the point on the ray which has distance $r = 1$ from the origin, then dropped a perpendicular. We know from the above formulas that $\sin(\theta) = y/r = y$ (since $r = 1$), so we really need to find y. Well, that little angle between our ray at $7\pi/6$ and the negative x-axis—which itself is at π—must be the difference between these two angles, $\pi/6$. The little angle is called the *reference angle*. In general, the reference angle for θ is the smallest angle between the ray which represents θ and the x-axis. It must be between 0 and $\pi/2$. In our example, the closest route to the x-axis is up, so the reference angle looks like this:

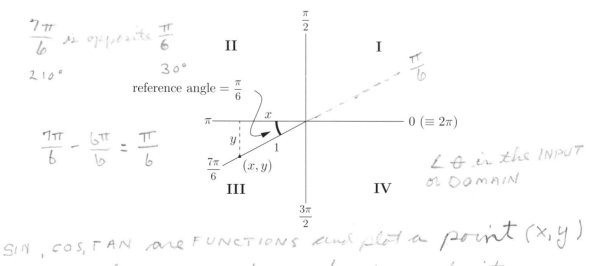

So in the little triangle, we know that $r = 1$ and the angle is $\pi/6$. It looks like $y = \sin(\pi/6) = 1/2$, except that can't be right! Since we're below the x-axis, the quantity y must be negative. That is, $y = -1/2$. Since $\sin(\theta) = y$, we have shown that $\sin(7\pi/6) = -1/2$. We can also repeat this with cosine instead of sine to see that $x = -\cos(\pi/6) = -\sqrt{3}/2$. After all, x has to be negative, since the point (x, y) is to the left of the y-axis. This shows that $\cos(7\pi/6) = -\sqrt{3}/2$ and we have identified our point (x, y) as $(-\sqrt{3}/2, -1/2)$.

TRIG FUNCTION

2.2.1 The ASTC method

210°

The key in the previous example is that $\sin(7\pi/6)$ is related to $\sin(\pi/6)$, where $\pi/6$ is the reference angle for $7\pi/6$. In fact, it's not hard to see that the sine of any angle is plus or minus the sine of the reference angle! This narrows it down to just two possibilities, and there's no need to mess around with x, y, or r. So in our example, we just needed to find that the reference angle for $7\pi/6$ is $\pi/6$; this immediately told us that $\sin(7\pi/6)$ is equal to either $\sin(\pi/6)$ or $-\sin(\pi/6)$ and we just had to make sure we got the correct one. We saw that it was the negative one because y was negative.

30°

Actually, the sine of anything in the third or fourth quadrant must be negative because y is negative there. Similarly, the cosine of anything in the second or third quadrant must be negative, since x is negative there. The tangent is the ratio y/x, which is negative in the second and fourth quadrants (since one, but not both, of x and y is then negative) but positive in the first and third quadrants.

Let's summarize these findings in words as well as with a picture. First, all three functions are positive in the first quadrant (I). In the second quadrant (II), only sin is positive; the other two functions are negative. In the third quadrant (III), only tan is positive; the other two functions are negative. And finally, in the fourth quadrant (IV), only cos is positive; the other two functions are negative. Here's what it all looks like:

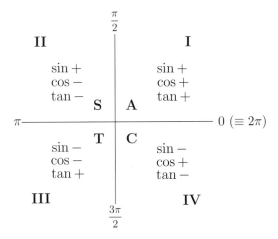

In fact, the letters ASTC on the diagram are all you need to remember. They show you which of the functions are positive in that quadrant. "A" stands for "All," meaning all the functions are positive in the first quadrant; the other letters obviously stand for sin, tan, and cos, respectively. In our example,

$7\pi/6$ is in the third quadrant, so only tan is positive there. In particular, sin is negative, so since we had narrowed the value of $\sin(7\pi/6)$ down to $1/2$ or $-1/2$, it must be the negative possibility: indeed, $\sin(7\pi/6) = -1/2$.

The only problem with the ASTC diagram is that it doesn't really tell you how to handle the angles 0, $\pi/2$, π, or $3\pi/2$, since they lie on the axes. In this case, it's best to forget all about the ASTC stuff and draw a graph of $y = \sin(x)$ (or $\cos(x)$ or $\tan(x)$, as appropriate) and read the value off the graph. We'll discuss this in Section 2.3 below.

Meanwhile, here's a summary of the ASTC method for finding trig functions of angles between 0 and 2π:

1. Draw the quadrant diagram, decide where in the picture the angle you care about is, and then mark that angle in the diagram.

2. If the angle you want is on the x- or y-axis (that is, not within any quadrant), draw a graph of the trig function and read the value off the graph (there are some examples in Section 2.3 below).

3. Otherwise, find the smallest angle between the one we want and the x-axis; this is called the reference angle.

4. If you can, use the important table to work out the value of the trig function of the reference angle. That's the answer you need, except that you might need a minus sign in front.

5. Use the ASTC diagram to decide whether or not you need a minus sign.

Let's look at a couple of examples. How would you find $\cos(7\pi/4)$ and $\tan(9\pi/13)$? We'll look at them one at a time. For $\cos(7\pi/4)$, we notice that $7/4$ is between $3/2$ and 2, so the angle must be in the fourth quadrant:

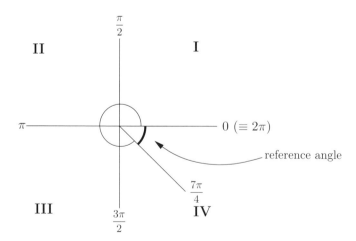

To work out the reference angle, notice that we have to go up to 2π (not down to 0, beware!) so the reference angle is the difference between 2π and $7\pi/4$, which is $(2\pi - 7\pi/4)$ or simply $\pi/4$. So, $\cos(7\pi/4)$ is plus or minus $\cos(\pi/4)$, which is $1/\sqrt{2}$ according to our table. Is it plus or minus? The ASTC picture says that cos is positive in the fourth quadrant, so it's plus: $\cos(7\pi/4) = 1/\sqrt{2}$.

Now let's look at $\tan(9\pi/13)$. We see that $9/13$ is between $1/2$ and 1, so the angle $9\pi/13$ is in the second quadrant:

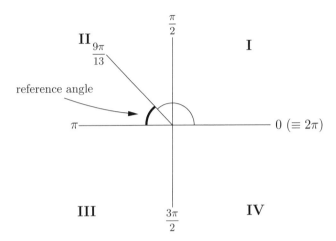

This time we have to go up to π to get to the x-axis, so the reference angle is the difference between π and $9\pi/13$, which is $\pi - 9\pi/13$ or simply $4\pi/13$. So, we know that $\tan(9\pi/13)$ is plus or minus $\tan(4\pi/13)$. Alas, the number $4\pi/13$ isn't in our table, so we can't simplify $\tan(4\pi/13)$. We also need to work out whether it's plus or minus. Well, the ASTC diagram shows that only sin is positive in the second quadrant, so tan must be negative there and we see that $\tan(9\pi/13) = -\tan(4\pi/13)$. That's as simplified as we can get without approximating. When solving calculus problems, I don't recommend approximating the answer unless you are explicitly asked to. A common misconception is that the number that comes out on the calculator when you calculate something like $-\tan(4\pi/13)$ is the actual answer. On the contrary, it's just an approximation! So you shouldn't write

$$-\tan(4\pi/13) = -0.768438861,$$

since it's just not true. Instead, just leave it as $-\tan(4\pi/13)$ unless you are specifically asked for an approximation. In that case, use the approximately-equal symbol and fewer decimal places, rounding appropriately (unless you are asked for more):

$$-\tan(4\pi/13) \cong -0.768.$$

By the way, you should rarely need to use a calculator—in fact, some colleges don't even allow them in exams! So you should try to avoid the temptation ever to use one.

2.2.2 Trig functions outside $[0, 2\pi]$

There's still the question of how to take trig functions of angles bigger than 2π or less than 0. In fact this isn't so bad: simply add or subtract multiples of 2π until you get between 0 and 2π. You see, it doesn't just stop at 2π. It just keeps on wrapping around. For example, if I asked you to stand on the spot facing due east and then turn around counterclockwise an angle of 450 degrees, it would be reasonable to assume that you'd turn a full revolution and then an extra 90 degrees. You'd be facing due north. Sure, you'd be a little dizzier than if you just did a 90-degree counterclockwise turn, but you'd be facing the same way. So 450 degrees is an equivalent angle to 90 degrees, and of course the same sort of thing is true in radians: in this case, $5\pi/2$ radians is an equivalent angle to $\pi/2$ radians. But why stop at one revolution? How about $9\pi/2$ radians? That's the same as going around 2π twice (which gets us up to 4π) and then an extra $\pi/2$, so we've done 2 useless revolutions before our final $\pi/2$ twist. The revolutions don't matter, so once again $9\pi/2$ is equivalent to $\pi/2$. This procedure can be extended indefinitely to get a whole family of angles which are equivalent to $\pi/2$:

$$\frac{\pi}{2}, \frac{5\pi}{2}, \frac{9\pi}{2}, \frac{13\pi}{2}, \frac{17\pi}{2}, \ldots$$

Of course, each angle is a full revolution, or 2π, more than the first one. Still, that's not the full story: if I'm going to insist that you do all these counterclockwise revolutions and get that dizzy, you might as well ask to be allowed to do a clockwise revolution or two to recover. This corresponds to a negative angle. In particular, if you were facing east and I asked you to turn -270 degrees counterclockwise, the only sane interpretation of my bizarre request is to turn 270 degrees (or $3\pi/2$) clockwise. Evidently you'll still end up facing due north, so -270 degrees must be equivalent to 90 degrees. Indeed, adding 360 degrees to -270 degrees just gives us 90 degrees. In radians, we see that $-3\pi/2$ is an equivalent angle to $\pi/2$. In addition, we could insist on more negative (clockwise) full revolutions. In the end, here is the complete set of angles which are equivalent to $\pi/2$:

$$\ldots, -\frac{15\pi}{2}, -\frac{11\pi}{2}, -\frac{7\pi}{2}, -\frac{3\pi}{2}, \frac{\pi}{2}, \frac{5\pi}{2}, \frac{9\pi}{2}, \frac{13\pi}{2}, \frac{17\pi}{2}, \ldots$$

The sequence has no beginning or end; when I say it's "complete," I'm glossing over the fact that there are infinitely many angles included in the dots at the beginning and the end. We can avoid the dots by writing the collection in set notation as $\{\pi/2 + 2\pi n\}$, where n runs over all the integers.

Let's see if we can apply this. How would you find $\sec(15\pi/4)$? The first thing to note is that if we can find $\cos(15\pi/4)$, all we need to do is take the reciprocal in order to get $\sec(15\pi/4)$. So let's find $\cos(15\pi/4)$ first. Since $15/4$ is more than 2, let's try lopping off 2 from it. Hmm, $15/4 - 2 = 7/4$, which is now between 0 and 2, so that looks promising. Restoring the π, we see that $\cos(15\pi/4)$ is the same as $\cos(7\pi/4)$ which we already saw is equal to $1/\sqrt{2}$. So, $\cos(15\pi/4) = 1/\sqrt{2}$. Taking reciprocals, we see that $\sec(15\pi/4)$ is just $\sqrt{2}$.

Finally, how about $\sin(-5\pi/6)$? There are several ways of doing this problem, but the way suggested above is to try to add multiples of 2π to $-5\pi/6$ until we are between 0 and 2π. In fact, adding 2π to $-5\pi/6$ gives $7\pi/6$, so $\sin(-5\pi/6) = \sin(7\pi/6)$, which we already saw is equal to $-1/2$. Alternatively, we could have drawn a diagram directly:

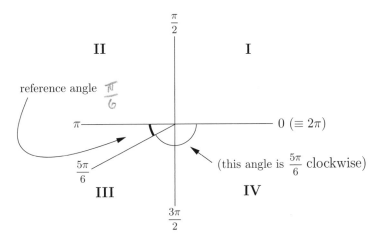

Now you have to work out the reference angle from the diagram, and it's not too hard to see that it is $\pi/6$ and continue as before.

2.3   The Graphs of Trig Functions

It's really useful to remember what the graphs of the sin, cos, and tan functions look like. These functions are all *periodic*, meaning that they repeat themselves over and over again from left to right. For example, consider $y = \sin(x)$. The graph from 0 to 2π looks like this:

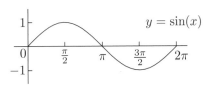

You should be able to produce this graph without thinking, including the positions of 0, $\pi/2$, π, $3\pi/2$, and 2π. Since $\sin(x)$ repeats every 2π units (we say that $\sin(x)$ is periodic in x with period 2π), we can extend the graph by repeating the pattern:

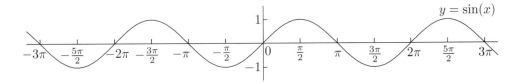

 Just reading values off the graph, we can see that $\sin(3\pi/2) = -1$ and $\sin(-\pi) = 0$. As noted earlier, this is how you should deal with multiples of $\pi/2$; no need to mess around with reference angles. Another thing to note is that the graph has 180° point symmetry about the origin, which means that $\sin(x)$ is an odd function of x. (We looked at odd and even functions in Section 1.4 of the previous chapter.)

The graph of $y = \cos(x)$ is similar to that of $y = \sin(x)$. When x ranges from 0 to 2π, it looks like this:

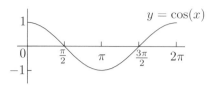

Now extend the graph using the fact that $\cos(x)$ is periodic with period 2π:

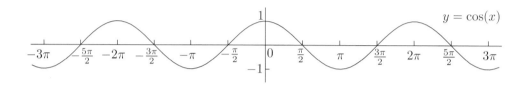

For example, if you want to find $\cos(\pi)$, you can see that it is -1 just by reading it off the graph. Furthermore, notice that this time the graph has mirror symmetry in the y-axis. This means that $\cos(x)$ is an even function of x.

Now $y = \tan(x)$ is a little different. It's best to graph it for x between $-\pi/2$ and $\pi/2$ first:

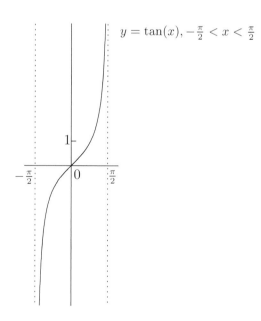

The tan function has vertical asymptotes, unlike the sin and cos functions. Also, its period is actually π, not 2π, so the above pattern can be repeated to obtain the full graph of $y = \tan(x)$:

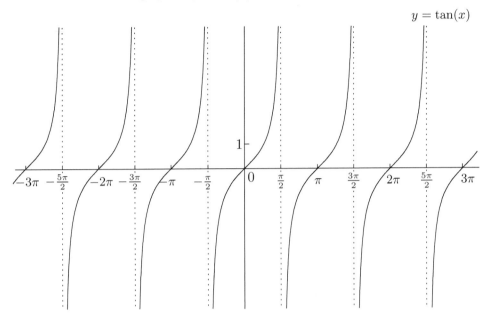

$y = \tan(x)$

It's clear that $y = \tan(x)$ has vertical asymptotes (and is consequently undefined) when x is any odd multiple of $\pi/2$. Also, the symmetry of the graph indicates that $\tan(x)$ is an odd function of x.

It's also worthwhile learning the graphs of $y = \sec(x)$, $y = \csc(x)$, and $y = \cot(x)$:

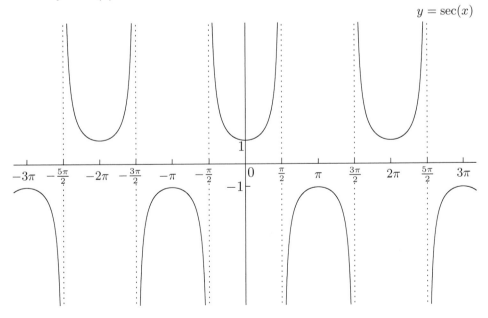

$y = \sec(x)$

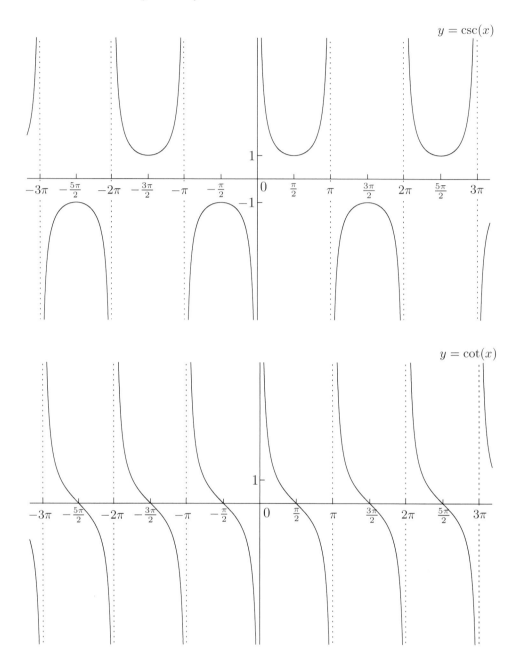

From their graphs, we can get the following symmetry properties of all six basic trig functions, which are worth learning:

$\sin(x)$, $\tan(x)$, $\cot(x)$, and $\csc(x)$ are odd functions of x.
$\cos(x)$ and $\sec(x)$ are even functions of x.

So $\sin(-x) = -\sin(x)$, $\tan(-x) = -\tan(x)$, and $\cos(-x) = \cos(x)$ for all real numbers x.

2.4 Trig Identities

There are relations between trig functions which will come in handy. First, note that tan and cot may be expressed in terms of sin and cos as follows:

$$\tan(x) = \frac{\sin(x)}{\cos(x)}, \qquad \cot(x) = \frac{\cos(x)}{\sin(x)}.$$

(Sometimes it's helpful to replace every instance of tan or cot by sin and cos using these identities, but you shouldn't really do this unless you're stuck.)

The most important of all the trig identities is Pythagoras' Theorem (written in trig form),

$$\cos^2(x) + \sin^2(x) = 1.$$

This is true for any x. (Why is this Pythagoras' Theorem? If the hypotenuse of a right-angled triangle is 1 and one of the angles is x, convince yourself that the other two sides of the triangle have lengths $\cos(x)$ and $\sin(x)$.)

Now divide this equation by $\cos^2(x)$. I want you to check that you end up with

$$1 + \tan^2(x) = \sec^2(x).$$

This also comes up a lot in calculus. Alternatively, you could have divided the Pythagorean equation above by $\sin^2(x)$ to get

$$\cot^2(x) + 1 = \csc^2(x).$$

This equation seems to come up less frequently than the others.

There are some more relationships between trig functions. Have you noticed that some of the names begin with the syllable "co"? This is short for the word "complementary." To say that two angles are complementary means that they add up to $\pi/2$ (or 90 degrees). It does not mean that they are nice to each other. All puns aside, the fact is that we have the following general relationship:

$$\text{trig function}(x) = \text{co-trig function}\left(\frac{\pi}{2} - x\right).$$

So in particular, we have

$$\sin(x) = \cos\left(\frac{\pi}{2} - x\right), \quad \tan(x) = \cot\left(\frac{\pi}{2} - x\right), \quad \text{and} \quad \sec(x) = \csc\left(\frac{\pi}{2} - x\right).$$

It even works when the trig function is already a "co"; you just have to realize that the complement of a complement is the original angle! For example, co-co-sine is really just sine, and co-co-tan is just tan. Basically this means that we can also say that

$$\cos(x) = \sin\left(\frac{\pi}{2} - x\right), \quad \cot(x) = \tan\left(\frac{\pi}{2} - x\right), \quad \text{and} \quad \csc(x) = \sec\left(\frac{\pi}{2} - x\right).$$

Finally, there's another group of identities which are worth learning. These are the identities involving sums of angles and the double-angle formulas.

Specifically, you should remember that

$$\begin{array}{l} \sin(A + B) = \sin(A)\cos(B) + \cos(A)\sin(B) \\ \cos(A + B) = \cos(A)\cos(B) - \sin(A)\sin(B). \end{array}$$

It's also useful to remember that you can switch all the pluses and minuses to get some related formulas:

$$\sin(A - B) = \sin(A)\cos(B) - \cos(A)\sin(B)$$
$$\cos(A - B) = \cos(A)\cos(B) + \sin(A)\sin(B).$$

A nice consequence of the $\sin(A + B)$ and $\cos(A + B)$ formulas in the box above is obtained by letting $A = B = x$. It's clear that the sine formula is $\sin(2x) = 2\sin(x)\cos(x)$, but let's take a closer look at the cosine formula. This becomes $\cos(2x) = \cos^2(x) - \sin^2(x)$; true as this is, it is more useful to use the Pythagorean identity $\cos^2(x) + \sin^2(x) = 1$ to express $\cos(2x)$ as either $2\cos^2(x) - 1$ or $1 - 2\sin^2(x)$ (convince yourself that these are both valid!). In summary, the double-angle formulas are

$$\begin{array}{l} \sin(2x) = 2\sin(x)\cos(x) \\ \cos(2x) = 2\cos^2(x) - 1 = 1 - 2\sin^2(x). \end{array}$$

So, how would you write $\sin(4x)$ in terms of $\sin(x)$ and $\cos(x)$? Well, think of $4x$ as double $2x$ and use the sine identity to write $\sin(4x) = 2\sin(2x)\cos(2x)$. Then use both identities to get

$$\sin(4x) = 2(2\sin(x)\cos(x))(2\cos^2(x) - 1) = 8\sin(x)\cos^3(x) - 4\sin(x)\cos(x).$$

Similarly,

$$\cos(4x) = 2\cos^2(2x) - 1 = 2(2\cos^2(x) - 1)^2 - 1 = 8\cos^4(x) - 8\cos^2(x) + 1.$$

You shouldn't memorize these last two formulas; instead, make sure you understand how to derive them using the double-angle formulas.

Now, if you can master all the trig in this chapter, you will be in very good shape indeed for the rest of the book. So don't leave it till too late—get cracking on a bunch of examples and make sure you learn the table and all the boxed formulas!

CHAPTER 3 _____

Introduction to Limits

Calculus wouldn't exist without the concept of limits. This means that we are going to spend a lot of time looking at them. It turns out that it's pretty tricky to define a limit properly, but you can get an intuitive understanding of limits even without going into the gory details. This will be enough to tackle differentiation and integration. So, this chapter contains only the intuitive version; check out Appendix A for the formal version. All in all, here's what we'll look at in this chapter:

- an intuitive idea of what a limit is;
- left-hand, right-hand, and two-sided limits, and limits at ∞ and $-\infty$;
- when limits fail to exist; and
- the sandwich principle (also known as the "squeeze principle").

3.1 Limits: The Basic Idea

Let's dive in. We start with some function f and a point on the x-axis, which we'll call a. Here is what we'd like to understand: what does $f(x)$ look like when x is really really close to a, **but not equal to a**? This is a pretty strange question to ask, which is probably why it took until relatively recently for humankind to develop calculus.

Here's an example showing why we might want to ask this question. Let f have domain $\mathbb{R}\backslash\{2\}$ (all real numbers except for 2), and set $f(x) = x - 1$ on this domain. Formally, you might write:

$$f(x) = x - 1 \qquad \text{when } x \neq 2.$$

This seems like a weird sort of function: after all, why on earth would we want to exclude 2 from the domain? Actually, in the next chapter, we'll see that f arises quite naturally as a rational function (see the second example in Section 4.1). In the meantime, let's just take f for what it is and sketch a graph of it:

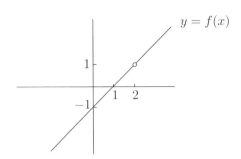

What is $f(2)$? Perhaps you'd like to say that $f(2) = 1$, but that would be a load of bull since 2 isn't even in the domain of f. The best you can do is to say that $f(2)$ is undefined. On the other hand, we can find the value of $f(x)$ when x is really really close to 2 and see what happens. For example, $f(2.01) = 1.01$, and $f(1.999) = 0.999$. If you think about it, you can see that when x is really really close to 2, the value of $f(x)$ is really really close to 1.

What's more, you can get as close as you want to 1, without actually getting to 1, by letting x be close enough to 2. For example, if you want $f(x)$ to be within 0.0001 of 1, you could take any x between 1.9999 and 2.0001 (except of course for $x = 2$, which is forbidden). If you instead wanted $f(x)$ to be within 0.000007 of 1, then you'd have to be a little more picky about your choice of x—this time you'd need to take x between 1.999993 and 2.000007 (except for $x = 2$, once again).

Anyway, these ideas are described in much greater detail in Section A.1 of Appendix A. Without getting bogged down, let's cut to the chase and just write

$$\lim_{x \to 2} f(x) = 1.$$

If you read this out loud, it should sound like "the limit, as x goes to 2, of $f(x)$ is equal to 1." Again, this means that when x is near 2 (but not equal to it), the value of $f(x)$ is near 1. How near? As near as your heart desires. Another way of writing the above statement is

$$f(x) \to 1 \quad \text{as } x \to 2.$$

This is harder to do computations with, but its meaning is quite clear: as x journeys along the number line from the left or the right toward the number 2, the value of $f(x)$ gets very very close to 1 (and stays close!).

Now, let's take the above function f and modify it slightly. Indeed, suppose that a new function g has the following graph:

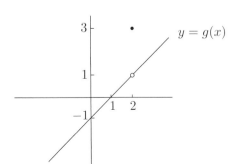

The domain of g is all real numbers, and $g(x)$ can be defined in piecewise fashion as

$$g(x) = \begin{cases} x - 1 & \text{if } x \neq 2, \\ 3 & \text{if } x = 2. \end{cases}$$

What is $\lim_{x \to 2} g(x)$? The trick here is that the value of $g(2)$ is irrelevant! It's only the values of $g(x)$ where x is close to 2, not actually **at** 2, which matter. Ignoring $x = 2$, the function g is identical to the function f we looked at earlier. So, $\lim_{x \to 2} g(x) = 1$ as before, even though $g(2) = 3$.

Here's an important point: when you write something like

$$\lim_{x \to 2} f(x) = 1,$$

the left-hand side isn't actually a function of x! Remember, the equation means that $f(x)$ is close to 1 when x is close to 2. We could actually replace x by any other letter and this would still be true. For example, $f(q)$ is close to 1 when q is close to 2, so we have

$$\lim_{q \to 2} f(q) = 1.$$

We can go nuts with this and also write

$$\lim_{b \to 2} f(b) = 1, \qquad \lim_{z \to 2} f(z) = 1, \qquad \lim_{\alpha \to 2} f(\alpha) = 1,$$

and so on until we run out of letters and symbols! The point is that in the limit

$$\lim_{x \to 2} f(x) = 1,$$

the variable x is just a *dummy variable*. It is a temporary label for some quantity that is (in this case) getting very close to 2. It can be replaced by any other letter, as long as you swap it out wherever else it appears; also, when you work out the value of the limit, the answer cannot include the dummy variable. So be smart about your dummy variables.

3.2 Left-Hand and Right-Hand Limits

We've seen that limits describe the behavior of a function near a certain point. Think about how you would describe the behavior of $h(x)$ near $x = 3$:

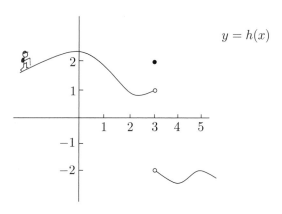

Of course, the fact that $h(3) = 2$ is irrelevant as far as the limiting behavior is concerned. Now, what happens when you approach $x = 3$ from the left? Imagine that you're the hiker in the picture, climbing up and down the hill. The value of $h(x)$ tells you how high up you are when your horizontal position is at x. So, if you walk rightward from the left of the picture, then when your horizontal position is close to 3, your height is close to 1. Sure, there's a sheer drop when you get to $x = 3$ (not to mention a weird little ledge floating in space above you!), but we don't care about this for the moment. Everything to the right of $x = 3$, including $x = 3$ itself, is irrelevant. So we've just seen that the *left-hand limit* of $h(x)$ at $x = 3$ is equal to 1.

On the other hand, if you are walking leftward from the right-hand side of the picture, your height becomes close to -2 as your horizontal position gets close to $x = 3$. This means that the *right-hand limit* of $h(x)$ at $x = 3$ is equal to -2. Now everything to the left of $x = 3$ (including $x = 3$ itself) is irrelevant!

We can summarize our findings from above by writing

$$\lim_{x \to 3^-} h(x) = 1 \quad \text{and} \quad \lim_{x \to 3^+} h(x) = -2.$$

The little minus sign after the 3 in the first limit above means that the limit is a left-hand limit, and the little plus sign in the second limit means that it's a right-hand limit. It's important to write the minus or plus sign **after** the 3, not before it! For example, if you write

$$\lim_{x \to -3} h(x),$$

then you are referring to the regular two-sided limit of $h(x)$ at $x = -3$, not the left-hand limit at $x = 3$. These are two very different animals indeed. By the way, the reason that you write $x \to 3^-$ under the limit sign for the left-hand limit is that this limit only involves values of x less than 3. That is, you need to take a little bit away from 3 to see what's going on. In a similar manner, when you write $x \to 3^+$ for the right-hand limit, this means that you only need to consider what happens when you add a little bit onto 3.

Now, limits don't always exist, as we'll see in the next section. But here's something important: the regular two-sided limit at $x = a$ exists **exactly when** both left-hand and right-hand limits at $x = a$ exist **and are equal to each other**! In that case, all three limits—two-sided, left-hand, and right-hand—are the same. In math-speak, I'm saying that

$$\lim_{x \to a^-} f(x) = L \quad \text{and} \quad \lim_{x \to a^+} f(x) = L$$

is the same thing as

$$\lim_{x \to a} f(x) = L.$$

If the left-hand and right-hand limits are not equal, as in the case of our function h from above, then the two-sided limit does not exist. We'd just write

$$\lim_{x \to 3} h(x) \text{ does not exist}$$

or you could even write "DNE" instead of "does not exist."

3.3   When the Limit Does Not Exist

We just saw that a two-sided limit doesn't exist when the corresponding left-hand and right-hand limits are different. Here's an even more dramatic example of this. Consider the graph of $f(x) = 1/x$:

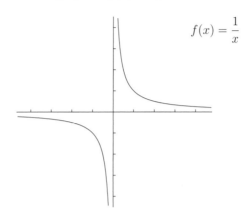

$$f(x) = \frac{1}{x}$$

What is $\lim_{x \to 0} f(x)$? It may be a bit much to expect the two-sided limit to exist here, so let's first try to find the right-hand limit, $\lim_{x \to 0^+} f(x)$. Looking at the graph, it seems as though $f(x)$ is very large when x is positive and close to 0. It doesn't really get close to any number in particular as x slides down to 0 from the right; it just gets larger and larger. How large? Larger than anything you can imagine! We say that the limit is infinity, and write

$$\lim_{x \to 0^+} \frac{1}{x} = \infty.$$

Similarly, the left-hand limit here is $-\infty$, since $f(x)$ gets arbitrarily more and more negative as x slides upward to 0. That is,

$$\lim_{x \to 0^-} \frac{1}{x} = -\infty.$$

The two-sided limit certainly doesn't exist, since the left-hand and right-hand limits are different. On the other hand, consider the function g defined by $g(x) = 1/x^2$. The graph looks like this:

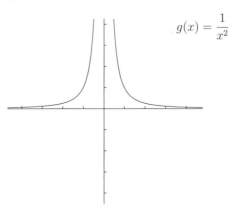

$$g(x) = \frac{1}{x^2}$$

Both the left-hand and right-hand limits at $x = 0$ are ∞, so you can say that $\lim_{x \to 0} 1/x^2 = \infty$ as well. By the way, we now have a formal definition of the term "vertical asymptote":

> "f has a vertical asymptote at $x = a$" means that at least one of $\lim_{x \to a^+} f(x)$ and $\lim_{x \to a^-} f(x)$ is equal to ∞ or $-\infty$.

Now, is it possible that even a left-hand or right-hand limit fails to exist? The answer is yes! For example, let's meet the funky function g defined by $g(x) = \sin(1/x)$. What does the graph of this function look like? Let's worry about the positive values of x first. Since $\sin(x)$ has zeroes at the values $x = \pi, 2\pi, 3\pi, \ldots$, then $\sin(1/x)$ will have zeroes when $1/x = \pi, 2\pi, 3\pi, \ldots$. Taking reciprocals, we see that $\sin(1/x)$ has zeroes when $x = \frac{1}{\pi}, \frac{1}{2\pi}, \frac{1}{3\pi}, \ldots$. These numbers are the x-intercepts of $\sin(1/x)$. On the number line, this is what they look like:

As you see, they really bunch up as you get close to 0. Now, $\sin(x)$ goes up to 1 or down to -1 between every x-intercept, so $\sin(1/x)$ does the same. Let's graph what we know so far:

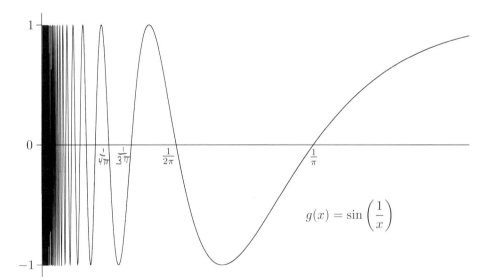

So what is $\lim_{x \to 0^+} \sin(1/x)$? The above graph is a real mess near $x = 0$. It oscillates infinitely often between 1 and -1, faster and faster as you move from the right toward $x = 0$. There is no vertical asymptote, but there's still no limit.* The function doesn't tend toward any one number as x goes to 0 from the right. So, we say that $\lim_{x \to 0^+} \sin(1/x)$ does not exist (DNE). We'll finish the graph of $y = \sin(1/x)$ in the next section.

---

*See Section A.3.4 of Appendix A for a real proof of this.

3.4 Limits at ∞ and $-\infty$

There is one more type of limit that we need to investigate. We've concentrated on the behavior of a function near a point $x = a$. However sometimes it is important to understand how a function behaves when x gets really huge. Another way of saying this is that we are interested in the behavior of a function as its argument x goes to ∞. We'd like to write something like

$$\lim_{x \to \infty} f(x) = L$$

and mean that $f(x)$ gets really close, and stays close, to the value L when x is large. (More details can be found in Section A.3.3 of Appendix A.) The important thing to realize is that writing "$\lim_{x \to \infty} f(x) = L$" indicates that the graph of f has a right-hand horizontal asymptote at $y = L$. There is a similar notion for when x heads toward $-\infty$: we write

$$\lim_{x \to -\infty} f(x) = L,$$

which means that $f(x)$ gets extremely close, and stays close, to L when x gets more and more negative (or more precisely, $-x$ gets larger and larger). This of course corresponds to the graph of $y = f(x)$ having a left-hand horizontal asymptote. You can turn these definitions around if you like and say:

> "f has a right-hand horizontal asymptote at $y = L$" means that $\lim_{x \to \infty} f(x) = L$.
>
> "f has a left-hand horizontal asymptote at $y = M$" means that $\lim_{x \to -\infty} f(x) = M$.

Of course, something like $y = x^2$ doesn't have any horizontal asymptotes: the values of y just go up and up as x gets larger. In symbols, we can write this as $\lim_{x \to \infty} x^2 = \infty$. Alternatively, the limit may not even exist. For example, what is $\lim_{x \to \infty} \sin(x)$? Well, what value is $\sin(x)$ getting closer and closer to (and staying close)? It's just oscillating back and forth between -1 and 1, so it never really gets anywhere. There's no horizontal asymptote, nor does the function wander off to ∞ or $-\infty$; the best you can do is to say that $\lim_{x \to \infty} \sin(x)$ does not exist (DNE). Again, see Section A.3.4 of Appendix A for a proof of this.

Let's return to the function f given by $f(x) = \sin(1/x)$ that we looked at in the previous section. What happens when x gets very large? Well, when x is large, $1/x$ is very close to 0. Since $\sin(0) = 0$, it should be true that $\sin(1/x)$ is also very close to 0. The larger x gets, the closer $\sin(1/x)$ is to 0. My argument has been a little sketchy but hopefully you're convinced that[*]

$$\lim_{x \to \infty} \sin(1/x) = 0.$$

So $\sin(1/x)$ has a horizontal asymptote at $y = 0$. This allows us to extend the graph of $y = \sin(1/x)$ that we drew above, at least to the right. We should

---

[*]If not, see Section A.4.1 of Appendix A!

still worry about what happens when $x < 0$. This isn't too bad, since f is an odd function. Here's why:

$$f(-x) = \sin\left(\frac{1}{-x}\right) = \sin\left(-\frac{1}{x}\right) = -\sin\left(\frac{1}{x}\right) = -f(x).$$

Note that we used the fact that $\sin(x)$ is an odd function of x to get from $\sin(-1/x)$ to $-\sin(1/x)$. So, since odd functions have that nice symmetry about the origin (see Section 1.4 in Chapter 1), we can complete the graph of $y = \sin(1/x)$ as follows:

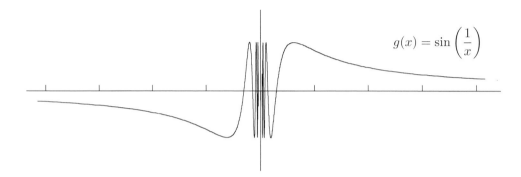

Again, it's hard to draw what happens for x near 0. The closer x is to 0, the more wildly the function oscillates, and of course the function is undefined at $x = 0$. In the above picture, I chose to avoid the black smudge in the middle and just leave the oscillations up to your imagination.

3.4.1 Large numbers and small numbers

I hope we can all agree that 1,000,000,000,000 is a large number. So how about $-1{,}000{,}000{,}000{,}000$? Perhaps controversially, I want you to think of this as a large negative number rather than a small number. An example of a small number would be 0.000000001, while -0.000000001 is a small number too— more precisely, a small negative number. Funnily enough, we're not going to think of 0 itself as being small: it's just zero. So our informal definition of large numbers and small numbers looks like this:

- A number is *large* if its absolute value is a really big number.
- A number is *small* if it is really close to 0 (but not actually equal to 0).

Although the above definition will serve us well in practice, it's a really lame definition. What do I mean by "really big" and "really close to 0"? Well, consider the limit equation

$$\lim_{x \to \infty} f(x) = L.$$

As we saw above, this means that when x is a large enough number, the value of $f(x)$ is almost L. The question is, how large is "large enough"? It depends on how close to L you want $f(x)$ to be! Still, from a practical point of view, a number x is large enough if the graph of $y = f(x)$ starts looking like it's getting serious about snuggling up to the horizontal asymptote at $y = L$. Of course, everything depends on what the function f is, as you can see from the following picture:

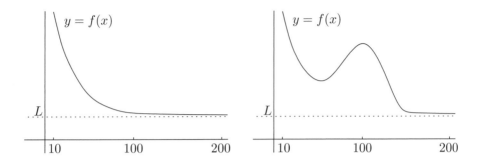

In both cases, $f(10)$ is nowhere near L. In the left-hand picture, it looks like $f(x)$ is pretty close to L when x is at least 100, so any number above 100 would be large. In the right-hand picture, $f(100)$ is far away from L, so now 100 isn't large enough. You probably need to go up to about 200 in this case. So can't you just pick a number like 1,000,000,000,000 and say that it's always large? Nope—a function might wander around until 5,000,000,000,000 before it starts getting close to its horizontal asymptote. The point is that the term "large" has to be taken in context, relative to some function or limit. Luckily, there's plenty of room up above—even a number like 1,000,000,000,000 is pretty puny compared to 10^{100} (a googol), which itself is chicken feed in comparison with $10^{1000000}$, and so on. By the way, we'll often use the term "near ∞" in place of "large and positive." (A number can't really be near ∞ in the literal sense, since ∞ is so far away from everything. The term "near ∞" makes sense, though, in the context of limits as $x \to \infty$.)

Of course, all this also applies to limits as $x \to -\infty$, except that you just stick a minus sign in front of all the large positive numbers above. In this case we'll sometimes say "near $-\infty$" to emphasize that we are referring to large negative numbers.

On the other hand, we'll often be looking at limit equations of the form

$$\lim_{x \to 0} f(x) = L, \qquad \lim_{x \to 0^+} f(x) = L \qquad \text{or} \qquad \lim_{x \to 0^-} f(x) = L.$$

In all three of these cases, we know that when x is close enough to 0, the value of $f(x)$ is almost L. (For the right-hand limit, x also has to be positive, while for the left-hand limit, x has to be negative.) Again, how close does x have to be to 0? It depends on the function f. So, when we say a number is "small" (or "near 0"), we'll have to take this in the context of some function or limit, just as in the case of "large."

Although this discussion really tightens up the above lame definition, it's still not perfect. If you want to learn more, you should really check out Sections A.1 and A.3.3 in Appendix A.

3.5 Two Common Misconceptions about Asymptotes

Now seems like a good time to correct a couple of common misconceptions about horizontal asymptotes. First, a function doesn't have to have the same horizontal asymptote on the left as on the right. In the graph of $f(x) = 1/x$ on page 45 above, there is a horizontal asymptote at $y = 0$ on both the right-hand side and the left-hand side—which means that

$$\lim_{x \to \infty} \frac{1}{x} = 0 \qquad \text{and} \qquad \lim_{x \to \infty} -\frac{1}{x} = 0.$$

However, consider the graph of $y = \tan^{-1}(x)$ (or if you prefer, $y = \arctan(x)$— this is the inverse tangent function and you can write it either way):

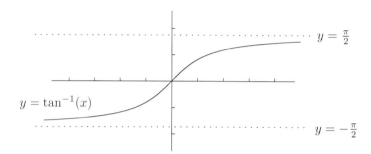

This function has a right-hand horizontal asymptote at $y = \pi/2$ and a left-hand horizontal asymptote at $y = -\pi/2$; these are not the same. We can also express this in terms of limits:

$$\boxed{\lim_{x \to \infty} \tan^{-1}(x) = \frac{\pi}{2}} \qquad \text{and} \qquad \boxed{\lim_{x \to -\infty} \tan^{-1}(x) = -\frac{\pi}{2}}$$

So a function can indeed have different right- and left-hand horizontal asymptotes, but there can be at most two horizontal asymptotes—one on the right and one on the left. It might have none or one: for example, $y = 2^x$ has a left-hand horizontal asymptote but not a right-hand one (see the graph on page 22). This is in contrast to vertical asymptotes: a function can have as many of those as it feels like (for example, $y = \tan(x)$ has infinitely many).

Another common misconception is that a function can't cross its asymptote. Perhaps you have learned that an asymptote is a line that a function gets closer and closer to without ever crossing. This just isn't true, at least when you're talking about horizontal asymptotes. For example, consider the function f given by $f(x) = \sin(x)/x$, where for the moment we only care about what happens when x is positive and large. The value of $\sin(x)$ oscillates between -1 and 1, so the value of $\sin(x)/x$ oscillates between the curves $y = -1/x$ and $y = 1/x$. Also, $\sin(x)/x$ has the same zeroes as $\sin(x)$ does, namely $\pi, 2\pi, 3\pi, \ldots$. Putting it all together, the graph looks like this:

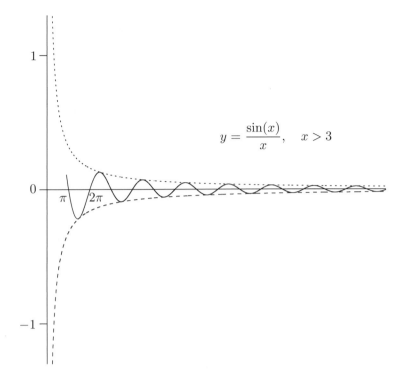

$$y = \frac{\sin(x)}{x}, \quad x > 3$$

The curves $y = 1/x$ and $y = -1/x$, which are drawn as dotted curves in the graph, form what's called the envelope of the sine wave. In any event, as you can see from the graph, if there's any justice in the world, then it should be true that

$$\lim_{x \to \infty} \frac{\sin(x)}{x} = 0.$$

This means that the x-axis is a horizontal asymptote for f, even though the graph of $y = f(x)$ crosses the axis over and over again. Now, to justify the above limit, we'll need to apply something called the sandwich principle. The justification is at the end of the next section.

3.6 The Sandwich Principle

The *sandwich principle*, also known as the *squeeze principle*, says that if a function f is sandwiched between two functions g and h that converge to the same limit L as $x \to a$, then f **also** converges to L as $x \to a$.

Here's a more precise statement of the principle. Suppose that for all x near a, we have $g(x) \le f(x) \le h(x)$. That is, $f(x)$ is sandwiched (or squeezed) between $g(x)$ and $h(x)$. Also, let's suppose that $\lim_{x \to a} g(x) = L$ and $\lim_{x \to a} h(x) = L$. Then we can conclude that $\lim_{x \to a} f(x) = L$; that is, all three functions have the same limit as $x \to a$. As usual, the picture tells the story:

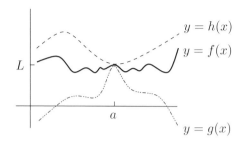

The function f, shown as a solid curve in the picture, is really squished between the other functions g and h; the values of $f(x)$ are forced to tend to L in the limit as $x \to a$. (See Section A.2.4 of Appendix A for a proof of the sandwich principle.)

There's a similar version of the sandwich principle for one-sided limits, except this time the inequality $g(x) \le f(x) \le h(x)$ only has to hold for x on the side of a that you care about. For example, what is

$$\lim_{x \to 0^+} x \sin\left(\frac{1}{x}\right)?$$

The graph of $y = x \sin(1/x)$ is similar to that of $y = \sin(1/x)$ but now there is the factor of x which causes the function to be trapped between the envelopes of $y = x$ and $y = -x$. Here's what the graph looks like for x between 0 and 0.3:

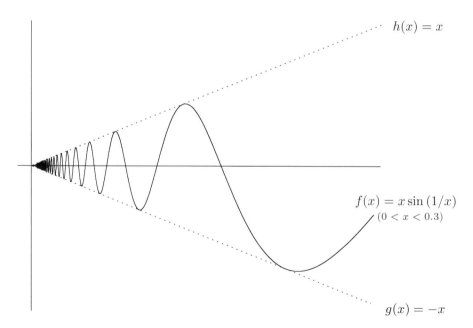

We still have the wild oscillations as x goes to 0 from above, but they are now damped by the envelope lines. In particular, finding the limit we want is a perfect application of the sandwich principle. The function g is the lower envelope line $y = -x$, and the function h is the upper envelope line $y = x$. We need to show that $g(x) \le f(x) \le h(x)$ for $x > 0$. We don't care about $x < 0$ since we only need the right-hand limit of $f(x)$ at $x = 0$. (Indeed, if you extend the lines to negative x, you can see that $g(x)$ is actually greater than $h(x)$ for $x < 0$, so the sandwich is the wrong way around!) So, how do we show that $g(x) \le f(x) \le h(x)$ when $x > 0$? We'll use the fact that the sine of any number (in our case, $1/x$) is between -1 and 1 inclusive:

$$-1 \le \sin\left(\frac{1}{x}\right) \le 1.$$

Now multiply this inequality through by x, which is cool because $x > 0$; we get

$$-x \le x \sin\left(\frac{1}{x}\right) \le x.$$

But this is precisely $g(x) \le f(x) \le h(x)$, which is what we need. Finally, note that

$$\lim_{x \to 0^+} g(x) = \lim_{x \to 0^+} (-x) = 0 \quad \text{and} \quad \lim_{x \to 0^+} h(x) = \lim_{x \to 0^+} x = 0.$$

So, since the values $g(x)$ and $h(x)$ of the sandwiching functions converge to the same number, 0, as $x \to 0^+$, so does $f(x)$. That is, we've shown that

$$\lim_{x \to 0^+} x \sin\left(\frac{1}{x}\right) = 0.$$

Remember, this certainly isn't true without the factor x out front; the limit of $\sin(1/x)$ as $x \to 0^+$ does not exist, as we saw in Section 3.3 above.

 We still haven't resolved the issue of justifying the limit from the end of the previous section! Remember, we wanted to show that

$$\lim_{x \to \infty} \frac{\sin(x)}{x} = 0.$$

To do this, we have to invoke a slightly different form of the sandwich principle, involving limits at ∞. In this case we need $g(x) \le f(x) \le h(x)$ to be true for all large x; then if we know that $\lim_{x \to \infty} g(x) = L$ and $\lim_{x \to \infty} h(x) = L$, we can also say that $\lim_{x \to \infty} f(x) = L$. This is almost the same as the sandwich principle for finite limits. To establish the above limit, we again use the fact that $-1 \le \sin(x) \le 1$ for all x, but this time we divide by x to get

$$-\frac{1}{x} \le \frac{\sin(x)}{x} \le \frac{1}{x}$$

for all $x > 0$. Now let $x \to \infty$; since both $-1/x$ and $1/x$ have 0 as their limit, the same must be true for $\sin(x)/x$. That is, since

$$\lim_{x \to \infty} -\frac{1}{x} = 0 \quad \text{and} \quad \lim_{x \to \infty} \frac{1}{x} = 0,$$

we must also have

$$\lim_{x \to \infty} \frac{\sin(x)}{x} = 0.$$

In summary, here's what the sandwich principle says:

> If $g(x) \leq f(x) \leq h(x)$ for all x near a,
> and $\lim_{x \to a} g(x) = \lim_{x \to a} h(x) = L$, then
> $$\lim_{x \to a} f(x) = L.$$

This also works for left-hand or right-hand limits; in that case, the inequality only has to be true for x on the appropriate side of a. It also works when a is ∞ or $-\infty$; in that case, the inequality has to be true for x really large (positively or negatively, respectively).

3.7 Summary of Basic Types of Limits

 We have looked at a whole bunch of different basic types of limits. Let's finish this chapter with some representative diagrams showing the most common possibilities:

1. The right-hand limit at $x = a$. Behavior of $f(x)$ to the left of $x = a$, and at $x = a$ itself, is irrelevant. (This means that it doesn't matter what values $f(x)$ takes for $x \leq a$, as far as the right-hand limit is concerned. In fact, $f(x)$ need not even be defined for $x \leq a$.)

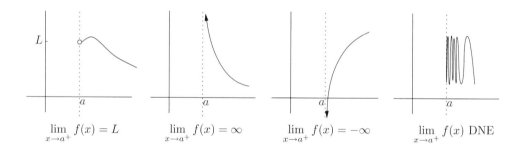

$$\lim_{x \to a^+} f(x) = L \qquad \lim_{x \to a^+} f(x) = \infty \qquad \lim_{x \to a^+} f(x) = -\infty \qquad \lim_{x \to a^+} f(x) \text{ DNE}$$

2. The left-hand limit at $x = a$. Behavior of $f(x)$ to the right of $x = a$, and at $x = a$ itself, is irrelevant.

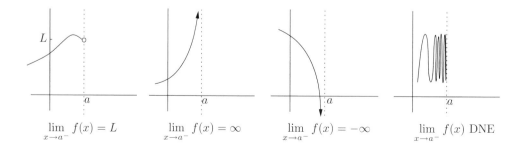

$$\lim_{x \to a^-} f(x) = L \qquad \lim_{x \to a^-} f(x) = \infty \qquad \lim_{x \to a^-} f(x) = -\infty \qquad \lim_{x \to a^-} f(x) \text{ DNE}$$

3. **The two-sided limit at $x = a$.** In the first picture below, the left- and right-hand limits exist but are different, so the two-sided limit does not exist. In the second picture, the left- and right-hand limits agree, so the two-sided limit exists and is equal to the common value. The value of $f(a)$ is irrelevant.

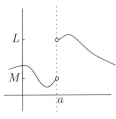

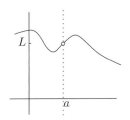

$$\left.\begin{array}{l} \lim_{x \to a^-} f(x) = M \\ \lim_{x \to a^+} f(x) = L \end{array}\right\} \lim_{x \to a} f(x) \text{ DNE}$$

$$\left.\begin{array}{l} \lim_{x \to a^-} f(x) = L \\ \lim_{x \to a^+} f(x) = L \end{array}\right\} \lim_{x \to a} f(x) = L$$

4. The limit as $x \to \infty$.

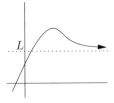

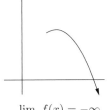

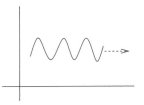

$$\lim_{x \to \infty} f(x) = L \qquad \lim_{x \to \infty} f(x) = \infty \qquad \lim_{x \to \infty} f(x) = -\infty \qquad \lim_{x \to \infty} f(x) \text{ DNE}$$

5. The limit as $x \to -\infty$.

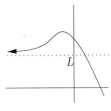

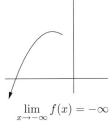

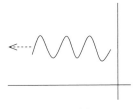

$$\lim_{x \to -\infty} f(x) = L \qquad \lim_{x \to -\infty} f(x) = \infty \qquad \lim_{x \to -\infty} f(x) = -\infty \qquad \lim_{x \to -\infty} f(x) \text{ DNE}$$

CHAPTER 4

How to Solve Limit Problems Involving Polynomials

In the previous chapter, we looked at limits from a mostly conceptual viewpoint. Now it's time to see some of the techniques used to evaluate limits. For the moment, we'll concentrate on limits involving polynomials; later on we'll see how to deal with trig functions, exponentials, and logarithms. As we'll see in the next chapter, differentiation involves taking limits of ratios, so most of our focus will be on this type of limit.

When you're taking the limit of a ratio of two polynomials, it's really important to notice where the limit is being taken. In particular, the techniques for dealing with $x \to \infty$ and $x \to a$ (for some finite a) are completely different. So, we'll split up our plan of attack into limits involving the following types of functions:

- rational functions as $x \to a$;
- functions involving square roots as $x \to a$;
- rational functions as $x \to \infty$;
- ratios of polynomial-like (or "poly-type") functions as $x \to \infty$;
- rational functions/poly-type functions as $x \to -\infty$; and
- functions involving absolute values.

4.1 Limits Involving Rational Functions as $x \to a$

Let's start off with limits that look like this:

$$\lim_{x \to a} \frac{p(x)}{q(x)},$$

where p and q are polynomials and a is a finite number. (Remember that the quotient $p(x)/q(x)$ of two polynomials is called a rational function.) The first thing you should always try is to substitute the value of a for x. If the denominator isn't 0, then you're in good shape—the value of the limit is just what you get when you substitute. For example, what is

$$\lim_{x \to -1} \frac{x^2 - 3x + 2}{x - 2}?$$

Simply plug $x = -1$ into the expression $(x^2 - 3x + 2)/(x - 2)$, and you get

$$\frac{(-1)^2 - 3(-1) + 2}{-1 - 2} = \frac{6}{-3} = -2.$$

The denominator isn't 0, so -2 is the value of the limit. (I know that I said in the previous chapter that the value of the function at the limit point, which is $x = -1$ in this case, is irrelevant; but in the next chapter we'll look at the concept of continuity, which will justify this "plugging-in" method.)

On the other hand, if you want to find

$$\lim_{x \to 2} \frac{x^2 - 3x + 2}{x - 2},$$

then plugging in $x = 2$ won't work so well: you get $(4 - 6 + 2)/(2 - 2)$, which simplifies down to $0/0$. This is called an *indeterminate form*. If you use the plugging in method and get zero divided by zero, then anything could happen: the limit might be finite, the limit might be ∞ or $-\infty$, or the limit might not exist. The above example can be solved by the important technique of factoring everything in sight. In particular, $x^2 - 3x + 2$ can be factored as $(x - 2)(x - 1)$, so we can write

$$\lim_{x \to 2} \frac{x^2 - 3x + 2}{x - 2} = \lim_{x \to -2} \frac{(x - 2)(x - 1)}{x - 2} = \lim_{x \to 2} (x - 1)$$

by canceling. Now there's no impediment to plugging $x = 2$ into the expression $(x - 1)$; you just get $2 - 1$, which equals 1. That's the value of our limit.

This brings us to a point which is often misunderstood: are the two functions f and g defined by

$$f(x) = \frac{x^2 - 3x + 2}{x - 2} \qquad \text{and} \qquad g(x) = x - 1$$

the same function? Why can't you say that

$$f(x) = \frac{x^2 - 3x + 2}{x - 2} = \frac{(x - 2)(x - 1)}{x - 2} = x - 1 = g(x)?$$

Well, you almost can! The only problem is when $x = 2$, because then the denominator $(x - 2)$ is equal to 0 and that doesn't make sense. So f and g are not the same function: the number 2 is not in the domain of f but it is in the domain of g. (We've actually encountered this function f before—check out the discussion and graph at the beginning of Chapter 3.) On the other hand, if you put limits in front of everything in the above chain of equations, it all becomes correct because the values of $f(x)$ and $g(x)$ at $x = 2$ don't matter—it's only the values of $f(x)$ and $g(x)$ **near** $x = 2$ that count. So the solution of the previous limit problem is indeed valid.

Let's look at another example of an indeterminate form. Again, the technique is to try to factor everything in sight. In addition to knowing how to factor quadratics, it's really useful to know the formula for the difference of two cubes:

$$a^3 - b^3 = (a - b)(a^2 + ab + b^2).$$

Here's a harder example where you need to use this formula: find

$$\lim_{x \to 3} \frac{x^3 - 27}{x^4 - 5x^3 + 6x^2}.$$

If you plug in $x = 3$, you indeed get $0/0$ (try it and see). So let's try to factor both the numerator and the denominator. The numerator is the difference between x^3 and 3^3, so we can use the boxed formula above. The denominator has an obvious factor of x^2, so it can be written as $x^2(x^2 - 5x + 6)$. The quadratic $x^2 - 5x + 6$ can also be factored; altogether, then, you should convince yourself that we have

$$\lim_{x \to 3} \frac{x^3 - 27}{x^4 - 5x^3 + 6x^2} = \lim_{x \to 3} \frac{(x - 3)(x^2 + 3x + 9)}{x^2(x - 3)(x - 2)}.$$

Substituting $x = 3$ doesn't work because of the factor of $(x - 3)$ in the denominator. On the other hand, since we are taking limits, we only need to see what happens when x is near 3; so we are perfectly justified in canceling out the factors of $(x - 3)$ from the numerator and denominator—they are never equal to 0. So, using the plugging-in technique after factoring and canceling, the whole solution looks like this:

$$\lim_{x \to 3} \frac{x^3 - 27}{x^4 - 5x^3 + 6x^2} = \lim_{x \to 3} \frac{(x - 3)(x^2 + 3x + 9)}{x^2(x - 3)(x - 2)} = \lim_{x \to 3} \frac{x^2 + 3x + 9}{x^2(x - 2)}$$

$$= \frac{3^2 + 3 \cdot 3 + 9}{3^2(3 - 2)} = 3.$$

What if the denominator is 0 but the numerator isn't 0? In that case, there's always a vertical asymptote involved; that is, the graph of the rational function will have a vertical asymptote at the value of x that you're interested in. The problem is that there are four types of behavior that could arise. In each of the following diagrams, f is the rational function we care about, and the various limits at $x = a$ are shown under the picture:

$$\lim_{x \to a^+} f(x) = \infty \qquad \lim_{x \to a^+} f(x) = \infty \qquad \lim_{x \to a^+} f(x) = -\infty \qquad \lim_{x \to a^+} f(x) = -\infty$$

$$\lim_{x \to a^-} f(x) = -\infty \qquad \lim_{x \to a^-} f(x) = \infty \qquad \lim_{x \to a^-} f(x) = \infty \qquad \lim_{x \to a^-} f(x) = -\infty$$

$$\lim_{x \to a} f(x) \text{ DNE} \qquad \lim_{x \to a} f(x) = \infty \qquad \lim_{x \to a} f(x) \text{ DNE} \qquad \lim_{x \to a} f(x) = -\infty$$

So, how do you tell which of the four cases you're dealing with? You just have to explore the sign of $f(x)$ on either side of $x = a$. If it's positive on both sides, for example, then you must be in the second case above. Here's an actual example: how would you find

$$\lim_{x \to 1} \frac{2x^2 - x - 6}{x(x-1)^3}?$$

First, plugging in $x = 1$ gives $-5/0$ (try it!). So we must be dealing with one of the four cases above. Which one? Let's set $f(x) = (2x^2 - x - 6)/(x(x-1)^3)$ and see what happens when we move x around near 1. The first thing to notice is that the numerator $2x^2 - x - 6$ is actually equal to -5 when $x = 1$, so when we wobble x around a little bit, the numerator will stay negative. How about the factor of x in the denominator? When $x = 1$, this factor is of course 1, which is positive—and it stays positive when you move x around a bit. The crucial factor is $(x - 1)^3$. This is positive when $x > 1$ but negative when $x < 1$. So we can summarize the situation like this (using $(+)$ and $(-)$ to denote positive and negative quantities, respectively, and of course using the fact that $(-) \cdot (-) = (+)$ and so on):

$$\text{when } x > 1: \frac{(-)}{(+) \cdot (+)} = (-); \qquad \text{when } x < 1: \frac{(-)}{(+) \cdot (-)} = (+).$$

That is, $f(x)$ is negative when x is a little greater than 1, but positive when x is a little less than 1. Look up at the four pictures above—the only one that works is the third figure. In particular, we can see that the two-sided limit

$$\lim_{x \to 1} \frac{2x^2 - x - 6}{x(x-1)^3}$$

does not exist, but the one-sided limits do (although they are infinite); in particular,

$$\lim_{x \to 1+} \frac{2x^2 - x - 6}{x(x-1)^3} = -\infty \qquad \text{and} \qquad \lim_{x \to 1-} \frac{2x^2 - x - 6}{x(x-1)^3} = \infty.$$

Now suppose we change the limit slightly to

$$\lim_{x \to 1} \frac{2x^2 - x - 6}{x(x-1)^2}.$$

How does that change anything? Well, the numerator is still negative when x is near 1, and the factor x is still positive, but how about $(x-1)^2$? Since it's a square, it must be positive when x is near but not equal to 1. So we now have the following situation:

$$\text{when } x > 1: \frac{(-)}{(+) \cdot (+)} = (-); \qquad \text{when } x < 1: \frac{(-)}{(+) \cdot (+)} = (-).$$

Now we have negative values on either side of $x = 1$, so we must have

$$\lim_{x \to 1} \frac{2x^2 - x - 6}{x(x-1)^2} = -\infty.$$

Of course, the left- and right-hand limits are both equal to $-\infty$ as well.

4.2 Limits Involving Square Roots as $x \to a$

Consider the following limit:

$$\lim_{x \to 5} \frac{\sqrt{x^2 - 9} - 4}{x - 5}.$$

If you plug in $x = 5$, you get the indeterminate form $0/0$ (try it and see!). Trying to factor everything in sight doesn't work so well—you can write $x^2 - 9$ as $(x-3)(x+3)$, but that doesn't really help because of that blasted -4 in the numerator. What you need to do is multiply and divide by $\sqrt{x^2 - 9} + 4$; this is called the *conjugate expression* of $\sqrt{x^2 - 9} - 4$. (You have probably already met conjugate expressions in your math studies, especially when rationalizing the denominator. The basic idea is that the conjugate expression of $a - b$ is $a + b$, and vice versa.) So, here's what we get when we do this multiplication and division:

$$\lim_{x \to 5} \frac{\sqrt{x^2 - 9} - 4}{x - 5} = \lim_{x \to 5} \frac{\sqrt{x^2 - 9} - 4}{x - 5} \times \frac{\sqrt{x^2 - 9} + 4}{\sqrt{x^2 - 9} + 4}.$$

This looks more complicated, but something nice happens: using the formula $(a - b)(a + b) = a^2 - b^2$, the numerator simplifies to $(\sqrt{x^2 - 9})^2 - 4^2$, or simply $x^2 - 25$. So the above limit is just

$$\lim_{x \to 5} \frac{x^2 - 25}{(x - 5)(\sqrt{x^2 - 9} + 4)}.$$

Factor $x^2 - 25$ as $(x - 5)(x + 5)$ and cancel to see that this limit becomes

$$\lim_{x \to 5} \frac{(x - 5)(x + 5)}{(x - 5)(\sqrt{x^2 - 9} + 4)} = \lim_{x \to 5} \frac{x + 5}{\sqrt{x^2 - 9} + 4}.$$

Now if you substitute $x = 5$, there are no problems: you simply get $10/8$, or $5/4$. The moral of the story is that if you have a square root plus or minus another quantity, try multiplying and dividing by its conjugate—you might be pleasantly surprised!

4.3 Limits Involving Rational Functions as $x \to \infty$

OK, back to rational functions, but this time we'll look at what happens as $x \to \infty$ instead of some finite value. In symbols, we are now trying to find limits of the form

$$\lim_{x \to \infty} \frac{p(x)}{q(x)},$$

where p and q are polynomials. Now, here's a very important property of a polynomial: **when x is large, the leading term dominates**. What this means is that if you have a polynomial p, then as x gets larger and larger, $p(x)$ behaves as if only its leading term were present. For example, let's say $p(x) = 3x^3 - 1000x^2 + 5x - 7$. Let's put $p_L(x) = 3x^3$, which is the leading term of p. Here's what I'm claiming: when x is really really large, $p(x)$ and $p_L(x)$ are relatively close to each other. More precisely, we have

$$\lim_{x \to \infty} \frac{p(x)}{p_L(x)} = 1.$$

Before we see why this is true, let's just look at the implications of what it is saying. Imagine that the limit wasn't there. This equation would say

$$\frac{p(x)}{p_L(x)} = 1,$$

which means that $p(x) = p_L(x)$. Well, that clearly isn't true (at least for most values of x), but the larger x is, the closer it is to being true. So why not just write

$$\lim_{x \to \infty} p(x) = \lim_{x \to \infty} p_L(x)?$$

This is actually true, but it's meaningless because both sides are ∞. So we have to settle for saying that $p(x)$ and $p_L(x)$ are very close to each other in the sense that their **ratio is close to** 1. As x gets large, the ratio approaches 1 without ever necessarily equaling 1.

Does this make sense? Why is it the leading term, anyway? Why not one of the other terms? If you want, you can skip to the next paragraph and see the mathematical proof; first, however, I'd like to get a feel for what happens in our example, $p(x) = 3x^3 - 1000x^2 + 5x - 7$, by testing it on actual large values of x. Let's start off with $x = 100$. In that case, $3x^3$ is 3 million, while $1000x^2$ is 10 million. The quantity $5x$ is only 500, and the 7 doesn't make much difference, so all together we can see that $p(100)$ is about -7 million. On the other hand, $p_L(100)$ is 3 million, so it's not looking so great: $p(100)$ and $p_L(100)$ are completely different. Let's not lose heart—after all, 100 isn't that large. Suppose we instead set x equal to 1,000,000—that's a million. Then $3x^3$ is freakin' huge: it's 3,000,000,000,000,000,000, or three million trillion! In comparison, $1000x^2$ is relatively puny at only a thousand trillion (that's 1,000,000,000,000,000) and $5x$ is only 5 million, which is a microscopic speck of dust in comparison to these numbers. The -7 term is just laughable and makes no noticeable difference. So, to calculate $p(1{,}000{,}000)$, we need to take 3 million trillion and take away a thousand trillion plus some spare change (a little under 5 million). Let's face it, it's still darned close to 3 million trillion! After all, how many trillions are we dealing with here? We have 3 million of them, and we're taking away a mere one thousand of them, so we still have almost 3 million trillions. That is, $p(1{,}000{,}000)$ is about 3 million trillion—but that is exactly the value of $p_L(1{,}000{,}000)$. The point is that the highest-degree term is growing much faster than the other terms as x gets large. Indeed, if you replace 1,000,000 with an even larger number, the difference between x^3 and the lower order terms like x^2 and x becomes even more pronounced.

Enough philosophical rambling. Let's try to give a real proof that

$$\lim_{x \to \infty} \frac{p(x)}{p_L(x)} = 1.$$

We have to do some actual math. Start off by writing

$$\lim_{x \to \infty} \frac{p(x)}{p_L(x)} = \lim_{x \to \infty} \frac{3x^3 - 1000x^2 + 5x - 7}{3x^3}$$

which simplifies to

$$\lim_{x\to\infty} \left(\frac{3x^3}{3x^3} - \frac{1000x^2}{3x^3} + \frac{5x}{3x^3} - \frac{7}{3x^3} \right) = \lim_{x\to\infty} \left(1 - \frac{1000}{3x} + \frac{5}{3x^2} - \frac{7}{3x^3} \right).$$

How do you handle this? Well, the first thing to note is that you can bust up this last expression into four separate limits. So if you know what happens to the four quantities 1, $-1000/3x$, $5/3x^2$, and $-7/3x^3$ as x becomes very large, then you can just add the four limits together to get the limit you want. Technically, this could be described in words as "the limit of the sum is equal to the sum of the limits"; this is true when all the limits are finite.*
So, we have four quantities to worry about. The first is 1, which is always 1 regardless of what happens to x. The second quantity is $-1000/3x$. What happens to this when x gets large? That is, what is

$$\lim_{x\to\infty} -\frac{1000}{3x}?$$

The trick here is to realize that you can take out a factor of $-1000/3$. In particular, the limit can be expressed as

$$\lim_{x\to\infty} -\frac{1000}{3}\frac{1}{x}.$$

The cool thing about something like $-1000/3$ is that it's constant. It doesn't change, no matter what x is, so it turns out that you can just go ahead and drag it out of the limit (see Section A.2.2 of Appendix A for more details). So we have

$$\lim_{x\to\infty} -\frac{1000}{3}\frac{1}{x} = -\frac{1000}{3}\lim_{x\to\infty}\frac{1}{x}.$$

We've already seen that the reciprocal of a very large number is a very small number (remember, this means a number very close to zero). So $\lim_{x\to\infty} 1/x = 0$, and $-1000/3$ times the limit is also 0. The conclusion is that

$$\lim_{x\to\infty} -\frac{1000}{3x} = 0.$$

In fact, you should just write that down without going into any more detail. More generally, you can use the following theorem:

$$\boxed{\lim_{x\to\infty} \frac{C}{x^n} = 0}$$

for any $n > 0$, as long as C is constant. This fact allows us to see that the other two pieces, $5/3x^2$ and $-7/3x^3$, also tend to 0 as x becomes very large. So the whole argument is

$$\lim_{x\to\infty} \frac{3x^3 - 1000x^2 + 5x - 7}{3x^3} = \lim_{x\to\infty} \left(1 - \frac{1000}{3x} + \frac{5}{3x^2} - \frac{7}{3x^3} \right)$$
$$= 1 - 0 + 0 + 0 = 1.$$

---

*It's not true if the limits aren't finite! Consider $\lim_{x\to\infty}(x + (1 - x))$. For any x, it's true that $(x + (1 - x)) = 1$, so this limit is just 1. On the other hand, the individual limits of the two pieces (x) and $(1 - x)$ are $\lim_{x\to\infty}(x)$ and $\lim_{x\to\infty}(1 - x)$. The first limit is ∞ and the second is $-\infty$, but it's not true that $\infty + (-\infty) = 1$. In fact, the expression $\infty + (-\infty)$ is meaningless.

So we have proved that

$$\lim_{x \to \infty} \frac{p(x)}{\text{leading term of } p(x)} = 1$$

in the special case where $p(x) = 3x^3 - 1000x^2 + 5x - 7$. Luckily the same method works for any polynomial, and we'll be using it over and over again during the rest of this chapter!

4.3.1 Method and examples

Here's the general idea: when you see $p(x)$ for some polynomial p with more than one term, replace it by

$$\frac{p(x)}{\text{leading term of } p(x)} \times (\text{leading term of } p(x)).$$

Do this for **every** polynomial around! Note that all we've done is to divide and multiply by the leading term, so we haven't changed the quantity $p(x)$. The point is that the fraction in the expression above has limit 1 as $x \to \infty$, and the leading term is much simpler. Let's see how this works in practice: for example, what is

$$\lim_{x \to \infty} \frac{x - 8x^4}{7x^4 + 5x^3 + 2000x^2 - 6}?$$

We have two polynomials: one on the top and one on the bottom. For the numerator, the leading term is $-8x^4$ (don't be fooled by the order in which the numerator is written—the leading term isn't always written first!). So we're going to replace the numerator by

$$\frac{x - 8x^4}{-8x^4} \times (-8x^4).$$

Similarly, the denominator has leading term $7x^4$, so we'll replace it by

$$\frac{7x^4 + 5x^3 + 2000x^2 - 6}{7x^4} \times (7x^4).$$

Making both these replacements leads to this:

$$\lim_{x \to \infty} \frac{x - 8x^4}{7x^4 + 5x^3 + 2000x^2 - 6} = \lim_{x \to \infty} \frac{\dfrac{x - 8x^4}{-8x^4} \times (-8x^4)}{\dfrac{7x^4 + 5x^3 + 2000x^2 - 6}{7x^4} \times (7x^4)}.$$

Looking at this, you should concentrate on the ratio

$$\frac{-8x^4}{7x^4},$$

because that's what's **really** going on here. The other fractions all have limit 1, but we have effectively squeezed all the important juice out of our two polynomials into the simple ratio of leading terms. Luckily that ratio just simplifies to $-8/7$, so that should be our answer. To nail that down, we have to prove that the other fractions have limit 1, but that's no problem. You see, in each of the little fractions, we can do the division and we see that our above limit can be written as

$$\lim_{x \to \infty} \frac{-\dfrac{1}{8x^3} + 1}{1 + \dfrac{5}{7x} + \dfrac{2000}{7x^2} - \dfrac{6}{7x^4}} \times \frac{-8x^4}{7x^4}.$$

Now we take limits; from the fact in the box in the previous section, any expression of the form C/x^n goes to 0 as $x \to \infty$ (provided that C is constant and $n > 0$). So most of the stuff goes away! We also cancel out the x^4 factor on the right to see that we are reduced to

$$\frac{0+1}{1+0+0-0} \times \frac{-8}{7} = \frac{1}{1} \times \frac{-8}{7} = \frac{-8}{7}$$

and we're all done.

Here's another example: find

$$\lim_{x \to \infty} \frac{(x^4 + 3x - 99)(2 - x^5)}{(18x^7 + 9x^6 - 3x^2 - 1)(x + 1)}.$$

We have four polynomials here, with leading terms x^4, $-x^5$, $18x^7$, and x. So we'll use our method for each one of them! Try it and see for yourself before reading further. Even if you don't, make sure you understand every step of the argument below:

$$\lim_{x \to \infty} \frac{(x^4 + 3x - 99)(2 - x^5)}{(18x^7 + 9x^6 - 3x^2 - 1)(x + 1)}$$

$$= \lim_{x \to \infty} \frac{\left(\dfrac{x^4 + 3x - 99}{x^4} \times (x^4)\right)\left(\dfrac{2 - x^5}{-x^5} \times (-x^5)\right)}{\left(\dfrac{18x^7 + 9x^6 - 3x^2 - 1}{18x^7} \times (18x^7)\right)\left(\dfrac{x + 1}{x} \times (x)\right)}$$

$$= \lim_{x \to \infty} \frac{\left(1 + \dfrac{3}{x^3} - \dfrac{99}{x^4}\right)\left(-\dfrac{2}{x^5} + 1\right)}{\left(1 + \dfrac{9}{18x} - \dfrac{3}{18x^5} - \dfrac{1}{18x^7}\right)\left(1 + \dfrac{1}{x}\right)} \times \frac{(x^4)(-x^5)}{(18x^7)(x)}$$

$$= \frac{(1 + 0 - 0)(0 + 1)}{(1 + 0 - 0 - 0)(1 + 0)} \times \lim_{x \to \infty} \frac{-x}{18} = \lim_{x \to \infty} \frac{-x}{18} = -\infty.$$

The main point is that we boiled out the leading terms into the ratio

$$\frac{(x^4)(-x^5)}{(18x^7)(x)},$$

which simplifies to $-x/18$. Everything else had no effect! Finally, when $x \to \infty$, the quantity $-x/18$ goes to $-\infty$, so that's the "value" of the limit we're looking for.

In the previous two examples, we've seen that the limit might be finite and nonzero (we got the answer $-8/7$) or it might be infinite (we got the answer $-\infty$). Let's look at the degree of the polynomials in these examples. In the first example, both the numerator and the denominator were of degree 4. In the second example, the numerator is the product of polynomials of degree 4 and degree 5, so if you multiply it out, you get a polynomial of degree 9. Similarly, the denominator is the product of polynomials of degree 7 and degree 1, so it has total degree 8. In this case, the numerator is of greater degree than the denominator. On the other hand, consider this limit:

$$\lim_{x \to \infty} \frac{2x+3}{x^2-7}.$$

Let's use our methods to solve it:

$$\lim_{x \to \infty} \frac{2x+3}{x^2-7} = \lim_{x \to \infty} \frac{\dfrac{2x+3}{2x} \times (2x)}{\dfrac{x^2-7}{x^2} \times (x^2)} = \lim_{x \to \infty} \left(\frac{1 + \dfrac{3}{2x}}{1 - \dfrac{7}{x^2}} \right) \times \frac{2x}{x^2}$$

$$= \frac{1+0}{1-0} \times \lim_{x \to \infty} \frac{2}{x} = 0.$$

Here, the denominator has degree 2, which is greater than the numerator's degree (which is 1). The result is that the denominator dominates, so the limit is 0. In general, here's what we can say considering the limit

$$\lim_{x \to \infty} \frac{p(x)}{q(x)}$$

where p and q are polynomials:

1. If the degree of p equals the degree of q, the limit is finite and nonzero.
2. If the degree of p is greater than the degree of q, the limit is ∞ or $-\infty$.
3. If the degree of p is less than the degree of q, the limit is 0.

(All this is also true when $x \to -\infty$, so that the limit is

$$\lim_{x \to -\infty} \frac{p(x)}{q(x)};$$

we'll consider this case in Section 4.5 below.) These facts are easily proved in general using the above methods. Useful as these facts are, you really don't need them to solve problems; you should use the dividing and multiplying method, then use the facts to check that your answer makes sense.

4.4 Limits Involving Poly-type Functions as $x \to \infty$

Consider functions f, g and h defined by

$$f(x) = x^3 + 4x^2 - 5x^{2/3} + 1, \quad g(x) = \sqrt{x^9 - 7x^2 + 2},$$

$$\text{and} \qquad h(x) = x^4 - \sqrt{x^3 + \sqrt[5]{x^2 - 2x + 3}}.$$

These aren't polynomials because they involve fractional powers or nth roots, but they look a little like polynomials. In fact, the methods of the previous section work on these objects as well, so I'll call them "poly-type functions."

The principles for poly-type functions are similar to those for polynomials, except that this time it may not be so clear what the leading term is. The presence of square roots (or cube roots, fourth roots, and so on) can have a big impact on this. For example, let's consider

$$\lim_{x \to \infty} \frac{\sqrt{16x^4 + 8} + 3x}{2x^2 + 6x + 1}.$$

The bottom is a polynomial with leading term $2x^2$, so we'll replace it by

$$\frac{2x^2 + 6x + 1}{2x^2} \times (2x^2).$$

How about the top? The part under the square root is a polynomial, $16x^4 + 8$, and its leading term is $16x^4$. If you take the square root of that, you get $4x^2$. So mentally you should think of the top as behaving like $4x^2 + 3x$. The leading term of that is $4x^2$, so that's what we're going to use. Specifically, we will replace the top by

$$\frac{\sqrt{16x^4 + 8} + 3x}{4x^2} \times (4x^2).$$

How do you simplify the first fraction? The answer is that you can drag the $4x^2$ under the square root, and it becomes $16x^4$:

$$\frac{\sqrt{16x^4 + 8} + 3x}{4x^2} = \frac{\sqrt{16x^4 + 8}}{4x^2} + \frac{3x}{4x^2} = \sqrt{\frac{16x^4 + 8}{16x^4}} + \frac{3x}{4x^2}.$$

Now if you split up more and cancel, you can reduce this to

$$\sqrt{1 + \frac{8}{16x^4}} + \frac{3}{4x}.$$

As $x \to \infty$, the parts with x on the bottom just go away, so this expression goes to

$$\sqrt{1 + 0} + 0 = 1.$$

So, let's put it all together and write out the solution to the original problem:

$$\lim_{x \to \infty} \frac{\sqrt{16x^4 + 8} + 3x}{2x^2 + 6x + 1} = \lim_{x \to \infty} \frac{\dfrac{\sqrt{16x^4 + 8} + 3x}{4x^2} \times (4x^2)}{\dfrac{2x^2 + 6x + 1}{2x^2} \times (2x^2)}$$

$$= \lim_{x \to \infty} \frac{\sqrt{\dfrac{16x^4 + 8}{16x^4}} + \dfrac{3x}{4x^2}}{\dfrac{2x^2 + 6x + 1}{2x^2}} \times \frac{4x^2}{2x^2} = \lim_{x \to \infty} \frac{\sqrt{1 + \dfrac{8}{16x^4}} + \dfrac{3}{4x}}{1 + \dfrac{6}{2x} + \dfrac{1}{2x^2}} \times \frac{4}{2}$$

$$= \frac{\sqrt{1 + 0} + 0}{1 + 0 + 0} \times 2 = 2.$$

Nice, huh? Messy, but nice. Now let's see what happens when we modify the situation very slightly. Consider

$$\lim_{x \to \infty} \frac{\sqrt{16x^4 + 8} + 3x^3}{2x^2 + 6x + 1}.$$

The only change is that the $3x$ term in the numerator in the previous example has become $3x^3$. How does this affect things? Well, we already said that the $\sqrt{16x^4 + 8}$ term behaves like $4x^2$ for large x, but this time it gets swamped by the higher-degree term $3x^3$. So now we have to replace the top by

$$\frac{\sqrt{16x^4 + 8} + 3x^3}{3x^3} \times (3x^3);$$

of course, when we drag $3x^3$ under the square root, it will become $9x^6$. All together, then, the solution looks like this:

$$\lim_{x \to \infty} \frac{\sqrt{16x^4 + 8} + 3x^3}{2x^2 + 6x + 1} = \lim_{x \to \infty} \frac{\dfrac{\sqrt{16x^4 + 8} + 3x^3}{3x^3} \times (3x^3)}{\dfrac{2x^2 + 6x + 1}{2x^2} \times (2x^2)}$$

$$= \lim_{x \to \infty} \frac{\sqrt{\dfrac{16x^4 + 8}{9x^6}} + \dfrac{3x^3}{3x^3}}{\dfrac{2x^2 + 6x + 1}{2x^2}} \times \frac{3x^3}{2x^2} = \lim_{x \to \infty} \frac{\sqrt{\dfrac{16}{x^2} + \dfrac{8}{9x^6}} + 1}{1 + \dfrac{6}{2x} + \dfrac{1}{2x^2}} \times \frac{3x}{2}$$

$$= \frac{\sqrt{0 + 0} + 1}{1 + 0 + 0} \times \lim_{x \to \infty} \frac{3x}{2} = \infty.$$

Make sure you understand each step of the last two solutions. In the first example, the leading term came from the $16x^4$ under the square root; even when you take the square root, the resulting term $4x^2$ still dominated the rest of the numerator ($3x$). In the second example, the rest of the numerator

($3x^3$) was the dominant force. But wait, you say—what if they are the same? For example, what is

$$\lim_{x \to \infty} \frac{\sqrt{4x^6 - 5x^5} - 2x^3}{\sqrt[3]{27x^6 + 8x}}?$$

The denominator isn't too nasty, actually, but let's just look at the numerator for a second. Under the square root, we have $4x^6 - 5x^5$, which behaves like its leading term $4x^6$ when x is large. So we should think that $\sqrt{4x^6 - 5x^5}$ behaves like $\sqrt{4x^6}$, which is just $2x^3$ (since x is positive). The problem is that we are taking away $2x^3$ in the numerator, so it looks like we're left with nothing! Crap. What do we do?

The solution is to use the same technique as described in Section 4.2 above: multiply top and bottom by the conjugate expression of the numerator. So before we even look at leading terms, we need to do some prep work:

$$\lim_{x \to \infty} \frac{\sqrt{4x^6 - 5x^5} - 2x^3}{\sqrt[3]{27x^6 + 8x}} = \lim_{x \to \infty} \frac{\sqrt{4x^6 - 5x^5} - 2x^3}{\sqrt[3]{27x^6 + 8x}} \times \frac{\sqrt{4x^6 - 5x^5} + 2x^3}{\sqrt{4x^6 - 5x^5} + 2x^3}.$$

Now the formula $(a - b)(a + b) = a^2 - b^2$ allows us to simplify this whole thing to

$$\lim_{x \to \infty} \frac{(4x^6 - 5x^5) - (2x^3)^2}{\sqrt[3]{27x^6 + 8x}(\sqrt{4x^6 - 5x^5} + 2x^3)}.$$

In fact we can even tidy up the numerator further and reduce the situation to

$$\lim_{x \to \infty} \frac{-5x^5}{\sqrt[3]{27x^6 + 8x}(\sqrt{4x^6 - 5x^5} + 2x^3)}.$$

There, that's not so bad! There's nothing we need to do on the numerator; let's just concentrate on the denominator. For $\sqrt[3]{27x^6 + 8x}$, we can actually just multiply and divide by the cube root of the leading term $27x^6$, giving

$$\frac{\sqrt[3]{27x^6 + 8x}}{\sqrt[3]{27x^6}} \times \sqrt[3]{27x^6},$$

which is just

$$\frac{\sqrt[3]{27x^6 + 8x}}{\sqrt[3]{27x^6}} \times (3x^2).$$

Of course, we'll combine the terms under the square root and cancel to get

$$\sqrt[3]{\frac{27x^6 + 8x}{27x^6}} \times (3x^2) = \sqrt[3]{1 + \frac{8}{27x^5}} \times (3x^2).$$

Note that the part involving the cube root just goes to 1 as $x \to \infty$.

As for the other term, $\sqrt{4x^6 - 5x^5} + 2x^3$, here we need to be a little careful. Under the square root, we have $4x^6 - 5x^5$, so the leading term is $4x^6$. The square root of this is $2x^3$. Now we have to add $2x^3$ to this, and the total "leading term" on the numerator is therefore $2x^3 + 2x^3$, or $4x^3$. Let's see how it works. We'll replace the numerator by

$$\frac{\sqrt{4x^6 - 5x^5} + 2x^3}{4x^3} \times (4x^3),$$

then split up the fraction and drag the $4x^3$ under the square root, where it becomes $16x^6$; we get

$$\left(\sqrt{\frac{4x^6 - 5x^5}{16x^6}} + \frac{2x^3}{4x^3} \right) \times (4x^3) = \left(\sqrt{\frac{1}{4} - \frac{5}{16x}} + \frac{1}{2} \right) \times (4x^3).$$

Now when you let $x \to \infty$, the first product goes to

$$\sqrt{\frac{1}{4} + 0} + \frac{1}{2} = \frac{1}{2} + \frac{1}{2} = 1,$$

which is what we want! (Note that the square root of $\frac{1}{4}$ is $\frac{1}{2}$.)

Now let's try to put it all together and solve this darned problem. We started off by multiplying the numerator by its conjugate, which reduced matters to

$$\lim_{x \to \infty} \frac{-5x^5}{\sqrt[3]{27x^6 + 8x}(\sqrt{4x^6 - 5x^5} + 2x^3)}.$$

Now we'll use the multiply and divide method on the bottom, giving

$$\lim_{x \to \infty} \frac{-5x^5}{\left(\dfrac{\sqrt[3]{27x^6 + 8x}}{\sqrt[3]{27x^6}} \times (3x^2) \right) \left(\dfrac{\sqrt{4x^6 - 5x^5} + 2x^3}{4x^3} \times (4x^3) \right)}.$$

Pull out the quantities $-5x^5$, $3x^2$, and $4x^3$ to get

$$\lim_{x\to\infty} \frac{1}{\left(\dfrac{\sqrt[3]{27x^6 + 8x}}{\sqrt[3]{27x^6}}\right)\left(\dfrac{\sqrt{4x^6 - 5x^5 + 2x^3}}{4x^3}\right)} \times \frac{-5x^5}{(3x^2)(4x^3)}.$$

Now all you have to do is cancel x^5 from the top and the bottom and use the arguments from above to show that the final answer is $-5/12$. I've left you with a bit of work, but you should try to assemble all the bits and pieces from above into a complete solution.

4.5 Limits Involving Rational Functions as $x \to -\infty$

Now let's spend a little time on limits of the form

$$\lim_{x\to-\infty} \frac{p(x)}{q(x)},$$

where p and q are polynomials or even poly-type functions. All the principles we've been using apply equally well here. When x is a very large negative number, the highest-degree term in any sum still dominates. Also, it's true that C/x^n still goes to 0 as $x \to -\infty$, provided that C is constant and n is a positive integer. (Can you see why?) This all means that the solutions are almost identical to what we've already seen. For example, consider some adaptations of two examples we've already looked at in Section 4.3.1 above:

$$\lim_{x\to-\infty} \frac{x - 8x^4}{7x^4 + 5x^3 + 2000x^2 - 6} \quad \text{and} \quad \lim_{x\to-\infty} \frac{(x^4 + 3x - 99)(2 - x^5)}{(18x^7 + 9x^6 - 3x^2 - 1)(x + 1)}.$$

All I've done is change ∞ to $-\infty$, so that we are now interested in what becomes of the two rational functions when x is a very large negative number. The solution to the first one is the same as it was when x tended to ∞; you just multiply and divide by the leading term of each polynomial:

$$\lim_{x\to-\infty} \frac{x - 8x^4}{7x^4 + 5x^3 + 2000x^2 - 6} = \lim_{x\to-\infty} \frac{\dfrac{x - 8x^4}{-8x^4} \times (-8x^4)}{\dfrac{7x^4 + 5x^3 + 2000x^2 - 6}{7x^4} \times (7x^4)}$$

$$= \lim_{x\to-\infty} \frac{-\dfrac{1}{8x^3} + 1}{1 + \dfrac{5}{7x} + \dfrac{2000}{7x^2} - \dfrac{6}{7x^4}} \times \frac{-8}{7} = -\frac{8}{7}.$$

The point here is that any term that looks like C/x^n for some positive n goes to 0 as $x \to -\infty$, just the same as it does when $x \to \infty$. On the other hand, the second example is not quite identical; the very last step is different from the previous version of the problem:

$$\lim_{x \to -\infty} \frac{(x^4 + 3x - 99)(2 - x^5)}{(18x^7 + 9x^6 - 3x^2 - 1)(x+1)}$$

$$= \lim_{x \to -\infty} \frac{\left(\dfrac{x^4 + 3x - 99}{x^4} \times (x^4)\right)\left(\dfrac{2 - x^5}{-x^5} \times (-x^5)\right)}{\left(\dfrac{18x^7 + 9x^6 - 3x^2 - 1}{18x^7} \times (18x^7)\right)\left(\dfrac{x+1}{x} \times (x)\right)}$$

$$= \lim_{x \to -\infty} \frac{\left(1 + \dfrac{3}{x^3} - \dfrac{99}{x^4}\right)\left(-\dfrac{2}{x^5} + 1\right)}{\left(1 + \dfrac{9}{18x} - \dfrac{3}{18x^5} - \dfrac{1}{18x^7}\right)\left(1 + \dfrac{1}{x}\right)} \times \frac{(x^4)(-x^5)}{(18x^7)(x)}$$

$$= \frac{(1 + 0 - 0)(-0 + 1)}{(1 + 0 - 0 - 0)(1 + 0)} \times \lim_{x \to -\infty} \frac{-x}{18} = \lim_{x \to -\infty} \frac{-x}{18} = \infty.$$

Only when we take the limit at the very end do we see anything different from when $x \to \infty$: as $x \to -\infty$, now $-x/18$ goes to ∞ rather than $-\infty$.

There's only one other thing you have to beware. We've been dragging factors into square roots without being too careful. To show you what I mean, try simplifying $\sqrt{x^2}$. Did you get x? That's not right if x is negative, unfortunately. For example, if you square -2 and then take the square root, you will get 2. So in fact $\sqrt{x^2} = -x$ when x is negative. This sort of thing comes up when you look at poly-type limits as $x \to -\infty$, for example:

$$\lim_{x \to -\infty} \frac{\sqrt{4x^6 + 8}}{2x^3 + 6x + 1}.$$

The denominator behaves like its leading term $2x^3$, but how about the numerator? The term in the square root, $4x^6 + 8$, behaves like $4x^6$, so $\sqrt{4x^6 + 8}$ behaves like $\sqrt{4x^6}$. Tempting as it is to simplify this as $2x^3$, it is simply not correct! Since $x \to -\infty$, we are interested in what happens when x is negative. This means that $2x^3$ is negative, but $\sqrt{4x^6}$ is positive, so we must simplify $\sqrt{4x^6}$ as $-2x^3$. Here's how the solution goes:

$$\lim_{x \to -\infty} \frac{\sqrt{4x^6 + 8}}{2x^3 + 6x + 1} = \lim_{x \to -\infty} \frac{\dfrac{\sqrt{4x^6 + 8}}{\sqrt{4x^6}} \times \sqrt{4x^6}}{\dfrac{2x^3 + 6x + 1}{2x^3} \times (2x^3)}$$

$$= \lim_{x \to -\infty} \frac{\sqrt{\dfrac{4x^6 + 8}{4x^6}}}{\dfrac{2x^3 + 6x + 1}{2x^3}} \times \frac{\sqrt{4x^6}}{2x^3} = \lim_{x \to -\infty} \frac{\sqrt{1 + \dfrac{8}{4x^6}}}{1 + \dfrac{6x}{2x^3} + \dfrac{1}{2x^3}} \times \frac{-2x^3}{2x^3}$$

$$= \frac{\sqrt{1 + 0}}{1 + 0 + 0} \times (-1) = -1.$$

You have to exercise similar care when you deal with fourth roots, sixth roots, and so on. For example,

$$\sqrt[4]{x^4} = -x \qquad \text{if } x \text{ is negative.}$$

The same would be true if you replaced every instance of 4 with any even number. On the other hand, it's not true if you replace 4 by an odd number; for example,

$$\sqrt[3]{x^3} = x \qquad \text{for all } x \text{ (positive, negative, or zero).}$$

One other point, though: it's still true that

$$\sqrt{x^4} = x^2$$

even if $x < 0$! Why? Because x^2 can't be negative, and $\sqrt{x^4}$ can't be negative by definition, so there can't possibly be a minus sign! Here's a summary of the situation:

> if $x < 0$ and you want to write $\sqrt[n]{x^{\text{something}}} = x^m$, the only time you need a minus sign in front of x^m is when n is even and m is odd.

4.6 Limits Involving Absolute Values

Sometimes you have to deal with functions involving absolute values. Consider this limit:

$$\lim_{x \to 0^-} \frac{|x|}{x}.$$

In order to answer this, let's set $f(x) = |x|/x$ and check it out some more. First, note that 0 can't be in the domain of f, since the denominator would then be 0. On the other hand, everything else is fine. Let's look at what happens when x is positive. The quantity $|x|$ is then just x, so we see that $f(x) = 1$ if x is any positive number. On the other hand, if x is negative, then $|x| = -x$, so $f(x) = -x/x = -1$ if $x < 0$. That is, writing $f(x) = |x|/x$ is just a fancy way of saying that $f(x) = 1$ if $x > 0$ and $f(x) = -1$ if $x < 0$. The graph of $y = f(x)$ looks like this:

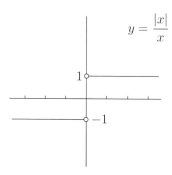

So, for the left-hand limit that we were looking at, you need to approach $x = 0$ from the left, and it's clear that

$$\lim_{x \to 0^-} \frac{|x|}{x} = -1,$$

and we may as well note that

$$\lim_{x \to 0^+} \frac{|x|}{x} = 1.$$

Since the left- and right-hand limits don't agree, the two-sided limit doesn't exist:

$$\lim_{x \to 0} \frac{|x|}{x} \text{ DNE.}$$

Most examples involving absolute values can be solved in a similar fashion by considering two or more different ranges of x, depending on the sign of what's inside the absolute value signs. A very slight variation of the above example is

$$\lim_{x \to (-2)^-} \frac{|x+2|}{x+2}.$$

Looking at the absolute value, we see that it matters whether $x + 2 \geq 0$ or $x + 2 < 0$. These conditions can be rewritten as $x \geq -2$ or $x < -2$. In the first case, $|x + 2| = x + 2$, whereas in the second case $|x + 2| = -(x + 2)$. The end result is that the quantity $|x + 2|/(x + 2)$ is equal to 1 when $x > -2$; whereas the quantity is just -1 when $x < -2$. In fact, the graph of $y = |x+2|/(x+2)$ is the same as the graph of $y = |x|/x$ shifted to the left by 2 units:

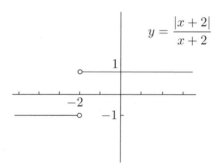

This means that the left-hand limit that we're looking for is equal to -1 (and the right-hand limit is 1, and the two-sided limit does not exist).

CHAPTER 5 _____

Continuity and Differentiability

In general, there's only one special thing about the graph of a function: it just has to obey the vertical line test. That's not particularly exclusive. The graph could be all over the place—a little bit here, a vertical asymptote there, or any number of individual disconnected points wherever the hell they feel like being. So now we're going to see what happens if we're a little more exclusive: we want to look at two types of *smoothness*. First, continuity: intuitively, this means that the graph now has to be drawn in one piece, without taking the pen off the page. Second, differentiability: the intuition here is that there are no sharp corners in the graph. In both cases, we'll do a lot better job with the definition, and we'll see some of the things you can expect to get from functions with these special properties. In detail, this is what we'll look at in this chapter:

- continuity at a point, and over an interval;
- some examples of continuous functions;
- the Intermediate Value Theorem for continuous functions;
- maxima and minima of continuous functions;
- displacement, average velocity, and instantaneous velocity;
- tangent lines and derivatives;
- second and higher-order derivatives; and
- the relationship between continuity and differentiability.

5.1   Continuity

We'll start off by looking at what it means for a function to be continuous. As I said above, the intuition is that you can draw the graph of the function in one piece, without lifting your pen off the page. This is all very well for something like $y = x^2$, which is all in one piece; but it's a little unfair for something like $y = 1/x$. This would have had a graph in one piece except for the vertical asymptote at $x = 0$, which breaks it into two. In fact, if $f(x) = 1/x$, then we want to say that f is continuous everywhere except at

$x = 0$. So we have to understand what it means to be continuous at a point, and then we'll worry about continuity over larger regions like intervals.

5.1.1 Continuity at a point

Let's start with a function f and a point a on the x-axis which is in the domain of f. When we draw the graph of $y = f(x)$, we don't want to lift up the pen as we pass through the point $(a, f(a))$ on the graph. It doesn't matter if we have to lift up our pen elsewhere, as long as we don't lift it up near $(a, f(a))$. This means that we want a stream of points $(x, f(x))$ which get closer and closer—arbitrarily close, in fact—to the point $(a, f(a))$. In other words, as $x \to a$, we need $f(x) \to f(a)$. Yes, ladies and gentlemen, we're dealing with limits here. We can now give a proper definition:

$$\boxed{\text{A function } f \text{ is } continuous \text{ at } x = a \quad \text{if } \lim_{x \to a} f(x) = f(a).}$$

Of course, for this last equation to make sense at all, both sides must be defined. If the limit doesn't exist, then f isn't continuous at $x = a$, whereas if $f(a)$ doesn't exist, then you're totally screwed: there isn't even a point $(a, f(a))$ to go through! So we can be a little more precise about the definition and explicitly require three things to be true:

1. The two-sided limit $\lim_{x \to a} f(x)$ exists (and is finite).
2. The function is defined at $x = a$; that is, $f(a)$ exists (and is finite).
3. The two above quantities are equal: that is,

$$\lim_{x \to a} f(x) = f(a).$$

 Let's see what happens if any of these properties fail. Consider the following graphs:

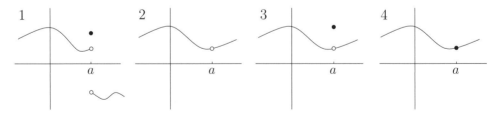

In diagram #1, the left- and right-hand limits aren't the same at $x = a$, so the two-sided limit doesn't exist there; therefore the function isn't continuous at $x = a$. In diagram #2, the left- and right-hand limits exist and are finite and equal to each other, so the two-sided limit exists; however the function isn't even defined at $x = a$, so it isn't continuous there. In diagram #3, the two-sided limit again exists, and the function is defined at $x = a$, but the limit isn't the same as the function value; once again, the function isn't continuous at $x = a$. On the other hand, the function in diagram #4 is indeed continuous at $x = a$, since the two-sided limit at $x = a$ exists, $f(a)$ exists, and the limit is the same as the value of the function. By the way, we say that the functions in the first three diagrams have a *discontinuity* at $x = a$.

5.1.2 Continuity on an interval

We now know what it means for a function to be continuous at a single point. Let's extend this definition and say that a function f is continuous on the interval (a, b) if it is continuous at every point in the interval. Notice that f doesn't actually have to be continuous at the endpoints $x = a$ or $x = b$. For example, if $f(x) = 1/x$, then f is continuous on the interval $(0, \infty)$ even though $f(0)$ isn't defined. This function is also continuous on $(-\infty, 0)$, but not on $(-2, 3)$, since 0 lies within that interval, and f isn't continuous there.

How about an interval like $[a, b]$? We have to be a little more flexible. For example, below is the graph of a function with domain $[a, b]$; we'd like to say that it's continuous on $[a, b]$:

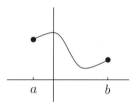

The problem is that the two-sided limits at the endpoints $x = a$ and $x = b$ don't exist: we only have a right-hand limit at $x = a$ and a left-hand limit at $x = b$. That's OK; we just modify our definition a bit by using the appropriate one-sided limits at the endpoints. So we say that a function f is continuous on $[a, b]$ if

1. the function f is continuous at every point in (a, b);
2. the function f is *right-continuous* at $x = a$. That is, $\lim_{x \to a^+} f(x)$ exists (and is finite), $f(a)$ exists, and these two quantities are equal; and
3. the function f is *left-continuous* at $x = b$. That is, $\lim_{x \to b^-} f(x)$ exists (and is finite), $f(b)$ exists, and these two quantities are equal.

Finally, we just say that a function is *continuous* if it is continuous at all the points in its domain, with the understanding that if its domain includes an interval with a left and/or right endpoint, then we only need one-sided continuity there.

5.1.3 Examples of continuous functions

Many common functions are continuous. For example, every polynomial is continuous. This seems a little hard to prove, since there are so many different polynomials, but actually it's not so bad. First, let's prove that the constant function f, defined by $f(x) = 1$ for all x, is continuous at any point a. Well, we need to show that

$$\lim_{x \to a} f(x) = f(a).$$

Since $f(x) = 1$ for any x, and $f(a) = 1$, then this means that we need to show that

$$\lim_{x \to a} 1 = 1.$$

Of course, this is obviously true, since nothing depends on x or a! Now, let's set $g(x) = x$. Is g continuous? Well, now we need

$$\lim_{x \to a} g(x) = g(a).$$

Since $g(x) = x$ and $g(a) = a$, this reduces to showing that

$$\lim_{x \to a} x = a.$$

This is also obviously true: as $x \to a$, well, $x \to a$! Now we just need to observe that a constant multiple of a continuous function is continuous; also, if you add, subtract, multiply or take the composition of two continuous functions, you get another continuous function (see Section A.4.1 of Appendix A for more info). The same is almost true if you divide one continuous function by another: the quotient function is continuous everywhere except where the denominator is 0. For example, $1/x$ is continuous except at $x = 0$, since we've seen that both the numerator and denominator are continuous functions of x.

Anyway, back to polynomials. Because $g(x) = x$ is continuous in x, we can multiply g by itself to see that x^2 is also continuous in x. You can keep multiplying by x as often as you like to prove the continuity of any power of x (as a function of x). Then you can multiply by constant coefficients and add different powers together to get any polynomial—and everything's still continuous!

It turns out that all exponentials and logarithms are continuous, as are all the trig functions (except where they have vertical asymptotes). We'll just take that for granted for the moment and return to this point in Section 5.2.11 below. Meanwhile, I want to look at a more exotic function. Consider the function f defined by $f(x) = x \sin(1/x)$. We looked at the graph of this (at least when $x > 0$) in Section 3.6 of Chapter 3. In fact, it's really easy to extend the graph to $x < 0$, because f is an even function. Why? Remembering that $\sin(x)$ is an odd function of x, we have

$$f(-x) = (-x) \sin\left(\frac{1}{-x}\right) = (-x)\left(-\sin\left(\frac{1}{x}\right)\right) = x \sin\left(\frac{1}{x}\right) = f(x).$$

So f is indeed even, and we can get the graph of all of f by reflecting the previous graph using the y-axis as our mirror (the graph only shows the domain $-0.3 < x < 0.3$):

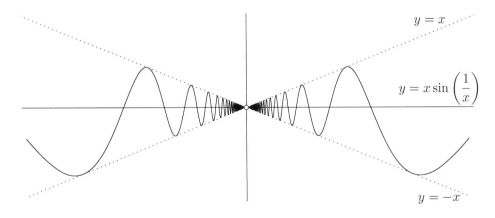

Now let's consider the continuity of the function. As a function of x, we know that $1/x$ is continuous away from $x = 0$; now compose this with the sine function, which is also continuous, and you can see that $\sin(1/x)$ is also continuous away from $x = 0$. Now you just have to multiply $\sin(1/x)$ by x (which is obviously a continuous function of x!) to see that f is continuous everywhere except at $x = 0$.

Now, what happens at $x = 0$? Clearly f is not continuous at $x = 0$, since it's not even defined there (there's a hole in the graph). Let's plug up this hole by defining a function g as follows:

$$g(x) = \begin{cases} x \sin\left(\dfrac{1}{x}\right) & \text{if } x \neq 0, \\ 0 & \text{if } x = 0. \end{cases}$$

So $g(x) = f(x)$ everywhere except at $x = 0$, where g equals 0 but f is undefined. As a result, g is automatically continuous everywhere f is—namely, everywhere except $x = 0$—but now we need to see what happens at $x = 0$. We have a hope because $g(0)$ is defined. Also, we used the sandwich principle in Section 3.6 of Chapter 3 to show that

$$\lim_{x \to 0+} g(x) = \lim_{x \to 0+} x \sin\left(\frac{1}{x}\right) = 0.$$

By symmetry (or the sandwich principle, again), we can see that the left-hand limit is also equal to 0. So in fact the two-sided limit is 0 as well:

$$\lim_{x \to 0} g(x) = \lim_{x \to 0} x \sin\left(\frac{1}{x}\right) = 0.$$

So we have shown that

$$\lim_{x \to 0} g(x) = g(0)$$

since both sides exist and are equal to 0. This means that g is actually continuous at $x = 0$, even though it was cobbled together in piecewise fashion.

We're almost ready to look at two nice facts involving continuity; first I want to return to a point I made at the beginning of Chapter 4. The first example we looked at was

$$\lim_{x \to -1} \frac{x^2 - 3x + 2}{x - 2},$$

which we solved by just substituting $x = -1$ to get the answer -2. Why is this justified? The argument seems to contradict the idea that the value of the above limit has nothing to do with what happens **at** $x = -1$, only what happens **near** $x = -1$. This is where continuity comes in: it connects the "near" with the "at." Specifically, if we let $f(x) = (x^2 - 3x + 2)/(x - 2)$, then since the numerator and denominator are polynomials, f is continuous everywhere except where the denominator is 0. That is, f is continuous everywhere except at $x = 2$. So f is continuous at $x = -1$, which means that

$$\lim_{x \to -1} f(x) = f(-1).$$

Replacing f by its definition, we have

$$\lim_{x \to -1} \frac{x^2 - 3x + 2}{x - 2} = \frac{(-1)^2 - 3(-1) + 2}{(-1) - 2} = -2.$$

That is the complete solution. In practice, few mathematicians would bother spelling it out in such gory detail, but it's worth understanding what you're doing whenever possible!

5.1.4   The Intermediate Value Theorem

Knowing that a function is continuous brings some benefits. We're going to look at two such benefits. The first is called the *Intermediate Value Theorem*, or *IVT* for short. Here's the idea: let's suppose that a function f is continuous on a closed interval $[a, b]$. Also suppose that $f(a) < 0$ and $f(b) > 0$. So in the graph of $y = f(x)$, we know that the point $(a, f(a))$ lies below the x-axis and that the point $(b, f(b))$ lies above the x-axis, like this:

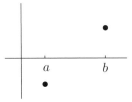

Now, if you have to connect those two points with a curve (which of course has to obey the vertical line test), and you're not allowed to lift your pen up, it's intuitively obvious that your pen will have to cross the x-axis somewhere between a and b, at least once. It could be close to a or close to b, or somewhere in the middle; you might cross back and forth many times; but the critical thing is that you have to cross at least once. That is, there is an x-intercept somewhere between a and b. It's crucial that the function f is continuous at every point in $[a, b]$; look what can happen if f is discontinuous at even one point:

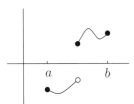

The discontinuity allows this function to jump over the x-axis without passing through it. So, we need continuity on the whole region $[a, b]$. All this is also true if we start above the axis and end below it; that is, if $f(a) > 0$ and $f(b) < 0$, we must have an x-intercept somewhere in $[a, b]$ if f is continuous on all of $[a, b]$. Since an x-intercept at c means that $f(c) = 0$, we can state the Intermediate Value Theorem as follows:

> **Intermediate Value Theorem:** if f is continuous on $[a, b]$, and $f(a) < 0$ and $f(b) > 0$, then there is at least one number c in the interval (a, b) such that $f(c) = 0$. The same is true if instead $f(a) > 0$ and $f(b) < 0$.

There's a proof of this theorem in Section A.4.2 of Appendix A. For now, let's look at a few examples of how to apply this theorem. First, suppose you want to show that the polynomial $p(x) = -x^5 + x^4 + 3x + 1$ has an x-intercept between $x = 1$ and $x = 2$. All you have to do is notice that p is continuous everywhere (including $[1, 2]$) because it's a polynomial; also, calculate $p(1) = 4 > 0$ and $p(2) = -9 < 0$. Since $p(1)$ and $p(2)$ have opposite signs and p is continuous on $[1, 2]$, we know that there is at least one number c in the interval $(1, 2)$ such that $p(c) = 0$. This number c is an x-intercept of the polynomial p.

Here's a slightly harder example. How would you show that the equation $x = \cos(x)$ has a solution? You don't have to find the solution, only to show that there is one. You could start by drawing the graphs of $y = x$ and $y = \cos(x)$ on the same axes. If you do, you'll find that the intersection of the graphs has x-coordinate somewhere around $\pi/4$. This graphical argument, while compelling, doesn't cut it so far as a mathematical proof is concerned. How can we do better?

The first step is to use a little trick: **put everything onto the left-hand side**. So, instead of solving $x = \cos(x)$, we try to solve $x - \cos(x) = 0$. Now we must take the initiative and set $f(x) = x - \cos(x)$. We'll be all done if we can show that there is a number c such that $f(c) = 0$. Let's check that this makes sense: if $f(c) = 0$, then $c - \cos(c) = 0$, so $c = \cos(c)$ and we have found a solution to the equation $x = \cos(x)$, namely $x = c$.

Now it's time to use the Intermediate Value Theorem. We need to find two numbers a and b such that one of $f(a)$ and $f(b)$ is negative and the other one is positive. Since we think (from the graph) that the answer is around $\pi/4$, we'll be conservative and take $a = 0$ and $b = \pi/2$. Let's check the values of $f(0)$ and $f(\pi/2)$. First, $f(0) = 0 - \cos(0) = 0 - 1 = -1$, which is negative, and second, $f(\pi/2) = \pi/2 - \cos(\pi/2) = \pi/2 - 0 = \pi/2$, which is positive. Since f is continuous (it is the difference of two continuous functions), we can conclude by the Intermediate Value Theorem that $f(c) = 0$ for some c in the interval $(0, \pi/2)$, and we have shown that $x = \cos(x)$ has a solution. We don't know where the solution is, nor how many solutions there are—only that there is at least one solution in the interval $(0, \pi/2)$. (Note that the solution is not really at $\pi/4$! It's not possible to find a nice expression for the answer, actually.)

Here's a small variation. So far, we have required that $f(a) < 0$ and $f(b) > 0$ (or the other way around), then concluded that there's a number c in (a, b) such that $f(c) = 0$. Instead, we can replace 0 by any number M and the result is still true. So, suppose f is continuous on $[a, b]$; if $f(a) < M$ and $f(b) > M$ (or the other way around), then there is some c in (a, b) such that $f(c) = M$. For example, if $f(x) = 3^x + x^2$, then does the equation $f(x) = 5$ have a solution? Certainly f is continuous; also we can guess to plug in 0 and 2, which leads to $f(0) = 1$ and $f(2) = 13$. Since the numbers 1 and 13 surround the target number 5 (one is smaller and the other is bigger), the Intermediate Value Theorem tells us that $f(c) = 5$ for some c in $(0, 2)$.

That is, $f(x) = 5$ does have a solution. Now, try to repeat the problem by starting with a new function g, where $g(x) = 3^x + x^2 - 5$. Convince yourself that if $f(x) = 5$ has a solution c, then this number c is also a solution of the equation $g(x) = 0$. Since $g(0) < 0$ and $g(2) > 0$, you can use the previous method instead of the variation! In fact, the variation doesn't really give us anything new—it just makes life a little easier sometimes.

5.1.5 A harder IVT example

One last example: let's show that any polynomial of odd degree has at least one root. That is, let p be a polynomial of odd degree; I claim that there is at least one number c such that $p(c) = 0$. (This isn't true for polynomials of even degree: for example, the quadratic $x^2 + 1$ doesn't have any roots—its graph doesn't cross the x-axis.) So, how do we prove my claim?

The key is actually found in the methods of Section 4.3 of the previous chapter. There we saw that if $p(x)$ is any polynomial and $a_n x^n$ is its leading term, then

$$\lim_{x \to \infty} \frac{p(x)}{a_n x^n} = 1 \qquad \text{and} \qquad \lim_{x \to -\infty} \frac{p(x)}{a_n x^n} = 1.$$

So when x gets very large, $p(x)$ and $a_n x^n$ are relatively close to each other (their ratio is near 1). This means that they at least have the same sign as each other! One can't be negative and the other positive, or else their ratio would be negative, not close to 1. The same is true when x is a very large negative number.

So let's suppose that A is a large negative number, so large that $p(A)$ and $a_n A^n$ have the same sign. Also, we'll pick some huge positive number B so that $p(B)$ and $a_n B^n$ have the same sign. Now let's compare the signs of $a_n A^n$ and $a_n B^n$. Since n is an odd number, these must have opposite signs! One is negative and one is positive. For example, if $a_n > 0$, then $a_n B^n$ is positive and $a_n A^n$ is negative. (This is only true because n is odd: if n were even then both quantities would be positive.) So here's the situation:

$$p(A) \quad \overset{\text{same sign as}}{\longleftrightarrow} \quad a_n A^n \quad \overset{\text{opposite sign to}}{\longleftrightarrow} \quad a_n B^n \quad \overset{\text{same sign as}}{\longleftrightarrow} \quad p(B).$$

So $p(A)$ and $p(B)$ have opposite signs. Since p is a polynomial, it is continuous; by the Intermediate Value Theorem, there is a number c between A and B such that $p(c) = 0$. That is, p has a root, although we really have no idea where it is. That makes sense since we knew virtually nothing about p to start with, only that its degree was odd.

5.1.6 Maxima and minima of continuous functions

Let's move on to the second benefit of knowing that a function is continuous. Suppose we have a function f which we know is continuous on the closed interval $[a, b]$. (It's very important that the interval is closed at both ends.) That means that we put our pen down at the point $(a, f(a))$ and draw a curve that ends up at $(b, f(b))$ without taking our pen off the paper. The question is, how high can we go? In other words, is there any limit to how high up this curve could go? The answer is yes: there must be a highest point, although the curve could reach that height multiple times.

In symbols, let's say that the function f defined on the interval $[a,b]$ has a *maximum* at $x = c$ if $f(c)$ is the highest value of f on the whole interval $[a,b]$. That is, $f(c) \geq f(x)$ for all x in the interval. The idea that I've been driving at is that a continuous function on $[a,b]$ has a maximum in the interval $[a,b]$. The same is true for the limbo question, "how low can you go?" We'll say that f has a *minimum* at $x = c$ if $f(c)$ is the lowest value of f on the whole interval; that is, that $f(c) \leq f(x)$ for all x in $[a,b]$. Once again, any continuous function on the interval $[a,b]$ has a minimum in that interval. These facts form a theorem, sometimes known as the *Max-Min Theorem*, which can be stated as follows:

> **Max-Min Theorem:** if f is continuous on $[a,b]$, then f has at least one maximum and one minimum on $[a,b]$.

Here are some examples of continuous functions on $[a,b]$ and their maxima and minima (these are the plurals of maximum and minimum, respectively, of course):

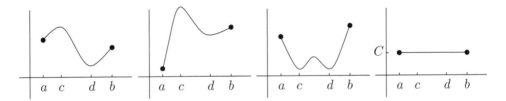

In the first graph, the function attains its maximum at $x = c$ and its minimum at $x = d$. In the second, the function has a maximum at $x = c$ but the minimum is at the left endpoint $x = a$. The third graph has a maximum at $x = b$ but the minimum is at both $x = c$ and $x = d$. This is acceptable—there are allowed to be multiple minima, as long as there is at least one. Finally, the fourth graph shows a constant function, which is continuous; in fact, every point in the interval $[a,b]$ is both a maximum and a minimum, since the function never goes above or below the constant value C.

So, why does the function f need to be continuous? And why can't it be an open interval, like (a,b)? The following diagrams show some potential problems:

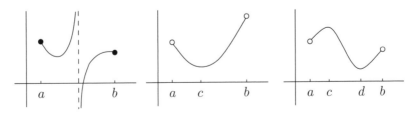

In the first figure, the function f has an asymptote in the middle of the interval $[a,b]$, which certainly creates a discontinuity. The function has no maximum value—it just keeps going up and up on the left side of the asymptote. Similarly, it has no minimum value either, since it just plummets way down on the right side of the asymptote.

The middle diagram on the previous page involves a more subtle situation. Here the function is only continuous on the open interval (a, b). It clearly has a minimum at $x = c$, but what is the maximum of this function? You might think that it occurs at $x = b$, but think again. The function isn't even defined at $x = b$! So it can't have a maximum there. If the function has a maximum, it must be somewhere near b. In fact, you'd like it to be the number less than b which is closest to b. Unfortunately, there is no such number! Whatever you think the closest number is, you can always take the average of this number and b to get an even closer number. So there is no maximum; this illustrates that the interval of continuity has to be closed in order to guarantee that the Max-Min Theorem works.

Of course, the conclusion of the theorem could still be true even if the interval isn't closed. For example, the function in the third diagram above is only continuous on the open interval (a, b), but it still has a maximum at $x = c$ and a minimum at $x = d$. This was just a lucky accident: you can only use the theorem to guarantee the existence of a maximum and minimum in an interval $[a, b]$ if you know the function is continuous on the entire closed interval.

5.2 Differentiability

We've spent a while looking at continuity. Now it's time to look at another degree of smoothness that a function can have: differentiability. This essentially means that the function has a derivative. So, we'll spend quite a bit of time looking at derivatives. One of the original inspirations for developing calculus came from trying to understand the relationship between speed, distance, and time for moving objects. So let's start there and work our way back to functions later on.

5.2.1 Average speed

Imagine looking at a photo of a car on a highway. The exposure time was very short, so it's not blurry—you can't even tell whether the car was moving or not. Now, I ask you this: how fast was the car moving when the picture was taken? No problem, you say—just use the classic formula

$$\text{speed} = \frac{\text{distance}}{\text{time}}.$$

The problem is that the photo conveys no sense of distance (the car hasn't moved) or time (the photo essentially captures an instant of time). So you can't answer my question.

Ah, but what if I tell you that a minute after the picture was taken, the car had traveled one mile? Then you could use the above formula to see that the car was going at a mile a minute, or 60 mph. Still, how do you know that the car was going the same speed for that whole minute? It might have accelerated and decelerated many times during that minute. You have no idea how fast it was actually going at the beginning of that minute. In fact, the above formula isn't really accurate: the left-hand side should say *average speed*, since that's all we've found.

OK, I'll take pity on you and tell you that the car went 0.25 miles in the first 10 seconds. Now you can use the formula and see that the average speed over the first 10 seconds is 1.5 miles per minute, or 90 mph. This helps, but the car could still have changed its speed over the 10 seconds—we don't really know how fast it was going at the beginning of the period. It's unlikely that it was too far away from 90 mph because the car can only accelerate and decelerate so much in such a short time.

It would be even better to know how far the car went in 1 second after the photo was taken, but it would still not be perfect. Even 0.0001 seconds might be enough for the car's speed to change, but not by much. If you sensed that we're heading toward whipping out a limit, you'd be quite right. We need to look at the concept of velocity first, though.

5.2.2 Displacement and velocity

Imagine that the car is driving down a long straight highway. The mile markers are a little weird—there's a 0 marker at some point, and to the left of it, the markers start at -1 and become more and more negative. To the right of the 0 marker, they go as normal. In fact, the whole situation looks exactly like a number line:

Suppose that the car starts at mile 2 and goes directly to mile 5. Then it has gone a distance of 3 miles. If instead it starts at mile 2 but goes left to mile -1, it's also gone a distance of 3 miles. We'd like to distinguish between these two cases, so we'll use *displacement* instead of distance. The formula for displacement is just

$$\text{displacement} = (\text{final position}) - (\text{initial position}).$$

If the car goes from position 2 to 5, then the displacement is $5 - 2 = 3$ miles. If instead the car goes from 2 to -1, the displacement is $(-1) - 2 = -3$ miles. So displacement can be negative, unlike distance. In fact, if the displacement is negative, then the car ends up to the left of where it began.

Another important difference between distance and displacement is that the displacement only involves the final and initial positions—what the car does in between is irrelevant. If it went from 2 to 11 and then back to 5, the distance is $9 + 6 = 15$ miles but the total displacement is still only 3 miles. If instead it went from 2 to -4 and then back to 2, the displacement is actually 0 miles even though the distance is 12 miles. It is true, however, that if the car just goes in one direction without backtracking, then the distance is the absolute value of the displacement.

As we saw in the last section, average speed is the distance traveled divided by the time taken. If you replace distance by displacement, you get the *average velocity* instead. That is,

$$\text{average velocity} = \frac{\text{displacement}}{\text{time}}.$$

Again, velocity can be negative while speed must be nonnegative. If the car has a negative average velocity over a certain time period, then it has ended to the left of where it began. If instead the average velocity is 0 over the time period, then the car has ended up exactly where it began. Notice that in this case the car might have a high average speed even though its average velocity is 0! In general, just like displacement, if the car is going in just one direction, then the average speed is just the absolute value of the average velocity.

5.2.3 Instantaneous velocity

We now revisit our crucial question in terms of velocity: how do you measure the velocity of the car at a given instant? The idea, as we saw above, is to take the average velocity of the car over smaller and smaller time periods beginning at the instant the photo was taken. Here's how it works in symbols.

Let t be the instant of time we care about. For example, if a race started at 2 p.m., you might decide to work in seconds with 0 representing the starting time; in that case, if the photo was taken at 2:03 p.m. then you'd want to take $t = 180$. Anyway, suppose that u is a short time later than t. Let's write $v_{t \leftrightarrow u}$ to mean the average velocity of the car during the time interval beginning at time t and ending at time u. Now we just push u closer and closer to t. How close? As close as we can! That's where the limit comes in. In fact,

$$\text{instantaneous velocity at time } t = \lim_{u \to t^+} v_{t \leftrightarrow u}.$$

Why neglect what happens before time t, though? We can make the above definition a little more general by allowing u to be before t; then we can replace the right-hand limit by a two-sided limit:

$$\text{instantaneous velocity at time } t = \lim_{u \to t} v_{t \leftrightarrow u}.$$

Now we need a few more formulas. Let's suppose we know exactly where on the highway the car is at any instant of time. In particular, suppose that at time t, the car is at position $f(t)$. That is, let

$$f(t) = \text{position of car at time } t.$$

We can now calculate the average velocity $v_{t \leftrightarrow u}$ exactly:

$$v_{t \leftrightarrow u} = \frac{\text{position at time } u - \text{position at time } t}{u - t} = \frac{f(u) - f(t)}{u - t}.$$

Notice that the denominator $u - t$ is the length of time involved (provided that u is after* t). Anyway, now we just take a limit as $u \to t$:

$$\text{instantaneous velocity at time } t = \lim_{u \to t} \frac{f(u) - f(t)}{u - t}.$$

Of course, you cannot just substitute $u = t$ in the previous limit, because then you get the indeterminate form $0/0$. You really do need to use limits.

---

*If u is before t, then the denominator should be $t - u$, but then the numerator should be $f(t) - f(u)$, so it all works out!

One more little variation. Let's define $h = u - t$. Then since u is very close to t, the difference h between the two times must be very small. Indeed, as $u \to t$, we can see that $h \to 0$. If we make this substitution in the above limit, then because $u = t + h$, we also have

$$\text{instantaneous velocity at time } t = \lim_{h \to 0} \frac{f(t+h) - f(t)}{h}.$$

There's no real difference between this formula and the previous one; it's just written a little differently.

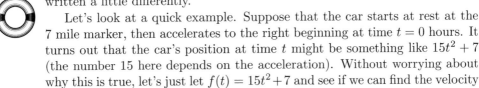

Let's look at a quick example. Suppose that the car starts at rest at the 7 mile marker, then accelerates to the right beginning at time $t = 0$ hours. It turns out that the car's position at time t might be something like $15t^2 + 7$ (the number 15 here depends on the acceleration). Without worrying about why this is true, let's just let $f(t) = 15t^2 + 7$ and see if we can find the velocity of the car at any time t.

Using the above formula, we have

$$\text{instantaneous velocity at time } t = \lim_{h \to 0} \frac{f(t+h) - f(t)}{h}$$
$$= \lim_{h \to 0} \frac{(15(t+h)^2 + 7) - (15t^2 + 7)}{h}.$$

Now expand $(t+h)^2 = t^2 + 2th + h^2$ and simplify a bit to see that the above expression is

$$\lim_{h \to 0} \frac{15t^2 + 30th + 15h^2 + 7 - 15t^2 - 7}{h} = \lim_{h \to 0} \frac{30th + 15h^2}{h} = \lim_{h \to 0} (30t + 15h).$$

It's particularly nice that the h gets canceled from the denominator in the last step, since that's what was giving us all the trouble. Now we can just put $h = 0$ to see that

$$\text{instantaneous velocity at time } t = \lim_{h \to 0} (30t + 15h) = 30t.$$

So at time 0, the car's velocity is $30 \times 0 = 0$ mph—the car is at rest. Half an hour later, at time $t = 1/2$, its velocity is $30 \times 1/2 = 15$ mph. One hour after the start time, the velocity is 30. In fact, the fact that the velocity is $30t$ at time t tells us that the car gets faster and faster at the constant rate of 30 mph every hour. That is, the car is constantly accelerating at 30 miles per hour per hour, or 30 miles per hour squared.

5.2.4 The graphical interpretation of velocity

It's time to look at a graph of the situation. Suppose that $f(t)$ again represents the position of the car at time t. If we want the instantaneous velocity at a particular time t, we need to pick a time u close to t. Let's draw the graph of $y = f(t)$ and mark in the points $(t, f(t))$ and $(u, f(u))$ as well as the line through them:

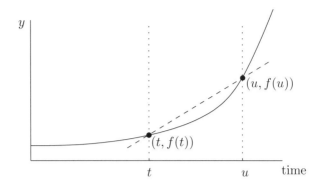

The slope of this line is given by

$$\text{slope} = \frac{f(u) - f(t)}{u - t},$$

which is exactly the formula for the average velocity $v_{t \leftrightarrow u}$ from the previous section. So we have a graphical interpretation for average velocity over the time period t to u: it's the slope of the line joining the points $(t, f(t))$ and $(u, f(u))$ on the graph of position versus time.

Let's try to find a similar interpretation for the instantaneous velocity. We need to take the limit as u goes to t, so let's repeat the previous graph a few times, each time with u closer and closer to the fixed value t:

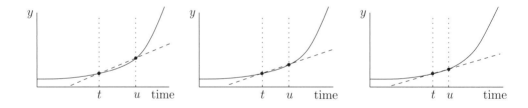

The lines seem to be getting closer to the tangent line at the point $(t, f(t))$. Since the instantaneous velocity is the limit of the slopes of these lines as $u \to t$, we'd like to say that the instantaneous velocity is exactly equal to the slope of the tangent line through $(t, f(t))$. Looks like we need to understand tangent lines better....

5.2.5 Tangent lines

Suppose we pick a number x in the domain of some function f. Then the point $(x, f(x))$ lies on the graph of $y = f(x)$. We want to try to draw a line through that point which is tangential to the curve—that is, we want to find a tangent line. Intuitively, this means that the line we're looking for just grazes the curve at our point $(x, f(x))$. The tangent line doesn't have to intersect the curve only once! For example, the tangent line through $(x, f(x))$ in the following picture hits the curve again, and that's not a problem:

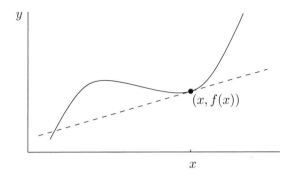

It's possible that there's no tangent line through a given point on a graph. For example, consider the graph of $y = |x|$:

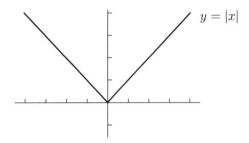

The graph passes through $(0,0)$, but there's no tangent line through that point. What could the tangent line possibly be, after all? No matter what you draw, you can't cuddle up to the graph there since it's got a sharp point at the origin. We'll return to this example in Section 5.2.10 below.

Even if the tangent line through $(x, f(x))$ exists, how on earth do you find it? Remember, to specify a line, you only need to provide two pieces of information: a point the line goes through and its slope. Then you can use the point-slope form to find the equation of the line. Well, we have one ingredient: we know the line passes through the point $(x, f(x))$. Now we just need to find the slope. To do this, we'll play a game similar to the one we played with instantaneous velocities in the previous section.

Start by picking a number z which is close to x (either to the right or to the left) and plot the point $(z, f(z))$ on the curve. Now draw the line through the points $(x, f(x))$ and $(z, f(z))$:

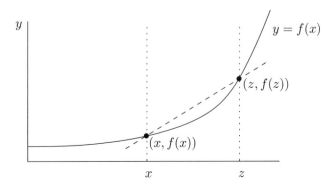

Since the slope is the rise over the run, the slope of the dashed line is

$$\frac{f(z) - f(x)}{z - x}.$$

Now, as the point z gets closer and closer to x, without ever actually getting to x itself, the slope of the above line should become closer and closer to the slope of the tangent we're looking for. So, if there's any justice in the world, then it should be true that

$$\text{slope of tangent line through } (x, f(x)) = \lim_{z \to x} \frac{f(z) - f(x)}{z - x}.$$

Let's set $h = z - x$; then we see that as $z \to x$, we have $h \to 0$, so we also have

$$\text{slope of tangent line through } (x, f(x)) = \lim_{h \to 0} \frac{f(x + h) - f(x)}{h}.$$

Of course, this only makes sense if the limit actually exists!

5.2.6 The derivative function

In the following picture, I've drawn in the tangent lines through three different points on the curve:

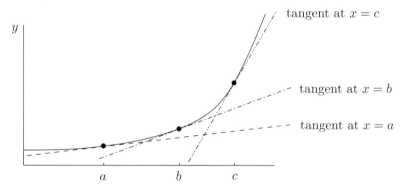

These lines have different slopes. That is, the slope of the tangent line **depends** on which value of x you start with. Another way of saying this is that the slope of the tangent line through $(x, f(x))$ is itself a function of x. This function is called the *derivative* of f and is written as f'. We say that we have *differentiated* the function f with respect to its variable x to get the function f'. By the formula at the end of the previous section, we see that

$$f'(x) = \lim_{h \to 0} \frac{f(x + h) - f(x)}{h}$$

provided that the limit exists. In this case, we say that f is *differentiable* at x. If the limit doesn't exist for some particular x, then that value of x is not in the domain of the derivative function f', so we say that f is *not differentiable* at x. The limit could fail to exist for a variety of reasons. In particular, there could be a sharp corner as in the example of $y = |x|$ above. On an even more basic level, if x isn't in the domain of f, then you can't even plot the point $(x, f(x))$, let alone draw a tangent line there!

Now let's recall the definition of instantaneous velocity in Section 5.2.3 above:

$$\text{instantaneous velocity at time } t = \lim_{h \to 0} \frac{f(t+h) - f(t)}{h},$$

where $f(t)$ is the position of the car at time t. This right-hand side of this the same as the definition of $f'(x)$ above, except with x replaced by t! That is, if $v(t)$ is the instantaneous velocity at time t, then $v(t) = f'(t)$. Velocity is precisely the derivative of position with respect to time.

Let's look at one example of finding a derivative. If $f(x) = x^2$, what is $f'(x)$? The computation is very similar to the one we did at the end of Section 5.2.3 above:

$$\begin{aligned} f'(x) &= \lim_{h \to 0} \frac{f(x+h) - f(x)}{h} = \lim_{h \to 0} \frac{(x+h)^2 - x^2}{h} \\ &= \lim_{h \to 0} \frac{x^2 + 2xh + h^2 - x^2}{h} = \lim_{h \to 0} \frac{2xh + h^2}{h} \\ &= \lim_{h \to 0} (2x + h) = 2x. \end{aligned}$$

So the derivative of $f(x) = x^2$ is given by $f'(x) = 2x$. This means that the slope of the tangent to the parabola $y = x^2$ at the point (x, x^2) is precisely $2x$. Let's draw the curve and a few tangent lines to check it out:

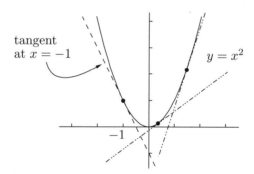

The slope of the tangent at $x = -1$ does indeed look like it's about -2, which is consistent with the formula $f'(x) = 2x$. (Twice -1 is -2!) The same is true with the other tangents—their slopes are all twice the corresponding x-coordinate.

5.2.7 The derivative as a limiting ratio

In our formula for the derivative $f'(x)$, we have to evaluate the quantity $f(x+h)$. What is this quantity? Well, if $y = f(x)$, and you change x into $x + h$, then $f(x+h)$ is simply the new value of y. The amount h represents how much you changed x, so let's replace it by the quantity Δx. Here the symbol Δ means "change in," so that Δx is just the change in x. (Don't think of Δx as the product of Δ and x—this is just plain wrong!) So, let's rewrite the formula for $f'(x)$ with h replaced by Δx:

$$f'(x) = \lim_{\Delta x \to 0} \frac{f(x + \Delta x) - f(x)}{\Delta x}.$$

OK, here's what happens. We start out with our pair (x, y), where $y = f(x)$. We now take a new value of x, which we'll call x_{new}. The value of y then changes as well to a new value y_{new}, which of course is just $f(x_{\text{new}})$. Now, the amount of change of any quantity is just the new value minus the old one, so we have two equations:

$$\Delta x = x_{\text{new}} - x \qquad \text{and} \qquad \Delta y = y_{\text{new}} - y.$$

The first equation says that $x_{\text{new}} = x + \Delta x$, so now the second equation can be transformed as follows:

$$\Delta y = y_{\text{new}} - y = f(x_{\text{new}}) - f(x) = f(x + \Delta x) - f(x).$$

But this is just the numerator of the fraction in the definition of $f'(x)$ above! What this means is that

$$f'(x) = \lim_{\Delta x \to 0} \frac{\Delta y}{\Delta x}.$$

An interpretation of this is that a small change in x produces approximately $f'(x)$ times as much change in y. Indeed, if $y = f(x) = x^2$, then we've seen in the previous section that $f'(x) = 2x$. Let's concentrate on what happens when $x = 6$, for example. First, note that our formula for $f'(x)$ shows us that $f'(6) = 2 \times 6 = 12$. So, if you take the equation $6^2 = 36$ and change the 6 a little bit, the 36 will change by about 12 times as much. For example, if we add 0.01 to 6, we should add 0.12 to 36. So I'm saying that $(6.01)^2$ should be about 36.12. In fact, the actual answer is 36.1201, so I was really close.

Now, why didn't I get the exact answer? The reason is that $f'(x)$ isn't actually equal to the ratio of Δy to Δx: it's equal to the **limit** of that ratio as Δx tends to 0. This means that if we don't move as far away from 6, we should do even better. Let's try to guess the value of $(6.0004)^2$. We have changed our original x-value 6 by 0.0004, so the y-value should change by 12 times this much, which is 0.0048. Our guess is therefore that $(6.0004)^2$ is approximately 36.0048. Not bad—the actual answer is 36.00480016, so we were very close! The smaller the change from 6, the better our method will work.

Of course, the magic number 12 only works when you start at $x = 6$. If instead you start at $x = 13$, the magic number is $f'(13)$, which equals $2 \times 13 = 26$. So, we know $13^2 = 169$; what is $(13.0002)^2$? To get from 13 to 13.0002, you have to add 0.0002; since the magic number is 26, we have to add 26 times as much to 169 to get our guess. That is, we add 0.0052 to 169 and come up with the guess 169.0052. Again, that's pretty darn good: $(13.0002)^2$ is actually exactly 169.00520004.

Anyway, we'll return to these ideas in Chapter 13 when we look at linearization. For now, let's look at the formula

$$f'(x) = \lim_{\Delta x \to 0} \frac{\Delta y}{\Delta x}.$$

once again. The right-hand side is the limit of the ratio of the change in y to the change in x, as the change in x becomes small. Suppose that x is so small that the change is barely noticeable. Instead of writing Δx, which means "change in x," we'd now like to write dx, which should mean "really really tiny change in x," and similarly for y. Unfortunately neither dx nor dy really means anything by itself;* nevertheless this provides the inspiration for writing the derivative in a different, more convenient way:

$$\text{if } y = f(x), \text{ then you can write } \frac{dy}{dx} \text{ instead of } f'(x).$$

For example, if $y = x^2$, then $\frac{dy}{dx} = 2x$. In fact, if you replace y by x^2, you get a variety of different ways of expressing the same thing:

$$f'(x) = \frac{dy}{dx} = \frac{d(x^2)}{dx} = \frac{d}{dx}(x^2) = 2x.$$

As another example, in Section 5.2.3 above, we saw that if the position of a car at time t is $f(t) = 15t^2 + 7$, then its velocity is $30t$. Remembering that velocity is just $f'(t)$, this means that $f'(t) = 30t$. If instead we decided to call the position p, so that $p = 15t^2 + 7$, we could write $\frac{dp}{dt} = 30t$. The point is that not everything comes in x's and y's—you have to be able to deal with other letters.

In summary, the quantity $\frac{dy}{dx}$ is the derivative of y with respect to x. If $y = f(x)$, then $\frac{dy}{dx}$ and $f'(x)$ are the same thing. Finally, remember that the quantity $\frac{dy}{dx}$ is not actually a fraction at all—it's the **limit** of the fraction $\frac{\Delta y}{\Delta x}$ as $\Delta x \to 0$.

5.2.8   The derivative of linear functions

Let's just pause for breath and go back to a simple case: suppose that f is linear. This means that $f(x) = mx + b$ for some m and b. What do you think that $f'(x)$ should be? Remember, this measures the slope of the tangent to the curve $y = f(x)$ at the point $(x, f(x))$. In our case, the graph of $y = mx + b$ is just a line of slope m and y-intercept equal to b. If there's any justice in the world, then the tangent at any point on the line is just the line itself! This means that the value of $f'(x)$ should be m no matter what x is: the curve $y = mx + b$ has constant slope m. Let's check it out using the formula:

$$f'(x) = \lim_{h \to 0} \frac{f(x+h) - f(x)}{h} = \lim_{h \to 0} \frac{(m(x+h) + b) - (mx + b)}{h}$$

$$= \lim_{h \to 0} \frac{mh}{h} = \lim_{h \to 0} m = m.$$

So there is justice in the world: $f'(x) = m$ regardless of what x is. That is, the derivative of a linear function is constant. As you might expect, only linear functions have constant slope (this is a consequence of the so-called Mean Value Theorem; see Section 11.3.1 in Chapter 11). By the way, if f is actually constant, so that $f(x) = b$, then the slope is always 0. In particular, $f'(x) = 0$ for all x. So we've proved that the derivative of a constant function is identically 0.

---

*There is a theory of "infinitesimals," but it's beyond the scope of this book!

5.2.9 Second and higher-order derivatives

Since you can start with a function f and take its derivative to get a new function f', you can actually take this new function and differentiate it again. You end up with the derivative of the derivative; this is called the *second derivative*, and it's written as f''.

For example, we've seen that if $f(x) = x^2$, then the derivative $f'(x) = 2x$. Now we want to differentiate this result. Let's put $g(x) = 2x$ and try to work out $g'(x)$. Since g is a linear function with slope 2, we know from the previous section above that $g'(x) = 2$. So the derivative of the derivative of f is the constant function 2, and we have shown that $f''(x) = 2$ for all x.

If $y = f(x)$, then we've seen that we can write $\frac{dy}{dx}$ instead of $f'(x)$. There's a similar sort of notation for the second derivative:

if $y = f(x)$, then you can write $\dfrac{d^2y}{dx^2}$ instead of $f''(x)$.

In the above example, if $y = f(x) = x^2$, then we've seen that

$$f''(x) = \frac{d^2y}{dx^2} = \frac{d^2(x^2)}{dx^2} = \frac{d^2}{dx^2}(x^2) = 2.$$

These are all valid ways of expressing that the second derivative of $f(x) = x^2$ (with respect to x) is the constant function 2.

Why stop at taking two derivatives? The third derivative of a function f is the derivative of the derivative of the derivative of f. That's a lot of derivatives! Realistically, you should think of the third derivative of f as being the derivative of the second derivative of f, and you can write it in any of the following ways:

$$f'''(x), \qquad f^{(3)}(x), \qquad \frac{d^3y}{dx^3}, \qquad \text{or} \qquad \frac{d^3}{dx^3}(y).$$

The notation $f^{(3)}(x)$ is particularly convenient for higher derivatives, because writing so many apostrophes is just plain stupid. So, for example, the fourth derivative, which is just the derivative of the third derivative, would be written $f^{(4)}(x)$ and not $f''''(x)$. That said, it will sometimes be convenient to go the other way and write $f^{(2)}(x)$ for the second derivative instead of $f''(x)$. It's even possible to write $f^{(1)}(x)$ instead of $f'(x)$, since we are only taking one derivative, and also $f^{(0)}(x)$ instead of just $f(x)$ itself (no derivatives!). That way, any derivative can be written in the form $f^{(n)}(x)$ for some integer n.

5.2.10 When the derivative does not exist

In Section 5.2.5 above, I mentioned that the graph of $f(x) = |x|$ has a sharp corner at the origin. This should mean that the derivative doesn't exist at $x = 0$. Now let's try to see why this is. Using the formula for the derivative, we have

$$f'(x) = \lim_{h \to 0} \frac{f(x+h) - f(x)}{h} = \lim_{h \to 0} \frac{|x+h| - |x|}{h}.$$

We are interested in what happens when $x = 0$, so let's replace x by 0 in the above chain of equations:

$$f'(0) = \lim_{h \to 0} \frac{f(0+h) - f(0)}{h} = \lim_{h \to 0} \frac{|0+h| - |0|}{h} = \lim_{h \to 0} \frac{|h|}{h}.$$

We have seen this limit before! In fact, in Section 4.6 of the previous chapter, we saw that the limit does not exist. This means that the value of $f'(0)$ is undefined: 0 is not in the domain of f'. We also saw, however, that the above limit does exist if you change it from a two-sided limit to a one-sided limit. In particular, the right-hand limit is 1 and the left-hand limit is -1. This motivates the idea of *right-hand* and *left-hand derivatives*, which are defined by the formulas

$$\lim_{h \to 0^+} \frac{f(x+h) - f(x)}{h} \quad \text{and} \quad \lim_{h \to 0^-} \frac{f(x+h) - f(x)}{h},$$

respectively. They look pretty similar to the definition of the ordinary derivative, except that the two-sided limit (that is, as $h \to 0$) is replaced by right-hand and left-hand limits, respectively. Just as in the case of limits, if the left- and right-hand derivatives both exist and are equal, then the actual derivative exists and is equal to the same thing. Also,* if the derivative exists then the left- and right-hand derivatives both exist and are equal to the derivative.

Anyway, the point is that if $f(x) = |x|$, at $x = 0$ the right-hand derivative is 1 and the left-hand derivative is -1. Do you believe this? Look at the graph again:

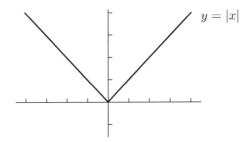

As you head from the origin along the curve to the right, it's definitely slope 1 (and in fact it stays at slope 1, that is, $f'(x) = 1$ if $x > 0$). Similarly, to the immediate left of the origin, the slope is -1 (and in fact $f'(x) = -1$ if $x < 0$). Since the left-hand slope doesn't equal the right-hand slope, there can be no derivative at $x = 0$.

OK, so we have come up with a continuous function which isn't differentiable everywhere in its domain. Still, it is clearly differentiable except at one measly little point. It turns out that you can have a continuous function which is so spiky and jittery that it effectively has a sharp corner at every single x, so it can't be differentiated anywhere! This sort of funky function is beyond the scope of this book, but I might as well mention that some of these sorts of functions are used to model stock prices—if you've ever seen the graph of a stock price, you'll know what I mean by "spiky and jittery." Anyway, my point is that there are continuous functions which are not differentiable. Are there any differentiable functions which aren't continuous? The answer is "no," and we're about to see why.

---

*You might say "conversely," but only if you know what a "converse" is!

5.2.11 Differentiability and continuity

Now it's time to relate the two big concepts in this chapter. I'm going to show that every differentiable function is also continuous. Another way of looking at this is that if you know a function is differentiable, you get the continuity of your function for free. More precisely, I will show:

> if a function f is differentiable at x, then it's continuous at x.

For example, we'll show in Chapter 7 that $\sin(x)$ is differentiable as a function of x. This will automatically imply that it's also continuous in x. The same goes for the other trig functions, exponentials, and logarithms (except at their vertical asymptotes).

So, how do we prove our big claim? Let's start by seeing what we want to prove. To show that f is continuous at x, we're going to need to show that

$$\lim_{u \to x} f(u) = f(x),$$

remembering from Section 5.1.1 above that this equation can only be true if both sides actually exist! Before we proceed farther, I want to substitute $h = u - x$ as we've done before. In that case, $u = x + h$, and as $u \to x$, we see that $h \to 0$. So the above equation can be replaced by

$$\lim_{h \to 0} f(x + h) = f(x).$$

We need to show that both sides exist and that equality holds—then we'll be all done.

Now that we are aware of our destination, let's start with what we actually know. Well, we know that f is differentiable at x; this means that $f'(x)$ exists, so by the definition of f', the limit

$$\lim_{h \to 0} \frac{f(x + h) - f(x)}{h}$$

exists. Let's first notice that $f(x)$ is involved in this formula, so it must exist or else the formula is all whacked. So we've already gotten somewhere: we know that $f(x)$ exists. We still need to do something clever. The trick is to start with another limit:

$$\lim_{h \to 0} \left(\frac{f(x + h) - f(x)}{h} \times h \right).$$

On the one hand, we can work out this limit exactly by splitting it into two factors:

$$\lim_{h \to 0} \left(\frac{f(x + h) - f(x)}{h} \times h \right) = \lim_{h \to 0} \frac{f(x + h) - f(x)}{h} \times \lim_{h \to 0} h = f'(x) \times 0 = 0.$$

This works just fine because all the limits involved exist. (That's where you need the fact that $f'(x)$ exists—otherwise it wouldn't work.) On the other hand, we could have taken the original limit and instead canceled out the factor of h to get

$$\lim_{h \to 0} \left(\frac{f(x + h) - f(x)}{h} \times h \right) = \lim_{h \to 0} (f(x + h) - f(x)).$$

Comparing these two previous equations, we just have

$$\lim_{h \to 0} (f(x + h) - f(x)) = 0.$$

Of course, the value of $f(x)$ doesn't depend on the limit at all, so we can pull it out and see that

$$\left(\lim_{h \to 0} f(x + h) \right) - f(x) = 0.$$

Now all we have to do is add $f(x)$ to both sides to get

$$\lim_{h \to 0} f(x + h) = f(x)$$

which is exactly what we wanted! In particular, the limit on the left exists and equality holds. So we have proved a nice result: **differentiable functions are automatically continuous**. Remember, though, that continuous functions aren't always differentiable!

CHAPTER 6 _____

How to Solve Differentiation Problems

Now we'll see how to apply some of the theory from the previous chapter to solve problems involving differentiation. Finding derivatives from the formula is possible but cumbersome, so we'll look at a few rules that make life a lot easier. All in all, here's what we'll tackle in this chapter:

- finding derivatives using the definition;
- using the product, quotient, and chain rules;
- finding equations of tangent lines;
- velocity and acceleration;
- finding limits which are derivatives in disguise;
- how to differentiate piecewise-defined functions; and
- using the graph of a function to draw the graph of its derivative.

6.1 Finding Derivatives Using the Definition

Let's say we want to differentiate $f(x) = 1/x$ with respect to x. We know from the previous chapter that the definition of the derivative is

$$f'(x) = \lim_{h \to 0} \frac{f(x+h) - f(x)}{h},$$

so in our case we have

$$f'(x) = \lim_{h \to 0} \frac{\dfrac{1}{x+h} - \dfrac{1}{x}}{h}.$$

If you just replace h by 0 in the fraction, you end up with the indeterminate form $\frac{0}{0}$. So you need to work a little. In this case, the idea is to simplify the numerator by taking a common denominator. You get

$$f'(x) = \lim_{h \to 0} \frac{\dfrac{x - (x+h)}{x(x+h)}}{h} = \lim_{h \to 0} \frac{-h}{hx(x+h)}.$$

Now cancel out a factor of h from top and bottom, then evaluate the limit by setting $h = 0$:

$$f'(x) = \lim_{h \to 0} \frac{-1}{x(x+h)} = \frac{-1}{x(x)} = -\frac{1}{x^2}.$$

That is,

$$\boxed{\frac{d}{dx}\left(\frac{1}{x}\right) = -\frac{1}{x^2}.}$$

On the other hand, to find the derivative of $f(x) = \sqrt{x}$, you have to employ the trick that we used in Section 4.2 of Chapter 4. Here's how it goes:

$$f'(x) = \lim_{h \to 0} \frac{f(x+h) - f(x)}{h} = \lim_{h \to 0} \frac{\sqrt{x+h} - \sqrt{x}}{h},$$

and we are again in $\frac{0}{0}$ territory. Let's multiply top and bottom by the conjugate of the numerator to get

$$f'(x) = \lim_{h \to 0} \frac{\sqrt{x+h} - \sqrt{x}}{h} \times \frac{\sqrt{x+h} + \sqrt{x}}{\sqrt{x+h} + \sqrt{x}} = \lim_{h \to 0} \frac{(x+h) - x}{h(\sqrt{x+h} + \sqrt{x})};$$

now we can cancel the x terms on the top, cancel a factor of h from top and bottom, and take the limit to see that

$$f'(x) = \lim_{h \to 0} \frac{h}{h(\sqrt{x+h} + \sqrt{x})} = \lim_{h \to 0} \frac{1}{\sqrt{x+h} + \sqrt{x}} = \frac{1}{\sqrt{x} + \sqrt{x}} = \frac{1}{2\sqrt{x}}.$$

In summary, we have shown that

$$\boxed{\frac{d}{dx}(\sqrt{x}) = \frac{1}{2\sqrt{x}}.}$$

Now how would you find the derivative of $f(x) = \sqrt{x} + x^2$ **using the definition of the derivative**? Even if you can just write down the answer, I've asked you to use the definition, so you must put all temptations aside and use the formula:

$$f'(x) = \lim_{h \to 0} \frac{f(x+h) - f(x)}{h} = \lim_{h \to 0} \frac{(\sqrt{x+h} + (x+h)^2) - (\sqrt{x} + x^2)}{h}.$$

This looks pretty messy, but if we split it up into the terms involving the square-root stuff and the terms involving the square stuff, we see that

$$f'(x) = \lim_{h \to 0} \frac{\sqrt{x+h} - \sqrt{x}}{h} + \lim_{h \to 0} \frac{(x+h)^2 - x^2}{h}.$$

We know how to do both of these limits; we have just seen that the first one is $1/2\sqrt{x}$, and we did the second one in Section 5.2.6 of the previous chapter and got the answer $2x$. You should try doing both of them without looking back at the previous work and make sure you get the answer

$$f'(x) = \frac{1}{2\sqrt{x}} + 2x.$$

It's now time to find the derivative of x^n with respect to x, where n is some positive integer. Set $f(x) = x^n$; then we have

$$f'(x) = \lim_{h \to 0} \frac{f(x+h) - f(x)}{h} = \lim_{h \to 0} \frac{(x+h)^n - x^n}{h}.$$

Somehow we have to deal with $(x+h)^n$. There are several ways of doing this; let's try the most direct approach, which is to write

$$(x+h)^n = (x+h)(x+h) \cdots (x+h).$$

There are n factors in the above product. This would be a real mess to multiply out, but it turns out we don't need to do the whole thing—we just need to get started. If you take the term x from each factor, there are n of them, so you get one term x^n in the product. That's the only way to get all x factors, so we already have

$$(x+h)^n = (x+h)(x+h) \cdots (x+h) = x^n + \text{stuff involving } h.$$

We need to do a little more work, though. What if you take the term h from the first factor and x from the others? Then you have one h and $(n-1)$ copies of x, so you get hx^{n-1} when you multiply them all together. There are other ways to choose one h and the rest x—you could take the h from the second factor and all others x, or the h from the third factor, and so on. In fact, there are n ways you could pick one h and the rest x, so you actually have n copies of hx^{n-1}. Together, this makes nhx^{n-1}. Every other term in the expansion has at least two copies of h, so every other term has a factor of h^2. All in all, we can write

$$(x+h)^n = (x+h)(x+h) \cdots (x+h) = x^n + nhx^{n-1} + \text{stuff with } h^2 \text{ as a factor.}$$

Let's tidy this up one little bit: we'll write the "stuff with h^2 as a factor" in the form $h^2 \times (\text{junk})$, where "junk" is just a polynomial in x and h. That is,

$$(x+h)^n = (x+h)(x+h) \cdots (x+h) = x^n + nhx^{n-1} + h^2 \times (\text{junk}).$$

Now we can substitute into the formula for the derivative:

$$f'(x) = \lim_{h \to 0} \frac{(x+h)^n - x^n}{h} = \lim_{h \to 0} \frac{x^n + nhx^{n-1} + h^2 \times (\text{junk}) - x^n}{h}.$$

The x^n terms cancel, and then we can cancel out a factor of h:

$$f'(x) = \lim_{h \to 0} \frac{nhx^{n-1} + h^2 \times (\text{junk})}{h} = \lim_{h \to 0} (nx^{n-1} + h \times (\text{junk})).$$

As $h \to 0$, the second term goes to 0 (since the junk is pretty benign and doesn't blow up!), but the first term remains as nx^{n-1}. So we conclude that

$$\frac{d}{dx}(x^n) = nx^{n-1}$$

when n is a positive integer. In fact, we'll show in Section 9.5.1 of Chapter 9 that

$$\boxed{\frac{d}{dx}(x^a) = ax^{a-1}}$$

when a is any real number at all. In words, you are simply taking the power, putting a copy of it out front as the coefficient, and then knocking the power down by 1.

Let's take a closer look at the above formula. First, when $a = 0$, then x^a is the constant function 1. The derivative is then $0x^{-1}$, which is just 0. This agrees with the computation we did in Section 5.2.8 of the previous chapter; in summary,

$$\text{if } C \text{ is constant, then } \frac{d}{dx}(C) = 0.$$

Now, if $a = 1$, then x^a is just x. According to the formula, the derivative is $1x^0$, which is the constant function 1. Again, this agrees with our results from Section 5.2.8 of the previous chapter; we have confirmed that

$$\frac{d}{dx}(x) = 1.$$

When $a = 2$, then we see that the derivative of x^2 with respect to x is $2x^1$, which is just $2x$. This agrees with what we found previously. Similarly, when $a = -1$, we can use our formula to see that the derivative of x^{-1} is $-1 \times x^{-2}$. In fact, this just says that the derivative of $1/x$ is $-1/x^2$, which we already knew from the beginning of this section! This example comes up so often that you should just learn it individually.

Now let's try some fractional powers. When $a = \frac{1}{2}$, you see that the derivative with respect to x of $x^{1/2}$ is $\frac{1}{2}x^{-1/2}$. By the exponential rules (see Section 9.1.1 in Chapter 9 for a review of these!), you can rewrite this and see that the derivative of $\sqrt{x}$ is $1/2\sqrt{x}$, which is exactly what we found above. Again, this comes up so often that it's worth learning it individually so that you don't have to mess around with powers of $\frac{1}{2}$ and $-\frac{1}{2}$. Finally, let's try $a = \frac{1}{3}$. Our formula says that

$$\frac{d}{dx}(x^{1/3}) = \frac{1}{3}x^{1/3-1} = \frac{1}{3}x^{-2/3}.$$

Using exponential rules (again, you can find these in Section 9.1.1 of Chapter 9), we can rewrite this whole thing as

$$\frac{d}{dx}(\sqrt[3]{x}) = \frac{1}{3\sqrt[3]{x^2}}.$$

This one is a little more esoteric, so I wouldn't worry about learning it. Just make sure you can derive it using the formula for the derivative of x^a with respect to x from the box above.

6.2 Finding Derivatives (the Nice Way)

All this messing about with limits in order to find derivatives is getting a bit tedious. Luckily, once you do it, you can build up other derivatives from the ones you've already found by means of simple rules. Let's define a function f as follows:

$$f(x) = \frac{3x^7 + x^4\sqrt{2x^5} + 15x^{4/3} - 23x + 9}{6x^2 - 4}.$$

The key to differentiating a function like this is to understand how it is synthesized from simpler functions. In Section 6.2.6 below, we'll see how to use simple operations—multiplication by a constant, adding and subtracting, multiplying, dividing, and composing functions—to build f from atoms of the form x^a, which we already know how to differentiate. First we need to see how taking derivatives is affected by each of these operations; then we'll come back and find $f'(x)$ for our nasty function f above. (See Section A.6 of Appendix A for proofs of the rules below, although there are intuitive justifications of some of them in Section 6.2.7.)

6.2.1 Constant multiples of functions

It's easy to deal with a constant multiple of a function: you just multiply by the constant after you differentiate. For example, we know the derivative of x^2 is $2x$; so the derivative of $7x^2$ is 7 times $2x$, or $14x$. The derivative of $-x^2$ is $-2x$, since you can think of the minus out front as multiplication by -1. There's actually an easy way to take the derivative of a constant multiple of x^a. Simply bring the power down, multiply it by the coefficient, and then knock the power down by one. So for the derivative of $7x^2$, bring the 2 down, mulitply it by 7 to get the coefficient 14, then knock the power of x down by one to get $14x^1$ or just $14x$. Similarly, to find the derivative of $13x^4$, multiply 13 by 4, giving a coefficient of 52, and then knock the power down by one to get $52x^3$.

6.2.2 Sums and differences of functions

It's even easier to differentiate sums and differences of functions: just differentiate each piece and then add or subtract. For example, what's the derivative with respect to x of

$$3x^5 - 2x^2 + \frac{7}{\sqrt{x}} + 2?$$

First write $1/\sqrt{x}$ as $x^{-1/2}$, so this means that we really have to differentiate $3x^5 - 2x^2 + 7x^{-1/2} + 2$. Using the method for constant multiples that we have just seen, the derivative of $3x^5$ is $15x^4$; similarly, the derivative of $-2x^2$ is $-4x$, and the derivative of $7x^{-1/2}$ is $-\frac{7}{2}x^{-3/2}$. Finally, the derivative of 2 is 0, since 2 is a constant. That is, the $+2$ at the end is irrelevant, as far as taking derivatives is concerned. So, we can just put the pieces together to see that

$$\frac{d}{dx}\left(3x^5 - 2x^2 + \frac{7}{\sqrt{x}} + 2\right) = \frac{d}{dx}(3x^5 - 2x^2 + 7x^{-1/2} + 2) = 15x^4 - 4x - \tfrac{7}{2}x^{-3/2}.$$

By the way, it's useful to realize that you can write $x^{3/2}$ as $x\sqrt{x}$, so you could also write the above derivative as

$$15x^4 - 4x - \frac{7}{2}\frac{1}{x\sqrt{x}}.$$

Similarly, $x^{5/2}$ is the same as $x^2\sqrt{x}$, and $x^{7/2}$ is the same as $x^3\sqrt{x}$, and so on.

6.2.3 Products of functions via the product rule

It's a little trickier dealing with products—you can't just multiply the two derivatives together. For example, let's say we want to find the derivative of

$$h(x) = (x^5 + 2x - 1)(3x^8 - 2x^7 - x^4 - 3x)$$

without expanding everything first (that would take way too long). Let's set $f(x) = x^5 + 2x - 1$ and $g(x) = 3x^8 - 2x^7 - x^4 - 3x$. The function h is the product of f and g. We can easily write down the derivatives of f and g: they are $f'(x) = 5x^4 + 2$ and $g'(x) = 24x^7 - 14x^6 - 4x^3 - 3$. As I said, it's not true that the derivative of the product h is the product of these two derivatives. That is, $h'(x) \neq (5x^4 + 2)(24x^7 - 14x^6 - 4x^3 - 3)$. It's no good saying what $h'(x)$ isn't—we need to say what it is!

It turns out that you have to mix and match. That is, you take the derivative of f and multiply it by g (not the derivative of g). Then you also have to take the derivative of g and multiply it by f. Finally, add the two things together. Here's the rule:

Product rule (version 1): if $h(x) = f(x)g(x)$, then
$$h'(x) = f'(x)g(x) + f(x)g'(x).$$

So, for our example of $h(x) = (x^5 + 2x - 1)(3x^8 - 2x^7 - x^4 - 3x)$, we write h as the product of f and g and then take their derivatives, as we did above. Let's summarize what we found, taking a column each for f and g:

$$f(x) = x^5 + 2x - 1 \qquad\qquad g(x) = 3x^8 - 2x^7 - x^4 - 3x$$
$$f'(x) = 5x^4 + 2 \qquad\qquad g'(x) = 24x^7 - 14x^6 - 4x^3 - 3.$$

Now we can use the product rule and do a sort of cross-multiplication. You see, we need to multiply $f'(x)$ on the bottom left by $g(x)$ on the top right, then add to this the product of $f(x)$ from the top left and $g'(x)$ from the bottom right. So we get

$$\begin{aligned}
h'(x) &= f'(x)g(x) + f(x)g'(x) \\
&= (5x^4 + 2)(3x^8 - 2x^7 - x^4 - 3x) \\
&\quad + (x^5 + 2x - 1)(24x^7 - 14x^6 - 4x^3 - 3).
\end{aligned}$$

You could multiply this out, but it would be even worse than multiplying out the original function h and then differentiating that. Just leave it as it is.

There's another way to write the product rule. Indeed, sometimes you have to deal with $y = $ stuff in x, instead of the $f(x)$ form. For example, suppose $y = (x^3 + 2x)(3x + \sqrt{x} + 1)$. What is dy/dx? In this case, it's easier to let $u = x^3 + 2x$ and $v = 3x + \sqrt{x} + 1$. Then we can take the above form of the product rule and make some replacements: first, u replaces $f(x)$, so that du/dx replaces $f'(x)$; we also do the same thing with v and $g(x)$. We get

---

Product rule (version 2): if $y = uv$, then
$$\frac{dy}{dx} = v\frac{du}{dx} + u\frac{dv}{dx}.$$

So, in our example, we have

$$u = x^3 + 2x \qquad\qquad v = 3x + \sqrt{x} + 1$$

$$\frac{du}{dx} = 3x^2 + 2 \qquad\qquad \frac{dv}{dx} = 3 + \frac{1}{2\sqrt{x}}.$$

This means that

$$\frac{dy}{dx} = v\frac{du}{dx} + u\frac{dv}{dx} = (3x + \sqrt{x} + 1)(3x^2 + 2) + (x^3 + 2x)\left(3 + \frac{1}{2\sqrt{x}}\right).$$

What if you have a product of **three** terms? For example, suppose

$$y = (x^2 + 1)(x^2 + 3x)(x^5 + 2x^4 + 7)$$

and you want to find dy/dx. You could multiply it all out and differentiate, or instead you could use the product rule for three terms:

> **Product rule (three variables):** if $y = uvw$, then
> $$\frac{dy}{dx} = \frac{du}{dx}vw + u\frac{dv}{dx}w + uv\frac{dw}{dx}.$$

Before we finish the example, here's a tip for remembering the above formula: just add up uvw three times, but put a d/dx in front of a different variable in each term. (The same trick works for four or more variables—every variable gets differentiated once!) Anyway, in our example, we'll let $u = x^2 + 1$, $v = x^2 + 3x$, and $w = x^5 + 2x^4 + 7$, so that y is the product uvw. We have $du/dx = 2x$, $dv/dx = 2x+3$, and $dw/dx = 5x^4 + 8x^3$. According to the above formula, we have

$$\begin{aligned}\frac{dy}{dx} &= \frac{du}{dx}vw + u\frac{dv}{dx}w + uv\frac{dw}{dx} \\ &= (2x)(x^2 + 3x)(x^5 + 2x^4 + 7) + (x^2 + 1)(2x + 3)(x^5 + 2x^4 + 7) \\ &\quad + (x^2 + 1)(x^2 + 3x)(5x^4 + 8x^3).\end{aligned}$$

Since we didn't multiply out and simplify the original expression for y above, I'm certainly not going to simplify the derivative! I do want to mention, however, that you can't always multiply everything out. Sometimes you just have to use the product rule. For example, after you learn how to differentiate trig functions in the next chapter, you'll want to be able to use the product rule to find derivatives of things like $x\sin(x)$. You can't really multiply this expression out—it's already as expanded as it can get. So if you want to differentiate it with respect to x, there's no easy way of avoiding using the product rule.

6.2.4 Quotients of functions via the quotient rule

Quotients are handled in a way similar to products, except that the rule is a little different. Let's say you want to differentiate

$$h(x) = \frac{2x^3 - 3x + 1}{x^5 - 8x^3 + 2}$$

with respect to x. You can let $f(x) = 2x^3 - 3x + 1$ and $g(x) = x^5 - 8x^3 + 2$; then you can write h as the quotient of f and g, or $h(x) = f(x)/g(x)$. Now here's the quotient rule:

> **Quotient rule (version 1):** if $h(x) = \dfrac{f(x)}{g(x)}$, then
> $$h'(x) = \frac{f'(x)g(x) - f(x)g'(x)}{(g(x))^2}.$$

Notice that the numerator of the right-hand fraction is the same as the numerator in the product rule, except with a minus instead of a plus. In our example, we need to differentiate f and g and summarize our results:

$$f(x) = 2x^3 - 3x + 1 \qquad\qquad g(x) = x^5 - 8x^3 + 2$$
$$f'(x) = 6x^2 - 3 \qquad\qquad g'(x) = 5x^4 - 24x^2.$$

By the quotient rule, since $h(x) = f(x)/g(x)$, we have

$$\begin{aligned}
h'(x) &= \frac{f'(x)g(x) - f(x)g'(x)}{(g(x))^2} \\
&= \frac{(6x^2 - 3)(x^5 - 8x^3 + 2) - (2x^3 - 3x + 1)(5x^4 - 24x^2)}{(x^5 - 8x^3 + 2)^2}.
\end{aligned}$$

There's also another version, just as there is in the case of the product rule. If instead you are given that

$$y = \frac{3x^2 + 1}{2x^8 - 7},$$

and you want to find dy/dx, then start by writing $u = 3x^2 + 1$ and $v = 2x^8 - 7$, so that $y = u/v$. Now we use:

> **Quotient rule (version 2):** if $y = \dfrac{u}{v}$, then
> $$\frac{dy}{dx} = \frac{v\dfrac{du}{dx} - u\dfrac{dv}{dx}}{v^2}.$$

Our summary box looks like this:

$$u = 3x^2 + 1 \qquad\qquad v = 2x^8 - 7$$
$$\frac{du}{dx} = 6x \qquad\qquad \frac{dv}{dx} = 16x^7.$$

By the quotient rule,

$$\frac{dy}{dx} = \frac{v\dfrac{du}{dx} - u\dfrac{dv}{dx}}{v^2} = \frac{(2x^8 - 7)(6x) - (3x^2 + 1)(16x^7)}{(2x^8 - 7)^2}.$$

As you can see, quotients aren't any harder than products (just a bit messier).

6.2.5 Composition of functions via the chain rule

Suppose $h(x) = (x^2 + 1)^{99}$ and you want to find $h'(x)$. It would be ridiculous to multiply it out—you'd have to multiply $x^2 + 1$ by itself 99 times and it would take days. It would also be crazy to use the product rule, since you'd need to use it too many times.

Instead, let's view h as the composition of two functions f and g, where $g(x) = x^2 + 1$ and $f(x) = x^{99}$. Indeed, if you take your x and hit it with g, you end up with $x^2 + 1$. If you now hit **that** with f, you get $(x^2 + 1)^{99}$, which is just $h(x)$. So we have written $h(x)$ as $f(g(x))$. (Check out Section 1.3 in Chapter 1 for more on how composition of functions works.) Now we can apply the *chain rule*:

> **Chain rule (version 1):** if $h(x) = f(g(x))$, then $h'(x) = f'(g(x))g'(x)$.

The formula looks a little tricky. Let's decompose it. The second factor is easy: it's just the derivative of g. How about the first factor? Well, you have to differentiate f, then evaluate the result at $g(x)$ instead of x.

In our example, we have $f(x) = x^{99}$, so $f'(x) = 99x^{98}$. We also have $g(x) = x^2 + 1$, so $g'(x) = 2x$. There's our second factor: just $2x$. How about the first one? Well, we take $f'(x)$, but instead of x, we put in $x^2 + 1$ (since that's what $g(x)$ is). That is, $f'(g(x)) = f'(x^2 + 1) = 99(x^2 + 1)^{98}$. Now we multiply our two factors together to get

$$h'(x) = f'(g(x))g'(x) = 99(x^2 + 1)^{98}(2x) = 198x(x^2 + 1)^{98}.$$

This might seem a little tortuous, to say the least. Here's another way to solve the same problem.

We start with $y = (x^2 + 1)^{99}$ and we want to find dy/dx. The $(x^2 + 1)$ term makes life difficult, so we'll just call it u. This means that $y = u^{99}$ where $u = x^2 + 1$. Now we can invoke the other version of the chain rule:

> **Chain rule (version 2):** if y is a function of u, and u is a function of x, then
> $$\frac{dy}{dx} = \frac{dy}{du}\frac{du}{dx}.$$

So in our case, we have

$$y = u^{99} \qquad\qquad u = x^2 + 1$$
$$\frac{dy}{du} = 99u^{98} \qquad\qquad \frac{du}{dx} = 2x.$$

Using the chain rule formula in the box above, we see that

$$\frac{dy}{dx} = \frac{dy}{du}\frac{du}{dx} = 99u^{98} \times 2x = 198xu^{98}.$$

Now you just need to tidy it up by replacing u by $x^2 + 1$ to see that we have $dy/dx = 198x(x^2 + 1)^{98}$, as we found above.

Here's another straightforward example. If $y = \sqrt{x^3 - 7x}$, what is dy/dx? Start by setting $u = x^3 - 7x$, so that $y = \sqrt{u}$. Our table looks like this:

$$y = \sqrt{u} \qquad\qquad u = x^3 - 7x$$
$$\frac{dy}{du} = \frac{1}{2\sqrt{u}} \qquad\qquad \frac{du}{dx} = 3x^2 - 7.$$

So by the chain rule, we have

$$\frac{dy}{dx} = \frac{dy}{du}\frac{du}{dx} = \frac{1}{2\sqrt{u}} \times (3x^2 - 7) = \frac{3x^2 - 7}{2\sqrt{u}}.$$

Now we just have to get rid of the u in the denominator; since $u = x^3 - 7x$, we see that

$$\frac{dy}{dx} = \frac{3x^2 - 7}{2\sqrt{x^3 - 7x}}.$$

Not so bad when you get the hang of it.

Two quick comments on the chain rule. First, why is it called the chain rule, anyway? Well, you start with x and it gives you u; then you take that u and get y. So there's a sort of chain from x to y through the extra variable u. Second, you might think that the chain rule is obvious. After all, in the formula in the box on the previous page, can't you just cancel out the factor of du? The answer is no—remember, expressions like dy/du and du/dx aren't actually fractions, they are limits of fractions (see Section 5.2.7 in the previous chapter for more on this). The nice thing is that they often behave as if they were fractions—they certainly do in this case.

The chain rule can actually be invoked multiple times all at once. For example, let

$$y = ((x^3 - 10x)^9 + 22)^8.$$

What is dy/dx? Simply let $u = x^3 - 10x$, and $v = u^9 + 22$, so that $y = v^8$. Then use a longer form of the chain rule:

$$\frac{dy}{dx} = \frac{dy}{dv}\frac{dv}{du}\frac{du}{dx}.$$

You can't get this wrong if you think about it: y is a function of v, which is a function of u, which is a function of x. So there's only one way the formula could possibly look! Anyway, we have

$$y = v^8 \qquad v = u^9 + 22 \qquad u = x^3 - 10x$$
$$\frac{dy}{dv} = 8v^7 \qquad \frac{dv}{du} = 9u^8 \qquad \frac{du}{dx} = 3x^2 - 10.$$

Plugging everything in, we have

$$\frac{dy}{dx} = \frac{dy}{dv}\frac{dv}{du}\frac{du}{dx} = (8v^7)(9u^8)(3x^2 - 10).$$

We're close, but we need to get rid of the u and v terms. First, replace v by $u^9 + 22$:

$$\frac{dy}{dx} = (8v^7)(9u^8)(3x^2 - 10) = (8(u^9 + 22)^7)(9u^8)(3x^2 - 10).$$

Now replace u by $x^3 - 10x$ and group the factors of 8 and 9 together to get the actual answer:

$$\frac{dy}{dx} = (8(u^9+22)^7)(9u^8)(3x^2-10) = 72((x^3-10x)^9+22)^7(x^3-10x)^8(3x^2-10).$$

We've mostly used the second version of the chain rule above, but there are times when the first version comes in useful. For example, if you know that $h(x) = \sqrt{g(x)}$ for some functions g and h, and all you know about g is that $g(5) = 4$ and $g'(5) = 7$, then you can still find $h'(5)$. Just set $f(x) = \sqrt{x}$ so that $h(x) = f(g(x))$, then use the formula $h'(x) = f'(g(x))g'(x)$ from above. Since $f(x) = \sqrt{x}$, we have $f'(x) = 1/2\sqrt{x}$; so

$$h'(x) = f'(g(x))g'(x) = \frac{1}{2\sqrt{g(x)}}g'(x).$$

Now substitute $x = 5$ to get

$$h'(5) = \frac{1}{2\sqrt{g(5)}}g'(5).$$

Since $g(5) = 4$ and $g'(5) = 7$, we have

$$h'(5) = \frac{1}{2\sqrt{4}}(7) = \frac{7}{4}.$$

One more example: suppose that $j(x) = g(\sqrt{x})$, where g is as above. What is $j'(25)$? Now we have $j(x) = g(f(x))$; here $f(x) = \sqrt{x}$ as before. This time, it works out that

$$j'(x) = g'(f(x))f'(x) = g'(\sqrt{x})\frac{1}{2\sqrt{x}}.$$

So if $x = 25$, we have

$$j'(25) = g'(\sqrt{25})\frac{1}{2\sqrt{25}} = g'(5)\frac{1}{10} = \frac{7}{10}$$

since $g'(5) = 7$. Compare these two examples: the order of composition makes a big difference!

6.2.6 A nasty example

Let's return to our function f from above:

$$f(x) = \frac{3x^7 + x^4\sqrt{2x^5 + 15x^{4/3} - 23x + 9}}{6x^2 - 4}.$$

To find $f'(x)$, we have to synthesize f from easier functions using the rules from the previous sections. It's not a bad idea to do this using the function notation (version 1 of all the rules above). Try this now!

Meanwhile, I'm going to use version 2 of all the rules. We'll set $y = f(x)$ and try to find dy/dx. The first thing to notice is that y is the quotient of two things: $u = 3x^7 + x^4\sqrt{2x^5 + 15x^{4/3} - 23x + 9}$ and $v = 6x^2 - 4$. We're going to use the quotient rule to deal with the fraction, so we'll need du/dx and dv/dx. The second of these is easy: it's just $12x$. The first is a bit harder. Let's summarize what we know so far:

$$u = 3x^7 + x^4\sqrt{2x^5 + 12x^{4/3} - 23x + 9} \qquad\qquad v = 6x^2 - 4$$

$$\frac{du}{dx} = ??? \qquad\qquad\qquad\qquad \frac{dv}{dx} = 12x.$$

If we just knew du/dx, we could use the quotient rule and we'd be done. So let's find du/dx.

First, note that u is the sum of $q = 3x^7$ and the nasty quantity r defined by $r = x^4\sqrt{2x^5 + 15x^{4/3} - 23x + 9}$. We need the derivatives of both pieces. The derivative of q is easy: it's just $21x^6$. Now, r is the product of $w = x^4$ and $z = \sqrt{2x^5 + 15x^{4/3} - 23x + 9}$, so we'll have to use the product rule to find dr/dx. We'll need to note the following:

$$w = x^4 \qquad\qquad\qquad z = \sqrt{2x^5 + 15x^{4/3} - 23x + 9}$$

$$\frac{dw}{dx} = 4x^3 \qquad\qquad\qquad \frac{dz}{dx} = ???$$

Darn, we don't know what dz/dx is. We're going to need to find that. Here we are taking the square root of a big expression, so let's call it t. Specifically, if $t = 2x^5 + 15x^{4/3} - 23x + 9$, then $z = \sqrt{t}$. Now we can actually differentiate everything! Let's set up one last table:

$$t = 2x^5 + 15x^{4/3} - 23x + 9 \qquad\qquad z = \sqrt{t}$$

$$\frac{dt}{dx} = 10x^4 + 20x^{1/3} - 23 \qquad\qquad \frac{dz}{dt} = \frac{1}{2\sqrt{t}}.$$

By the chain rule (changing the variables to the letters we need),

$$\frac{dz}{dx} = \frac{dz}{dt}\frac{dt}{dx} = \frac{1}{2\sqrt{t}}\left(10x^4 + 20x^{1/3} - 23\right).$$

Replacing t by its definition, $2x^5 + 15x^{4/3} - 23x + 9$, we see that

$$\frac{dz}{dx} = \frac{10x^4 + 20x^{1/3} - 23}{2\sqrt{2x^5 + 15x^{4/3} - 23x + 9}}.$$

Great—we finally know dz/dx. Now we can fill in the question marks from above:

$$w = x^4 \qquad\qquad\qquad z = \sqrt{2x^5 + 15x^{4/3} - 23x + 9}$$

$$\frac{dw}{dx} = 4x^3 \qquad\qquad \frac{dz}{dx} = \frac{10x^4 + 20x^{1/3} - 23}{2\sqrt{2x^5 + 15x^{4/3} - 23x + 9}}.$$

Now look back above—we were trying to find dr/dx, where $r = wz$. Let's use the product rule:

$$\frac{dr}{dx} = z\frac{dw}{dx} + w\frac{dz}{dx}.$$

Again, notice that you have to be flexible with the variables—they're not always u and v! Anyway, if you substitute from the table above, you find that

$$\frac{dr}{dx} = \left(\sqrt{2x^5 + 15x^{4/3} - 23x + 9}\right)(4x^3) + (x^4)\frac{10x^4 + 20x^{1/3} - 23}{2\sqrt{2x^5 + 15x^{4/3} - 23x + 9}}.$$

Taking a common denominator and simplifying reduces this (check it!) to

$$\frac{dr}{dx} = \frac{26x^8 + 140x^{13/3} - 207x^4 + 72x^3}{2\sqrt{2x^5 + 15x^{4/3} - 23x + 9}}.$$

Now go back to u. We saw that $u = q + r$, where we have $q = 3x^7$ and $r = x^4\sqrt{2x^5 + 15x^{4/3} - 23x + 9}$. We know that $dq/dx = 21x^6$, and we just worked out the messy formula for dr/dx, so we just add them together to get

$$\frac{du}{dx} = 21x^6 + \frac{26x^8 + 140x^{13/3} - 207x^4 + 72x^3}{2\sqrt{2x^5 + 15x^{4/3} - 23x + 9}}.$$

Finally, we can come back to our original quotient rule computation from the top of the previous page, and fill in du/dx:

$$u = 3x^7 + x^4\sqrt{2x^5 + 15x^{4/3} - 23x + 9} \qquad\qquad v = 6x^2 - 4$$

$$\frac{du}{dx} = 21x^6 + \frac{26x^8 + 140x^{13/3} - 207x^4 + 72x^3}{2\sqrt{2x^5 + 15x^{4/3} - 23x + 9}} \qquad \frac{dv}{dx} = 12x.$$

Since $y = u/v$, we just use the standard quotient rule

$$\frac{dy}{dx} = \frac{v\dfrac{du}{dx} - u\dfrac{dv}{dx}}{v^2}$$

to see (after splitting up and canceling) that

$$\frac{dy}{dx} = \frac{21x^6 + \dfrac{26x^8 + 140x^{13/3} - 207x^4 + 72x^3}{2\sqrt{2x^5 + 15x^{4/3} - 23x + 9}}}{6x^2 - 4}$$
$$- \frac{\left(3x^7 + x^4\sqrt{2x^5 + 15x^{4/3} - 23x + 9}\right)(12x)}{(6x^2 - 4)^2}.$$

We're finally done! It's certainly not pretty, but it's certainly effective.

6.2.7 Justification of the product rule and the chain rule

You can find formal proofs of the product rule and chain rule in Sections A.6.3 and A.6.5 of Appendix A, but it's not a bad idea to get an intuitive idea for why these rules work. So let's take a quick look.

In the case of the product rule, we'll use version 2 of the rule from Section 6.2.3 above. We start off with two quantities, u and v, which both depend on some variable x. We want to see how the product uv changes when we change x by a tiny amount Δx. Well, u will change to $u + \Delta u$, and v will change to $v + \Delta v$, so the product changes to $(u + \Delta u)(v + \Delta v)$. We can visualize this by thinking of a rectangle with side lengths u and v units. The rectangle changes shape a little bit so that its new dimensions are $u + \Delta u$ and $v + \Delta v$ units, like this:

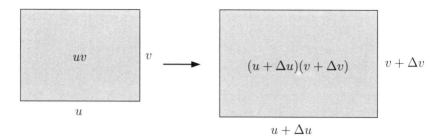

The products uv and $(u+\Delta u)(v+\Delta v)$ are just the areas of the two rectangles in square units, respectively. So how much does the area change? Let's see by superimposing the two rectangles:

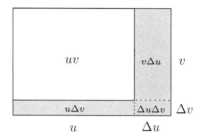

The change in areas is precisely the area of the shaded L-shaped region. This region is made up of two thin rectangles (of areas $v\Delta u$ and $u\Delta v$ square units) and one little one (of area $(\Delta u)(\Delta v)$ square units). Since the change of areas is $\Delta(uv)$ square units, we have shown that

$$\Delta(uv) = v\Delta u + u\Delta v + (\Delta u)(\Delta v).$$

When the quantities Δu and Δv are very small, the little area is very very small indeed, so we can basically ignore it. Here's what we're saying:

$$\Delta(uv) \cong v\Delta u + u\Delta v.$$

If you divide by Δx and take limits, the approximation becomes perfect and we get the product rule

$$\frac{d}{dx}(uv) = v\frac{du}{dx} + u\frac{dv}{dx}.$$

This is actually pretty close to the real proof!

Before we move onto the chain rule, let's prove the product rule for three functions, which is (as we saw above) given by

$$\frac{d}{dx}(uvw) = \frac{du}{dx}vw + u\frac{dv}{dx}w + uv\frac{dw}{dx}.$$

The trick is to let $z = vw$, so that uvw is just uz. We can use the product rule on $z = vw$ first:

$$\frac{dz}{dx} = w\frac{dv}{dx} + v\frac{dw}{dx}.$$

Now use the product rule on uz to get

$$\frac{d}{dx}(uvw) = \frac{d}{dz}(uz) = z\frac{du}{dx} + u\frac{dz}{dx}.$$

All that's left is to replace z by vw and dz/dx by the above expression to see that

$$\frac{d}{dx}(uvw) = z\frac{du}{dx} + u\frac{dz}{dx} = vw\frac{du}{dx} + u\left(w\frac{dv}{dx} + v\frac{dw}{dx}\right).$$

If you expand this, you get the desired formula.

Finally, let's think about the chain rule for a little bit. Suppose $y = f(u)$ and $u = g(x)$. This means that u is a function of x, and y is a function of u. If we change x by a little bit, as a result u will also change by a little bit. As a result of **that**, y will change too. By how much will y change?

Well, let's start off by concentrating on u and seeing how it reacts to a small change in x. Remember that $u = g(x)$; so as we discussed in Section 5.2.7 in the previous chapter, the change in u will be approximately $g'(x)$ times the change in x. You can think of $g'(x)$ as a sort of stretching factor. (For example, if you stand in front of one of those amusement park mirrors that make you look twice as tall and skinny as you are, then stand on your toes, your reflection will rise by twice as much as you do.) Here's an equation that describes this:

$$\Delta u \cong g'(x)\,\Delta x.$$

Now we can repeat the exercise with y in terms of u. Since $y = f(u)$, a change in u will produce approximately $f'(u)$ times as much change in y:

$$\Delta y \cong f'(u)\,\Delta u.$$

Putting these two equations together, we get

$$\Delta y \cong f'(u)g'(x)\,\Delta x.$$

So the change in x is first stretched by a factor of $g'(x)$, then again by a factor of $f'(u)$. The overall effect is to stretch by the **product** of the two stretching factors $f'(u)$ and $g'(x)$. (After all, if you stretch a piece of chewing gum by a factor of 2, then stretch **that** by a factor of 3, this would be the same as stretching the original piece of gum by a factor of 6.) This last equation suggests that

$$\frac{dy}{dx} = \lim_{\Delta x \to 0}\frac{\Delta y}{\Delta x} = f'(u)g'(x).$$

From here, you can get to either of the two versions of the chain rule without too much difficulty. To get version 1, remember that $u = g(x)$ and $y = f(u)$, so that $y = f(g(x))$; then let $y = h(x)$ and rewrite the above equation as

$$h'(x) = f'(u)g'(x) = f'(g(x))g'(x).$$

To get version 2, just interpret $f'(u)$ as dy/du and also $g'(x)$ as du/dx, so that the above equation for dy/dx says that

$$\frac{dy}{dx} = \frac{dy}{du}\frac{du}{dx}.$$

The above explanation isn't a formal proof, but it's pretty close.

6.3   Finding the Equation of a Tangent Line

What's the use of finding derivatives, anyway? Well, one benefit is that you can use derivatives to find the equation of a tangent line to a given curve. Suppose you have a curve $y = f(x)$ and a particular point $(x, f(x))$ on the curve. Then the tangent line through that point has slope $f'(x)$ and passes through the point $(x, f(x))$. Now you can just use the point-slope form to find the equation of the tangent line. In gory detail:

1. **find the slope**, by finding the derivative and plugging in the given value of x;

2. **find a point on the line**, by substituting the value of x into the function itself to get the y-coordinate. Put the coordinates together and call the resulting point (x_0, y_0). Finally,

3. **use the point-slope form** $y - y_0 = m(x - x_0)$ to find the equation.

Here's an example. Let $y = (x^3 - 7)^{50}$. What is the equation of the tangent line to the graph of this function at $x = 2$? First we need the derivative. We'll have to use the chain rule, as follows: let $u = x^3 - 7$, so $y = u^{50}$. Then we have $dy/du = 50u^{49}$ and $du/dx = 3x^2$. By the chain rule,

$$\frac{dy}{dx} = \frac{dy}{du}\frac{du}{dx} = 50u^{49} \times 3x^2 = 150x^2(x^3 - 7)^{49}.$$

(Remember, we have to replace u by $x^3 - 7$ in order to get everything in terms of x.) Now we need to plug in $x = 2$; for this value of x, we have

$$\frac{dy}{dx} = 150(2)^2(2^3 - 7)^{49} = 150 \times 4 \times 1^{49} = 600.$$

Great—we've found the slope of the tangent line we're looking for. Now we need to find a point it goes through: just put $x = 2$ and see what y is. In fact, $y = (2^3 - 7)^{50} = 1^{50} = 1$. So the tangent line passes through $(2, 1)$. Using the point-slope form, we see that the equation of the tangent line is $(y - 1) = 600(x - 2)$, which you can rewrite as $y = 600x - 1199$ if you like. And that's all there is to finding tangent lines!

6.4   Velocity and Acceleration

Another application of finding derivatives is to compute velocities and accelerations of moving objects. In Section 5.2.2 of the previous chapter, we imagined that an object moves along a number line. We saw that if its position at time t is x, then its velocity* at time t is given by

$$\boxed{\text{velocity} = v = \frac{dx}{dt}.}$$

Now, just as the velocity is the instantaneous rate at which the position changes, the *acceleration* of the object is the instantaneous rate at which the

---

*From now on, we'll drop the word "instantaneous"; the term "velocity" will always refer to instantaneous velocity unless we actually say "average velocity."

velocity changes. That is, the acceleration is the derivative of the velocity with respect to time t. Since the velocity is the derivative of the position, we see that the acceleration is actually the second derivative of the position. So we have

$$\text{acceleration} = a = \frac{dv}{dt} = \frac{d^2x}{dt^2}.$$

For example, let's say that we know that the position of an object at time t is given by $x = 3t^3 - 6t^2 + 4t - 2$, where x is in feet and t is in seconds. What are the object's velocity and acceleration at time $t = 3$? Well, we get the velocity by differentiating the position with respect to time, just like this: $v = dx/dt = 9t^2 - 12t + 4$. Now we differentiate this new expression with respect to time to get the acceleration: $a = dv/dt = 18t - 12$. Now plug in $t = 3$ to get $v = 9(3)^2 - 12(3) + 4 = 49$ ft/sec, and $a = 18(3) - 12 = 42$ ft/sec^2.

Why is the acceleration given in feet per second squared? Well, when you ask what the acceleration of an object is, you are really asking how fast the object's speed is changing. If the speed changes from 15 ft/sec to 25 ft/sec over a time period of 2 seconds, then it has (on average) changed by 5 ft/sec per second. So acceleration should be written in feet per second per second, or just feet per second squared. In general, you always have to square the time unit when you are dealing with acceleration.

6.4.1　Constant negative acceleration

Suppose you throw a ball directly up in the air. It goes up and comes back down (unless it hits something or someone else catches it!). This is because the Earth's gravitational pull exerts a force on the ball, pulling it toward the Earth. Newton—one of the pioneers of calculus—realized that the effect of the force is that the ball moves downward with constant acceleration. (We'll assume that there's no air resistance.)

Since the ball is going up and down, we'd better reorient our number line so that it points up and down. Let's set the 0 point as being on the ground, and we'll make upward positive. Since the acceleration is downward, it must be a negative quantity, and since it's constant, we can just call it $-g$. On Earth, g is about 9.8 meters per second squared, but it's a lot less on the moon. Anyway, if we're going to understand how this ball moves, we need to know its position and its velocity at time t.

Let's start off with velocity. We know that $a = dv/dt$. In the example in the previous section, we knew what v was, so we differentiated it to find a. Unfortunately, this time we know what a is (it's the constant $-g$) and we need to find v; so we're all topsy-turvy here. The same thing happens for x, once we know v. In both cases, we need to reverse the process of differentiation. Unfortunately, we're not ready for this yet—that's part of what integration is all about. So I'm just going to tell you the answer, then verify it by differentiating:

> An object thrown at time $t = 0$ from initial height h with initial velocity u satisfies the equations
>
> $$a = -g, \quad v = -gt + u, \quad \text{and} \quad x = -\frac{1}{2}gt^2 + ut + h.$$

It's not hard to check that these equations are consistent. Differentiating with respect to t, we see that $dv/dt = -g$, which is equal to a; and that $dx/dt = -gt + u$, which is just v. So $a = dv/dt$ and $v = dx/dt$ after all. Also, when $t = 0$, we see that $v = u$ and $x = h$. This means that the initial velocity is u and the initial height is h. Everything checks out.

Now, let's look at an example of how to use the above formulas. Suppose you throw a ball up from a height of 2 meters above the ground with a speed of 3 meters per second. Taking g to be 10 meters per second squared, we want to know five things:

1. How long does it take for the ball to hit the ground?
2. How fast is the ball moving when it hits the ground?
3. How high does the ball go?
4. If instead you throw the ball at the same speed but downward, how long does the ball take to hit the ground?
5. In that case, how fast does it hit the ground?

In the original situation, we know that $g = 10$, the initial height $h = 2$, and the initial velocity $u = 3$. This means that the above formulas become

$$a = -10, \quad v = -10t + 3, \quad \text{and} \quad x = -\frac{1}{2}(10)t^2 + 3t + 2 = -5t^2 + 3t + 2.$$

For part 1, we need to find how long it takes for the ball to get to the ground. This surely happens when its height is 0. So set $x = 0$ and let's find t; we get $0 = -5t^2 + 3t + 2$. If you factor this quadratic as $-(5t + 2)(t - 1)$, you can see that the solution of our equation is $t = 1$ or $t = -2/5$. Clearly the second answer is unrealistic—the ball can't hit the ground before you even throw it! So the answer must be $t = 1$. That is, the ball hits the ground 1 second after we throw it.

For part 2, we need to find the speed at the time when the ball hits the ground. No problem—we know that $v = -10t + 3$, and that the ball hits the ground when $t = 1$. Plugging that in, we see that $v = -10 + 3 = -7$. So the velocity of the ball when it hits the ground is -7 meters per second. Why negative? Because the ball is going downward when it hits, and downward is negative. The speed of the ball is just the absolute value of the velocity, or 7 meters per second.

To solve the third part, you have to realize that the ball reaches the top of its path when its velocity is exactly 0. On the way up, the velocity is positive; on the way down, the velocity is negative; it must be 0 when it's changing from up to down. So, when is v equal to 0? We just need to solve $-10t + 3 = 0$. The answer is $t = 3/10$. That is, the ball reaches the top of its trajectory three-tenths of a second after we release it. How high is it then? Just plug $t = 3/10$ into the formula $x = -5t^2 + 3t + 2$ to see that

$$x = -5\left(\frac{3}{10}\right)^2 + 3\left(\frac{3}{10}\right) + 2 = \frac{49}{20}.$$

That is, the ball reaches a height of 49/20 meters above the ground.

For the last two parts, you're throwing the ball downward instead. We still have $g = 10$ and the initial height $h = 2$, but what is the starting velocity u? Don't make the mistake of thinking that u is still 3! Since you are throwing the ball downward, the initial velocity is **negative**. A speed of 3 meters per second downward translates into an initial velocity $u = -3$. Omitting this minus sign is a common mistake, so be warned. Anyway, our equations are now

$$a = -10, \quad v = -10t - 3, \quad \text{and} \quad x = -\frac{1}{2}(10)t^2 - 3t + 2 = -5t^2 - 3t + 2.$$

Notice how similar they are to the equations for the scenario when we threw the ball upward. To solve part 4 of the problem, we need to find the time the ball hits the ground. Just as we did in part 1, set $x = 0$; then we have $0 = -5t^2 - 3t + 2 = -(5t - 2)(t + 1)$. So $t = 2/5$ or $t = -1$. This time we reject $t = -1$, since it's before we threw the ball, so we must have $t = 2/5$. That is, the ball hits the ground two-fifths of a second after we throw it. It makes sense that it's less than the time taken when we threw the ball up (which was 1 second), since the ball doesn't have to go up and then down. For the final part, we want to see how fast the ball is moving when it hits the ground; so put $t = 2/5$ in the formula for velocity. We get $v = -10(2/5) - 3 = -4 - 3 = -7$. Once again, the ball hits the ground with a speed of 7 meters per second. Interesting that it doesn't matter whether you throw the ball up or down (as long as it's from the same height and with the same speed in both cases): it still hits the ground with the same speed, although the time taken to hit the ground is different.

6.5 Limits Which Are Derivatives in Disguise

That's enough motion for now. Consider how you'd find the following limit:

$$\lim_{h \to 0} \frac{\sqrt[5]{32 + h} - 2}{h}.$$

It looks pretty hopeless. Even the trick of multiplying by the conjugate-type expression $\sqrt[5]{32 + h} + 2$ doesn't work because it's a 5th root, not a square root (try it and see for yourself!). So let's take a break from this and consider a related limit:

$$\lim_{h \to 0} \frac{\sqrt[5]{x + h} - \sqrt[5]{x}}{h}.$$

Note that h, not x, is the dummy variable here. Now this limit looks pretty difficult too, but perhaps it rings a bell. It's pretty similar to the limit in our formula

$$\lim_{h \to 0} \frac{f(x + h) - f(x)}{h} = f'(x).$$

All you have to do is set $f(x) = \sqrt[5]{x}$, and note that $f'(x) = \frac{1}{5}x^{-4/5}$. (Here we wrote $\sqrt[5]{x}$ as $x^{1/5}$ in order to find the derivative.) The derivative equation becomes

$$\lim_{h \to 0} \frac{\sqrt[5]{x + h} - \sqrt[5]{x}}{h} = \frac{1}{5}x^{-4/5}.$$

So the limit on the left is a derivative in disguise! We had to create a function f and differentiate it to solve the limit.

Now we can return to the original limit

$$\lim_{h \to 0} \frac{\sqrt[5]{32 + h} - 2}{h}.$$

This is actually a special case of the limit

$$\lim_{h \to 0} \frac{\sqrt[5]{x + h} - \sqrt[5]{x}}{h} = \frac{1}{5}x^{-4/5},$$

which we just worked out. If you set $x = 32$ in this limit, you get

$$\lim_{h \to 0} \frac{\sqrt[5]{32 + h} - \sqrt[5]{32}}{h} = \frac{1}{5} \times 32^{-4/5}.$$

Since $\sqrt[5]{32} = 2$ and $32^{-4/5} = 1/16$, we have shown that

$$\lim_{h \to 0} \frac{\sqrt[5]{32 + h} - 2}{h} = \frac{1}{5} \times 32^{-4/5} = \frac{1}{5} \times \frac{1}{16} = \frac{1}{80}.$$

Make no mistake: this is hard. There is a double disguise here: not only are we dealing with a derivative, we're actually evaluating the derivative at a particular point (32 in this case). You're better off generalizing the situation first, then substituting the specific value of x. Here's another example:

$$\lim_{h \to 0} \frac{\sqrt{(4 + h)^3 - 7(4 + h)} - 6}{h}.$$

This one **could** be done by multiplying top and bottom by the conjugate, but it's also a derivative in disguise. Since we are dealing with $4+h$, let's try replacing 4 by x. The first term in the numerator becomes $\sqrt{(x + h)^3 - 7(x + h)}$. This suggests that we might try setting $f(x) = \sqrt{x^3 - 7x}$. In Section 6.2.5 above, we saw that $f'(x) = (3x^2 - 7)/2\sqrt{x^3 - 7x}$, so the equation

$$\lim_{h \to 0} \frac{f(x + h) - f(x)}{h} = f'(x)$$

becomes

$$\lim_{h \to 0} \frac{\sqrt{(x + h)^3 - 7(x + h)} - \sqrt{x^3 - 7x}}{h} = \frac{3x^2 - 7}{2\sqrt{x^3 - 7x}}.$$

Finally, if you put $x = 4$, and simplify everything (noticing that you have $\sqrt{x^3 - 7x} = \sqrt{64 - 28} = \sqrt{36} = 6$), you get

$$\lim_{h \to 0} \frac{\sqrt{(4 + h)^3 - 7(4 + h)} - 6}{h} = \frac{3(4)^2 - 7}{2(6)} = \frac{41}{12}.$$

If you get stuck on a limit, it might be a derivative in disguise. Telltale signs are that the dummy variable is by itself in the denominator, and the numerator is the difference of two quantities. Even if this doesn't happen, you could still be dealing with a derivative in disguise; for example,

$$\lim_{h \to 0} \frac{h}{(x+h)^6 - x^6}$$

has the dummy variable in the numerator. No matter—just flip it over and find this limit first:

$$\lim_{h \to 0} \frac{(x+h)^6 - x^6}{h}.$$

To do this, set $f(x) = x^6$, so that $f'(x) = 6x^5$. We have

$$\lim_{h \to 0} \frac{(x+h)^6 - x^6}{h} = \lim_{h \to 0} \frac{f(x+h) - f(x)}{h} = f'(x) = 6x^5.$$

Now just flip it over again and get

$$\lim_{h \to 0} \frac{h}{(x+h)^6 - x^6} = \frac{1}{6x^5}.$$

We'll see a few other examples of limits which are derivatives in disguise in the future (in Chapters 9 and 17, to be precise). Keep your eyes peeled: many limits are derivatives in disguise, and your job is to unmask them.*

6.6 Derivatives of Piecewise-Defined Functions

Consider the following piecewise-defined function f:

$$f(x) = \begin{cases} 1 & \text{if } x \leq 0, \\ x^2 + 1 & \text{if } x > 0. \end{cases}$$

Is this function differentiable? Let's graph it and see:

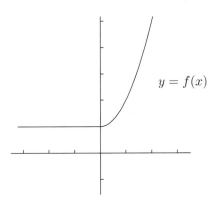

$$y = f(x)$$

---

* Actually, if you use l'Hôpital's Rule (see Chapter 14), you often don't even need to recognize when a limit is a derivative in disguise.

Looks pretty smooth—no sharp corners. In fact, it's pretty obvious that the function f is differentiable everywhere except perhaps at $x = 0$. To the left of $x = 0$, the function f inherits its differentiability from the constant function 1, and to the right of $x = 0$, it inherits its differentiability from $x^2 + 1$. The question is, what happens at $x = 0$, the interface between the two pieces?

The first thing to check is that the function is actually continuous there. You can't have differentiability without continuity, as we saw in Section 5.2.11 of the previous chapter. To see that f is continuous at $x = 0$, we need to show that $\lim_{x \to 0} f(x) = f(0)$. Well, we can see from the definition of f that $f(0) = 1$. As for the limit, let's break it up into left-hand and right-hand limits. For the left-hand limit, we have

$$\lim_{x \to 0^-} f(x) = \lim_{x \to 0^-} (1) = 1,$$

since $f(x) = 1$ when x is to the left of 0. As for the right-hand limit,

$$\lim_{x \to 0^+} f(x) = \lim_{x \to 0^+} (x^2 + 1) = 0^2 + 1 = 1,$$

since $f(x) = x^2 + 1$ when x is to the right of 0. So the left-hand limit equals the right-hand limit, which means that the two-sided limit exists and is 1. This agrees with $f(0)$, so we have proved that f is continuous at $x = 0$. (Notice that for both the left-hand and right-hand limits, you effectively just have to substitute $x = 0$ into the appropriate piece of f to get the limit.)

We still need to show that f is differentiable at $x = 0$. To do this, we have to show that the left-hand and right-hand derivatives match at $x = 0$ (see Section 5.2.10 in the previous chapter to refresh your memory of left-hand and right-hand derivatives). To the left of 0, we have $f(x) = 1$, so $f'(x) = 0$ in this case. It turns out that we can push it all the way up to $x = 0$ like this:

$$\lim_{x \to 0^-} f'(x) = \lim_{x \to 0^-} 0 = 0.$$

This shows that the left-hand derivative of f at $x = 0$ is 0. (See Section A.6.10 of Appendix A for more details.) To the right of 0, we have $f(x) = x^2 + 1$, so $f'(x) = 2x$ there. Again, we can push this down to $x = 0$:

$$\lim_{x \to 0^+} f'(x) = \lim_{x \to 0^+} 2x = 2 \times 0 = 0.$$

So the right-hand derivative of f at $x = 0$ is $2 \times 0 = 0$. Since the left-hand and right-hand derivatives at $x = 0$ match, the function is differentiable there.

 So, to check that a piecewise-defined function is differentiable at a point where the pieces join together, you need to check that the pieces agree at the join point (for continuity) **and** that the derivatives of the pieces agree at the join point. Otherwise it's not differentiable at the join point.* If you have more than two pieces, you have to check continuity and differentiability at all the join points.

---

*Actually, this is only true if the left- and right-hand limits of the derivatives at the join points exist and are finite. See Section 7.2.3 in the next chapter for an example of this.

Let's look at one more example of differentiating a piecewise-defined function. Suppose that

$$g(x) = \begin{cases} |x^2 - 4| & \text{if } x \leq 1, \\ -2x + 5 & \text{if } x > 1. \end{cases}$$

Where is g differentiable? You might think that the only issue is at the join point $x = 1$, but actually the absolute value makes life more complicated. Remember, the absolute value function is really a piecewise-defined function in disguise! In particular, $|x| = x$ when $x \geq 0$, but $|x| = -x$ when $x < 0$. It follows that

$$|x^2 - 4| = \begin{cases} x^2 - 4 & \text{if } x^2 - 4 \geq 0, \\ -(x^2 - 4) & \text{if } x^2 - 4 < 0. \end{cases}$$

In fact, the inequality $x^2 - 4 < 0$ can be rewritten as $x^2 < 4$, which means that $-2 < x < 2$. (Be careful to include the $-2 < x$ bit as well as the more obvious $x < 2$ bit!) So we can simplify this a little to get

$$|x^2 - 4| = \begin{cases} x^2 - 4 & \text{if } x \geq 2 \text{ or } x \leq -2, \\ -x^2 + 4 & \text{if } -2 < x < 2. \end{cases}$$

Now, in the definition of $g(x)$ above, the term $|x^2 - 4|$ only appears when $x \leq 1$. So, we can throw everything together and remove the absolute values for once and for all, rewriting $g(x)$ as follows:

$$g(x) = \begin{cases} x^2 - 4 & \text{if } x \leq -2, \\ -x^2 + 4 & \text{if } -2 < x \leq 1, \\ -2x + 5 & \text{if } x > 1. \end{cases}$$

So actually there are two join points: $x = -2$ and $x = 1$. Since the three pieces making up g are differentiable everywhere, we know that g itself is differentiable everywhere except perhaps at the join points. Let's check the join points one at a time, starting with $x = -2$. First, continuity. From the left, we have

$$\lim_{x \to (-2)^-} g(x) = \lim_{x \to (-2)^-} x^2 - 4 = (-2)^2 - 4 = 0,$$

while from the right, we have

$$\lim_{x \to (-2)^+} g(x) = \lim_{x \to (-2)^+} -x^2 + 4 = -(-2)^2 + 4 = 0.$$

Since the limits are equal, g is continuous at $x = -2$. Now, let's check the derivatives: for the left-hand derivative, we have

$$\lim_{x \to (-2)^-} g'(x) = \lim_{x \to (-2)^-} 2x = 2(-2) = -4,$$

whereas for the right-hand derivative, we have

$$\lim_{x \to (-2)^+} g'(x) = \lim_{x \to (-2)^+} -2x = 2(-2) = 4.$$

Since these don't match, the function g is not differentiable at $x = -2$.

How about at the other join point, $x = 1$? We repeat the exercise as follows: left-continuity:

$$\lim_{x \to 1^-} g(x) = \lim_{x \to 1^-} -x^2 + 4 = -(1)^2 + 4 = 3.$$

Right-continuity:

$$\lim_{x \to 1^+} g(x) = \lim_{x \to 1^+} -2x + 5 = -2(1) + 5 = 3.$$

So they match, and g is continuous at $x = 1$. Now, left-differentiability:

$$\lim_{x \to 1^-} g'(x) = \lim_{x \to 1^-} -2x = -2(1) = -2.$$

As for right-differentiability:

$$\lim_{x \to 1^+} g'(x) = \lim_{x \to 1^+} -2 = -2.$$

Since they match, the function g is in fact differentiable at $x = 1$.

We've answered the original question, but let's draw a graph anyway and see what's going on. To sketch the graph of $y = |x^2 - 4|$, let's first graph $y = x^2 - 4$. This is a parabola with x-intercepts at 2 and -2 (that's where $y = 0$) and y-intercept at -4. To get the absolute value, we take everything below the x-axis and reflect it in the x-axis. The bit that we flip over is part of the curve $y = -x^2 + 4$. Finally, the line $y = -2x + 5$ has y-intercept 5 and x-intercept $5/2$, so that graph is not hard to draw either. In the following two graphs, the left-hand graph shows all the functions that are ingredients for making $g(x)$, and the right-hand graph takes only what we need and is purely the graph of $y = g(x)$:

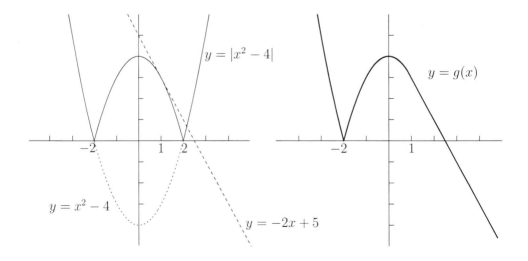

It actually looks continuous everywhere and differentiable everywhere except where there's that sharp corner at $(-2, 0)$. In particular, everything's nice at the join point $x = 1$, just as we calculated.

6.7 Sketching Derivative Graphs Directly

Suppose you have a graph of a function but not its equation, and you want to sketch the graph of its derivative. Formulas and rules aren't going to help you here: instead, you need a good understanding of differentiation.

Here's the basic idea. Imagine the graph of the function as a mountain, and imagine that there is a little mountain-climber walking up and down the graph from left to right. At each point of the climb, the climber calls out how difficult he or she thinks the climb is. If the terrain is flat, the climber calls out the number 0 for the degree of difficulty. If the terrain goes uphill, the climber calls out a positive number; the steeper the climb, the higher the number. If the terrain goes downhill, then the climb is actually easy, so the degree of difficulty is negative. That is, the climber will call out a negative number. The more downhill the terrain, the easier it is, so the number will be more negative. (If it's really steep going downhill, it might be difficult to climb down safely, but it's certainly easy to descend quickly!)

One important point: the height of the mountain itself isn't relevant. It's only the steepness that matters. In particular, you could shift the whole graph upward, and the climber would still be calling out the same degree of difficulty. A consequence of this is that if you are drawing the graph of a derivative from the graph of a function, the x-intercepts of the function are not important!

Let's look at an example: sketch the derivative of the following fearsome-looking function:

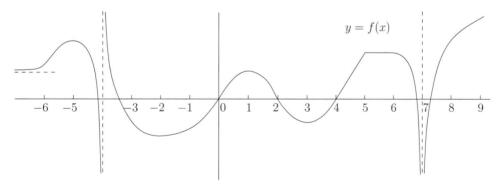

Don't panic. Just draw a little mountain-climber at a whole bunch of different points and imagine the climber shouting out the degree of difficulty at each point. Then all you have to do is plot these degrees of difficulty on another set of axes. Of particular interest are the points where the path is flat; this can occur in a long flat section (such as between $x = 5$ and $x = 6$ in the above graph), or at the top of a crest (such as at $x = -5$ or $x = 1$) or at the bottom of a valley (such as at $x = -2$ or $x = 3$). You definitely want to draw the mountain-climber there. Here's what the graph of f looks like with the climber in a bunch of positions:

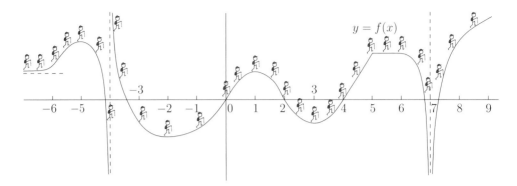

Now let's draw a set of axes for our graph of the derivative. Label the y-axis as "degree of difficulty," ranging from hard, down to flat at the origin, down to easy. Then you should be able to pencil in some points based on what the various copies of the little climber have shouted out. Remember, the climber doesn't care how high the mountain is, **only how steep** the climb is! Based on this, you get the following points:

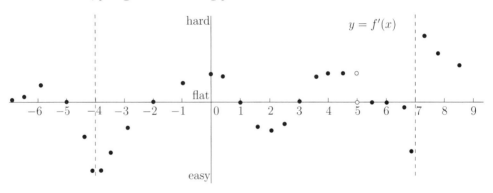

Here's a detailed explanation of how we came to these conclusions:

- At the far left of the graph of $y = f(x)$, the climber starts out going only slightly uphill. So we'll plot points of height a little above 0.
- Moving along to $x = -6$, the climber starts to go uphill, so the difficulty has gone up, so the points get higher (more difficult).
- Then it starts getting a little easier, until when $x = -5$, the climber has reached the top of the crest and it's now flat. In particular, when $x = -5$, the derivative has an x-intercept.
- After $x = -5$, the original curve starts to go downhill, first gently and then more and more steely. This means that it's getting easier and easier, until it gets ridiculously easy. So the derivative also has a vertical asymptote at $x = -4$.
- On the other side of the asymptote, the climb is also really easy—the climber is going downhill, starting very steeply and then leveling out at the valley when $x = -2$. So the vertical asymptote on the derivative curve actually starts at $-\infty$ (really easy) and climbs up to 0 at $x = -2$. (The fact that there are x-intercepts between -5 and -4 and also between -4 and -3 is irrelevant. The x-intercepts of the original function don't matter.)

- After the bottom of the valley at $x = -2$, the climber has to go uphill for a while, so it gets harder. After $x = 0$, though, it gets a little easier, until he or she reaches the top of the hill at $x = 1$. This means that the derivative curve goes up until $x = 0$, then comes back down to an x-intercept at $x = 1$.

- The reverse happens on the way to the bottom of the valley at $x = 3$: it gets easier and easier until $x = 2$, then it flattens out while still being downhill. So the derivative curve goes down, reaches a minimum at $x = 2$, then comes back up for an x-intercept at $x = 3$.

- From the bottom of the valley at $x = 3$, the climb gets steadily harder until $x = 4$. Between $x = 4$ and $x = 5$, however, the climb is of uniform difficulty, since the slope is constant. So the derivative curve increases from $x = 3$ until $x = 4$, but then it stays at the same height (degree of difficulty) between $x = 4$ and $x = 5$.

- At $x = 5$, the slope abruptly changes—it becomes flat without any warning, then stays flat until $x = 6$. So the derivative curve must jump down to 0 and stay there until $x = 6$. The derivative will have a discontinuity at $x = 5$.

- After $x = 6$, the climber finds things easier and easier as the curve dips down to the vertical asymptote at $x = 7$. The derivative curve also has a vertical asymptote there.

- To the right of the vertical asymptote, the climb is extremely difficult, but it does get a little easier as x moves up to 9. So the derivative curve will start very high on the right side of $x = 7$ and then get a little lower as the climb becomes easier.

Now, just connect the dots! Here are the graphs of $y = f(x)$ and $y = f'(x)$:

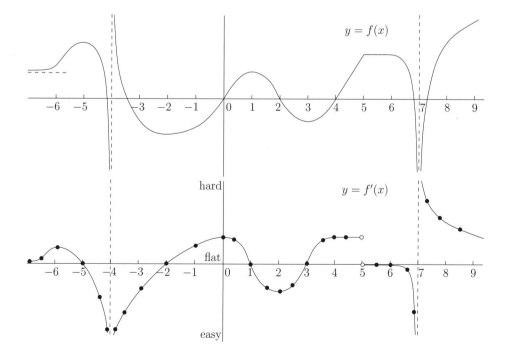

Let's just summarize the ideas that we used:

- When the original graph is flat, the derivative graph has an x-intercept. In the above example, this occurs at $x = -5$, $x = -2$, $x = 1$, $x = 3$, and everywhere in the interval $[5, 6]$.
- When a portion of the original graph is a straight line, the derivative graph is constant (this occurs in the interval $[4, 5]$ in our example).
- If the original has a horizontal asymptote, it's often true that the derivative also has one, but in that case it will be at $y = 0$ instead of the original height of the asymptote (as at the left-hand edge of our example).
- Vertical asymptotes in the original usually lead to vertical asymptotes in the derivative at the same place,* although the directions may change. For example, in our graph above, at $x = 7$ the original curve goes to $-\infty$ on both sides of the asymptote, but the derivative has opposite signs. The vertical asymptote at $x = -4$ is similarly affected.

When in doubt, use the trusty mountain-climber!

---

*It's not actually true in general that if a function has a vertical asymptote, then its derivative also has a vertical asymptote at the same place. An example is $y = 1/x + \sin(1/x)$ at $x = 0$. Can you see why?

CHAPTER 7 _____

Trig Limits and Derivatives

So far, most of our limits and derivatives have involved only polynomials or poly-type functions. Now let's expand our horizons by looking at trig functions. In particular, we'll focus on the following topics:

- the behavior of trig functions at small, large, and other argument values;
- derivatives of trig functions; and
- simple harmonic motion.

7.1 Limits Involving Trig Functions

Consider the following two limits:

$$\lim_{x \to 0} \frac{\sin(5x)}{x} \qquad \text{and} \qquad \lim_{x \to \infty} \frac{\sin(5x)}{x}.$$

They look almost the same. The only difference is that the first limit is taken as $x \to 0$ while the second is taken as $x \to \infty$. What a difference, though! As we'll soon see, the answers and the techniques used have almost nothing in common. So, it's really important to note whether your limit involves taking the sine—or cosine or tangent—of really small numbers (as in the first limit above) or really large numbers (as in the second limit). We'll look at these two cases separately, then see what happens when neither case applies.

Before we do, it's important to note that you can't tell what case you're dealing with just by looking at whether $x \to 0$ or $x \to \infty$. You need to see where you are evaluating your trig functions. For example, consider the following pair of limits:

$$\lim_{x \to 0} x \sin\left(\frac{5}{x}\right) \qquad \text{and} \qquad \lim_{x \to \infty} x \sin\left(\frac{5}{x}\right).$$

In the first limit, you are taking the sine of $5/x$, which is actually a huge number (positive or negative, depending on the sign of x) when x is near 0. So the first limit isn't covered by the small case at all—it belongs to the large case! Similarly, in the second limit, the quantity $5/x$ is very small when x is large, so that's really the small case. We'll solve all four of the above limits in the next few sections.

7.1.1 The small case

We know $\sin(0) = 0$. OK, so what does $\sin(x)$ look like when x is **near** 0? Sure, $\sin(x)$ is near 0 as well in that case, but **how near** to 0 is it? It turns out that $\sin(x)$ is approximately the same as x itself!

For example, if you take your calculator, put it in radian mode, and find $\sin(0.1)$, you get about 0.0998, which is very close to 0.1. Try it with a number even closer to 0 and you'll see that taking the sine of your number leaves you with something very close to your original number.

It's always good to look at a picture of the situation. Here's a graph of $y = \sin(x)$ and $y = x$ on the same set of axes, concentrating only on the values of x between -1 and 1 (approximately):

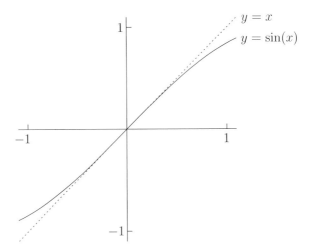

The graphs are very similar, especially when x is close to 0. (Of course, if we graph a little more of $y = \sin(x)$, it starts making the familiar waves; it's only when we zoom in like this that we see how close $\sin(x)$ is to x.) So there's good justification for making the statement that $\sin(x)$ is close to x when x is small. If $\sin(x)$ were actually equal to x, then

$$\frac{\sin(x)}{x} = 1$$

would be true. In fact, the above equation is never true, but it **is** true in the limit as $x \to 0$:

$$\lim_{x \to 0} \frac{\sin(x)}{x} = 1.$$

This is very important. It's basically the key to doing calculus involving trig functions. We'll use it in Section 7.2 to find the derivatives of the trig functions, and we'll actually prove it in Section 7.1.5 below.

How about $\cos(x)$? Well, $\cos(0) = 1$, so things are very different in this case. For the moment, let's just say that the cosine of a small number is very

close to 1. We write

$$\lim_{x \to 0} \cos(x) = 1$$

taking special care to notice that there's no factor of x in the denominator as there is in the previous formula involving $\sin(x)$. What if you do put a factor of x in the denominator? We'll see very soon, but first I want to look at $\tan(x)$.

The key is to write $\tan(x)$ as $\sin(x)/\cos(x)$. The numerator is $\sin(x)$, which is close to x when x is small. On the other hand, the denominator is close to 1. If there's any justice in the world, then the ratio should behave like $x/1$, which is just x. In fact, this is true, as we can see by isolating the harmless factor $\cos(x)$ in the denominator:

$$\lim_{x \to 0} \frac{\tan(x)}{x} = \lim_{x \to 0} \frac{\dfrac{\sin(x)}{\cos(x)}}{x} = \lim_{x \to 0} \left(\frac{\sin(x)}{x} \right) \left(\frac{1}{\cos(x)} \right) = (1)\left(\frac{1}{1} \right) = 1.$$

So we have shown that

$$\lim_{x \to 0} \frac{\tan(x)}{x} = 1.$$

This means that $\sin(x)$ and $\tan(x)$ behave in a similar way when x is small, but $\cos(x)$ is the odd one out. Let's take a look at what happens to $\cos(x)/x$ as $x \to 0$. So we are trying to understand

$$\lim_{x \to 0} \frac{\cos(x)}{x}.$$

If you just substitute $x = 0$, then you get $1/0$. This means that the graph of $y = \cos(x)/x$ has a vertical asymptote at $x = 0$. It looks very much like $1/x$ for small x; in particular, you should try to convince yourself that

$$\lim_{x \to 0^+} \frac{\cos(x)}{x} = \infty, \qquad \lim_{x \to 0^-} \frac{\cos(x)}{x} = -\infty, \qquad \text{so} \qquad \lim_{x \to 0} \frac{\cos(x)}{x} \text{ DNE}.$$

(Remember, "DNE" stands for "does not exist.") This is really different from what happens with sin or tan in place of cos.

7.1.2 Solving problems—the small case

Here's a simple example: find

$$\lim_{x \to 0} \frac{\sin(x^2)}{x^2}.$$

First note that when x is near 0, so is x^2, so we are indeed taking the sine of a small number. Now, we know that the following limit holds:

$$\lim_{x \to 0} \frac{\sin(x)}{x} = 1.$$

If you replace x by x^2 (which is a continuous function of x), then you get the following valid limit:

$$\lim_{x^2 \to 0} \frac{\sin(x^2)}{x^2} = 1.$$

This is almost the limit we want. In fact, the only thing we need to note is that $x^2 \to 0$ when $x \to 0$, so we can finally evaluate our limit as

$$\lim_{x \to 0} \frac{\sin(x^2)}{x^2} = 1.$$

Of course, there's nothing special about x^2; any other continuous function of x that is 0 when $x = 0$ will do. In particular, we know all the following limits automatically:

$$\lim_{x \to 0} \frac{\sin(5x)}{5x} = 1; \quad \lim_{x \to 0} \frac{\sin(3x^7)}{3x^7} = 1; \quad \text{and even} \quad \lim_{x \to 0} \frac{\sin(\sin(x))}{\sin(x)} = 1.$$

These are all true with "sin" replaced by "tan," but not by "cos"! Anyway, we can summarize the whole situation by noting that

$$\boxed{\lim_{x \to 0} \frac{\sin(\text{small})}{\text{same small}} = 1} \quad \text{and} \quad \boxed{\lim_{x \to 0} \frac{\tan(\text{small})}{\text{same small}} = 1.}$$

It's vital that the denominator **matches** the argument of sin or tan in the numerator, and also that this quantity is small when x is small. Of course, for cosine, the best we can say is

$$\boxed{\lim_{x \to 0} \cos(\text{small}) = 1.}$$

There's no need to worry about matching anything in this case!

Now let's return to one of the examples from the beginning of the chapter:

$$\lim_{x \to 0} \frac{\sin(5x)}{x}.$$

The problem is that we are taking the sine of $5x$, but we only have x in the denominator. These two quantities don't match. Never mind—we'll take that $\sin(5x)$ term and divide it by $5x$, which does match, then multiply it again to make it work out. That is, we'll rewrite $\sin(5x)$ as

$$\frac{\sin(5x)}{5x} \times (5x).$$

This is almost the same trick as the one we used in Section 4.3 of Chapter 4 for limits involving rational functions! Let's see how it works in this case:

$$\lim_{x \to 0} \frac{\sin(5x)}{x} = \lim_{x \to 0} \frac{\dfrac{\sin(5x)}{5x} \times (5x)}{x}.$$

Now keep the $\sin(5x)/5x$ part together, but cancel out a factor of x from the other two factors to get

$$\lim_{x \to 0} \frac{\sin(5x)}{x} = \lim_{x \to 0} \frac{\sin(5x)}{5x} \times 5,$$

As we saw above, since we have matched the $5x$ terms—once in the denominator and once in the argument of sin—we know that the fraction has limit 1, so the total limit is 5. In one line, the solution looks like this:

$$\lim_{x \to 0} \frac{\sin(5x)}{x} = \lim_{x \to 0} \frac{\dfrac{\sin(5x)}{5x} \times (5x)}{x} = \lim_{x \to 0} \frac{\sin(5x)}{5x} \times 5 = 1 \times 5 = 5.$$

Now let's check out a harder example. What is

$$\lim_{x \to 0} \frac{\sin^3(2x) \cos(5x^{19})}{x \tan(5x^2)}?$$

Let's look at the four factors of this expression one at a time. First, consider $\sin^3(2x)$. This is just another way of writing $(\sin(2x))^3$. To deal with $\sin(2x)$, we'd divide and multiply by $2x$; so to deal with its cube, we divide and multiply by $(2x)^3$ instead. That is, we'll replace $(\sin(2x))^3$ by

$$\frac{(\sin(2x))^3}{(2x)^3} \times (2x)^3.$$

How about the $\cos(5x^{19})$ factor? Well, when x is small, so is $5x^{19}$, so we are just taking the cosine of a small number. This should be 1 in the limit, so we don't touch this second factor.

In the denominator, we have a factor x, which we can't do anything with (nor do we want to—it's really easy to deal with already!). That leaves the $\tan(5x^2)$ factor. We simply divide and multiply by $(5x^2)$, so that we will be replacing $\tan(5x^2)$ by

$$\frac{\tan(5x^2)}{5x^2} \times (5x^2).$$

Putting all of this together, we have

$$\lim_{x \to 0} \frac{\sin^3(2x) \cos(5x^{19})}{x \tan(5x^2)} = \lim_{x \to 0} \frac{\left[\dfrac{(\sin(2x))^3}{(2x)^3} \times (2x)^3 \right] \cos(5x^{19})}{x \left[\dfrac{\tan(5x^2)}{5x^2} \times (5x^2) \right]}.$$

Now let's pull out all the powers of x that don't match the trig functions: the $(2x)^3$ term on the numerator and the x and $(5x^2)$ terms in the denominator. Then we rewrite the fraction $(\sin(2x))^3/(2x)^3$ as $(\sin(2x)/2x)^3$ and simplify to see that the limit becomes

$$\lim_{x \to 0} \frac{\dfrac{(\sin(2x))^3}{(2x)^3} \cdot \cos(5x^{19})}{\dfrac{\tan(5x^2)}{5x^2}} \times \frac{(2x)^3}{x(5x^2)} = \lim_{x \to 0} \frac{\left(\dfrac{\sin(2x)}{2x} \right)^3 \cos(5x^{19})}{\dfrac{\tan(5x^2)}{5x^2}} \times \frac{8x^3}{5x^3}.$$

Finally, we can cancel out x^3 from top and bottom, and take limits. Since the sin and tan terms have matching numerators and denominators, and $\cos(\text{small}) \to 1$, the limit is

$$\frac{(1)^3(1)}{1} \times \frac{8}{5} = \frac{8}{5}.$$

Here's another example from the beginning of the chapter: what is

$$\lim_{x \to \infty} x \sin\left(\frac{5}{x}\right)?$$

As we saw, this example does belong in this section, because when x is large, the quantity $5/x$ is small. So we use the same method, in this case dividing and multiplying $\sin(5/x)$ by $5/x$, to write:

$$\lim_{x \to \infty} x \sin\left(\frac{5}{x}\right) = \lim_{x \to \infty} x \cdot \frac{\sin\left(\dfrac{5}{x}\right)}{\dfrac{5}{x}} \times \frac{5}{x}.$$

Now we can cancel out a factor of x to simplify this down to

$$\lim_{x \to \infty} 5 \times \frac{\sin(5/x)}{5/x}.$$

Thinking of "small" as $5/x$, we can immediately see that the limit of the big fraction as $x \to \infty$ is 1, and so the overall answer is 5.

It's also possible to have trig limits involving sec, csc, or cot. For example, what is

$$\lim_{x \to 0} \sin(3x) \cot(5x) \sec(7x)?$$

To do this, the best bet is to write it in terms of cos, sin, or tan, as follows:

$$\lim_{x \to 0} (\sin(3x)) \left(\frac{1}{\tan(5x)}\right) \left(\frac{1}{\cos(7x)}\right).$$

Now we can do our standard trick of multiplying and dividing for the sin and tan terms, but ignoring the cos term, to see that the limit is equal to

$$\lim_{x \to 0} \left(\frac{\sin(3x)}{3x} \times (3x)\right) \left(\frac{1}{\dfrac{\tan(5x)}{5x} \times (5x)}\right) \left(\frac{1}{\cos(7x)}\right).$$

Now the $(3x)$ and the $(5x)$ terms cancel to leave $3/5$, and all the other fractions tend to 1 in the limit, so you can see that the overall limit is $3/5$.

There is one thing you have to be very careful of: when you say that $\sin(x)$ behaves like x when x is small, you should only use this fact in the context of products or quotients. For example,

$$\lim_{x \to 0} \frac{x - \sin(x)}{x^3}$$

cannot be done by the methods of this chapter. It is a mistake to say that $\sin(x)$ behaves like x, so $x - \sin(x)$ behaves like 0. (In fact, nothing behaves like 0 except for the constant function 0 itself!) In order to solve the above limit, you need l'Hôpital's Rule (see Chapter 14) or Maclaurin series (see Chapter 24). On the other hand, here's a limit which has a similar difficulty that we can nevertheless solve now:

$$\lim_{x \to 0} \frac{1 - \cos^2(x)}{x^2}.$$

Again, you can't just say that $\cos(x)$ behaves like 1 when x is small, so $1 - \cos^2(x)$ behaves like $1 - 1^2 = 0$. So we just use $\cos^2(x) + \sin^2(x) = 1$ to rewrite the numerator as $\sin^2(x)$:

$$\lim_{x \to 0} \frac{1 - \cos^2(x)}{x^2} = \lim_{x \to 0} \frac{\sin^2(x)}{x^2}.$$

Since $\sin^2(x)$ is another way of writing $(\sin(x))^2$, we can rewrite the limit as

$$\lim_{x \to 0} \frac{(\sin(x))^2}{x^2} = \lim_{x \to 0} \left(\frac{\sin(x)}{x} \right)^2.$$

This limit is simply $1^2 = 1$. So

$$\lim_{x \to 0} \frac{1 - \cos^2(x)}{x^2} = 1.$$

 In effect, we're saying that $1 - \cos^2(x)$ behaves like x^2 when x is small, not like 0 after all. Anyway, let's use the same idea to solve some other limits:

$$\lim_{x \to 0} \frac{1 - \cos(x)}{x^2} \qquad \text{and} \qquad \lim_{x \to 0} \frac{1 - \cos(x)}{x}.$$

We'll do both of these limits with the same clever trick. The idea is to multiply top and bottom by $1 + \cos(x)$ so that the numerator becomes $1 - \cos^2(x)$, which we write as $\sin^2(x)$. In the first case, we have

$$\lim_{x \to 0} \frac{1 - \cos(x)}{x^2} = \lim_{x \to 0} \frac{1 - \cos(x)}{x^2} \times \frac{1 + \cos(x)}{1 + \cos(x)}$$

$$= \lim_{x \to 0} \frac{1 - \cos^2(x)}{x^2} \times \frac{1}{1 + \cos(x)} = \lim_{x \to 0} \frac{\sin^2(x)}{x^2} \times \frac{1}{1 + \cos(x)}$$

$$= \lim_{x \to 0} \left(\frac{\sin(x)}{x} \right)^2 \times \frac{1}{1 + \cos(x)} = 1^2 \times \frac{1}{1 + 1} = \frac{1}{2}.$$

Here we used the fact that $\cos(0) = 1$. The second example is similar:

$$\lim_{x \to 0} \frac{1 - \cos(x)}{x} = \lim_{x \to 0} \frac{1 - \cos(x)}{x} \times \frac{1 + \cos(x)}{1 + \cos(x)}$$

$$= \lim_{x \to 0} \frac{1 - \cos^2(x)}{x} \times \frac{1}{1 + \cos(x)} = \lim_{x \to 0} \frac{\sin^2(x)}{x} \times \frac{1}{1 + \cos(x)}.$$

At this point, we could divide and multiply the $\sin^2(x)$ term by x^2, but here's a simpler way to handle the limit: simply write $\sin^2(x)$ as $\sin(x) \times \sin(x)$, and group one of the $\sin(x)$ factors with the x in the denominator. The limit becomes

$$\lim_{x \to 0} \left(\sin(x) \times \frac{\sin(x)}{x} \times \frac{1}{1 + \cos(x)} \right) = 0 \times 1 \times \frac{1}{1 + 1} = 0,$$

since $\sin(0) = 0$. This last limit will be useful in Section 7.2 below, so let's summarize it as something to keep in mind:

$$\boxed{\lim_{x \to 0} \frac{1 - \cos(x)}{x} = 0.}$$

Enough of the small case—let's see how to deal with limits involving trig functions evaluated at large numbers.

7.1.3 The large case

Consider the limit

$$\lim_{x \to \infty} \frac{\sin(x)}{x}.$$

As we just saw, if $x \to 0$ instead of ∞, then the limit is 1. This is because $\sin(x)$ behaves like x when x is small. How does $\sin(x)$ behave when x gets larger and larger? It just keeps on oscillating between -1 and 1. So it doesn't really "behave" like anything when x is large. Often one is forced to resort to one of the simplest things you can say about $\sin(x)$ (and also $\cos(x)$):

$$\boxed{-1 \le \sin(x) \le 1} \qquad \text{and} \qquad \boxed{-1 \le \cos(x) \le 1} \qquad \text{for any } x.$$

This is pretty darn handy for applying the sandwich principle (see Section 3.6 in Chapter 3). In fact, we saw on page 53 that

$$\lim_{x \to \infty} \frac{\sin(x)}{x} = 0.$$

Take a look back at the proof right now to refresh your memory.

Remember how $\cos(x)$ is the odd one out when x is small? Unlike $\sin(x)$ and $\tan(x)$, it doesn't behave like x itself. When x is large, on the other hand, $\tan(x)$ is the odd one out. There are no inequalities for $\tan(x)$ similar to the boxed inequalities for $\sin(x)$ and $\cos(x)$ above; this is because $\tan(x)$ keeps on having vertical asymptotes and never settles down when x becomes large (see page 37 for the graph of $y = \tan(x)$).

Here's a much harder example using the sandwich principle: find

$$\lim_{x \to \infty} \frac{x \sin(11x^7) - \frac{1}{2}}{2x^4}.$$

The gut feeling is that the $\sin(11x^7)$ term isn't doing much, so the top is really of size about x. The x^4 on the bottom should overwhelm the numerator, so the whole thing should go to 0 as $x \to \infty$. In order to show this, let's look at the numerator first. We know that the sine of any number is between -1 and 1, so it's true that

$$-1 \le \sin(11x^7) \le 1.$$

The numerator isn't just $\sin(11x^7)$, though: we need to multiply by x and then subtract $1/2$. We can in fact multiply by x and then subtract $1/2$ from all three "sides" of the above inequality to get

$$-x - \frac{1}{2} \le x \sin(11x^7) - \frac{1}{2} \le x - \frac{1}{2}$$

for any $x > 0$. (If instead $x < 0$, which would be the case if the limit were as $x \to -\infty$, then multiplying by the negative number x would just mean that you'd have to flip those less-than-or-equal signs around to become greater-than-or-equal signs. Otherwise the solution would be identical.) Anyway, that takes care of the numerator. We still need to divide by the denominator. Since $2x^4 > 0$, we can divide the above inequality by $2x^4$ to get

$$\frac{-x - \frac{1}{2}}{2x^4} \le \frac{x \sin(11x^7) - \frac{1}{2}}{2x^4} \le \frac{x - \frac{1}{2}}{2x^4}.$$

This is all we need. I leave it to you to use the methods of Section 4.3 of Chapter 4 to show that the limits of the outside terms are both 0 as $x \to \infty$, that is,

$$\lim_{x \to \infty} \frac{-x - \frac{1}{2}}{2x^4} = 0 \qquad \text{and} \qquad \lim_{x \to \infty} \frac{x - \frac{1}{2}}{2x^4} = 0.$$

(Don't be lazy! These are pretty easy limits, but you should try to justify them now.) Now we invoke the sandwich principle; since our original function is trapped between two functions which tend to 0 as $x \to \infty$, it also tends to 0 then. That is,

$$\lim_{x \to \infty} \frac{x \sin(11x^7) - \frac{1}{2}}{2x^4} = 0.$$

Another consequence of the inequality $-1 \le \sin(x) \le 1$ (and the similar one for $\cos(x)$) is that you can treat sin(anything) or cos(anything) as being of lower degree than any positive power of x, so long as you are only adding or subtracting. More precisely, if you are solving a problem of the form

$$\lim_{x \to \infty} \frac{p(x)}{q(x)},$$

where p and q are polynomials or poly-type functions but with some sines and cosines added on, then the degrees of the top and bottom are the same as they would be without the sines and cosines added on. The only exception is when p or q has degree 0; then the trig part could be significant.

Here's an example of how adding sines and cosines doesn't make much of a difference: what is

$$\lim_{x \to \infty} \frac{3x^2 + 2x + 5 + \sin(3000x^9)}{2x^2 - 1 - \cos(22x)}?$$

In the numerator, the dominant term is still $3x^2$, since the $\sin(3000x^9)$ term is only between -1 and 1 and is insignificant in comparison. Compare this to the previous example, where we **multiplied** the highest-degree term x by $\sin(11x^7)$; there the sine factor matters. In our current example, the sine term is **added** instead.

How about the denominator? Well, the cosine term is much smaller than the dominant term $2x^2$. All up, we'll multiply and divide the numerator by $3x^2$ and the denominator by $2x^2$:

$$\lim_{x \to \infty} \frac{3x^2 + 2x + 5 + \sin(3000x^9)}{2x^2 - 1 - \cos(22x)} = \lim_{x \to \infty} \frac{\dfrac{3x^2 + 2x + 5 + \sin(3000x^9)}{3x^2} \times (3x^2)}{\dfrac{2x^2 - 1 - \cos(22x)}{2x^2} \times (2x^2)}$$

$$= \lim_{x \to \infty} \frac{1 + \dfrac{2}{3x} + \dfrac{5}{3x^2} + \dfrac{\sin(3000x^9)}{3x^2}}{1 - \dfrac{1}{2x^2} - \dfrac{\cos(22x)}{2x^2}} \times \frac{3x^2}{2x^2}.$$

Now what happens? We certainly know that $2/3x$, $5/3x^2$, and $1/2x^2$ go to 0 in the limit, but how about the $\sin(3000x^9)/3x^2$ and $\cos(22x)/2x^2$ terms? If you want to give a complete solution, you need to use the sandwich principle (once for each term) to show that they both go to 0. I suggest you try it as an exercise now. In practice, most mathematicians would automatically write down the answer 0, having established the general principle that

$$\lim_{x \to \infty} \frac{\sin(\text{anything})}{x^\alpha} = 0$$

for any positive exponent α, and similarly when sine is replaced by cosine. In any case, the above limit works out to be

$$\frac{1 + 0 + 0 + 0}{1 - 0 - 0} \times \frac{3}{2} = \frac{3}{2}.$$

Finally, let's return to the example

$$\lim_{x \to 0} x \sin \left(\frac{5}{x} \right),$$

which was mentioned at the beginning of this chapter. As we saw then, this does belong to the large case even though the limit is taken as $x \to 0$, because $5/x$ is a large number (positive or negative) when x is near 0. So the best we can do is to use the sandwich principle, combined with the fact that the sine of any number is between -1 and 1. In particular, we have

$$-1 \leq \sin \left(\frac{5}{x} \right) \leq 1$$

for any x. Now the temptation is to multiply by x:

$$-x \leq x \sin \left(\frac{5}{x} \right) \leq x.$$

Unfortunately, this is only true for $x > 0$. For example, if $x = -2$, then the leftmost part of the inequality would be 2 and the rightmost part would be -2, which is crazy. So let's worry about the right-hand limit first:

$$\lim_{x \to 0^+} x \sin \left(\frac{5}{x} \right).$$

Now we can use the above inequalities and note that both $-x$ and x go to 0 as $x \to 0^+$, so the sandwich principle applies and the above limit is 0. As for the left-hand limit (as $x \to 0^-$), now we start off with the same inequality for $\sin(5/x)$ and multiply it by x, but this time we have to reverse the inequalities since x is negative. In particular, when $x < 0$, we have

$$-x \geq x \sin \left(\frac{5}{x} \right) \geq x.$$

It doesn't matter much, though—the outer quantities still go to 0 as $x \to 0^-$, so the middle quantity also goes to 0. Since the left-hand and right-hand limits are both 0, so is the two-sided limit; we have proved that

$$\lim_{x \to 0} x \sin \left(\frac{5}{x} \right) = 0.$$

(This example is very similar to the one on page 52.)

7.1.4 The "other" case

Consider the limit

$$\lim_{x \to \pi/2} \frac{\cos(x)}{x - \frac{\pi}{2}}.$$

The trig function, cosine in this case, is being evaluated near $\pi/2$. This is neither small nor large, so apparently the previous cases don't apply. If you just plug in $x = \pi/2$, you get the indeterminate form $0/0$, which sucks. If you know your trig identities, though, you're golden. Here's why.

A good general principle when dealing with a limit involving $x \to a$ for some $a \neq 0$ is to **shift the problem to 0 by substituting $t = x - a$.** So in the above limit, set $t = x - \pi/2$. Then when $x \to \pi/2$, you can see that $t \to 0$. Also, $x = t + \pi/2$, so we have

$$\lim_{x \to \pi/2} \frac{\cos(x)}{x - \frac{\pi}{2}} = \lim_{t \to 0} \frac{\cos\left(t + \frac{\pi}{2}\right)}{t}.$$

Notice that we still need to know the behavior of cosine near $\pi/2$ (as you can see by setting t near 0 and looking what you're taking cosine of!); the substitution hasn't changed that fact. Now, this is where you need to know the following trig identity from Section 2.4 of Chapter 2:

$$\cos\left(\frac{\pi}{2} - x\right) = \sin(x).$$

In our limit, we have $\cos(\frac{\pi}{2} + t)$, so we need to apply the above trig identity with x replaced by $-t$ in order for it to be useful. We get

$$\cos\left(\frac{\pi}{2} + t\right) = \sin(-t).$$

The other thing we need to remember is that sine is an odd function, so in fact

$$\cos\left(\frac{\pi}{2} + t\right) = \sin(-t) = -\sin(t).$$

Now we can put this into the limit and finish the problem. All in all,

$$\lim_{x \to \pi/2} \frac{\cos(x)}{x - \frac{\pi}{2}} = \lim_{t \to 0} \frac{\cos\left(t + \frac{\pi}{2}\right)}{t} = \lim_{t \to 0} \frac{-\sin(t)}{t} = -1.$$

Not so easy, but knowing the trig identities certainly helps in situations like these.

7.1.5 Proof of an important limit

We've been using the following limit over and over again in this chapter, and now it's time to prove it:

$$\lim_{x \to 0} \frac{\sin(x)}{x} = 1.$$

The proof has to rely on the geometry of right-angled triangles, since that's where the sine function comes from. Let's start with the right-hand limit (as $x \to 0^+$). Once we get that, we'll see that the two-sided limit is pretty easy. So, we'll start off by assuming that x is near 0 but positive. Let's draw a wedge OAB of a circle of radius 1 unit with angle x:

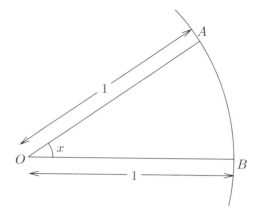

We're going to doctor this figure a little, but first, a question: what is the area of this wedge? Imagine that the wedge is a slice from a big pizza. The pizza has radius 1 unit, so its area is $\pi r^2 = \pi$ square units. Now, how much of the pizza do we have in our slice? The whole pizza has 2π radians of angle, while the slice has x radians, so the slice accounts for $x/2\pi$ of the pizza. The area is therefore $(x/2\pi) \times \pi$, or simply $x/2$ square units. That is,

$$\text{area of wedge } OAB \;=\; \frac{x}{2} \text{ square units.}$$

(This is a special case of the general formula: the area of a wedge of angle x radians in a circle of radius r units is simply $xr^2/2$ square units.)

Now let's do a few things to the figure. First, we'll draw in the line AB. Then we'll drop a perpendicular from A down to the line OB; call the base point C. We'll also extend the line OA out a little bit, and finally draw the tangent line to the circle at the point B. That tangent line intersects the extended line OA at a point D. After we do all that, we get the following picture:

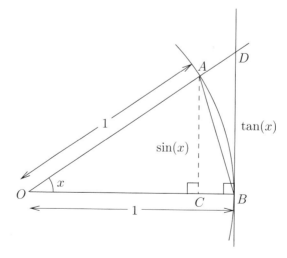

I marked the lengths of AC and DB on the diagram. To see how I worked out these lengths, note that $\sin(x) = \frac{|AC|}{|OA|}$ (remember, $|AC|$ means "the length of the segment AC"). Since $|OA| = 1$, we have $|AC| = \sin(x)$. Also, we have $\tan(x) = \frac{|DB|}{|OB|}$, and $|OB| = 1$, so $|DB| = \tan(x)$.

I want to focus attention on three objects. One is the original wedge; we already found that the area of this is $x/2$ square units. Let's also look at the triangles $\triangle OAB$ and $\triangle OBD$. The base of $\triangle OAB$ is OB, which has length 1 unit. The height is AC, which has length $\sin(x)$ units. So the area of $\triangle OAB$ is half the base times the height, or $\sin(x)/2$ square units. As for $\triangle OBD$, its base OB has length 1 unit and its height DB has length $\tan(x)$ units, so the area of $\triangle OBD$ is $\tan(x)/2$ square units. The crucial observation is that

$\triangle OAB$ is contained in the wedge OAB which is contained in $\triangle OBD$.

This means that the area of $\triangle OAB$ is less than the area of the wedge OAB, which itself is less than the area of $\triangle OBD$:

$$\text{area of } \triangle OAB < \text{area of wedge } OAB < \text{area of } \triangle OBD.$$

We know all three of these quantities in terms of the variable x; substituting them in, we have

$$\frac{\sin(x)}{2} < \frac{x}{2} < \frac{\tan(x)}{2}.$$

Multiplying this by 2, we get a really nice inequality which is worth remembering:

$$\boxed{\sin(x) < x < \tan(x) \qquad \text{for } 0 < x < \frac{\pi}{2}.}$$

Now we can find our limit. Let's first take reciprocals of the nice inequality. Remember, this forces us to switch the less-than signs to greater-than signs. Writing $\tan(x) = \sin(x)/\cos(x)$, the reciprocal inequality is

$$\frac{1}{\sin(x)} > \frac{1}{x} > \frac{\cos(x)}{\sin(x)}.$$

Finally, multiply by the positive quantity $\sin(x)$ to see that

$$1 > \frac{\sin(x)}{x} > \cos(x).$$

If it creeps you out to write it backward like this, you can always rewrite it as

$$\cos(x) < \frac{\sin(x)}{x} < 1.$$

(Remember, this is true for any x between 0 and $\pi/2$.) Now we use the sandwich principle: since $\cos(0) = 1$ and $y = \cos(x)$ is continuous, we know that $\lim_{x \to 0^+} \cos(x) = 1$. Also, $\lim_{x \to 0^+} (1) = 1$; so the quantity $\sin(x)/x$ is squished between $\cos(x)$ and 1, both of which tend to 1 as $x \to 0^+$. By the sandwich principle,

$$\lim_{x \to 0^+} \frac{\sin(x)}{x} = 1$$

as well. So we've got our right-hand limit.

We still have to deal with the left-hand limit and show that

$$\lim_{x \to 0^-} \frac{\sin(x)}{x} = 1.$$

If we can do it, then we will have proved that both the left-hand and right-hand limits are 1, so the two-sided limit is also 1 and we'll be done.

To prove that the left-hand limit is 1, set $t = -x$. Then when x is a small negative number, t is a small positive number. In math symbols, we can say that as $x \to 0^-$, we have $t \to 0^+$. So the above limit can be written as

$$\lim_{t \to 0^+} \frac{\sin(-t)}{-t}.$$

Now we know that $\sin(-t) = -\sin(t)$ (since sine is an odd function), so we can simplify the limit down to

$$\lim_{t \to 0^+} \frac{-\sin(t)}{-t} = \lim_{t \to 0^+} \frac{\sin(t)}{t}.$$

We've already seen that this limit is 1 (well, with x instead of t, but so what?), so we're all done.

Before we move on to differentiating trig functions, I want to consider the graph of $f(x) = \sin(x)/x$. The argument for the left-hand limit has in fact shown that f is an even function (can you see why?). This means that the y-axis acts as a mirror for the graph of $y = f(x)$. If you look back at page 51, you can see that we have already drawn the graph of $y = f(x)$ when $x > 3$. We didn't do $x \le 3$ since we didn't know what happens. Now we know: as $x \to 0$, the quantity $f(x) = \sin(x)/x \to 1$. In fact, we have shown that $\sin(x)/x$ lies between $\cos(x)$ and 1. This allows us to extend the graph down to $x > 0$. Finally we use the evenness of f to give the complete graph of $y = \sin(x)/x$ in all its glory (note the different scales on the x- and y-axes):

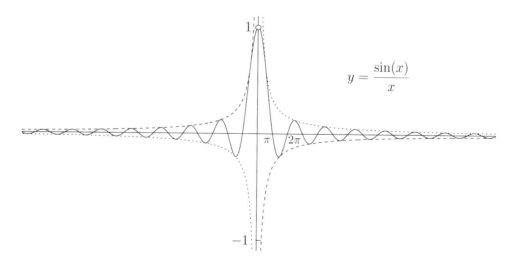

The graphs of the envelope functions $y = 1/x$ and $y = -1/x$ are shown as dotted curves. Also, the x-intercepts are at all the multiples of π except for

0. Finally, as you can see, the function isn't continuous at $x = 0$ since it isn't defined there. However, if we define the function g by $g(x) = \sin(x)/x$ if $x \neq 0$ and $g(0) = 1$, then we have effectively filled in the open circle at $(0, 1)$ in the above picture, and the function g is continuous.

7.2 Derivatives Involving Trig Functions

Now, time to differentiate some functions. Let's start off by differentiating $\sin(x)$ with respect to x. To do this, we're going to use two of the limits from Section 7.1.2 above:

$$\lim_{h \to 0} \frac{\sin(h)}{h} = 1 \quad \text{and} \quad \lim_{h \to 0} \frac{1 - \cos(h)}{h} = 0.$$

(OK, so I changed x to h, but no matter—the h is a dummy variable anyway and could be replaced by any letter at all.) Anyway, with $f(x) = \sin(x)$, let's differentiate:

$$f'(x) = \lim_{h \to 0} \frac{f(x + h) - f(x)}{h} = \lim_{h \to 0} \frac{\sin(x + h) - \sin(x)}{h}.$$

Now what? Well, you should remember the formula

$$\sin(A + B) = \sin(A)\cos(B) + \cos(A)\sin(B);$$

if not, you'd better look at Chapter 2 again. Anyway, we want to replace A by x and B by h, so we have

$$\sin(x + h) = \sin(x)\cos(h) + \cos(x)\sin(h).$$

Inserting this in the above limit, we get

$$f'(x) = \lim_{h \to 0} \frac{\sin(x)\cos(h) + \cos(x)\sin(h) - \sin(x)}{h}.$$

All that's left is to group the terms a little differently and do a bit of factoring; we get

$$
\begin{aligned}
f'(x) &= \lim_{h \to 0} \frac{\sin(x)(\cos(h) - 1) + \cos(x)\sin(h)}{h} \\
&= \lim_{h \to 0} \left(\sin(x) \left(\frac{\cos(h) - 1}{h} \right) + \cos(x) \left(\frac{\sin(h)}{h} \right) \right).
\end{aligned}
$$

Notice that we separated as much x-stuff as we could from h-stuff. Now we actually have to take the limit as $h \to 0$ (not as $x \to 0$!). Using the two limits from the beginning of this section, we get

$$f'(x) = \sin(x) \times 0 + \cos(x) \times 1 = \cos(x).$$

That is, the derivative of $f(x) = \sin(x)$ is $f'(x) = \cos(x)$, or in other words,

$$\boxed{\frac{d}{dx}\sin(x) = \cos(x).}$$

Now you should try to repeat the argument but this time with $f(x) = \cos(x)$. You just need the identity

$$\cos(A + B) = \cos(A)\cos(B) - \sin(A)\sin(B)$$

from Chapter 2. It's a really good exercise, so try to do it now. If you've done it correctly, you should see that

$$\boxed{\frac{d}{dx}\cos(x) = -\sin(x).}$$

Anyway, it's a piece of cake to get the derivatives of the other trig functions now; you don't need to use any limits. You can just use the quotient rule and the chain rule. Let's start with the derivative of $y = \tan(x)$. We can write $\tan(x)$ as $\sin(x)/\cos(x)$, so if we set $u = \sin(x)$ and $v = \cos(x)$, then $y = u/v$. We just worked out that $du/dx = \cos(x)$ and $dv/dx = -\sin(x)$. Using the quotient rule, we get

$$\frac{dy}{dx} = \frac{v\dfrac{du}{dx} - u\dfrac{dv}{dx}}{v^2} = \frac{\cos(x)(\cos(x)) - \sin(x)(-\sin(x))}{\cos^2(x)}.$$

The numerator of this last fraction is just $\cos^2(x) + \sin^2(x)$, which is always equal to 1; so the derivative is just

$$\frac{dy}{dx} = \frac{1}{\cos^2(x)} = \sec^2(x).$$

We've just shown that

$$\boxed{\frac{d}{dx}\tan(x) = \sec^2(x).}$$

Now let's calculate the derivative of $y = \sec(x)$. Here we are able to write $y = 1/\cos(x)$, so you might think that the quotient rule is best. Indeed, you can do it by using the quotient rule, but the chain rule is nicer. If $u = \cos(x)$, then $y = 1/u$. We can differentiate both of these things: $dy/du = -1/u^2$, and $du/dx = -\sin(x)$. By the chain rule,

$$\frac{dy}{dx} = \frac{dy}{du}\frac{du}{dx} = \left(-\frac{1}{u^2}\right)(-\sin(x)) = \frac{\sin(x)}{\cos^2(x)},$$

where we had to replace u by $\cos(x)$ in the last step. Actually, you can tidy up the answer as follows:

$$\frac{\sin(x)}{\cos^2(x)} = \frac{1}{\cos(x)}\frac{\sin(x)}{\cos(x)} = \sec(x)\tan(x),$$

so we've shown that

$$\boxed{\frac{d}{dx}\sec(x) = \sec(x)\tan(x).}$$

As for $y = \csc(x)$, that should be written as $1/\sin(x)$. Once again, it's best to use the chain rule, letting $u = \sin(x)$ and writing $y = 1/u$. But I know you want to use the quotient rule, since it's a quotient, even though it's inferior. You just don't believe me. Well, check this. To use the quotient rule on $y = 1/\sin(x)$, we'll actually let $u = 1$ and $v = \sin(x)$. Then $du/dx = 0$ and $dv/dx = \cos(x)$. By the quotient rule,

$$\frac{dy}{dx} = \frac{v\dfrac{du}{dx} - u\dfrac{dv}{dx}}{v^2} = \frac{\sin(x)(0) - 1(\cos(x))}{\sin^2(x)} = -\frac{\cos(x)}{\sin^2(x)}.$$

OK, it wasn't that bad, but the chain rule is still nicer. Anyway, by splitting up the answer as we just did for the derivative of $y = \sec(x)$, you should be able to simplify it to get

$$\boxed{\frac{d}{dx}\csc(x) = -\csc(x)\cot(x).}$$

Finally, consider $y = \cot(x)$, which of course can be written as either $y = \cos(x)/\sin(x)$ or $y = 1/\tan(x)$. You could use the quotient rule on $y = \cos(x)/\sin(x)$, or now that we know the derivative of $\tan(x)$, you could use the chain rule (or even the quotient rule) on $y = 1/\tan(x)$. You could even write $\cot(x)$ as the product $\cos(x)\csc(x)$ and use the product rule. Whichever way you do it, you should get

$$\boxed{\frac{d}{dx}\cot(x) = -\csc^2(x).}$$

You should learn all six boxed formulas by heart. Notice that the three cofunctions (cos, csc, cot) all have minus signs in front of them, and the derivatives are the co- versions of the regular ones. For example, the derivative of $\sec(x)$ is $\sec(x)\tan(x)$, so throwing a "co" in front of everything and also putting in a minus sign, we get that the derivative of $\csc(x)$ is $-\csc(x)\cot(x)$. The same is true for cos and cot, remembering (in the case of cos) that co-co-sine is just the original sine function.

By the way, what is the second derivative of $f(x) = \sin(x)$? We know that $f'(x) = \cos(x)$, and so $f''(x)$ is the derivative of $\cos(x)$, which we saw is $-\sin(x)$. That is,

$$\frac{d^2}{dx^2}(\sin(x)) = -\sin(x).$$

The second derivative of the function is just the negative of the original function. The same is true for $g(x) = \cos(x)$. This sort of thing doesn't happen at all with (nonzero) polynomials, since the derivative of a polynomial is a new polynomial whose degree is one less than the original one.

7.2.1 Examples of differentiating trig functions

Now that you have some more functions to differentiate, you'd better make sure you still know how to use the product rule, the quotient rule, and the chain rule. For example, how would you find the following derivatives:

$$\frac{d}{dx}(x^2\sin(x)), \qquad \frac{d}{dx}\left(\frac{\sec(x)}{x^5}\right) \qquad \text{and} \qquad \frac{d}{dx}(\cot(x^3))?$$

Let's take them one at a time. If $y = x^2 \sin(x)$, then we can write $y = uv$ where $u = x^2$ and $v = \sin(x)$. Now we just need to set up our table:

$$u = x^2 \qquad\qquad\qquad v = \sin(x)$$

$$\frac{du}{dx} = 2x \qquad\qquad\qquad \frac{dv}{dx} = \cos(x).$$

Using the product rule (see Section 6.2.3 in the previous chapter), we get

$$\frac{dy}{dx} = v\frac{du}{dx} + u\frac{dv}{dx} = \sin(x) \cdot (2x) + x^2 \cos(x).$$

This would normally be written as $2x \sin(x) + x^2 \cos(x)$. Anyway, let's do the second example. If $y = \sec(x)/x^5$, this time we set $u = \sec(x)$ and $v = x^5$ so that $y = u/v$. Our table looks like this:

$$u = \sec(x) \qquad\qquad\qquad v = x^5$$

$$\frac{du}{dx} = \sec(x)\tan(x) \qquad\qquad\qquad \frac{dv}{dx} = 5x^4.$$

Whipping out the quotient rule leads to

$$\frac{dy}{dx} = \frac{v\dfrac{du}{dx} - u\dfrac{dv}{dx}}{v^2} = \frac{x^5 \sec(x)\tan(x) - \sec(x) \cdot 5x^4}{(x^5)^2} = \frac{\sec(x)(x\tan(x) - 5)}{x^6}.$$

Note that we canceled out a factor of x^4 at the end. Now, moving on to the third example, set $y = \cot(x^3)$. Here we are dealing with a composition of two functions, so we'd better use the chain rule. The first thing that happens to x is that it gets cubed, so let $u = x^3$. Then $y = \cot(u)$. Our table is

$$y = \cot(u) \qquad\qquad\qquad u = x^3$$

$$\frac{dy}{du} = -\csc^2(u) \qquad\qquad\qquad \frac{du}{dx} = 3x^2.$$

By the chain rule, we have

$$\frac{dy}{dx} = \frac{dy}{du}\frac{du}{dx} = -\csc^2(u) \cdot 3x^2.$$

We can't just leave that u term lying around—we need to replace it by x^3. Altogether, then, our derivative is $-3x^2 \csc^2(x^3)$.

Before we move on, I want to show you a neat trick. Suppose you have $y = \sin(8x)$ and you want to find dy/dx. You could do it by using the chain rule, setting $u = 8x$, so that $y = \sin(u)$. It's an easy exercise (try it!) to show that $dy/dx = 8\cos(8x)$. Of course, there's nothing special about the number 8; it could have been anything. So the general rule is that

$$\frac{d}{dx}(\sin(ax)) = a\cos(ax)$$

for any constant a. Basically, **if x is replaced by ax, then there is an extra factor of a out front when you differentiate**. This also works

for the other trig functions. For example, the derivative with respect to x of $\tan(x)$ is $\sec^2(x)$, so the derivative of $\tan(2x)$ is $2\sec^2(2x)$. In the same way, the derivative of $\csc(x)$ is $-\csc(x)\cot(x)$, so the derivative of $\csc(19x)$ is $-19\csc(19x)\cot(19x)$. This saves you the trouble of using the chain rule in this easy case.

7.2.2 Simple harmonic motion

One place where trig functions appear naturally is in describing the motion of a weight on a spring bouncing up and down. It turns out that if x is the position of a weight on a spring at time t, taking upward as positive, then a possible equation for x is something like $x = 3\sin(4t)$. The numbers 3 and 4 might change, and the "sin" might be a "cos," but that's the basic idea. The equation is reasonable—after all, cosine keeps bouncing back and forth, and so does the weight. This sort of motion is called *simple harmonic motion*.

So, if $x = 3\sin(4t)$ is the displacement of the weight from its starting point, what are the velocity and the acceleration of the weight at time t? All we have to do is differentiate. We know that $v = dx/dt$, so we just have to differentiate $3\sin(4t)$ with respect to t. We could use the chain rule, but it's simpler to use the observation at the end of the previous section. Indeed, to differentiate $\sin(4t)$ with respect to t, we just observe that the derivative of $\sin(t)$ would be $\cos(t)$, so the derivative of $\sin(4t)$ is $4\cos(4t)$. (Don't forget that factor of 4 out front!) All in all, we have

$$v = \frac{d}{dt}(3\sin(4t)) = 3 \times 4\cos(4t) = 12\cos(4t).$$

Now we can repeat the exercise for acceleration, which is given by dv/dt, using the same technique:

$$a = \frac{dv}{dt} = \frac{d}{dt}(12\cos(4t)) = -12 \times 4\sin(4t) = -48\sin(4t).$$

Notice that the acceleration—which of course is the second derivative of the displacement—is basically the same as the displacement itself, except that there's a minus out front and the coefficient is different (48 instead of 3). The minus means that the acceleration is in the opposite direction from the displacement. In fact, we have shown that

$$a = -16x,$$

since $48 = 3 \times 16$. Now let's interpret this equation by examining the motion of the weight a little more closely.

The position x is given by $x = 3\sin(4t)$, with the understanding that the rest position of the weight is at $x = 0$. Now, if we multiply the inequality $-1 \le \sin(4t) \le 1$ (which is true for all t) by 3, we get $-3 \le 3\sin(4t) \le 3$. That is, $-3 \le x \le 3$. So we can see that x is oscillating between -3 and 3. When x is positive, the weight is above its rest position; then a is negative, which is good: the acceleration is downward, as it should be. As x gets bigger and bigger, the spring compresses even more, causing the weight to experience a greater force and acceleration downward. Eventually the weight starts going down, and after a little while x becomes negative. Then the weight is below its rest position, so the spring is expanded and tries to pull the weight back up. Indeed, when x is negative, a is positive, so the force is upward. The following picture shows what's going on:

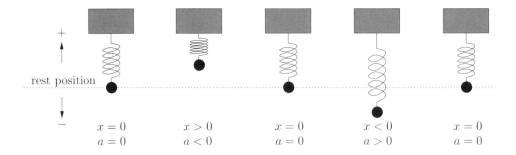

When the weight is at the top of its motion, the velocity is 0. Since we have $v = -12\cos(4t)$, this occurs whenever $4t$ is an odd multiple of $\pi/2$, that is, when $t = (2n + 1)\pi/8$ for some integer n. Now, enough about simple harmonic motion—let's just look at one more example of trig differentiation before moving on to implicit differentiation in the next chapter.

7.2.3 A curious function

Consider the function f given by

$$f(x) = x^2 \sin\left(\frac{1}{x}\right).$$

What is its derivative? We'd better not worry about $x = 0$, since f isn't defined there, but we'll be fine for other values of x. Set $y = f(x)$; then y is the product of $u = x^2$ and $v = \sin(1/x)$. It's easy to differentiate u with respect to x (the answer is just $2x$), but v is a little harder. The best bet is to set $w = 1/x$, so that $v = \sin(w)$. Then we can draw up our standard table:

$$v = \sin(w) \qquad\qquad w = \frac{1}{x}$$

$$\frac{dv}{dw} = \cos(w) \qquad\qquad \frac{dw}{dx} = -\frac{1}{x^2}.$$

Now we can use the chain rule:

$$\frac{dv}{dx} = \frac{dv}{dw}\frac{dw}{dx} = \cos(w)\left(-\frac{1}{x^2}\right) = -\frac{\cos(1/x)}{x^2}.$$

Now that we have du/dx and dv/dx, we can finally use the product rule on $y = uv$:

$$\frac{dy}{dx} = v\frac{du}{dx} + u\frac{dv}{dx} = \sin\left(\frac{1}{x}\right)(2x) + x^2\left(-\frac{\cos(1/x)}{x^2}\right) = 2x\sin\left(\frac{1}{x}\right) - \cos\left(\frac{1}{x}\right),$$

 and we're done.

It turns out that the function f is pretty curious. Let's see why. (If you don't feel like it, I guess you can go on to the next chapter and come back to it later.) Anyway, to investigate further, we'll need the following three limits:

$$\lim_{x\to 0} x^2 \sin\left(\frac{1}{x}\right) = 0, \quad \lim_{x\to 0} x \sin\left(\frac{1}{x}\right) = 0, \quad \text{and} \quad \lim_{x\to 0^+} \cos\left(\frac{1}{x}\right) \text{ DNE}.$$

You can do the first two of these limits using the sandwich principle and the fact that sine or cosine of anything (even $1/x$) is between -1 and 1. The third limit is a little trickier, but we did it for $\sin(1/x)$ in Section 3.3 of Chapter 3, and changing sin to cos doesn't make any difference. The issue (as you may recall) is that the oscillations of $\cos(1/x)$ between -1 and 1 become more and more wild as $x \to 0^+$, so the limit doesn't exist.

Anyway, the first limit says that $\lim_{x \to 0} f(x) = 0$, even though $f(0)$ is undefined. This means that we can extend f to be continuous by filling in the point $f(0) = 0$. So we'll throw away the old f from above and define a new one by the following formula:

$$f(x) = \begin{cases} x^2 \sin\left(\dfrac{1}{x}\right) & \text{if } x \neq 0, \\ 0 & \text{if } x = 0. \end{cases}$$

We have just shown that this new, improved f is continuous everywhere. We have already found its derivative when $x \neq 0$:

$$f'(x) = 2x \sin\left(\frac{1}{x}\right) - \cos\left(\frac{1}{x}\right).$$

So, what's the derivative of f at $x = 0$? None of our fancy-shmancy rules will help here: we have to use the formula for the derivative:

$$f'(0) = \lim_{h \to 0} \frac{f(0+h) - f(0)}{h} = \lim_{h \to 0} \frac{h^2 \sin(1/h) - 0}{h} = \lim_{h \to 0} h \sin\left(\frac{1}{h}\right).$$

Now this last limit is the middle of our three limits from above (with h replacing x), and it exists and equals 0. This means that f is actually differentiable at $x = 0$, and in fact $f'(0) = 0$. Can you tell that from the graph of $y = f(x)$? Here's what it looks like for $-0.1 < x < 0.1$, along with the envelope functions $y = x^2$ and $y = -x^2$:

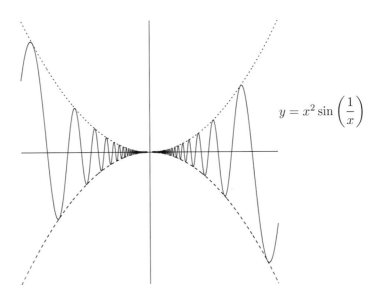

$$y = x^2 \sin\left(\frac{1}{x}\right)$$

It looks pretty wobbly at $x = 0$ to me, so it's not clear at all that the derivative should even exist there—but we've just shown that it does! This leads to the following question: what is

$$\lim_{x \to 0^+} f'(x)?$$

Since we know that $f'(0) = 0$, you might think that the above limit is just 0. Let's check it out by using the above formula for $f'(x)$ when $x \neq 0$:

$$\lim_{x \to 0^+} f'(x) = \lim_{x \to 0^+} \left(2x \sin \left(\frac{1}{x} \right) - \cos \left(\frac{1}{x} \right) \right).$$

We have two terms to deal with here. The first term $(2x \sin(1/x))$ goes to 0 in the limit, since it's just twice the middle of our three limits from above. On the other hand, the second term $(\cos(1/x))$ has no limit as $x \to 0$; this is exactly what the third limit from above says. The conclusion is that $\lim_{x \to 0^+} f'(x)$ doesn't exist. By symmetry (check that f is an odd function), neither does $\lim_{x \to 0^-} f'(x)$.

Now let's summarize what we've found. Our function f is continuous everywhere and also differentiable everywhere, even at $x = 0$. Indeed, at $x = 0$, the derivative $f'(0)$ equals 0, but near 0, the derivative $f'(x)$ oscillates wildly: $\lim_{x \to 0} f'(x)$ doesn't exist even though $f'(0)$ does. In particular, we have now shown that the derivative function f' is not itself a continuous function. So, there are functions out there which are differentiable, yet their derivatives aren't continuous. That's pretty darned curious!

CHAPTER 8 _____

Implicit Differentiation and Related Rates

Let's take a break from trying to work out how to differentiate everything in sight. It's time to look at implicit differentiation, which is a nice generalization of regular differentiation. We'll then see how to use this technique to solve word problems involving changing quantities. Knowing how fast one quantity is changing allows us to find how fast a different, but related, quantity is changing too. Anyway, the summary for this chapter is the same as the title:

- implicit differentiation; and
- related rates.

8.1 Implicit Differentiation

Consider the following two derivatives:

$$\frac{d}{dx}(x^2) \qquad \text{and} \qquad \frac{d}{dx}(y^2).$$

The first is just $2x$, as we've seen. So isn't the second one $2y$? That would be the answer if the differentiation were with respect to y, but it isn't: the dx in the denominator tells us that the differentiation is with respect to x. How do we unravel this?

The best way is to say to yourself that the first of the derivatives above is asking how much the quantity x^2 changes when we change x a little bit. As we saw in Section 5.2.7 of Chapter 5, if we do change x by a little bit, then x^2 changes by approximately $2x$ times as much.

On the other hand, if you change x by a little bit, what does that do to y^2? This is what we need to know in order to find the second of our above derivatives, $d(y^2)/dx$. Think of it this way: if you change x, then y will change a little bit; this change in y will cause y^2 to change. (All this is true only if y depends on x, of course—if not, then when you change x, nothing at all will happen to y.)

If you think that it sounds as if I'm hinting at the chain rule here, you're quite right. Here's how it actually works. Let $u = y^2$, so that $du/dy = 2y$.

By the chain rule,

$$\frac{d}{dx}(y^2) = \frac{du}{dx} = \frac{du}{dy} \cdot \frac{dy}{dx} = 2y\frac{dy}{dx}.$$

So if you change x by a little bit, then y^2 changes by $2y(dy/dx)$ times as much. Now you might complain that the answer contains dy/dx in it, but what did you expect? If you want to know how the quantity y^2 changes when you change x a little, then first you need to know something about how y changes! (Again, if y doesn't depend on x, then dy/dx equals 0 for all x, so $d(y^2)/dx$ is also 0 for all x. That is, y^2 doesn't depend on x either.)

8.1.1 Techniques and examples

 Now it's time to get practical. Consider the following equation:

$$x^2 + y^2 = 4.$$

The quantity y isn't a function of x. In fact, when $-2 < x < 2$, there are two values of y satisfying this equation. On the other hand, the graph of the above relation is the circle of radius 2 units centered at the origin. This circle has nice tangents everywhere, and we should be able to find their slopes without having to write $y = \pm\sqrt{4 - x^2}$ and differentiating. In fact, all we have to do is whack a d/dx in front of both sides:

$$\frac{d}{dx}(x^2 + y^2) = \frac{d}{dx}(4).$$

As we know, the left-hand side can be split into two pieces without any problem. In fact, normally one would just automatically start by writing

$$\frac{d}{dx}(x^2) + \frac{d}{dx}(y^2) = \frac{d}{dx}(4).$$

To simplify this, note that we have already identified the two quantities on the left-hand side in the previous section, and the right-hand side is 0 because 4 is constant. Be careful not to write 4 instead—this is a very common mistake! Anyway, here's what we get:

$$2x + 2y\frac{dy}{dx} = 0.$$

Dividing by 2 and rearranging leads to

$$\frac{dy}{dx} = -\frac{x}{y}.$$

This formula says that at the point (x, y) on the circle, the slope of the tangent is $-x/y$. If the point isn't on the circle, then the formula doesn't say anything (at least as far as we're concerned). Now, let's use the formula to find the equation of the tangent to the circle at the point $(1, \sqrt{3})$. This point certainly does lie on the circle, since $x^2 + y^2 = 1^2 + (\sqrt{3})^2 = 4$. By the above formula, the slope is given by $dy/dx = -1/\sqrt{3}$. So, the tangent line has slope $-1/\sqrt{3}$

and goes through $(1, \sqrt{3})$. Using the point-slope formula, we see that the equation of the line is

$$y - \sqrt{3} = -\frac{1}{\sqrt{3}}(x - 1).$$

This can be simplified slightly to $y = (4 - x)/\sqrt{3}$, if you like.

Here's another example: if

$$5\sin(x) + 3\sec(y) = y - x^2 + 3,$$

what is the equation of the tangent at the origin? Unlike the previous example, it's impossible to solve this equation for y (or x). So we have to use implicit differentiation. Let's first verify that the origin actually lies on the curve. Plugging in $x = 0$ and $y = 0$ gives $5\sin(0) + 3\sec(0)$ on the left-hand side, which is just 3 (remember that $\sec(0) = 1/\cos(0) = 1$). The right-hand side is also 3, so the origin is on the curve. Now let's differentiate the above equation, splitting it up as we do so:

$$\frac{d}{dx}(5\sin(x)) + \frac{d}{dx}(3\sec(y)) = \frac{dy}{dx} - \frac{d}{dx}(x^2) + \frac{d}{dx}(3).$$

The only one of these quantities that's hard to simplify is the second one on the left-hand side. It's not too bad, though: let $u = 3\sec(y)$. Then $du/dy = 3\sec(y)\tan(y)$, so by the chain rule, we have

$$\frac{d}{dx}(3\sec(y)) = \frac{du}{dx} = \frac{du}{dy} \cdot \frac{dy}{dx} = 3\sec(y)\tan(y)\frac{dy}{dx}.$$

So we can return to the previous equation and differentiate everything, getting

$$5\cos(x) + 3\sec(y)\tan(y)\frac{dy}{dx} = \frac{dy}{dx} - 2x.$$

Note that when you differentiate the constant 3, you get 0. In any case, we could solve for dy/dx here: just throw all the stuff involving dy/dx on one side and everything else on the other side:

$$\frac{dy}{dx} - 3\sec(y)\tan(y)\frac{dy}{dx} = 2x + 5\cos(x).$$

Now factor—

$$\frac{dy}{dx}(1 - 3\sec(y)\tan(y)) = 2x + 5\cos(x)$$

—and then divide to get

$$\frac{dy}{dx} = \frac{2x + 5\cos(x)}{1 - 3\sec(y)\tan(y)}.$$

Finally, plug in $x = 0$ and $y = 0$ to see that

$$\frac{dy}{dx} = \frac{2(0) + 5\cos(0)}{1 - 3\sec(0)\tan(0)} = \frac{2(0) + 5(1)}{1 - 2(1)(0)} = 5.$$

Since the tangent line has slope 5 and goes through the origin, its equation is just $y = 5x$, and we're done. But do you see how we might have saved a little effort? Go back to the equation

$$5\cos(x) + 3\sec(y)\tan(y)\frac{dy}{dx} = \frac{dy}{dx} - 2x$$

from above. We manipulated this to find the general expression for dy/dx, but actually we only care about what happens at the origin. So we could have saved a little time by plugging $x = 0$ and $y = 0$ into the above equation. We would have gotten

$$5\cos(0) + 3\sec(0)\tan(0)\frac{dy}{dx} = \frac{dy}{dx} - 2(0).$$

This easily reduces to $dy/dx = 5$. So a good rule of thumb is that **if you only need the derivative at a certain point, substitute before rearranging**—it often saves time.

So far, we've only used the chain rule. Sometimes you might need to use the product rule or the quotient rule. For example, if

$$y\cot(x) = 3\csc(y) + x^7,$$

then you'll need the product rule and the chain rule to find dy/dx. Indeed, if we differentiate, we get

$$\frac{d}{dx}(y\cot(x)) = \frac{d}{dx}(3\csc(y)) + \frac{d}{dx}(x^7).$$

The left-hand side is the product of y and $\cot(x)$. We should give it a name—I'll call it s, so that $s = y\cot(x)$. If we also set $v = \cot(x)$, then $s = yv$, and we can use the product rule to differentiate s with respect to x:

$$\frac{ds}{dx} = v\frac{dy}{dx} + y\frac{dv}{dx} = \cot(x)\frac{dy}{dx} + y(-\csc^2(x)).$$

(Remember that the derivative of $\cot(x)$ with respect to x is $-\csc^2(x)$.) Now, let's worry about the right-hand side of our original equation from above. For the first term, $3\csc(y)$, we'll use the chain rule. Let's call the term u, so $u = 3\csc(y)$. We can see that $du/dy = -3\csc(y)\cot(y)$, so by the chain rule we have

$$\frac{du}{dx} = \frac{du}{dy}\frac{dy}{dx} = -3\csc(y)\cot(y)\frac{dy}{dx}.$$

Finally, the derivative of the last term, x^7, with respect to x is just $7x^6$. Putting it all together, we see that when we differentiate both sides of our original equation

$$y\cot(x) = 3\csc(y) + x^7$$

with respect to x, we get

$$\cot(x)\frac{dy}{dx} - y\csc^2(x) = -3\csc(y)\cot(y)\frac{dy}{dx} + 7x^6.$$

Let's throw everything involving dy/dx on the left and everything else on the right:

$$\cot(x)\frac{dy}{dx} + 3\csc(y)\cot(y)\frac{dy}{dx} = y\csc^2(x) + 7x^6.$$

Now factor the left-hand side and divide to solve for dy/dx:

$$\frac{dy}{dx} = \frac{y\csc^2(x) + 7x^6}{\cot(x) + 3\csc(y)\cot(y)},$$

and we're done.

Finally, consider the equation

$$x - y\cos\left(\frac{y}{x^4}\right) = \pi + 1.$$

What is the equation of the tangent to the point $(1, \pi)$ on the curve? I leave it to you to substitute $x = 1$ and $y = \pi$, and make sure that the left- and right-hand sides agree, so that the point is indeed on the curve. Now we have to differentiate. We get

$$\frac{d}{dx}(x) - \frac{d}{dx}\left(y\cos\left(\frac{y}{x^4}\right)\right) = \frac{d}{dx}(\pi + 1).$$

The first term is easy: it's just 1. Also, the right-hand side is 0, since $\pi + 1$ is constant. This leaves us with an awful mess in the middle. Suppose we set

$$s = y\cos\left(\frac{y}{x^4}\right).$$

Then s is the product of y and v, where $v = \cos(y/x^4)$. By the product rule, we have

$$\frac{ds}{dx} = v\frac{dy}{dx} + y\frac{dv}{dx}.$$

There's no escape: we are going to have to differentiate v. Suppose we set $t = y/x^4$. Then $v = \cos(t)$, so $dv/dt = -\sin(t)$, and the chain rule tells us that

$$\frac{dv}{dx} = \frac{dv}{dt}\cdot\frac{dt}{dx} = -\sin(t)\frac{dt}{dx} = -\sin\left(\frac{y}{x^4}\right)\frac{dt}{dx}.$$

We're not out of the woods yet, though—we need to find dt/dx. Now $t = y/x^4$, so set $U = y$ and $V = x^4$. (I already used a little v, so I'll use the capital letter here.) The quotient rule says that

$$\frac{dt}{dx} = \frac{V\dfrac{dU}{dx} - U\dfrac{dV}{dx}}{V^2} = \frac{x^4\dfrac{dy}{dx} - y\dfrac{d}{dx}(x^4)}{(x^4)^2} = \frac{x^4\dfrac{dy}{dx} - 4x^3 y}{x^8} = \frac{x\dfrac{dy}{dx} - 4y}{x^5}.$$

Now we just need to unwind everything. Working backward, we can now finish the calculation of dv/dx:

$$\frac{dv}{dx} = -\sin\left(\frac{y}{x^4}\right)\frac{dt}{dx} = -\sin\left(\frac{y}{x^4}\right) \times \frac{x\dfrac{dy}{dx} - 4y}{x^5}.$$

This in turn allows us to find ds/dx:

$$\frac{ds}{dx} = v\frac{dy}{dx} + y\frac{dv}{dx} = \cos\left(\frac{y}{x^4}\right)\frac{dy}{dx} - y\sin\left(\frac{y}{x^4}\right) \times \frac{x\dfrac{dy}{dx} - 4y}{x^5}.$$

Finally, we can go back to our original differentiated equation

$$\frac{d}{dx}(x) - \frac{d}{dx}\left(y\cos\left(\frac{y}{x^4}\right)\right) = \frac{d}{dx}(\pi + 1)$$

from above and simplify this down to

$$1 - \cos\left(\frac{y}{x^4}\right)\frac{dy}{dx} + y\sin\left(\frac{y}{x^4}\right) \times \frac{x\dfrac{dy}{dx} - 4y}{x^5} = 0.$$

Don't bother solving for dy/dx! We only need to know what happens when $x = 1$ and $y = \pi$. So plug those in. Noting that $\cos(\pi) = -1$ and $\sin(\pi) = 0$, you should check that the whole darn thing simplifies to

$$1 - (-1)\frac{dy}{dx} + \pi \times 0 \times \text{irrelevant junk} = 0,$$

or just $dy/dx = -1$. Our tangent line therefore has slope -1 and goes through $(1, \pi)$, so its equation is $y - \pi = -(x-1)$; you can rewrite this as $y = -x + \pi + 1$ if you like.

We still have to look at how to find the second derivative using implicit differentiation. Just before we do that, here's a brief summary of the above methods:

- in your original equation, differentiate everything and simplify using the chain, product, and quotient rules;
- if you want to find dy/dx, rearrange and divide to solve for dy/dx; but
- if instead you want to find the slope or equation of the tangent at a particular point on the curve, first substitute the known values of x and y, then rearrange to find dy/dx. Then use the point-slope formula to find the equation of the tangent, if needed.

8.1.2 Finding the second derivative implicitly

It's also possible to differentiate twice to get the second derivative. For example, if

$$2y + \sin(y) = \frac{x^2}{\pi} + 1,$$

then what is the value of d^2y/dx^2 at the point $(\pi, \pi/2)$ on the curve? Once again, you should verify that the point does lie on the curve by plugging in these values of x and y and seeing that the equation checks out. Now, if you want to differentiate twice, you have to start by differentiating once! You should get

$$2\frac{dy}{dx} + \cos(y)\frac{dy}{dx} = \frac{2x}{\pi},$$

having used the chain rule to tackle the $\sin(y)$ term. Now we need to differentiate again. Do not substitute first! In order to differentiate, we need to see what happens when x and y are varying. This can't happen if we fix them at certain values (like π and $\pi/2$). Instead, differentiate the above equation with respect to x:

$$\frac{d}{dx}\left(2\frac{dy}{dx}\right) + \frac{d}{dx}\left(\cos(y)\frac{dy}{dx}\right) = \frac{d}{dx}\left(\frac{2x}{\pi}\right).$$

The right-hand side is just $2/\pi$, and the first term on the left-hand side is just $2(d^2y/dx^2)$. The tricky bit is the second term on the left. We'll need to use the product rule: set $s = \cos(y)(dy/dx)$, and also $u = \cos(y)$ and $v = dy/dx$, so that $s = uv$. By the product rule,

$$\frac{ds}{dx} = v\frac{du}{dx} + u\frac{dv}{dx} = \frac{dy}{dx}\cdot\frac{du}{dx} + \cos(y)\frac{d}{dx}\left(\frac{dy}{dx}\right) = \frac{dy}{dx}\cdot\frac{du}{dx} + \cos(y)\frac{d^2y}{dx^2}.$$

We still need to find du/dx, where $u = \cos(y)$. This is just the chain rule once again:

$$\frac{du}{dx} = \frac{du}{dy}\cdot\frac{dy}{dx} = -\sin(y)\frac{dy}{dx}.$$

Putting it all together, we see that

$$\frac{ds}{dx} = \frac{dy}{dx}\cdot\frac{du}{dx} + \cos(y)\frac{d^2y}{dx^2} = \frac{dy}{dx}\cdot\left(-\sin(y)\frac{dy}{dx}\right) + \cos(y)\frac{d^2y}{dx^2}$$

$$= -\sin(y)\left(\frac{dy}{dx}\right)^2 + \cos(y)\frac{d^2y}{dx^2}.$$

Beware: the quantities

$$\left(\frac{dy}{dx}\right)^2 \qquad \text{and} \qquad \frac{d^2y}{dx^2}$$

are completely different! The left one is the square of the first derivative, while the right one is the second derivative. Anyway, let's put everything together. Starting from

$$\frac{d}{dx}\left(2\frac{dy}{dx}\right) + \frac{d}{dx}\left(\cos(y)\frac{dy}{dx}\right) = \frac{d}{dx}\left(\frac{2x}{\pi}\right),$$

we can now write this as

$$2\frac{d^2y}{dx^2} - \sin(y)\left(\frac{dy}{dx}\right)^2 + \cos(y)\frac{d^2y}{dx^2} = \frac{2}{\pi}.$$

Phew. That was exhausting. We're not done yet, though: we still need to find d^2y/dx^2 when $x = \pi$ and $y = \pi/2$. So plug that in to the above equation: you get

$$2\frac{d^2y}{dx^2} - \sin\left(\frac{\pi}{2}\right)\left(\frac{dy}{dx}\right)^2 + \cos\left(\frac{\pi}{2}\right)\frac{d^2y}{dx^2} = \frac{2}{\pi}.$$

This simplifies down to

$$2\frac{d^2y}{dx^2} - \left(\frac{dy}{dx}\right)^2 = \frac{2}{\pi}.$$

The problem is, we still need dy/dx! No problem: in our equation

$$2\frac{dy}{dx} + \cos(y)\frac{dy}{dx} = \frac{2x}{\pi}$$

from way above, put $x = \pi$ and $y = \pi/2$ (I didn't let you do this before!) and you get

$$2\frac{dy}{dx} + 0\frac{dy}{dx} = \frac{2\pi}{\pi} = 2,$$

so $dy/dx = 1$. Put that into our second derivative equation and we get

$$2\frac{d^2y}{dx^2} - (1)^2 = \frac{2}{\pi}.$$

This means that

$$\frac{d^2y}{dx^2} = \frac{1}{\pi} + \frac{1}{2}$$

when $x = \pi$ and $y = \pi/2$, so we're finally done!

8.2 Related Rates

Consider two quantities—they can measure anything you like—that are related to each other. If you know one, you can find the other. For example, if you keep your eyes on an airplane that passes over your head, then the angle that your line of sight makes with the ground depends on the position of the plane. In this case, the two quantities are the position of the plane and the angle I just described.

Of course, as one of the two quantities changes, so does the other. Suppose that we know how fast one of the quantities is changing. Then how fast is the other one changing? That is exactly what we mean by the term *related rates*. You see, a *rate of change* is the speed at which a quantity is changing over time. We have two quantities which are related to each other, and we want to know how their rates of change are related to each other. (By the way, sometimes we'll abbreviate and say "rate" instead of "rate of change.")

The above definition of a rate of change was a little sketchy. If you want to know how fast something is changing over time, you simply have to differentiate with respect to time. So, here's the real definition: **the rate of change of a quantity Q is the derivative of Q with respect to time**. That is,

> if Q is some quantity, then the rate of change of Q is $\dfrac{dQ}{dt}$.

When you see the word "rate," you should automatically think "d/dt."

So, how do you go from an equation relating two quantities to an equation relating the rates of change of these quantities? You differentiate, of course!

If you differentiate both sides implicitly with respect to t, you'll find that the rates just pop out, giving you a new equation. The same thing works if you are dealing with three or more quantities which are related (for example, the length, width, and area of a rectangle). Just differentiate implicitly with respect to t and you'll relate the rates of change.

So, let's look at a general overview of how to solve problems involving related rates. Then we'll use it to solve a bunch of examples.

1. Read the question. Identify all the quantities and note which one you need to find the rate of. Draw a picture if you need to!

2. Write down an equation (sometimes you need more than one) that relates all the quantities. To do this, you may need to do some geometry, possibly involving similar triangles. If you have more than one equation, try to solve them simultaneously to eliminate unnecessary variables.

3. Differentiate your remaining equation(s) implicitly with respect to time t. That is, whack both sides of each equation with a $\frac{d}{dt}$. You end up with one or more equations relating the rates of change.

4. Finally, substitute values for everything you know into all the equations you have. Solve the equations simultaneously to find the rate you need.

The only difference between these types of problems and the word problems you've already seen is that step 3 was absent. Here, it makes all the difference. Just one more thing before we look at examples: it's vital that you **substitute values at the end, after differentiating!** That is, don't switch steps 3 and 4. If you substitute values first, denying the quantities the ability to change, then your rates will all be 0. That's what you get for freezing everything in place. . . .

8.2.1 A simple example

Here's a relatively simple example to illustrate the above method. Suppose that a perfectly spherical balloon is being inflated by a pump. Air is entering the balloon at the constant rate of 12π cubic inches per second. At what rate does the radius of the balloon change at the instant when the radius itself is 2 inches? Also, at what rate does the radius change when the volume is 36π cubic inches?

OK, let's write down our quantities (step 1). These are the volume and the radius of the balloon. Let's call the volume V (in cubic inches) and the radius r (in inches). We need to find the rate of change of the radius r. Now, we need an equation relating V and r (step 2). Here's where the geometry comes in. Since the balloon is a sphere, we know that

$$V = \frac{4}{3}\pi r^3.$$

This relates the quantities. Now we need to relate the rates (step 3). Differentiate both sides implicitly with respect to t:

$$\frac{d}{dt}(V) = \frac{d}{dt}\left(\frac{4}{3}\pi r^3\right).$$

The left-hand side is just dV/dt; to handle the right-hand side, let $s = r^3$, so $ds/dr = 3r^2$. By the chain rule,

$$\frac{ds}{dt} = \frac{ds}{dr}\frac{dr}{dt} = 3r^2\frac{dr}{dt}.$$

Now we can put this in our above equation and get

$$\frac{dV}{dt} = \frac{4}{3}\pi\left(3r^2\frac{dr}{dt}\right) = 4\pi r^2\frac{dr}{dt}.$$

So we have an equation relating the rate of V with the rate of r. Finally, we're ready to substitute (step 4). In both parts of the question, the rate of change of volume is 12π cubic inches per second. In symbols, we have $dV/dt = 12\pi$. Plugging this into the above equation, we get

$$12\pi = 4\pi r^2\frac{dr}{dt}.$$

Rearranging leads to

$$\frac{dr}{dt} = \frac{3}{r^2}.$$

Great—that means that if we know the radius r, then we can find the rate at which the radius is changing, which of course is dr/dt. Notice that the rate of change of the radius is itself a changing quantity: it depends on the radius. You've probably noticed that when you blow up a balloon, it grows in size (or radius) faster at the beginning, and then starts to slow down, even though you're blowing the same amount of air into the balloon all the time. This is consistent with the above formula for dr/dt, which is decreasing in r.

Armed with the formula, we can quickly do both parts of the question. In the first part, we know that the radius is 2 inches, so set $r = 2$ in our formula from above:

$$\frac{dr}{dt} = \frac{3}{2^2} = \frac{3}{4}.$$

So the answer is $\frac{3}{4}$. But $\frac{3}{4}$ what? It's important to write a sentence summarizing the situation, as well as including the **units** of measurement. In this case, we'd say that when the radius is 2 inches, the rate of change of the radius is $\frac{3}{4}$ inches per second.

Now, for the second part of the question, we know that the volume is 36π cubic inches. That means that $V = 36\pi$. The problem is that we need to know what r is in order to find dr/dt. Now we need to go back to the equation relating V and r, which was $V = \frac{4}{3}\pi r^3$. If you put $V = 36\pi$ and solve for r, you should be able to see that $r = 3$ inches. Finally, substituting into the equation for dr/dt gives

$$\frac{dr}{dt} = \frac{3}{r^2} = \frac{3}{3^2} = \frac{1}{3}.$$

So when the volume is 36π cubic inches, the rate of change of the radius is $\frac{1}{3}$ inches per second.

8.2.2 A slightly harder example

Let's look at another relatively straightforward example, this time involving three quantities. Suppose there are two cars, A and B. Car A is driving on a road heading directly north away from your house, and car B is driving on a different road heading directly west toward your house. Car A travels at 55 miles per hour and car B travels at 45 miles per hour. At what rate is the distance between the cars changing when A is 21 miles north of your house and car B is 28 miles east of your house?

To answer this question, we'd better draw a picture (step 1). Draw your house H and the cars A and B. Let the distance between H and A be given by a; let the distance between H and B be called b; and let the distance between the cars be called c. The diagram looks like this:

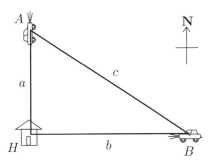

Note that it would be wrong to mark in 21 instead of a or 28 instead of b. You need to see what happens as a and b change, not when they are fixed at a certain number, so they need to have the flexibility of being variable. Also note that c is the quantity we need the rate of, since it's the distance between the cars.

Time for step 2. The equation relating a, b, and c is nothing other than Pythagoras' Theorem:

$$a^2 + b^2 = c^2.$$

Moving on to step 3, we differentiate implicitly with respect to time t. Make sure you agree that we get

$$2a\frac{da}{dt} + 2b\frac{db}{dt} = 2c\frac{dc}{dt}.$$

Now, we know that car A is moving at 55 miles an hour away from your house. This means that the distance a is increasing by 55 miles per hour, so $da/dt = 55$. As for B, it is moving at 45 miles an hour toward your house. This means that b is **decreasing** by 45 miles an hour, so $db/dt = -45$. You need that negative sign in there! Otherwise you'll screw the whole thing up. Plugging these values in to the above equation leads to

$$2a(55) + 2b(-45) = 2c\frac{dc}{dt},$$

which can be simplified to

$$c\frac{dc}{dt} = 55a - 45b.$$

Finally, we can see what happens at the instant of time we're interested in. This is when $a = 21$ and $b = 28$. At that instant, we know that $c^2 = 21^2 + 28^2$, which works out to be $c = \pm 35$. Since c is positive (it's the distance between the two cars!), we have $c = 35$. Put those numbers into the above equation and you get

$$(35)\frac{dc}{dt} = 55(21) - 45(28).$$

You can compute this easily by canceling a factor of 5 and a factor of 7 from both sides. The end result is that $dc/dt = -3$. This means that the distance between the cars is **decreasing** at a rate of 3 miles per hour (at the moment in time we are considering).

That's the answer we need. Notice that the cars are actually getting closer together at the moment of time we're considering, even though A is moving away from the house faster than B is coming toward it. If we wait a little bit, car A will be farther away from the house and car B will be closer; by staring at the equation for dc/dt, you might convince yourself that this quantity eventually becomes positive (although this observation isn't required to complete the question).

8.2.3 A much harder example

Here's a tougher example involving similar triangles: suppose there's a freakin' huge water tank in the shape of a cone (with the point at the bottom). The height of the cone is twice the radius of the cone. Water is being pumped into the tank at the rate of 8π cubic feet per second. At what rate is the water level changing when the volume of water in the tank is 18π cubic feet?

There's a second part as well: assume that the tank develops a little hole at the bottom that causes water to flow out at a rate of one cubic foot per second for every cubic foot of water in the tank. I want to know the same thing as before: at what rate is the water level changing when the volume of water in the tank is 18π cubic feet, but now with the leak in the tank?

Let's start with the first part. Here's a diagram of the situation:

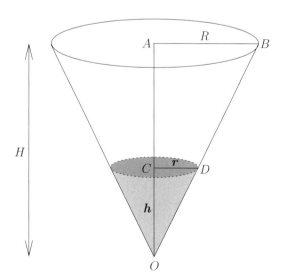

We have marked some quantities on the diagram. The height of the tank is H and its radius is R. The height of the water level is h and the radius of the top of the water surface is r. All these quantities are measured in feet. Let's also let v be the volume of water in the tank, measured in cubic feet. (You could let V be the volume of the whole tank, but we'll never need that quantity since the tank will never be full—it's that huge!). Anyway, that takes care of step 1.

For step 2, we have to start relating some of these quantities. We are given that the tank's height is twice the radius, so we have $H = 2R$. We're more interested in relating h and r, though. There are some similar triangles in the diagram: in fact, $\triangle ABO$ is similar to $\triangle CDO$, so $H/R = h/r$. Since $H = 2R$, we have $2R/R = h/r$, which means that $h = 2r$. So the water is like a mini-copy of the whole tank. Anyway, we still need to find the volume of water in the tank in terms of h and r. The volume of a cone of height h units and radius r units is given by $v = \frac{1}{3}\pi r^2 h$ cubic units. It would be nice to eliminate one of h and r at this point; since we're more interested in the water level h than the radius r (read the question and see why!), it makes sense to eliminate r. Using the equation $r = h/2$, we have

$$v = \frac{1}{3}\pi r^2 h = \frac{1}{3}\pi \left(\frac{h}{2}\right)^2 h = \frac{\pi h^3}{12}, \qquad \text{so} \qquad v = \frac{\pi h^3}{12}.$$

Now, for step 3, let's differentiate this with respect to time t. By the chain rule,

$$\frac{dv}{dt} = \frac{\pi}{12} \times 3h^2 \frac{dh}{dt} = \frac{\pi h^2}{4} \frac{dh}{dt}, \qquad \text{so} \qquad \frac{dv}{dt} = \frac{\pi h^2}{4} \frac{dh}{dt}.$$

Great—now for step 4, substitute in everything we know into the two equations above. We know that $dv/dt = 8\pi$ and we're interested in what happens when $v = 18\pi$. Substituting, we get

$$18\pi = \frac{\pi h^3}{12} \qquad \text{and} \qquad 8\pi = \frac{\pi h^2}{4} \frac{dh}{dt}.$$

The first equation tells us that $h^3 = 18 \times 12 = 216$, so $h = 6$. That is, when the water volume is 18π cubic feet, the water level is at 6 feet. Putting that into the second equation, we get

$$8\pi = \frac{\pi}{4} \times 6^2 \frac{dh}{dt},$$

which means that $dh/dt = 8/9$. That is, the water height is increasing at a rate of 8/9 feet per second at the moment we care about (when the volume is 18π cubic feet).

The second part is almost the same. In fact, the only difference occurs at step 4. We still want to substitute in $v = 18\pi$, which will mean that $h = 6$ once again. On the other hand, it's wrong to put $dv/dt = 8\pi$, since that doesn't take into account the leak. We know that 8π cubic feet of water is entering into the tank per second, but one cubic foot is leaving per second for every cubic foot of water in the tank. Since there are v cubic feet of water in the tank (by definition!), the rate of outflow from the leak is v cubic feet per second. So the rate of inflow is 8π and the rate of outflow is v (both in cubic feet per second), which means that

$$\frac{dv}{dt} = 8\pi - v.$$

Now, when $v = 18\pi$, we have $dv/dt = 8\pi - 18\pi = -10\pi$. So we need to substitute $dv/dt = -10\pi$ and $h = 6$ into our previous equation

$$\frac{dv}{dt} = \frac{\pi h^2}{4} \frac{dh}{dt}.$$

The answer works out to be $dh/dt = -10/9$, which means that the water level in the tank is falling at a rate of $10/9$ feet per second at the time we're considering. Even though we're pumping water in, the leak is letting even more water out and so the level is falling.

8.2.4 A really hard example

Here's one more problem. Now that you have seen a number of related rate problems, perhaps you should try to solve it before reading the solution.

Suppose that a plane is flying eastward directly away from you at a height of 2000 feet above your head. The plane moves at a constant speed of 500 feet per second. Meanwhile, some time ago a parachutist jumped out of a helicopter (which has since flown away). The parachutist is floating directly downward, 1000 feet due east of you, at a constant speed of 10 feet per second. The situation is summarized in the following picture:

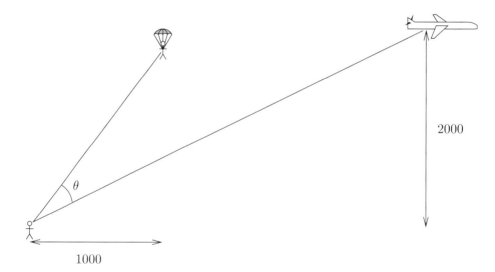

In the picture, what you might call the inter-azimuthal angle between the parachutist and the plane (with respect to you) is marked as θ. The question is, at what rate is θ changing when the plane and the parachutist have the same height but the plane is 8000 feet due east of you?

We have two objects to worry about, the plane and the parachutist. We know that the height of the plane is always 2000 feet (relative to your head), but we don't know how far east the plane is—the distance keeps on changing. Let the plane be p feet to the east of you. As for the parachutist, this time we know exactly how far east the parachutist is: 1000 feet. The problem is, how high is the parachutist? Let the height be h feet. By drawing a few extra lines, we can recast the above diagram as follows:

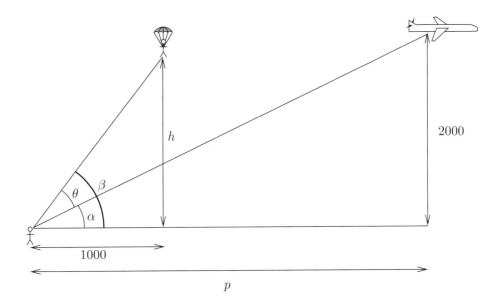

Notice that the quantities 1000 and 2000 never change, but the quantities p and h do change. In particular, the plane is heading to the right, so p is getting bigger; and the parachutist is heading down, so h is getting smaller. Even though the question asks us to concentrate on the moment when $p = 8000$ and $h = 2000$ (the same height as the plane), it's important that we allow p and h to vary so that we can work out the rate of change. After all, if p and h stay the same, then the plane and the parachutist just stay suspended in space in the same spot, and of course the angle θ wouldn't change. That's hardly realistic—so we need to let p and h vary, in which case θ varies and we can work out how fast it varies. That completes step 1.

Speaking of θ, it's clear from the diagram that it is simply the difference between the angle β the parachutist makes with the ground and the angle α the plane makes with the ground. (Let's assume that you have no height, or if you prefer, you are lying on the ground.) So we know that $\theta = \beta - \alpha$. Actually, we should probably write $\theta = |\beta - \alpha|$, just in case the parachutist is much lower than the plane. At around the time we're interested in, the heights are the same but the plane is much farther to the east than the parachutist, so β must be bigger than α and we don't need the absolute values.

Now, let's do some trig. We have two right-angled triangles. From one of them (the one with the plane), we get $\tan(\alpha) = 2000/p$. From the other one, we have $\tan(\beta) = h/1000$. Let's write these equations down in one place:

$$\tan(\alpha) = \frac{2000}{p} \quad \text{and} \quad \tan(\beta) = \frac{h}{1000}.$$

Step 2 is finally done, and we can move on to step 3, differentiating these two relations implicitly with respect to time. Starting with the first one, let $u = \tan(\alpha)$ and $v = 2000/p$, so our equation just becomes $u = v$. This means that $du/dt = dv/dt$. Let's find these two quantities using the chain rule. First, du/dt:

$$\frac{du}{dt} = \frac{du}{d\alpha}\frac{d\alpha}{dt} = \sec^2(\alpha)\frac{d\alpha}{dt}.$$

And now for dv/dt:

$$\frac{dv}{dt} = \frac{dv}{dp}\frac{dp}{dt} = -\frac{2000}{p^2}\frac{dp}{dt}.$$

Since $du/dt = dv/dt$, we have

$$\sec^2(\alpha)\frac{d\alpha}{dt} = -\frac{2000}{p^2}\frac{dp}{dt}.$$

That's just the first of our two trig equations. We need to repeat the exercise for the second one involving $\tan(\beta)$. The left-hand side is handled exactly the same way as we did $\tan(\alpha)$, but the right-hand side is much easier. Make sure you agree that we get

$$\sec^2(\beta)\frac{d\beta}{dt} = \frac{1}{1000}\frac{dh}{dt}.$$

Remember, we also know that $\theta = \beta - \alpha$, so we can differentiate this also with respect to time t and get $d\theta/dt = d\beta/dt - d\alpha/dt$. Since there are so many equations, let's write all six of them down in one place:

$$\tan(\alpha) = \frac{2000}{p} \qquad\qquad \sec^2(\alpha)\frac{d\alpha}{dt} = -\frac{2000}{p^2}\frac{dp}{dt}$$

$$\tan(\beta) = \frac{h}{1000} \qquad\qquad \sec^2(\beta)\frac{d\beta}{dt} = \frac{1}{1000}\frac{dh}{dt}$$

$$\theta = \beta - \alpha \qquad\qquad \frac{d\theta}{dt} = \frac{d\beta}{dt} - \frac{d\alpha}{dt}.$$

Now we'd better make some substitutions and get to the bottom of this mess. What do we know? Well, the speed of the plane is 500 feet per second, which means that $dp/dt = 500$. The speed of the parachutist is 10 feet per second, but the height is decreasing, so $dh/dt = -10$. If you forget the minus sign, you'll get the answer wrong! So be very careful. For example, if the plane were coming toward you, then p would be decreasing, so dp/dt would be negative. Anyway, we're interested in what happens when the plane is 8000 feet away, so $p = 8000$, and when the parachutist is at height 2000 feet (the same as the plane), so set $h = 2000$. The first four of our equations become a lot simpler:

$$\tan(\alpha) = \frac{2000}{8000} = \frac{1}{4} \qquad \sec^2(\alpha)\frac{d\alpha}{dt} = -\frac{2000}{8000^2} \times 500 = -\frac{1}{64}$$

$$\tan(\beta) = \frac{2000}{1000} = 2 \qquad \sec^2(\beta)\frac{d\beta}{dt} = \frac{1}{1000} \times (-10) = -\frac{1}{100}.$$

From the top right equation, we could find $d\alpha/dt$ if only we knew what $\sec^2(\alpha)$ was. But wait a second—we do know that $\tan(\alpha) = 1/4$, so surely we can find $\sec^2(\alpha)$. Remembering our trig identities (see Section 2.4 in Chapter 2), we get

$$\sec^2(\alpha) = 1 + \tan^2(\alpha) = 1 + \left(\frac{1}{4}\right)^2 = \frac{17}{16}.$$

So the top right equation becomes

$$\frac{17}{16}\frac{d\alpha}{dt} = -\frac{1}{64},$$

which works out to be

$$\frac{d\alpha}{dt} = -\frac{1}{68}.$$

This rocks—we now need to do the same with β and we'll be done. Here we know that $\tan(\beta) = 2$, so

$$\sec^2(\beta) = 1 + \tan^2(\beta) = 1 + 2^2 = 5.$$

Substituting into the bottom right equation above, we have

$$5\frac{d\beta}{dt} = -\frac{1}{100},$$

which means that

$$\frac{d\beta}{dt} = -\frac{1}{500}.$$

So we know $d\alpha/dt$ and $d\beta/dt$; from the final one of our original six equations above,

$$\frac{d\theta}{dt} = \frac{d\beta}{dt} - \frac{d\alpha}{dt} = \left(-\frac{1}{500}\right) - \left(-\frac{1}{68}\right) = \frac{-17 + 125}{8500} = \frac{27}{2125}.$$

So the angle θ is increasing at a rate of $27/2125$ radians per second (at the moment we're considering), and we're finally done.

CHAPTER 9 _____

Exponentials and Logarithms

Here's a big old chapter on exponentials and logarithms. After we review the properties of these functions, we need to do some calculus with them. It turns out that there's a special base, the number e, that works out particularly nicely. In particular, doing calculus with e^x and $\log_e(x)$ is a little easier than dealing with 2^x or $\log_3(x)$, for example. So we need to spend some time looking at e. There are other things we want to look at as well; all in all, the plan is to check out the following topics:

- review of the basics of exponentials and logs, and how they are related to each other;
- the definition and properties of e;
- how to differentiate exponentials and logs;
- how to solve limit problems involving exponentials and logs;
- logarithmic differentiation;
- exponential growth and decay; and
- hyperbolic functions.

9.1 The Basics

Before you start doing calculus with exponentials and logarithms, you really need to understand their properties. Basically, in addition to the actual definition of logs, you need to know three things: the exponential rules, the relationship between logs and exponentials, and the log rules.

9.1.1 Review of exponentials

The rough idea is that we'll take a positive number, called the *base*, and raise it to a power called the *exponent*:

$$\text{base}^{\text{exponent}}.$$

For example, the number $2^{-5/2}$ is an exponential with base 2 and exponent $-5/2$. It's essential that you know the so-called exponential rules, which effectively tell you how exponentials work. You've seen these before, no doubt, but here they are again to remind you. For any base $b > 0$ and real numbers x and y:

1. $\boxed{b^0 = 1.}$ The zeroth power of any nonzero number is 1.

2. $\boxed{b^1 = b.}$ The first power of a number is just the number itself.

3. $\boxed{b^x b^y = b^{x+y}.}$ When you multiply two exponentials with the same base, you **add** the exponents.

4. $\boxed{\dfrac{b^x}{b^y} = b^{x-y}.}$ When you divide two exponentials with the same base, you **subtract** the bottom exponent from the top one.

5. $\boxed{(b^x)^y = b^{xy}.}$ When you take the exponential of the exponential, you **multiply** the exponents.

You should also know what the graphs of exponentials look like. We looked at this a little in Section 1.6 in Chapter 1, but in any case we'll revisit the graph shortly.

9.1.2   Review of logarithms

Logarithms—a word that strikes fear into the hearts of many students. Watch carefully, and we'll see how to deal with these beasts. Suppose that you want to solve the following equation for x:

$$2^x = 7.$$

The way you can bring x down from the exponent is to hit both sides with a logarithm. Since the base on the left-hand side is 2, the base of the logarithm is 2. Indeed, by definition, the solution of the above equation is

$$x = \log_2(7).$$

In other words, to what power do you have to raise 2 in order to get 7? The answer is $\log_2(7)$. This particular number can't be simplified, but how about $\log_2(8)$? Ask yourself, to what power do you raise the base 2 in order to get 8? Since $2^3 = 8$, the power we need is 3. So $\log_2(8) = 3$.

Let's go back to the equation $2^x = 7$. We know that this means that $x = \log_2(7)$. If we now plug that value of x into the original equation, we get the bizarre looking formula

$$2^{\log_2(7)} = 7.$$

In more generality, $\log_b(y)$ **is the power you have to raise the base b to in order to get y.** This means that $x = \log_b(y)$ is the solution of the equation $b^x = y$ for given b and y. Plugging this value of y in, we get the formula

$$\boxed{b^{\log_b(y)} = y}$$

which is true for any $y > 0$ and $b > 0$ (except $b = 1$). Hey, why do I insist that b and y be positive? First, if b is negative, then many weird things can happen. The quantity b^x may not be defined. For example, if $b = -1$ and $x = 1/2$, then b^x is $(-1)^{1/2}$, which is $\sqrt{-1}$ (urk). So we avoid all this by requiring $b > 0$. Then there's no problem taking any power b^x. On the other hand, b^x is always positive! So if $y = b^x$ then $y > 0$ by necessity. This means that it's nonsense to take the log of a negative number or 0. After all, if $\log_b(y)$ is the power that you raise b to in order to get y, and you can't ever raise b to a power and get a negative number or 0, then y can't be negative or 0. **You can only take the logarithm of a positive number.**

You might also have noticed that I mentioned that $b = 1$ is bad. If you put $b = 1$ in the formula $b^{\log_b(y)} = y$ from above, you get $1^{\log_1(y)} = y$. The problem is, 1 raised to any power still equals 1, but y may not be 1, so the equation doesn't make sense. There just isn't any base 1 logarithm. How about base $1/2$? That's OK, but there's rarely any need for a base $1/2$ logarithm, since it turns out that $\log_{1/2}(y) = -\log_2(y)$ for any number y. (You can prove this by setting $y = (1/2)^x$ and noting that y also equals 2^{-x}.) The same sort of thing is true for any base b between 0 and 1: $\log_b(y) = -\log_{1/b}(y)$ for all y, and $1/b$ is greater than 1. So from now on, we'll always assume that our base b is greater than 1.

9.1.3 Logarithms, exponentials, and inverses

We can describe everything we've seen above in a more sophisticated manner by using inverse functions. Fix a base $b > 1$ and set $f(x) = b^x$. The function f has domain $\mathbb{R}$ and range $(0, \infty)$. Since it satisfies the horizontal line test, it has an inverse, which we'll call g. The domain of g is the range of f, which is $(0, \infty)$, while the range of g is the domain of f, which is $\mathbb{R}$. We say that g is the *logarithm of base b*; in fact, $g(x) = \log_b(x)$ by definition. Remembering that the graph of the inverse function is the reflection of the original function in the mirror line $y = x$, we can draw the graphs of $f(x) = b^x$ and its inverse $g(x) = \log_b(x)$ on the same axes:

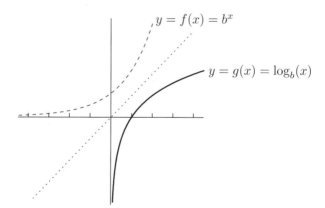

Since f and g are inverses of each other, we know that $f(g(x)) = x$ and $g(f(x)) = x$. (The first fact is only true for $x > 0$, as we will see.) Let's interpret these two facts, one at a time.

1. We'll start off with $f(g(x)) = x$. Since g is the logarithm function, x had better be positive—remember, you can only take the log of a positive number. Now let's take a close look at the quantity $f(g(x))$. You start out with a positive number x, and hit it with g, which is the base b logarithm. You then exponentiate the result: that is, you raise b to the power of $g(x)$. You end up with your original number! In fact, since $f(x) = b^x$ and $g(x) = \log_b(x)$, the formula $f(g(x)) = x$ just says that

$$b^{\log_b(x)} = x,$$

which was one of our formulas from the previous section (with y replaced by x). **The exponential of the logarithm is the original number**— provided that the bases match!

2. Our other fact is that $g(f(x)) = x$, which is true for all x. Now we take a number x, raise b to the power of our number x, then take the base b logarithm. Once again, we get the original number x back. It's sort of like taking a positive number, squaring it and then taking the square root: you get the original number back. Since $f(x) = b^x$ and $g(x) = \log_b(x)$, the equation $g(f(x)) = x$ becomes

$$\boxed{\log_b(b^x) = x} \qquad \text{for any real } x \text{ and } b > 1.$$

For example, when we looked at the equation $2^x = 7$ in the previous section, you can take $\log_2$ of both sides to get

$$\log_2(2^x) = \log_2(7).$$

The left-hand side is just x, because **the logarithm of the exponential is the original number** (provided that the bases match!). One more quick example: to solve

$$3^{x^2-1} = 19,$$

simply take $\log_3$ of both sides:

$$\log_3\left(3^{x^2-1}\right) = \log_3(19).$$

The left-hand side is just $x^2 - 1$, so we have $x^2 - 1 = \log_3(19)$. This means that $x = \pm\sqrt{\log_3(19) + 1}$.

9.1.4 Log rules

The exponential rules from Section 9.1.1 above all have log versions, which are (strangely enough) called log rules. There's actually an extra log rule—the change of base rule—that doesn't have a corresponding exponential rule (see #6 below).* So, here are the rules, which are valid for any base $b > 1$ and positive real numbers x and y:

---

*Actually, there is a change of base rule for exponentials too: $b^x = c^{x\log_c(b)}$ for $b > 0$, $c > 1$, and $x > 0$. This isn't normally included in the list of exponential rules because it involves logarithms!

1. $\boxed{\log_b(1) = 0.}$

2. $\boxed{\log_b(b) = 1.}$

3. $\boxed{\log_b(xy) = \log_b(x) + \log_b(y).}$ **The log of the product is the sum of the logs.**

4. $\boxed{\log_b(x/y) = \log_b(x) - \log_b(y).}$ **The log of the quotient is the difference of the logs.**

5. $\boxed{\log_b(x^y) = y\log_b(x).}$ **The log moves the exponent down in front of the log.** In this equation, y can be any real number (positive, negative or zero).

6. **Change of base rule:**

$$\boxed{\log_b(x) = \frac{\log_c(x)}{\log_c(b)}}$$

for any bases $b > 1$ and $c > 1$ and any number $x > 0$. This means that all the log functions with different bases are really constant multiples of each other. Indeed, the above equation says that

$$\log_b(x) = K\log_c(x),$$

where K is constant (it happens to be equal to $1/\log_c(b)$). When I say "constant," I mean it doesn't depend on x. We can conclude that the graphs of $y = \log_b(x)$ and $y = \log_c(x)$ are very similar—you just stretch the second one vertically by a factor of K to get the first one.

Now, let's see why these rules are all true. If you want, you can skip to the next section, but believe me, you'll understand logs a whole lot better if you read on. Anyway, #1 above is pretty easy: because $b^0 = 1$ for any base $b > 1$, we have $\log_b(1) = 0$. The same sort of thing works for #2: since $b^1 = b$ for any $b > 1$, we can just write down $\log_b(b) = 1$.

The third rule is harder. We must show that $\log_b(xy) = \log_b(x) + \log_b(y)$, where x and y are positive and $b > 1$. Let's start off with our important fact, which we've noted a couple of times above (with A replacing the previous variable):

$$b^{\log_b(A)} = A$$

for any $A > 0$. If we apply this three times with A replaced by x, y, and xy, respectively, we get

$$b^{\log_b(x)} = x, \qquad b^{\log_b(y)} = y, \quad \text{and} \quad b^{\log_b(xy)} = xy.$$

Now you can just multiply the first and second of these equations together, then compare with the third equation to get

$$b^{\log_b(x)}b^{\log_b(y)} = xy = b^{\log_b(xy)}.$$

So what? Well, use exponential rule #3 on the left-hand side; since we have to add the exponents, the equation becomes

$$b^{\log_b(x)+\log_b(y)} = b^{\log_b(xy)}.$$

Now hit both sides with a base b log to kill the base b on both sides; we're left with our log rule $\log_b(x) + \log_b(y) = \log_b(xy)$. Not so bad!

As for rule #4 above, I leave it to you to show this; the proof is almost identical to the one we just did for #3. So, let's go on to #5. We want to show that $\log_b(x^y) = y \log_b(x)$, where $x > 0$, $b > 1$, and y is any number at all. To do this, start with the important fact from above but with A replaced by x^y. We get

$$b^{\log_b(x^y)} = x^y.$$

This gives us a weird way of expressing x^y. We could also replace A instead by x to get

$$b^{\log_b(x)} = x,$$

then raise both sides to the power y:

$$(b^{\log_b(x)})^y = x^y.$$

The left-hand side of this is just $b^{y \log_b(x)}$ by exponential rule #5 (see Section 9.1.1 above). So we have two different expressions for x^y, which must be equal to each other:

$$b^{\log_b(x^y)} = b^{y \log_b(x)}.$$

Again, hitting both sides with a logarithm base b reduces everything to our log rule

$$\log_b(x^y) = y \log_b(x).$$

Finally, we just need to prove the change of base rule. We're actually going to show that

$$\log_b(x) \log_c(b) = \log_c(x).$$

You see, if that's true, then just divide both sides by $\log_c(b)$ to get the rule as it's described in #6 above. Anyway, let's take the equation above and raise c to the power the left-hand side and right-hand side separately. We get

$$c^{\log_b(x) \log_c(b)} \qquad \text{and} \qquad c^{\log_c(x)},$$

respectively. The right-hand side is easy: it's just x because of our important fact. How about the left-hand side? We use exponential rule #5 again in a tricky way to write

$$c^{\log_b(x) \log_c(b)} = c^{\log_c(b) \times \log_b(x)} = \left(c^{\log_c(b)} \right)^{\log_b(x)}.$$

Since $c^{\log_c(b)} = b$ and $b^{\log_b(x)} = x$ by our important fact (twice), we conclude that

$$c^{\log_b(x) \log_c(b)} = \left(c^{\log_c(b)} \right)^{\log_b(x)} = b^{\log_b(x)} = x.$$

So both of the quantities

$$c^{\log_b(x) \log_c(b)} \qquad \text{and} \qquad c^{\log_c(x)}$$

from above simplify down to just x! They must be equal to each other, then, and if we knock out the base of c (using a base c logarithm), we get our desired equation

$$\log_b(x) \log_c(b) = \log_c(x).$$

Well done if you took the trouble to understand all these proofs.

9.2 Definition of e

So far, we haven't done any calculus involving exponentials or logs. Let's start doing some. We'll begin with limits and then move on to derivatives. Along the way, we need to introduce a new constant e, which is a special number in the same sort of way that π is a special number—it just pops up when you start exploring math deeply enough. One way of seeing where e comes from involves a bit of a finance lesson.

9.2.1 A question about compound interest

A long time ago, a dude named Bernoulli answered a question about compound interest. Here's the setup for his question. Let's suppose you have a bank account at a bank that pays interest at a generous rate of 12% annually, compounded once a year. You put in an initial deposit; every year, your fortune increases by 12%. This means that after n years, your fortune has increased by a factor of $(1 + 0.12)^n$. In particular, after one year, your fortune is just $(1 + 0.12) = 1.12$ times the original amount. If you started with \$100, you'd finish the year with \$112.

Now suppose you find another bank that also offers an annual interest rate of 12%, but now it compounds twice a year. Of course you aren't going to get 12% for half a year; you have to divide that by 2. Basically this means that you are getting 6% interest for every 6 months. So, if you put money into this bank account, then after one year it has compounded twice at 6%; the result is that your fortune has expanded by a factor of $(1 + 0.06)^2$, which works out to be 1.1236. So if you started with \$100, you'd finish with \$112.36.

The second account is a little better than the first. It makes sense when you think about it—compounding is beneficial, so compounding more often at the same annual rate should be better. Let's try 3 times a year at the annual rate of 12%. We take 12% and divide by 3 to get 4%, then compound three times; our fortune has increased by $(1 + 0.04)^3$, which works out to be 1.124864. This is a little higher still. How about 4 times a year? That'd be $(1 + 0.03)^4$, which is approximately 1.1255. That's even higher. Now, the question is, where does it stop? If you compound more and more often at the same annual rate, do you get wads and wads of cash after a year, or is there some limitation on all this?

9.2.2 The answer to our question

To answer our question, let's turn to some symbols. First, let's suppose that we are compounding n times a year at an annual rate of 12%. This means that each time we compound, the amount of compounding is $0.12/n$. After this happens n times in one year, our original fortune has grown by a factor of

$$\left(1 + \frac{0.12}{n}\right)^n.$$

We want to know what happens if we compound more and more often; in fact, let's allow n to get larger and larger. That is, we'd like to know what happens in the limit as $n \to \infty$: what on earth is

$$\lim_{n \to \infty} \left(1 + \frac{0.12}{n}\right)^n ?$$

It would also be nice to know what happens at interest rates other than 12%. So let's replace 0.12 by r and worry about the more general limit

$$L = \lim_{n \to \infty} \left(1 + \frac{r}{n}\right)^n.$$

If this limit (which I called L) turns out to be infinite, then by compounding more and more often, you could get more and more money in a single year. On the other hand, if it turns out to be finite, we'll have to conclude that there is a limitation on how much we can increase our fortune with an annual interest rate of r, no matter how often we compound. There would be a sort of "speed limit," or more accurately, a "fortune-increase limit." Given a fixed annual interest rate r and one year to play with, you can never increase your fortune by a factor of more than the value of the above limit (assuming it's finite) no matter how often you compound.

The quantity $(1 + r/n)^n$ which occurs in the limit is a special case of the formula for compound interest. In general, suppose you start with \$$A$ in cash and you put it in a bank account at an annual interest rate of r, compounded n times a year. Then over t years, the compounding will occur nt times at a rate of r/n each time; so your fortune after t years will be given by the following formula:

$$\boxed{\text{fortune after } t \text{ years, compounded } n \text{ times a year at a rate of } r \text{ per year} = A \left(1 + \frac{r}{n}\right)^{nt}.}$$

So we are just starting with \$1 (so $A = 1$) and seeing what happens after one year (so $t = 1$), then seeing what happens in the limit if we compound more and more times a year.

Now let's attack our limit:

$$L = \lim_{n \to \infty} \left(1 + \frac{r}{n}\right)^n.$$

First, let's set $h = r/n$, so that $n = r/h$. Then as $n \to \infty$, we see that $h \to 0^+$ (since r is constant), so

$$L = \lim_{h \to 0^+} (1 + h)^{r/h}.$$

Now we can use our exponential rule to write

$$L = \lim_{h \to 0^+} \left((1 + h)^{1/h}\right)^r.$$

Let's pull a huge rabbit out of the hat and set

$$e = \lim_{h \to 0^+} (1 + h)^{1/h}.$$

Where is the trickery? Well, the limit might not exist. It turns out that it does; see Section A.5 of Appendix A if you want to know why. In any case, we have a special number e, which we'll look at in more detail very soon. Back to our limit, though; we now have

$$L = \lim_{h \to 0^+} ((1+h)^{1/h})^r = e^r.$$

That's the answer we're looking for! Let's put all the above steps together to see how it flows. With $h = r/n$, we have

$$L = \lim_{n \to \infty} \left(1 + \frac{r}{n}\right)^n = \lim_{h \to 0^+} (1+h)^{r/h} = \lim_{h \to 0^+} ((1+h)^{1/h})^r = e^r.$$

This means that if you compound more and more frequently at an annual rate of r, your fortune will increase by an amount very close to e^r, but never more than that. The quantity e^r is the "fortune-increase limit" we've been looking for. The only way you get this rate of increase is if you compound continuously—that is, all the time!

So, suppose you start with \$$A$ in cash and put it in a bank account which compounds continuously at an annual interest rate of r. After 1 year, you'll have \$$Ae^r$. After two years, you'll have \$$Ae^r \times e^r = Ae^{2r}$. It's easy to keep repeating this and see that after t years, you'll have \$$Ae^{rt}$. It actually works for partial years as well, because of the exponential rules. So, starting with \$$A$, we have:

> fortune after t years, compounded **continuously** at a rate of r per year $= Ae^{rt}$.

Compare this to the formula for compounding n times a year on the previous page. The quantities $A(1+r/n)^{nt}$ and Ae^{rt} look quite different, but for large n they're almost the same.

9.2.3 More about e and logs

Let's take a closer look at our number e. Remembering that

$$\lim_{n \to \infty} \left(1 + \frac{r}{n}\right)^n = e^r,$$

we can replace r by 1 to get

$$\lim_{n \to \infty} \left(1 + \frac{1}{n}\right)^n = e.$$

Of course, $r = 1$ corresponds to an interest rate of 100% per year. Let's draw up a little table of values of $(1 + 1/n)^n$ **to three decimal places** for some different values of n:

n	1	2	3	4	5	10	100	1000	10000	100000
$\left(1+\frac{1}{n}\right)^n$	2	2.25	2.353	2.441	2.488	2.594	2.705	2.717	2.718	2.718

Even compounding once a year at this humongous interest rate doubles your money (that's the "2" in the bottom row of the second column). Still, it seems as if we can't do much better than about 2.718, even if we compound many many times a year. Our number e, which is the limit as $n \to \infty$ of the numbers in the second row of the above table, turns out to be an irrational number whose decimal expansion begins like this:

$$e = 2.71828182845904523\ldots$$

It looks like there's a pattern near the beginning, with the repeated string "1828," but that's just a coincidence. In practice, just knowing that e is a little over 2.7 will be more than enough.

Now if $x = e^r$, then $r = \log_e(x)$. It turns out that taking logs base e is such a common thing to do that we can even write it a different way: $\ln(x)$ instead of $\log_e(x)$. The expression "$\ln(x)$" is **not** pronounced "lin x" or anything like that—just say "log x," or perhaps "ell en x," or if you're feeling particularly geeky, "the natural logarithm of x." In fact, most mathematicians write $\log(x)$ without a base to mean the same thing as $\log_e(x)$ or $\ln(x)$. The base e logarithm is called the *natural logarithm*. We'll see one of the reasons why it's so natural when we differentiate $\log_b(x)$ with respect to x in the next section.

Since we have a new base e, and a new way of writing logarithms in that base, let's take another look at the log rules and formulas we've seen so far. See if you can convince yourself that the following formulas are all true for $x > 0$ and $y > 0$:

$$\boxed{e^{\ln(x)} = x} \qquad \boxed{\ln(e^x) = x} \qquad \boxed{\ln(1) = 0} \qquad \boxed{\ln(e) = 1}$$

$$\boxed{\ln(xy) = \ln(x) + \ln(y)} \qquad \boxed{\ln\left(\frac{x}{y}\right) = \ln(x) - \ln(y)} \qquad \boxed{\ln(x^y) = y\ln(x)}$$

(Actually, in the second formula, x can even be negative or 0, and in the last formula, y can be negative or 0.) In any case, it's really worth knowing these formulas in this form, since we will almost always be working with natural logarithms from now on.

One more point before we move on to differentiating logs and exponentials. Suppose you take the important limit

$$\lim_{n \to \infty} \left(1 + \frac{r}{n}\right)^n = e^r,$$

and this time substitute $h = 1/n$. As we noticed in the previous section, when $n \to \infty$, we have $h \to 0^+$. So, replacing n by $1/h$, we get

$$\lim_{h \to 0^+} (1 + rh)^{1/h} = e^r.$$

This is a right-hand limit. In fact, you can replace $h \to 0^+$ by $h \to 0$ and the two-sided limit is still true. All we need to show is that the left-hand limit is e^r, and then, since both the left-hand and right-hand limits are the same, the two-sided limit equals e^r as well. So consider

$$\lim_{h \to 0^-} (1 + rh)^{1/h} = ?$$

Replace h by $-t$; then $t \to 0^+$ as $h \to 0^-$. (When h is a small negative number, $t = -h$ is a small positive number.) So

$$\lim_{h \to 0^-} (1 + rh)^{1/h} = \lim_{t \to 0^+} (1 - rt)^{-1/t}.$$

Since $A^{-1} = 1/A$ for any $A \neq 0$, we can rewrite the limit as

$$\lim_{t \to 0^+} \frac{1}{(1 + (-r)t)^{1/t}}.$$

The denominator is just the classic limit but with interest rate $-r$ instead of r. This means that in the limit as $t \to 0^+$, the denominator goes to e^{-r}. So altogether we have

$$\lim_{h \to 0^-} (1 + rh)^{1/h} = \lim_{t \to 0^+} (1 - rt)^{-1/t} = \lim_{t \to 0^+} \frac{1}{(1 + (-r)t)^{1/t}} = \frac{1}{e^{-r}} = e^r.$$

The last step works because $e^{-r} = 1/e^r$. So we have shown what we want to show. Let's change r to x in all our formulas (why not?) and summarize what we've found:

$$\boxed{\lim_{n \to \infty} \left(1 + \frac{x}{n}\right)^n = e^x} \qquad \text{and} \qquad \boxed{\lim_{h \to 0} (1 + xh)^{1/h} = e^x.}$$

When $x = 1$, we get two formulas for e:

$$\boxed{\lim_{n \to \infty} \left(1 + \frac{1}{n}\right)^n = e} \qquad \text{and} \qquad \boxed{\lim_{h \to 0} (1 + h)^{1/h} = e.}$$

These are important! We'll look at some examples at how to use them in Section 9.4.1 below. We'll also use one of them to differentiate the log function, right now.

9.3 Differentiation of Logs and Exponentials

Now the plot thickens. Let $g(x) = \log_b(x)$. What is the derivative of g? Using the definition,

$$g'(x) = \lim_{h \to 0} \frac{g(x + h) - g(x)}{h} = \lim_{h \to 0} \frac{\log_b(x + h) - \log_b(x)}{h}.$$

How do we simplify this mess? We use the log rules, of course! First, use rule #4 from Section 9.1.4 above to turn the difference of logs into the log of the quotient:

$$g'(x) = \lim_{h \to 0} \frac{1}{h} \log_b \left(\frac{x + h}{x}\right).$$

We can simplify the fraction down to $(1 + h/x)$, but we also need to use log rule #5 to pull the factor $1/h$ up to be an exponent. So

$$g'(x) = \lim_{h \to 0} \log_b \left(1 + \frac{h}{x}\right)^{1/h}.$$

Forget about the $\log_b$ for the moment. What happens to

$$\left(1 + \frac{h}{x}\right)^{1/h}$$

as h goes to 0? That is, what is

$$\lim_{h \to 0} \left(1 + \frac{h}{x}\right)^{1/h} ?$$

In the previous section, we saw that

$$\lim_{h \to 0} (1 + hr)^{1/h} = e^r;$$

so if we replace r by $1/x$, then this leads to

$$\lim_{h \to 0} \left(1 + \frac{h}{x}\right)^{1/h} = e^{1/x}.$$

So, if we go back to our expression for $g'(x)$, we see that

$$g'(x) = \lim_{h \to 0} \log_b \left(1 + \frac{h}{x}\right)^{1/h} = \log_b(e^{1/x}).$$

In fact we can even make the expression simpler by using log rule #5 again—the power $1/x$ comes down out front and we have shown that

$$\frac{d}{dx} \log_b(x) = \frac{1}{x} \log_b(e).$$

Now, let's set $b = e$, so that we are taking the derivative of the log function of base e. We get

$$\frac{d}{dx} \log_e(x) = \frac{1}{x} \log_e(e).$$

But wait a second—by log rule #2, $\log_e(e)$ is simply equal to 1. So this means that

$$\frac{d}{dx} \log_e(x) = \frac{1}{x}.$$

That's pretty nice. It's actually really really nice. Kind of amazing, really. Who would have thought that the derivative of $\log_e(x)$ is just $1/x$? This is one of the reasons why the logarithm base e is called the natural logarithm. Writing $\log_e(x)$ as $\ln(x)$ (we made this definition in the previous section), we get the important formula

$$\boxed{\frac{d}{dx} \ln(x) = \frac{1}{x}.}$$

Also, the above expression $\frac{1}{x} \log_b(e)$ for the derivative of $\log_b(x)$ can be written in terms of natural logarithms by using the change of base formula (that's #6 in Section 9.1.4 above). You see, by changing to base e, we get

$$\log_b(e) = \frac{\log_e(e)}{\log_e(b)} = \frac{1}{\ln(b)}.$$

So we have

$$\frac{d}{dx}\log_b(x) = \frac{1}{x\ln(b)}.$$

This is the nicest way to express the derivative of a logarithm of a base other than e. Now watch this: if $y = b^x$, then we know that $x = \log_b(y)$. Now differentiate with respect to y; using the above formula with x replaced by y, we get

$$\frac{dx}{dy} = \frac{1}{y\ln(b)}.$$

By the chain rule, we can flip both sides to get

$$\frac{dy}{dx} = y\ln(b).$$

Since $y = b^x$, we have proved the nice formula

$$\frac{d}{dx}(b^x) = b^x\ln(b).$$

In particular, if $b = e$, then $\ln(b) = \ln(e) = 1$. (That is just log rule #1 in disguise—remember, $\ln(e) = \log_e(e) = 1$.) So if $b = e$, this formula becomes

$$\frac{d}{dx}(e^x) = e^x.$$

This is a pretty freaky formula. If $h(x) = e^x$, then $h'(x) = e^x$ as well—the function h is its own derivative! Of course, the second derivative of e^x (with respect to x) is also e^x, as are the third derivative, the fourth derivative, and so on.

9.3.1 Examples of differentiating exponentials and logs

Now let's look at how to apply some of the above formulas. First, if $y = e^{-3x}$, what is dy/dx? Well, if $u = -3x$, then $y = e^u$. We have

$$\frac{dy}{du} = \frac{d}{du}(e^u) = e^u \qquad \text{and} \qquad \frac{du}{dx} = \frac{d}{dx}(-3x) = -3.$$

By the chain rule,

$$\frac{dy}{dx} = \frac{dy}{du}\frac{du}{dx} = e^u(-3) = -3e^{-3x};$$

notice that we replaced u by $-3x$ in the last step. In fact, this is a special case of a nice rule: if a is constant, then

$$\frac{d}{dx}e^{ax} = ae^{ax}.$$

This formula can be proved in the same way by letting $u = ax$. In fact, it's exactly the same as the principle we saw at the end of Section 7.2.1 in Chapter 7: **if x is replaced by ax, then there is an extra factor of**

a out front when you differentiate. So it should be no problem, for example, to differentiate $\ln(8x)$ with respect to x. In fact,

$$\frac{d}{dx}(\ln(8x)) = 8 \times \frac{1}{8x},$$

since the derivative of $\ln(x)$ with respect to x is $1/x$. Now, the factors of 8 cancel and we see that

$$\frac{d}{dx}(\ln(8x)) = 8 \times \frac{1}{8x} = \frac{1}{x}.$$

That's weird—the derivative of $\ln(8x)$ is the same as the derivative of $\ln(x)$! Not so weird when you think about it: $\ln(8x) = \ln(8) + \ln(x)$, so in fact the quantities $\ln(8x)$ and $\ln(x)$ just differ by a constant and therefore have the same derivative with respect to x.

Here's a harder example:

$$\text{if } y = e^{x^2} \log_3(5^x - \sin(x)), \text{ what is } \frac{dy}{dx}?$$

Let's use the product rule and the chain rule. Start off by setting $u = e^{x^2}$ and $v = \log_3(5^x - \sin(x))$, so $y = uv$. For the product rule, we need to differentiate u and v (with respect to x), so let's do them one at a time. Starting with $u = e^{x^2}$, let $t = x^2$ so that $u = e^t$; then, using the chain rule, we have

$$\frac{du}{dx} = \frac{du}{dt}\frac{dt}{dx} = e^t(2x) = 2xe^{x^2}.$$

As for v, let $s = 5^x - \sin(x)$ so that $v = \log_3(s)$. By the chain rule,

$$\frac{dv}{dx} = \frac{dv}{ds}\frac{ds}{dx} = \frac{1}{s\ln(3)}(5^x\ln(5) - \cos(x)) = \frac{5^x\ln(5) - \cos(x)}{\ln(3)(5^x - \sin(x))}.$$

Here we've used the formulas from the previous section for the derivatives of $\log_b(x)$ (with $b = 3$) and b^x (with b now equal to 5). Anyway, since $y = uv$, we have

$$\frac{dy}{dx} = v\frac{du}{dx} + u\frac{dv}{dx} = \log_3(5^x - \sin(x))2xe^{x^2} + e^{x^2}\frac{5^x\ln(5) - \cos(x)}{\ln(3)(5^x - \sin(x))}.$$

As usual, it's a bit of mess, but the example does illustrate the main points involved; as long as you know the basic formulas for differentiating exponentials and logs (they are the boxed equations in the previous section), then you'll be all set.

9.4 How to Solve Limit Problems Involving Exponentials or Logs

Now it's time to see how to solve a bunch of limit problems. As in the case of all the previous limits we've looked at, it's really important to note whether you are evaluating functions near 0 (that is, at small arguments), near ∞ or $-\infty$ (large arguments), or somewhere else that's neither small nor large. We'll examine some of these cases in some detail with respect to exponentials and logarithms. Let's start off, though, with limits involving the definition of e.

9.4.1 Limits involving the definition of e

Consider the following limit:

$$\lim_{h \to 0} (1 + 3h^2)^{1/3h^2}.$$

It looks pretty similar to the limit involving e from Section 9.2.3 above:

$$\lim_{h \to 0} (1 + h)^{1/h} = e.$$

If we take this limit, and replace h by $3h^2$ everywhere we see it, then we get

$$\lim_{3h^2 \to 0} (1 + 3h^2)^{1/3h^2} = e.$$

This is almost exactly what we want. All we have to do is note that $3h^2 \to 0$ as $h \to 0$, so

$$\lim_{h \to 0} (1 + 3h^2)^{1/3h^2} = e.$$

Using the same logic, we can show (for example) that

$$\lim_{h \to 0} (1 + \sin(h))^{1/\sin(h)} = e.$$

Indeed, if you replace h by any quantity that goes to 0 as $h \to 0$, such as $3h^2$ or $\sin(h)$, then the limit is still e. So how about

$$\lim_{h \to 0} (1 + \cos(h))^{1/\cos(h)}?$$

You can't just repeat the previous argument, since $\cos(h) \to 1$ as $h \to 0$. In fact, if you just substitute $h = 0$ into the expression $(1 + \cos(h))^{1/\cos(h)}$ then you get $(1 + 1)^1 = 2$, so the above limit is in fact equal to 2.

Now consider

$$\lim_{h \to 0} (1 + h^2)^{1/3h^2}.$$

There is a mismatch between an h^2 term and a $3h^2$ term. They are similar, but the coefficients aren't the same. We need to write the exponent $1/3h^2$ as $(1/h^2) \times (1/3)$ and use an exponential rule:

$$\lim_{h \to 0} (1 + h^2)^{1/3h^2} = \lim_{h \to 0} (1 + h^2)^{(1/h^2) \times (1/3)} = \lim_{h \to 0} \left((1 + h^2)^{1/h^2} \right)^{1/3}.$$

Since the h^2 terms match up, the part inside the big parentheses goes to e, and the whole limit is therefore $e^{1/3}$.

Here's a slightly harder example: what is the value of

$$\lim_{h \to 0} (1 - 5h^3)^{2/h^3}?$$

It's annoying, but the small quantities $-5h^3$ and h^3 don't quite match, and there's also that 2. We need to match them up by fiddling with the exponent $2/h^3$ to match the $-5h^3$ term. The best way to look at it is to see how nice everything would be if we instead wanted to find

$$\lim_{h \to 0} (1 - 5h^3)^{1/(-5h^3)},$$

because this limit is just e. Yup, the $-5h^3$ terms match and so this is nothing more than our classic limit

$$\lim_{h \to 0} (1+h)^{1/h} = e,$$

with h replaced by $-5h^3$. Unfortunately, we have to do a little more work. Somehow we need to turn $1/(-5h^3)$ into $2/h^3$. To do that, we have to multiply by -5 to get rid of the -5 in the denominator, and then multiply again by 2 to fix up the numerator. The overall effect is that we should multiply by -10. So, we have

$$\lim_{h \to 0} (1 - 5h^3)^{2/h^3} = \lim_{h \to 0} (1 - 5h^3)^{(1/(-5h^3)) \times (-10)}$$
$$= \lim_{h \to 0} \left((1 - 5h^3)^{1/(-5h^3)} \right)^{-10} = e^{-10}.$$

9.4.2 Behavior of exponentials near 0

We'd like to understand how e^x behaves when x is really close to 0. In fact, since $e^0 = 1$, we know that

$$\lim_{x \to 0} e^x = e^0 = 1.$$

Of course, you can replace x by another quantity that goes to 0 when $x \to 0$ and get the same limit. For example,

$$\lim_{x \to 0} e^{x^2} = e^{0^2} = 1$$

as well. So, we can find

$$\lim_{x \to 0} \frac{e^{x^2} \sin(x)}{x}$$

by splitting up like this:

$$\lim_{x \to 0} \frac{e^{x^2} \sin(x)}{x} = \lim_{x \to 0} \left(e^{x^2} \right) \left(\frac{\sin(x)}{x} \right).$$

Both factors tend to 1 as $x \to 0$, so the overall limit is $1 \times 1 = 1$. Now, here's a trickier example:

$$\lim_{x \to \infty} \frac{2x^2 + 3x - 1}{e^{1/x}(x^2 - 7)}.$$

As x gets very large, $1/x$ gets very close to 0; so $e^{1/x}$ is very close to 1 and can be ignored. Your best bet is to write the limit as

$$\lim_{x \to \infty} \frac{1}{e^{1/x}} \times \frac{2x^2 + 3x - 1}{x^2 - 7}.$$

The first fraction goes to 1, and using the techniques from Section 4.3 of Chapter 4, you can show that the second factor goes to 2; so the limit is 2.

This sort of approach works well if your exponential term appears in a product or a quotient, but it fails miserably with something like this:

$$\lim_{h \to 0} \frac{e^h - 1}{h}.$$

It's tempting to replace the e^h by 1, which is fair enough, except that you get a useless $0/0$ case. The problem is that we have a difference between e^h and 1, which gets very small when h is near 0. So what do we do? As we saw in Section 6.5 in Chapter 6, when the dummy variable is by itself on the bottom, your limit might be a derivative in disguise. Try setting $f(x) = e^x$, so that $f'(x) = e^x$ as well (as we saw in Section 9.3 above). In this case, the standard formula

$$\lim_{h \to 0} \frac{f(x + h) - f(x)}{h} = f'(x)$$

becomes

$$\lim_{h \to 0} \frac{e^{x+h} - e^x}{h} = e^x.$$

Now all we need to do is replace x by 0. Since $e^0 = 1$, we get the useful fact that

$$\boxed{\lim_{h \to 0} \frac{e^h - 1}{h} = 1.}$$

Once again, you can replace h by any small quantity. For example,

$$\lim_{s \to 0} \frac{e^{3s^5} - 1}{s^5} = \lim_{s \to 0} \frac{e^{3s^5} - 1}{3s^5} \times 3 = 1 \times 3 = 3.$$

The standard matching trick works; this is really the same trick we used in poly-type limits (Chapter 4), trig limits where the arguments are small (Chapter 7), and the limits in Section 9.4.1 above.

9.4.3 Behavior of logarithms near 1

Now let's look at how logs behave near 1. It turns out that the situation is pretty similar to the case of exponentials near 0. We know that $\ln(1) = 0$, but what is

$$\lim_{h \to 0} \frac{\ln(1 + h)}{h}?$$

Believe it or not, this is another example of a limit which is a derivative in disguise (see Section 6.5 in Chapter 6). Set $f(x) = \ln(x)$ and note that $f'(x) = 1/x$, as we saw in Section 9.3. Now the equation

$$\lim_{h \to 0} \frac{f(x + h) - f(x)}{h} = f'(x)$$

becomes

$$\lim_{h \to 0} \frac{\ln(x + h) - \ln(x)}{h} = \frac{1}{x}$$

for any x. All that's left is to put $x = 1$ and get

$$\lim_{h \to 0} \frac{\ln(1 + h) - \ln(1)}{h} = \frac{1}{1}.$$

Since $\ln(1) = 0$, this simplifies to

$$\lim_{h \to 0} \frac{\ln(1+h)}{h} = 1.$$

Once again, h can be replaced by any quantity which goes to 0 as $h \to 0$ and the limit will still be 1. For example, to find

$$\lim_{h \to 0} \frac{\ln(1-7h^2)}{5h^2},$$

you have to mess with the denominator to make it look like $-7h^2$ as follows:

$$\lim_{h \to 0} \frac{\ln(1-7h^2)}{5h^2} = \lim_{h \to 0} \frac{\ln(1-7h^2)}{-7h^2} \times \frac{-7h^2}{5h^2}.$$

It's just our old trick of multiplying and dividing by a useful quantity ($-7h^2$ in this case). Anyway, the first fraction has limit 1 since the small quantity $-7h^2$ matches, and the second fraction just cancels down to be $-7/5$. So the limit is $-7/5$.

9.4.4 Behavior of exponentials near ∞ or $-\infty$

Now we want to understand what happens to e^x when $x \to \infty$ or $x \to -\infty$. Let's take another look at the graph of $y = e^x$:

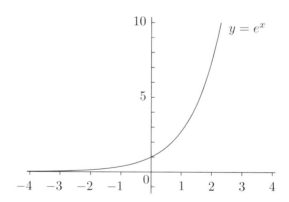

Beware: the curve above looks as if it touches the x-axis at the left side of the graph, but **it doesn't**; remember, $e^x > 0$ for all x, so there are no x-intercepts. (This is a good argument against relying too strongly on graphing calculators in order to understand what's going on!) In any case, it seems that we should at least have

$$\lim_{x \to \infty} e^x = \infty \qquad \text{and} \qquad \lim_{x \to -\infty} e^x = 0.$$

What if the base e is replaced by some other base? For example, consider

$$\lim_{x \to \infty} 2^x \qquad \text{and} \qquad \lim_{x \to \infty} \left(\frac{1}{3}\right)^x.$$

To handle the first one, let's use the identity $A = e^{\ln(A)}$ with $A = 2^x$ to write

$$2^x = e^{\ln(2^x)} = e^{x \ln(2)}.$$

Now as $x \to \infty$, we also have $x \ln(2) \to \infty$, so the first limit is ∞. As for the second limit, this time we can use the same trick to write

$$\left(\frac{1}{3}\right)^x = \frac{1}{3^x} = \frac{1}{e^{x \ln(3)}}.$$

As $x \to \infty$, we see that $e^{x \ln(3)} \to \infty$, so the reciprocal goes to 0. We have proved that

$$\lim_{x \to \infty} 2^x = \infty \qquad \text{and} \qquad \lim_{x \to \infty} \left(\frac{1}{3}\right)^x = 0.$$

These are special cases of the following important limit:

$$\lim_{x \to \infty} r^x = \begin{cases} \infty & \text{if } r > 1, \\ 1 & \text{if } r = 1, \\ 0 & \text{if } 0 \le r < 1. \end{cases}$$

The middle case, when $r = 1$, is obvious, since $1^x = 1$ for all $x \ge 0$. The other two cases can be shown in the same way as we handled the limits of 2^x and $(1/3)^x$ above—just write r^x as $e^{x \ln(r)}$.

This is not the whole story. The limit

$$\lim_{x \to \infty} e^x = \infty$$

says that e^x gets larger and larger—as large as you want—when x gets larger; but how fast does this happen? After all,

$$\lim_{x \to \infty} x^2 = \infty$$

as well. Which one grows faster, x^2 or e^x? The answer is that e^x kicks butt over x^2 when x is large. After all, when $x = 100$, the quantity x^2 is only 100×100, while

$$e^{100} = e \times e \times \cdots \times e.$$

There are a hundred factors of e but only two factors of 100, so e^{100} is much bigger than 100^2. The situation is even more in favor of e^x when x is larger still. Since e^x is so much bigger than x^2, when you divide x^2 by e^x you should get a tiny number. In fact,

$$\lim_{x \to \infty} \frac{x^2}{e^x} = 0.$$

We won't prove this until we look at l'Hôpital's Rule in Chapter 14. For the moment, I want to point out that the above limit is also true if you replace x^2 by any power of x. Even x^{999} can't compete with e^x. When x is a billion, x^{999} is 999 copies of a billion, multiplied together—but e^x is a **billion** copies of e, multiplied together! Even though e is a lot smaller than a billion, e^x just walks all over x^{999} in terms of size when x is large. So in general we have the following principle:

Exponentials grow quickly: $\boxed{\lim_{x \to \infty} \dfrac{x^n}{e^x} = 0}$ no matter how large n is.

In fact, by tweaking this a little, you can get a more general statement:

$$\lim_{x \to \infty} \frac{\text{poly-type stuff}}{\text{exponential of large, positive poly-type stuff}} = 0.$$

For example,

$$\lim_{x \to \infty} \frac{x^8 + 100x^7 - 4}{e^x} = 0.$$

To see why, simply split up the fraction into three pieces, each of which goes to 0 because exponentials grow quickly. More subtly,

$$\lim_{x \to \infty} \frac{x^{10000} + 300x^9 + 32}{e^{2x^3 - 19x^2 - 100}} = 0.$$

Here the crucial fact is that $2x^3 - 19x^2 - 100$ behaves like $2x^3$ when x is large, so the exponential is indeed of large, positive poly-type stuff.* In fact, the base e can be replaced by any other base greater than 1. For example,

$$\lim_{x \to \infty} \frac{x^{10000} + 300x^9 + 32}{2^{2x^3 - 19x^2 - 100}} = 0$$

as well. Another variation involves the fact that e^{-x} is just another way of writing $1/e^x$. Here's an example of this:

$$\lim_{x \to \infty} (x^5 + 3)^{101} e^{-x}.$$

We can just write this as

$$\lim_{x \to \infty} (x^5 + 3)^{101} e^{-x} = \lim_{x \to \infty} \frac{(x^5 + 3)^{101}}{e^x} = 0;$$

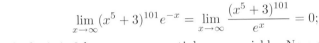

here the limit is 0 because exponentials grow quickly. Now consider the very similar limit

$$\lim_{x \to -\infty} (x^5 + 3)^{101} e^x.$$

This of course involves the behavior of e^x near $-\infty$, but you can just throw the situation over to $+\infty$ by setting $t = -x$. We can see that as $x \to -\infty$, we have $t \to +\infty$. So

$$\lim_{x \to -\infty} (x^5 + 3)^{101} e^x = \lim_{t \to \infty} ((-t)^5 + 3)^{101} e^{-t} = \lim_{t \to \infty} \frac{(-t^5 + 3)^{101}}{e^t} = 0.$$

Once again, the limit is 0 because the numerator is a polynomial (it doesn't matter that its leading coefficient is negative). So you can deal with e^x as $x \to -\infty$ by substituting $t = -x$; this means that you now have to deal with e^{-t} as $t \to \infty$, and you just handle that by writing e^{-t} as $1/e^t$.

---

*If you really want to nail it, you must write something clever like $2x^3 - 19x^2 - 100 > x^3$ for large enough x. After all, if $2x^3 - 19x^2 - 100$ behaves like $2x^3$, then clearly it must eventually be larger than x^3. So our denominator is bigger than e^{x^3}. Now replace x^3 by u, so that the denominator is just e^u and the numerator is some easy-to-deal-with mess. Finally, use the sandwich principle.

9.4.5 Behavior of logs near ∞

The saga continues. Let's look at what happens to $\ln(x)$ when x is a large positive number. (Remember, you can't take the log of any negative number, so there's no point in studying the behavior of logs near $-\infty$!) Here's the graph of $y = \ln(x)$ once again:

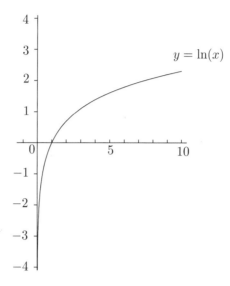

Again, it's important to note that the curve never touches the y-axis, even though it looks as if it does. It just gets very, very close. In any event, it seems as if

$$\lim_{x \to \infty} \ln(x) = \infty.$$

This is actually easy to show directly. Do you believe that $\ln(x)$ ever makes it up to 1000? Sure, it does: $\ln(e^{1000}) = 1000$. The same trick works for any number N. Just take $x = e^N$ and you will find that $\ln(x) = \ln(e^N) = N$. So there's no limit to how big $\ln(x)$ gets: it goes to ∞ as $x \to \infty$ ... but how fast?

It's pretty easy to see that it must be quite slow. As we just noted, $\ln(e^{1000}) = 1000$. The number e^{1000} is positively humongous—much greater than the number of atoms in the universe—yet its log is only 1000. Talk about cutting things down to size!

More precisely, it turns out that $\ln(x)$ goes to infinity much more slowly than **any positive power** of x, even something like $x^{0.0001}$. So if you take the ratio of $\ln(x)$ to any positive power of x, the ratio should be small (at least, when x is very large). In symbols, we have

Logs grow slowly: $\quad \boxed{\text{if } a > 0, \ \lim_{x \to \infty} \dfrac{\ln(x)}{x^a} = 0}$ $\quad$ no matter how small a is.

Just as in the case of exponentials, it's not too hard to extend this to a more general form:

$$\lim_{x \to \infty} \frac{\text{log of any positive poly-type stuff}}{\text{poly-type stuff of positive ``degree''}} = 0.$$

This works for logs of any base $b > 1$, not just the natural logarithm. (That's because of the change of base rule.) For example,

$$\lim_{x \to \infty} \frac{\log_7(x^3 + 3x - 1)}{x^{0.1} - 99} = 0$$

even though the power $x^{0.1}$ is very small.

Actually, we shouldn't be surprised that logs grow slowly, once we know that exponentials grow quickly. After all, logs and exponentials are inverses of each other. More precisely, if you take $\ln(x)/x^a$ and replace x by e^t, you get

$$\lim_{x \to \infty} \frac{\ln(x)}{x^a} = \lim_{t \to \infty} \frac{\ln(e^t)}{(e^t)^a} = \lim_{t \to \infty} \frac{t}{e^{at}} = 0.$$

The last limit is 0 because the exponential e^{at} on the bottom grows much more quickly than the polynomial t on the top. So we have shown that the fact that exponentials grow quickly automatically leads to the fact that logs grow slowly.

9.4.6 Behavior of logs near 0

It's tempting to write $\ln(0) = -\infty$, but it's just not true: $\ln(0)$ is not defined. On the other hand, the graph of $y = \ln(x)$ above suggests that

$$\boxed{\lim_{x \to 0^+} \ln(x) = -\infty.}$$

You need to use the right-hand limit here, since $\ln(x)$ isn't even defined for $x < 0$. Once again, though, we need to say more. Sure, $\ln(x)$ goes to $-\infty$ as $x \to 0^+$, but how quickly? For example, consider the limit

$$\lim_{x \to 0^+} x \ln(x).$$

If you just plug in 0, it doesn't work at all, since $\ln(0)$ doesn't exist. When x is a little bigger than 0, the quantity x is small and $\ln(x)$ is a large negative number. What happens when you multiply a small number by a large one? It could be anything at all, depending on how small and how large the numbers are.

Here's one way to solve the above problem. Replace x by $1/t$. Then as $x \to 0^+$, we can see that $t \to \infty$. So we have

$$\lim_{x \to 0^+} x \ln(x) = \lim_{t \to \infty} \frac{1}{t} \ln\left(\frac{1}{t}\right).$$

Of course, $\ln(1/t)$ is just $\ln(1) - \ln(t)$, which equals $-\ln(t)$, since $\ln(1) = 0$. So we get

$$\lim_{x \to 0^+} x \ln(x) = \lim_{t \to \infty} \frac{1}{t} \ln\left(\frac{1}{t}\right) = \lim_{t \to \infty} \frac{-\ln(t)}{t} = 0,$$

where the limit is 0 because logs grow slowly.

The trick of replacing x by $1/t$ to transfer the behavior near 0 to behavior near ∞ works because $\ln(1/t) = -\ln(t)$. You can use it to show the following principle, of which the above example is a special case:

Logs "grow" slowly at 0: $\quad$ if $a > 0$, $\displaystyle\lim_{x \to 0^+} x^a \ln(x) = 0$ $\quad$ no matter how small a is.

(I put "grow" in quotation marks because $\ln(x)$ really grows downward to $-\infty$ as $x \to 0^+$.) Once again, you can replace x^a by poly-type stuff, as long as it becomes small when $x \to 0^+$, and "ln" can be replaced by "$\log_b$" for any other base $b > 1$ (that is, not just the base e).

9.5 Logarithmic Differentiation

Logarithmic differentiation is a useful technique for dealing with derivatives of things like $f(x)^{g(x)}$, where both the base and the exponent are functions of x. After all, how on earth would you find

$$\frac{d}{dx}(x^{\sin(x)})$$

with what we have seen already? It doesn't fit any of the rules. Still, we have these nice log rules which cut exponents down to size. If we let $y = x^{\sin(x)}$, then

$$\ln(y) = \ln(x^{\sin(x)}) = \sin(x)\ln(x)$$

by log rule #5 from Section 9.1.4 above. Now let's differentiate both sides (implicitly) with respect to x:

$$\frac{d}{dx}(\ln(y)) = \frac{d}{dx}(\sin(x)\ln(x)).$$

Let's look at the right-hand side first. This is just a function of x and requires the product rule; you should check that the derivative works out to be $\cos(x)\ln(x) + \sin(x)/x$. Now let's look at the left-hand side. To differentiate $\ln(y)$ with respect to x (not y!), we should use the chain rule. Set $u = \ln(y)$, so that $du/dy = 1/y$. We need to find du/dx; by the chain rule,

$$\frac{du}{dx} = \frac{du}{dy}\frac{dy}{dx} = \frac{1}{y}\frac{dy}{dx}.$$

So, implicitly differentiating the equation $\ln(y) = \sin(x)\ln(x)$ produces

$$\frac{1}{y}\frac{dy}{dx} = \cos(x)\ln(x) + \frac{\sin(x)}{x}.$$

Now we just have to multiply both sides by y and then replace y by $x^{\sin(x)}$:

$$\frac{dy}{dx} = \left(\cos(x)\ln(x) + \frac{\sin(x)}{x}\right)y = \left(\cos(x)\ln(x) + \frac{\sin(x)}{x}\right)x^{\sin(x)}.$$

That's the answer we're looking for. (By the way, there is another way we could have done this problem. Instead of using the variable y, we could just have used our formula $A = e^{\ln(A)}$ to write

$$x^{\sin(x)} = e^{\ln(x^{\sin(x)})} = e^{\sin(x)\ln(x)}.$$

Now I leave it to you to differentiate the right-hand side of this with respect to x by using the product and chain rules. When you've finished, you should replace $e^{\sin(x)\ln(x)}$ by $x^{\sin(x)}$ and check that you get the same answer as the original one above.)

Let's review the main technique. Suppose you want to find the derivative with respect to x of

$$y = f(x)^{g(x)},$$

where both the base f and the exponent g involve the variable x. Here's what you do:

1. Let y be the function of x you want to differentiate. Take (natural) logs of both sides. The exponent g comes down on the right-hand side, so you should get

$$\ln(y) = g(x)\ln(f(x)).$$

2. Differentiate both sides implicitly with respect to x. The right-hand side often requires the product rule and the chain rule (at least). The left-hand side **always** works out to be $(1/y)(dy/dx)$. So you get

$$\frac{1}{y}\frac{dy}{dx} = \text{nasty stuff in } x.$$

3. Multiply both sides by y to isolate dy/dx, then replace y by the original expression $f(x)^{g(x)}$, and you're done.

Here's another example: what is

$$\frac{d}{dx}\left((1 + x^2)^{1/x^3}\right)?$$

According to the first step, we let $y = (1 + x^2)^{1/x^3}$, then take logs of both sides, bringing the exponent down; we get

$$\ln(y) = \ln\left((1 + x^2)^{1/x^3}\right) = \frac{1}{x^3}\ln(1 + x^2) = \frac{\ln(1 + x^2)}{x^3}.$$

The second step is to differentiate both sides implicitly with respect to x. The left-hand side, as always, becomes $(1/y)(dy/dx)$, but we'll have to use the quotient rule on the right-hand side. First, differentiate $z = \ln(1 + x^2)$ using the chain rule: if $u = 1 + x^2$, then $z = \ln(u)$, so

$$\frac{dz}{dx} = \frac{dz}{du}\frac{du}{dx} = \frac{1}{u}(2x) = \frac{2x}{1 + x^2}.$$

Now you can use the quotient rule; you should check that when you implicitly differentiate the equation $\ln(y) = \ln(1 + x^2)/x^3$ from above, you get (after simplifying)

$$\frac{1}{y}\frac{dy}{dx} = \frac{x^3\dfrac{2x}{1 + x^2} - 3x^2\ln(1 + x^2)}{(x^3)^2} = \frac{2x^2 - 3(1 + x^2)\ln(1 + x^2)}{x^4(1 + x^2)}.$$

Finally, multiply through y and replace y by $(1+x^2)^{1/x^3}$ to get

$$\frac{dy}{dx} = \frac{(2x^2 - 3(1+x^2)\ln(1+x^2))y}{x^4(1+x^2)}$$

$$= \frac{(2x^2 - 3(1+x^2)\ln(1+x^2))(1+x^2)^{1/x^3}}{x^4(1+x^2)}$$

$$= \frac{(2x^2 - 3(1+x^2)\ln(1+x^2))}{x^4(1+x^2)^{1-1/x^3}}$$

and we're all done.

Even if the base and exponent are not both functions of x, logarithmic differentiation can still come in handy. If your function is really nasty and involves lots of products and quotients of powers (like x^2) and exponentials (like e^x), you might want to try logarithmic differentiation. For example,

$$\text{if} \quad y = \frac{(x^2 - 3)^{100} 3^{\sec(x)}}{2x^5(\log_7(x) + \cot(x))^9}, \quad \text{what is} \quad \frac{dy}{dx}?$$

I must be joking, right? How can you be expected to differentiate something so foul? By logarithmic differentiation, that's how. Just take natural logs of both sides, and you'll find that the right-hand side becomes much more manageable (provided that you remember your log rules), like this:

$$\ln(y) = \ln\left(\frac{(x^2 - 3)^{100} 3^{\sec(x)}}{2x^5(\log_7(x) + \cot(x))^9}\right)$$

$$= \ln((x^2 - 3)^{100}) + \ln(3^{\sec(x)}) - \ln(2) - \ln(x^5) - \ln((\log_7(x) + \cot(x))^9)$$

$$= 100\ln(x^2 - 3) + \sec(x)\ln(3) - \ln(2) - 5\ln(x) - 9\ln(\log_7(x) + \cot(x)).$$

Make sure you understand these log manipulations before reading on. Anyway, now we can differentiate this expression implicitly with respect to x without too much drama:

$$\frac{d}{dx}(\ln(y)) = \frac{d}{dx}\left(100\ln(x^2 - 3) + \sec(x)\ln(3)\right.$$
$$\left. - \ln(2) - 5\ln(x) - 9\ln(\log_7(x) + \cot(x))\right).$$

The left-hand side is $(1/y)(dy/dx)$ as usual, so let's take a look at the right-hand side, term by term.

- The first term is $100\ln(x^2 - 3)$; it's a straightforward chain rule exercise to see that the derivative is $100 \times (2x)/(x^2 - 3)$, which is of course $200x/(x^2 - 3)$.
- The second term is $\sec(x)\ln(3)$. Before you whip out the product rule, remember that $\ln(3)$ is a constant, so in fact you can just take the derivative of $\sec(x)$ and then multiply by $\ln(3)$ to get $\ln(3)\sec(x)\tan(x)$.
- The third term is $-\ln(2)$, which is a constant, so its derivative is just 0.
- The fourth term is $-5\ln(x)$, which has derivative $-5/x$.
- The fifth term, $-9\ln(\log_7(x) + \cot(x))$, which I'll call z, requires the chain rule. Here are the details, although you should be able to work

this out for yourself. Let $u = \log_7(x) + \cot(x)$, so $z = -9\ln(u)$. Then we have

$$\frac{dz}{dx} = \frac{dz}{du}\frac{du}{dx} = -\frac{9}{u}\left(\frac{1}{x\ln(7)} - \csc^2(x)\right)$$

$$= \frac{9}{\log_7(x) + \cot(x)}\left(\csc^2(x) - \frac{1}{x\ln(7)}\right).$$

Let's put it all together to get

$$\frac{1}{y}\frac{dy}{dx} = \frac{200x}{x^2 - 3} + \ln(3)\sec(x)\tan(x) - \frac{5}{x}$$

$$+ \frac{9}{\log_7(x) + \cot(x)}\left(\csc^2(x) - \frac{1}{x\ln(7)}\right).$$

Now multiply by y to get

$$\frac{dy}{dx} = \left(\frac{200x}{x^2 - 3} + \ln(3)\sec(x)\tan(x) - \frac{5}{x}\right.$$

$$\left.+ \frac{9}{\log_7(x) + \cot(x)}\left(\csc^2(x) - \frac{1}{x\ln(7)}\right)\right) \times y.$$

Finally, replace y by the original (horrible) expression to get

$$\frac{dy}{dx} = \left(\frac{200x}{x^2 - 3} + \ln(3)\sec(x)\tan(x) - \frac{5}{x}\right.$$

$$\left.+ \frac{9}{\log_7(x) + \cot(x)}\left(\csc^2(x) - \frac{1}{x\ln(7)}\right)\right) \times \frac{(x^2 - 3)^{100}3^{\sec(x)}}{2x^5(\log_7(x) + \cot(x))^9}.$$

It seems nasty, but just imagine trying to do it without logarithmic differentiation!

9.5.1 The derivative of x^a

Now we can finally show something that we've been taking for granted:

$$\boxed{\frac{d}{dx}(x^a) = ax^{a-1}}$$

for **any** number a, not just integers as we've seen before. Let's suppose $x > 0$. Now use logarithmic differentiation: set $y = x^a$, so that $\ln(y) = a\ln(x)$. If you differentiate both sides implicitly, you get

$$\frac{1}{y}\frac{dy}{dx} = \frac{a}{x}.$$

Now multiply both sides by y and replace y by x^a:

$$\frac{dy}{dx} = \frac{ay}{x} = \frac{ax^a}{x} = ax^{a-1}.$$

This is exactly what we want, at least when $x > 0$. When $x \le 0$, we have a bit of a problem. For example, you can't even take $(-1)^{1/2}$ because this is the square root of a negative number. So what on earth should $(-1)^{\sqrt{2}}$ be? In fact, without using complex numbers (after all, we won't look at these until Chapter 28), you can only make sense of x^a for $x < 0$ when a is a rational number with an odd denominator (after canceling out common factors). For example, $x^{5/3}$ makes sense for negative x since you can always take a cube root—we're OK because 3 is odd. In the case where x^a makes sense for $x < 0$, it turns out that it's either an even or an odd function of x; you can use that fact to show that the derivative is still ax^{a-1}.

 Here are a couple of simple examples of using the formula. Working on the domain $(0, \infty)$, what is the derivative of $x^{\sqrt{2}}$ with respect to x? How about x^{π}? Just use the formula to show that

$$\frac{d}{dx}(x^{\sqrt{2}}) = \sqrt{2}x^{\sqrt{2}-1} \qquad \text{and} \qquad \frac{d}{dx}(x^{\pi}) = \pi x^{\pi-1}$$

for $x > 0$. It's not really any different from what we've done before—just that we can handle non-integer exponents now.

9.6 Exponential Growth and Decay

We've seen that bank accounts with continuous compounding grow exponentially. We don't need to look to such human-made devices to find exponential growth, though: it occurs in nature too. For example, under certain circumstances, populations of animals, like rabbits (and humans!), grow exponentially. There's also exponential decay, where a quantity gets smaller and smaller in an exponential fashion (we'll see what this means very soon). This occurs in radioactive decay, allowing scientists to find out how old some ancient artifacts, fossils, or rocks are.

Here's the basic idea. Suppose $y = e^{kx}$. Then, as we saw at the beginning of Section 9.3.1 above, $dy/dx = ke^{kx}$. The right-hand side of this equation can be written as ky, since $y = e^{kx}$. That is,

$$\frac{dy}{dx} = ky.$$

This is an example of a *differential equation*. After all, it's an equation involving derivatives. We'll look at many more differential equations in Chapter 30, but let's just focus on this one for the moment. What other functions satisfy the above equation? We know that $y = e^{kx}$ does, but there must be others. For example, if $y = 2e^{kx}$, then $dy/dx = 2ke^{kx}$, which is once again equal to ky. More generally, if $y = Ae^{kx}$, then $dy/dx = Ake^{kx}$, which is once again equal to ky. It turns out that this is the **only** way you can have $dy/dx = ky$:

$$\boxed{\text{if} \ \ \frac{dy}{dx} = ky, \ \ \text{then} \ \ y = Ae^{kx} \ \text{for some constant } A.}$$

We'll see why in Section 30.2 of Chapter 30. In the meantime, let's take a closer look at the differential equation $dy/dx = ky$. The first thing we'll do is change the variable x to t, so that we are looking at

$$\frac{dy}{dt} = ky.$$

This means that the rate of change of y is equal to ky. Interesting! The rate that the quantity is changing depends on how much of the quantity you have. If you have more of the quantity, then it grows faster (assuming $k > 0$). This makes sense in the case of population growth: the more rabbits you have, the more they can breed. If you have twice as many rabbits, they also **produce** twice as many rabbits in any given time period. The number k, which is called the *growth constant*, controls how fast the rabbits are breeding in the first place. The hornier they are, the higher k is!

9.6.1 Exponential growth

So, suppose we have a population which grows exponentially. In symbols, let P (or $P(t)$, if you prefer) be the population at time t, and let k be the growth constant. The differential equation for P is

$$\frac{dP}{dt} = kP.$$

This is the same as the differential equation in the box above, except that some symbols have changed. Instead of y, we have P; and instead of x, we have t. Never mind, we're good at adapting to these situations; we'll just make the same changes in the solution $y = Ae^{kx}$. We end up with $P = Ae^{kt}$ for some constant A. Now, when $t = 0$, we have $P = Ae^{k(0)} = Ae^0 = A$, since $e^0 = 1$. This means that A is the initial population, that is, the population at time 0. It's customary to relabel this variable as well. Instead of A, we'll write P_0 to indicate that it represents the population at time 0. Altogether, we have found the

> exponential growth equation: $\qquad P(t) = P_0 e^{kt}.$

Remember, P_0 is the initial population and k is the growth constant.

This formula is easy to apply in practice, provided that you know your exponential and log rules (see Sections 9.1.1 and 9.1.4 above). For example, if you know that a population of rabbits started 3 years ago at 1000, but now has grown to 64,000, then what will the population be one year from now? Also, what is the total time it will take for the population to grow from 1000 to 400,000?

Well, we have $P_0 = 1000$, since that's the initial population. So the equation in the box above becomes $P(t) = 1000e^{kt}$. The problem is, we don't know what k is. We do know that $P = 64000$ when $t = 3$, so let's plug this in:

$$64000 = 1000e^{3k}.$$

This means that $e^{3k} = 64$. Take logs of both sides to get $3k = \ln(64)$, so $k = \frac{1}{3}\ln(64)$. Actually, if you write $\ln(64) = \ln(2^6) = 6\ln(2)$, then you can simplify down to $k = 2\ln(2)$. This means that

$$P(t) = 1000e^{2\ln(2)t}$$

for any time t. Now we can solve both parts of the problem. For the first part, we want to know what happens a year from now. This is actually 4 years from the initial time, so set $t = 4$. We get

$$P(4) = 1000e^{2\ln(2)\times 4} = 1000e^{8\ln(2)}.$$

Now we get a little tricky: write $8\ln(2)$ as $\ln(2^8) = \ln(256)$, so

$$P(4) = 1000e^{\ln(256)} = 1000 \times 256 = 256000.$$

Here we have used the crucial formula $e^{\ln(A)} = A$ for any number $A > 0$. The conclusion is that the population will be 256,000 a year from now. Now let's tackle the second part of the problem. We want to see how long it will take for the population to get up to 400,000, so set $P = 400000$ to get

$$400000 = 1000e^{2\ln(2)t}.$$

This becomes $e^{2\ln(2)t} = 400$. To solve this, take logs of both sides; we get $2\ln(2)t = \ln(400)$, which means that

$$t = \frac{\ln(400)}{2\ln(2)}.$$

This is the number of years it takes for the population to grow from 1000 to 400,000, but it's not very intuitive. You could use a calculator to get an approximation; but suppose you don't have one handy. You just have to know that $\ln(5)$ is approximately 1.6 and $\ln(2)$ is approximately 0.7. Start off by writing $400 = 20^2$, so $\ln(400) = \ln(20^2) = 2\ln(20)$. We can do even better, though: $\ln(20) = \ln(4 \times 5) = \ln(4) + \ln(5) = 2\ln(2) + \ln(5)$. All told, we get

$$t = \frac{\ln(400)}{2\ln(2)} = \frac{2(2\ln(2) + \ln(5))}{2\ln(2)} = 2 + \frac{\ln(5)}{\ln(2)}.$$

Using our approximations, we get

$$t \cong 2 + \frac{1.6}{0.7} = 2 + \frac{16}{7} = 4\tfrac{2}{7}.$$

So although it takes 4 years to get up to a population of 256,000, it only takes approximately two-sevenths of a year more—about $3\tfrac{1}{2}$ months—to get up to 400,000. That's the power of exponential growth....

9.6.2 Exponential decay

Let's turn things upside-down and look at exponential decay. To set the scene, let me tell you that there are certain atoms which are radioactive. They are like little time bombs: after awhile they break apart into different atoms, emitting energy at the same time. The only problem is that you never know when they are going to break apart (we'll say "decay" instead of "break apart"). All you know is that over a given time, there's a certain chance that the decay will happen.

For example, you might have a certain type of atom which has a 50% chance of decaying within any 7-year period. So if you have one of these atoms in a box, close the box, and then open it up in 7 years, there's a 50-50 chance that it will have decayed. Of course, it's pretty difficult to see an individual atom! So let's suppose, a little more realistically, that you have a trillion atoms. (That's still a tiny speck of material, by the way.) You put them in the box and come back 7 years later. What do you expect to find? Well, about half the atoms should have decayed, while the other half remain intact. So you should have about half a trillion of the original atoms. What if you come back in another 7 years? Then half the remaining original atoms will be left, leaving you with a quarter of a trillion of the original atoms. Every 7 years, you lose half of your remaining sample.

So let's try to write down an equation to model the situation. If $P(t)$ is the number (population?) of atoms at time t, then I claim that

$$\frac{dP}{dt} = -kP$$

for some constant k. This says that the rate of change of P is a negative multiple of P. That is, P decays at a rate proportional to P. The more atoms you have, the faster the decay. This agrees with our above example: in the first 7 years, we lost half a trillion atoms, but in the next 7 years, we only lost a quarter of a trillion. In another 7 years, we'll only lose one-eighth of a trillion atoms. The more we have, the more we lose. Anyway, the solution to the above differential equation is

$$P(t) = P_0 e^{-kt},$$

where P_0 is the original number of atoms (at time $t = 0$). This is exactly the same as the equation for exponential growth from the previous section, except that we have replaced the growth constant k by a negative constant $-k$, which is called the *decay constant*.

In the above example, we know that it takes 7 years for any sample of atoms to halve in size. This length of time is called the *half-life* of the atom (or material). In the above equation, this means that if you start with P_0 atoms, then in 7 years, you'll have $\frac{1}{2}P_0$ atoms. So, setting $t = 7$ and $P(7) = \frac{1}{2}P_0$ in the above equation, we have

$$\frac{1}{2}P_0 = P_0 e^{-k(7)}.$$

Now cancel out the factor of P_0 from both sides and take the log of both sides to get

$$\ln\left(\frac{1}{2}\right) = -7k.$$

Since $\ln(1/2) = \ln(1) - \ln(2) = -\ln(2)$, the above equation becomes

$$k = \frac{\ln(2)}{7}.$$

This means that

$$P(t) = P_0 e^{-t(\ln(2)/7)}$$

in this case.

Now let's generalize a little. Suppose you have some other radioactive material with a half-life of $t_{1/2}$ years. This means that half of any size sample of the material will decay in $t_{1/2}$ years. It doesn't mean that the whole sample will decay in twice that many years! Anyway, by the same reasoning as in the previous paragraph, we can show that $k = \ln(2)/t_{1/2}$. In summary,

$$\text{for radioactive decay with half-life } t_{1/2}, \quad P(t) = P_0 e^{-kt} \quad \text{with } k = \frac{\ln(2)}{t_{1/2}}.$$

For example, if the half-life of the material is still 7 years, and you start off with 50 pounds of the material, how much do you have after 10 years, and how long does it take before you are down to 1 pound of the material? We know $t_{1/2} = 7$, so $k = \ln(2)/7$, as we saw before. Since $P_0 = 50$ (in pounds), the decay equation $P(t) = P_0 e^{-kt}$ becomes

$$P(t) = 50e^{-t(\ln(2)/7)}.$$

So when $t = 10$, we have

$$P(10) = 50e^{-10\ln(2)/7}.$$

That is, we are down to $50e^{-10\ln(2)/7}$ pounds. If we use our approximation $\ln(2) \cong 0.7$ from above, then we see that we have approximately $50e^{-1}$ pounds, which we can further approximate to about 18.4 pounds.

As for the second part of the question, now we need to find out how long it takes before we are down to one pound of material, so set $P(t) = 1$ in the above equation for $P(t)$ to get

$$1 = 50e^{-t(\ln(2)/7)}.$$

Divide both sides by 50 and take logs to get

$$\ln\left(\frac{1}{50}\right) = -\frac{t\ln(2)}{7}.$$

Since $\ln(1/50) = -\ln(50)$, we have $-7\ln(50) = -t\ln(2)$; that is,

$$t = \frac{7\ln(50)}{\ln(2)}.$$

We can estimate this using our previous approximations $\ln(5) \cong 1.6$ and $\ln(2) \cong 0.7$. Write $\ln(50) = \ln(2 \times 5 \times 5) = \ln(2) + 2\ln(5)$ to see that

$$t = \frac{7\ln(50)}{\ln(2)} = \frac{7(\ln(2) + 2\ln(5))}{\ln(2)} = 7 + \frac{14\ln(5)}{\ln(2)} \cong 7 + \frac{14(1.6)}{0.7},$$

which works out to be 39 years. So it takes approximately 39 years for the sample to decay from 50 pounds down to a single pound. By the way, 39 years is a little more than $5\frac{1}{2}$ half-lives (since one half-life is 7 years). So if you have 50 pounds of a different radioactive material with a half-life of 10 years, then this material will take a little more than 55 years to decay to 1 pound. (The actual number is $10\ln(50)/\ln(2)$ years, which is closer to $56\frac{1}{2}$ years.)

9.7 Hyperbolic Functions

Let's change course and look at the so-called *hyperbolic functions*. These are actually exponential functions in disguise, but they are similar to trig functions in many ways. We won't be using them much but they do come up occasionally, so it's good to be familiar with them.

We'll start by defining the hyperbolic cosine and hyperbolic sine functions:

$$\cosh(x) = \frac{e^x + e^{-x}}{2} \qquad \sinh(x) = \frac{e^x - e^{-x}}{2}$$

No triangles needed! This isn't trigonometry, after all.* These functions behave somewhat like their ordinary cousins, but not exactly. For example, if you square $\cosh(x)$ and $\sinh(x)$, you find that

$$\cosh^2(x) = \left(\frac{e^x + e^{-x}}{2}\right)^2 = \frac{e^{2x} + e^{-2x} + 2}{4},$$

and

$$\sinh^2(x) = \left(\frac{e^x - e^{-x}}{2}\right)^2 = \frac{e^{2x} + e^{-2x} - 2}{4}.$$

(We used the fact that $e^x e^{-x} = 1$.) Anyway, let's take the difference of these two quantities:

$$\cosh^2(x) - \sinh^2(x) = \frac{e^{2x} + e^{-2x} + 2}{4} - \frac{e^{2x} + e^{-2x} - 2}{4} = \frac{4}{4} = 1.$$

So we've proved that

$$\boxed{\cosh^2(x) - \sinh^2(x) = 1}$$

for any x. Not quite the same as the regular old trig identity—the minus makes all the difference. (Indeed, $x^2 - y^2 = 1$ is the equation of a hyperbola.)

How about calculus properties? Well, let's differentiate $y = \sinh(x)$; we'll need the fact that the derivative of e^{-x} is $-e^{-x}$:

$$\frac{d}{dx}\sinh(x) = \frac{d}{dx}\left(\frac{e^x - e^{-x}}{2}\right) = \frac{e^x + e^{-x}}{2} = \cosh(x).$$

So the derivative of hyperbolic sine is hyperbolic cosine. That's just like what happens with regular old sine and cosine. On the other hand,

$$\frac{d}{dx}\cosh(x) = \frac{d}{dx}\left(\frac{e^x + e^{-x}}{2}\right) = \frac{e^x - e^{-x}}{2} = \sinh(x).$$

If these were ordinary trig functions, then the derivative would be negative hyperbolic sine, but we don't have a negative here. In any case, we have shown that

$$\boxed{\frac{d}{dx}\sinh(x) = \cosh(x)} \qquad \text{and} \qquad \boxed{\frac{d}{dx}\cosh(x) = \sinh(x).}$$

---

*There is actually a branch of geometry called *hyperbolic geometry*, in which the triangles have wacky properties that lead to hyperbolic functions.

Now let's look at the graphs of these functions. First, you should try to convince yourself that $\cosh(x)$ is an even function of x and that $y = \sinh(x)$ is an odd function of x. (Just plug in $-x$ and see what happens.) Furthermore, $\cosh(0) = 1$ and $\sinh(0) = 0$ (check this too). Finally, let's note that

$$\lim_{x \to \infty} \cosh(x) = \lim_{x \to \infty} \frac{e^x + e^{-x}}{2}.$$

The term e^x goes to ∞, but e^{-x} goes to 0. The overall effect is that the limit is ∞. The same thing works for $\sinh(x)$, so our graphs must look something like this:

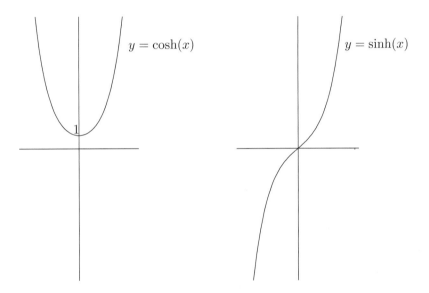

Of course you can define $\tanh(x)$ as $\sinh(x)/\cosh(x)$, as well as the reciprocals $\operatorname{sech}(x)$, $\operatorname{csch}(x)$, and $\coth(x)$. Each of the functions sech, csch, and coth can be differentiated by replacing them with the appropriate exponentials—for example,

$$\operatorname{sech}(x) = \frac{1}{\cosh(x)} = \frac{1}{\dfrac{e^x + e^{-x}}{2}} = \frac{2}{e^x + e^{-x}},$$

which you can then differentiate using the chain rule or the quotient rule. There are also identities connecting the functions, the most important of which is

$$1 - \tanh^2(x) = \operatorname{sech}^2(x).$$

This follows directly from the identity $\cosh^2(x) - \sinh^2(x) = 1$ by dividing both sides by $\cosh^2(x)$. Now I'm just going to list the derivatives of the other hyperbolic functions and display their graphs—I leave it to you to check that the derivatives all work out and that the graphs at least make sense. First, the derivatives:

$$\frac{d}{dx}\tanh(x) = \text{sech}^2(x)$$

$$\frac{d}{dx}\text{sech}(x) = -\text{sech}(x)\tanh(x)$$

$$\frac{d}{dx}\text{csch}(x) = -\text{csch}(x)\coth(x)$$

$$\frac{d}{dx}\coth(x) = -\text{csch}^2(x).$$

Now the graphs:

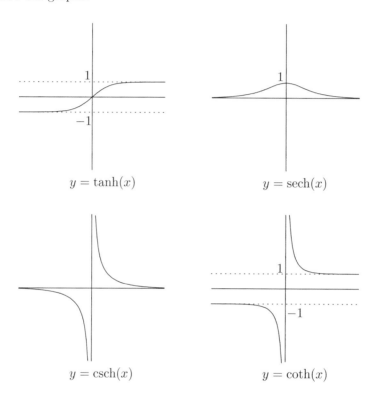

$$y = \tanh(x)$$

$$y = \text{sech}(x)$$

$$y = \text{csch}(x)$$

$$y = \coth(x)$$

From the definitions of the functions, you can see that all the hyperbolic trig functions are odd functions except for cosh and sech, which are even. This is the same as in the case of regular old trig functions! Also, $y = \tanh(x)$ and $y = \coth(x)$ both have horizontal asymptotes at $y = 1$ and $y = -1$, whereas $y = \text{sech}(x)$ and $y = \text{csch}(x)$ both have a horizontal asymptote at $y = 0$.

CHAPTER 10 _____

Inverse Functions and Inverse Trig Functions

In the previous chapter, we looked at exponentials and logarithms. We got a lot of mileage out of the fact that e^x and $\ln(x)$ are inverses of each other. In this chapter, we'll look at some more general properties of inverse functions, then examine inverse trig functions (and their hyperbolic cousins) in greater detail. Here's the game plan:

- using the derivative to show that a function has an inverse;
- finding the derivative of inverse functions;
- inverse trig functions, one by one; and
- inverse hyperbolic functions.

10.1   The Derivative and Inverse Functions

In Section 1.2 of Chapter 1, we reviewed the basics of inverse functions. I strongly suggest you take a quick look over that section before reading further, familiarizing yourself with the general idea. Now that we know some calculus, we can say more. In particular, we're going to explore two connections between derivatives and inverse functions.

10.1.1   Using the derivative to show that an inverse exists

Suppose that you have a differentiable function f whose derivative is always positive. What do you think the graph of this function looks like? Well, the slope of the tangent has to be positive everywhere, so the function can't dip up and down: it has to go upward as we look from left to right. In other words, the function must be **increasing**.

We'll prove this fact in the next chapter (see Section 11.3.1 and also Section 11.2), but it at least seems clear that it should be true. In any case, if our function f is always increasing, then it must satisfy the horizontal line test. No horizontal line could possibly hit the graph of $y = f(x)$ twice. Since the horizontal line test is satisfied by f, we know that f has an inverse. This has given us a nice strategy for showing that a function has an inverse: show that its derivative is always positive on its domain.

For example, suppose that

$$f(x) = \frac{1}{3}x^3 - x^2 + 5x - 11$$

on the domain $\mathbb{R}$ (the whole real line). Does f has an inverse? It would be a real mess to switch x and y in the equation $y = \frac{1}{3}x^3 - x^2 + 5x - 11$ and then try to solve for y. (Try it and see!) A much better way to show that f has an inverse is to find the derivative. We get

$$f'(x) = x^2 - 2x + 5.$$

So what? Well, f' is just a quadratic. Its discriminant is -16, which is negative, so the equation $f'(x) = 0$ has no solutions. (See Section 1.6 in Chapter 1 for a review of the discriminant.) That means that $f'(x)$ must be always positive or negative: its graph can't cross the x-axis. Well, which is it—positive or negative? Since $f'(0) = 5$, it must be positive;* that is, $f'(x) > 0$ for all x. This means that f is increasing. In particular, f satisfies the horizontal line test, so it has an inverse.

We've seen that if $f'(x) > 0$ for all x in the domain, then f has an inverse. There are some variations. For example, if $f'(x) < 0$ for all x, then the graph $y = f(x)$ is decreasing. The horizontal line test still works, though—the graph is just going down and down, so it can't come back up and hit the same horizontal line twice. Another variation is that the derivative might be 0 for an instant but positive everywhere else. This is OK as long as the derivative doesn't stay at 0 for a long time. Here's a summary of the situation:

Derivatives and inverse functions: if f is differentiable on its domain (a, b) and any of the following are true:

1. $f'(x) > 0$ for all x in (a, b);
2. $f'(x) < 0$ for all x in (a, b);
3. $f'(x) \geq 0$ for all x in (a, b) and $f'(x) = 0$ for only a finite number of x; or
4. $f'(x) \leq 0$ for all x in (a, b) and $f'(x) = 0$ for only a finite number of x,

then f has an inverse. If instead the domain is of the form $[a, b]$, or $[a, b)$, or $(a, b]$, and f is continuous on the whole domain, then f still has an inverse if any of the above four conditions are true.

Here's another example. Suppose $g(x) = \cos(x)$ on the domain $(0, \pi)$. Does g have an inverse? Well, $g'(x) = -\sin(x)$. We know that $\sin(x) > 0$ on the interval $(0, \pi)$—just look at its graph if you don't believe this. Since $g'(x) = -\sin(x)$, we see that $g'(x) < 0$ for all x in $(0, \pi)$. This means that g has an inverse. In fact, we know that g has an inverse on all of $[0, \pi]$, since g is continuous there. The idea is that $g(0) = 1$, so g starts out at height 1; then, since $g'(x) < 0$ when $0 < x < \pi$, we know that g immediately gets lower than 1. Since $g(\pi) = -1$, the values of $g(x)$ go down to -1 without ever hitting

---

*Another way to show this is to complete the square: $x^2 - 2x + 5 = (x-1)^2 + 4 > 0$, since all squares (such as $(x-1)^2$) are nonnegative.

the same value twice. So g has an inverse on all of $[0, \pi]$. We'll come back to this particular function in Section 10.2.2 below.

Finally, let $h(x) = x^3$ on all of $\mathbb{R}$. We know that $h'(x) = 3x^2$, which can't be negative. So $h'(x) \geq 0$ for all x. Luckily, $h'(x) = 0$ only when $x = 0$, so there's just one little point where $h'(x) = 0$. That's OK, so h still has an inverse; in fact, $h^{-1}(x) = \sqrt[3]{x}$.

10.1.2 Derivatives and inverse functions: what can go wrong

We noticed that the derivative of our function is allowed to be 0 occasionally and the function can still have an inverse. Why can't $f'(x) = 0$ a little more often? For example, suppose that f is defined by

$$f(x) = \begin{cases} -x^2 + 1 & \text{if } x < 0, \\ 1 & \text{if } 0 \leq x < 1, \\ x^2 - 2x + 2 & \text{if } x \geq 1. \end{cases}$$

When $x < 0$, we have $f'(x) = -2x$, which is positive (since x is negative!). When $0 < x < 1$, we have $f'(x) = 0$; and when $x > 1$, we can see that $f'(x) = 2x - 2 = 2(x - 1)$, which is certainly positive. Also, the function values and derivatives both match at the join points $x = 0$ and $x = 1$, so we've shown that f is differentiable and $f'(x) \geq 0$ for all x. (See Section 6.6 in Chapter 6 to review why this works.) Unfortunately the horizontal line test fails, and there is no inverse! Check out the graph:

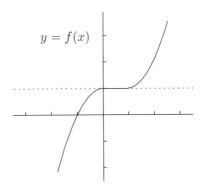

$y = f(x)$

The horizontal line $y = 1$ hits this graph infinitely often—everywhere between $x = 0$ and $x = 1$ inclusive. The function f is constant on $[0, 1]$, which is consistent with the fact that $f'(x) = 0$ for these x.

Here's another potential problem. The four conditions on the previous page all require that the domain be an interval like (a, b). What if the domain isn't in one piece? Unfortunately, then the conclusion can totally fail to hold. For example, if $f(x) = \tan(x)$, then $f'(x) = \sec^2(x)$, which can't be negative; however, you can see from the graph that $y = \tan(x)$ fails the horizontal line test pretty miserably. (See Section 10.2.3 below to remind yourself about the graph of $y = \tan(x)$.) So the methods of the previous section won't work, in general, when your function has discontinuities or vertical asymptotes.

10.1.3 Finding the derivative of an inverse function

If you know that a function f has an inverse, which we'll call f^{-1} as usual, then what's the derivative of that inverse? Here's how you find it. Start off with the equation $y = f^{-1}(x)$. You can rewrite this as $f(y) = x$. Now differentiate implicitly with respect to x to get

$$\frac{d}{dx}(f(y)) = \frac{d}{dx}(x).$$

The right-hand side is easy: it's just 1. To find the left-hand side, we use implicit differentiation (see Chapter 8). If we set $u = f(y)$, then by the chain rule (noting that $du/dy = f'(y)$), we have

$$\frac{d}{dx}(f(y)) = \frac{d}{dx}(u) = \frac{du}{dy}\frac{dy}{dx} = f'(y)\frac{dy}{dx}.$$

Now divide both sides by $f'(y)$ to get the following principle:

$$\text{if } y = f^{-1}(x), \qquad \text{then} \quad \frac{dy}{dx} = \frac{1}{f'(y)}.$$

If you want to express everything in terms of x, then you have to replace y by $f^{-1}(x)$ to get

$$\frac{d}{dx}(f^{-1}(x)) = \frac{1}{f'(f^{-1}(x))}.$$

In words, this means that the derivative of the inverse is basically the reciprocal of the derivative of the original function, except that you have to evaluate this latter derivative at $f^{-1}(x)$ instead of x.

For example, set $f(x) = \frac{1}{3}x^3 - x^2 + 5x - 11$. We saw in Section 10.1.1 above that f has an inverse on all of $\mathbb{R}$. If we set $y = f^{-1}(x)$, then what is dy/dx in general? What is its value when $x = -11$? To do the first part, all you have to do is to see that $f'(x) = x^2 - 2x + 5$, so

$$\frac{dy}{dx} = \frac{1}{f'(y)} = \frac{1}{y^2 - 2y + 5}.$$

Note that it's important to replace x by y here. Anyway, now we can solve the second part. We know that $x = -11$, but what is y? Since $y = f^{-1}(x)$, we know that $f(y) = x$. By the definition of f, we have

$$\frac{1}{3}y^3 - y^2 + 5y - 11 = -11.$$

Now clearly $y = 0$ is a solution to this equation, and it **must** be the only solution because the inverse exists. So, when $x = -11$, we have $y = 0$, and then

$$\frac{dy}{dx} = \frac{1}{y^2 - 2y + 5} = \frac{1}{(0)^2 - 2(0) + 5} = \frac{1}{5}.$$

More formally, one can write $(f^{-1})'(-11) = 1/5$.

Now suppose that $h(x) = x^3$ as in Section 10.1.1 above. We saw there that h has an inverse, and we even have a way to write it: $h^{-1}(x) = x^{1/3}$. Of course, we could just use the rule for differentiating x^a with respect to x, but let's try the above method. We know that $h'(x) = 3x^2$; if $y = h^{-1}(x)$, then

$$\frac{dy}{dx} = \frac{1}{h'(y)} = \frac{1}{3y^2}.$$

Now we can solve the equation $x = y^3$ for y to get $y = x^{1/3}$, and substitute into the above equation to get

$$\frac{dy}{dx} = \frac{1}{3(x^{1/3})^2} = \frac{1}{3x^{2/3}}.$$

This is all pretty silly, because we could just have differentiated $y = x^{1/3}$ and gotten the same answer without nearly so much work. Nevertheless it's nice to know that it all works out.

Before we move on to another example, let's just note that the derivative of the inverse function doesn't exist when $x = 0$, since the denominator $3x^{2/3}$ vanishes. So even though the original function is differentiable everywhere, the inverse isn't differentiable everywhere: its derivative doesn't exist at $x = 0$. This is true in general, not just for the function h from above. If you have any function which has an inverse, and it has slope 0 at the point (x, y), the inverse function will have infinite slope at the point (y, x), as the following picture illustrates:

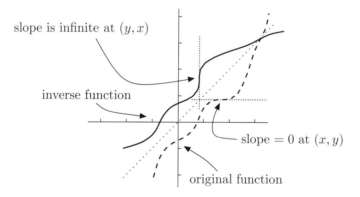

Sometimes you don't know much about a function, but you can still find out something about the derivative of the inverse function. For example, suppose you know that $g(x) = \sin(f^{-1}(x))$ for some invertible function f, but all you know about f is that $f(\pi) = 2$ and $f'(\pi) = 5$. That's actually enough information to find the values of $g(2)$ and $g'(2)$. In particular, since $f(\pi) = 2$ and f is invertible, we have $f^{-1}(2) = \pi$, so $g(2) = \sin(f^{-1}(2)) = \sin(\pi) = 0$. Also, by the chain rule and the above boxed formula for $(f^{-1})'(x)$, we have

$$g'(x) = \cos(f^{-1}(x)) \times (f^{-1})'(x) = \cos(f^{-1}(x)) \times \frac{1}{f'(f^{-1}(x))}.$$

Putting $x = 2$ and using the facts that $f^{-1}(2) = \pi$ and $f'(\pi) = 5$, we get

$$g'(2) = \cos(f^{-1}(2)) \times \frac{1}{f'(f^{-1}(2))} = \cos(\pi) \times \frac{1}{f'(\pi)} = -1 \times \frac{1}{5} = -\frac{1}{5}.$$

Make sure you know both the above versions of the formula for the derivative of an inverse function!

10.1.4 A big example

Let's finish off with an example that involves most of the theory we've looked at so far in this chapter. Suppose that

$$f(x) = x^2(x-5)^3 \qquad \text{on the domain } [2, \infty).$$

Here's what we want to do:

1. show that f is invertible;
2. find the domain and range of the inverse f^{-1};
3. check that $f(4) = -16$; and finally,
4. compute $(f^{-1})'(-16)$.

For #1, use the product rule and the chain rule to see that

$$f'(x) = 2x(x-5)^3 + 3x^2(x-5)^2.$$

Noticing that x and $(x-5)^2$ are factors of both terms on the right, we can rewrite this as

$$f'(x) = x(x-5)^2(2(x-5) + 3x) = x(x-5)^2(5x - 10) = 5x(x-5)^2(x-2).$$

When $x > 2$ (remember, the domain of f is $[2, \infty)$), all three of the factors $5x$, $(x-5)^2$, and $(x-2)$ are nonnegative, so their product is as well. We have now shown that $f'(x) \geq 0$ on $(2, \infty)$. Also, the only place in this domain where $f'(x) = 0$ is $x = 5$. Since f is continuous on $[2, \infty)$, the methods of Section 10.1.1 above show that f has an inverse.

Let's move on to #2. The range of the inverse f^{-1} is just the domain of f, which of course is $[2, \infty)$. Alas, the domain of f^{-1} is harder to find. Indeed, the domain of f^{-1} is precisely the range of f, so we need to do some work and find this range. It's not such a big deal, though. We know that f is always increasing, so this means that $f(2)$ is the lowest point. That is, the function starts at height $f(2)$, which works out to be $2^2(-3)^3 = -108$, and increases. How high does it get? Well, as x gets larger and larger, f does as well—there's no limit to how much it increases. This means that f covers all the numbers from -108 upward, so the domain of f^{-1} is the same as the range of f, which is $[-108, \infty)$.

We still have to do the last two parts of the problem. For #3, it's an easy calculation to show that $f(4) = -16$, which means that $f^{-1}(-16) = 4$. Moving on to #4, if $y = f^{-1}(x)$, then we know that

$$\frac{dy}{dx} = \frac{1}{f'(y)} = \frac{1}{5y(y-5)^2(y-2)}.$$

When $x = -16$, we know from part #3 that $y = 4$. Plugging this in, we get

$$\frac{dy}{dx} = \frac{1}{5(4)(4-5)^2(4-2)} = \frac{1}{40}.$$

We've finished all the parts of the question, but it's really useful to sketch the graph of $y = x^2(x-5)^3$ to get an idea what on earth we've accomplished here. In Section 12.3.3 of Chapter 12, we'll return to this example and do a thorough job of sketching the graph, but meanwhile we can still get a great idea of what the graph looks like. Let's work on the domain $\mathbb{R}$, then restrict ourselves to $[2, \infty)$ at the end. Here's what we know:

- To find the y-intercept, put $x = 0$; we get $y = 0^2(0-5)^3 = 0$. So the y-intercept is at 0.
- To find the x-intercepts, set $x^2(x-5)^3 = 0$; we find that $x = 0$ or $x = 5$. These are the x-intercepts.
- When x is near 0, the quantity $(x-5)^3$ is very close to $(-5)^3 = -125$, so $x^2(x-5)^3$ should be pretty close to $-125x^2$. The graph should convey this fact.
- When x is near 5, we see that x^2 is also near 25, so the curve behaves like $25(x-5)^3$. The graph of $y = 25(x-5)^3$ is just like the graph of x^3, except shifted to the right by 5 units and stretched vertically by a factor of 25. So we'll build that into our graph as well.

All in all, it's not surprising that the graph looks something like this (I have ghosted out the part of the graph where $x < 2$; also note that the axes have different scales):

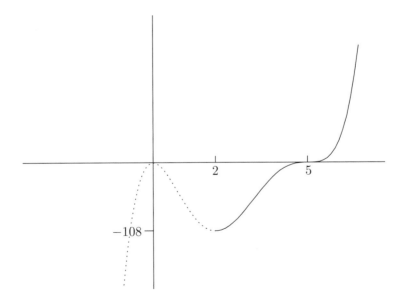

The graph is consistent with the fact that the function f is invertible on the restricted domain $[2, \infty)$, and also that the range of f on this restricted domain is indeed $[-108, \infty)$.

10.2 Inverse Trig Functions

Now it's time to investigate the inverse trig functions. We'll see how to define them, what their graphs look like, and how to differentiate them. Let's look at them one at a time, beginning with inverse sine.

10.2.1 Inverse sine

Let's start by looking at the graph of $y = \sin(x)$ once again:

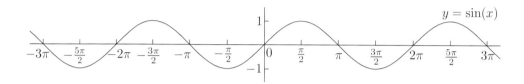

Does the sine function have an inverse? You can see from the above graph that the horizontal line test fails pretty miserably. In fact, every horizontal line of height between -1 and 1 intersects the graph **infinitely** many times, which is a lot more than the zero or one time we can tolerate. So, using the tactic described in Section 1.2.3 in Chapter 1, we throw away as little of the domain as possible in order to pass the horizontal line test. There are many options, but the sensible one is to restrict the domain to the interval $[-\pi/2, \pi/2]$. Here's the effect of this:

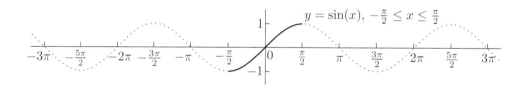

The solid portion of the curve is all we have left after we restrict the domain. Clearly we can't go to the right of $\pi/2$ or else we'll start repeating the values immediately to the left of $\pi/2$ as the curve dips back down. A similar thing happens at $-\pi/2$. So, we're stuck with our interval.

OK, if $f(x) = \sin(x)$ with domain $[-\pi/2, \pi/2]$, then it satisfies the horizontal line test, so it has an inverse f^{-1}. We'll write $f^{-1}(x)$ as $\sin^{-1}(x)$ or $\arcsin(x)$. (Beware: the first of these notations is a little confusing at first, since $\sin^{-1}(x)$ does **not** mean the same thing as $(\sin(x))^{-1}$, even though $\sin^2(x) = (\sin(x))^2$ and $\sin^3(x) = (\sin(x))^3$.)

So, what is the domain of the inverse sine function? Well, since the range of $f(x) = \sin(x)$ is $[-1, 1]$, the domain of the inverse function is $[-1, 1]$. And since the domain of our function f is $[-\pi/2, \pi/2]$ (since that's how we restricted the domain), the range of the inverse is $[-\pi/2, \pi/2]$.

How about the graph of $y = \sin^{-1}(x)$? We just have to take the restricted graph of $y = \sin(x)$ and reflect it in the mirror line $y = x$; it looks like this:

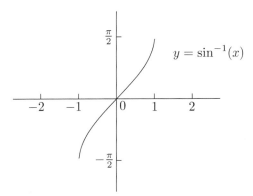

Here's a neat way to remember how to draw this graph. Start by reflecting all of $y = \sin(x)$ in the line $y = x$, then throw away all but the correct part of it. This graph shows how the above graph of $y = \sin^{-1}(x)$ is just part of the tipped-over graph of $y = \sin(x)$:

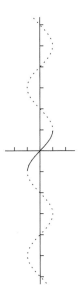

Note that since $\sin(x)$ is an odd function of x, so is $\sin^{-1}(x)$. This is consistent with the above graphs.

Now let's differentiate the inverse sine function. Set $y = \sin^{-1}(x)$; we want to find dy/dx. The snazziest way to do this is to write $x = \sin(y)$ and then differentiate both sides implicitly with respect to x:

$$\frac{d}{dx}(x) = \frac{d}{dx}(\sin(y)).$$

The left-hand side is just 1, but the right-hand side needs the chain rule. You should check that you get $\cos(y)(dy/dx)$. So we have

$$1 = \cos(y)\frac{dy}{dx}$$

which simplifies to

$$\frac{dy}{dx} = \frac{1}{\cos(y)}.$$

Actually, we could have written this down immediately using the formula from Section 10.1.3 above. Now, we really want the derivative in terms of x, not y. No problem—we know that $\sin(y) = x$, so it shouldn't be too hard to find $\cos(y)$. In fact, $\cos^2(y) + \sin^2(y) = 1$, which means that $\cos^2(y) + x^2 = 1$. This leads to the equation $\cos(y) = \pm\sqrt{1 - x^2}$, so we have

$$\frac{dy}{dx} = \pm\frac{1}{\sqrt{1 - x^2}}.$$

But which is it? Plus or minus? If you look at the graph of $y = \sin^{-1}(x)$ above, you can see that the slope is always positive. This means that we have to take the positive square root:

$$\boxed{\frac{d}{dx}\sin^{-1}(x) = \frac{1}{\sqrt{1 - x^2}} \qquad \text{for } -1 < x < 1.}$$

Note that $\sin^{-1}(x)$ is not differentiable, even in the one-sided sense, at the endpoints $x = 1$ and $x = -1$, since the denominator $\sqrt{1 - x^2}$ is 0 in both these cases.

In addition to the derivative formula and the above graph, here's a summary of the important facts about the inverse sine function:

$$\boxed{\sin^{-1} \text{ is odd; it has domain } [-1, 1] \text{ and range } [-\tfrac{\pi}{2}, \tfrac{\pi}{2}].}$$

Now that you have a new derivative formula, you should become comfortable using the product, quotient, and chain rules in association with it. For example, what are

$$\frac{d}{dx}(\sin^{-1}(7x)) \qquad \text{and} \qquad \frac{d}{dx}(x\sin^{-1}(x^3))?$$

For the first one, you could use the chain rule, setting $t = 7x$, or you could use the principle from the end of Section 7.2.1 in Chapter 7: when you replace x by ax, you have to multiply the derivative by a. So we have

$$\frac{d}{dx}(\sin^{-1}(7x)) = 7 \times \frac{1}{\sqrt{1 - (7x)^2}} = \frac{7}{\sqrt{1 - 49x^2}}.$$

For the second question, start by setting $y = x\sin^{-1}(x^3)$; also put $u = x$ and $v = \sin^{-1}(x^3)$, so that $y = uv$. We'll need to use the product rule:

$$\frac{dy}{dx} = v\frac{du}{dx} + u\frac{dv}{dx} = \sin^{-1}(x^3) \times 1 + x\frac{dv}{dx}.$$

To finish it off, we must find dv/dx. Since $v = \sin^{-1}(x^3)$, if we set $t = x^3$ then $v = \sin^{-1}(t)$. By the chain rule,

$$\frac{dv}{dx} = \frac{dv}{dt}\frac{dt}{dx} = \frac{1}{\sqrt{1 - t^2}}(3x^2) = \frac{3x^2}{\sqrt{1 - (x^3)^2}} = \frac{3x^2}{\sqrt{1 - x^6}}.$$

Plug this into the previous equation to see that

$$\frac{dy}{dx} = \sin^{-1}(x^3) \times 1 + x\frac{dv}{dx} = \sin^{-1}(x^3) + \frac{3x^3}{\sqrt{1-x^6}},$$

and we're all done.

10.2.2 Inverse cosine

We're going to repeat the procedure from the previous section in order to understand the inverse cosine function. Start with the graph of $y = \cos(x)$:

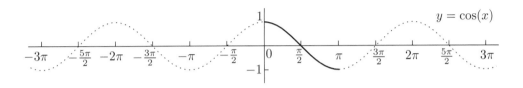

Once again, no inverse. This time, restricting the domain to $[-\pi/2, \pi/2]$ won't work, since the horizontal line test would fail and also we'd be throwing away part of the range that would be useful. Already on the above graph, you can see that the section between $[0, \pi]$ is highlighted and obeys the horizontal line test, so that's what we'll use. We get an inverse function which we write as $\cos^{-1}$ or arccos. Like inverse sine, the domain of inverse cosine is $[-1, 1]$, since that's the range of cosine. On the other hand, the range of inverse cosine is $[0, \pi]$, since that's the restricted domain of cosine that we're using. The graph of $y = \cos^{-1}(x)$ is formed by reflecting the graph of $y = \cos(x)$ in the mirror $y = x$:

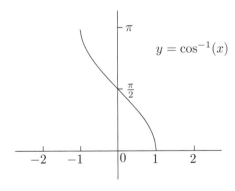

Notice that the graph shows that $\cos^{-1}$ is neither even nor odd. This is despite the fact that $\cos(x)$ is an even function of x! In any case, if you have trouble drawing the above graph from memory, just draw the graph of $\cos(x)$ on its side and pick out the bit with range $[0, \pi]$, like this:

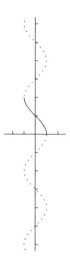

Now it's time to differentiate $y = \cos^{-1}(x)$ with respect to x. We do exactly the same thing we did in the previous section. Start by writing $x = \cos(y)$ and differentiating implicitly with respect to x:

$$\frac{d}{dx}(x) = \frac{d}{dx}(\cos(y)).$$

The left-hand side is 1 and the right-hand side is $-\sin(y)(dy/dx)$. This can be rearranged into

$$\frac{dy}{dx} = -\frac{1}{\sin(y)}.$$

Since $\cos^2(y) + \sin^2(y) = 1$, and also $x = \cos(y)$, we have $\sin(y) = \pm\sqrt{1 - x^2}$. This means that

$$\frac{dy}{dx} = -\frac{1}{\pm\sqrt{1 - x^2}} = \pm\frac{1}{\sqrt{1 - x^2}}.$$

Unlike the case of inverse sine, the graph of inverse cosine is all downhill, which means that the slope is always negative, so we get

$$\boxed{\frac{d}{dx}\cos^{-1}(x) = -\frac{1}{\sqrt{1 - x^2}} \qquad \text{for } -1 < x < 1.}$$

Here are the other facts about inverse cosine that we collected above:

$$\boxed{\cos^{-1} \text{ is neither even nor odd; it has domain } [-1, 1] \text{ and range } [0, \pi].}$$

Before we move on to the inverse tangent function, let's just look at the derivatives of inverse sine and inverse cosine side by side:

$$\frac{d}{dx}\sin^{-1}(x) = \frac{1}{\sqrt{1 - x^2}} \qquad \text{and} \qquad \frac{d}{dx}\cos^{-1}(x) = -\frac{1}{\sqrt{1 - x^2}}.$$

The derivatives are negatives of each other! Let's try to see why this makes sense. If you plot $y = \sin^{-1}(x)$ and $y = \cos^{-1}(x)$ on the same set of axes, here's what you get:

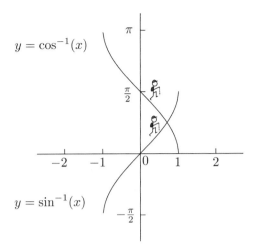

The two mountain-climbers in the above picture experience exactly opposite conditions at the same horizontal point, so it makes sense that the derivatives should be negatives of each other. Indeed, we now know that

$$\frac{d}{dx}(\sin^{-1}(x) + \cos^{-1}(x)) = \frac{1}{\sqrt{1-x^2}} - \frac{1}{\sqrt{1-x^2}} = 0.$$

So $y = \sin^{-1}(x) + \cos^{-1}(x)$ has constant slope 0, which means that it's flat as a pancake. In fact, if you add up the heights of the function values in the two graphs above, you can see that you get $\pi/2$ for any value of x. We've just used calculus to prove the following identity:

$$\sin^{-1}(x) + \cos^{-1}(x) = \frac{\pi}{2}$$

for any x in the interval $[-1, 1]$. When you think about it, this makes sense, though! Look at the following diagram:

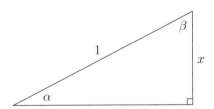

Since $\sin(\alpha) = x$, we have $\alpha = \sin^{-1}(x)$. Similarly, $\cos(\beta) = x$ which means that $\beta = \cos^{-1}(x)$. But $\alpha + \beta = \pi/2$, which means that

$$\sin^{-1}(x) + \cos^{-1}(x) = \frac{\pi}{2}$$

once again. Kind of nice how the calculus agrees with the geometry, huh?

10.2.3 Inverse tangent

Here we go again. Let's remember the graph of $y = \tan(x)$:

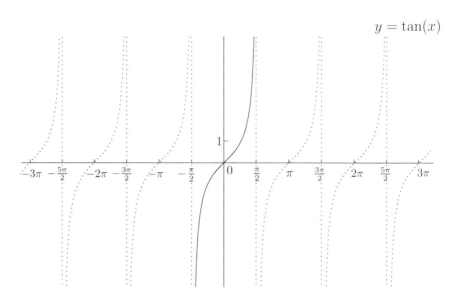

$$y = \tan(x)$$

We'll restrict the domain to $(-\pi/2, \pi/2)$ so that we can get an inverse function $\tan^{-1}$, also written as arctan. The domain of this function is the range of the tangent function, which is all of $\mathbb{R}$. The range of the inverse function is $(\pi/2, \pi/2)$, which of course is the restricted domain of $\tan(x)$ that we're using. The graph of $y = \tan^{-1}(x)$ looks like this:

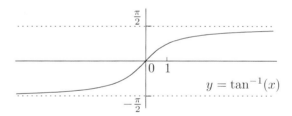

$$y = \tan^{-1}(x)$$

Now $\tan^{-1}(x)$ is an odd function of x, as you can see from the graph—it inherits its oddness from that of $\tan(x)$, in fact. Once again, you can remember the graph by drawing $y = \tan(x)$ on its side and throwing most of it away:

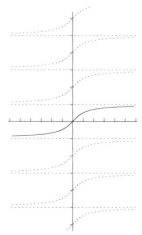

Now let's differentiate $y = \tan^{-1}(x)$ with respect to x. Write $x = \tan(y)$ and differentiate implicitly with respect to x. Check to make sure that you believe that

$$\frac{dy}{dx} = \frac{1}{\sec^2(y)}.$$

Since $\sec^2(y) = 1 + \tan^2(y)$, and $\tan(y) = x$, we see that $\sec^2(y) = 1 + x^2$. This means that

$$\boxed{\frac{d}{dx}\tan^{-1}(x) = \frac{1}{1 + x^2} \qquad \text{for all real } x.}$$

We also have the following facts from above:

$$\boxed{\tan^{-1} \text{ is odd; it has domain } \mathbb{R} \text{ and range } \left(-\tfrac{\pi}{2}, \tfrac{\pi}{2}\right).}$$

Unlike inverse sine and inverse cosine, the inverse tangent function has horizontal asymptotes. (The first two functions don't have a chance, since their domains are both $[-1, 1]$.) As you can see from the graph above, $\tan^{-1}(x)$ tends to $\pi/2$ as $x \to \infty$, and it tends to $-\pi/2$ as $x \to -\infty$. In fact, the vertical asymptotes $x = \pi/2$ and $x = -\pi/2$ of the tangent function have become horizontal asymptotes of the inverse tan function. This means that we have the following useful limits:

$$\boxed{\lim_{x \to \infty} \tan^{-1}(x) = \frac{\pi}{2}} \qquad \text{and} \qquad \boxed{\lim_{x \to -\infty} \tan^{-1}(x) = -\frac{\pi}{2}.}$$

By the way, we've seen these limits before, in Section 3.5 of Chapter 3. In any case, these limits can come up in conjunction with other limits at $\pm\infty$; for example, to find

$$\lim_{x \to -\infty} \frac{x^2 - 6x + 4}{(2x^2 + 7x - 8)\tan^{-1}(3x)},$$

first separate the fraction to get

$$\lim_{x \to -\infty} \frac{x^2 - 6x + 4}{2x^2 + 7x - 8} \times \frac{1}{\tan^{-1}(3x)}.$$

The first fraction has limit $1/2$ (check it!), but what happens to the second fraction? Well, as x becomes very negatively large, $3x$ also does, so $\tan^{-1}(3x)$ tends to $-\pi/2$. So the whole limit is

$$\frac{1}{2} \times \frac{1}{-\frac{\pi}{2}} = -\frac{1}{\pi}.$$

However, suppose that we replace the $3x$ term by $3x^2$, like this:

$$\lim_{x \to -\infty} \frac{x^2 - 6x + 4}{(2x^2 + 7x - 8)\tan^{-1}(3x^2)}.$$

Now $\tan^{-1}(3x^2)$ has limit $\pi/2$ even when $x \to -\infty$, because then $3x^2$ tends to ∞, not $-\infty$. So the overall limit in this case is $1/\pi$.

10.2.4 Inverse secant

The saga continues. Here's the graph of $y = \sec(x)$:

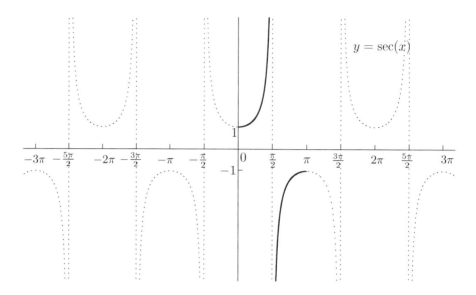

The situation is (unsurprisingly) very similar to the one we faced when we inverted the cosine function. The domain has to be restricted to $[0, \pi]$, except for the point $\pi/2$, which isn't even in the original domain of $\sec(x)$. The range of secant is the union of the two intervals $(-\infty, -1]$ and $[1, \infty)$, so this becomes the domain of the inverse function $\sec^{-1}$ (alternatively arcsec). As for the range of $\sec^{-1}$, it's the same as the restricted domain: $[0, \pi]$ minus the point $\pi/2$. The graph looks like this:

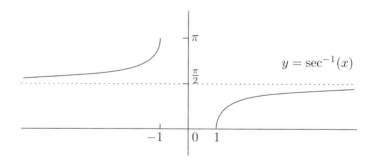

Note that there's a two-sided horizontal asymptote at $y = \pi/2$, so

$$\lim_{x \to \infty} \sec^{-1}(x) = \frac{\pi}{2} \qquad \text{and} \qquad \lim_{x \to -\infty} \sec^{-1}(x) = \frac{\pi}{2}.$$

Let's find the derivative. If $y = \sec^{-1}(x)$ then $x = \sec(y)$, so

$$\frac{d}{dx}(x) = \frac{d}{dx}(\sec(y)).$$

Make sure you see why this leads to

$$\frac{dy}{dx} = \frac{1}{\sec(y)\tan(y)}.$$

Now $x = \sec(y)$, so since $\sec^2(y) = 1 + \tan^2(y)$, we can rearrange and take square roots to show that $\tan(y) = \pm\sqrt{x^2 - 1}$. This means that

$$\frac{dy}{dx} = \frac{1}{\pm x\sqrt{x^2 - 1}}.$$

Is it plus or minus? Looking at the graph of $y = \sec^{-1}(x)$ above, you can see that the slope is always positive. So in fact we need to be a little more clever—instead of the plus or minus, we can simply put $|x|$ instead of x and we always get something positive. That is,

$$\boxed{\frac{d}{dx}\sec^{-1}(x) = \frac{1}{|x|\sqrt{x^2 - 1}} \qquad \text{for } x > 1 \text{ or } x < -1.}$$

We can summarize the other facts about inverse secant like this:

$$\boxed{\sec^{-1} \text{ is neither odd nor even; it has domain } (-\infty, -1] \cup [1, \infty) \text{ and range } [0, \pi]\backslash\{\tfrac{\pi}{2}\}.}$$

(Here I used the standard abbreviations of $\cup$ to mean the union of two intervals, and $\backslash$ to mean "not including.")

10.2.5 Inverse cosecant and inverse cotangent

Let's just wrap the last two inverse trig functions up quickly. You can repeat the above analyses to find the domain, range, and graphs of $y = \csc^{-1}(x)$ and $y = \cot^{-1}(x)$:

$$\boxed{\csc^{-1} \text{ is odd; it has domain } (-\infty, -1] \cup [1, \infty) \text{ and range } [-\tfrac{\pi}{2}, \tfrac{\pi}{2}]\backslash\{0\}.}$$

$$\boxed{\cot^{-1} \text{ is neither odd nor even; it has domain } \mathbb{R} \text{ and range}(0, \pi).}$$

This is what the graphs look like:

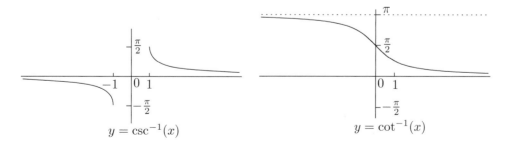

$$y = \csc^{-1}(x) \qquad\qquad y = \cot^{-1}(x)$$

Both functions have horizontal asymptotes: $y = \csc^{-1}(x)$ has a two-sided horizontal asymptote at $y = 0$, and $y = \cot^{-1}(x)$ has a left-hand horizontal asymptote at $y = \pi$ and a right-hand one at $y = 0$. We can summarize the limits as follows:

$$\lim_{x \to \infty} \csc^{-1}(x) = 0 \qquad \text{and} \qquad \lim_{x \to -\infty} \csc^{-1}(x) = 0$$

$$\lim_{x \to \infty} \cot^{-1}(x) = 0 \qquad \text{and} \qquad \lim_{x \to -\infty} \cot^{-1}(x) = \pi.$$

Of course, if you know the above graphs, you can reconstruct the limits without having to remember them. Notice that the graphs of $y = \csc^{-1}(x)$ and $y = \sec^{-1}(x)$ from above are very similar; in fact, you can get one from the other by flipping about the line $y = \pi/4$. This is exactly the same relation as the one that $y = \sin^{-1}(x)$ and $y = \cos^{-1}(x)$ have with each other. So it's not surprising that the derivative of $\csc^{-1}(x)$ is just the negative of the derivative of $\sec^{-1}(x)$:

$$\frac{d}{dx} \csc^{-1}(x) = -\frac{1}{|x|\sqrt{x^2 - 1}} \qquad \text{for } x > 1 \text{ or } x < -1.$$

The same thing happens with $\cot^{-1}(x)$ and $\tan^{-1}(x)$, so that

$$\frac{d}{dx} \cot^{-1}(x) = -\frac{1}{1 + x^2} \qquad \text{for all real } x.$$

10.2.6 Computing inverse trig functions

We've completed a pretty thorough survey of the inverse trig functions. Since you have a few more derivative rules, it's a great idea to practice differentiating functions involving inverse trig functions. Meanwhile, let's not neglect some basic computations involving inverse trig functions which don't involve any calculus. For one thing, you should try to make sure that you can compute quantities like $\sin^{-1}(1/2)$, $\cos^{-1}(1)$, and $\tan^{-1}(1)$ without stretching your brain. For example, to find $\sin^{-1}(1/2)$, remember that you're looking for an angle in $[-\pi/2, \pi/2]$ whose sine is $1/2$. Of course—it's $\pi/6$. Similarly, it should be almost second nature to write down $\cos^{-1}(1) = 0$ and $\tan^{-1}(1) = \pi/4$. All the common values are in the table near the beginning of Chapter 2.

Now, here's a more interesting question: how would you simplify

$$\sin^{-1}\left(\sin\left(\frac{13\pi}{10}\right)\right)?$$

The knee-jerk reaction is to cancel out the inverse sine and the sine, leaving only $13\pi/10$. This can't be correct, though—the range of inverse sine is $[-\pi/2, \pi/2]$, as we saw in Section 10.2.1 above. What we really need to do is find an angle in that range which has the same sine as $13\pi/10$. Well, note that $13\pi/10$ is in the third quadrant, since it's greater than π but less than $3\pi/2$, so its sine is negative. Furthermore, the reference angle is $3\pi/10$. The possible angles in the range $[\pi/2, \pi/2]$ with the same reference angle are $3\pi/10$ and $-3\pi/10$. The first one has a positive sine, while the second has a negative sine. We need a negative sine, so we've proved that

$$\sin^{-1}\left(\sin\left(\frac{13\pi}{10}\right)\right) = -\frac{3\pi}{10}.$$

Now, how about finding

$$\cos^{-1}\left(\cos\left(\frac{13\pi}{10}\right)\right)?$$

The previous answer $-3\pi/10$ can't be correct here, since the range of inverse cosine is $[0, \pi]$. Man, why does this stuff have to be so messy? Nothing I can do about it, unfortunately ... so let's deal with it like this: once again, $13\pi/10$ is in the third quadrant, so its cosine is negative. The reference angle is $3\pi/10$; the only angles in $[0, \pi]$ with the same reference angle are $3\pi/10$ and $7\pi/10$. The cosines of these two angles are positive and negative, respectively; since we want a negative cosine, we must have

$$\cos^{-1}\left(\cos\left(\frac{13\pi}{10}\right)\right) = \frac{7\pi}{10}.$$

I now leave it to you to show that

$$\tan^{-1}\left(\tan\left(\frac{13\pi}{10}\right)\right) = \frac{3\pi}{10}.$$

Just remember that tan is positive in the third quadrant! In any case, those are all difficult examples, so I wouldn't blame you if you also thought that finding

$$\sin\left(\sin^{-1}\left(-\frac{1}{5}\right)\right)$$

would be hard as well. Luckily, it's not: the answer is just $-1/5$. In general, $\sin(\sin^{-1}(x)) = x$, provided that x is in the domain $[-1, 1]$ of inverse sine. (Otherwise, $\sin(\sin^{-1}(x))$ doesn't even make sense!) The trouble comes when you try to write $\sin^{-1}(\sin(x)) = x$. This just isn't true, as the above example where $x = 13\pi/10$ shows. Of course, the same observations apply to all the other inverse trig functions. (See also the discussion at the end of Section 1.2 in Chapter 1.)

Two more examples: consider how you would find

$$\sin\left(\cos^{-1}\left(\frac{\sqrt{15}}{4}\right)\right) \quad \text{and} \quad \sin\left(\cos^{-1}\left(-\frac{\sqrt{15}}{4}\right)\right).$$

The trick in both cases is to use the trig identity $\cos^2(x) + \sin^2(x) = 1$. For the first problem, let

$$x = \cos^{-1}\left(\frac{\sqrt{15}}{4}\right)$$

and note that we want to find $\sin(x)$. We actually know $\cos(x)$:

$$\cos(x) = \cos\left(\cos^{-1}\left(\frac{\sqrt{15}}{4}\right)\right) = \frac{\sqrt{15}}{4}.$$

Remember, there's no problem taking the cosine of an inverse cosine: it's only the other way around that poses a problem. Anyway, we know $\cos(x)$, so by rearranging the identity $\cos^2(x) + \sin^2(x) = 1$, we must have

$$\sin(x) = \pm\sqrt{1 - \cos^2(x)} = \pm\sqrt{1 - \left(\frac{\sqrt{15}}{4}\right)^2} = \pm\sqrt{\frac{1}{16}} = \pm\frac{1}{4}.$$

So the answer we want is either $1/4$ or $-1/4$. Which one is it? Well, since $\sqrt{15}/4$ is positive, inverse cosine of it must lie in $[0, \pi/2]$. That is, x is in the first quadrant, so its sine is positive. We've finally shown that

$$\sin\left(\cos^{-1}\left(\frac{\sqrt{15}}{4}\right)\right) = \frac{1}{4}.$$

As for

$$\sin\left(\cos^{-1}\left(-\frac{\sqrt{15}}{4}\right)\right),$$

you can repeat the above argument to show that

$$\sin(x) = \pm\sqrt{1 - \cos^2(x)} = \pm\sqrt{1 - \left(-\frac{\sqrt{15}}{4}\right)^2} = \pm\sqrt{\frac{1}{16}} = \pm\frac{1}{4}.$$

You might guess that the answer this time is $-1/4$, but that's no good. You see, $-\sqrt{15}/4$ is negative, so its inverse cosine must lie in the interval $[\pi/2, \pi]$. That is, x is in the second quadrant. The thing is, sine is positive in the second quadrant as well! So $\sin(x)$ must be positive, and we've shown that

$$\sin\left(\cos^{-1}\left(-\frac{\sqrt{15}}{4}\right)\right) = \frac{1}{4}$$

as well. In fact, we've noticed that $\sin(\cos^{-1}(A))$ must always be nonnegative, even if A is negative (note that A has to lie in $[-1, 1]$, since that's the domain of inverse cosine). This is because $\cos^{-1}(A)$ is in the interval $[0, \pi]$, and sine is nonnegative on that interval.

We'll actually look at another method of finding things like $\sin(\cos^{-1}(A))$ when we see how to do trig substitutions in Section 19.3 of Chapter 19. For now, let's take a well-deserved rest from inverse trig functions and take a quick look at inverse hyperbolic functions.

10.3 Inverse Hyperbolic Functions

The situation is a little different for hyperbolic functions, which we looked at in Section 9.7 of the previous chapter. Look back now and remind yourself what the graphs of these functions look like. In particular, you can see that the graph of $y = \cosh(x)$ is sort of like the graph of $y = x^2$, except shifted up by 1 and shaped a little differently. If you want an inverse for this function, you have to throw away the left half of the graph, just as you do when you take the positive square root (and throw away the negative one). On the other hand, $y = \sinh(x)$ already satisfies the horizontal line test, so there's nothing that needs to be done. So we get two inverse functions with the following properties:

$\cosh^{-1}$ is neither odd nor even; it has domain $[1, \infty)$ and range $[0, \infty)$.

$\sinh^{-1}$ is odd; its domain and range are all of $\mathbb{R}$.

The graphs are obtained by reflecting the original graphs in the line $y = x$ as usual:

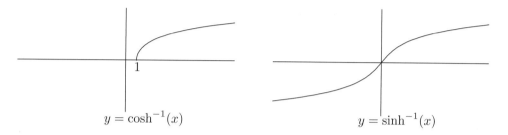

$$y = \cosh^{-1}(x) \qquad\qquad y = \sinh^{-1}(x)$$

The derivatives are obtained in the same way that we got the derivatives of the inverse trig functions. In particular, if $y = \cosh^{-1}(x)$, then $x = \cosh(y)$; differentiating implicitly with respect to x, we get

$$1 = \sinh(y)\frac{dy}{dx}.$$

(Remember that the derivative of $\cosh(x)$ with respect to x is $\sinh(x)$, not $-\sinh(x)$.) Now $\cosh^2(y) - \sinh^2(y) = 1$, so we can rearrange and take square roots to see that $\sinh(y) = \pm\sqrt{\cosh^2(y) - 1} = \pm\sqrt{x^2 - 1}$. Since $\cosh^{-1}(x)$ is clearly increasing in x, we end up with

$$\frac{d}{dx}\cosh^{-1}(x) = \frac{1}{\sqrt{x^2 - 1}} \qquad \text{for } x > 1.$$

In exactly the same way, you should be able to check that

$$\frac{d}{dx}\sinh^{-1}(x) = \frac{1}{\sqrt{x^2 + 1}} \qquad \text{for all real } x.$$

Now, let's forget about the calculus for a few seconds and recall the definitions of $\cosh(x)$ and $\sinh(x)$:

$$\cosh(x) = \frac{e^x + e^{-x}}{2} \qquad \text{and} \qquad \sinh(x) = \frac{e^x - e^{-x}}{2}.$$

Since we can write $\cosh(x)$ and $\sinh(x)$ in terms of exponentials, we should be able to write the inverse functions in terms of logarithms. After all, exponentials and logarithms are inverses of each other. Let's see how it works. For example, if $y = \cosh^{-1}(x)$, then $x = \cosh(y) = (e^y + e^{-y})/2$. Now you can solve for y by using a little trick. Let $u = e^y$; then $e^{-y} = 1/u$. The equation then looks like this:

$$x = \frac{u + 1/u}{2}.$$

Multiply both sides by $2u$ and rearrange; we get a quadratic equation in u, which is $u^2 - 2xu + 1 = 0$. By the quadratic formula,

$$e^y = u = x \pm \sqrt{x^2 - 1},$$

so taking logs of both sides,

$$y = \ln(x \pm \sqrt{x^2 - 1}).$$

Well, is it plus or minus? After a bit of gymnastics, you can actually see that $x - \sqrt{x^2 - 1} < 1$ if $x > 1$. This means that the logarithm of it is negative (remember, the log of a number between 0 and 1 is negative!). That's not what we want. So it's the positive square root, and we just showed that

$$\cosh^{-1}(x) = \ln(x + \sqrt{x^2 - 1})$$

when $x \geq 1$. In a similar way, you can show that

$$\sinh^{-1}(x) = \ln(x + \sqrt{x^2 + 1})$$

for all x. As an exercise, you should try differentiating the right-hand sides of these last two equations and check that your answers agree with the derivatives of $\cosh^{-1}(x)$ and $\sinh^{-1}(x)$ we found above.

10.3.1 The rest of the inverse hyperbolic functions

So far, we've only looked at hyperbolic sine and cosine. If you repeat the analysis for the other four hyperbolic functions, you should be able to conclude that:

$\tanh^{-1}$ is odd; its domain is $(-1, 1)$; its range is all of $\mathbb{R}$.

sech^{-1} is neither even nor odd; its domain is $(0, 1]$; its range is $[0, \infty)$.

csch^{-1} is odd; its domain and range are both $\mathbb{R}\backslash\{0\}$.

$\coth^{-1}$ is odd; its domain is $(-\infty, -1) \cup (1, \infty)$; its range is $\mathbb{R}\backslash\{0\}$.

Note that we've restricted the domain of sech to $[0, \infty)$ in order to get an inverse, just as we did for cosh.

Now, here are the graphs, which you should compare with the graphs of the original (non-inverse) functions in Section 9.7 of the previous chapter:

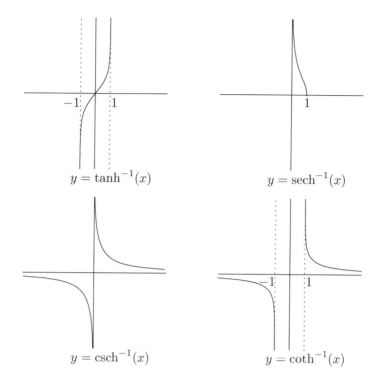

$y = \tanh^{-1}(x)$

$y = \operatorname{sech}^{-1}(x)$

$y = \operatorname{csch}^{-1}(x)$

$y = \coth^{-1}(x)$

Finally, you can find the derivatives using the standard trick of solving for x and differentiating implicitly with respect to x. Here's what the derivatives turn out to be:

$$\frac{d}{dx}\tanh^{-1}(x) = \frac{1}{1-x^2} \qquad (-1 < x < 1)$$

$$\frac{d}{dx}\coth^{-1}(x) = \frac{1}{1-x^2} \qquad (x > 1 \text{ or } x < -1)$$

$$\frac{d}{dx}\operatorname{sech}^{-1}(x) = -\frac{1}{x\sqrt{1-x^2}} \qquad (0 < x < 1)$$

$$\frac{d}{dx}\operatorname{csch}^{-1}(x) = -\frac{1}{|x|\sqrt{1+x^2}} \qquad (x \neq 0).$$

Remember, all these derivatives only hold when x is in the domain of the relevant function itself. This explains why the derivatives of $\tanh^{-1}(x)$ and $\coth^{-1}(x)$ are the same even though the graphs look very different. In particular, $\tanh^{-1}(x)$ is only defined on $(-1, 1)$, whereas $\coth^{-1}(x)$ is defined only **outside** the interval $[-1, 1]$. There's no overlap, therefore it's no problem that both functions have the same derivative. And that's quite enough about inverse functions for now!

CHAPTER 11 _____

The Derivative and Graphs

We have seen how to differentiate functions from several different families: polynomials and poly-type functions, trig and inverse trig functions, exponentials and logs, and even hyperbolic functions and their inverses. Now we can use this knowledge to help us sketch graphs of functions in general. We'll see how the derivative helps us understand the maxima and minima of functions, and how the second derivative helps us to understand the so-called concavity of functions. All in all, we have the following agenda:

- global and local maxima and minima (that is, extrema) of functions, and how to find them using the derivative;
- Rolle's Theorem and the Mean Value Theorem, and their implications for sketching graphs;
- the graphical interpretation of the second derivative; and
- classifying points where the derivative vanishes.

Then in the next chapter, we'll look a comprehensive method of sketching graphs of functions using the above methods.

11.1 Extrema of Functions

If we say that $x = a$ is an *extremum* of a function f, this means that f has a maximum or minimum at $x = a$. (The plural of "extremum" is "extrema," of course.) We've already looked a little bit at maxima and minima in Section 5.1.6 of Chapter 5; I strongly suggest taking a peek back at that before you read on. In any event, we need to go a little deeper and distinguish between two types of extrema: global and local.

11.1.1 Global and local extrema

The basic idea of a maximum is that it occurs when the function value is highest. Think about where the maximum of the following function on its domain $[0, 7]$ should be:

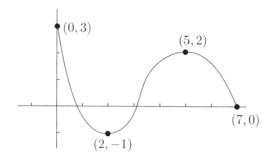

Certainly the maximum value that this function gets to is 3, which occurs when $x = 0$, so it's true that the function has a maximum at $x = 0$. On the other hand, imagine the graph is a hill (in cross-section) and you're climbing up it. Suppose you start at the point $(2, -1)$ and walk up the hill to the right. Eventually you reach the peak at $(5, 2)$, and then you start going back down again. It sure feels as if the peak is some sort of maximum—it's the top of the mountain, at height 2, even though there's a neighboring peak to the left that's taller. If the high ground near $x = 0$ were covered in fog, you couldn't even see it when you climbed the peak at $(5, 2)$, so you'd really feel as if you were at a maximum. In fact, if we restrict the domain to $[2, 7]$, then the point $x = 5$ is **actually** a maximum.

We need a way of clarifying the situation. Let's say that a *global maximum* (or *absolute maximum*) occurs at $x = a$ if $f(a)$ is the highest value of f on the **entire** domain of f. In symbols, we want $f(a) \geq f(x)$ for any value x in the domain of f. This is exactly the same definition we used before when we looked at maxima in general; we're simply being more precise and saying "global maxima" instead of just "maxima."

As we noted before, there could be multiple global maxima; for example, $\cos(x)$ has a maximum value of 1, but this occurs for infinitely many values of x. (These values are all the integer multiples of 2π, as you can see from the graph of $y = \cos(x)$.)

How about that other type of maximum? Let's say that a *local maximum* (or *relative maximum*) occurs at $x = a$ if $f(a)$ is the highest value of f **on some small interval containing a.** You can think of this as throwing away most of the domain, just concentrating on values of x close to a, then insisting that the function is at its maximum out of only those values.

Let's see how this works in the case of our above graph. We see that $x = 5$ is a local maximum, since $(5, 2)$ is the highest point around if you only concentrate on the function near $x = 5$. For example, if you cover up the part of the graph to the left of $x = 3$, then the point $(5, 2)$ is the highest point remaining. On the other hand, $x = 5$ isn't a global maximum, since the point $(0, 3)$ is higher up. This means that $x = 0$ is a global maximum. It's also a local maximum; in fact, it's pretty obvious that **every global maximum is also a local maximum**.

In the same way, we can define global and local minima. In the above graph, you can see that $x = 2$ is a global minimum (with value -1), since the height is at its lowest. On the other hand, $x = 7$ is actually a local minimum (with value 0). Indeed, if you just look at the function to the right of $x = 5$, you can see that the lowest height occurs at the endpoint $x = 7$.

11.1.2 The Extreme Value Theorem

In Chapter 5, we looked at the Max-Min Theorem. This says that a **continuous** function on a **closed** interval $[a, b]$ must have a global maximum somewhere in the interval and also a global minimum somewhere in the interval. We also saw that if the function isn't continuous, or even if it is continuous but the domain isn't a closed interval, then there might not be a global maximum or minimum. For example, the function f given by $f(x) = 1/x$ on the domain $[-1, 1]\backslash\{0\}$ doesn't have a global maximum or minimum on that domain. (Draw it and see why!)

The problem with the Max-Min Theorem is that it doesn't tell you anything about where these global maxima and minima are. That's where the derivative comes in. Let's say that $x = c$ is a *critical point* for the function f if either $f'(c) = 0$ or if $f'(c)$ does not exist. Then we have this nice result:*

> **Extreme Value Theorem:** suppose that f is defined on (a, b) and c is in (a, b). If c is a local maximum or minimum of f, then c must be a critical point for f. That is, either $f'(c) = 0$ or $f'(c)$ does not exist.

So local maxima and minima in an open interval occur only at critical points. But it's not true that a critical point must be a local maximum or minimum! For example, if $f(x) = x^3$, then $f'(x) = 3x^2$, and you can see that $f'(0) = 0$. This means that $x = 0$ is a critical point for f. On the other hand, $x = 0$ is neither a local maximum nor a local minimum, as you can see by drawing the graph of $y = x^3$.

The above theorem applies to open intervals. How about when the domain of your function is a closed interval $[a, b]$? Then the endpoints a and b might be local maxima and minima; they aren't covered by the theorem. So in the case of a closed interval, local maxima and minima can occur only at critical points **or** at the endpoints of the interval. For example, let's take a closer look at our graph from the previous section:

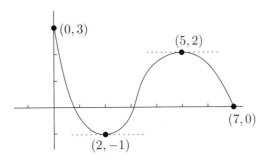

As we saw, the local maxima are at $x = 0$ and $x = 5$, while the local minima are at $x = 2$ and $x = 7$. The points $x = 5$ and $x = 2$ are critical points, because the slope is 0 there, while the points $x = 0$ and $x = 7$ are endpoints.

You might like to think about why the theorem makes sense. Suppose you have a local minimum at some value $x = a$. Then when you are to the

---

*The Max-Min Theorem is often called the Extreme Value Theorem, sometimes in conjunction with the above version of the Extreme Value Theorem.

immediate left of $x = a$, you must be going downhill, so the slope (if it exists) is negative. When you are to the immediate right of $x = a$, you are going uphill, so the slope is positive. If you are to get from a negative to a positive slope, you would think that you have to go through 0. On the other hand, if $f(x) = |x|$, then f goes from a slope of -1 to a slope of 1 without passing through 0. This is because $f'(0)$ doesn't exist (as we saw in Section 5.2.10 in Chapter 5). That's OK, though—the point $x = 0$ is still a critical point, because the derivative doesn't exist there. It's also a local minimum. (Can you see why?) By the way, the above logic doesn't constitute a proof of the theorem; a real proof is in Section A.6.6 of Appendix A.

11.1.3 How to find global maxima and minima

The Extreme Value Theorem really makes finding global extrema pretty easy, since it narrows down where they can be. Here's the idea: every global extremum is also a local extremum. Local extrema can only occur at critical points. So just find all the critical points and look at the corresponding function values. The biggest one gives the global maximum, while the smallest gives the global minimum! In gory detail, here's how to find the global maximum and minimum of the function f with domain $[a, b]$:

1. Find $f'(x)$. Make a list of all the points in (a, b) where $f'(x)$ does not exist or $f'(x) = 0$. That is, make a list of all the critical points in the interval (a, b).
2. Add the endpoints $x = a$ and $x = b$ to the list.
3. For each of the points in the list, find the y-coordinates by substituting into the equation $y = f(x)$.
4. Pick the highest y-coordinate and note all the values of x from the list corresponding to that y-coordinate. These are the global maxima.
5. Do the same for the lowest y-coordinate to find the global minima.

We'll worry about local extrema in Section 11.5 below. For now, let's look at an example of how to apply this method. Suppose that

$$f(x) = 12x^5 + 15x^4 - 40x^3 + 1$$

on the domain $[-1, 2]$. What are the global maxima and minima of f on this domain?

Let's follow the above program. For step 1, we need to find $f'(x)$. No problem: you should check that $f'(x) = 60x^4 + 60x^3 - 120x^2$. Clearly $f'(x)$ exists for all x in $(-1, 2)$, so we just need to find all the values of x satisfying $f'(x) = 0$. That's not so bad if you factor $f'(x)$ as $f'(x) = 60x^2(x - 1)(x + 2)$. So we can see that if $f'(x) = 0$, we must have $x = 0$, $x = 1$ or $x = -2$. The last of these is irrelevant since -2 is not in the interval $(-1, 2)$. So our list just contains $x = 0$ and $x = 1$. Step 2 tells us to add the endpoints $x = -1$ and $x = 2$ to the list.

So, we arrive at step 3 armed with the following list of candidates for global maxima and minima: -1, 0, 1, and 2. We need to find the corresponding function values. This is just a matter of plugging them in and calculating that $f(-1) = 44$, $f(0) = 1$, $f(1) = -12$, and $f(2) = 305$. As for the last two steps, all we have to do is select the highest and lowest values from this list.

The highest is 305, which occurs when $x = 2$, so $x = 2$ is a global maximum for f. The lowest function value is -12, which occurs when $x = 1$, so $x = 1$ is a global minimum for f, and we're all done!

Before we start lounging around after our efforts, let's take a closer look at the function f. First, note that if we made the domain larger, the situation could change for two reasons: the new endpoints would be different, and also the critical point at $x = -2$ could come into play. Second, we should look at what happens at the critical point $x = 0$ a little more closely. Is this a local maximum, a local minimum, or neither? One way to tell is to inspect the graph, which must look something like this:

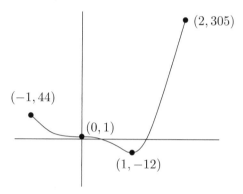

The point $(-1, 44)$ is higher than $(0, 1)$, which is in turn higher than $(1, -12)$. So we can't possibly have a local maximum or a local minimum at 0. But wait, you say—perhaps the graph looks something like this:

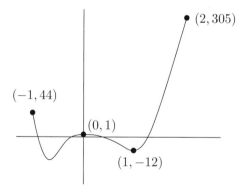

In this picture, $x = 0$ is a local maximum. The problem is that we've had to introduce another local minimum somewhere between -1 and 0. After all, if the curve is supposed to get from $(-1, 44)$ to $(0, 1)$ while still being on a plateau at $(0, 1)$, it's got to go down below a height of 1. This means there has to be a valley as well, which means a local minimum somewhere between $x = -1$ and $x = 0$! That can't happen, though, since there are no critical points between $x = -1$ and $x = 0$. So the graph must look more like the first picture above, and the conclusion is that $x = 0$ is neither a local maximum nor a local minimum.

If the domain isn't bounded, then the situation is a little more complicated. For example, consider the two functions f and g, both with domain $[0, \infty)$, whose graphs look like this:

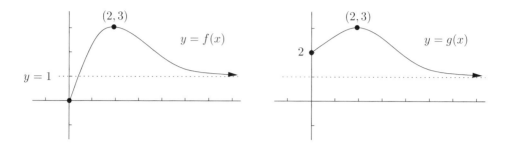

In both cases, $x = 2$ is obviously a critical point, while the endpoints are 0 and ∞. Wait a second, ∞ isn't really an endpoint, since it doesn't really exist! Let's add it to the list anyway, so that the list is 0, 2, and ∞; note that the same list works for both f and g.

Let's take a look at f first. We see that $f(0) = 0$, $f(2) = 3$, while $f(\infty)$ only makes sense if you think of it as

$$\lim_{x \to \infty} f(x).$$

This limit is 1, since $y = 1$ is a horizontal asymptote for f. The highest of these function values is 3, which occurs at $x = 2$, so $x = 2$ is a global maximum for f. The lowest function value is at $x = 0$, so $x = 0$ is a global minimum for f. The right-hand "endpoint" at ∞ doesn't even come into it.

How about g? Well, this time $g(0) = 2$, $g(2) = 3$, and the right-hand endpoint is covered by the observation that

$$\lim_{x \to \infty} g(x) = 1.$$

The highest value is still 3, which occurs at $x = 2$, so $x = 2$ is also a global maximum for g. How about the lowest value? Well, that value, which is 1, occurs as $x \to \infty$. Does this mean that ∞ is a global minimum for g? Of course not, because ∞ isn't even a number; the function g has no global minimum.*

11.2 Rolle's Theorem

Imagine you're driving down a long straight highway. I watch you stop at a gas station. Then you proceed, always facing the same direction, although you can put the car in reverse if you want. Later on, I see you at the gas station again, without watching what you did in the meantime. I make the following conclusion: at some point when I wasn't looking, your car had velocity equal to zero.

---

*On the other hand, g does have a global *infimum*. This concept is a little beyond our scope, though. Check out a book on real analysis if you want to learn more.

How can I be so confident about this? Well, it's possible that you never even left the gas station, in which case your velocity was zero the whole time. If you did leave the gas station and went forward, well, you must have eventually have gone backward or else you wouldn't be back at the gas station again. So what happened when you ceased going forward and started going backward? You must have stopped, even for an instant! You can't just change from going forward to backward without coming to rest. It's similar to the situation we saw in Section 6.4.1 of Chapter 6 when we studied the motion of a ball being thrown up in the air. At the instant the ball reaches the top of its path, its velocity is 0.

On the other hand, you might actually have started backing up from the gas station. In that case, you would have switched some time from backward to forward motion, and the effect would be the same: you still stopped somewhere. Regardless of which way you set out, you might have stopped many times; but I know you stopped at least once. This is the content of Rolle's Theorem,* which says:

> **Rolle's Theorem:** suppose that f is continuous on $[a, b]$ and differentiable on (a, b). If $f(a) = f(b)$, then there must be at least one number c in (a, b) such that $f'(c) = 0$.

In terms of your journey, we are supposing that $f(t)$ is the position of your car at time t. This means that $f'(t)$ is your velocity at time t. The times a and b are when I observed you at the gas station; the equation $f(a) = f(b)$ means that you were in the same place at time a as at time b, which of course was the gas station. Finally, the number c is a time that you stopped, since $f'(c) = 0$. Rolle's Theorem is telling me that you must have stopped at least once. I don't know when, because I wasn't watching, but I know it happened. (I am assuming that your car's motion is differentiable, which is pretty reasonable in most circumstances. On the other hand, if you consider the point of view of a crash test dummy, perhaps the car's motion isn't differentiable at the moment the car hits the wall....)

Now, let's look at some pictures of a few possibilities of functions where Rolle's Theorem applies:

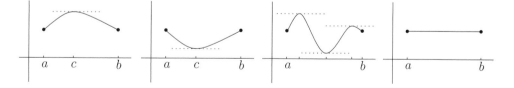

In the first two diagrams, there is only one possible value of c such that $f'(c) = 0$. In the third diagram, there are three potential candidates for c, but that's OK—Rolle's Theorem says that there must be at least one. The fourth diagram shows a constant function, so its derivative is always 0. This means that c could be **any** number between a and b. Now, let's look at some pictures where Rolle's Theorem does **not** apply:

---

*See Section A.6.7 of Appendix A for a proof of Rolle's Theorem.

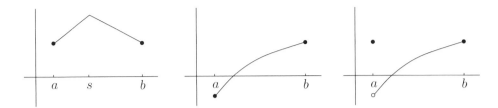

In all three cases, the derivative is never 0. That's OK, because Rolle's Theorem doesn't apply in any of these cases. In the first picture, the function isn't differentiable on all of (a, b) because of that spike at s. Yes, even one point where the function isn't differentiable is enough to screw everything up. In the middle picture, the function **is** differentiable, but $f(a) \neq f(b)$, so Rolle's Theorem cannot be used. In the right-hand picture, $f(a) = f(b)$ and the function is differentiable on (a, b), but it isn't continuous on all of $[a, b]$: the point $x = a$ spoils everything. Once again, no Rolle's Theorem allowed.

    Here's an example of an application of Rolle's Theorem. Suppose that you have a function f satisfying $f'(x) > 0$ for all x. In Section 10.1.1 in the previous chapter, we claimed that f must satisfy the horizontal line test. Let's prove this using Rolle's Theorem, arguing by contradiction. Start off by supposing that f does **not** satisfy the horizontal line test. Then there's some horizontal line, say $y = L$, which intersects the graph of $y = f(x)$ twice (or more). Suppose that two of these intersection points have x-coordinates a and b. So we know that $f(a) = L$ and $f(b) = L$. In particular, $f(a) = f(b)$, and we can use Rolle's Theorem (we already know that f is differentiable everywhere, so it must be continuous everywhere as well). The theorem says that there is some c between a and b such that $f'(c) = 0$. This is impossible because $f'(x)$ is always supposed to be positive! So the horizontal line test does not fail.

    Now, let's look at an even harder example. Suppose now that the second derivative of f exists everywhere and that $f''(x) > 0$ for all real x. The problem is to show that f has at most two x-intercepts. Before we tackle the problem itself, let's just think about what it means for a second or two. Can you think of a function f with $f''(x) > 0$ for all x that has no x-intercepts? How about one x-intercept? Two x-intercepts? If you can do all these, then try to find one with **three** x-intercepts. Don't spend too long on this one, though, because it's impossible! Indeed, our problem is to show that you can't have more than two x-intercepts.

    In fact, here's the key idea: if there are more than two x-intercepts, then there must be at least three! Let's suppose that there are more than two; call any three of them you like a, b, and c, where we choose the variables so that $a < b < c$. Since they are all x-intercepts, we have $f(a) = f(b) = f(c) = 0$. So, start off by applying Rolle's Theorem to the interval $[a, b]$. Since f is continuous and differentiable everywhere, and $f(a) = f(b)$, we know that $f'(p) = 0$ for some p in the interval (a, b). Why do I use p? Because c is already taken!

    Now let's move on to the interval $[b, c]$. Again, since $f(b) = f(c)$, we can use Rolle's Theorem to show that there must be some number q in (b, c) such that $f'(q) = 0$. Don't forget that we also have $f'(p) = 0$. Hey, now we can use Rolle's Theorem on the interval $[p, q]$, but instead of taking the function as f, we'll use f'. After all, we know that $f'(p) = f'(q)$, since both of these quantities are 0. So by Rolle's Theorem, we have some point r where $(f')'(r) = 0$. Wait a second, $(f')'$ is just the second derivative f''. So we know that $f''(r) = 0$ for some r between p and q. This is a big problem because we had supposed that $f''(x) > 0$ for all x. The only way out is that our idea that there are more than two x-intercepts is all out of whack. There can't be more than two, and we've solved the problem.

Tricky stuff. By the way, did you find some functions satisfying $f''(x) > 0$ for all x which have 0, 1 and 2 x-intercepts? If not, check out $f(x) = x^2 + C$, where C is positive, zero, or negative, respectively.

11.3 The Mean Value Theorem

Suppose you go on another journey, and I find out that you have traveled 100 miles in 2 hours. Your average velocity was 50 miles per hour. This doesn't mean that you were going at exactly 50 miles per hour the whole time. Now, here's my question: were you **ever** going at 50 miles per hour, even for an instant?

The answer is yes. Even if you go at 45 mph for the first hour and 55 mph for the second hour, you still have to accelerate from the slow velocity to the fast velocity. Along the way, your velocity will pass through 50 mph for an instant. You can't avoid it! No matter how you do your journey, if your average velocity is 50 mph, then your instantaneous velocity must be 50 mph at least once.* Of course, you might be going at 50 mph more than just once—there might be several times, or you can even go at 50 mph the whole time. This leads to the Mean Value Theorem, which says:

> **The Mean Value Theorem:** suppose that f is continuous on $[a, b]$ and differentiable on (a, b). Then there's at least one number c in (a, b) such that
> $$f'(c) = \frac{f(b) - f(a)}{b - a}.$$

It seems a little weird, but it actually makes sense. You see, if $f(t)$ is your position at time t, and you start and finish at times a and b, respectively, then what is your average velocity? The displacement is $f(b) - f(a)$, while the time taken is $b - a$, so the quantity on the right-hand side of the above equation is just your average velocity. On the other hand, $f'(c)$ is your instantaneous velocity at time c. The Mean Value Theorem says that there is at least one time c where your instantaneous velocity equals your average velocity over the whole journey.

Let's look at a picture of the situation. Suppose your function f looks like this:

---

*Again, all this assumes—very reasonably—that your car's motion is differentiable!

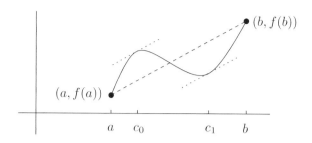

The dashed line joining $(a, f(a))$ and $(b, f(b))$ has slope

$$\frac{f(b) - f(a)}{b - a}.$$

According to the Mean Value Theorem, there is some tangent whose slope equals this quantity; that is, some tangent is parallel to the dashed line. In the above picture, there are actually two tangents that work—the x-coordinates are at c_0 and c_1. Either one would be an acceptable candidate for the number c in the theorem.

The Mean Value Theorem looks a lot like Rolle's Theorem. In fact, the conditions for applying the two theorems are almost the same. In both cases, your function f has to be continuous on a closed interval $[a, b]$ and differentiable on (a, b). Rolle's Theorem also requires that $f(a) = f(b)$, but the Mean Value Theorem doesn't require that. In fact, if you apply the Mean Value Theorem to a function f satisfying $f(a) = f(b)$, you'll see that $f(b) - f(a) = 0$, so you get a number c in (a, b) satisfying $f'(c) = 0$. So the Mean Value Theorem reduces to Rolle's Theorem!

Now let's look at a couple of examples of how to use the theorem. First, how would you show that the equation

$$2xe^{x^2} - e + 1 = 0$$

has a solution? One way is to use the Intermediate Value Theorem (see Section 5.1.4 in Chapter 5)—try it now and see. Suppose instead that I give you a nudge by suggesting that you apply the Mean Value Theorem to $f(x) = e^{x^2}$ on the domain $[0, 1]$. That's acceptable because f is continuous and differentiable everywhere. The theorem says that there's a number c in $[0, 1]$ such that

$$f'(c) = \frac{f(1) - f(0)}{1 - 0}.$$

Clearly, we'll need to find $f'(x)$. Using the chain rule, you should be able to show that $f'(x) = 2xe^{x^2}$. So the above equation becomes

$$2ce^{c^2} = \frac{e^{1^2} - e^{0^2}}{1 - 0} = e - 1.$$

So we have $2ce^{c^2} - e + 1 = 0$, and we have shown that our original equation above does have a solution. In fact, we've shown that there's a solution between 0 and 1.

Here's a harder example. Suppose that a function f is differentiable everywhere and that $f'(x) > 4$ for all values of x. The problem is to show that the graph $y = f(x)$ intersects the line $y = 3x - 2$ at most once. Try it and see if you can solve it before reading on.

So, how on earth do we do this problem? Actually it's pretty similar to the Rolle's Theorem examples from the previous section. First, note that if (x, y) is a point lying on both $y = f(x)$ and $y = 3x - 2$, then we must have $f(x) = 3x - 2$. That equation is **not** true for most x! It's only true at the intersection points. So, suppose that there's more than one intersection point. Pick any two and call them a and b, where they are arranged so that $a < b$. Since they are intersection points, we know that $f(a) = 3a - 2$ and $f(b) = 3b - 2$. Now since f is continuous and differentiable everywhere, we can use the Mean Value Theorem to show that there's a number c between a and b such that

$$f'(c) = \frac{f(b) - f(a)}{b - a}.$$

Plug in $f(b) = 3b - 2$ and $f(a) = 3a - 2$ to get

$$f'(c) = \frac{(3b - 2) - (3a - 2)}{b - a} = \frac{3(b - a)}{b - a} = 3.$$

This can't be right, since $f'(x) > 4$ for all x. So there can't be more than one intersection point.

That completes the solution, but you might like to consider another interpretation of it. Indeed, imagine a car going at a constant speed of 3 mph, starting at position -2. Its position at time t is therefore $3t - 2$. If your position at time t is $f(t)$, then the condition that $f'(t) > 4$ means that you're always going faster than 4 mph (in the same direction as the other car). So all the problem says is that you can't pass the other car more than once. If you were alongside the other car more than once, then since it's going at a constant speed of 3 mph, you'd have to be going at 3 mph for at least one instant. This is impossible because you're always going faster than 4 mph. It makes a lot of sense if you think about it like that!

11.3.1 Consequences of the Mean Value Theorem

We've been taking a few things about the derivative for granted. For example, if a function has derivative equal to 0 everywhere, it must be constant. Facts like this seem obvious but they actually deserve to be proved. Let's use the Mean Value Theorem to show three useful facts about derivatives:

1. Suppose that a function f has derivative $f'(x) = 0$ for **every** x in some interval (a, b). This means that the function is pretty darn flat. In fact, it's intuitively obvious that the function should be constant on the whole interval. How do we prove it? First, fix some special number S in the interval, and then pick any other number x in the interval. We know from the Mean Value Theorem that there's some number c between S and x such that

$$f'(c) = \frac{f(x) - f(S)}{x - S}.$$

Now we have assumed that f' is always equal to 0, the quantity $f'(c)$ must be 0. So the above equation says that

$$\frac{f(x) - f(S)}{x - S} = 0,$$

which means that $f(x) = f(S)$. If we now let $C = f(S)$, we have shown that $f(x) = C$ for all x in the interval (a, b), so f is constant! In summary,

> if $f'(x) = 0$ for all x in (a, b), then f is constant on (a, b).

Actually, we've already used this fact in Section 10.2.2 of the previous chapter. There we saw that if $f(x) = \sin^{-1}(x) + \cos^{-1}(x)$, then $f'(x) = 0$ for all x in the interval $(-1, 1)$. We concluded that f is constant on that interval, and in fact since $f(0) = \pi/2$, we have $\sin^{-1}(x) + \cos^{-1}(x) = \pi/2$ for all x in $(-1, 1)$.

2. Suppose that two differentiable functions have exactly the same derivative. Are they the same function? Not necessarily. They could differ by a constant; for example, $f(x) = x^2$ and $g(x) = x^2 + 1$ have the same derivative, $2x$, but f and g are clearly not the same function. Is there any other way that two functions could have the same derivative everywhere? The answer is no. Differing by a constant is the only way:

> if $f'(x) = g'(x)$ for all x, then $f(x) = g(x) + C$ for some constant C.

It turns out to be quite easy to show this using #1 above. Suppose that $f'(x) = g'(x)$ for all x. Now set $h(x) = f(x) - g(x)$. Then we can differentiate to get $h'(x) = f'(x) - g'(x) = 0$ for all x, so h is constant. That is, $h(x) = C$ for some constant C. This means that $f(x) - g(x) = C$, or $f(x) = g(x) + C$. The functions f and g do indeed differ by a constant. This fact will be very useful when we look at integration in a few chapters' time.

3. If a function f has a derivative that's always positive, then it must be *increasing*. This means that if $a < b$, then $f(a) < f(b)$. In other words, take two points on the curve; the one on the left is lower than the one on the right. The curve is getting higher as you look from left to right. Why is it so? Well, suppose $f'(x) > 0$ for all x, and also suppose that $a < b$. By the Mean Value Theorem, there's a c in the interval (a, b) such that

$$f'(c) = \frac{f(b) - f(a)}{b - a}.$$

This means that $f(b) - f(a) = f'(c)(b - a)$. Now $f'(c) > 0$, and $b - a > 0$ since $b > a$, so the right-hand side of this equation is positive. So we have $f(b) - f(a) > 0$, hence $f(b) > f(a)$, and the function is indeed increasing. On the other hand, if $f'(x) < 0$ for all x, the function is always *decreasing*; this means that if $a < b$ then $f(b) < f(a)$. The proof is basically the same.

11.4 The Second Derivative and Graphs

So far, we haven't paid much attention to the second derivative. We've only used it to define acceleration, and that's about all. Actually, the second derivative can tell you a lot about what the graph of your function looks like. For example, suppose that you know that $f''(x) > 0$ for all x in some interval (a, b). If you think of the second derivative f'' as the derivative of the derivative, then you can write $(f')'(x) > 0$. This means that the derivative $f'(x)$ is always increasing.

So what? Well, if you know that the derivative is increasing, this means that it's getting more and more difficult to "climb up" the function. The situation could look like this:

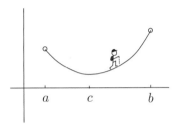

Just to the right of $x = a$, the mountain-climber has it nice and easy: the slope is negative. It's getting harder all the time, though; first it gets flatter, until the climber reaches the flat part at $x = c$; then the going keeps on getting tougher as the slope increases up to $x = b$. The important thing is that the slope is increasing all the way from $x = a$ up to $x = b$. This is exactly what is implied by the equation $f''(x) > 0$.

We need a way to describe this sort of behavior. We'll say a function is *concave up* on an interval (a, b) if its slope is always increasing on that interval, or equivalently if its second derivative is always positive on the interval (assuming that the second derivative exists). Here are some other examples of graphs of functions which are concave up on their whole domains:

They all look like part of a bowl. Notice that you can't tell anything about the sign of the first derivative $f'(x)$ just by knowing that $f''(x) > 0$. Indeed, the middle two graphs have negative first derivative; the rightmost graph has positive first derivative; while the leftmost graph has a first derivative that is negative and then positive.

If instead the second derivative $f''(x)$ is negative, then everything is reversed. You end up with something more like an upside-down bowl, saying that f is *concave down* on any interval where its second derivative is always negative.* Here are some examples of functions which are concave down on their entire domain:

In this case, the derivative is always decreasing: it's getting easier and easier to climb as you go along in each case. If you're going uphill, this means it's getting less and less steep, but if you're going downhill, it's getting steeper and steeper downhill (as you go from left to right).

Of course, the concavity doesn't have to be the same everywhere: it can change:

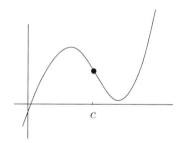

To the left of $x = c$, the curve is concave down, while to the right of $x = c$, the curve is concave up. We'll say that the point $x = c$ is a *point of inflection* for f because the concavity changes as you go from left to right through c.

11.4.1 More about points of inflection

In the above picture, we see that $f''(x) < 0$ to the left of c and $f''(x) > 0$ to the right of c. What about $f''(c)$ itself? It must be 0, since everything is nice and smooth. In general, if c is a point of inflection, then the sign of $f''(x)$ must be different on either side of $x = c$, assuming of course that $f''(x)$ actually exists when x is near c. In that case, it must be true that

if $x = c$ is a point of inflection for f, then $f''(c) = 0$.

On the other hand, if $f''(c) = 0$, then c may or may not be an inflection point! That is,

if $f''(c) = 0$, then it's not always true that $x = c$ is a point of inflection for f.

---

*If you have trouble remembering which one is concave up and which is concave down, the following rhyme might help: "like a cup, concave up; like a frown, concave down."

For example, suppose that $f(x) = x^4$. Then $f'(x) = 4x^3$ and $f''(x) = 12x^2$. At $x = 0$, the second derivative vanishes, because $f''(0) = 12(0)^2 = 0$. So is $x = 0$ a point of inflection? The answer is no. Here's a miniature graph of $y = x^4$:

You can see that f is always concave up; so the concavity doesn't change around $x = 0$. That is, $x = 0$ is not a point of inflection, despite the fact that $f''(0) = 0$.

On the other hand, if you want to find points of inflection, you do need to find where the second derivative vanishes. That at least narrows down the list of potential candidates, which you can check one by one. For example, suppose that $f(x) = \sin(x)$. We have $f'(x) = \cos(x)$ and $f''(x) = -\sin(x)$. The second derivative $-\sin(x)$ vanishes whenever x is a multiple of π. Let's focus on what happens at $x = 0$. We have $f''(0) = -\sin(0) = 0$. Is $x = 0$ an inflection point? Let's take a look at the graph:

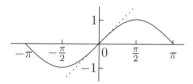

Yes, $x = 0$ is a point of inflection: $\sin(x)$ is concave up immediately to the left of 0 but concave down to the right of 0. Notice that the tangent line at $x = 0$ passes through the curve $y = \sin(x)$. This is typical of points of inflection: the curve must be above the tangent line on one side and below the tangent line on the other side.

11.5 Classifying Points Where the Derivative Vanishes

It's time to apply some of the above theory to a practical problem. Suppose that you have a function f and a number c such that $f'(c) = 0$. You can say for sure that c is a critical point for f, but what else can you say? It turns out that there are only three common possibilities: $x = c$ could be a local maximum; it could be a local minimum; or it could be a horizontal point of inflection, which means that it is a point of inflection with a horizontal tangent line.* (It's also possible that $f(x)$ is constant for all x near c, but in that case c is both a local maximum and a local minimum.) In any case, here are some pictures of the common possibilities:

---

*Another possibility is that the concavity isn't even well-defined near the critical point. For example, if $f(x) = x^4 \sin(1/x)$, then the sign of $f''(x)$ oscillates wildly as x approaches the critical point 0 from either above or below, so the concavity keeps switching between up and down!

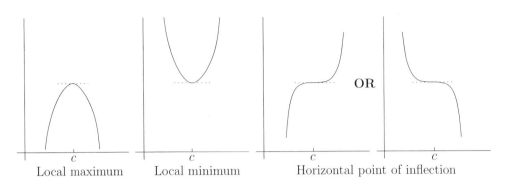

Local maximum   Local minimum   Horizontal point of inflection

In each case, the tangent line is horizontal; that's all you can tell if you only know that $f'(c) = 0$. How do you tell which case applies? There are two methods, one involving only the first derivative, and the other involving the second derivative. When you use the first derivative, you have to look at the sign (positive or negative) of the first derivative **near** $x = c$. On the other hand, if you use the second derivative, then you need to consider its sign **at** $x = c$. Let's look at these methods one at a time.

11.5.1 Using the first derivative

Let's take another look at the above cases, but this time we'll draw in some tangent lines near $x = c$:

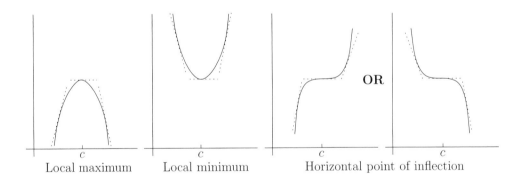

Local maximum   Local minimum   Horizontal point of inflection

In the first case, we have a local maximum at $x = c$. To the left of c, the slope is positive. This means that the function is increasing in that portion of the domain (as we saw in Section 11.3.1 above). On the other hand, to the right of c, the slope is negative: the function is decreasing there. It's clear that whenever the slope changes from positive to negative as you move from left to right, the point where the slope is 0 must be a local maximum.

In the second case, the situation is reversed. If the slope changes from negative to positive as you go from left to right, the point where the slope is 0 must be a local minimum. In the third case, the slope is always positive (except at $x = c$), while in the fourth case, the slope is always negative (except at $x = c$). Both cases give a point of inflection: the derivative doesn't change sign.

Here's a summary of what we have just observed. Suppose that $f'(c) = 0$. Then:

- if $f'(x)$ changes sign from positive to negative as you pass from left to right through $x = c$, then $x = c$ is a local maximum;
- if $f'(x)$ changes sign from negative to positive as you pass from left to right through $x = c$, then $x = c$ is a local minimum;
- if $f'(x)$ doesn't change sign as you pass through $x = c$ from left to right, then $x = c$ is a horizontal point of inflection.

For example, if $f(x) = x^3$, then we have $f'(x) = 3x^2$. This is 0 when $x = 0$, so $x = 0$ must be a local maximum, local minimum, or horizontal point of inflection. Which is it? Well, $f'(x)$ is always positive when $x \neq 0$, so the derivative doesn't change sign as you pass through $x = 0$ from left to right. So $x = 0$ must be a point of inflection. Draw the graph and check that this makes sense! (You can also find the graph in Section 11.5.2 below.)

Here's another example. If we now set $f(x) = x \ln(x)$, then where are the local maxima, minima, and horizontal points of inflection of f? Well, you should use the product rule to find that $f'(x) = \ln(x) + 1$. (Check that you believe this!) We are looking for solutions to the equation $f'(x) = 0$, which means that $\ln(x) + 1 = 0$. Rearranging, we get $\ln(x) = -1$; now exponentiate both sides to get $x = e^{-1}$, otherwise known as $1/e$. This is the only potential candidate. But what sort of critical point is it?

Well, let's look at the sign of $f'(x) = \ln(x) + 1$ when x is near $1/e$. The easiest way to do this is to draw a quick graph of $y = f'(x)$. All we have to do is take our graph for $\ln(x)$ and shift it up by 1. Here's what we get:

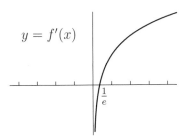

You can see from the graph that $f'(x)$ goes from negative to positive as we pass through $1/e$. So $x = 1/e$ must be a local minimum for f. Now, what is the value of $f(1/e)$? We can plug it in and get $f(1/e) = (1/e) \ln(1/e) = -1/e$, noting that $\ln(1/e) = \ln(e^{-1}) = -\ln(e) = -1$. So the graph of $y = f(x)$ has a local minimum at the point $(1/e, -1/e)$. It must look something like this:

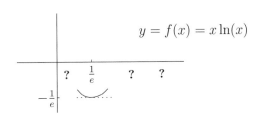

As you can see, we don't know much about the graph yet! We'll finish it off in Section 12.3.2 of the next chapter.

11.5.2 Using the second derivative

Take another look at the common possibilities which arise when $f'(c) = 0$:

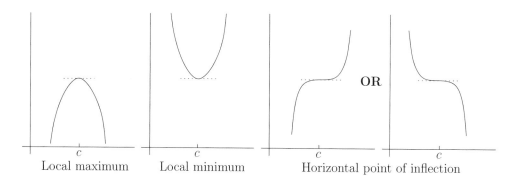

| Local maximum | Local minimum | Horizontal point of inflection |

Imagine that $f''(c) > 0$. We saw in Section 11.4 above that this means that the curve $y = f(x)$ is concave up near $x = c$. The only one of the above four graphs which is concave up is the second one, that is, the case of a local minimum at $x = c$. Similarly, if $f''(c) < 0$, then the curve is concave down, and we must be in the first case above: c is a local maximum in that case.

This is pretty useful, but there's a catch: if $f''(c) = 0$, then you could be in any one of the four cases! For example, suppose that $f(x) = x^3$ and $g(x) = x^4$. We have $f'(x) = 3x^2$, so $f'(0) = 0$. Let's find $f''(0)$ to try to classify the critical point. Since $f''(x) = 6x$, we have $f''(0) = 0$.

On the other hand, what about g? As we saw in Section 11.4.1 above, we have $g'(x) = 4x^3$, so $g'(0) = 0$. What sort of critical point is $x = 0$? Let's check the second derivative: $g''(x) = 12x^2$, so $g''(0) = 0$.

In both cases, at the critical point $x = 0$, the second derivative is 0. As you can see from the miniature graphs below, f has a point of infection at $x = 0$ while g has a local minimum there:

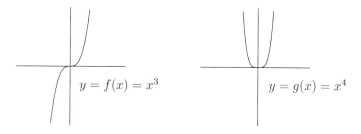

$y = f(x) = x^3$ $y = g(x) = x^4$

So much for using the second derivative to distinguish between these two cases. When the second derivative is 0, you are so in the dark, you might as well be in an underground room with your eyes closed and one of those really thick blindfolds on. You just can't tell whether you're dealing with a local maximum, a local minimum, or a horizontal point of inflection. So, here's the summary of the situation. Suppose that $f'(c) = 0$. Then:

- if $f''(c) < 0$, then $x = c$ is a local maximum;
- if $f''(c) > 0$, then $x = c$ is a local minimum;
- if $f''(c) = 0$, then **you can't tell what happens!** Use the first derivative test from the previous section.

Yes, the first derivative test is better, although it's a little more cumbersome to use. It always works, while the second derivative test sometimes lets you down. Here's an example where things do work out, though: suppose that $f(x) = x \ln(x)$. Hey, this is the same example as one from the previous section! There we saw using the first derivative test that $1/e$ is a local minimum for f. Let's try using the second derivative test instead.

First, recall that $f'(x) = \ln(x) + 1$, so $f'(1/e) = 0$. We can easily see that $f''(x) = 1/x$. When $x = 1/e$, we have $f''(1/e) = e$, which is positive. So the concavity is upward at $1/e$, which means that we're dealing with the bowl shape; indeed, according to the above summary, $1/e$ is indeed a local minimum.

CHAPTER 12 _____

Sketching Graphs

Now it's time to look at a general method for sketching the graph of $y = f(x)$ for some given function f. When we sketch a graph, we're not looking for perfection; we just want to illustrate the main features of the graph. Indeed, we're going to use the calculus tools we've developed: limits to understand the asymptotes, the first derivative to understand maxima and minima, and the second derivative to investigate the concavity. Here's what we'll look at:

- the useful technique of making a table of signs;
- a general method for sketching graphs; and
- five examples of how to use the method.

12.1 How to Construct a Table of Signs

Suppose you want to sketch the graph of $y = f(x)$. For any number x, the quantity $f(x)$ could be positive, negative, zero, or undefined. Luckily, if f is continuous except for maybe a few points, and you can find all of the zeroes and discontinuities of f, then it's easy to see where $f(x)$ is positive and where it's negative by using a table of signs.

 Here's how it works: start off by making a list of all the zeroes and discontinuities of f in ascending order. For example, if

$$f(x) = \frac{(x-3)(x-1)^2}{x^3(x+2)},$$

then the zeroes of f are 3 and 1, and f is discontinuous at 0 and -2. So our list, in order, is -2, 0, 1, 3. Now, draw a table with three rows and plenty of columns. We'll label the first two rows x and $f(x)$; the third row will actually be blank. Now, write the values in your list of zeroes and discontinuities across the top row so that there's one space on either side of each number. In our example, the table would look like this:

x	-2	0	1	3
$f(x)$				

Now you can fill in some of the second row—just put a 0 where $f(x)$ is 0 and a star where f is discontinuous:

x	-2	0	1	3
$f(x)$	$\star$	$\star$	0	0

Next, pick your favorite number between each of the special numbers on the top, as well as one at the beginning and one at the end. In our example, you might pick -3 as being to the left of -2; and -1 as being between -2 and 0; and so on, until the table looks something like this:

x	-3	-2	-1	0	$\frac{1}{2}$	1	2	3	4
$f(x)$		$\star$		$\star$		0		0	

We could have chosen -4 instead of -3, or $\frac{1}{3}$ instead of $\frac{1}{2}$—it wouldn't have made any difference. We can pick **any** number between the special numbers. Now, the next thing is to find whether $f(x)$ is positive or negative for each of the values we just chose. In our example, consider $x = -3$; then

$$f(-3) = \frac{(-3-3)(-3-1)^2}{(-3)^3(-3+2)} = -\frac{32}{9}.$$

So we can put a minus sign in the box under -3. Now we didn't actually need to work that hard, since we could care less about the value of $f(-3)$: we only care whether it's positive or negative. We should just have looked at each factor to see whether it's positive or negative. In particular, when $x = -3$, you can see that $(x-3)$ is negative, $(x-1)^2$ is positive (it can't be negative since it's a square!), x^3 is negative, and $(x+2)$ is negative as well. The overall effect is

$$\frac{(-)(+)}{(-)(-)} = -,$$

so $f(-3)$ is negative. Now try it for each of our other numbers, and verify that you can fill in the whole table like this:

x	-3	-2	-1	0	$\frac{1}{2}$	1	2	3	4
$f(x)$	$-$	$\star$	$+$	$\star$	$-$	0	$-$	0	$+$

The main point is not that $f(-3)$ is negative, but that $f(x)$ is negative for **all** $x < -2$. The number -3 is just a representative sample point for the region $(-\infty, -2)$. Whatever sign $f(-3)$ is, $f(x)$ has the same sign on the whole region. Similarly, since $f(-1)$ is positive, $f(x)$ is positive on the entire interval $(-2, 0)$. Already this gives us lots of information about the graph of $y = f(x)$, which we'll look at in Section 12.3.1 below.

Here's another example. Suppose that

$$f(x) = x^2(x - 5)^3.$$

We've actually already looked at this function f a little bit in Section 10.1.4 of Chapter 10. Let's take a closer look, starting with a table of signs. The zeroes of f clearly occur at $x = 0$ and $x = 5$ only, and there are no discontinuities. So our special points are at 0 and 5. We need to fill in the gaps. To the left of 0, I'll choose -1; in between I'll choose 2; and to the right, I'll choose 6. So our table of signs looks like this:

x	-1	0	2	5	6
$f(x)$	$-$	0	$-$	0	$+$

Here's how I came up with the signs at -1, 2, and 6:

- When $x = -1$, both x and $(x - 5)$ are negative. The sign of $f(-1)$ is therefore $(-)^2(-)^3 = (+)(-) = (-)$.
- When $x = 2$, now x is positive and $(x - 5)$ is negative. The sign of $f(2)$ is $(+)^2(-)^3$ which is still $(-)$.
- When $x = 6$, now both x and $(x - 5)$ are positive, so $f(6)$ has sign $(+)^2(+)^3 = (+)$.

We'll use this table to help us sketch the graph of $y = f(x)$ in Section 12.3.3 below. For now, let's see how to make a table of signs for the derivative and the second derivative.

12.1.1 Making a table of signs for the derivative

As we saw in Section 11.3.1 of the previous chapter, the sign of the derivative of a function tells you a lot about the function. Whenever the derivative is positive, the function is increasing; when the derivative is negative, the function is decreasing; and when the derivative is 0, the function has a local maximum, a local minimum, or a horizontal point of inflection. A table of signs for the derivative can summarize all this information in a compact, simple way.

The method is the same as for the table of signs for $f(x)$ that we looked at above, except that now you apply it to $f'(x)$ instead. The only other difference is that when $f'(x)$ is zero, we'll put a little flat line in the third row; when $f'(x)$ is positive, the line will slope upward; and when $f'(x)$ is negative, the line will slope downward.

Let's see how it works for our previous example where $f(x) = x^2(x - 5)^3$. In Section 10.1.4 of Chapter 10, we calculated that $f'(x) = 5x(x - 5)^2(x - 2)$.

(Try it yourself if you don't want to look back!) This means that $f'(x) = 0$ when $x = 0$, $x = 2$ or $x = 5$. Let's pick some points in between: we'll choose -1 to the left of 0; between 0 and 2, we'll pick 1; between 2 and 5, we'll choose 3; finally, we'll select 6 to the right of 5. Our table of signs looks like this, so far:

x	-1	0	1	2	3	5	6
$f'(x)$		0		0		0	

Now we need to find the sign of $f'(x)$ at the new points we chose. For example, when $x = -1$, we see that $5x$ is negative, $(x-5)$ is negative, and $(x-2)$ is also negative, so $f'(-1)$ has sign $(-)(-)^2(-) = (+)$. I leave it to you to repeat this exercise with the other values and verify that the filled in table looks like this:

x	-1	0	1	2	3	5	6
$f'(x)$	$+$	0	$-$	0	$+$	0	$+$
	╱	─	╲	─	╱	─	╱

Notice how I drew the little lines in the third row: upward-sloping when $f'(x)$ has sign $(+)$, downward when its sign is $(-)$, and flat when its sign is 0. We immediately know that f is increasing when $x < 0$ and when $x > 2$, while it's decreasing for $0 < x < 2$. The table also reveals that $x = 0$ is a local maximum, $x = 2$ is a local minimum, and $x = 5$ is a horizontal point of inflection. We'll use the above table again when we sketch the graph of $y = f(x)$ in Section 12.3.3 below.

A word of warning: the lines in the third row of the table are meant only to guide you as you sketch the graph of $y = f(x)$. The graph probably doesn't look like a collection of lines tacked together! Instead, just use the information in that third row to understand where the graph is increasing, decreasing or temporarily flat.

12.1.2 Making a table of signs for the second derivative

We've also seen that the sign of the second derivative is important (check out Section 11.4 of the previous chapter). When the sign is positive, the curve is concave up; when the sign is negative, the curve is concave down; and when it's 0, you may or may not get a point of inflection. The table of signs for the second derivative tells all.

The method is the same as for the function or the derivative, except that the third row is now used to show whether the function is concave up or concave down. Put a little upward parabola-like curve whenever the sign is $(+)$, a downward version when the sign is $(-)$, and a dot when the sign is 0.

If we return to our example $f(x) = x^2(x-5)^3$ from above, we have already seen that $f'(x) = 5x(x-5)^2(x-2)$. To differentiate this, let's combine the x

and $(x-2)$ factors to write $f'(x) = 5(x-5)^2(x^2 - 2x)$. Now we can use the product rule:

$$f''(x) = 5\left((x^2 - 2x) \times (2(x-5)) + (x-5)^2(2x - 2)\right).$$

Taking a common factor of $(x-5)$ and rearranging, we find that we have $f''(x) = 10(x-5)(2x^2 - 8x + 5)$. Actually, you can use the quadratic formula to see that the solutions of $2x^2 - 8x + 5 = 0$ are $2 \pm \frac{1}{2}\sqrt{6}$. So we can completely factor $f''(x)$ as

$$f''(x) = 20\left(x - (2 - \tfrac{1}{2}\sqrt{6})\right)\left(x - (2 + \tfrac{1}{2}\sqrt{6})\right)(x - 5).$$

This means that $f''(x)$ has sign 0 at $x = 2 - \frac{1}{2}\sqrt{6}$, $x = 2 + \frac{1}{2}\sqrt{6}$ and $x = 5$. Let's start on our table of signs for $f''(x)$:

x	$2 - \frac{1}{2}\sqrt{6}$		$2 + \frac{1}{2}\sqrt{6}$		5	
$f''(x)$	0		0		0	

Now we have to fill in the gaps. It would be nice to know something more about $2 \pm \frac{1}{2}\sqrt{6}$, so let's try to estimate it without resorting to a calculator! You see, $\sqrt{6}$ is between 2 and 3 (since 6 is between 4 and 9), so $\frac{1}{2}\sqrt{6}$ is between 1 and $\frac{3}{2}$. This means that $2 - \frac{1}{2}\sqrt{6}$ is somewhere between $2 - \frac{3}{2} = \frac{1}{2}$ and $2 - 1 = 1$, and also that $2 + \frac{1}{2}\sqrt{6}$ is between $2 + 1 = 3$ and $2 + \frac{3}{2} = 3\frac{1}{2}$. So we can choose 0 to the left of $2 - \frac{1}{2}\sqrt{6}$; between $2 - \frac{1}{2}\sqrt{6}$ and $2 + \frac{1}{2}\sqrt{6}$, we'll pick 2; between $2 + \frac{1}{2}\sqrt{6}$ and 5, we'll choose 4; finally, we'll pick 6 to the right of 5. Here's what we get:

x	0	$2 - \frac{1}{2}\sqrt{6}$	2	$2 + \frac{1}{2}\sqrt{6}$	4	5	6
$f''(x)$	$-$	0	$+$	0	$-$	0	$+$
	$\frown$	$\bullet$	$\smile$	$\bullet$	$\frown$	$\bullet$	$\smile$

Make sure you agree with all the signs I've filled in. For example, when $x = 0$, all three factors of $f''(x)$ are negative, so the product is negative. Also, notice how I drew in the little curves in the third row. You can clearly see that f is concave up when $2 - \frac{1}{2}\sqrt{6} < x < 2 + \frac{1}{2}\sqrt{6}$ or $x > 5$; and that f is concave down when $x < 2 - \frac{1}{2}\sqrt{6}$ or $2 + \frac{1}{2}\sqrt{6} < x < 5$. All three points $2 - \frac{1}{2}\sqrt{6}$, $2 + \frac{1}{2}\sqrt{6}$ and 5 are points of inflection, since the concavity is opposite to the left and the right of these points. Once again, we'll return to the table in Section 12.3.3 below.

Let's look at one more example. Suppose that

$$g(x) = x^9 - 9x^8.$$

You can easily calculate that $g'(x) = 9x^8 - 72x^7$ and that

$$g''(x) = 72x^7 - 72 \times 7x^6 = 72x^6(x - 7).$$

So $g''(x) = 0$ when $x = 0$ or $x = 7$. Let's pick $x = -1$, $x = 3$ and $x = 8$ as our fill-in points. I leave it to you to show that $g''(-1) < 0$, $g''(3) < 0$ and $g''(8) > 0$. So the table of signs for $g''(x)$ looks like this:

x	-1	0	3	7	8
$g''(x)$	$-$	0	$-$	0	$+$
	$\frown$	$\bullet$	$\frown$	$\bullet$	$\smile$

So we see that $x = 0$ is **not** a point of inflection for g: the function is concave down on both sides of $x = 0$. On the other hand, the point $x = 7$ is a point of inflection, since g is concave down to the left of 7 and concave up to the right of 7.

As we noted in the case of the first derivative in the previous section, the pictures in the third row are meant only as a guide to sketching the graph. They show where the original function is concave up and concave down, but they won't necessarily give anything more than a rough idea of what the curve $y = f(x)$ actually looks like. That's why we're going to look at a big method for sketching curves. The three types of tables of signs we've looked at above will be used in the process, but that's not the whole story. Now, fasten your seatbelts. . . .

12.2   The Big Method

Here is an eleven-step method for sketching the graph of $y = f(x)$. Before you start, draw up a set of axes so you can start putting some of the information you gather on the graph.

1. **Symmetry:** check whether the function is even, odd, or neither by replacing x by $-x$ and seeing whether you get back the original function or its negative. If the function is even or odd, you only need to sketch it for $x \geq 0$, then use the symmetry to sketch the left half of the graph. This could save you a lot of time.

2. **y-intercept:** find the y-intercept (if it exists) by setting $x = 0$. Mark it on the graph.

3. **x-intercepts:** find the x-intercepts by setting $y = 0$ and solving for x. This is sometimes difficult or impossible! For example, if you have to factor a polynomial of degree 3 or higher, you may have to scrabble around to find a root, then do a polynomial division to continue factoring. Mark the x-intercepts on your graph.

4. **Domain:** find the domain of f. If it's specified in the definition of f, there's nothing to do; otherwise, the domain is assumed to be as much of the real line as possible. Remember, you have to avoid numbers which lead to 0 in the denominator, or the square root of a negative number, or the log of a negative number or 0. If inverse trig functions are involved, the situation is more complicated—so I suggest you learn the domains of all the inverse trig functions. (For example, you can't take the inverse sine of a number outside the interval $[-1, 1]$.)

5. **Vertical asymptotes:** these generally occur where the denominator is zero (if there is a denominator!). Beware: if the numerator is zero too, then you might have a removable discontinuity* instead of a vertical asymptote. Also, you may have a vertical asymptote due to a log factor. Mark all the vertical asymptotes as dotted vertical lines on your graph.

6. **Sign of the function:** at this point, draw up a table of signs for $f(x)$, as described in Section 12.1. We already know where f is zero from #3 above, and we know where it's discontinuous from #4 and #5. The table tells you exactly where the curve is above or below the x-axis.

7. **Horizontal asymptotes:** find the horizontal asymptotes by calculating

$$\lim_{x \to \infty} f(x) \qquad \text{and} \qquad \lim_{x \to -\infty} f(x).$$

Even if the limits are $\pm\infty$, it may be that you can still work out what $f(x)$ behaves like for large (or negatively large) x and thereby get a sort of "diagonal" asymptote. In any case, draw dashed horizontal lines on your graph to remind you about the horizontal asymptotes, if there are any. At this point, you can fill in little bits of the function near both the horizontal and vertical asymptotes, using the table of signs for $f(x)$ to tell which side of each of the asymptotes the function lies on.

8. **Sign of the derivative:** now, time for calculus. Find the derivative, then find all the critical points—remember, these are points where the derivative is 0 or does not exist. Now draw up a table of signs for $f'(x)$, as described in Section 12.1.1 above. Use the third row of the table to tell where the function is increasing, decreasing, or flat.

9. **Maxima and minima:** from the table of signs, you can find all the local maxima or minima—remember, these only occur at critical points. For each maximum or minimum x, you need to find the value of y by substituting the value of x into the equation $y = f(x)$. Make sure you label all these points on your graph.

10. **Sign of the second derivative:** find the second derivative, then find all the points where the second derivative is zero or does not exist. Now you should draw up a table of signs for $f''(x)$, as described in Section 12.1.2 above. The pictures in the third row of the table indicate where the curve is concave up and where it's concave down.

11. **Points of inflection:** use the table of signs for the second derivative to identify the inflection points. Remember, the second derivative at an inflection point has to be zero, and the sign of the second derivative has to be different on either side of the inflection point. For each inflection point x, you need to find the y-coordinate by substituting into the equation $y = f(x)$. Make sure these points are labeled on your graph.

Now, using all the information you've gathered, complete the sketch of the graph. If anything looks inconsistent, then you might have made a mistake! All the information you gather should work nicely together.

---

*For example, if $f(x) = (x^2 - 3x + 2)/(x - 2)$, then by factoring the numerator as $(x - 1)(x - 2)$, you can easily see that $f(x) = x - 1$ except at $x = 2$, where f is undefined. The graph is on page 42.

By the way, remember that you can also find the local maxima and minima in step 9 above by looking at the sign of the second derivative (see Section 11.5.2 in the previous chapter). This method doesn't always work, though—that's why I recommend using the table of signs for $f'(x)$.

12.3 Examples

We'll start with an example of sketching a curve without using the first or second derivatives, then look at four more examples of the complete method.

12.3.1 An example without using derivatives

At the beginning of Section 12.1 above, we looked at

$$f(x) = \frac{(x-3)(x-1)^2}{x^3(x+2)}.$$

Let's sketch $y = f(x)$ using only the first seven steps of our program:

1. **Symmetry:** you can plug in $-x$ instead of x, and play around with it, but it's a lost cause: the function is neither odd nor even.
2. **y-intercept:** set $x = 0$; then the denominator vanishes and the numerator doesn't. So the function blows up at $x = 0$ and there's no y-intercept.
3. **x-intercepts:** set $y = 0$; then we must have $x - 3 = 0$ or $x - 1 = 0$, so the x-intercepts are at 1 and 3.
4. **Domain:** clearly we're fine for all x except $x = 0$ and $x = 2$.
5. **Vertical asymptotes:** the denominator vanishes when $x = 0$ or when $x = -2$; the numerator doesn't also vanish there, so these are the vertical asymptotes.
6. **Sign of the function:** we already investigated this thoroughly, and found that the function is positive on $(-2, 0)$ and $(3, \infty)$ and negative everywhere else (except at the x-intercepts and vertical asymptotes). For reference, here's the table we saw in Section 12.1:

x	-3	-2	-1	0	$\frac{1}{2}$	1	2	3	4
$f(x)$	$-$	$\star$	$+$	$\star$	$-$	0	$-$	0	$+$

7. **Horizontal asymptotes:** we need to look at

$$\lim_{x \to \infty} \frac{(x-3)(x-1)^2}{x^3(x+2)} \quad \text{and} \quad \lim_{x \to -\infty} \frac{(x-3)(x-1)^2}{x^3(x+2)}.$$

I leave it to you to show that both these limits are 0 (using the methods of Section 4.3 in Chapter 4), so there's a two-sided horizontal asymptote at $y = 0$.

Now we can sketch the graph. Let's first mark in what we know:

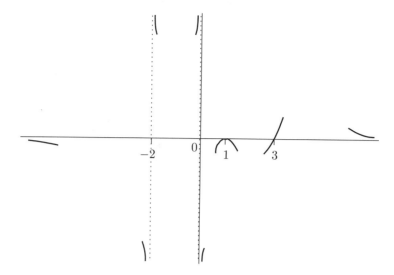

The horizontal asymptotes are both at $y = 0$. On the left-hand side of the graph, the curve is below the x-axis since the function values are negative when $x < -2$. On the right-hand side of the graph, the curve is above the axis since the function values are positive when $x > 3$ (we know this from the table of signs). As for the vertical asymptotes, the one at $x = -2$ must be negative on the left and positive on the right, using the table of signs once again. The asymptote at $x = 0$ is analyzed in the same way. Now consider the x-intercepts. The intercept at $x = 1$ must touch the curve, since the sign of $f(x)$ is negative on either side of 1. On the other hand, the function changes sign on either side of $x = 3$, so the intercept there passes through the axis. Now we can join the curve pieces and get something like this:

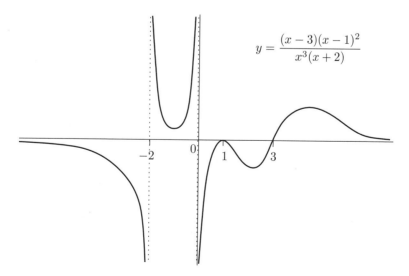

$$y = \frac{(x-3)(x-1)^2}{x^3(x+2)}$$

This is a pretty good approximation to the shape of the graph. The problem is, we don't know where the local maxima and minima are except for the local maximum at $x = 1$. Certainly there's at least one local minimum between $x = -2$ and $x = 0$, at least one local minimum between $x = 1$ and $x = 3$, and at least one local maximum greater than $x = 3$. There could be more, though—the graph might have a lot more wobbles than shown. We can't tell without using the derivative.

So why not use the derivative? For this function, it's too difficult to deal with! If you go to the trouble of calculating it, you will find that

$$f'(x) = \frac{-x^4 + 10x^3 - 11x^2 - 16x + 18}{x^4(x + 2)^2}.$$

Actually, we know $x = 1$ is a local maximum, so $f'(1)$ should be 0. You can check and see that the numerator does indeed vanish at $x = 1$. This means that $(x - 1)$ is a factor of the numerator, and you can do a long division to see that the numerator is $(x - 1)(-x^3 + 9x^2 - 2x - 18)$. That still leaves a nasty cubic to deal with. At least we do know one thing: the cubic has at most three solutions. This means that, in addition to $x = 1$, there are at most three other critical points. In particular, our graph doesn't have extra wobbles—just the four critical points you can see in the picture above.

As for using the second derivative to find the concavity and points of inflection, well, suffice it to say that it's even worse than the first derivative! On the other hand, not every function has such difficult derivatives—let's look at four more examples where we can use the full method.

12.3.2 The full method: example 1

At the end of Section 11.5.1 in the previous chapter, we saw that if

$$f(x) = x \ln(x),$$

then f has a local minimum at $x = 1/e$. We even started to sketch its graph. Let's use the full method to complete the graph of $y = f(x)$:

1. **Symmetry:** the function isn't even defined for $x \le 0$, so it certainly can't be odd or even.

2. **y-intercept:** set $x = 0$; then $f(x)$ is undefined, so there can't be any y-intercept.

3. **x-intercepts:** set $y = 0$; then we must have $x = 0$ or $\ln(x) = 0$. We can't have $x = 0$, since f isn't defined there, and $\ln(x) = 0$ only when $x = 1$. So the only x-intercept is at $x = 1$.

4. **Domain:** because of the $\ln(x)$ factor, the domain must be $(0, \infty)$.

5. **Vertical asymptotes:** the $\ln(x)$ factor might actually introduce a vertical asymptote at $x = 0$. Let's check it out. Since $f(x)$ is only defined when $x > 0$, the best we can do is to consider the right-hand limit

$$\lim_{x \to 0^+} x \ln(x).$$

Actually, we know from Section 9.4.6 that this limit is 0, as logs grow slowly (to $-\infty$) as $x \to 0^+$. So there are no vertical asymptotes; there's just a (right-hand) removable discontinuity at the origin.

6. **Sign of the function:** we know that the function is undefined for $x \leq 0$, and the only x-intercept is at $x = 1$. So we need to fill in the gaps with something like $x = 1/2$ and $x = 2$. When $x = 1/2$, we have $\ln(1/2) = -\ln(2)$, which is negative, so f has sign $(-)$. When $x = 2$, you can easily see that f has sign $(+)$. So the table of signs looks like this:

x	≤ 0	$\frac{1}{2}$	1	2
$f(x)$	$\star$	$-$	0	$+$

7. **Horizontal asymptotes:** we only need to look at

$$\lim_{x \to \infty} x \ln(x)$$

since the limit as $x \to -\infty$ doesn't even make sense. The above limit is clearly ∞, since both x and $\ln(x)$ go to ∞ as $x \to \infty$. So there are no horizontal asymptotes.

8. **Sign of the derivative:** by the product rule, we have $f'(x) = \ln(x) + 1$ (as we saw in Section 11.5.1 of the previous chapter). So $f'(x) = 0$ when $\ln(x) = -1$, that is, when $x = e^{-1} = 1/e$. We just need to pick a point between $x = 0$ and $x = 1/e$, and some other point greater than $x = 1/e$. Let's choose $x = 1/10$ for the first and $x = 1$ for the second. Note that $f'(1/10) = \ln(1/10) + 1 = -\ln(10) + 1$, which is clearly negative; and $f'(1) = \ln(1) + 1 = 1$, which is positive. Our table of signs for $f'(x)$ looks like this:

x	≤ 0	$\frac{1}{10}$	$\frac{1}{e}$	1
$f'(x)$	$\star$	$-$	0	$+$
		$\searrow$	$-$	$\nearrow$

9. **Maxima and minima:** looking at the table of signs, we see that we only have a local minimum at $x = 1/e$. We just need to calculate the y-value there: we have $y = e^{-1} \ln(e^{-1}) = -e^{-1} = -1/e$. So there is a local minimum at $(1/e, -1/e)$, as we already observed in Section 11.5.1 of the previous chapter.

10. **Sign of the second derivative:** since $f'(x) = \ln(x) + 1$, we have $f''(x) = 1/x$. Since f is only defined when $x > 0$, we see that $f''(x) > 0$ for all relevant x. This means that f is always concave up.

11. **Points of inflection:** since $f''(x)$ is never 0, there aren't any!

Now, let's put the information we've gathered on a graph. We have a removable discontinuity at the origin, a local minimum at $(1/e, -1/e)$, an x-intercept at 1, and no horizontal or vertical asymptotes. The graph is below the x-axis when $x < 1$ and above it when $x > 1$. Also, the function is decreasing for $0 < x < 1/e$ and increasing when $x > 1/e$, and is always concave up. Its graph must look something like this:

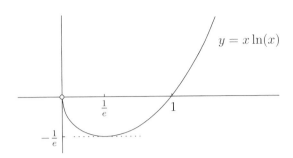

It's not perfect, but it's a heck of a lot better than our first attempt on page 241, since we have a lot more information.

12.3.3 The full method: example 2

Let's look at another function we've already investigated somewhat:

$$f(x) = x^2(x-5)^3.$$

In Section 10.1.4 of Chapter 10, we already made a rough sketch of the graph of $y = f(x)$; we've also made tables of signs for $f(x)$, $f'(x)$, and $f''(x)$ in Section 12.1 above. This means that we can step on the gas and rip right through our method:

1. **Symmetry:** if you replace x by $(-x)$, you get $(-x)^2(-x-5)^3$, which simplifies to $-x^2(x+5)^2$. This is neither $f(x)$ nor $-f(x)$, so f is neither odd nor even. Oh well, you can't win them all.

2. **y-intercept:** when $x = 0$, we see that $y = f(0) = 0$. So the y-intercept is at $y = 0$.

3. **x-intercepts:** if $y = 0$, then we must have $x^2 = 0$ or $(x-5)^3 = 0$. So the x-intercepts are at $x = 0$ and $x = 5$.

4. **Domain:** there are no problems taking $f(x)$ for any x, so the domain is the set of all real numbers $\mathbb{R}$.

5. **Vertical asymptotes:** since the domain is all of $\mathbb{R}$, there aren't any vertical asymptotes.

6. **Sign of the function:** as we saw in Section 12.1, the table of signs looks like this:

x	-1	0	2	5	6
$f(x)$	$-$	0	$-$	0	$+$

So the graph is only above the x-axis when $x > 5$.

7. **Horizontal asymptotes:** it's pretty easy to see that

$$\lim_{x \to \infty} x^2(x-5)^3 = \infty \qquad \text{and} \qquad \lim_{x \to -\infty} x^2(x-5)^3 = -\infty.$$

After all, when $x \to \infty$, both x^2 and $(x-5)^3$ also go to ∞, so their product does as well. When $x \to -\infty$, the x^2 term goes to ∞ and the

$(x-5)^3$ term goes to $-\infty$, so the product goes to $-\infty$. We might note that when x is large (positive or negative), the quantity $(x-5)$ behaves like its highest-degree term x; so $x^2(x-5)^3$ behaves like x^5 near the edges of the graph, but not near the origin!

8. **Sign of the derivative:** as we saw in Section 12.1.1, the table of signs for $f'(x)$ is as follows:

x	-1	0	1	2	3	5	6
$f'(x)$	$+$	0	$-$	0	$+$	0	$+$
	↗	—	↘	—	↗	—	↗

This tells us where the function is increasing, decreasing or flat.

9. **Maxima and minima:** we see from the above table that $x=0$ is a local maximum, $x=2$ is a local minimum, and $x=5$ is a horizontal point of inflection. Now we need to calculate the corresponding y-coordinates by using the formula $y=f(x)=x^2(x-5)^3$. This isn't too bad: $f(0)=0$, $f(2)=(2)^2(-3)^3=-108$, and $f(5)=0$. So there's a local maximum at the origin, a local minimum at $(2,-108)$ and a horizontal point of inflection at $(5,0)$.

10. **Sign of the second derivative:** we already found this in Section 12.1.2:

x	0	$2-\frac{1}{2}\sqrt{6}$	2	$2+\frac{1}{2}\sqrt{6}$	4	5	6
$f''(x)$	$-$	0	$+$	0	$-$	0	$+$
	⌢	•	⌣	•	⌢	•	⌣

We can use this to see where the function is concave up and where it's concave down. Notice that $f''(0)<0$, which confirms that the critical point $x=0$ is a local maximum; and also that $f''(2)>0$, confirming that the critical point $x=2$ is a local minimum.

11. **Points of inflection:** from the above table, we have points of inflection at $x=2-\frac{1}{2}\sqrt{6}$, $x=2+\frac{1}{2}\sqrt{6}$ and $x=5$. Actually, we already knew about this last one, since we saw in step 9 above that $(5,0)$ is a horizontal point of inflection. The other two are a lot messier. We need to substitute $x=2-\frac{1}{2}\sqrt{6}$ and $x=2+\frac{1}{2}\sqrt{6}$, one at a time, into the original equation $y=x^2(x-5)^3$. Unfortunately, you get a bit of a mess. Let's cheat a little and define $\alpha=f(2-\frac{1}{2}\sqrt{6})$ and $\beta=f(2+\frac{1}{2}\sqrt{6})$. This means that

$$\alpha=(2-\tfrac{1}{2}\sqrt{6})^2(-3-\tfrac{1}{2}\sqrt{6})^3 \qquad \text{and} \qquad \beta=(2+\tfrac{1}{2}\sqrt{6})^2(-3+\tfrac{1}{2}\sqrt{6})^3.$$

Actually, if you go to the trouble of multiplying everything out, you can simplify these expressions, but it's no fun at all. We might also make a rare use of a calculator to see that α is approximately -45.3 and β is approximately -58.2. These are approximations only! The calculator can never give you the true value of an irrational number such as α or β. Anyway, we have found points of inflection at $(2-\frac{1}{2}\sqrt{6},\alpha)$ and $(2+\frac{1}{2}\sqrt{6},\beta)$ as well as $(5,0)$.

Now let's put everything together. Starting with a set of axes, mark in the y-intercept at the origin, the x-intercepts at 0 and 5, the local maximum at the origin, the local minimum at $(-2, 108)$, the horizontal inflection point at $(5, 0)$, and the nonhorizontal inflection points at $(2 - \frac{1}{2}\sqrt{6}, \alpha)$ and $(2 + \frac{1}{2}\sqrt{6}, \beta)$. We also know that $y \to \infty$ as $x \to \infty$, and $y \to -\infty$ as $x \to -\infty$, so we can put a small piece of curve to indicate this. Altogether, here's what we get:

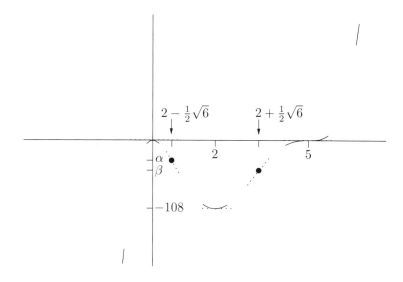

Note that we know from the table of signs for $f'(x)$ that the slope at the inflection point $(2 - \frac{1}{2}\sqrt{6})$ is negative and that the slope at $(2 + \frac{1}{2}\sqrt{6})$ is positive. Now all we have to do is join the pieces:

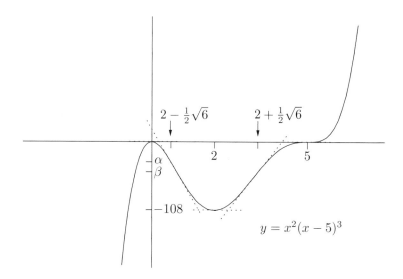

Again, this is better than our previous attempt at sketching this graph on page 207, because it shows the inflection points as well.

12.3.4 The full method: example 3

Let's sketch the graph of $y = f(x)$, where

$$f(x) = xe^{-3x^2/2}.$$

1. **Symmetry:** replace x by $(-x)$ and we get $-xe^{-3(-x)^2/2} = -xe^{-3x^2/2}$, which is just $-f(x)$. This means that the function is odd, which is a major bonus: we only have to graph it for $x \geq 0$, then it'll be easy to get the other half.

2. **y-intercept:** if $x = 0$, then $y = 0e^{-3(0)^2/2} = 0$. So the y-intercept is at $y = 0$.

3. **x-intercepts:** if $y = 0$, then we have $0 = xe^{-3x^2/2}$. So either $x = 0$ or $e^{-3x^2/2} = 0$. The latter equation has no solution, since exponentials are always positive! So the only x-intercept is at $x = 0$. So far, all we know is that the function is odd and the only place it crosses the axes is at the origin.

4. **Domain:** clearly you can make x equal to anything and never have a problem—there are no square roots or logs, and even if you write the function as

$$y = \frac{x}{e^{3x^2/2}},$$

the denominator can't be zero since exponentials are always positive. So the domain is the real line $\mathbb{R}$.

5. **Vertical asymptotes:** there aren't any, since the domain is $\mathbb{R}$.

6. **Sign of the function:** we know that the only place $f(x) = 0$ is when $x = 0$, so the table of signs is ridiculously simple:

x	-1	0	1
$f(x)$	$-$	0	$+$

The function is positive when $x > 0$ and negative when $x < 0$.

7. **Horizontal asymptotes:** we need to find

$$\lim_{x \to \infty} \frac{x}{e^{3x^2/2}} \qquad \text{and} \qquad \lim_{x \to -\infty} \frac{x}{e^{3x^2/2}}.$$

Note that $3x^2/2$ is a large positive number in either case, so the denominator is a large exponential. Since exponentials grow quickly (see Section 9.4.4 in Chapter 9), both the above limits are 0. So there is a two-sided horizontal asymptote at $y = 0$.

8. **Sign of the derivative:** now we have to differentiate. By the product rule and the chain rule, you can check that

$$f'(x) = x(-3x)e^{-3x^2/2} + e^{-3x^2/2} = (1 - 3x^2)e^{-3x^2/2}.$$

This is defined everywhere, but where is it 0? Since exponentials are positive, it is only 0 when $1 - 3x^2 = 0$, that is, when $x = 1/\sqrt{3}$ or

$x = -1/\sqrt{3}$. Let's choose the points -1, 0, and 1 to fill in the gaps; our table of signs for the derivative looks like this:

x	-1	$\frac{-1}{\sqrt{3}}$	0	$\frac{1}{\sqrt{3}}$	1
$f'(x)$	$-$	0	$+$	0	$-$
	$\searrow$	$-$	$\nearrow$	$-$	$\searrow$

We see that the function is increasing between $-1/\sqrt{3}$ and $1/\sqrt{3}$, and decreasing elsewhere. Notice that the oddness of f (as in step 1 above) is clearly apparent from the third row of the above table.

9. **Maxima and minima:** looking at the table of signs, it's pretty evident that $x = 1/\sqrt{3}$ is a local maximum and $x = -1/\sqrt{3}$ is a local minimum. The only thing left is to substitute these values of x into the equation for y. When $x = 1/\sqrt{3}$, we have

$$y = \frac{1}{\sqrt{3}} e^{-3(1/\sqrt{3})^2/2} = \frac{e^{-1/2}}{\sqrt{3}}.$$

So there's a local maximum at $(1/\sqrt{3}, e^{-1/2}/\sqrt{3})$. Since the function is odd, we don't even need to substitute $x = -1/\sqrt{3}$ to see that there must be a local minimum at $(-1/\sqrt{3}, -e^{-1/2}/\sqrt{3})$.

10. **Sign of the second derivative:** now we have to differentiate again, using the product rule and chain rule once more. We find that

$$f''(x) = (1 - 3x^2)(-3x)e^{-3x^2/2} + (-6x)e^{-3x^2/2} = 9x(x^2 - 1)e^{-3x^2/2}.$$

Once again, since exponentials are positive, the only way that $f''(x)$ can equal 0 is if $x = 0$ or $x^2 - 1 = 0$, that is, if $x = 0$, $x = 1$ or $x = -1$. The table of signs looks like this:

x	-2	-1	$-\frac{1}{2}$	0	$\frac{1}{2}$	1	2
$f''(x)$	$-$	0	$+$	0	$-$	0	$+$
	$\frown$	$\bullet$	$\smile$	$\bullet$	$\frown$	$\bullet$	$\smile$

For $x = 1/2$, the factor $9x$ is positive whereas $(x^2 - 1)$ is negative, and the exponential is positive, so the whole thing is negative. When $x = 2$, it's just as easy to see that the second derivative is positive. The situation for $x = -1/2$ and $x = -2$ is just as easy and in fact follows by symmetry. (Since the original function is odd, its derivative is even and its second derivative is odd. You may have to think about this point a little!) The third row indicates that the graph is concave down when $x < -1$ or $0 < x < 1$, and concave up when $x > 1$ or $-1 < x < 0$. By the way, notice that at the critical point $x = 1/\sqrt{3}$, the second derivative is negative—this confirms that we have a local maximum there. Similarly, when $x = -1/\sqrt{3}$, the second derivative is positive, so we do indeed have a local minimum there.

11. **Points of inflection:** from the above table, we can see that the concavity clearly changes at $x = 1$, $x = -1$, and $x = 0$; so these are all points

of inflection and we just need to find the y-coordinates. By substituting in the equation $y = xe^{-3x^2/2}$, it's easy to see that the points of inflection should be displayed on the graph as $(1, e^{-3/2})$, $(-1, -e^{-3/2})$ and $(0,0)$.

If you've been really good, you would have been plotting what we already know on a set of axes, and you should have something like this:

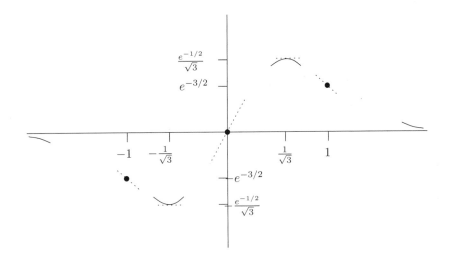

On this graph, you can see the x- and y-intercepts (at the origin), the horizontal asymptote (the x-axis), the maximum at $(1/\sqrt{3}, e^{-1/2}/\sqrt{3})$, the minimum at $(-1/\sqrt{3}, -e^{-1/2}/\sqrt{3})$, and the inflection points at $(1, e^{-3/2})$, $(-1, -e^{-3/2})$, and $(0,0)$ (shown as dotted lines for now). Because we know the sign of $f(x)$ from step 6, we've even diagnosed the behavior near the horizontal asymptotes and displayed this information on the graph. Anyway, all that's left is to connect the dots:

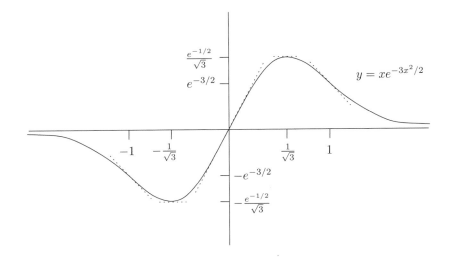

This sketch really illustrates all the important features of the graph.

12.3.5 The full method: example 4

Now let's do it all over again: we'll sketch the graph of $y = f(x)$, where f is the fearsome-looking function defined by

$$f(x) = \frac{x^3 - 6x^2 + 13x - 8}{x}.$$

1. **Symmetry:** replacing x by $-x$, we get $(-x^3 - 6x^2 - 13x - 8)/(-x)$, which is neither $f(x)$ nor $-f(x)$, so there's no symmetry. Bummer.

2. **y-intercept:** put $x = 0$, and you get $-8/0$ which is undefined. So there's no y-intercept.

3. **x-intercepts:** now things get nasty. We need to set $y = 0$, which means that $x^3 - 6x^2 + 13x - 8 = 0$. This is a cubic equation, so factoring might be a pain in the butt. The best bet is to try to guess a solution. Try $x = 1$. Well, you get $1 - 6 + 13 - 8 = 0$, and it works! (Basically, the only nice solutions would be factors of the constant term -8, so if ± 1, ± 2, ± 4 and ± 8 don't work, you're screwed.) Luckily our first try worked and we know that $(x - 1)$ is a factor. Now we have to divide:

$$x - 1 \,\overline{\big)\, x^3 - 6x^2 + 13x - 8}$$

I leave it to you to do this division and show that the other factor is $x^2 - 5x + 8$. Can you factor this quadratic? The discriminant is $(-5)^2 - 4(8) = -7$, which is negative, so you can't factor the quadratic. That is, we have $x^3 - 6x^2 + 13x - 8 = (x - 1)(x^2 - 5x + 8)$, and the second factor is always positive, so the only x-intercept is $x = 1$.

4. **Domain:** the only problem is at $x = 0$, so the domain is $\mathbb{R} \backslash \{0\}$.

5. **Vertical asymptotes:** there's one at $x = 0$, since the denominator vanishes there but the numerator doesn't. There can't be any other vertical asymptotes because the function is defined everywhere else.

6. **Sign of the function:** write $f(x)$ as

$$f(x) = \frac{(x - 1)(x^2 - 5x + 8)}{x}.$$

The only x-intercept is at $x = 1$, and the only discontinuity is at $x = 0$, so our table of signs looks like this:

x	-1	0	$\frac{1}{2}$	1	2
$f(x)$	$+$	$\star$	$-$	0	$+$

(Make sure you believe the signs at $x = -1$, $x = 1/2$, and $x = 2$.)

7. **Horizontal asymptotes:** consider

$$\lim_{x \to \infty} \frac{x^3 - 6x^2 + 13x - 8}{x} \qquad \text{and} \qquad \lim_{x \to -\infty} \frac{x^3 - 6x^2 + 13x - 8}{x}.$$

These can be written as

$$\lim_{x \to \infty} \left(x^2 - 6x + 13 - \frac{8}{x} \right) \qquad \text{and} \qquad \lim_{x \to -\infty} \left(x^2 - 6x + 13 - \frac{8}{x} \right).$$

It's quite clear that both these limits are infinity, so there are no horizontal asymptotes. On the other hand, when x is large (or negatively large), $f(x)$ acts like its dominant term, which is x^2. So the curve should look pretty similar to the parabola $y = x^2$ but only when x is large. Anyway, we've taken no derivatives but we still know a lot about the function:

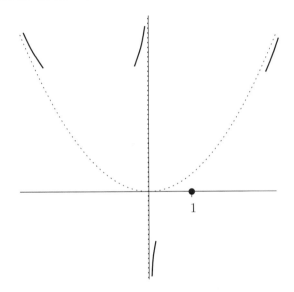

Notice that we used the table of signs for $f(x)$ to see how the graph looks near the vertical asymptote at $x = 0$. In particular, when x is a little less than zero, $f(x)$ is positive, so the curve goes up to ∞ on the left side of the asymptote. Similarly, when x is a little larger than zero, $f(x)$ is negative, which means that the curve goes down to $-\infty$ on the right side of the asymptote.

8. **Sign of the derivative:** we have three forms for $f(x)$ that we've already used:

$$f(x) = \frac{x^3 - 6x^2 + 13x - 8}{x} = \frac{(x-1)(x^2 - 5x + 8)}{x} = x^2 - 6x + 13 - \frac{8}{x}.$$

We need $f'(x)$, and you can take your pick which form of $f(x)$ you want to use. I vote for the last one since it doesn't require any use of the product rule or the quotient rule. We have

$$f'(x) = 2x - 6 + \frac{8}{x^2},$$

which we can now write as

$$f'(x) = \frac{2x^3 - 6x^2 + 8}{x^2}.$$

So where is the derivative equal to 0, and where does it not exist? It's pretty obvious that the only place it doesn't exist is when $x = 0$. On the other hand, if $f'(x) = 0$, then we must have $2x^3 - 6x^2 + 8 = 0$. Once again we need a solution to a cubic equation; this time, $x = 1$ doesn't work, so try $x = -1$. Hey, it does work! After you do the long division, you find that you can factor the cubic as $2(x+1)(x-2)^2$. That is,

$$f'(x) = \frac{2(x+1)(x-2)^2}{x^2}.$$

So the derivative is undefined at $x = 0$ and it equals zero when $x = -1$ or $x = 2$. Now we can draw up a table of signs for $f'(x)$:

x	-2	-1	$-\frac{1}{2}$	0	1	2	3
$f'(x)$	$-$	0	$+$	$\star$	$+$	0	$+$
	$\searrow$	$-$	$\nearrow$	$\vdots$	$\nearrow$	$-$	$\nearrow$

Make sure you check the details of this table! In any case, we can see that the function is increasing when $x > -1$ (except at the critical points $x = 0$ and $x = 2$) and the function is decreasing when $x < -1$.

9. **Maxima and minima:** looking at the table of signs, we see that $x = -1$ is a local minimum and $x = 2$ is a horizontal point of inflection. We need the y-coordinates; it's not too hard to see that $f(-1) = 28$ and $f(2) = 1$. So $(-1, 28)$ is a local minimum and $(2, 1)$ is a horizontal point of inflection.

10. **Sign of the second derivative:** we know that $x = 2$ is a point of inflection, but are there any others? Let's find out. Use the form

$$f'(x) = 2x - 6 + \frac{8}{x^2}$$

to find that

$$f''(x) = 2 - \frac{16}{x^3} = \frac{2(x^3 - 8)}{x^3}.$$

So the second derivative is undefined at $x = 0$ and it's zero only when $x^3 - 8 = 0$, so $x = 2$. There aren't any other points of inflection! Let's draw up the table of signs:

x	-1	0	1	2	3
$f''(x)$	$+$	$\star$	$-$	0	$+$
	$\smile$	$\vdots$	$\frown$	$\bullet$	$\smile$

You can see that the graph is concave up when $x < 0$ and $x > 2$, and concave down when $0 < x < 2$. By the way, at the critical point $x = -1$, we have $f''(x) > 0$, so we indeed have a local minimum there; on the other hand, at the critical point $x = 2$, we see that $f''(2) = 0$, which by itself wouldn't have been enough information to confirm the inflection point. The best way to nail that down is to show that the sign of the derivative is the same on either side of $x = 2$. This information is nicely conveyed by the table of signs.

11. **Points of inflection:** we know that $x = 2$ is the only one, and we've already seen that this leads to the inflection point $(2, 1)$.

Let's complete our sketch of the graph, based on our newfound knowledge in the last few steps. We need to put in the minimum at $(-1, 28)$ and the horizontal point of inflection at $(2, 1)$. Unfortunately 28 is a big number, so we'll need to squish the y-axis (compared to our rough draft above) to get the scale right. We end up with this:

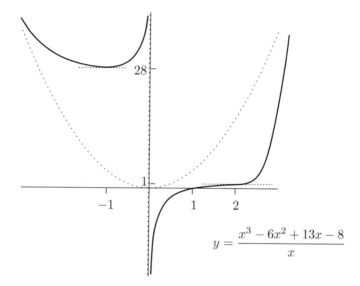

$$y = \frac{x^3 - 6x^2 + 13x - 8}{x}$$

The dotted curve is supposed to be $y = x^2$, although the scale isn't right. Also, on the right-hand side of the graph, the solid curve is supposed to get close to $y = x^2$, but I didn't do a great job of it. Unfortunately, if you get this sort of behavior right, you end up missing the detailed behavior at the inflection point. Indeed, here's what the output from a graphing calculator might look like:

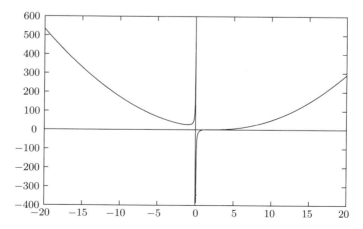

So you can see that the curve looks roughly like $y = x^2$ with some strange stuff going on near $x = 0$, but you can't really make out the details. This really illustrates the difference between "plotting" and "sketching" a graph. After all, the graphing calculator has just plotted enough points to make the curve look smooth, but it doesn't emphasize the interesting features of the graph. You might get a better idea if you zoom in, but then you wouldn't see the behavior for large x. Even though it's inaccurate, our rough sketch above is much more useful for understanding what's really going on, especially as far as turning points and points of inflection are concerned: it shows exactly where all these features are.

CHAPTER 13 _____

Optimization and Linearization

We're now going to look at two practical applications of calculus: optimization and linearization. Believe it or not, these techniques are used every day by engineers, economists, and doctors, for example. Basically, optimization involves finding the best situation possible, whether that be the cheapest way to build a bridge without it falling down or something as mundane as finding the fastest driving route to a specific destination. On the other hand, linearization is a useful technique for finding approximate values of hard-to-calculate quantities. It can also be used to find approximate values of zeroes of functions; this is called Newton's method. In summary, we'll look at

- how to solve optimization problems, and three examples of such problems;
- using linearization and the differential to estimate certain quantities;
- how good our estimates are; and
- Newton's method for estimating zeroes of functions.

13.1 Optimization

To "optimize" something means to make it as good as possible. This being math, we're going for quantity over quality here. Suppose there is a certain quantity we care about. It could be a number, a length, an angle, an area, a cost, an amount of money earned, or one of oodles of other possibilities. If it's a good thing, like amount of money earned, then we'd like to make the quantity as large as possible; if it's a bad thing, like cost, then we'd like to make it as small as possible. In a nutshell, we want to maximize or minimize the quantity. So in our context, the term "optimize" just means "maximize or minimize, as appropriate."

13.1.1 An easy optimization example

In the last few chapters, we've spent quite a lot of time learning how to find maxima and minima of functions. So far as optimization is concerned, normally we would be interested in finding global maxima and minima. In

Section 11.1.3 of Chapter 11, we looked at a nice method for doing this. I urge you to go back and read this section now to refresh your memory.

In any case, to use our method, we need to express the quantity as a function of one other quantity that we can control. For example, suppose that two real numbers add up to 10, but neither number is greater than 8. How large could the product of the two numbers possibly be, and how small could it be?

Before we bust out our method, let's just explore the situation first. If one of the numbers is 8, which is as large as either number can be, then the other number is 2 and the product is 16. At the other extreme, the numbers are both equal to 5 and the product is 25, which is certainly larger than 16. Can we make the product larger than 25 or smaller than 16? How about if the numbers are $4\frac{1}{2}$ and $5\frac{1}{2}$? Try it and see.

Now let's get serious and choose some variables. Suppose that the numbers are x and y, and that their product is P. Well, we know that $P = xy$. The quantity we want to optimize is P, but it's a function of two variables: x and y. This doesn't suit us at all. We really need P to be a function of one variable—it doesn't matter which one. Luckily we have one other piece of information: we know that $x + y = 10$. This means that we can eliminate y by writing $y = 10 - x$. If we do that, then $P = x(10 - x)$. This expresses P as a function of x alone.

One important point, though: what is the domain of P? Sure, you could plug any x into the formula $x(10 - x)$ and get a meaningful answer, but we know something about x that we haven't expressed in math terms yet: it can't be more than 8. Actually, it can't be less than 2 either, or else y would be bigger than 8. So x must lie in the interval $[2, 8]$. We should consider this to be the domain of P.

So we have rewritten our word problem as follows: maximize $P = x(10-x)$ on the domain $[2, 8]$. Not so bad! We just write $P = 10x - x^2$, so we have $dP/dx = 10 - 2x$. This is 0 when $x = 5$, so that's the only critical point. We also could have a maximum or minimum at the endpoints $x = 2$ and $x = 8$. Our list of potential maxima and minima is therefore 2, 5, and 8. When $x = 2$ or $x = 8$, we see that $P = 16$, and when $x = 5$, we have $P = 25$. The conclusion is that the maximum value of the product is indeed 25, and this occurs when both numbers are 5. The minimum value is 16, which occurs when one number is 8 and the other is 2. Notice that when I stated this conclusion, I didn't mention P, x, or y, since those were variables that I introduced. If the variables aren't actually given in the problem, then you not only have to identify them and pick names for them; you also have to write your final conclusion without mentioning them!

It doesn't hurt to verify that $x = 5$ is indeed a maximum by looking at a table of signs* for $P'(x)$, using the formula $P'(x) = 10 - 2x$:

x	4	5	6
$P'(x)$	$+$	0	$-$
	$\nearrow$	$-$	$\searrow$

*See Section 12.1.1 in the previous chapter.

Yup, it's a maximum. We could also verify that $x = 5$ is a maximum by looking at the sign of the second derivative, as described in Section 11.5.2 of Chapter 11. Indeed, $P''(x) = -2$, so $P''(5) = -2$ as well. Since that's negative, we again see that $x = 5$ is a local maximum (which is also a global maximum). Neither of these methods works on the endpoints, though—they only work for critical points.

13.1.2 Optimization problems: the general method

Here's a way to tackle optimization problems in general:

1. Identify all the variables you might possibly need. One of them should be the quantity you want to maximize or minimize—make sure you know which one! Let's call it Q for now, although of course it might be another letter like P, m, or α.

2. Get a feel for the extremes of the situation, seeing how far you can push your variables. (For example, in the problem from the previous section, we saw that x had to be between 2 and 8.)

3. Write down equations relating the variables. One of them should be an equation for Q.

4. Try to make Q a function of only one variable, using all your equations to eliminate the other variables.

5. Differentiate Q with respect to that variable, then find the critical points; remember, these occur where the derivative is 0 or the derivative doesn't exist.

6. Find the values of Q at all the critical points and at the endpoints. Pick out the maximum and minimum values. As a verification, use a table of signs or the sign of the second derivative to classify the critical points.

7. Write out a summary of what you've found, identifying the variables in words rather than symbols (wherever possible).

Actually, sometimes step 4 can be quite difficult, but you might be able to avoid it altogether by using implicit differentiation. We'll see how to do this in Section 13.1.5 below.

13.1.3 An optimization example

Let's see how to apply the method. Suppose that the border of a farm is a long, straight fence, and that the farmer wants to fence off a little enclosure for some horses to graze in. The farmer is a little eccentric and would like to make the enclosure in the shape of a right-angled triangle with the existing fence as one of the sides which is not the hypotenuse, like this:

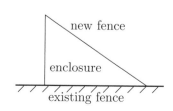

Assuming that only 300 feet of fencing are available, and that the farmer wants the enclosure to have the largest possible area, what are the dimensions and area of the enclosure?

Let's pick some variables. We'll let the base of the triangle be b, the height be h, the hypotenuse be H (all in feet), and the area be A (in square feet), like this:

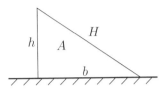

Note that the fence is of length $h + H$, and we want to maximize A. That completes step 1. Moving on to the second step, consider some extreme shapes that you can make out of 300 feet of fencing:

In the first case, h is nearly 0, while b and H are both almost 300, but the area is tiny! In the second case, b is nearly 0, while h and H are both almost 150. The area is still very small. So we can do better by some middle-of-the-road solution. We have at least determined that b and H are between 0 and 300, and that h is between 0 and 150.

Moving on to step 3, we see that $A = \frac{1}{2}bh$ and also that $h + H = 300$. We still need one more equation, since we have to condense the three variables b, h, and H down to one. In fact, we can use Pythagoras' Theorem to say that $b^2 + h^2 = H^2$.

Now we should try to eliminate some variables. We can take square roots and write $H = \sqrt{b^2 + h^2}$, since we know $H > 0$; substituting into $h + H = 300$, we get the equation $h + \sqrt{b^2 + h^2} = 300$. Let's try to eliminate b from this. Subtract h from both sides and square to get

$$b^2 + h^2 = (300 - h)^2 = 90000 - 600h + h^2.$$

This means that $b = \sqrt{90000 - 600h} = 10\sqrt{900 - 6h}$, again since b is positive (that is, it can't be the negative square root!). Finally, the equation $A = \frac{1}{2}bh$ can be rewritten as

$$A = \frac{1}{2} \times 10\sqrt{900 - 6h} \times h = 5h\sqrt{900 - 6h},$$

where h lies in the interval $[0, 150]$. That's step 4. As for step 5, you can use the product rule and the chain rule to see that

$$\frac{dA}{dh} = 5\left(\sqrt{900 - 6h} + h\frac{-6}{2\sqrt{900 - 6h}}\right) = \frac{45(100 - h)}{\sqrt{900 - 6h}}.$$

This equals 0 when $100 - h = 0$, that is, when $h = 100$. Moving on to step 6, we substitute $h = 100$ into the equation for A above, and we get

$$A = 5(100)\sqrt{900 - 6(100)} = 500\sqrt{300} = 5000\sqrt{3}.$$

On the other hand, at the endpoint $h = 0$, we see that $A = 0$; similarly, when $h = 150$, the quantity $900 - 6h$ vanishes, so $A = 0$ once again. The conclusion is that A is maximized when $h = 100$. We can check this with a table of signs. This isn't so bad, since the numerator of dA/dh is just $45(100 - h)$, while the denominator is always positive. The table of signs for dA/dh looks like this:

h	99	100	101
dA/dh	+	0	−
	╱	−	╲

So $h = 100$ is indeed a local maximum, as we suspected.

Now we just have to finish it off. The question asks for the dimensions, and we only have one: $h = 100$. We'd better find b and H. Just look back at the equations: we know that $h + H = 300$, so we immediately get $H = 200$. Also, we know that $b^2 + h^2 = H^2$; plugging in $h = 100$ and $H = 200$, we can see that $b = 100\sqrt{3}$. Finally, we already found that the maximum value of A is $5000\sqrt{3}$. So our concluding sentence could go something like this: the enclosure of maximal area has base $100\sqrt{3}$ feet, height 100 feet, and hypotenuse 200 feet, and the area is then $5000\sqrt{3}$ square feet.

13.1.4 Another optimization example

Here's a nice problem. Suppose that you are manufacturing closed, hollow cylindrical metal cans. You can choose their dimensions, but the volume of a can must be 16π cubic inches. You'd like to use as little metal as possible, since the metal costs 2 cents per square inch. What dimensions should the cans be to make your costs as low as possible, and how much does each can cost in that case?

As a follow-up problem, how does the situation change if we now take into account that the top and bottom of each can have to be welded onto the curved bit, and it costs 14 cents an inch to weld?

Let's start with the first part. Here's a diagram of the situation:

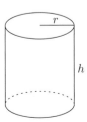

To describe the cylinder, we only need to say what its radius and height are, so let's call them r and h (in inches). We'll also need the volume V (in cubic inches), since the question mentions it. Also, the cost depends on how much metal we use, which is basically the surface area of the cylinder. Let's call a can's surface area A (in square inches) and its cost C (in cents). The quantity C is the one we want to minimize, although it's pretty obvious that it will be minimized if we can also minimize A. (This won't be true for the follow-up question!)

Now, moving on to step 2 of our method, what happens when the radius r is really really small? The height h then has to be large so we can have our volume of 16π cubic inches. We'd get a really tall, skinny cylinder like the first picture below. On the other hand, if r is really large, then h has to be small, and you get a wide, squat cylinder like the second picture:

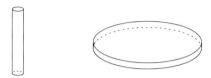

Even though they look pretty extreme, actually they can get weirder. In fact, r can be any positive number at all! So there aren't really endpoints; both r and h have to lie in the open interval $(0, \infty)$ and we'll have to be careful. In either of the above pictures, it looks like there's a whole lot of metal involved, so the low-cost solution probably looks more like the nicely proportioned cylinder above than either of the two extreme ones.

Now it's time for step 3: we have to find some equations. We know $V = 16\pi$; also, since $V = \pi r^2 h$ for a cylinder, we have our first useful equation:

$$16\pi = \pi r^2 h.$$

We can rewrite this as $16 = r^2 h$ or

$$h = \frac{16}{r^2}.$$

On the other hand, the surface area of a closed cylinder is

$$A = 2\pi r h + 2\pi r^2,$$

where the first term in the sum comes from the curved part and the second term is from the top and bottom. (If there were no top, the second term would just be πr^2 without the factor 2.) Finally, the cost is 2 cents times the total area, so we have

$$C = 2A = 4\pi r h + 4\pi r^2.$$

For step 4, notice that both terms on the right-hand side above involve r, so it's easier to get rid of h. Since we saw that $h = 16/r^2$, we can just substitute and get

$$C = 4\pi r \left(\frac{16}{r^2}\right) + 4\pi r^2 = 4\pi \left(\frac{16}{r} + r^2\right).$$

Great—we've expressed C in terms of r, and now the question is to minimize C when r lies in the interval $(0, \infty)$. We have

$$\frac{dC}{dr} = 4\pi \left(-\frac{16}{r^2} + 2r\right),$$

which exists for all r in $(0, \infty)$ and is zero precisely when

$$-\frac{16}{r^2} + 2r = 0,$$

or $2r^3 = 16$. This means that $r^3 = 8$, so $r = 2$ is the only critical point. How about the endpoints? We can't substitute $r = 0$ into the formula for C, but we can take a limit:

$$\lim_{r \to 0^+} C = \lim_{r \to 0^+} 4\pi \left(\frac{16}{r} + r^2 \right) = \infty.$$

The limit is infinite because the $16/r$ term blows up as $r \to 0^+$. This means that as the radius goes down to 0, our costs get larger and larger. This isn't what we want at all! So we'll stay away from that endpoint. How about the other endpoint of our interval $(0, \infty)$? Once again, we can't just set $r = \infty$, so we'll take a limit:

$$\lim_{r \to \infty} C = \lim_{r \to \infty} 4\pi \left(\frac{16}{r} + r^2 \right) = \infty.$$

This time it's the r^2 term that blows up. No matter, we still need to avoid this endpoint. So our conclusion is that $r = 2$ gives a local and global minimum. We can check this by using a table of signs for dC/dr or by looking at the sign of the second derivative. Let's use the second derivative:

$$\frac{d^2 C}{dr^2} = 4\pi \left(\frac{32}{r^3} + 2 \right).$$

This is always positive when r is in the domain $(0, \infty)$; in particular, when $r = 2$, it's positive, so we must have a local minimum there.

All that's left is to find the other variables when $r = 2$ and write up our conclusion. Indeed, when $r = 2$, we can see that $h = 16/r^2 = 4$, and $C = 4\pi rh + 4\pi r^2 = 48\pi$. This means that the cheapest shape occurs when the radius is 2 inches and the height is 4 inches; each can costs 48π cents, which is about \$1.50 (pretty expensive for a lousy can!). Notice that the diameter and the height of the can are the same in this case.

Now let's do the follow-up problem. Everything is the same as it was in the original problem, except that we now have to add on the welding cost of 14 cents per inch, so our formula for C will change. How much welding is there per can? Well, we need to weld on the top and the bottom, so we're dealing with twice the circumference of each of these circles. That means we need to weld twice $2\pi r$, or $4\pi r$, inches per can. This adds a cost of $14 \times 4\pi r$ cents per can, so our new formula for C is

$$C = 4\pi \left(\frac{16}{r} + r^2 \right) + 14 \times 4\pi r = 4\pi \left(\frac{16}{r} + r^2 + 14r \right).$$

(Factoring out that pesky 4π is a good idea.) Anyway, now we differentiate to find that

$$\frac{dC}{dr} = 4\pi \left(-\frac{16}{r^2} + 2r + 14 \right),$$

which equals 0 when

$$-\frac{16}{r^2} + 2r + 14 = 0.$$

To solve this equation, multiply through by r^2, divide by 2, and switch the sign of everything to get

$$r^3 + 7r^2 - 8 = 0.$$

(Make sure you check that this is right!) Great. Now we have to solve a cubic equation. Luckily, something simple works: $r = 1$. So you can do a long division and see that the other factor is $(r^2 + 8r + 8)$ (check this!). So we have

$$(r - 1)(r^2 + 8r + 8) = 0,$$

and either $r = 1$ or $r^2 + 8r + 8 = 0$. The solutions of the quadratic equation are

$$\frac{-8 \pm \sqrt{32}}{2},$$

both of which are negative since $\sqrt{32}$ is only about 6. So the only critical point when r is positive is $r = 1$. Once again, this is a minimum because the costs are infinite at the endpoints (for the same reason as before—the welding certainly doesn't make it cheaper). Alternatively, we have

$$\frac{d^2 C}{dr^2} = 4\pi \left(\frac{32}{r^3} + 2 \right),$$

which is actually the same as it was before. So it's positive, the curve is concave up and we do have a minimum when $r = 1$.

Now we just need to substitute. We find that $h = 16/r^2 = 16$, and $C = 4\pi(16/1 + 1^2 + 14 \times 1) = 124\pi$ cents, which is nearly \$4! Looks like we have to cut costs somehow. In any case, the ideal can now has radius 1 inch and height 16 inches, and it costs 124π cents to make. Notice that the optimal radius is now less than it was in the first part of the question, which makes sense since a smaller radius really cuts down on those expensive welding costs.

13.1.5 Using implicit differentiation in optimization

Before we move on to our final example, let's just take another look at the first part of the previous example. There we knew that

$$C = 4\pi(rh + r^2) \qquad \text{and} \qquad r^2 h = 16,$$

and we minimized C by eliminating the variable h. Another way of doing the minimization is to differentiate both sides implicitly with respect to the variable r, which is the one we wanted to keep anyway. (See Section 8.1 in Chapter 8 for a review of implicit differentiation.) Here's what we get:

$$\frac{dC}{dr} = 4\pi \left(h + r\frac{dh}{dr} + 2r \right) \qquad \text{and} \qquad 2rh + r^2 \frac{dh}{dr} = 0.$$

Check to make sure you agree with this! Anyway, if we solve the second equation for dh/dr, then since $r \neq 0$, we get

$$\frac{dh}{dr} = -\frac{2rh}{r^2} = -\frac{2h}{r}.$$

Put this into the first equation:

$$\frac{dC}{dr} = 4\pi \left(h + r \times -\frac{2h}{r} + 2r \right) = 4\pi(h - 2h + 2r) = 4\pi(2r - h).$$

So $dC/dr = 0$ precisely when $2r = h$, which is what we found before! To see that the critical point here is a minimum, differentiate the above equation with respect to r once more to get

$$\frac{d^2C}{dr^2} = 4\pi \left(2 - \frac{dh}{dr} \right) = 4\pi \left(2 + \frac{2h}{r} \right).$$

(Here we used the fact from above that $dh/dr = -2h/r$.) The main thing to notice is that the right-hand side of the above equation is always positive, so the graph of C against r is concave up and we do have a minimum. Of course, knowing that $2r = h$ at the minimum doesn't tell you what either variable actually is! To find that, substitute into the equation $16 = r^2h$ to get $16 = r^2(2r) = 2r^3$, so $r = 2$ and $h = 4$ as before.

Now, see if you can redo the follow-up part of the question using implicit differentiation and make sure you get the same answer as the one we found above.

13.1.6 A difficult optimization example

Suppose that an oil-drilling platform in the sea is 8 miles due east of a lighthouse on the shore. The backup power generator for the platform is 2 miles due north of the lighthouse. You need to run a cable from the generator to the platform. The water is quite shallow for the first mile east of the lighthouse, but gets much deeper after that. It takes your crew only 1 day per mile to run the cable in the shallow water, but it takes 5 days per mile to run it in the deep water. Show that the quickest way to run the cable is as in the following diagram (in which all measurements are in miles), and find out how long it takes to run the cable in that case:

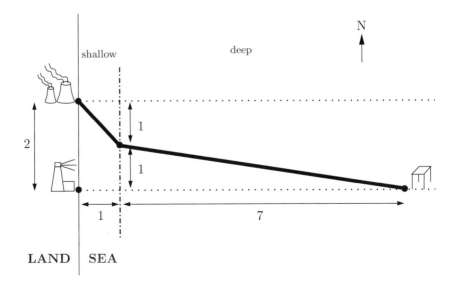

Urk. This seems hard. First, let's note that the diagram is at least somewhat realistic. It would be crazy to position the cable with lots of curves, since that would just add to the length. On the other hand, we need to think carefully about where the cable should hit the interface between the shallow and the deep water. Once we know where that is, it makes a lot of sense to run the cable in straight lines from the generator to the point on the interface, and from the point on the interface to the platform. Once again, it would be crazy to have the point on the interface to the north of the generator or south of the platform—that would have to take longer. Here are some reasonable possibilities:

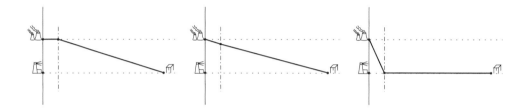

In the first picture, there's quite a lot of cable in the deep water, so it probably won't be great. The second picture shows the least possible amount of total cable, but that doesn't mean it takes the shortest time: there's still a quite a lot of cable in the deep water. The third picture shows a scenario with the least possible amount of cable in the deep water, but this comes at the expense of having a lot of cable in the shallow water. These explorations have confirmed that the quickest solution is probably somewhere between the situations from the second and third pictures, as the problem suggests.

It's time to introduce some variables. Let y, z, s, and t be as shown in the following diagram:

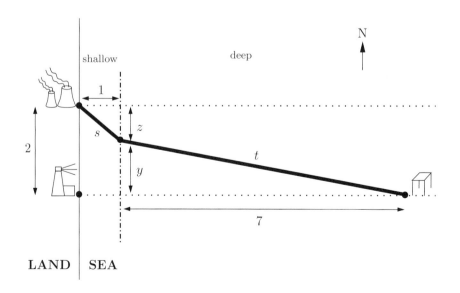

So s is the length of cable in the shallow water and t is the length in the deep water.* Also, y is how far north the interface point is, in miles, from the line joining the lighthouse and the platform, and z makes up the rest of the distance to the east-west line through the generator, so $y + z = 2$. We want to show that the quickest way to lay the cable is when both y and z are equal to 1. We've already seen that y and z should lie in the interval $[0, 2]$, but actually we don't even need to assume this.

We also want to work out the total time taken. Since it takes 1 day per mile for the shallow water, and we have s miles of cable, it takes $1 \times s = s$ days to run the part of the cable in the shallow water. Similarly, it takes 5 days per mile for the deep water, for a total of $5t$ days. Letting T be the total time taken, we see that

$$T = s + 5t.$$

This is the quantity we want to minimize. Now, we need to find equations for s and t. To do this, we use Pythagoras' Theorem twice to get two equations:

$$s^2 = z^2 + 1,$$
$$t^2 = y^2 + 49.$$

Now take the square root of both equations and substitute the results into the equation for T above; you should get

$$T = \sqrt{z^2 + 1} + 5\sqrt{y^2 + 49}.$$

Since $y + z = 2$, we can replace z by $2 - y$ and get

$$T = \sqrt{(2 - y)^2 + 1} + 5\sqrt{y^2 + 49}.$$

 I leave it to you to differentiate this and check that

$$\frac{dT}{dy} = -\frac{2 - y}{\sqrt{(2 - y)^2 + 1}} + \frac{5y}{\sqrt{y^2 + 49}}.$$

We want to show that the shortest time occurs when $y = 1$. Let's substitute that into the above equation and see what we get:

$$\frac{dT}{dy} = -\frac{1}{\sqrt{1^2 + 1}} + \frac{5}{\sqrt{1 + 49}} = -\frac{1}{\sqrt{2}} + \frac{5}{\sqrt{50}} = -\frac{1}{\sqrt{2}} + \frac{5}{5\sqrt{2}} = 0.$$

Hey, $y = 1$ is a critical point after all! So at least there's a hope that it's the global minimum. Unfortunately, we still need to prove this. One way to do this is to take the second derivative. After a lot of grunt-work, you can show that

$$\frac{d^2T}{dy^2} = \frac{1}{((2 - y)^2 + 1)^{3/2}} + \frac{245}{(y^2 + 49)^{3/2}}.$$

So the second derivative is always positive, so the curve is concave up and $y = 1$ is indeed a local minimum. In fact, it must be the only local minimum!

---

*I guess the length of cable in the deep water should be called d, but how weird does dd/dx look? Don't use d as a variable in calculus problems!

Indeed, if there were other critical points, then they would all be local minima as the second derivative is positive. You just can't have lots of local minima without local maxima in between, so there aren't any. This means that $y = 1$ is also the **global** minimum, which is what we want.

We have nearly finished: just substitute $y = 1$ into the equation for T to see that

$$T = \sqrt{(2-1)^2 + 1} + 5\sqrt{1^2 + 49} = \sqrt{2} + 5\sqrt{50} = \sqrt{2} + 25\sqrt{2} = 26\sqrt{2},$$

so it takes $26\sqrt{2}$ days in total (or approximately 36.75 days).

Before we move on to our next topic, let's just look at one other way to see that $y = 1$ is a minimum. The trick is to take the expression

$$\frac{dT}{dy} = -\frac{2-y}{\sqrt{(2-y)^2 + 1}} + \frac{5y}{\sqrt{y^2 + 49}}$$

and rewrite it in a clever way. In the second term on the right, we divide top and bottom by y, while in the first term, we divide by $(2-y)$. Making the reasonable assumption that y and $(2-y)$ are both positive, we can write

$$\frac{dT}{dy} = -\frac{1}{\sqrt{1 + \dfrac{1}{(2-y)^2}}} + \frac{5}{\sqrt{1 + \dfrac{49}{y^2}}}.$$

What happens when y gets bigger? Well, $(2-y)$ gets smaller, as does $(2-y)^2$, so $1/(2-y)^2$ gets bigger. This means that the denominator in the first term gets bigger, so its reciprocal gets smaller, but its negative gets bigger. If you have chased this around properly, you'll have to conclude that when y gets bigger, so does the first term above. In the same way, $49/y^2$ gets smaller, so the denominator of the second term gets smaller, but the term itself gets bigger.

What we've just shown, without too much work, is that dT/dy is an increasing function of y, at least on the interval $(0, 2)$. Since dT/dy is increasing, its derivative d^2T/dy^2 is positive! So we have managed to show that the second derivative is positive without actually having to calculate it, and we conclude that $y = 1$ is a minimum, once again.

13.2 Linearization

Now we're going to use the derivative to estimate certain quantities. For example, suppose you want to get a decent estimate of $\sqrt{11}$ without using a calculator. We know that $\sqrt{11}$ is a little bigger than $\sqrt{9} = 3$, so you could certainly say that $\sqrt{11}$ is approximately 3-and-a-bit. That's OK, but you can actually do a better job without too much work. Here's how it's done.

Start off by setting $f(x) = \sqrt{x}$ for any $x \geq 0$. We want to estimate the value of $f(11) = \sqrt{11}$, since we don't know the actual value. On the other hand, we know exactly what $f(9)$ is—it's just $\sqrt{9} = 3$. Inspired by our knowledge of $f(x)$ when $x = 9$, let's sketch the graph of $y = f(x)$, and draw in the tangent line through the point $(9, 3)$, like this:

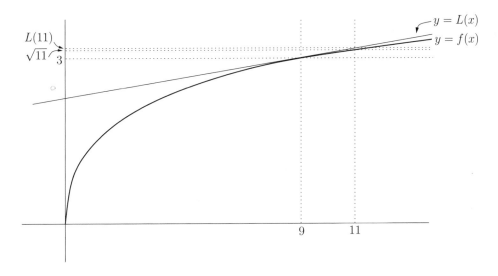

The tangent line, which I've written as $y = L(x)$, is very close to the curve $y = f(x)$ when x is near 9. It's not so close when x is near 0. That's not important, since we want to approximate $f(11)$, and 11 is pretty close to 9. In the above picture, the line and the curve are close to each other at $x = 11$. This means that the value of $L(11)$ is a good approximation to $f(11) = \sqrt{11}$. Indeed, look how close the two values are on the y-axis in the picture above!

All this is irrelevant unless we can actually calculate $L(11)$. So let's do it. The linear function $L(x)$ passes through the point $(9, 3)$, and since it's the tangent to the curve $y = f(x)$ at $x = 9$, the slope of $L(x)$ is exactly $f'(9)$. Now $f'(x) = 1/2\sqrt{x}$, so $f'(9) = 1/2\sqrt{9} = 1/6$. So, $L(x)$ has slope $1/6$ and passes through $(9, 3)$. Its equation is therefore

$$y - 3 = \frac{1}{6}(x - 9),$$

which simplifies to $y = x/6 + 3/2$. That is,

$$L(x) = \frac{x}{6} + \frac{3}{2}.$$

Now all we need to do is calculate $L(11)$ by substituting $x = 11$ into the above equation. We get

$$L(11) = \frac{11}{6} + \frac{3}{2} = \frac{10}{3} = 3\tfrac{1}{3}.$$

We conclude that

$$\sqrt{11} \cong 3\tfrac{1}{3}.$$

That's a lot better than 3-and-a-bit! In fact, you can use a calculator to see that $\sqrt{11}$ is 3.317 (to three decimal places), so the approximation $3\tfrac{1}{3}$ is pretty good.

13.2.1 Linearization in general

Let's generalize the above example. If you want to estimate some quantity, try to write it as $f(x)$ for some nice function f. In the above example, we wanted to estimate $\sqrt{11}$, so we set $f(x) = \sqrt{x}$ and realized that we were interested in the value of $f(11)$.

Next, we pick some number a, close to x, such that $f(a)$ is really nice. In our example, we couldn't deal with $f(11)$, but $f(9)$ was nice because we can take the square root of 9 without any problems. We could have chosen $a = 25$ instead, since we understand $\sqrt{25}$, but this isn't as good because 25 is quite far away from 11.

So, given our function f and our special number a, we find the tangent to the curve $y = f(x)$ at the point $(a, f(a))$. This tangent has slope $f'(a)$, so its equation is

$$y - f(a) = f'(a)(x - a).$$

If the tangent line is $y = L(x)$, then by adding $f(a)$ to both sides in the above equation, we get

$$L(x) = f(a) + f'(a)(x - a).$$

The linear function L is called the *linearization* of f at $x = a$. Remember, we're going to use $L(x)$ as an approximation to $f(x)$. So we have

$$f(x) \cong L(x) = f(a) + f'(a)(x - a),$$

with the understanding that the approximation is very good when x is close to a. In fact, if x actually equals a, the approximation is perfect! Both sides of the above equation become $f(a)$. This isn't helpful, though, since we already understand $f(a)$. The benefit is that we now have an approximation for $f(x)$ for x **near** a.

Let's check that our formula works for the example in the previous section. We have $f(x) = \sqrt{x}$ and $a = 9$. Clearly $f(a) = f(9) = 3$; and since $f'(x) = 1/2\sqrt{x}$, we have $f'(9) = 1/2\sqrt{9} = 1/6$. According to the formula, the linearization is given by

$$L(x) = f(a) + f'(a)(x - a) = 3 + \frac{1}{6}(x - 9).$$

This agrees with our formula $L(x) = x/6 + 3/2$ from above, which we used to find the estimate $\sqrt{11} \cong 3\frac{1}{3}$. Now, how would you estimate $\sqrt{8}$? We see that 8 is also close to 9, so we can just use the same linearization:

$$\sqrt{8} = f(8) \cong L(8) = 3 + \frac{1}{6}(8 - 9) = \frac{17}{6}.$$

So the formula $L(x) = 3 + (x - 9)/6$ gives a good approximation to $\sqrt{x}$ for **any** x near 9, not just 11.

On the other hand, suppose you also want to estimate $\sqrt{62}$. It wouldn't be ideal to use $L(62)$ as an approximation. Let's see what happens if we do:

$$L(62) = 3 + \frac{62 - 9}{6} = 11\tfrac{5}{6}.$$

Wait a second, $\sqrt{62}$ should be a little less than $\sqrt{64}$, which is 8. The value of $L(62)$, which is $11\frac{5}{6}$, is way too high. The problem is that our linearization is at $x = 9$, while 62 is a long way from 9; so the approximation isn't very good. To estimate $\sqrt{62}$, you're much better off using the linearization at $x = 64$

instead. So, set $a = 64$; we now have $f(a) = 8$ and $f'(a) = 1/2\sqrt{64} = 1/16$. This means that our new linearization is given by

$$L(x) = f(a) + f'(a)(x - a) = 8 + \frac{1}{16}(x - 64).$$

When $x = 62$, we have

$$\sqrt{62} = f(62) \cong L(62) = 8 + \frac{1}{16}(62 - 64) = 7\tfrac{7}{8}.$$

This approximation makes a lot more sense than $11\tfrac{5}{6}$ does!

13.2.2 The differential

Let's take a look at the general situation once more. We saw that

$$f(x) \cong f(a) + f'(a)(x - a).$$

Let's define Δx to be $x - a$, so that $x = a + \Delta x$. The above formula becomes

$$\boxed{f(a + \Delta x) \cong f(a) + f'(a)\Delta x.}$$

Here's a graph of the situation:

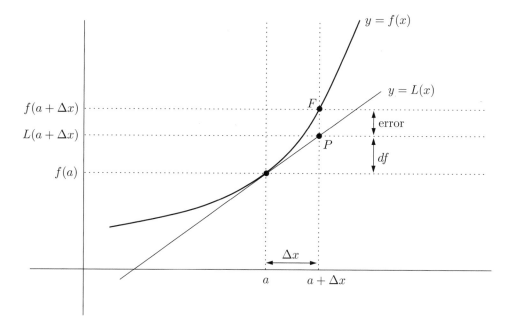

The graph shows the curve $y = f(x)$ and the linearization $y = L(x)$, which is the tangent line to the curve at $x = a$. We want to estimate the value of $f(a + \Delta x)$. That's the height of the point F in the above picture. As an approximate value, we're actually using $L(a + \Delta x)$, which is the height of P in the picture. The difference between the two quantities is labeled "error"; we'll come back to this in Section 13.2.4 below.

In the above graph, there's one more quantity marked: this is df, which is the difference between the height of P and $f(a)$. It is the amount we needed to add to $f(a)$ in order to get our estimate. Since $L(a+\Delta x) = f(a)+f'(a)\Delta x$, we see that

$$df = f'(a)\Delta x.$$

The quantity df is called the *differential of f at $x = a$*. It is an approximation to the amount that f changes when x moves from a to $a + \Delta x$.

We've actually touched on these ideas before. In Section 5.2.7 of Chapter 5, we saw that if $y = f(x)$, then

$$f'(x) = \lim_{\Delta x \to 0} \frac{\Delta y}{\Delta x}.$$

This means that a small change in x produces approximately $f'(x)$ times the change in y. This is exactly what the equation $df = f'(a)\Delta x$ says, taking into account that this time we are starting at $x = a$.

For example, suppose we want to estimate $(6.01)^2$. Set $f(x) = x^2$ and $a = 6$; then you can easily see that $f'(x) = 2x$, so that $f(6) = 12$. We want to know what happens when we shift x from 6 over by the amount 0.01; so we should set $\Delta x = 0.01$. We have

$$df = f'(a)\Delta x = f'(6)(0.01) = 12(0.01) = 0.12.$$

So if we add 0.12 to the value of $f(a)$, we should get a good approximation. Since $f(a) = f(6) = 6^2 = 36$, this means that $(6.01)^2 \cong 36.12$. Now look at back at Section 5.2.7 in Chapter 5 again: we solved the same example there, using basically the same method—we just have some nicer formulas now, that's all!

Here's another example of how to use the differential. Suppose that you use a ruler to measure the diameter of a round ball and get 6 inches, but this measurement is only accurate to 0.5%. If we use our measurement to calculate the volume of the ball, how accurate is our result? Let's use the differential to work this out, at least approximately. If the ball has radius r, diameter D, and volume V, then $r = D/2$, so

$$V = \frac{4}{3}\pi r^3 = \frac{4}{3}\pi \left(\frac{D}{2}\right)^3 = \frac{\pi D^3}{6}.$$

When $D = 6$, we have $V = \pi(6)^3/6 = 36\pi$. So we've calculated the volume to be 36π cubic inches, but the true answer might be a little more or a little less. To find out how much more or less, let's use the above boxed formula, $df = f'(a)\Delta x$. We need to change f to V, a to 6, and x to D to get the appropriate formula for this case:

$$dV = V'(6)\Delta D.$$

Differentiating the previous formula for V with respect to D, we find that

$$V'(D) = \frac{\pi(3D^2)}{6} = \frac{\pi D^2}{2}.$$

This means that $V'(6) = 18\pi$, so

$$dV = 18\pi\Delta D.$$

This equation means that if you change the diameter D from 6 to $6 + \Delta D$, the volume V changes by about $18\pi\Delta D$. In our case, the true diameter might be 0.5% more or less than 6 inches, which is $0.005 \times 6 = 0.03$ inches. So ΔD might be as high as 0.03 or as low as -0.03; in this worst case scenario, we have

$$dV = 18\pi \times (\pm 0.03) = \pm 0.54\pi.$$

This is a good approximation to the true error in the measurement, so we can say that the volume of the ball is 36π cubic inches, accurate to about 0.54π cubic inches. Since the original error in the diameter was expressed as a percentage of the diameter, we should probably do the same for the volume. In percentage terms, an approximate error of $dV = \pm 0.54\pi$ on a quantity $V = 36\pi$ is

$$\frac{dV}{V} \times 100\% = \frac{\pm 0.54\pi}{36\pi} \times 100\% = \pm 1.5\%.$$

In other words, the relative (percentage) error in the volume measurement is about three times the relative error in the original diameter measurement. That's what happens when you compound the error in a one-dimensional measurement in the calculation of a three-dimensional quantity.

13.2.3 Linearization summary and examples

Here's the basic strategy for estimating, or approximating, a nasty number:

1. Write down the main formula

$$\boxed{f(x) \cong L(x) = f(a) + f'(a)(x - a).}$$

2. Choose a function f, and a number x such that the nasty number is equal to $f(x)$. Also, choose a close to x such that $f(a)$ can easily be computed.

3. Differentiate f to find f'.

4. In the above formula, replace f and f' by the actual functions, and a by the actual number you've chosen.

5. Finally, plug in the value of x from step 2 above. Also note that the differential df is the quantity $f'(a)(x - a)$.

Let's look at a few examples. First, how would you estimate $\sin(11\pi/30)$? Start off with the standard formula

$$f(x) \cong L(x) = f(a) + f'(a)(x - a).$$

We need to take the sine of something, so let's set $f(x) = \sin(x)$. We are interested in what happens when $x = 11\pi/30$. Now, we need some number a which is close to $11\pi/30$, such that $f(a)$ is nice. Of course, $f(a)$ is just $\sin(a)$. What number close to $11\pi/30$ has a manageable sine? How about $10\pi/30$? After all, that's just $\pi/3$, and we certainly understand $\sin(\pi/3)$. So, set $a = \pi/3$.

We've completed the first two steps. Moving on to the third step, we find that $f'(x) = \cos(x)$, so the linearization formula becomes

$$f(x) \cong L(x) = \sin\left(\frac{\pi}{3}\right) + \cos\left(\frac{\pi}{3}\right)\left(x - \frac{\pi}{3}\right).$$

Since $f(x) = \sin(x)$, this simplifies to

$$\sin(x) \cong L(x) = \frac{\sqrt{3}}{2} + \frac{1}{2}\left(x - \frac{\pi}{3}\right).$$

Finally, put $x = 11\pi/30$ to get

$$\sin\left(\frac{11\pi}{30}\right) \cong L\left(\frac{11\pi}{30}\right) = \frac{\sqrt{3}}{2} + \frac{1}{2}\left(\frac{11\pi}{30} - \frac{\pi}{3}\right) = \frac{\sqrt{3}}{2} + \frac{\pi}{60}.$$

This may still seem bad, but at least the estimate doesn't involve any trig functions—only the numbers π and $\sqrt{3}$, which are not too hard to deal with.

Now, consider this example: find an approximation for $\ln(0.99)$ using a linearization. Well, this time we set $f(x) = \ln(x)$ and note that we are interested in what happens when $x = 0.99$. A number near 0.99 that is nice, so far as taking the log of it is concerned, is 1; so we set $a = 1$. Since $f(x) = \ln(x)$ and $f'(x) = 1/x$, the formula $f(x) \cong L(x) = f(a) + f'(a)(x-a)$ becomes

$$\ln(x) \cong L(x) = \ln(1) + \frac{1}{1}(x-1).$$

Since $\ln(1) = 0$, we have shown that

$$\ln(x) \cong x - 1.$$

Replacing x by 0.99, we get

$$\ln(0.99) \cong L(0.99) = 0.99 - 1 = -0.01,$$

and we're done.

More generally, how would you find an approximation for $\ln(1+h)$, where h is **any** small number? In fact, you can use the linearization that we just found, $f(x) \cong L(x) = x - 1$, to approximate $\ln(1+h)$. Just replace x by $1+h$ and we see that $\ln(1+h) \cong L(1+h) = (1+h) - 1$. That is,

$$\ln(1+h) \cong h$$

when h is small. Actually, this shouldn't be a surprise! In Section 9.4.3 of Chapter 9, we saw that

$$\lim_{h \to 0} \frac{\ln(1+h)}{h} = 1,$$

so we already knew that $\ln(1+h)$ is approximately equal to h when h is small. Finally, how about an approximation for $\ln(e+h)$ when h is small? We now need a different linearization, as the quantity $(e+h)$ is close to e, not

1. So let's set $a = e$ and start again, once again using $f(x) = \ln(x)$ and $f'(x) = 1/x$. We get

$$f(x) \cong L(x) = f(a) + f'(a)(x - a) = \ln(e) + \frac{1}{e}(x - e).$$

Since $\ln(e) = 1$, we get

$$\ln(x) \cong L(x) = 1 + \frac{x}{e} - 1 = \frac{x}{e}.$$

When $x = e + h$, we have

$$\ln(e + h) \cong L(e + h) = \frac{e + h}{e} = 1 + \frac{h}{e}.$$

That is, $\ln(e + h) \cong 1 + h/e$ when h is small. This answer is quite different from the answer in previous example, where we saw that $\ln(1 + h) \cong h$ for small h. Everything depends on the value of a.

13.2.4 The error in our approximation

We've been using $L(x)$ as an approximation for $f(x)$. They are not the same thing, though. The question is, how wrong could we be to use $L(x)$ instead of $f(x)$? The way to find out is to consider the difference between the two quantities. The smaller that distance, the better the approximation. So, set

$$r(x) = f(x) - L(x),$$

where $r(x)$ is the error[*] in using the linearization at $x = a$ in order to estimate $f(x)$. It turns out that if the second derivative of f exists, at least between x and a, then there's a nice formula[†] for $r(x)$:

$$r(x) = \frac{1}{2}f''(c)(x - a)^2 \qquad \text{for some number } c \text{ between } x \text{ and } a.$$

The problem is, we don't know what c is, only that it's between x and a. The above formula is related to the Mean Value Theorem, which we looked at in Section 11.3 of Chapter 11. Since that theorem tells you about a number c without telling you much about it, we shouldn't be surprised to see it popping up here.

We can use the above formula to tell us two things. First, note that the quantity $(x - a)^2$ is always positive. This means that the sign of $r(x)$ is the same as the sign of $f''(c)$. So if we know that the curve is concave up, at least between a and x, then $r(x)$ is positive. Since $r(x) = f(x) - L(x)$, we see that $f(x) > L(x)$. This means that our estimate $L(x)$ is lower than $f(x)$, so we have made an underestimate. This situation is shown in the graph in Section 13.2.2 above. On the other hand, if the curve is concave down, then

---

[*]The letter r in $r(x)$ stands for "remainder," since it's what's left when you remove the linearization.

[†]See Section A.6.9 of Appendix A for a proof.

$f''(c)$ must be negative; so you can chase it around and see that $L(x) > f(x)$. This means that our approximation is an overestimate.

For example, when we estimated $\sqrt{11}$ at the beginning of Section 13.2 above, we used $f(x) = \sqrt{x}$. If you calculate that $f'(x) = 1/2\sqrt{x}$ and that $f''(x) = -1/4x\sqrt{x}$, you can see that the curve is always concave down. Or you can just see it from the graph. In any case, we see that the estimate that we found $(3\frac{1}{3})$ must be an overestimate.

In summary,

- if f'' is **positive** between a and x, then using the linearization leads to an **underestimate**;
- if f'' is **negative** between a and x, then using the linearization leads to an **overestimate**.

Now look back at the equation for the error $r(x)$ above. If we take absolute values of both sides of the equation, then we get

$$|\text{error}| = \frac{1}{2}|f''(c)||x - a|^2.$$

Suppose we know that the biggest $|f''(t)|$ could be, as t ranges between a and x, is some number M. Then even though we don't know what c is, we do know that $|f''(c)| \leq M$, so we get the following formula:

$$\boxed{|\text{error}| \leq \frac{1}{2}M|x - a|^2.}$$

Again, M is the largest value of $|f''(t)|$ for all t between x and a. Actually, the important factor in the above equation isn't the M; it's the $|x - a|^2$ factor. You see, when x is close to a, the quantity $|x - a|$ is small, but when you square it, it becomes tiny. (For example, when you square 0.01, you get the tiny number 0.0001.) This means that the error is small, so our approximation is good!

Let's see how this applies to our above example of estimating $\sqrt{11}$. We set $f(x) = \sqrt{x}$, $f'(x) = 1/2\sqrt{x}$ and $f''(x) = -1/4x\sqrt{x}$. We also took $a = 9$ and $x = 11$. The question is, how big could the value of $|f''(t)|$ be for t between 9 and 11? Clearly

$$|f''(t)| = \frac{1}{4t\sqrt{t}}.$$

The right-hand side is a decreasing function of t, so it's biggest when t is smallest, that is, when $t = 9$. So $M = |f''(9)|$, which turns out to be $1/108$. The conclusion is that

$$|\text{error}| \leq \frac{1}{2}M|x - a|^2 = \frac{1}{2}\frac{1}{108}|11 - 9|^2 = \frac{1}{54}.$$

So when we said earlier that $\sqrt{11} \cong 3\frac{1}{3}$, now we have confidence that we're pretty close. In fact, we are within $\pm 1/54$ of the correct answer. More precisely, we actually know that

$$3\tfrac{1}{3} - \tfrac{1}{54} \leq \sqrt{11} \leq 3\tfrac{1}{3} + \tfrac{1}{54}.$$

In fact, since we discovered earlier that $3\tfrac{1}{3}$ is an overestimate for $\sqrt{11}$, we can say more:

$$3\tfrac{1}{3} - \tfrac{1}{54} \leq \sqrt{11} \leq 3\tfrac{1}{3}.$$

Now, let's repeat this for the example of estimating $\ln(0.99)$, which we looked at in Section 13.2.3 above. There we saw that $\ln(0.99) \cong -0.01$. How good is this approximation? With $f(x) = \ln(x)$, we have $f'(x) = 1/x$ and $f''(x) = -1/x^2$. Since the second derivative is negative, we again have an overestimate. Now, when t ranges between $a = 1$ and $x = 0.99$, how big could $|f''(t)| = 1/t^2$ be? Again, this is decreasing in t, so the biggest value occurs when $t = 0.99$. So we have $M = 1/(0.99)^2$, and our error estimate looks like this:

$$|\text{error}| \leq \frac{1}{2}M|x - a|^2 = \frac{1}{2}\frac{1}{0.99^2}|0.99 - 1|^2 = \frac{1}{20000(0.99)^2}.$$

This simplifies to about 0.000051, which is really tiny. This means that -0.01 is a very good approximation to $\ln(0.99)$. More precisely, we've proved the inequalities

$$-0.01 - \frac{1}{20000(0.99)^2} \leq \ln(0.99) \leq -0.01 + \frac{1}{20000(0.99)^2}.$$

In fact, since -0.01 is an overestimate, we can once again tighten up the right-hand side and write that

$$-0.01 - \frac{1}{20000(0.99)^2} \leq \ln(0.99) \leq -0.01.$$

We've narrowed down the value of $\ln(0.99)$ to lie in a really tiny interval.

We're going to return to the topic of finding approximations and estimating errors when we look at Taylor series in Chapter 24. There we'll use not only the first derivative, but the second and higher derivatives to get even better approximations.

13.3 Newton's Method

Here's another useful application of linearization. Suppose that you have an equation of the form $f(x) = 0$ that you'd like to solve, but you just can't solve the darned thing. So you do the next best thing: you take a guess at a solution, which you call a. The situation might look something like this:

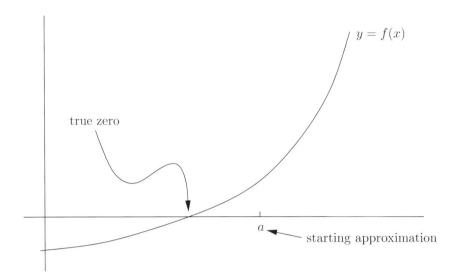

As you can see, $f(a)$ isn't actually equal to 0, so a isn't really a solution; it's just an approximation, or an estimate, of the solution. Think of it as a first stab at an approximation, which is why it's labeled "starting approximation" in the picture above. Now, the idea of Newton's method is that you can (hopefully) improve upon your estimate by using the linearization of f about $x = a$. (This means that f needs to be differentiable at $x = a$, of course!) Anyway, let's see what this looks like:

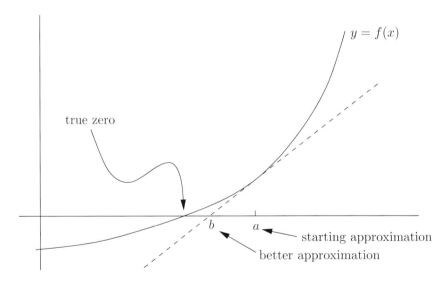

The x-intercept of the linearization is labeled b, and it's clearly a better approximation to the true zero than a is. Starting with one guess, we've gotten a better one. So what is the value of b? Well, it's just the x-intercept of the linearization L, which is given by

$$L(x) = f(a) + f'(a)(x - a),$$

as in Section 13.2.1 above. To find the x-intercept, set $L(x) = 0$; then we have $f(a) + f'(a)(x - a) = 0$. Solving for x, we get

$$x = a - \frac{f(a)}{f'(a)}.$$

Since we called the x-intercept b, we have found the following formula:

Newton's method: suppose that a is an approximation to a solution of $f(x) = 0$. If you set

$$b = a - \frac{f(a)}{f'(a)},$$

then a lot of the time b is a better approximation than a.

It doesn't work all the time, so I put in the phrase "a lot of the time" to cover my ass. We'll come back to this detail on the next page. First, let's look at some examples. Suppose that

$$f(x) = x^5 + 2x - 1$$

and we'd like to find a solution to the equation $f(x) = 0$. Does it even have one? Since f is continuous, $f(0) = -1$ (negative), and $f(1) = 2$ (positive), the Intermediate Value Theorem (see Section 5.1.4 in Chapter 5) shows that f has at least one solution. On the other hand, $f'(x) = 5x^4 + 2$, which is always positive; so f is always increasing, which means that the equation $f(x) = 0$ has at most one solution. (See Section 10 in Chapter 10.1.1 to remind yourself about this.) We have shown that f has exactly one solution. Let's approximate the solution as 0. We know that $f(0) = -1$, which isn't very close to 0. No problem, just use Newton's method with $a = 0$:

$$b = a - \frac{f(a)}{f'(a)} = 0 - \frac{f(0)}{f'(0)} = 0 - \frac{0^5 + 2(0) - 1}{5(0)^4 + 2} = \frac{1}{2}.$$

So $b = 1/2$ should be a better approximation than 0. Indeed, you can calculate that $f(1/2) = 1/32$, which is quite close to 0. What's to stop us repeating the method and getting an even better solution? Nothing! Let's now take $a = 1/2$ instead, and repeat:

$$b = a - \frac{f(a)}{f'(a)} = \frac{1}{2} - \frac{f(1/2)}{f'(1/2)} = \frac{1}{2} - \frac{1/32}{37/16} = \frac{18}{37}.$$

(Here we used the calculation $f'(1/2) = 5(1/2)^4 + 2 = 37/16$.) Anyway, this means that $18/37$ should be an even better approximation to the true zero of f. If you calculate $f(18/37)$, you'll get something close to 0.0002, which is pretty darned small. The number $18/37$ is really a pretty good approximation to the true zero of f.

It might seem confusing to reuse a and b like this. A way around it is to use x_0 as the initial guess and x_1 as the first improvement; then x_2 is the second improvement, starting with x_1; and so on. The formula can now be written like this:

$$x_1 = x_0 - \frac{f(x_0)}{f'(x_0)}, \quad x_2 = x_1 - \frac{f(x_1)}{f'(x_1)}, \quad x_3 = x_2 - \frac{f(x_2)}{f'(x_2)}, \quad \text{and so on.}$$

Here's another example. To find an approximate solution of the equation $x = \cos(x)$, first set $f(x) = x - \cos(x)$. If we can estimate the zero of f, then the same number will be an approximate solution of $x = \cos(x)$. (We already used this trick in Section 5.1.4 of Chapter 5.) Let's make the guess $x_0 = \pi/2$; then $f(\pi/2) = \pi/2 - \cos(\pi/2) = \pi/2$. That's a pretty lousy guess. Never mind; since $f'(x) = 1 + \sin(x)$, we have $f'(\pi/2) = 1 + \sin(\pi/2) = 2$. This means that

$$x_1 = x_0 - \frac{f(x_0)}{f'(x_0)} = \frac{\pi}{2} - \frac{\pi/2}{2} = \frac{\pi}{4}.$$

So $x_1 = \pi/4$ is a better approximation; indeed, $f(\pi/4)$ works out to be the quantity $\pi/4 - 1/\sqrt{2}$, which is about 0.08. Now repeat:

$$x_2 = x_1 - \frac{f(x_1)}{f'(x_1)} = \frac{\pi}{4} - \frac{f(\pi/4)}{f'(\pi/4)} = \frac{\pi}{4} - \frac{\pi/4 - 1/\sqrt{2}}{1 + 1/\sqrt{2}},$$

since $f'(\pi/4) = 1 + \sin(\pi/4) = 1 + 1/\sqrt{2}$. The above equation simplifies to

$$x_2 = \frac{1 + \pi/4}{1 + \sqrt{2}} = (1 + \pi/4)(\sqrt{2} - 1),$$

which is actually a little less than $\pi/4$. Also, $f(x_2)$ turns out to be about 0.0008. This means that $x - \cos(x_2)$ is about 0.0008, so the number x_2 above is a pretty good approximation to the solution of the equation $x = \cos(x)$. Of course, we could repeat the method to find an even better approximation x_3, but the calculations become horrible. Computers and calculators are very good at it, though, and in fact often use Newton's method to give good approximations. (Remember, a calculator **only** gives approximations! Even 10 or 12 decimal places is still not exact, although it's close enough in most situations.)

As we noted before (but never explained), sometimes Newton's method doesn't work. Here are four different things that could go wrong:

1. **The value of $f'(a)$ could be near 0.** Clearly, if

$$b = a - \frac{f(a)}{f'(a)},$$

then $f'(a)$ can't be 0 or else b isn't even defined. In that case, the tangent line at $x = a$ doesn't even intersect the x-axis, since it's horizontal! Even if $f'(a)$ is close, but not equal to 0, Newton's method can still give a whacked-out result; for example, check out this picture:

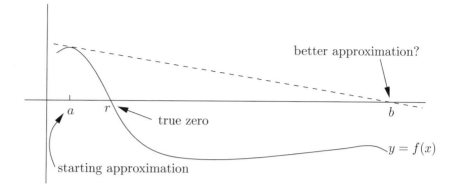

Even though we started with a pretty good approximation a of the actual zero r, the result of Newton's method (b) is really far away from r. So we didn't get a better approximation after all. To get around this, make sure that your initial approximation is not near a critical point of your function f.

2. **If $f(x) = 0$ has more than one solution, you might not get the right one.** For example, in the following picture, if you are trying to estimate the left-hand root r, and you guess to start at a, you'll end up estimating s instead:

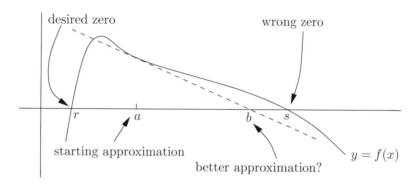

So you should make some effort to start with an estimate a which is close to the zero you want, unless you're sure there's only one zero!

3. **The approximations might get worse and worse.** For example, if $f(x) = x^{1/3}$, the only solution to the equation $f(x) = 0$ is $x = 0$. If you try to use Newton's method (for reasons best known to yourself, I guess!), then something weird happens. You see, unless you start with $a = 0$, this is what you find:

$$b = a - \frac{f(a)}{f'(a)} = a - \frac{a^{1/3}}{a^{-2/3}/3} = -2a.$$

So the next approximation is always -2 times the one you started with. For example, if you start with $a = 1$, then the next approximation will be -2. If you keep on repeating the process, you'll get 4, then -8, then

16, and so on. These are just getting farther and farther away from the correct value 0. There's not much you can do with Newton's method if this sort of thing happens.

4. **You might get stuck in a loop.** It's possible that your estimate a leads to another estimate b, which then leads back to a again. This means that there's no point in repeating the process, as you just keep going around in circles! Here's how the situation might look:

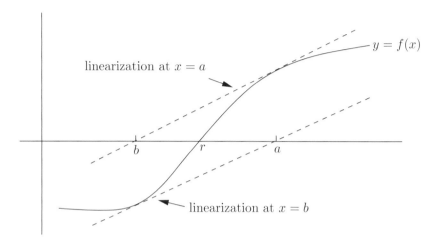

The linearization at $x = a$ has x-intercept b, and the linearization at $x = b$ has x-intercept a, so Newton's method just doesn't work. A concrete (but messy) example is

$$f(x) = \left(x^2 - \frac{4 + 3\pi}{4 - \pi}\right)\tan^{-1}(x).$$

If you start with $a = 1$, I leave it to you to show that $b = -1$. Since f is an odd function, it's now clear that restarting with -1 leads to 1 again. It's pretty unlucky to encounter a loop! Just try some other starting guess. (By the way, the study of these sorts of loops leads to a nice type of fractal that you might have seen as a screensaver on someone's computer....)

CHAPTER 14 _____

L'Hôpital's Rule and Overview of Limits

We've used limits to find derivatives. Now we'll turn things upside-down and use derivatives to find limits, by way of a nice technique called l'Hôpital's Rule. After looking at various varieties of the rule, we'll give a summary, followed by an overview of all the methods we've used so far to evaluate limits. So, we'll look at:

- l'Hôpital's Rule, and four types of limits which naturally lead to using the rule; and
- a summary of limit techniques from earlier chapters.

14.1 L'Hôpital's Rule

Most of the limits we've looked at are naturally in one of the following forms:

$$\lim_{x \to a} \frac{f(x)}{g(x)}, \qquad \lim_{x \to a} (f(x) - g(x)), \qquad \lim_{x \to a} f(x)g(x), \qquad \text{and} \qquad \lim_{x \to a} f(x)^{g(x)}.$$

Sometimes you can just substitute $x = a$ and evaluate the limit directly, effectively using the continuity of f and g. This method doesn't always work, though—for example, consider the limits

$$\lim_{x \to 3} \frac{x^2 - 9}{x - 3}, \quad \lim_{x \to 0} \left(\frac{1}{\sin(x)} - \frac{1}{x} \right), \quad \lim_{x \to 0^+} x \ln(x), \quad \text{and} \quad \lim_{x \to 0} (1 + 3 \tan(x))^{1/x}.$$

In the first case, replacing x by 3 gives the indeterminate form $0/0$. The second limit involves the difference between two terms which become infinite as $x \to 0$. Actually, they both go to ∞ as $x \to 0^+$ and $-\infty$ as $x \to 0^-$, so you can think of the form in this case as $\pm(\infty - \infty)$. As for the third limit above (involving $x \ln(x)$), this leads to the form $0 \times (-\infty)$, remembering that $\ln(x) \to -\infty$ as $x \to 0^+$. Finally, the fourth limit looks like 1^∞, which is also problematic. Luckily, all four types can often be solved using l'Hôpital's Rule.

It turns out that the first type, involving the ratio $f(x)/g(x)$, is the most suitable for applying the rule, so we'll call it "Type **A**." The next two types, involving $f(x) - g(x)$ and $f(x)g(x)$, both reduce directly to Type **A**, so we'll call them Type **B1** and Type **B2**, respectively. Finally, we'll say that limits involving exponentials like $f(x)^{g(x)}$ are Type **C**, since you can solve them by reducing them to Type **B2** and then back to Type **A**. Let's look at all these types individually, then summarize the whole situation in Section 14.1.6 below.

14.1.1 Type **A**: 0/0 case

Consider limits of the form

$$\lim_{x \to a} \frac{f(x)}{g(x)},$$

where f and g are nice differentiable functions. If $g(a) \neq 0$, everything's great—you just substitute $x = a$ to see that the limit is $f(a)/g(a)$. If $g(a) = 0$ but $f(a) \neq 0$, then you're dealing with a vertical asymptote at $x = a$ and the above limit is either ∞, $-\infty$ or it doesn't exist. (See page 59 for graphs of the four possibilities that can arise in this case.)

The only other possibility is that $f(a) = 0$ and $g(a) = 0$. That is, the fraction $f(a)/g(a)$ is the indeterminate form $0/0$. The majority of the limits we've seen have been of this form. In fact, every derivative is of this form! After all,

$$f'(x) = \lim_{h \to 0} \frac{f(x + h) - f(x)}{h},$$

and if you put $h = 0$ in the fraction, you get $0/0$. So let's concentrate on the case where $f(a) = 0$ and $g(a) = 0$.

Here's the basic idea. Since f and g are differentiable, we can find the linearization of both of them at $x = a$. In fact, as we saw in the previous chapter, if x is close to a, then

$$f(x) \cong f(a) + f'(a)(x - a) \qquad \text{and} \qquad g(x) \cong g(a) + g'(a)(x - a).$$

Now, we're assuming that $f(a)$ and $g(a)$ are both zero. This means that

$$f(x) \cong f'(a)(x - a) \qquad \text{and} \qquad g(x) \cong g'(a)(x - a).$$

If you divide the first equation by the second one, then assuming that $x \neq a$, we get

$$\frac{f(x)}{g(x)} \cong \frac{f'(a)(x - a)}{g'(a)(x - a)} = \frac{f'(a)}{g'(a)}.$$

The closer x is to a, the better the approximation. This leads* us to one version of l'Hôpital's Rule:

$$\boxed{\text{if } f(a) = g(a) = 0, \text{ then} \qquad \lim_{x \to a} \frac{f(x)}{g(x)} = \lim_{x \to a} \frac{f'(x)}{g'(x)}}$$

provided that the limit on the right-hand side exists. (Actually, there's another condition as well: $g'(x)$ can't be 0 when x is close to, but not equal

---

*We haven't actually proved l'Hôpital's Rule here; see Section A.6.11 Appendix A for a real proof.

to, a. You have to be really unlucky for this to be a problem, though!) It's really important that $f(a)$ and $g(a)$ are both zero, or else everything could get screwed up.

Let's apply the rule to an example from the beginning of the chapter:

$$\lim_{x \to 3} \frac{x^2 - 9}{x - 3}.$$

Notice that if you put $x = 3$, then both top and bottom of the fraction are 0. This means we can use l'Hôpital's Rule. All you have to do is differentiate the top and bottom separately—don't use the quotient rule! The solution looks like this:

$$\lim_{x \to 3} \frac{x^2 - 9}{x - 3} \overset{\text{l'H}}{=} \lim_{x \to 3} \frac{2x}{1} = 6.$$

Notice how there's a little "l'H" above the equal sign to show that we're using l'Hôpital's Rule. By the way, you don't need to use l'Hôpital's Rule here—you can just factor $x^2 - 9$ as $(x - 3)(x + 3)$, like this:

$$\lim_{x \to 3} \frac{x^2 - 9}{x - 3} = \lim_{x \to 3} \frac{(x - 3)(x + 3)}{x - 3} = \lim_{x \to 3} (x + 3) = 3 + 3 = 6.$$

Hey, we got the same answer! That's a relief.

Here's a harder example where the factoring trick doesn't work:

$$\lim_{x \to 0} \frac{x - \sin(x)}{x^3}.$$

If you put $x = 0$, then both top and bottom are 0. The principle that $\sin(x)$ behaves like x for small x is useless in this case, since we're taking the difference of the two quantities. So let's apply l'Hôpital's Rule, differentiating $x - \sin(x)$ and x^3 separately:

$$\lim_{x \to 0} \frac{x - \sin(x)}{x^3} \overset{\text{l'H}}{=} \lim_{x \to 0} \frac{1 - \cos(x)}{3x^2}.$$

We actually saw how to solve the right-hand limit (without the 3 on the bottom) in Section 7.1.2 of Chapter 7. There we used the trick of multiplying top and bottom by $1 + \cos(x)$. There's an easier way: just notice that the right-hand limit is also of the form 0/0 when you replace x by 0 (since $\cos(0) = 1$), so we can use l'Hôpital's Rule again! We get

$$\lim_{x \to 0} \frac{x - \sin(x)}{x^3} \overset{\text{l'H}}{=} \lim_{x \to 0} \frac{1 - \cos(x)}{3x^2} \overset{\text{l'H}}{=} \lim_{x \to 0} \frac{\sin(x)}{6x}.$$

We could actually use l'Hôpital's Rule once more to find the final limit, but a better way is to write

$$\lim_{x \to 0} \frac{\sin(x)}{6x} = \frac{1}{6} \lim_{x \to 0} \frac{\sin(x)}{x} = \frac{1}{6} \times 1 = \frac{1}{6}.$$

(Here we used our classic trig limit which we proved in Section 7.1.5 of Chapter 7.) All in all, we have proved that

$$\lim_{x \to 0} \frac{x - \sin(x)}{x^3} = \frac{1}{6}.$$

Before we move on to the next variation, here's a nice observation. Way back in Section 6.5 of Chapter 6, we saw that some limits can be thought of as derivatives in disguise. For example, we worked out

$$\lim_{h \to 0} \frac{\sqrt[5]{32 + h} - 2}{h}$$

by the tricky technique of setting $f(x) = \sqrt[5]{x}$, then finding $f'(x)$, writing it as a limit, and finally putting $x = 32$. (Check back to see the details.) The point is that l'Hôpital's Rule actually makes all these shenanigans unnecessary! For example, since the above limit is of the indeterminate form $0/0$, we can find it by differentiating the top and bottom with respect to h. First write $\sqrt[5]{32 + h}$ as $(32 + h)^{1/5}$; then we have

$$\lim_{h \to 0} \frac{\sqrt[5]{32 + h} - 2}{h} = \lim_{h \to 0} \frac{(32 + h)^{1/5} - 2}{h} \stackrel{\text{l'H}}{=} \lim_{h \to 0} \frac{\frac{1}{5}(32 + h)^{-4/5}}{1} = \frac{1}{5}(32)^{-4/5},$$

which works out to be $1/80$. This agrees with the answer we found previously. Now you should go back and look at the other examples in Section 6.5 of Chapter 6 and try using l'Hôpital's Rule on them instead.

14.1.2 Type A: $\pm\infty/\pm\infty$ case

L'Hôpital's Rule also works in the case where $\lim_{x \to a} f(x) = \infty$ and $\lim_{x \to a} g(x) = \infty$. That is, when you try to put $x = a$, the top and bottom both look infinite, so you are dealing with the indeterminate form ∞/∞. For example, to find

$$\lim_{x \to \infty} \frac{3x^2 + 7x}{2x^2 - 5},$$

you could note that both top and bottom go to ∞ as $x \to \infty$, then use l'Hôpital's Rule:

$$\lim_{x \to \infty} \frac{3x^2 + 7x}{2x^2 - 5} \stackrel{\text{l'H}}{=} \lim_{x \to \infty} \frac{6x + 7}{4x} = \lim_{x \to \infty} \left(\frac{6}{4} + \frac{7}{4x} \right).$$

The term $7/4x$ goes to 0 as $x \to \infty$, so the limit is $6/4$, which is just $3/2$. Of course, you could just have used the methods of Section 4.3 of Chapter 4 to find the limit; try checking that you still get $3/2$ using those methods.

Here's another example. To find

$$\lim_{x \to 0^+} \frac{\csc(x)}{1 - \ln(x)},$$

notice that as $x \to 0^+$, both the numerator and the denominator tend to ∞. Why? Well, $\sin(x)$ goes to 0 as $x \to 0$, so $\csc(x)$ blows up; and also $\ln(x) \to -\infty$ as $x \to 0^+$, so $1 - \ln(x) \to \infty$. Now use l'Hôpital's Rule:

$$\lim_{x \to 0^+} \frac{\csc(x)}{1 - \ln(x)} \stackrel{\text{l'H}}{=} \lim_{x \to 0^+} \frac{-\csc(x)\cot(x)}{-1/x} = \lim_{x \to 0^+} x\csc(x)\cot(x).$$

To find the limit, write it as

$$\lim_{x \to 0^+} \frac{x}{\sin(x)} \frac{1}{\tan(x)}.$$

We have

$$\lim_{x\to 0^+} \frac{x}{\sin(x)} = \frac{1}{\displaystyle\lim_{x\to 0^+}\frac{\sin(x)}{x}} = \frac{1}{1} = 1,$$

but for the other factor we have

$$\lim_{x\to 0^+} \frac{1}{\tan(x)} = \infty,$$

since $\tan(x) \to 0^+$ as $x \to 0^+$. So we have proved that

$$\lim_{x\to 0^+} \frac{\csc(x)}{1 - \ln(x)} = \infty.$$

The rule also applies as $x \to \infty$, as we saw above. Here's another example:

$$\lim_{x\to\infty} \frac{x}{e^x} \overset{\text{l'H}}{=} \lim_{x\to\infty} \frac{1}{e^x} = 0.$$

The last limit is 0 because $e^x \to \infty$ as $x \to \infty$. Also, the justification for using l'Hôpital's Rule is that both x and e^x go to ∞ as $x \to \infty$. Notice that the denominator e^x was unscathed by the differentiation, but the numerator x was knocked down to 1. This is even clearer when you consider the example

$$\lim_{x\to\infty} \frac{x^3}{e^x}.$$

Just use l'Hôpital's Rule three times, noting that in each case we are dealing with the indeterminate form ∞/∞:

$$\lim_{x\to\infty} \frac{x^3}{e^x} \overset{\text{l'H}}{=} \lim_{x\to\infty} \frac{3x^2}{e^x} \overset{\text{l'H}}{=} \lim_{x\to\infty} \frac{6x}{e^x} \overset{\text{l'H}}{=} \lim_{x\to\infty} \frac{6}{e^x} = 0.$$

Of course, the same technique applies to any power of x; you just have to apply the rule enough times, knocking the power down by 1 each time, while the e^x just sits there like some immovable lump. So we have proved the principle that exponentials grow quickly, which is discussed in some detail in Section 9.4.4 of Chapter 9.

Now, a gentle reminder: please, please, please check that you have an indeterminate form! The only acceptable forms for a quotient are $0/0$ or $\pm\infty/\pm\infty$. For example, if you try to use l'Hôpital's Rule on the limit

$$\lim_{x\to 0} \frac{x^2}{\cos(x)},$$

you'll get into a real tangle. Let's see what happens:

$$\lim_{x\to 0} \frac{x^2}{\cos(x)} \overset{\text{l'H?}}{=} \lim_{x\to 0} \frac{2x}{-\sin(x)} = -2\lim_{x\to 0}\frac{x}{\sin(x)} = -2.$$

This is clearly wrong, since x^2 and $\cos(x)$ are both positive when x is near 0. In fact, the correct solution is

$$\lim_{x\to 0} \frac{x^2}{\cos(x)} = \frac{0^2}{\cos(0)} = \frac{0}{1} = 0.$$

L'Hôpital's Rule can't be used here since the form is $0/1$, which is not indeterminate. So be careful!

14.1.3 Type **B1** ($\infty - \infty$)

Here's a limit from the beginning of this chapter:

$$\lim_{x \to 0} \left(\frac{1}{\sin(x)} - \frac{1}{x} \right).$$

As $x \to 0^+$, both $1/\sin(x)$ and $1/x$ go to ∞. As $x \to 0^-$, both quantities go to $-\infty$. Either way, you're looking at the difference of two huge (positive or negative) quantities, so we can express the indeterminate form as $\pm(\infty - \infty)$.

Luckily, it's pretty easy to reduce this to Type **A**. Just take a common denominator:

$$\lim_{x \to 0} \left(\frac{1}{\sin(x)} - \frac{1}{x} \right) = \lim_{x \to 0} \frac{x - \sin(x)}{x \sin(x)}.$$

Now you can put $x = 0$ and see that we are in the $0/0$ case. So we can apply l'Hôpital's Rule:

$$\lim_{x \to 0} \left(\frac{1}{\sin(x)} - \frac{1}{x} \right) = \lim_{x \to 0} \frac{x - \sin(x)}{x \sin(x)} \stackrel{\text{l'H}}{=} \lim_{x \to 0} \frac{1 - \cos(x)}{\sin(x) + x \cos(x)}.$$

Notice that we used the product rule to differentiate the denominator. In any case, we are again in $0/0$ territory—just put $x = 0$ and see that the top and bottom both become 0. So we use l'Hôpital's Rule (and the product rule) once more:

$$\lim_{x \to 0} \frac{1 - \cos(x)}{\sin(x) + x \cos(x)} \stackrel{\text{l'H}}{=} \lim_{x \to 0} \frac{\sin(x)}{\cos(x) + \cos(x) - x \sin(x)}.$$

Don't use l'Hôpital's Rule again! At this stage, just put $x = 0$; the numerator is 0 and the denominator is 2, so the overall limit is 0. Putting everything together, we have shown that

$$\lim_{x \to 0} \left(\frac{1}{\sin(x)} - \frac{1}{x} \right) = 0.$$

Taking a common denominator doesn't always work. Sometimes you might not even have a denominator at all, so you have to create one out of thin air. For example, to find

$$\lim_{x \to \infty} \left(\sqrt{x + \ln(x)} - \sqrt{x} \right),$$

first note that as $x \to \infty$, both $\sqrt{x + \ln(x)}$ and $\sqrt{x}$ go to ∞; so we are in the $\infty - \infty$ case. There's no denominator, so let's make one by multiplying and dividing by the conjugate expression:

$$\lim_{x \to \infty} \left(\sqrt{x + \ln(x)} - \sqrt{x} \right) = \lim_{x \to \infty} \left(\sqrt{x + \ln(x)} - \sqrt{x} \right) \times \frac{\sqrt{x + \ln(x)} + \sqrt{x}}{\sqrt{x + \ln(x)} + \sqrt{x}}.$$

Using the difference of squares formula $(a - b)(a + b)$, this becomes

$$\lim_{x \to \infty} \frac{x + \ln(x) - x}{\sqrt{x + \ln(x)} + \sqrt{x}} = \lim_{x \to \infty} \frac{\ln(x)}{\sqrt{x + \ln(x)} + \sqrt{x}}.$$

Now we are in the ∞/∞ case of Type **A**, so we just differentiate top and bottom (using the chain rule on the bottom) to see that

$$\lim_{x \to \infty} \frac{\ln(x)}{\sqrt{x + \ln(x)} + \sqrt{x}} \stackrel{\text{l'H}}{=} \lim_{x \to \infty} \frac{1/x}{\dfrac{1 + 1/x}{2\sqrt{x + \ln(x)}} + \dfrac{1}{2\sqrt{x}}}.$$

If you multiply the top and bottom of the fraction by x, you get

$$\lim_{x \to \infty} \frac{1}{\dfrac{x + 1}{2\sqrt{x + \ln(x)}} + \dfrac{\sqrt{x}}{2}}.$$

We're almost done, but we do need to take a little look at what happens to the first fraction in the denominator as $x \to \infty$:

$$\lim_{x \to \infty} \frac{x + 1}{2\sqrt{x + \ln(x)}}.$$

This is also an ∞/∞ indeterminate form, so whack out another application of ye olde l'Hôpital's Rule:

$$\lim_{x \to \infty} \frac{x + 1}{2\sqrt{x + \ln(x)}} \stackrel{\text{l'H}}{=} \lim_{x \to \infty} \frac{1}{\dfrac{2(1 + 1/x)}{2\sqrt{x + \ln(x)}}} = \lim_{x \to \infty} \frac{\sqrt{x + \ln(x)}}{1 + 1/x}.$$

As $x \to \infty$, the denominator $1 + 1/x$ goes to 1 but the numerator $\sqrt{x + \ln(x)}$ goes to ∞. This means that

$$\lim_{x \to \infty} \frac{x + 1}{2\sqrt{x + \ln(x)}} = \infty.$$

Returning to our original problem, we have already found that

$$\lim_{x \to \infty} \left(\sqrt{x + \ln(x)} - \sqrt{x} \right) = \lim_{x \to \infty} \frac{1}{\dfrac{x + 1}{2\sqrt{x + \ln(x)}} + \dfrac{\sqrt{x}}{2}}.$$

Both fractions in the denominator go to ∞ as $x \to \infty$, so the limit is 0.

Unfortunately, it's not always possible to use l'Hôpital's Rule on type **B1** limits. In fact, the only time it can actually work is when you're able to manipulate the original expression to be a ratio of two quantities, as in the above examples.

14.1.4 Type **B2** $(0 \times \pm\infty)$

Here's a limit we've looked at before, in Section 9.4.6 of Chapter 9 as well as at the beginning of this chapter:

$$\lim_{x \to 0^+} x \ln(x).$$

The limit has to be as $x \to 0^+$ since $\ln(x)$ isn't even defined when $x \leq 0$. Now, as $x \to 0^+$, we see that $x \to 0$ while $\ln(x) \to -\infty$, so we are dealing with the indeterminate form $0 \times (-\infty)$. Let's turn the limit into Type **A** by manufacturing a denominator. The idea is to move x into a new denominator by putting it there as $1/x$:

$$\lim_{x \to 0^+} x \ln(x) = \lim_{x \to 0^+} \frac{\ln(x)}{1/x}.$$

Now the form is $-\infty/\infty$, so we can use l'Hôpital's Rule:

$$\lim_{x \to 0^+} x \ln(x) = \lim_{x \to 0^+} \frac{\ln(x)}{1/x} \stackrel{\text{l'H}}{=} \lim_{x \to 0^+} \frac{1/x}{-1/x^2}.$$

We can simplify the fraction on the right to $-x$, so that the overall limit is

$$\lim_{x \to 0^+} (-x) = 0.$$

We've solved the problem, but let's just check out something: why did we move x into the denominator and not $\ln(x)$? It's true that

$$\lim_{x \to 0^+} x \ln(x) = \lim_{x \to 0^+} \frac{x}{1/\ln(x)}.$$

Now you have to differentiate $1/\ln(x)$ instead, which is much harder. If you try it, you'll see that

$$\lim_{x \to 0^+} x \ln(x) = \lim_{x \to 0^+} \frac{x}{1/\ln(x)} \stackrel{\text{l'H}}{=} \lim_{x \to 0^+} \frac{1}{(1/x)(-1/(\ln(x))^2)} = \lim_{x \to 0^+} -x(\ln(x))^2.$$

This is actually worse than the original limit! So, take care when you choose which factor to move down the bottom. As you can see from the above example, moving a log term can be a bad idea—so avoid doing that.

Here's another example:

$$\lim_{x \to \pi/2} \left(x - \frac{\pi}{2} \right) \tan(x).$$

When you put $x = \pi/2$, the first factor $(x - \pi/2)$ is 0, while the $\tan(x)$ factor is either ∞ (as $x \to (\pi/2)^-$) or $-\infty$ (as $x \to (\pi/2)^+$). Sketch the graph of $y = \tan(x)$ to make sure you believe this. In any case, we can move the $\tan(x)$ factor down to a new denominator by putting it there as $1/\tan(x)$, or $\cot(x)$. That is,

$$\lim_{x \to \pi/2} \left(x - \frac{\pi}{2} \right) \tan(x) = \lim_{x \to \pi/2} \frac{x - \pi/2}{\cot(x)}.$$

This is a lot easier than putting the $(x - \pi/2)$ term in the denominator—in fact, that doesn't even work. Anyway, the above limit is now in 0/0 form, so you can use l'Hôpital's Rule:

$$\lim_{x \to \pi/2} \left(x - \frac{\pi}{2} \right) \tan(x) = \lim_{x \to \pi/2} \frac{x - \pi/2}{\cot(x)} \stackrel{\text{l'H}}{=} \lim_{x \to \pi/2} \frac{1}{(-\csc^2(x))}.$$

Since $\sin(\pi/2) = 1$, we see that also $\csc(\pi/2) = 1$, so the above limit is -1.

14.1.5 Type C ($1^{\pm\infty}$, 0^0, or ∞^0)

Finally, the trickiest type involves limits like

$$\lim_{x\to 0^+} x^{\sin(x)},$$

where both the base and exponent involve the dummy variable (x in this case). If you just put $x = 0$, you get 0^0, which is another indeterminate form. To find the limit, we'll use a technique very similar to logarithmic differentiation (see Section 9.5 in Chapter 9). The idea is to take the logarithm of the quantity $x^{\sin(x)}$ first, and work out its limit as $x \to 0^+$:

$$\lim_{x\to 0^+} \ln(x^{\sin(x)}).$$

By our log rules (see Section 9.1.4 of Chapter 9), the exponent $\sin(x)$ comes down out front of the logarithm:

$$\lim_{x\to 0^+} \ln(x^{\sin(x)}) = \lim_{x\to 0^+} \sin(x)\ln(x).$$

As $x \to 0^+$, we have $\sin(x) \to 0$ and $\ln(x) \to -\infty$, so now we're dealing with a Type **B2** problem. We can put the $\sin(x)$ into a new denominator as $1/\sin(x)$, which is just $\csc(x)$, then use l'Hôpital's Rule on the resulting Type **A** problem:

$$\lim_{x\to 0^+} \sin(x)\ln(x) = \lim_{x\to 0^+} \frac{\ln(x)}{\csc(x)} \overset{\text{l'H}}{=} \lim_{x\to 0^+} \frac{1/x}{-\csc(x)\cot(x)}.$$

This can be rearranged to

$$\lim_{x\to 0^+} -\frac{\sin(x)}{x} \times \tan(x) = -1 \times 0 = 0.$$

Are we done? Not quite. We now know that

$$\lim_{x\to 0^+} \ln(x^{\sin(x)}) = 0;$$

so now we just have to exponentiate both sides to see that

$$\lim_{x\to 0^+} x^{\sin(x)} = e^0 = 1.$$

(The exponentiation works because e^x is a continuous function of x.)

Let's review what we just did. Instead of finding the original limit, we took logarithms and then found **that** limit, using the Type **B2** technique. Finally, we exponentiated at the end.

In fact, sometimes you don't even have to go through the Type **B2** step on your way to Type **A**. For example, to do

$$\lim_{x\to 0}(1 + 3\tan(x))^{1/x}$$

from the beginning of the chapter, first note that we are dealing with the form $1^{\pm\infty}$. So take logarithms:

$$\lim_{x\to 0} \ln\left((1 + 3\tan(x))^{1/x}\right) = \lim_{x\to 0} \frac{1}{x}\ln(1 + 3\tan(x)) = \lim_{x\to 0} \frac{\ln(1 + 3\tan(x))}{x}.$$

This is now of the form $0/0$, so it's already a Type **A** limit. By the chain rule, we have

$$\lim_{x\to 0}\frac{\ln(1+3\tan(x))}{x}\overset{\text{l'H}}{=}\lim_{x\to 0}\frac{\dfrac{3\sec^2(x)}{1+3\tan(x)}}{1}=\frac{3(1)^2}{1+3(0)}=3.$$

We have now shown that

$$\lim_{x\to 0}\ln\left((1+3\tan(x))^{1/x}\right)=3.$$

Exponentiate both sides to get

$$\lim_{x\to 0}(1+3\tan(x))^{1/x}=e^3.$$

There is one more indeterminate form of this type, ∞^0. An example is

$$\lim_{x\to\infty}x^{-1/x},$$

since $-1/x\to 0$ as $x\to\infty$. The same trick still works: take logarithms and use the Type **A** methodology to get

$$\lim_{x\to\infty}\ln(x^{-1/x})=\lim_{x\to\infty}\frac{\ln(x)}{-x}\overset{\text{l'H}}{=}\lim_{x\to\infty}\frac{1/x}{-1}=0.$$

Now exponentiate to get

$$\lim_{x\to\infty}x^{-1/x}=e^0=1.$$

It's not really necessary to learn that the only indeterminate forms involving exponentials are $1^{(\pm\infty)}$, 0^0, and ∞^0. You see, if you have any limit involving exponentials, you can always use the above logarithmic method to convert everything to a product or quotient, then work out the new limit L. The actual limit will just be e^L. The only exceptions are that if $L=\infty$, then you have to interpret e^∞ as ∞; and if $L=-\infty$, then you need to recognize $e^{-\infty}$ as 0. This is consistent with our limits

$$\lim_{x\to\infty}e^x=\infty\qquad\text{and}\qquad\lim_{x\to-\infty}e^x=0$$

from Section 9.4.4 of Chapter 9.

14.1.6 Summary of l'Hôpital's Rule types

Here are all the techniques we've looked at:

• Type **A**: if the limit involves a fraction, like

$$\lim_{x\to a}\frac{f(x)}{g(x)},$$

check that the form is indeterminate. It must be $0/0$ or $\pm\infty/\pm\infty$. Use the rule

$$\boxed{\lim_{x\to a}\frac{f(x)}{g(x)}\overset{\text{l'H}}{=}\lim_{x\to a}\frac{f'(x)}{g'(x)}.}$$

Do not use the quotient rule here! Now, solve the new limit, perhaps even using l'Hôpital's Rule again.

- Type **B1**: if the limit involves a difference, like

$$\lim_{x \to a} (f(x) - g(x)),$$

where the form is $\pm(\infty - \infty)$, try taking a common denominator or multiplying by a conjugate expression to reduce to a Type **A** form.

- Type **B2**: if the limit involves a product, like

$$\lim_{x \to a} f(x)g(x),$$

where the form is $0 \times \pm\infty$, pick the simplest of the two factors and put it on the bottom as its reciprocal. (Avoid picking a log term—keep that on the top.) You get something like

$$\lim_{x \to a} f(x)g(x) = \lim_{x \to a} \frac{g(x)}{1/f(x)}.$$

This is now a Type **A** form.

- Type **C**: if the limit involves an exponential where both base and exponent involve the dummy variable, like

$$\lim_{x \to a} f(x)^{g(x)},$$

then first work out the limit of the logarithm:

$$\lim_{x \to a} \ln(f(x)^{g(x)}) = \lim_{x \to a} g(x) \ln(f(x)).$$

This should be either Type **B2** or Type **A** (or else it's not indeterminate and you can just substitute). Once you've solved it, you can rewrite the equation as something like

$$\lim_{x \to a} \ln(f(x)^{g(x)}) = L,$$

then exponentiate both sides to get

$$\lim_{x \to a} f(x)^{g(x)} = e^L.$$

Now all that's left is for you to practice doing as many l'Hôpital's Rule problems as you can get your hands on!

14.2   Overview of Limits

It's time to consolidate. Here's a brief summary of all the techniques we've seen so far involving evaluating limits. The following techniques apply to limits of the form

$$\lim_{x \to a} F(x),$$

where F is a function which is at least continuous for x near a, but maybe not at $x = a$ itself. Also, a could be ∞ or $-\infty$. So, here's the summary:

- **Try substituting first.** You might be able to evaluate the limit.
- If your substitution leads to b/∞ or $b/(-\infty)$, where b is some finite number, then the limit is 0.
- If the substitution gives $b/0$, where $b \neq 0$, then you're dealing with a **vertical asymptote.** The left-hand and right-hand limits must be ∞ or $-\infty$, and the two-sided limit either doesn't exist (if the left-hand and right-hand limits are different) or is one of ∞ and $-\infty$. Use a table of signs around $x = a$ to investigate the left-hand and right-hand limits. (Also see Section 4.1 in Chapter 4.)
- If none of the above points are relevant, and your limit is of the form $0/0$, try seeing if it is a **derivative in disguise.** If you can rewrite it in the form

$$\lim_{h \to 0} \frac{f(x + h) - f(x)}{h}$$

for some particular function and possibly a specific number x, then the limit is just $f'(x)$. As we saw in Section 14.1.1 above, these sorts of problems can also be done by using l'Hôpital's Rule. (See also Section 6.5 in Chapter 6.)
- If **square roots** are involved, multiplication by a conjugate expression might help. (See Section 4.2 in Chapter 4.)
- If **absolute values** are involved, convert them into piecewise-defined functions using the formula

$$|A| = \begin{cases} A & \text{if } A \geq 0, \\ -A & \text{if } A < 0. \end{cases}$$

Remember to replace all five occurrences of A above with the actual expression you're taking the absolute value of! (See Section 4.6 in Chapter 4.)
- Otherwise, you can use the properties of various functions which can pop up as ingredients in your main function. Remember that "small" means "near 0," and "large" can mean large positive or negative numbers. (See Section 3.4.1 in Chapter 3.) Beware: if your limit is as $x \to \infty$, it doesn't necessarily mean that you are in the large case. For example, $\sin(1/x)$ involves the sine of a small number as $x \to \infty$, since $1/x \to 0$. The same warning applies to limits as $x \to 0$, which need not be in the small case. Anyway, here's the deal for polynomials, trig functions, exponentials, and logs:

  1. **Polynomials and poly-type functions:**

     - **General tip:** try factoring, then cancel common factors. (See Section 4.1 in Chapter 4.)

     - **Large arguments:** the **largest-degree term dominates**, so divide and multiply by that term. (See Section 4.3 in Chapter 4.)

2. **Trig and inverse trig functions:**

   - **General tip:** know the graphs of all the trig and inverse trig functions, and their values at some common arguments. All the stuff in Chapter 2 and Chapter 10 is helpful in this regard.

   - **Small arguments:** $\sin(A)$ behaves like A when A is small, so divide and multiply by A. The same goes for $\tan(A)$, but **not** $\cos(A)$: that just behaves like 1. This technique is useful when only products and quotients are involved. It probably won't work when the trig function is added to or subtracted from some other quantity. (See Section 7.1.2 in Chapter 7.)

   - **Large arguments:** for sine or cosine, use the facts that

     $$|\sin(\text{anything})| \le 1 \qquad \text{and} \qquad |\cos(\text{anything})| \le 1$$

     in conjunction with the sandwich principle. (See Section 7.1.3 in Chapter 7.) Some other useful facts are

     $$\lim_{x \to \infty} \tan^{-1}(x) = \frac{\pi}{2} \qquad \text{and} \qquad \lim_{x \to -\infty} \tan^{-1}(x) = -\frac{\pi}{2}.$$

     (Informally, you can think of these as $\tan^{-1}(\infty) = \pi/2$ and $\tan^{-1}(-\infty) = -\pi/2$, but make sure you understand that these are just crude ways of expressing the limits above.)

3. **Exponentials:**

   - **General tip:** know the graph of $y = e^x$, and learn the limits

     $$\lim_{h \to 0} (1 + hx)^{1/h} = e^x \qquad \text{and} \qquad \lim_{n \to \infty} \left(1 + \frac{x}{n}\right)^n = e^x.$$

     (See Section 9.4.1 in Chapter 9.)

   - **Small arguments:** since $e^0 = 1$, you can normally just isolate any factors which involve the exponential of a small number and replace them by 1 when you take the limit. The exception is when sums or differences occur; then you might want to use l'Hôpital's Rule, or perhaps the limit is actually a derivative in disguise. (See Section 9.4.2 in Chapter 9.)

   - **Large arguments:** learn the important limits

     $$\lim_{x \to \infty} e^x = \infty \qquad \text{and} \qquad \lim_{x \to -\infty} e^x = 0.$$

     (For substitution purposes only, you can think of these limits as $e^\infty = \infty$ and $e^{-\infty} = 0$, even though these equations aren't formally true.) Also remember that **exponentials grow quickly** as $x \to \infty$. This means that

     $$\lim_{x \to \infty} \frac{\text{poly}}{e^x} = 0.$$

     The base e could instead be any number bigger than 1, and the exponent x could instead be some other polynomial with positive leading coefficient. (See Section 9.4.4 in Chapter 9.)

4. **Logarithms:**

– **General tip:** know the graph of $y = \ln(x)$ and the log rules, which are in Section 9.1.4 of Chapter 9.

– **Small arguments:** a really important limit is

$$\lim_{x \to 0^+} \ln(x) = -\infty$$

(or, as a memory aid only, $\ln(0) = -\infty$). Also, logs "grow" slowly down to $-\infty$ as $x \to 0^+$:

$$\lim_{x \to 0^+} x^a \ln(x) = 0$$

for any $a > 0$, no matter how small. (See Section 9.4.6 in Chapter 9.)

– **Large arguments:** we have

$$\lim_{x \to \infty} \ln(x) = \infty,$$

which has the informal abbreviation $\ln(\infty) = \infty$. Nevertheless **logs grow slowly**, that is, more slowly than any polynomial:

$$\lim_{x \to \infty} \frac{\ln(x)}{\text{poly}} = 0$$

for any polynomial of positive degree. (See Section 9.4.5 in Chapter 9.)

– **Behavior near 1:** we have $\ln(1) = 0$. L'Hôpital's Rule can be very useful in this regard, or the limit might be a derivative in disguise. (See Section 9.4.3 in Chapter 9.)

• If none of the above techniques work, consider using l'Hôpital's Rule (see Section 14.1.6 above for a summary). If you do, you'll always get a new limit to solve, which you can attack using any of the above principles or l'Hôpital's Rule once again.

All these facts and methods above are just tools to help you solve limits. They may not work on every limit you see—in fact, we'll be looking at a completely different type of limit problem in Chapter 17—but they should help with a heck of a lot of them. There's an art to knowing which tool to use, and of course, practice makes perfect. So go forth and evaluate limits!

CHAPTER 15 _____

Introduction to Integration

So far as calculus is concerned, differentiation is only half the story. The other half concerns integration. This powerful tool enables us to find areas of curved regions, volumes of solids, and distances traveled by objects moving at variable speeds. In this chapter, we'll spend some time developing the theory we need to define the definite integral. Then, in the next chapter, we'll give the definition and see how to apply it. So here's the plan for the preliminaries on integration:

- sigma notation and telescoping sums;
- the relationship between displacement and area; and
- using partitions to find areas.

15.1 Sigma Notation

Consider the sum
$$\frac{1}{1} + \frac{1}{4} + \frac{1}{9} + \frac{1}{16} + \frac{1}{25} + \frac{1}{36}.$$
This is not just a sum of random numbers: there's a definite pattern. The terms in the sum are reciprocals of the squares from 1^2 through 6^2. Here's a more convenient way to write the sum:

$$\sum_{j=1}^{6} \frac{1}{j^2}.$$

To read it out loud, say "the sum, from $j = 1$ to 6, of $1/j^2$." Now, here's how it actually works. The idea is that you plug $j = 1$, $j = 2$, $j = 3$, $j = 4$, $j = 5$, and finally $j = 6$ into the expression $1/j^2$, one at a time, and then add everything up. We can tell that we're supposed to start at $j = 1$ and end up at $j = 6$ by the symbols below and above the big Greek letter Σ (which is a capital sigma, hence the term "sigma notation"). So we have

$$\sum_{j=1}^{6} \frac{1}{j^2} = \frac{1}{1^2} + \frac{1}{2^2} + \frac{1}{3^2} + \frac{1}{4^2} + \frac{1}{5^2} + \frac{1}{6^2}.$$

Notice that we haven't actually worked out the value of the sum! All we've done is abbreviate it.

Now consider the following series (that's another word for "sum") in sigma notation:

$$\sum_{j=1}^{1000} \frac{1}{j^2}.$$

The only difference between this sum and the previous one is that now we have to go to 1000, not 6. So

$$\sum_{j=1}^{1000} \frac{1}{j^2} = \frac{1}{1^2} + \frac{1}{2^2} + \frac{1}{3^2} + \cdots + \frac{1}{999^2} + \frac{1}{1000^2}.$$

In this case, the sigma notation is particularly nice, avoiding the "$\cdots$" altogether (unlike the right-hand side of the above equation). Here's another variation:

$$\sum_{j=5}^{30} \frac{1}{j^2} = \frac{1}{5^2} + \frac{1}{6^2} + \frac{1}{7^2} + \cdots + \frac{1}{29^2} + \frac{1}{30^2}.$$

This sum starts at $j = 5$, not $j = 1$, so the first term is $1/5^2$.

Sigma notation is also really useful when you want to vary where the sum stops (or starts). For example, consider the series

$$\sum_{j=1}^{n} \frac{1}{j^2}.$$

This starts at $j = 1$ and finishes at $j = n$, so we have

$$\sum_{j=1}^{n} \frac{1}{j^2} = \frac{1}{1^2} + \frac{1}{2^2} + \frac{1}{3^2} + \cdots + \frac{1}{(n-2)^2} + \frac{1}{(n-1)^2} + \frac{1}{n^2}.$$

Notice that the second-to-last term occurs when $j = n - 1$, and the third-to-last term occurs when $j = n - 2$; I wrote those terms, along with the first three and the last term, on the right-hand side of the above equation. The other terms are all absorbed into the "$\cdots$" in the middle.

In the sum

$$\sum_{j=1}^{n} \frac{1}{j^2},$$

it looks as if there are two variables, j and n, but in reality there is only one: it's n. You can easily see this by looking at the expanded form

$$\frac{1}{1^2} + \frac{1}{2^2} + \frac{1}{3^2} + \cdots + \frac{1}{(n-2)^2} + \frac{1}{(n-1)^2} + \frac{1}{n^2}.$$

There's no j at all! In fact, j is a dummy variable—it's just a temporary placeholder, called the *index of summation*, that runs through the integers from 1 to n. So we could even change it to another letter without affecting anything. For example, the following sums are all the same:

$$\sum_{j=1}^{6} \frac{1}{j^2} = \sum_{k=1}^{6} \frac{1}{k^2} = \sum_{a=1}^{6} \frac{1}{a^2} = \sum_{\alpha=1}^{6} \frac{1}{\alpha^2}.$$

By the way, this isn't the first time we've seen dummy variables: limits also use them, so there's nothing new here. (See the end of Section 3.1 of Chapter 3.)

Let's look at some more examples. What is

$$\sum_{m=1}^{200} 5?$$

Don't fall into the trap of saying that it's equal to 5. Let's look a little closer. When $m = 1$, we have a term 5. When $m = 2$, we again have 5. The same goes for $m = 3$, $m = 4$ and so on until $m = 200$. So in fact

$$\sum_{m=1}^{200} 5 = 5 + 5 + 5 + \cdots + 5 + 5 + 5,$$

where there are 200 terms in the sum. So the value works out to be 200×5, or 1000. Similarly, consider the series

$$\sum_{q=100}^{1000} 1 = 1 + 1 + 1 + \cdots + 1 + 1 + 1.$$

How many terms of 1 are there in this sum? You might be tempted to say that there are $1000 - 100$, or 900, but actually there's one more. The answer is 901. In general, the number of integers between A and B, **including** A and B, is $B - A + 1$.

How would you write

$$\sin(1) + \sin(3) + \sin(5) + \cdots + \sin(2997) + \sin(2999) + \sin(3001)$$

in sigma notation? You might try

$$\sum_{j=1}^{3001} \sin(j),$$

but that's no good: that would be

$$\sin(1) + \sin(2) + \sin(3) + \cdots + \sin(2999) + \sin(3000) + \sin(3001).$$

We don't want the even numbers. Here's how you get rid of them. First, imagine that j steps through the numbers 1, 2, 3, and so on. Then the quantity $(2j - 1)$ goes through all the odd numbers 1, 3, 5, and so on. So for our second try, let's guess

$$\sum_{j=1}^{3001} \sin(2j - 1).$$

This is better, but there's still a problem. When j gets to the end of its run, it's at 3001, but $(2j - 1)$ is then $2(3001) - 1 = 6001$. This means that

$$\sum_{j=1}^{3001} \sin(2j-1) = \sin(1) + \sin(3) + \sin(5) + \cdots + \sin(5997) + \sin(5999) + \sin(6001).$$

We have too many terms! How do you know when to stop? At the end, we need $\sin(2j-1)$ to be $\sin(3001)$, not $\sin(6001)$. So, just set $2j - 1 = 3001$, which means that $j = 1501$. Finally, we have

$$\sin(1)+\sin(3)+\sin(5)+\cdots+\sin(2997)+\sin(2999)+\sin(3001) = \sum_{j=1}^{1501} \sin(2j-1).$$

This is the correct answer. Make sure you agree with it by plugging in the values $j = 1$, $j = 2$, $j = 3$, and also $j = 1499$, $j = 1500$, and $j = 1501$. You should get the terms written out on the left-hand side above. On the other hand, the sum

$$\sum_{j=1}^{1501} \sin(2j)$$

expands as

$$\sin(2) + \sin(4) + \sin(6) + \cdots + \sin(2998) + \sin(3000) + \sin(3002).$$

So you get the even numbers using $2j$ instead of $(2j - 1)$. Of course, if you wanted multiples of 3, you'd use $3j$. The possibilities are endless!

15.1.1 A nice sum

Consider the sum

$$\sum_{j=1}^{100} j.$$

First, let's expand the sum. When $j = 1$, we get 1. When $j = 2$, we get 2. This continues until $j = 100$; then we just add up all these quantities. So

$$\sum_{j=1}^{100} j = 1 + 2 + 3 + \cdots + 98 + 99 + 100.$$

Yup, it's the sum of the numbers from 1 to 100. Now, how about the sum

$$\sum_{j=0}^{99} (j + 1)?$$

When $j = 0$, we get 1; when $j = 1$, we get 2; and so on until $j = 99$, in which case we get 100. So in fact

$$\sum_{j=0}^{99} (j + 1) = 1 + 2 + 3 + \cdots + 98 + 99 + 100.$$

This is the same sum as before! What we've done is shift the index of summation j down by 1. Now, consider this sum:

$$\sum_{j=1}^{100} (101 - j).$$

When $j = 1$, we get 100; when $j = 2$, we get 99; and so on until $j = 100$, in which case we get 1. That is, the numbers $101 - j$ march down from 100 to 1, so

$$\sum_{j=1}^{100}(101 - j) = 100 + 99 + 98 + \cdots + 3 + 2 + 1.$$

This is the same sum as before, just written backward. There are many ways of expressing any sum in sigma notation.

In fact, this last way of writing the sum isn't just a curiosity—we can actually use it to find the value of the sum. Suppose that we let S be the sum $1 + 2 + \cdots + 99 + 100$; then we have seen that

$$S = \sum_{j=1}^{100} j \quad \text{and also} \quad S = \sum_{j=1}^{100}(101 - j).$$

If you add up these two expressions, you get

$$2S = \sum_{j=1}^{100} j + \sum_{j=1}^{100}(101 - j).$$

In the first sum, the numbers increase from 1 to 100; in the second sum they decrease from 100 to 1. The nice thing is that you can add the numbers in any order and still get the same result. So we can combine the sums and write

$$2S = \sum_{j=1}^{100}\big(j + (101 - j)\big).$$

Since $j + (101 - j) = 101$, this just works out to be

$$2S = \sum_{j=1}^{100} 101.$$

There are 100 copies of the number 101, so we have $2S = 101 \times 100 = 10100$. This means that $S = 10100/2 = 5050$. We have shown that the sum of the numbers from 1 to 100 is 5050. Believe it or not, the great mathematician Gauss worked this out (using the same method) at the age of 10!

15.1.2 Telescoping series

Check out the following sum:

$$\sum_{j=1}^{5}\big(j^2 - (j - 1)^2\big).$$

This expands fully to

$$\big(1^2 - 0^2\big) + \big(2^2 - 1^2\big) + \big(3^2 - 2^2\big) + \big(4^2 - 3^2\big) + \big(5^2 - 4^2\big).$$

You can cancel a lot of the terms here. In fact, if you take a close look, you'll see that everything cancels out except $5^2 - 0^2$, so the sum is just $5^2 = 25$. The same sort of thing happens even if you have a lot more terms. For example,

$$\sum_{j=1}^{200}\left(j^2 - (j-1)^2\right)$$

expands as

$$\left(1^2 - 0^2\right) + \left(2^2 - 1^2\right) + \left(3^2 - 2^2\right) + \cdots + \left(198^2 - 197^2\right) + \left(199^2 - 198^2\right) + \left(200^2 - 199^2\right).$$

Once again, everything cancels except for $200^2 - 0^2$, so the sum is 40000. Wait a second, there doesn't seem to be anything to cancel out the 3^2 or -197^2 terms! Well, there are -3^2 and 197^2 terms hidden inside the "$\cdots$", so the cancelation does work.

This sort of series is called a *telescoping series*. You can compact it down to a much simpler expression, just like collapsing one of those old spyglasses. In general, we have

$$\boxed{\sum_{j=a}^{b}(f(j) - f(j-1)) = f(b) - f(a-1).}$$

For example, we have

$$\sum_{j=10}^{100}\left(e^{\cos(j)} - e^{\cos(j-1)}\right) = e^{\cos(100)} - e^{\cos(10-1)}$$

which is simply $e^{\cos(100)} - e^{\cos(9)}$. You just have to take the $e^{\cos(j)}$ part and replace j by the last number (100), then subtract the $e^{\cos(j-1)}$ part with the j replaced by the first number (10). You should try expanding the sum and check that the cancelation works.

Here's another example. To find

$$\sum_{j=1}^{n}(j^2 - (j-1)^2),$$

notice that the sum telescopes; so you just take $(j^2 - (j-1)^2)$ and replace the first j by n, and the second j by 1, to see that

$$\sum_{j=1}^{n}(j^2 - (j-1)^2) = n^2 - (1-1)^2 = n^2.$$

On the other hand, the quantity $j^2 - (j-1)^2$ works out to be $j^2 - (j^2 - 2j + 1)$, or just $2j - 1$. So we have actually shown that

$$\sum_{j=1}^{n}(2j - 1) = n^2.$$

If you think about it, the left-hand side is just the sum of the first n odd numbers. For example, when $n = 5$, the left-hand side is $1 + 3 + 5 + 7 + 9$, which works out to be 25. Hey, that's 5^2 exactly! If instead you take $n = 6$, then the left-hand side is $1 + 3 + 5 + 7 + 9 + 11$, which is 36. This is 6^2, so once again the formula works. We have proved that the sum of the first n odd numbers is n^2.

We can say even more, though. We can split up the sum like this:

$$\sum_{j=1}^{n}(2j) - \sum_{j=1}^{n} 1 = n^2.$$

If you're a little skeptical about this, then check out how it works for the first five terms. Instead of writing $1 + 3 + 5 + 7 + 9$, we're expressing the sum as $(2 - 1) + (4 - 1) + (6 - 1) + (8 - 1) + (10 - 1)$, then rearranging to get $(2 + 4 + 6 + 8 + 10) - (1 + 1 + 1 + 1 + 1)$. In fact, we can take out a factor of 2 from the first sum and express it as $2(1 + 2 + 3 + 4 + 5)$. In terms of our equation above, this means that we can pull out the constant 2 from the first sum and get

$$2\sum_{j=1}^{n} j - \sum_{j=1}^{n} 1 = n^2.$$

Stick the second sum on the right and divide by 2 to get

$$\sum_{j=1}^{n} j = \frac{1}{2}\left(n^2 + \sum_{j=1}^{n} 1\right).$$

The sum on the right-hand side is just n copies of 1, so it's actually equal to n. So the right-hand side is $(n^2 + n)/2$, which can be written as $n(n + 1)/2$. We have proved the useful formula

$$\sum_{j=1}^{n} j = \frac{n(n + 1)}{2}.$$

When $n = 100$, this formula specializes to

$$\sum_{j=1}^{100} j = \frac{100(100 + 1)}{2} = 5050,$$

agreeing with what we saw in the previous section.

Instead of starting with squares as we did in the previous example, let's try starting with cubes:

$$\sum_{j=1}^{n}(j^3 - (j - 1)^3) = n^3 - (1 - 1)^3 = n^3.$$

Once again, finding the value of the sum is easy because it's a telescoping series. In any case, you can do some algebra and see that $j^3 - (j - 1)^3$ simplifies to $3j^2 - 3j + 1$. So the above sum becomes

$$\sum_{j=1}^{n}(3j^2 - 3j + 1) = n^3.$$

Let's break the sum into three pieces and pull out some constants:

$$3\sum_{j=1}^{n} j^2 - 3\sum_{j=1}^{n} j + \sum_{j=1}^{n} 1 = n^3.$$

Now put the last two sums on the right-hand side and divide by 3 to get

$$\sum_{j=1}^{n} j^2 = \frac{1}{3}\left(n^3 + 3\sum_{j=1}^{n} j - \sum_{j=1}^{n} 1\right).$$

The previous example shows that the first sum on the right-hand side works out to be $n(n+1)/2$, while the second sum is again n copies of 1, which is n. So we have

$$\sum_{j=1}^{n} j^2 = \frac{1}{3}\left(n^3 + \frac{3n(n+1)}{2} - n\right).$$

A little algebra shows that the polynomial on the right-hand side can be simplified to $(2n^3 + 3n^2 + n)/6$, which factors to $n(n+1)(2n+1)/6$. So we have proved that

$$\sum_{j=1}^{n} j^2 = \frac{n(n+1)(2n+1)}{6}.$$

Now we know how to add up the first n square numbers. For example,

$$1^2 + 2^2 + 3^2 + \cdots + 99^2 + 100^2 = \frac{(100)(101)(201)}{6} = 338350.$$

Even Gauss might have had to wait until he was 11 years old to find that sum!

15.2 Displacement and Area

Let's move on from sigma notation, and spend some time investigating the following question:

> If you know the velocity of a car at every moment during some time interval, what is its total displacement over that time interval?

In symbols, this means that we know the velocity $v(t)$ at every time t in some interval $[a, b]$, and we want to find the displacement $x(t)$. We already know how to do this the other way around: if we know $x(t)$, then $v(t)$ is just $x'(t)$. That is, velocity is the derivative (with respect to time) of displacement. In order to answer the reverse question, let's look at some simple cases first.

15.2.1 Three simple cases

Consider three cars going in the forward direction along a long straight highway. Since the cars are always going forward, we can work with speed and distance instead of velocity and displacement (respectively)—there's no difference in this case. Each of the cars leaves from the same gas station at 3 p.m. and finishes the journey at 5 p.m.

The first car goes at a speed of 50 miles per hour the whole time. So $v(t) = 50$ for all t in the interval $[3, 5]$. To work out the distance traveled in this case, just use the fact that distance = average speed × time. Luckily, the average speed v_{av} and the instantaneous speed v are both equal to 50, since the speed never changes. So we get

$$\text{distance} = v \times t = 50 \times 2 = 100.$$

That is, the car has gone 100 miles. Now, if we draw the graph of v against t, it looks like this:

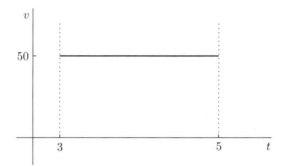

You can see a rectangle marked off between the solid line of the velocity at $v = 50$, the t-axis, and the vertical lines $t = 3$ and $t = 5$. The height of the rectangle is the speed 50 (mph), while its base is the time taken, 2 (hours). The quantity 50×2 is the **area** of the rectangle (in miles, but let's not get too bogged down about units for the moment). So in this case, the distance traveled is the area under the graph of v versus t.

As for the second car, it goes at a speed of 40 mph for the first hour; then at 4 p.m. it starts going 60 mph. Ignoring the few seconds that it takes to accelerate, the graph of the situation looks like this:

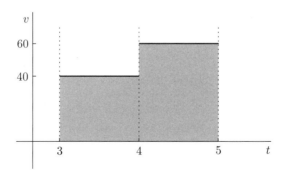

I've already shaded the area under the graph down to the t-axis between the lines $t = 3$ and $t = 5$, expecting this to be the distance. Let's check it out. During the first hour, the car travels at 40 mph, so the distance traveled is $40 \times 1 = 40$ miles. This is the area under the left-hand rectangle, which has height 40 (mph) and base 1 (hour). The same thing works for the second hour, where the distance traveled is $60 \times 1 = 60$ miles—the same as the area under the right-hand rectangle. The total distance traveled is again 100 miles.

The important thing to note is that we broke up the journey into pieces of time where the car was going at a constant speed, found the distance traveled for each piece, and then added them all up. Using a formula like $d = v_{av} \times t$ is no good on the whole journey unless you know the average speed. Wait, you say—the average speed here is obviously 50 mph, so there's no problem! OK, that's true, but let's look at the third car and then see if you still feel the same way.

The third car travels at 20 mph for the first 15 minutes, then goes 40 mph until 4 p.m. At that time, it switches to 60 mph for half an hour, before shifting to the slower speed of 50 mph for the rest of the journey. Once again ignoring the short accelerations and decelerations when the speed changes, the graph of v against t looks like this:

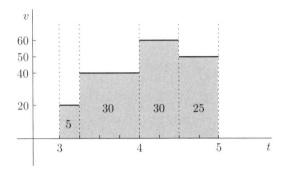

The average speed isn't obvious from looking at the graph. On the other hand, we can work out the distance by breaking the 2-hour time interval into smaller pieces corresponding to the four rectangles in the above graph:

- From 3 to 3.25 (which is the way to write 3:15 p.m. in decimal hours), the car traveled at 20 mph, so the distance traveled was $20 \times 0.25 = 5$ miles. That's the area of the first rectangle above, since its height is 20 mph and its base is 0.25 hours.
- From 3.25 to 4, the speed was 40 mph, so the distance was 40×0.75, or 30 miles. That's the area of the second rectangle.
- From 4 to 4.5 (that is, 4:30 p.m.), the car's speed was 60 mph, so the distance was $60 \times 0.5 = 30$ miles—the area of the third rectangle.
- Finally, from 4.5 to 5, the speed was 50 mph, so the distance traveled during that time was $50 \times 0.5 = 25$ miles, precisely the area of the fourth rectangle.

So, during the four time periods, the car went 5, 30, 30, and 25 miles, respectively, as shown on the above graph; the total is therefore $5 + 30 + 30 + 25 = 90$ miles. Finally, we've found the distance the third car traveled! This means that its average speed was actually $90/2 = 45$ mph, which isn't even one of the four speeds that the car went at. (This doesn't violate the Mean Value Theorem because the function in the above graph isn't differentiable.)

15.2.2   A more general journey

Let's look at a general framework to describe the sort of journey that the three cars made. Suppose that the time interval involved is $[a, b]$; also, suppose that we can chop up this interval into smaller intervals so that the car is going at a constant speed on each interval. We don't want to fix the number of intervals, so let's call it n. We also need to have some way of describing the beginning and end of each small interval:

- The first interval begins at time a and finishes at some later time t_1. Since a is earlier than t_1, we can say that $a < t_1$. In fact, it will be useful to also let $t_0 = a$, so that we have $a = t_0 < t_1$.
- The second interval begins at time t_1 and finishes at some later time t_2, so that $t_1 < t_2$.
- The third interval goes from t_2 to t_3, where $t_2 < t_3$.
- Keep going in the same way, so that the jth time interval starts at time t_{j-1} and ends at time t_j.
- The second-to-last interval goes from t_{n-2} to t_{n-1}, where $t_{n-2} < t_{n-1}$.
- Finally, the last interval goes from t_{n-1} to t_n, which is the same as the very end time b. So we have $t_{n-1} < t_n = b$.

All together, we can summarize the situation by saying that

$$a = t_0 < t_1 < t_2 < t_3 < \cdots < t_{n-2} < t_{n-1} < t_n = b.$$

We have chopped up the time interval $[a, b]$ into smaller intervals, which together are called a *partition* of the interval. On the number line, it looks something like this:

The dots in the middle are supposed to show that we don't want to fix the number of smaller intervals in the partition.

That takes care of the time aspect, but we need to talk about velocities. Let's suppose that the car goes at velocity v_1 during the first small time interval (t_0, t_1). This means that the graph of v against t will have a line segment above (t_0, t_1) at height v_1. As for the second interval, the velocity will then be v_2, so we get a different line segment at height v_2 above (t_1, t_2). This keeps on going until the last time interval (t_{n-1}, t_n), where the velocity will be v_n. Overall, the picture looks like this (for example):

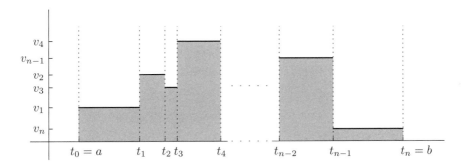

Now we're ready to calculate the total displacement. During the first small time interval (t_0, t_1), the car has gone at velocity v_1. The length of time is $(t_1 - t_0)$, so the displacement will be $v_1 \times (t_1 - t_0)$. Let's quickly repeat this for the second interval (t_1, t_2). The speed is v_2 and the length of time is $(t_2 - t_1)$, so the displacement is $v_2 \times (t_2 - t_1)$. Keep doing this all the way up to the last time interval (t_{n-1}, t_n). Finally, we add up all the displacements to see that

$$\text{total displacement} = v_1(t_1 - t_0) + v_2(t_2 - t_1) + \cdots$$
$$+ v_{n-1}(t_{n-1} - t_{n-2}) + v_n(t_n - t_{n-1}).$$

This is a perfect time to whip out the sigma notation that we looked at in Section 15.1 above. Check that you believe that we can write the above formula as

$$\text{total displacement} = \sum_{j=1}^{n} v_j(t_j - t_{j-1}).$$

Of course, this is also the shaded area in the above graph.

Let's see how the three examples from the previous sections fit into the framework. In each case, we know that $a = 3$ and $b = 5$.

- For the first car, we just have one interval $[3, 5]$, so set $n = 1$, $t_0 = 3$, and $t_1 = 5$. We also know that the velocity is $v_1 = 50$; so

$$\text{displacement} = \sum_{j=1}^{n} v_j(t_j - t_{j-1}) = v_1(t_1 - t_0) = 50(5 - 3) = 100.$$

- The second car needs two time intervals; set $n = 2$, $t_0 = 3$, $t_1 = 4$, and $t_2 = 5$, so that our partition looks like $3 < 4 < 5$. On the first interval, the velocity is $v_1 = 40$, while on the second interval, we have $v_2 = 60$. So

$$\text{displacement} = \sum_{j=1}^{n} v_j(t_j - t_{j-1}) = v_1(t_1 - t_0) + v_2(t_2 - t_1)$$
$$= 40(4 - 3) + 60(5 - 4) = 100.$$

- I'll let you fill in the full details for the third car. Suffice it to say that $n = 4$, the partition is $3 < 3.25 < 4 < 4.5 < 5$, and the velocities are $v_1 = 20$, $v_2 = 40$, $v_3 = 60$, and $v_4 = 50$, so

$$\text{displacement} = \sum_{j=1}^{n} v_j(t_j - t_{j-1})$$
$$= v_1(t_1 - t_0) + v_2(t_2 - t_1) + v_3(t_3 - t_2) + v_4(t_4 - t_3)$$
$$= 20(3.25 - 3) + 40(4 - 3.25) + 60(4.5 - 4) + 50(5 - 4.5)$$
$$= 5 + 30 + 30 + 25 = 90.$$

Notice that the calculations are identical to the ones we did in the previous section—only the notation has changed.

15.2.3   Signed area

What if our car goes backward? For example, suppose that the car goes forward at 40 mph between 3 and 4 p.m., then backward at 30 mph until 6 p.m. The graph looks like this:

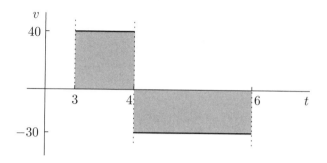

Now it's really important to distinguish between distance and displacement. Between 3 and 4 p.m., the distance and displacement are both 40 miles. From 4 to 6 p.m., the car travels a total of $30 \times 2 = 60$ miles, so the total distance traveled from 3 p.m. to 6 p.m. is $40 + 60 = 100$ miles. On the other hand, the displacement is $40 + (-60) = -20$ miles, since the second part of the journey is backward. This means that the car finishes up 20 miles back from where it started.

Now look at the above graph. The rectangle on the left has area 40 (miles), no problem, but the right-hand rectangle is interesting. Its base has length 2 (hours), and if you consider its height as 30 (mph), then sure enough, the area is 60 (miles). Adding the two areas gives $40 + 60 = 100$ miles, which is the distance.

On the other hand, take another look at that second rectangle. Suppose that we say that its "height" is actually -30 mph, since the rectangle goes below the horizontal axis. Of course, a rectangle can't actually have a negative height, but nevertheless it would be good to distinguish between rectangles above and below the axis. So if the "height" is -30 mph, then the "area" is $2 \times (-30) = -60$ miles. Let's drop the quotation marks and correctly refer to this as the *signed area*. Our convention, then, is that areas below the axis count as negative toward the total. If we do that, then the total signed area is 40 miles (from the first piece) plus -60 miles (from the second), giving a total of -20 miles. Hey, the displacement is -20 miles!

In terms of our formulas from the previous section, we have a partition of the total time interval $[3, 6]$ that looks like $3 < 4 < 6$. The first velocity is $v_1 = 40$ while the second is $v_2 = -30$. So we have

$$\text{displacement} = \sum_{j=1}^{n} v_j(t_j - t_{j-1}) = v_1(t_1 - t_0) + v_2(t_2 - t_1)$$

$$= 40(4 - 3) + (-30)(6 - 4) = -20.$$

If instead we take $v_2 = 30$, which is the speed (not the velocity!) during the second part of the journey, then the last sum is $40(4 - 3) + 30(6 - 4) = 100$, which gives the distance in miles. Of course, the speed 30 mph is the absolute value of the velocity -30 mph. So instead of adding up the actual (unsigned) area in the graph above to get the distance, we could graph $|v|$ against t:

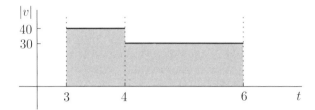

Now it's irrelevant whether the area is signed or not because there's nothing below the horizontal axis! So, we'll make the convention that **all areas are signed**. If we want the unsigned area, we'll take absolute values first. See Section 16.4.1 in the next chapter for some more on this point.

15.2.4 Continuous velocity

We've seen that if a car (or other object) moves along a straight line so that the velocity is constant on a finite number of intervals in a partition of $[a, b]$, then the displacement is the signed area between the graph of v versus t, the t-axis, and the lines $t = a$ and $t = b$. The distance is the same thing, except that you start with the graph of $|v|$ versus t instead.

What if the velocity **isn't** constant on a finite number of intervals? Unless you never turn off the cruise control, you'll be speeding up from time to time to pass another car, slowing down when you see a cop, and so on. Even getting from 40 to 60 mph requires some acceleration—you can't just change speeds instantaneously. So, let's consider the situation where velocity v is a **continuous** function of time t, for example:

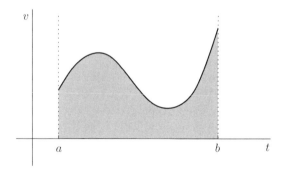

So the car is speeding up, then slowing down, and finally speeding up even more. The displacement should still be the shaded signed area—actually, since everything's above the t-axis, this is also the distance traveled. How on earth do we find the area?

Here's the idea. Our velocity is changing quite a lot over the whole period from a to b, but it doesn't change as much during a shorter period. Let's take some little interval of time, which we'll call $[p, q]$, and just focus on what happens during that interval. Even on this little interval, the velocity is changing, but let's pretend that it doesn't. Let's **sample** the velocity by picking some instant of time c during $[p, q]$, and seeing what the velocity is then. We'll pretend that the sampled velocity is the actual velocity for the whole interval $[p, q]$. If we write the velocity v as $v(t)$ to emphasize that v is a function of t, then the velocity at time c is $v(c)$. So, here's a graphical interpretation of what we're doing:

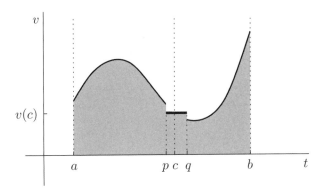

We've flattened out the curve above $[p, q]$ at a height of $v(c)$. The advantage of this is that we can get some idea about the displacement over the time interval $[p, q]$. The area of the little rectangle of height $v(c)$ and base $q - p$ is $v(c) \times (q - p)$. Now, this isn't **actually** the correct displacement over that time period, but it's mighty close.

Why stop at just one little interval like $[p, q]$? Let's repeat the process on an entire partition of $[a, b]$. Starting with the partition

$$a = t_0 < t_1 < t_2 < \cdots < t_{n-2} < t_{n-1} < t_n = b,$$

let's sample the velocity during **each** time period. The first time interval is from t_0 to t_1, so let's pick some time c_1 in that interval and pretend that the velocity is equal to $v(c_1)$ for the whole period. The number c_1 could be equal to the beginning number t_0 or the end number t_1, or some number in between, as long as it lies in $[t_0, t_1]$. Now, repeat this for the second interval: pick c_2 in the interval $[t_1, t_2]$, and use $v(c_2)$ as the sample velocity for that period. Keep doing this for every interval, up until c_n in the interval $[t_{n-1}, t_n]$. Here's an example of what this could look like with $n = 6$:

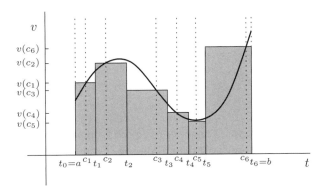

All we've done is approximate the nice smooth velocity curve using some staircase-like function, where each step intersects the curve. We can use the techniques from the previous sections to work out the shaded (signed) area, which will be an approximation to the actual area under the curve. We get

$$\text{area under velocity curve} \cong \sum_{j=1}^{n} v(c_j)(t_j - t_{j-1}).$$

Unfortunately, the approximation is pretty lousy. That big rectangle on the right in the picture at the bottom of the previous page doesn't really do a great job of approximating the area under the part of the curve above $[t_5, t_6]$, since there's so much of the rectangle above the curve. So let's take a different partition with more intervals which are smaller, for example:

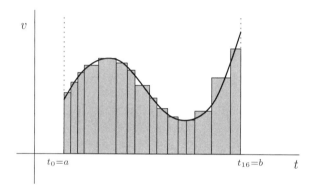

Here we used 16 partitions instead of 6, and it looks as if the shaded area is a much better approximation to the actual area than our previous attempt yielded. This wouldn't have been true if we used a lot of intervals in our partition, but some of the little intervals were still quite wide. For example, check out this picture:

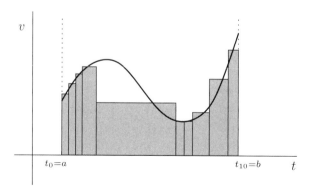

Even though most of the rectangles are pretty narrow, that one big-ass rectangle screws up the approximation. So somehow we need to make **all** the little time intervals small. The way to do this is to let the *mesh* of the partition be the longest of all the time intervals, then insist that the mesh get smaller and smaller, eventually down to 0 in the limit. That way, all the time intervals will become small and you won't have a huge rectangle like the one in the above picture.

Formally, the mesh is defined by

$$\text{mesh} = \text{maximum of } (t_1 - t_0), (t_2 - t_1), \ldots, (t_{n-1} - t_{n-2}), (t_n - t_{n-1}).$$

For example, if you have the partition $3 < 3.25 < 4 < 4.5 < 5$ of $[3, 5]$ (which was the partition that we used for the third car in Section 15.2.1 above), then the lengths of the little intervals are 0.25 (which is $3.25 - 3$), 0.75 (that's $4 - 3.25$), 0.5 ($4.5 - 4$), and 0.5 ($5 - 4.5$). The largest of the quantities 0.25, 0.75, 0.5, 0.5 is 0.75, so the mesh of the partition is 0.75.

Now we can try to replace the approximation

$$\sum_{j=1}^{n} v(c_j)(t_j - t_{j-1})$$

by a limit to get the actual answer. Suppose we repeat the above procedure over and over again, each time taking a partition which has a smaller mesh than the previous one, so that the meshes go down to 0 in the limit. Then the approximations should get better and better. This is what we're trying to achieve in the following formula:

$$\text{actual area under velocity curve} = \lim_{\text{mesh} \to 0} \sum_{j=1}^{n} v(c_j)(t_j - t_{j-1}).$$

For the mesh to go to 0, we need the number of small intervals in the partition to get larger and larger, so the limit automatically includes the idea that $n \to \infty$ as well.

15.2.5 Two special approximations

The above formula leaves a lot to be desired. How do you know that you get the same answer no matter what partitions you take and no matter how you choose the sampling times c_j? It's actually a theorem that if v is a continuous function of t, then the above limit is independent of the partitions and sampling times. The proof of the theorem is a little advanced for this book, but can be found in most textbooks on real analysis. On the other hand, we can get an idea of the flavor of the proof by investigating two special approximations: the upper sum and the lower sum.

Starting with a partition, we are allowed to pick sample points in each of the little intervals. Suppose that we always pick a point where the velocity is the greatest possible. For example, we'll choose c_1 in the interval $[t_0, t_1]$ so that $v(c_1)$ is the maximum possible value of v on that interval. We'll do the same for each of the intervals. This means that all our steps lie above the curve. Here's an example of what this looks like:

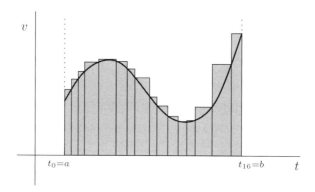

The area of the rectangles, which is called an *upper sum*, is clearly bigger than the area under the curve. On the other hand, if we always sample the lowest possible velocity, then we get a situation like this:

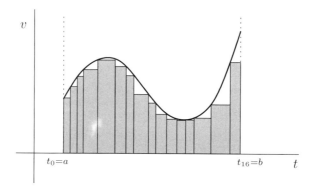

The partition is the same, but the sampling times are different. Because of the way they're chosen, all the steps lie below the curve; the area of all the rectangles, which is called a *lower sum*, is less than the area under the curve.

Combining these observations, we have

$$\text{lower sum} \leq \text{actual area under curve} \leq \text{upper sum}.$$

In fact, for the same partition, any choice of the sampling times c_j will lead to an area between the lower sum and the upper sum. If you use a sequence of partitions with smaller and smaller meshes, then the lower sum and the upper sum have the same limit (that's what I'm not going to prove). The sandwich principle then shows that the formula at the end of the previous section makes sense. It doesn't matter what values of c_j you choose—your sums are trapped, along with the actual area, between the lower and upper sums. As the mesh goes to zero, the sandwich principle ensures that your sums converge to the correct area.

We now have all the tools we need to define the definite integral. This is the subject of the next chapter....

CHAPTER 16 _____

Definite Integrals

Now it's time to get some facts straight about definite integrals. First we'll give an informal definition in terms of areas; then we'll use our ideas about partitions from the previous chapter to tighten up the definition. After one (exhausting) example of applying the tightened-up definition, we'll see what else we can say about definite integrals. More precisely, we'll look at the following topics:

- signed areas and definite integrals;
- the definition of the definite integral;
- an example using this definition;
- basic properties of definite integrals;
- using integrals to find unsigned areas, the area between two curves, and areas between a curve and the y-axis;
- estimating definite integrals;
- average values of functions and the Mean Value Theorem for integrals; and
- an example of a nonintegrable function.

16.1 The Basic Idea

We start off with some function f and an interval $[a, b]$. Take the graph of $y = f(x)$, and consider the region between the curve, the x-axis, and the two vertical lines $x = a$ and $x = b$:

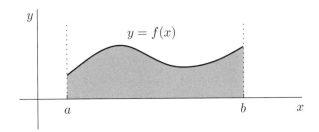

It would be nice to have a compact way to express the area of the shaded region. Since there aren't actually any units of length in the above picture, we'll just call them "units," so that area is measured in "square units." (If the above picture actually had some units, like inches, marked on it, the area would be given in square inches instead.) In any case, let's say that the area of the shaded region above, in square units, is

$$\int_a^b f(x)\,dx.$$

This is a *definite integral*. You would read it out loud as "the integral from a to b of $f(x)$ with respect to x." The expression $f(x)$ is called the *integrand*, and tells you what the curved part looks like. The a and b tell you where the two vertical lines go, and are called the *limits of integration* (not to be confused with regular old limits!) or the *endpoints of integration*. Finally, the dx tells you that x is the variable on the horizontal axis. Actually, x is a dummy variable—you can change it to any other letter, provided that you change it everywhere. So all the following are equal to each other:

$$\int_a^b f(x)\,dx = \int_a^b f(t)\,dt = \int_a^b f(q)\,dq = \int_a^b f(\beta)\,d\beta.$$

In fact, they are all equal to the same number, which is the shaded area (in square units) in the above picture; the only difference is that we are renaming the x-axis to be the t-axis, q-axis, or β-axis. This doesn't affect the value of the area!

What if the function dips below the x-axis? The situation could look like this:

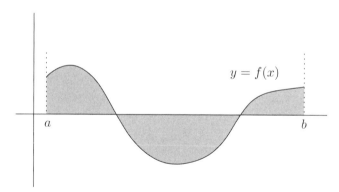

As we saw in Section 15.2.3 of the previous chapter, it makes sense for the part of the area below the x-axis to count as negative area. If all of the curve $y = f(x)$ between $x = a$ and $x = b$ actually lies below the x-axis, then the integral must be negative. In general, the integral gives the total amount of signed area. More precisely,

$$\int_a^b f(x)\,dx \text{ is the signed area (in square units) of the region between the}$$
curve $y = f(x)$, the lines $x = a$ and $x = b$, and the x-axis.

Note that the integral is a number, but the area is in square units.

We saw in the previous chapter that the displacement of an object between time $t = a$ and $t = b$ is the signed area between the graph of $y = v(t)$, the t-axis, and the lines $t = a$ and $t = b$. We also saw that the distance the object traveled is found in exactly the same way, except that you deal with $y = |v(t)|$ instead. Using our new notation, we can say that

$$\text{displacement} = \int_a^b v(t)\, dt \qquad \text{and} \qquad \text{distance} = \int_a^b |v(t)|\, dt.$$

Here the understanding is that our clock starts at $t = a$ and ends at $t = b$. Notice that the dummy variable here is t, and that the integrands are the velocity $v(t)$ and speed $|v(t)|$, respectively.

16.1.1   Some easy examples

Now, let's look at a few simple examples of definite integrals. First, consider

$$\int_0^1 x\, dx \qquad \text{and} \qquad \int_0^2 x\, dx.$$

In both cases, the integrand is x, so we should start off by drawing the graph of $y = x$. In the first case, the area is from $x = 0$ to $x = 1$, while the second case goes from $x = 0$ to $x = 2$. So we are looking for the following two areas (respectively):

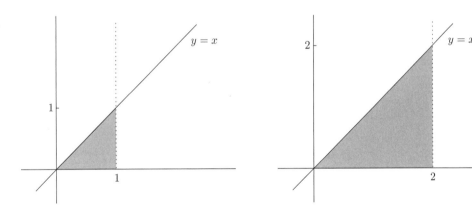

These areas are easy to find: both regions are triangles. The first has base and height both equal to 1 unit, so its area turns out to be $\frac{1}{2}(1)(1) = \frac{1}{2}$ square units; while the second has base and height both equal to 2 units, so the area is $\frac{1}{2}(2)(2) = 2$ square units. We have shown that

$$\int_0^1 x\, dx = \frac{1}{2} \qquad \text{and} \qquad \int_0^2 x\, dx = 2.$$

Now, let's use these formulas to solve a practical problem. Suppose that a car starts at rest, then accelerates at a constant rate of 1 yard per second squared; its speed (in yards per second) will be given by $v(t) = t$. So how fast does the car go in one second? How about two seconds? The answer is given by the above integrals! Just replace x by t and you're golden. First, note that the displacement and distance are the same thing, since the car's going in the positive direction all the time. So, in the first second, we have

$$\text{displacement} = \int_0^1 v(t)\,dt = \int_0^1 t\,dt = \frac{1}{2},$$

while for the first two seconds, we have

$$\text{displacement} = \int_0^2 v(t)\,dt = \int_0^2 t\,dt = 2.$$

These displacements are in yards, of course.

Now, let's take a look at another definite integral:

$$\int_{-2}^5 1\,dx.$$

To find the value of this integral, we need to draw a graph of $y = 1$, then put in the vertical lines $x = -2$ and $x = 5$. The area we're looking for looks like this:

So it's a rectangle of height 1 unit and base 7 units, which has area 7 square units. This means that

$$\int_{-2}^5 1\,dx = 7.$$

In fact, the more general integral

$$\int_a^b 1\,dx$$

represents the area of this region:

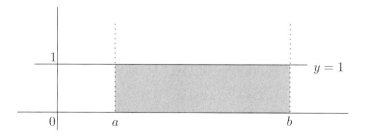

The rectangle has height 1 unit and base $b - a$ units (even if a and/or b are negative), so we have

$$\int_a^b 1\, dx = b - a$$

in general. This could also be written as simply

$$\int_a^b dx = b - a,$$

since we can think of $1\, dx$ as being just dx.

Finally, what is

$$\int_{-\pi}^{\pi} \sin(x)\, dx?$$

Let's draw a graph and see what area we're trying to find:

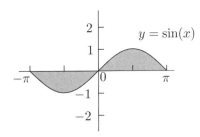

Luckily, we need the signed area, not the actual area. By symmetry, the area above the axis (between 0 and π) exactly matches the area below the axis (between $-\pi$ and 0), so they cancel out and the signed area is 0 square units. That is,

$$\int_{-\pi}^{\pi} \sin(x)\, dx = 0.$$

If you want the **actual** area, not the signed area, then you have to be more careful and chop up the integral into two pieces. We'll see how to do this in general in Section 16.4.1 below, then return to the above example in Section 17.6.3 of the next chapter.

Before we move on, I'd like to point out a generalization of the previous example. It turns out that the reason the integral is 0 is that the integrand $\sin(x)$ is an odd function of x, and also that the region of integration $[-\pi, \pi]$ is symmetric about 0. We could replace $\sin(x)$ by any other odd function of x, and we could change the bounds to $-a$ and a for any number a, and the integral would still be 0. That is,

if f is an odd function, then $\qquad \int_{-a}^{a} f(x)\, dx = 0 \qquad$ for any a.

This is true by symmetry: every bit of area above the x-axis has a corresponding bit of area below the x-axis, just as in the above picture. This fact can actually save you a lot of time, since it means that you don't have to do any calculations if your integral happens to fit the above template. We'll give a more formal proof of the above fact at the end of Section 18.1.1 of Chapter 18.

16.2   Definition of the Definite Integral

We have a nice working definition of the definite integral in terms of area, but that doesn't really help us to calculate specific integrals. Sure, we got by in the last few examples, but only because we already know how to find the area of any triangle or rectangle. We also got lucky with that last example involving $\sin(x)$, because everything canceled out. In general, we won't be so lucky.

Actually, we've been in this situation before in the case of derivatives. We could have defined $f'(x)$ to be the slope of the tangent to $y = f(x)$ at the point $(x, f(x))$, but that wouldn't have told us how to find the slope. Instead, we defined $f'(x)$ by the formula

$$f'(x) = \lim_{h \to 0} \frac{f(x+h) - f(x)}{h},$$

provided that the limit exists. As we've observed, this limit is of the indeterminate form $0/0$, but we can still work it out in many cases. Anyway, once we've made the above definition, the **interpretation** is that $f'(x)$ represents the slope of the tangent we're interested in.

Unfortunately, the definition of the definite integral is a lot nastier than the above definition of the derivative. The good news is that we've already done the grunt work in the previous chapter, and we can just state the definition:

$$\int_a^b f(x)\, dx = \lim_{\text{mesh} \to 0} \sum_{j=1}^n f(c_j)(x_j - x_{j-1}),$$

where $a = x_0 < x_1 < \cdots < x_{n-1} < x_n = b$ and c_j is in $[x_{j-1}, x_j]$ for each $j = 1, \ldots, n$.

Even though that definition is wordy, it **still** doesn't tell the full story! You also need to be aware of the following points:

- The expression $a = x_0 < x_1 < \cdots < x_{n-1} < x_n = b$ means that the points $x_0, x_1, x_2, \ldots, x_{n-1}$, and x_n form a partition of the interval $[a, b]$, with $x_0 = a$ on the left and $x_n = b$ on the right. The partition creates n smaller intervals $[x_0, x_1]$, $[x_1, x_2]$, and so on up to $[x_{n-1}, x_n]$.

- The mesh of the partition is the maximum length of these smaller intervals; so we have

  mesh = maximum of $(x_1 - x_0), (x_2 - x_1), \ldots, (x_{n-1} - x_{n-2}), (x_n - x_{n-1})$.

- The numbers c_j can be chosen anywhere in their corresponding smaller intervals, one for each smaller interval. This is what is meant by saying that c_j is in $[x_{j-1}, x_j]$.

- The above limit is taken by repeating the calculation of the sum for different partitions with smaller and smaller mesh, and consequently more and more smaller intervals; that is, as mesh $\to 0$, we must also have $n \to \infty$. Each partition involves a choice of all the numbers c_j.

- If f is continuous, then it doesn't matter what partitions are used, nor which c_j are chosen, as long as the mesh goes to 0. In fact, this is also true if f has a finite number of discontinuities, as long as f is bounded. Such functions are referred to as *integrable*, since they can be integrated. There are functions which are integrable even though they might have infinitely many discontinuities, but that's a little advanced for this book. On the other hand, if f is unbounded, which would happen (for example) if it has a vertical asymptote, then the integral is called improper; see Chapters 20 and 21 for how to deal with this sort of thing.

- The sum

$$\sum_{j=1}^{n} f(c_j)(x_j - x_{j-1})$$

which appears in the definition is called a *Riemann sum*. It gives an approximate value for the integral. If the mesh of the partition is very small, the approximation should be pretty good.

See, I told you it was nasty! Now we'll see how to use the definition to calculate a definite integral.

16.2.1 An example of using the definition

 Let's use the above formula to find the following integral:

$$\int_0^2 x^2 \, dx.$$

So we are looking for the following area:

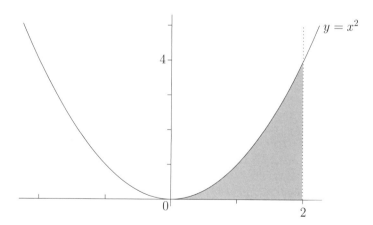

This isn't a triangle or a rectangle, and nothing cancels out since the area is entirely above the x-axis. So let's set $f(x) = x^2$ and use the definition of the definite integral to find the area.

We need to take partitions with smaller and smaller meshes. By far the easiest way to do this is to use small intervals of equal size. So, we want to chop up our interval $[0, 2]$ into n pieces, each the same length. Since the total length is 2, and we're using n pieces, each piece must have length $2/n$ units. The first piece goes from 0 to $2/n$; the second piece goes from $2/n$ to $4/n$; and so on. Zooming in on the region of interest, here's a picture of what we've done:

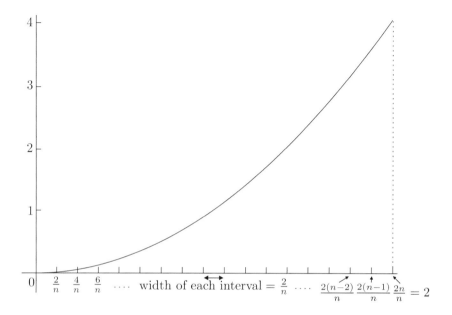

In this case, the general partition

$$a = x_0 < x_1 < x_2 < \cdots < x_{n-1} < x_n = b$$

specializes to

$$0 = \frac{0}{n} < \frac{2}{n} < \frac{4}{n} < \cdots < \frac{2(n-1)}{n} < \frac{2n}{n} = 2.$$

The mesh of this partition is $2/n$, since **every** smaller interval has width $2/n$. It's also pretty clear that the formula for a general x_j in this partition is $2j/n$. Now, we need to choose our numbers c_j. For example, c_0 could be anywhere in the interval $[0, 2/n]$, c_1 could be anywhere inside $[2/n, 4/n]$, and so on. We'll make life simple by always choosing the right endpoint of each smaller interval, so that $c_j = x_j = 2j/n$. That is,

$$c_j = \frac{2j}{n} \quad \text{is our choice for the smaller interval} \quad [x_{j-1}, x_j] = \left[\frac{2(j-1)}{n}, \frac{2j}{n}\right].$$

This will lead to the following rectangles:

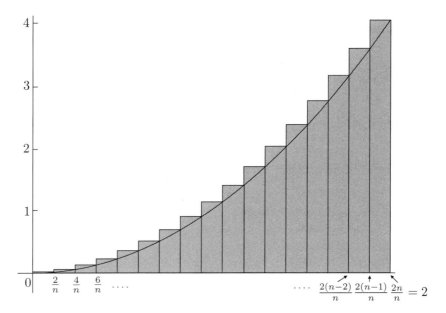

So we're actually dealing with an upper sum here—all the rectangles lie above the curve. (See Section 15.2.5 in the previous chapter for a more general discussion of upper sums.)

Now, we're finally ready to use the formula. Consider the Riemann sum

$$\sum_{j=1}^{n} f(c_j)(x_j - x_{j-1}).$$

We know that $f(x) = x^2$, $c_j = 2j/n$, $x_j = 2j/n$ as well, and $x_{j-1} = 2(j-1)/n$. So the sum becomes

$$\sum_{j=1}^{n} \left(\frac{2j}{n}\right)^2 \left(\frac{2j}{n} - \frac{2(j-1)}{n}\right).$$

The right-hand factor simplifies to $2/n$. This isn't surprising, since that's the width of each of the rectangles. On the other hand, the rectangles all have different heights, which are given by the first factor $(2j/n)^2$, as j ranges from 1 to n. In any case, the above sum simplifies to

$$\sum_{j=1}^{n} \frac{4j^2}{n^2} \times \frac{2}{n} = \sum_{j=1}^{n} \frac{8j^2}{n^3}.$$

Now what? Well, the denominator n^3 doesn't depend on the dummy variable j, so we can take it out as a common factor to write the sum as

$$\frac{8}{n^3} \sum_{j=1}^{n} j^2.$$

In Section 15.1.2 of the previous chapter, we actually found that the value of the above sum is $n(n + 1)(2n + 1)/6$. This means that

$$\frac{8}{n^3} \sum_{j=1}^{n} j^2 = \frac{8}{n^3} \times \frac{n(n + 1)(2n + 1)}{6} = \frac{4(n + 1)(2n + 1)}{3n^2}.$$

All in all, we have shown that the shaded area of the rectangles in the above picture (in square units) is given by

$$\sum_{j=1}^{n} f(c_j)(x_j - x_{j-1}) = \frac{4(n+1)(2n+1)}{3n^2}.$$

This is only an approximation to the area we're looking for. Since the mesh of the partition is $2/n$, we can force the mesh to go to 0 by letting $n \to \infty$. The rectangles become smaller and smaller, but there are more and more of them which hug the curve $y = x^2$ better and better. So we have

$$\int_0^2 x^2 \, dx = \lim_{\text{mesh} \to 0} \sum_{j=1}^{n} f(c_j)(x_j - x_{j-1}) = \lim_{n \to \infty} \frac{4(n+1)(2n+1)}{3n^2}.$$

All that's left is to find the last limit. You can use the techniques from Section 4.3 of Chapter 4 to show that the limit is $8/3$, so we have finally shown that

$$\int_0^2 x^2 \, dx = \frac{8}{3}.$$

The area we're looking for is $8/3$ square units. Now you should try to repeat the above method to show that

$$\int_0^1 x^2 \, dx = \frac{1}{3}.$$

As you can tell, this method is a pain in the butt. Not only is it long and involved, but you also need to know how to find the sum

$$\sum_{j=1}^{n} j^2.$$

If the integrand was x^3 instead of x^2, you'd need to deal with

$$\sum_{j=1}^{n} j^3.$$

Things would be even worse if the integrand happened to be $\sin(x)$ or something similar. So we need another method in order to avoid all these rectangles and sums. That will have to wait until we look at the Second Fundamental Theorem of Calculus in the next chapter. In the meantime, let's look at some nice properties of definite integrals.

16.3 Properties of Definite Integrals

Let's extend our definition of the definite integral a little bit. What do you think of

$$\int_2^0 x^2 \, dx?$$

The only difference between this integral and the one we calculated in the previous section is that the integral goes from 2 to 0, instead of 0 to 2. So what on earth is a partition of the interval $[2, 0]$? That's not even a legitimate interval, since 2 is greater than 0. The best we can do is to take a backward partition, like this:

$$2 = x_0 > x_1 > x_2 > \cdots > x_{n-1} > x_n = 0.$$

Now the quantity $(x_j - x_{j-1})$, which appears in the definition of the definite integral, is always negative. Our rectangles effectively have negative base length! The end result is that

$$\int_2^0 x^2 \, dx = -\frac{8}{3}.$$

So **if you reverse the limits of integration, you need to put in a minus sign out front**. In general, for an integrable function f and numbers a and b, we have

$$\boxed{\int_b^a f(x) \, dx = -\int_a^b f(x) \, dx.}$$

Another way of looking at this is that if you go backward in time, then the displacement is reversed. For example, if you make a movie of a car which is going forward, then play it in reverse, the car will appear to have gone backward, so the displacement should be negative.

Now, what if the limits of integration are equal? For example, consider

$$\int_3^3 x^2 \, dx.$$

This isn't much of an area. After all, there's no area between $x = 3$ and $x = 3$ at all! So the answer must be 0. In fact, it's generally true that

$$\boxed{\int_a^a f(x) \, dx = 0}$$

for any number a and function f defined at a. Again, this makes sense in terms of the physical interpretation: between times a and a, which is no time at all, an object can't move at all, so it has no displacement.

Moving on, let's consider the following picture:

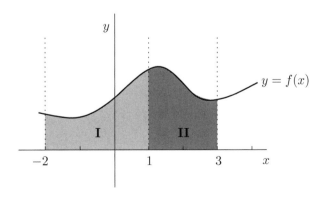

The whole area, from $x = -2$ to $x = 3$, is clearly the sum of the two areas labeled **I** and **II**. By definition, we have

$$\text{area of } \mathbf{I} = \int_{-2}^{1} f(x)\,dx \qquad \text{and} \qquad \text{area of } \mathbf{II} = \int_{1}^{3} f(x)\,dx,$$

respectively; the conclusion is that

$$\int_{-2}^{3} f(x)\,dx = \int_{-2}^{1} f(x)\,dx + \int_{1}^{3} f(x)\,dx.$$

All we've done is split up the area into two pieces and express this in terms of integrals. Of course, we could have split up the integral using any number in the interval $[-2, 3]$, as long as we replaced both the 1s in the above formula by the same number. In fact it even works when the number is outside the interval $[-2, 3]$. For example, the following formula is true:

$$\int_{-2}^{3} f(x)\,dx = \int_{-2}^{4} f(x)\,dx + \int_{4}^{3} f(x)\,dx.$$

Here's a picture of what's going on:

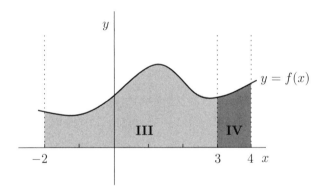

This time, we have

$$\text{area of } \mathbf{III} = \int_{-2}^{3} f(x)\,dx \qquad \text{and} \qquad \text{area of } \mathbf{IV} = \int_{3}^{4} f(x)\,dx.$$

So we can add them up to see that

$$\int_{-2}^{4} f(x)\,dx = \int_{-2}^{3} f(x)\,dx + \int_{3}^{4} f(x)\,dx.$$

Now reverse the limits of integration in the final integral above to get

$$\int_{-2}^{4} f(x)\,dx = \int_{-2}^{3} f(x)\,dx - \int_{4}^{3} f(x)\,dx.$$

It's pretty easy to rearrange this and get the desired formula

$$\int_{-2}^{3} f(x)\,dx = \int_{-2}^{4} f(x)\,dx + \int_{4}^{3} f(x)\,dx.$$

In general, for any integrable function f and numbers a, b, and c, we have

$$\int_a^b f(x)\,dx = \int_a^c f(x)\,dx + \int_c^b f(x)\,dx.$$

You can split an integral into two pieces, even if the break point c is outside the original interval $[a, b]$, as long as in both pieces the integrand f is still integrable.

For example, to find

$$\int_1^2 x^2\,dx,$$

we can use two facts that we've already found in the previous section:

$$\int_0^2 x^2\,dx = \frac{8}{3} \qquad \text{and} \qquad \int_0^1 x^2\,dx = \frac{1}{3}.$$

All you have to do is split up the first integral at $x = 1$, like this:

$$\int_0^2 x^2\,dx = \int_0^1 x^2\,dx + \int_1^2 x^2\,dx.$$

Using the above facts, this becomes

$$\frac{8}{3} = \frac{1}{3} + \int_1^2 x^2\,dx,$$

so we have

$$\int_1^2 x^2\,dx = \frac{8}{3} - \frac{1}{3} = \frac{7}{3}.$$

I now leave it to you to show that

$$\int_1^2 x\,dx = \frac{3}{2}$$

using the following facts from Section 16.1.1 above:

$$\int_0^2 x\,dx = 2 \qquad \text{and} \qquad \int_0^1 x\,dx = \frac{1}{2}.$$

There are two more simple properties of integrals which are even more useful. The first is that **constants move through integral signs**. That is, for any integrable f and numbers a, b, and C,

$$\int_a^b Cf(x)\,dx = C\int_a^b f(x)\,dx.$$

This is not true if C depends on x! C has to be constant. It's actually quite easy to prove this. Just write

$$\int_a^b Cf(x)\,dx = \lim_{\text{mesh}\to 0} \sum_{j=1}^n Cf(c_j)(x_j - x_{j-1})$$

and pull the constant C out of the sum **and** the limit:

$$\int_a^b Cf(x)\,dx = C \lim_{\text{mesh}\to 0} \sum_{j=1}^n f(c_j)(x_j - x_{j-1}) = C \int_a^b f(x)\,dx.$$

For example, to find

$$\int_0^2 7x^2\,dx,$$

just drag the 7 outside the integral:

$$\int_0^2 7x^2\,dx = 7\int_0^2 x^2\,dx = 7\left(\frac{8}{3}\right) = \frac{56}{3}.$$

The second property is that **integrals respect sums and differences**. That is, if f and g are both integrable functions, and a and b are two numbers, then

$$\boxed{\int_a^b (f(x) + g(x))\,dx = \int_a^b f(x)\,dx + \int_a^b g(x)\,dx.}$$

The same is true if you change both plus signs to minus signs. Either version is easy to show using partitions. All you have to do is break up the sum and limit, like this:

$$\int_a^b (f(x) + g(x))\,dx = \lim_{\text{mesh}\to 0} \sum_{j=1}^n (f(c_j) + g(c_j))(x_j - x_{j-1})$$

$$= \lim_{\text{mesh}\to 0} \sum_{j=1}^n f(c_j)(x_j - x_{j-1}) + \lim_{\text{mesh}\to 0} \sum_{j=1}^n g(c_j)(x_j - x_{j-1})$$

$$= \int_a^b f(x)\,dx + \int_a^b g(x)\,dx.$$

The same thing works with minus signs instead of plus signs.
For example, to find

$$\int_0^2 (3x^2 - 5x)\,dx,$$

split up the integral and also drag the constants through the integral signs. We get

$$\int_0^2 (3x^2 - 5x)\,dx = 3\int_0^2 x^2\,dx - 5\int_0^2 x\,dx = 3\left(\frac{8}{3}\right) - 5(2) = -2.$$

Here we have used the facts from above that

$$\int_0^2 x^2\,dx = \frac{8}{3} \qquad \text{and} \qquad \int_0^2 x\,dx = 2.$$

16.4   Finding Areas

If $y = f(x)$, then we can write

$$\int_a^b y \, dx$$

instead of using $f(x)$ as the integrand. This has a nice geometrical interpretation: if we look at one of our thin rectangles, or strips, arising from the partition method, we can think of it as having height y units and width equal to some small length dx units:

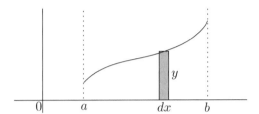

The area of the strip is the height times the width, or $y \, dx$ square units. Now draw in more strips so that the bases form a partition of $[a, b]$. If we were to add up the areas of all these strips, we'd get an approximating sum. The beauty of the integral sign is that it not only adds up the areas of all the strips, it also takes the limit as all the strip widths go to 0 (in the limit).

This idea is useful in helping to understand how to use the integral to find areas. Now, let's spend a little time looking at how to find three specific types of areas: unsigned area, the area between two curves, and the area between a curve and the y-axis.

16.4.1   Finding the unsigned area

We've seen that definite integrals deal with signed areas. Sure, if your curve is always above the x-axis, then it doesn't matter whether the area is signed or unsigned. But what if some of the curve lies below the axis? For example, suppose that $f(x) = -x^2 - 2x + 3$ and the region of interest is between $x = 0$ and $x = 2$. Since $f(0) = 3$ and $f(2) = -5$, the curve $y = f(x)$ looks like this:

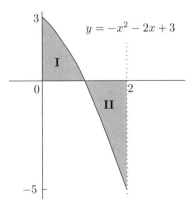

If you treat the shaded area as signed, so that the area of the region labeled **II** counts as negative, then we have

$$\text{signed area} = \int_0^2 (-x^2 - 2x + 3)\, dx = -\int_0^2 x^2\, dx - 2\int_0^2 x\, dx + 3\int_0^2 1\, dx.$$

Here we've broken up the integral using the principles from the previous section. We also know what all three integrals are, having found them above. We get

$$\text{shaded signed area} = -\frac{8}{3} - 2(2) + 3(2) = -\frac{2}{3} \text{ square units.}$$

This is clearly not the unsigned area, since it's negative! So, how do you find the unsigned area? The trick is to break up the integral into pieces to isolate the bits of area above and below the axis, then add up their absolute values. In the above example, we need to know where the curve hits the x-axis. So just solve $-x^2 - 2x + 3 = 0$ and you will see that $x = 1$ or $x = -3$. Clearly $x = 1$ is what we're looking for here, since it's between 0 and 2, while -3 isn't.

Now we can write down two integrals:

$$\int_0^1 (-x^2 - 2x + 3)\, dx \qquad \text{and} \qquad \int_1^2 (-x^2 - 2x + 3)\, dx.$$

These represent the signed areas of regions **I** and **II**, respectively, in the above picture. To calculate the integrals, you'll need some formulas that we've developed earlier in this chapter:

$$\int_0^1 x^2\, dx = \frac{1}{3}; \qquad \int_0^1 x\, dx = \frac{1}{2}; \qquad \int_0^1 1\, dx = 1;$$

$$\int_1^2 x^2\, dx = \frac{7}{3}; \qquad \int_1^2 x\, dx = \frac{3}{2}; \qquad \int_1^2 1\, dx = 1.$$

I leave it to you to work out that

$$\int_0^1 (-x^2 - 2x + 3)\, dx = \frac{5}{3} \qquad \text{and} \qquad \int_1^2 (-x^2 - 2x + 3)\, dx = -\frac{7}{3}.$$

As expected, the first integral is positive since region **I** is above the axis, and the second is negative since region **II** lies below the axis. Also, the sum of the two integrals is $-2/3$, which is the signed area (in square units). Now, here's the important point: we can get the actual area of region **II** just by ignoring the minus sign! This works because the region is entirely below the x-axis. So the actual area of region **II** is $7/3$ square units, while region **I** has area $5/3$ square units. The total area is therefore $5/3 + 7/3 = 4$ square units. Effectively, we just took the absolute value of each of the two pieces $5/3$ and $-7/3$, then added them up.

Incidentally, we have actually just proved that

$$\int_0^2 |-x^2 - 2x + 3| \, dx = 4.$$

To see why taking absolute values of the integrand gives the unsigned area, just look at the graph of $y = |-x^2 - 2x + 3|$:

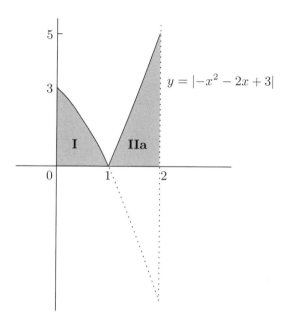

The region labeled **IIa** is just the reflection in the x-axis of the old region **II**, so it has the same unsigned area. The total shaded area is the same as the total unsigned area in the original picture above.

Let's summarize the method for finding the value of the unsigned area between $y = f(x)$, the x-axis, and the lines $x = a$ and $x = b$. The same method works for either of the following integrals, because they are both equal to the unsigned area:

$$\int_a^b |f(x)| \, dx \qquad \text{or} \qquad \int_a^b |y| \, dx.$$

So, here's the method:

- Find all the zeroes of f lying in the interval $[a, b]$.
- Write down a bunch of integrals with integrand $f(x)$, not $|f(x)|$. The first integral starts at a and goes up to the lowest of the zeroes you just found. The next one starts at that lowest zero and goes up until the next one. Keep going until you run out of zeroes. The last integral starts at this final zero and goes up to b.
- Work out each integral separately.
- Add up the absolute values of the numbers from the previous step to get the unsigned area.

We'll look at another example of this in Section 17.6.3 of the next chapter. Note that you should use the above method in order to find the distance an object travels, as opposed to the displacement. Indeed, as we saw in Section 16.1 above,

$$\text{distance} = \int_a^b |v(t)|\, dt,$$

so absolute values are involved and the above method applies.

16.4.2 Finding the area between two curves

Suppose you have two curves, one above the other, and you want to find the area of the region between the curves and the lines $x = a$ and $x = b$. If the curves are $y = f(x)$ and $y = g(x)$, where the first is above the second, then the situation looks like this:

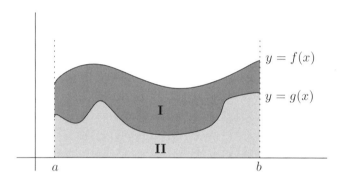

The actual region we want to find the area of is labeled **I**. On the other hand, the region **II** lies under the curve $y = g(x)$, so it has signed area

$$\int_a^b g(x)\, dx.$$

So what is

$$\int_a^b f(x)\, dx?$$

That must be the signed area below the top curve all the way to the x-axis, so it is actually the area of both regions put together. So we have

$$\int_a^b f(x)\, dx = \int_a^b g(x)\, dx + \text{signed area of region } \mathbf{I}.$$

We can rearrange this and stick the two integrals together into one integral, getting

$$\text{signed area of region } \mathbf{I} = \int_a^b (f(x) - g(x))\, dx.$$

So you just take the top curve's function and subtract the bottom curve's function, then integrate. For example, let's find the following shaded area:

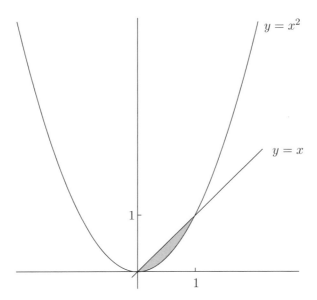

The region lies below $y = x$ and above $y = x^2$. The intersection points are at $x = 0$ and $x = 1$, so we have

$$\text{shaded area} = \int_0^1 (x - x^2)\, dx = \int_0^1 x\, dx - \int_0^1 x^2\, dx = \frac{1}{2} - \frac{1}{3} = \frac{1}{6} \text{ square units.}$$

What about going from 0 to 2 instead? Here's the picture:

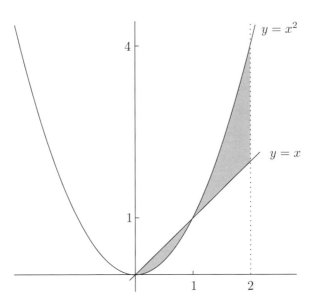

It would be wrong to express this area as

$$\int_0^2 (x - x^2)\, dx.$$

If you try it, you'll actually get the answer $-2/3$ again, which can't be a real area. The problem is that $y = x$ is above $y = x^2$ only when x is between 0 and 1. To the right of $x = 1$, the curve $y = x^2$ is on top. The quantity $x - x^2$ is no good, then—we should really use $|x - x^2|$ instead. That way, we'll make sure we're always using the actual area, no matter which curve is on top. So we have to apply the method from the previous section to find

$$\int_0^2 |x - x^2| \, dx.$$

No problem. First, notice that $x - x^2 = 0$ when $x = 0$ or $x = 1$, so we consider the integrals

$$\int_0^1 (x - x^2) \, dx \qquad \text{and} \qquad \int_1^2 (x - x^2) \, dx.$$

The first integral is $1/6$, but the second works out to be $3/2 - 7/3 = -5/6$. It makes sense that the second integral is negative, since $y = x$ is not above $y = x^2$ when x is in the interval $[1, 2]$. Never mind—we just add up the absolute values of the two integrals:

$$\int_0^2 |x - x^2| \, dx = \left|\frac{1}{6}\right| + \left|-\frac{5}{6}\right| = \frac{1}{6} + \frac{5}{6} = 1.$$

So the area we want is 1 square unit.

In summary, the area of the region bounded by $y = f(x)$, $y = g(x)$, $x = a$, and $x = b$ is given by the following formula:

$$\boxed{\text{area between } f \text{ and } g \text{ (in square units)} = \int_a^b |f(x) - g(x)| \, dx.}$$

If $f(x)$ is always greater than or equal to $g(x)$ on the interval $[a, b]$, then the absolute value signs aren't needed. Otherwise, use the method from Section 16.4.1 above to handle the absolute value in the above integral. We'll look at another example of this technique in Section 17.6.3 of the next chapter.

16.4.3 Finding the area between a curve and the y-axis

Let's try to find the area of the region enclosed by the curve $y = \sqrt{x}$, the y-axis, and the line $y = 2$. Here's a picture of the region:

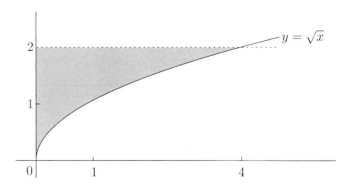

It would be a mistake to write the above area as

$$\int_0^2 \sqrt{x}\,dx \qquad \text{or even} \qquad \int_0^4 \sqrt{x}\,dx.$$

Both of the above integrals represent areas down to the x-axis, not the y-axis; in fact, they are equal to the following areas (respectively):

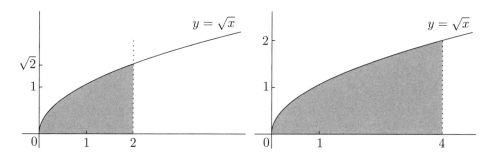

The second one is a little bit better, because $x = 4$ actually corresponds to $y = 2$. On the other hand, neither is correct! To find the correct area, the best way is to integrate with respect to y, not x. When we do this, we're effectively chopping up the region we want into horizontal strips, not the vertical ones we've used before. Here's an example of how this might look:

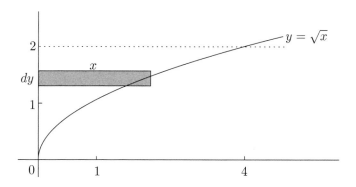

If you focus on any one of these strips, you can think of the dimensions as being dy and x:

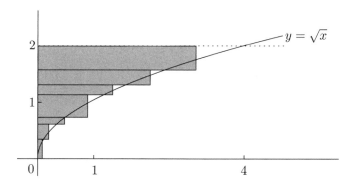

So the area of a little strip is $x\,dy$ square units, and you get the total by integrating. In our case, y ranges from 0 to 2 (not 4), so the area we want (in square units) is

$$\int_0^2 x\,dy.$$

Since $y = \sqrt{x}$, we know that $x = y^2$. So the above integral becomes

$$\int_0^2 y^2\,dy.$$

This is none other than our old integral

$$\int_0^2 x^2\,dx,$$

but with the dummy variable changed from x to y. This change has no effect: the value is still 8/3, so the area we want is 8/3 square units. Now, if you want to be clever about it, look back at the original area, and notice that all you have to do is flip the whole picture in the mirror line $y = x$ and you get the area under $y = x^2$ from $x = 0$ to $x = 2$ instead. That's all we're doing here—switching x and y. Of course, if $y = f(x)$, then $x = f^{-1}(y)$, provided that the inverse function exists. So, we can summarize the situation as follows:

$\int_A^B f^{-1}(y)\,dy$	is the signed area (in square units) of the region between the curve $y = f(x)$, the lines $y = A$ and $y = B$, and the y-axis, if f is invertible.

If you prefer, you can write the above integral as

$$\int_A^B x\,dy$$

instead. This is because $x = f^{-1}(y)$ when $y = f(x)$. Also, notice that I used capital letters A and B for the limits of integration—I did this to emphasize that these numbers are on the y-axis, not the x-axis. So in our above example, the integral has to be from 0 to 2, not the 0 to 4 that you might think by looking at the x-axis. Since $f(x) = \sqrt{x}$, we know that $f^{-1}(x) = x^2$. So the above formula does indeed give our integral

$$\int_A^B f^{-1}(y)\,dy = \int_0^2 y^2\,dy,$$

which is 8/3, as we saw above.

16.5 Estimating Integrals

Here's a very simple but important principle: **when one function is always larger than another, its integral is also larger.** Take a look at the following picture:

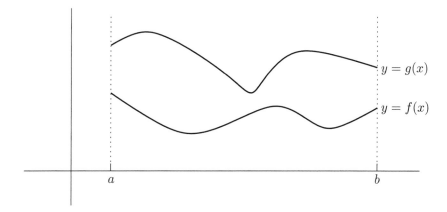

On the interval $[a, b]$, the function g always lies above f. (I know, I had them the other way around in Section 16.4.1 above!) In any case, the area under $y = f(x)$ (down to the x-axis) is clearly less than the area under $y = g(x)$ (down to the x-axis). In symbols:

$$\text{if } f(x) \leq g(x) \text{ for all } x \text{ in } [a, b], \text{ then } \int_a^b f(x)\,dx \leq \int_a^b g(x)\,dx.$$

This is true even if one or both of the curves go below the x-axis, thanks to the fact that we're using signed areas. For example, if f is always below the x-axis and g is always above the x-axis, then $\int_a^b f(x)\,dx$ is negative while $\int_a^b g(x)\,dx$ is positive, and the above inequality is still true.

The proof of the statement in the box above is quite easy using Riemann sums. Without getting into the gory details, you just have to take a partition and note that $f(c_j) \leq g(c_j)$ for all j, so the whole Riemann sum for f is less than the corresponding sum for g. I leave it to you to take it from there.

There's also a nice interpretation of the above fact in terms of velocity and displacement. Suppose that there are two cars starting at the same place. The first one travels with velocity $f(t)$ at time t, while the second goes at a velocity of $g(t)$ at time t. Since the integral of velocity is the displacement, the statement in the box above means that if the first car's velocity is always less than the second car's velocity, then the first car's displacement is less than the second car's displacement. This makes a lot of sense if you think about it! The first car will always be more to the left of the second car on our mythical number line, because it just doesn't have as much rightward oomph as the second car does.

16.5.1 A simple type of estimation

We can use the above inequality to get a feel for how big or small a definite integral is, without actually finding the integral. For example, suppose we'd like to estimate $\int_a^b f(x)\,dx$, which is the value of the following area:

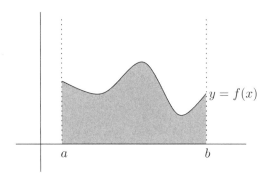

Let's set M equal to the maximum value of $f(x)$ on $[a, b]$, and we'll do the same thing with the minimum value, except we'll call it m instead. If we draw in the lines $y = M$ and $y = m$, then the situation looks like this:

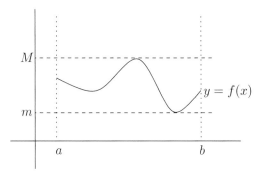

Notice that the area we want is less than the area under $y = M$, but greater than the area under $y = m$. This is easy to see by drawing some more pictures:

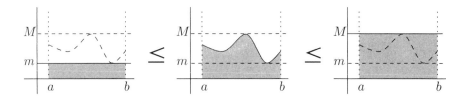

It's not hard to find the area of the two rectangles in the left-hand and right-hand pictures above. In the left-hand case, the base is $(b - a)$ units and the height is m units, so the area is $m(b - a)$ square units. In the right-hand case, the base is still $(b - a)$ units but the height is now M units, so the area is $M(b - a)$ square units. So the above graphs indicate the following principle:

$$
\boxed{
\begin{array}{c}
\text{if } m \leq f(x) \leq M \text{ for all } x \text{ in } [a, b], \text{ then} \\[2mm]
m(b - a) \leq \displaystyle\int_a^b f(x)\, dx \leq M(b - a).
\end{array}
}
$$

Of course, this is exactly the principle from the previous section applied twice. Let's look at an example of how to use it. Suppose we want to get some idea about the value of

$$\int_0^{1/2} e^{-x^2}\, dx.$$

The graph of $y = e^{-x^2}$ is a variety of the famous bell-shaped curve, which pops up all over the place, especially in probability theory and statistics. We are looking for the following area:

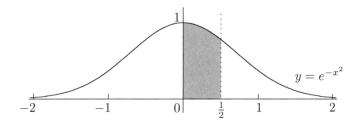

Even with all the techniques for finding integrals that we'll develop in the next three chapters, we still won't be able to find the exact value of the above integral. In fact, there **isn't** any nice way to express the value without using an integral sign or a sum which goes on forever or some other trick. We can at least estimate the value of the integral by using the above principle.

We need to find the maximum and minimum values of $y = e^{-x^2}$ on the interval $[0, \frac{1}{2}]$. The chain rule shows that $dy/dx = -2xe^{-x^2}$, which is 0 at the endpoint 0 and is negative otherwise. This confirms that e^{-x^2} is decreasing in x on the interval $[0, \frac{1}{2}]$; so the maximum value occurs when $x = 0$, and the minimum value occurs when $x = \frac{1}{2}$. Plugging these values in, we find that the maximum value is $e^{-0^2} = 1$, and the minimum value is $e^{-(1/2)^2} = e^{-1/4}$. That is, on the interval $[0, \frac{1}{2}]$, we have

$$e^{-1/4} \le e^{-x^2} \le 1.$$

By our principle from the box above with $a = 0$ and $b = \frac{1}{2}$, we have

$$e^{-1/4}\left(\frac{1}{2} - 0\right) \le \int_0^{1/2} e^{-x^2}\, dx \le 1\left(\frac{1}{2} - 0\right).$$

So the value of the integral we're looking for lies between $\frac{1}{2}e^{-1/4}$ and $\frac{1}{2}$. Again, you can clearly see this by looking at the following graphs, which show the underestimate and overestimate, respectively:

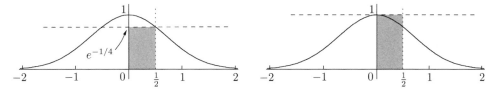

The areas of the two rectangles are $\frac{1}{2}e^{-1/4}$ and $\frac{1}{2}$ square units, respectively.

The above estimates are pretty crude. We can do a better job by using more rectangles, or even more exotic shapes like trapezoids or parabola-topped strips. See Appendix B for more details.

16.6 Averages and the Mean Value Theorem for Integrals

At last, we can return to average velocities. Yes, once upon a time, we thought nothing of saying that speed equals distance over time, or better still, velocity equals displacement over time. That's fine as long as the velocity is constant; otherwise, as we saw in Section 5.2.3 in Chapter 5, we really need to say **average** velocity.

Then we learned how to use differentiation to find the **instantaneous** velocity, knowing what the displacement is at all times during the time interval of interest. Using integration, we can find the displacement, knowing what the instantaneous velocity is at all times during our time interval. This last fact also allows us to find the average velocity, knowing the instantaneous velocity at all times. All you have to do is find the displacement and divide it by the total time. If the time interval goes from a to b, and the velocity at time t is $v(t)$, then we've already seen that

$$\text{displacement} = \int_a^b v(t)\, dt.$$

Since the total time is $b - a$, we have

$$\text{average velocity} = \frac{\text{displacement}}{\text{total time}} = \frac{1}{b-a} \int_a^b v(t)\, dt.$$

More generally, we can define the *average value* of an integrable function f on the interval $[a, b]$ as follows:

$$\boxed{\text{average value of } f \text{ on } [a, b] = \frac{1}{b-a} \int_a^b f(x)\, dx.}$$

For example, what is the average value of f on the interval $[0, 2]$, where $f(x) = x^2$? No problem:

$$\text{average value} = \frac{1}{2-0} \int_0^2 x^2\, dx = \frac{1}{2} \times \frac{8}{3} = \frac{4}{3}.$$

All you have to do is divide the integral by the difference between the limits of integration.

Let's look at a geometrical interpretation of this. Let's write the average value of f on $[a, b]$ as f_{av} for short. Here's an example of what the graphs of $y = f(x)$ and $y = f_{\text{av}}$ might look like:

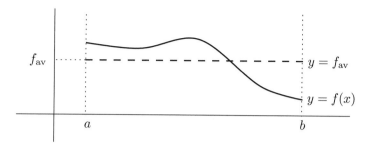

Notice that f_{av} is just a constant number, so the graph of $y = f_{av}$ is a horizontal line. Now, by the above boxed formula, we have

$$f_{av} = \frac{1}{b-a} \int_a^b f(x)\,dx.$$

Multiplying by $(b-a)$, we see that

$$\int_a^b f(x)\,dx = f_{av} \times (b-a).$$

This actually says that the following two areas are equal:

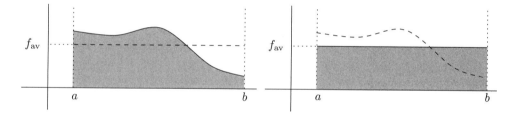

After all, the rectangle in the right-hand picture has height f_{av} units and base $(b - a)$ units, so its area is $f_{av} \times (b - a)$ square units. You can think of it this way: if you disturb the water in a thin long fish tank so that the water surface looks like $y = f(x)$ for an instant, then after the water stabilizes, the surface will look like the horizontal line $y = f_{av}$.

16.6.1 The Mean Value Theorem for integrals

In the above graphs, observe that the horizontal line $y = f_{av}$ intersects the graph of $y = f(x)$. Let's label the corresponding point on the x-axis as c, like this:

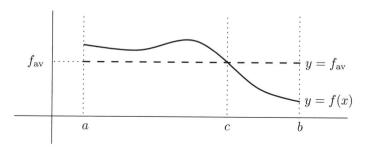

So we have $f(c) = f_{av}$. It turns out that if f is continuous, then there is always such a number c:

> **Mean Value Theorem for integrals:** if f is continuous on $[a, b]$, then there exists c in (a, b) such that $\qquad f(c) = \dfrac{1}{b-a} \displaystyle\int_a^b f(x)\,dx.$

In words, you could say that "a continuous function attains its average value at least once." For example, we saw in the previous section that the average value of $f(x) = x^2$ on $[0, 2]$ is $4/3$. According to the above theorem, we must have $f(c) = 4/3$ for some c in $[0, 2]$. Since $f(c) = c^2$, we can see that $c = \sqrt{4/3}$ is a solution which does indeed lie in $[0, 2]$ (unlike the other possible solution, $c = -\sqrt{4/3}$).

If you think of the above theorem in terms of velocities, it just says that $v(c) = v_{av}$ for some c in $[a, b]$. This means that for any journey, there is some point in time (c) such that the velocity at that time $(v(c))$ equals the average velocity (v_{av}). No matter how hard you try, during any journey you make, there must be at least one instant of time where your instantaneous velocity equals your average velocity. There might be more than one such instant, but there can't be none. Even if you go at 45 mph for an hour and 55 mph for an hour, for an average velocity of 50 mph, you will still have to go at 50 mph for an instant while you're accelerating from 45 to 55.

So, why is the above theorem also called the Mean Value Theorem? After all, we already have a Mean Value Theorem. If you look back at our discussion of the original theorem in Section 11.3 of Chapter 11, you'll see that we reached the same conclusion as we did above: the instantaneous velocity has to equal the average velocity at some point during any journey. The difference between the two versions of the theorem is that in the regular version, the conclusion was interpreted in terms of slopes on the graph of displacement versus time; whereas now we have interpreted it in terms of areas on the graph of velocity versus time.

Now let's see why the theorem is true. As we did in Section 16.5 above, we'll let M be the maximum value of f on $[a, b]$, and m be the minimum value of f on $[a, b]$. Could f_{av} possibly be greater than M? If so, the situation would look like this:

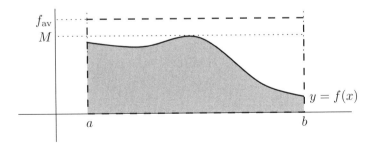

There's no way that the area of the dashed rectangle equals the area of the shaded region under $y = f(x)$, since the rectangle contains the region! So that situation can't happen. In a similar way, f_{av} can't be below the minimum m.

It must lie between m and M. The Intermediate Value Theorem implies that f takes every value between m and M (can you see why?), so in particular, f takes on the value f_{av} somewhere. That is, $f(c) = f_{av}$ for some c and the theorem is true. We'll use the theorem in Section 17.8 of the next chapter when we prove the First Fundamental Theorem of Calculus.

16.7 A Nonintegrable Function

In Section 16.2 above, I mentioned that if f is bounded and has only a finite number of discontinuities in $[a, b]$, then f is integrable. That is, the integral $\int_a^b f(x)\,dx$ exists. By the way, recall that discontinuities are a deal-breaker as far as differentiability is concerned—if f is discontinuous at $x = a$, then it can't be differentiable there. (See Section 5.2.11 of Chapter 5.) Integration is a little more forgiving, since it can deal with some discontinuities, as long as there aren't too many of them. Now, let's look at an example of a function where there **are** too many discontinuities.

First, remember that a rational number is a number that can be written in the form p/q where p and q are integers (with no common factor). An irrational number can't be written in that form. Now, for x in the domain $[0, 1]$, let

$$f(x) = \begin{cases} 1 & \text{if } x \text{ is rational,} \\ 2 & \text{if } x \text{ is irrational.} \end{cases}$$

This is a pretty weird function. There are lots and lots of rational and irrational numbers between 0 and 1. In fact, between every two rational numbers, there's an irrational number, and between every two irrational numbers, there's a rational number! So if you try to sketch a graph of $y = f(x)$, you might come up with the following picture:

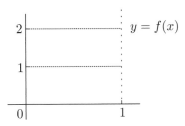

The values of $f(x)$ jump between heights 1 and 2 faster than you can imagine. There's no connectivity whatsoever in the above line segments at heights 1 and 2: they are full of holes. The function is actually discontinuous everywhere. So what on earth should

$$\int_0^1 f(x)\,dx$$

be? Let's try taking upper and lower Riemann sums and see what we get. Pick any partition of $[0, 1]$. No matter how narrow they are, your strips will pick up some irrational point. So the upper sum must look something like this:

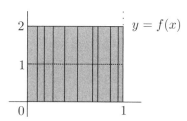

Every rectangle has to reach height 2 in order to create an upper sum, even if the rectangles are really really thin. Notice that the area of all the rectangles above is 2 square units, no matter how many there are, since they fill out a 1-by-2 unit rectangle. In particular,

$$\lim_{\text{mesh}\to 0}(\text{upper Riemann sum}) = \lim_{\text{mesh}\to 0} 2 = 2.$$

Similarly, in the lower sum for the same partition, every rectangle has to be of height 1 unit. After all, no matter how thin a rectangle is, its base (on the x-axis) will still contain a rational number, and the function has height 1 at all rational numbers. So a lower sum must look like this:

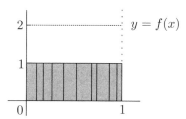

Now the area is 1 square unit, since the total rectangle filled in by all the little strips is 1-by-1 unit. So we have shown that

$$\lim_{\text{mesh}\to 0}(\text{lower Riemann sum}) = \lim_{\text{mesh}\to 0} 1 = 1.$$

The limits, as the mesh goes to 0, for the upper and lower Riemann sums are different. This doesn't happen for continuous functions, but it does happen for this crazy function! The only conclusion is that f cannot be integrated on its domain $[0, 1]$. We say that f is *nonintegrable*. Actually, there **is** a way to integrate this function, but it's called Lebesgue integration (as opposed to Riemann integration) and it's way beyond the scope of this book. So, let's not worry about these sorts of pathological examples and concentrate instead on finding a nice way to find definite integrals of well-behaved, continuous functions.

CHAPTER 17 _____

The Fundamental Theorems of Calculus

Here it is: the big kahuna. I'm talking about the Fundamental Theorems of Calculus, which not only provide the key for finding definite integrals without using messy Riemann sums, but also show how differentiation and integration are connected to each other. Without further ado, here's the roadmap for the chapter: we'll investigate

- functions which are based on integrals of other functions;
- the First Fundamental Theorem, and the basic idea of antiderivatives;
- the Second Fundamental Theorem; and
- indefinite integrals and their properties.

After all this theoretical stuff, we'll look at a lot of different examples in the following categories:

- problems based on the First Fundamental Theorem;
- finding indefinite integrals; and
- finding definite integrals and areas using the Second Fundamental Theorem.

17.1 Functions Based on Integrals of Other Functions

In the previous chapter, we used Riemann sums to show that

$$\int_0^1 x^2 \, dx = \frac{1}{3} \qquad \text{and} \qquad \int_0^2 x^2 \, dx = \frac{8}{3}.$$

(Actually, we only did the second one; I left the first one to you!) Unfortunately, the method of Riemann sums was really nasty. It would be nice to have an easier method to find the above integrals. Why stop there, though? Let's try to find

$$\int_0^{\text{any number}} x^2 \, dx.$$

So we want to allow the right-hand limit of integration to be **variable**. Everyone's favorite variable is x, but you can't write down

$$\int_0^x x^2\, dx$$

unless you want to be really confusing. After all, x is the dummy variable, so it can't be a real variable too. So let's start over, this time using t as the dummy variable. First, we have

$$\int_0^1 t^2\, dt = \frac{1}{3} \quad \text{and} \quad \int_0^2 t^2\, dt = \frac{8}{3}.$$

Remember, the letter we use for the dummy variable is irrelevant—we've just renamed the x-axis to be the t-axis. The actual area doesn't change. Now we want to consider the quantity

$$\int_0^x t^2\, dt.$$

If you substitute $x = 1$ into this quantity, you get $\int_0^1 t^2\, dt$, which is equal to $1/3$; if instead you substitute $x = 2$, you get $\int_0^2 t^2\, dt$, which is $8/3$. Why stop there? You can substitute **any** number in place of x and get a different integral. That is, the above quantity is a **function** of the right-hand limit of integration, x. Let's call the function F, so that

$$F(x) = \int_0^x t^2\, dt.$$

We have seen that $F(1) = 1/3$ and $F(2) = 8/3$. How about $F(0)$? Well,

$$F(0) = \int_0^0 t^2\, dt.$$

In Section 16.3 of the previous chapter, we saw that an integral with the same left-hand and right-hand limits of integration must be 0. That is, we know that $F(0) = 0$. Unfortunately, it's not so easy to find many other values of F, such as $F(9)$, $F(-7)$ or $F(1/2)$. We'll return to this point in the next section. In the meantime, how would you describe $F(x)$ in words? It's precisely the signed area (in square units) between the curve $y = t^2$, the t-axis, and the vertical line $t = x$.

There are two ways we can make this whole thing more general. First, the left-hand endpoint doesn't have to be 0. You could define another function G by setting

$$G(x) = \int_2^x t^2\, dt.$$

The quantity $G(x)$ is the area (in square units) of the region bounded by $y = t^2$, the t-axis, and the lines $t = 2$ and $t = x$. So what is $G(2)$? Well,

$$G(2) = \int_2^2 t^2\, dt = 0,$$

since the left-hand and right-hand limits of integration are the same. How about $G(0)$? We have

$$G(0) = \int_2^0 t^2 \, dt.$$

To handle this, remember from Section 16.3 of the previous chapter that you can switch the limits of integration as long as you put a minus sign out front. So

$$G(0) = \int_2^0 t^2 \, dt = -\int_0^2 t^2 \, dt = -\frac{8}{3}.$$

In fact, there's a really nice relationship between F and G. First, let's remind ourselves what these functions are:

$$F(x) = \int_0^x t^2 \, dt \qquad \text{and} \qquad G(x) = \int_2^x t^2 \, dt.$$

Let's split up the first of these integrals up at $t = 2$; see Section 16.3 in the previous chapter to remind yourself how to split up an integral. We get

$$\int_0^x t^2 \, dt = \int_0^2 t^2 \, dt + \int_2^x t^2 \, dt.$$

The left-hand side is $F(x)$. Meanwhile, the first term on the right-hand side is just $8/3$, while the second term is $G(x)$. Altogether, we have shown that

$$F(x) = \frac{8}{3} + G(x).$$

That is, F and G differ by the constant $8/3$. We can be even more general, though. Suppose that a is **any** fixed number, and set

$$H(x) = \int_a^x t^2 \, dt.$$

If you split the integral in the definition of F at $t = a$ instead of $t = 2$, you get this:

$$F(x) = \int_0^x t^2 \, dt = \int_0^a t^2 \, dt + \int_a^x t^2 \, dt.$$

The second term on the right-hand side is exactly $H(x)$, so we've shown that

$$F(x) = \int_0^a t^2 \, dt + H(x).$$

So what? Well, the integral $\int_0^a t^2 \, dt$ is actually a constant—it doesn't depend on x at all! Even though we didn't specify the value of a, we did say it was constant, so the integral must also be constant. We've shown that

$$F(x) = H(x) + C,$$

where C is some constant that depends on a but not on x. The moral of the story is that changing the left-hand endpoint from one constant to another doesn't make too much difference.

Our second generalization is that the integrand doesn't have to be t^2. It can be any continuous function of t. Let's suppose the integrand is $f(t)$. If a is some constant number, then let's define

$$F(x) = \int_a^x f(t)\, dt.$$

For example, if $a = 0$ and $f(t) = t^2$, you get the original function F from above. In general, for any number x, the value $F(x)$ is the signed area (in square units) between the curve $y = f(t)$, the t-axis, and the vertical lines $t = a$ and $t = x$. Here is an example of what this might look like for three different values of x:

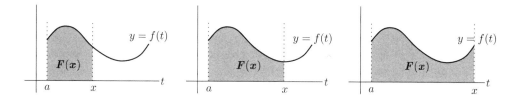

The above pictures are reminiscent of a curtain with fixed left-hand edge, while the right-hand edge slides back and forth. The only unrealistic aspect is that the curtain rod at the top is pretty warped, unless the function f is constant! In any case, note that the function F comes directly from the choice of the integrand $f(t)$ and the number a. By splitting up the integral, you can show that changing the number a just changes the function F by a constant. All these ideas will be very important in the next couple of sections. . . .

17.2 The First Fundamental Theorem

Here's the goal: find

$$\int_a^b f(x)\, dx$$

without using Riemann sums. Let's do three things which are not really obvious at all:

1. First, let's change the dummy variable to t and write the above integral as

$$\int_a^b f(t)\, dt.$$

   As we saw in the previous section, this doesn't make any difference—the name of the dummy variable doesn't matter.

2. Now, let's replace b by a variable x to get a new function F, defined like this:

$$F(x) = \int_a^x f(t)\, dt.$$

   This is exactly the sort of function that we looked at in the previous section. Eventually we're going to want the value of $F(b)$, which is exactly the integral in step 1 above, but first let's see what we can understand about F in general.

3. So we have this new function F. It's like a brand new shiny toy to play with. Since we've spent so much time differentiating functions, let's try differentiating this one with respect to the variable x. That is, we consider

$$F'(x) = \frac{d}{dx} \int_a^x f(t)\, dt.$$

Understanding the nature of $F'(x)$ will allow us to find $F(x)$ in general. Once we've done that, we can find $F(b)$, which is exactly the integral we want.

The expression

$$\frac{d}{dx} \int_a^x f(t)\, dt$$

might just about be the weirdest thing we've looked at so far in this book. Let's see how to unravel it. Pick your favorite number x and find $F(x)$. Then wobble x a little bit—let's move it to $x + h$, where h is a small number. So now our function value is $F(x + h)$. Here's a picture of the situation:

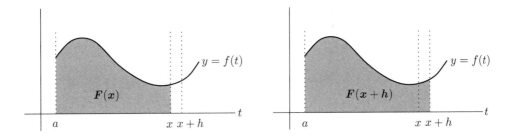

As you can see, x and $x + h$ are pretty close to each other. The values of $F(x)$ and $F(x+h)$ are pretty close to each other too—they represent the two shaded areas above (respectively). Now, to differentiate F, we have to find

$$\lim_{h \to 0} \frac{F(x+h) - F(x)}{h}.$$

The difference $F(x + h) - F(x)$ is just the difference between the two shaded areas, which is itself just the area of the thin little region (with curved top) between $t = x$ and $t = x + h$:

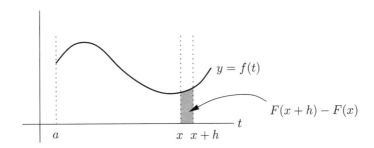

You can see this in symbols by splitting up the integral for $F(x+h)$ at $t = x$, like this:

$$F(x+h) = \int_a^{x+h} f(t)\,dt = \int_a^x f(t)\,dt + \int_x^{x+h} f(t)\,dt = F(x) + \int_x^{x+h} f(t)\,dt.$$

Rearranging, we get

$$F(x+h) - F(x) = \int_x^{x+h} f(t)\,dt,$$

which is exactly the shaded area (in square units) of the thin strip above. Actually, it's not a strip, since the top is curved, but it's **almost** a strip when h is small. The height of the strip at the left-hand side is $f(x)$ units, so we can approximate the thin region by a rectangle with base going from x to $x + h$ and height from 0 to $f(x)$, like this:

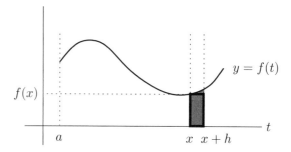

The base of the rectangle is h units, and the height is $f(x)$ units, so the area is $hf(x)$ square units. If h is small, then this is a good approximation to the integral we want. That is,

$$F(x+h) - F(x) = \int_x^{x+h} f(t)\,dt \cong hf(x).$$

Dividing by h, we have

$$\frac{F(x+h) - F(x)}{h} \cong f(x).$$

The approximation gets really good when h is really close to 0. It should be true, then, that the approximation is perfect in the limit as $h \to 0$:

$$\lim_{h \to 0} \frac{F(x+h) - F(x)}{h} = f(x).$$

As we'll see in Section 17.8 below, the above formula is indeed true; we conclude that

$$F'(x) = f(x).$$

Let's summarize our conclusion as follows:

> **First Fundamental Theorem of Calculus:** for f continuous on $[a, b]$, define a function F by
> $$F(x) = \int_a^x f(t)\,dt \qquad \text{for } x \text{ in } [a, b].$$
> Then F is differentiable on (a, b) and $F'(x) = f(x)$.

In short, you can write the whole thing as

$$\frac{d}{dx} \int_a^x f(t)\, dt = f(x).$$

So our weird expression simplifies down to $f(x)$!

A common concern with this last formula is that a appears on the left-hand side but not on the right-hand side. This actually makes sense, believe it or not. Suppose that A is some other number in (a, b), and set

$$F(x) = \int_a^x f(t)\, dt \qquad \text{and} \qquad H(x) = \int_A^x f(t)\, dt.$$

Then, as we saw in Section 17.2 above, F and H differ by a constant:

$$F(x) = H(x) + C$$

for some constant C. If we differentiate, the constant goes away and we see that $F'(x) = H'(x)$ for all x in (a, b). So the actual choice of a doesn't affect the derivative. In terms of the curtain, we only care how fast it's being pulled and how high the rail is at the right-hand point. Where it happens to be attached at the fixed left-hand end doesn't affect the **rate** of area being swept out all the way over at the right-hand part of the curtain.

17.2.1 Introduction to antiderivatives

Now, let's pause for breath. We started with some function f of the variable t, as well as some number a; then we constructed a new function F of the variable x. Differentiating F gives us back the original function f, except now we evaluate it at x instead of t. Weird!

OK, weird, but really useful. It actually solves our whole darn problem. Let's see how. Suppose that $f(t) = t^2$ and $a = 0$, so that

$$F(x) = \int_0^x t^2\, dt.$$

The First Fundamental Theorem tells us that $F'(x) = f(x)$. Since $f(t) = t^2$, we have $f(x) = x^2$; this means that $F'(x) = x^2$. In other words, F is a function whose derivative is x^2. We say that F is an *antiderivative* of x^2 (with respect to x). Can you think of any other function whose derivative is x^2? Here are a few:

$$G(x) = \frac{x^3}{3}, \qquad H(x) = \frac{x^3}{3} + 7, \qquad \text{and} \qquad J(x) = \frac{x^3}{3} - 2\pi.$$

In each case, you can check that the derivative is x^2. In fact, any function of x of the form

$$\frac{x^3}{3} + C \qquad \text{for some constant } C$$

is an antiderivative of x^2. Are there any others? The answer is no! We actually saw this in Section 11.3.1 of Chapter 11. If two functions have the same derivative, they differ by a constant. This means that all the antiderivatives of x^2 differ by a constant. Since one of the antiderivatives is $x^3/3$, then any other antiderivative must be $x^3/3 + C$, where C is constant. Wait a second—the weird function F above is also an antiderivative of x^2. This means that

$$F(x) = \int_0^x t^2 \, dt = \frac{x^3}{3} + C$$

for some constant C. Now all we have to do is find C. We know that

$$F(0) = \int_0^0 t^2 \, dt = 0.$$

So we have

$$0 = \frac{0^3}{3} + C.$$

This means that $C = 0$. We now have the formula we've been looking for:

$$\int_0^x t^2 \, dt = \frac{x^3}{3}.$$

Finally, we can integrate t^2 from 0 to any number! In particular, if we replace t by 1 and then by 2, we get our well-worn formulas

$$\int_0^1 t^2 \, dt = \frac{1^3}{3} = \frac{1}{3} \quad \text{and} \quad \int_0^2 t^2 \, dt = \frac{2^3}{3} = \frac{8}{3}.$$

This can be made even simpler—we'll do that in the next section. First, I'd like to make one more point. We now have a way of constructing an antiderivative of any continuous function. For example, what is an antiderivative of e^{-x^2}? Just change x to t, pick your favorite number as a left-hand limit of integration (let's say 0 for the moment), and integrate to see that

$$F(x) = \int_0^x e^{-t^2} \, dt \quad \text{is an antiderivative of } e^{-x^2}.$$

The number 0 could be replaced by any number you choose, and the same statement would be true. Of course, you get a different antiderivative for each potential choice of left-hand limit of integration.

17.3 The Second Fundamental Theorem

The example with $f(t) = t^2$ in the previous section points the way to finding $\int_a^b f(t) \, dt$ in general. First, we know that the function F defined by

$$F(x) = \int_a^x f(t) \, dt$$

is an antiderivative of f (with respect to x). We really want to find $F(b)$, since

$$F(b) = \int_a^b f(t) \, dt.$$

We know one more thing:

$$F(a) = \int_a^a f(t)\,dt = 0,$$

because the left-hand and right-hand limits of integration are equal.

Now, suppose we have some other antiderivative of f: let's call it G. Then F and G differ by a constant, so that $G(x) = F(x) + C$. Put $x = a$ and you see that $G(a) = F(a) + C$; since $F(a) = 0$ from above, we have $G(a) = C$. This means that

$$F(x) = G(x) - C = G(x) - G(a).$$

If you replace x by b, you get

$$F(b) = G(b) - G(a).$$

In other words,

$$\int_a^b f(t)\,dt = G(b) - G(a).$$

This is true for **any** antiderivative G. Notice that we've gotten rid of x altogether. So the convention now is to change the dummy variable back to x and also change the letter G to F, arriving at the

---

Second Fundamental Theorem of Calculus: if f is continuous on $[a, b]$, and F is any antiderivative of f (with respect to x), then

$$\int_a^b f(x)\,dx = F(b) - F(a).$$

In practice, the right-hand side is normally written as $F(x)\Big|_a^b$. That is, we set

$$F(x)\Big|_a^b = F(b) - F(a).$$

So, for example, to evaluate

$$\int_1^2 x^2\,dx,$$

start by finding an antiderivative of x^2. We have seen that $x^3/3$ is one antiderivative, so

$$\int_1^2 x^2\,dx = \frac{x^3}{3}\Big|_1^2.$$

Now just plug $x = 2$ and $x = 1$ into $x^3/3$, and take the difference:

$$\int_1^2 x^2\,dx = \frac{x^3}{3}\Big|_1^2 = \left(\frac{2^3}{3}\right) - \left(\frac{1^3}{3}\right),$$

which works out to be 7/3. Now, here's another example. Suppose you want to find

$$\int_{\pi/6}^{\pi/2} \cos(x)\,dx.$$

We need an antiderivative of $\cos(x)$. Luckily, we have one at hand: it's $\sin(x)$. After all, the derivative with respect to x of $\sin(x)$ is $\cos(x)$. So, we get

$$\int_{\pi/6}^{\pi/2} \cos(x)\,dx = \sin(x)\Big|_{\pi/6}^{\pi/2} = \sin\left(\frac{\pi}{2}\right) - \sin\left(\frac{\pi}{6}\right) = 1 - \frac{1}{2} = \frac{1}{2}.$$

We'll look at more examples of this sort in Section 17.6 below.

17.4 Indefinite Integrals

So far, we've used two different techniques to find definite integrals: limits of Riemann sums (what a pain) and antiderivatives (not so bad). It's quite clear that we're going to have to become pretty adept at finding antiderivatives—in fact, that's going to occupy us for the next couple of chapters after this one. So, we might as well have a shorthand way of expressing antiderivatives without having to write the long word "antiderivative." Inspired by the First Fundamental Theorem, we'll write

$$\int f(x)\,dx$$

to mean "the family of all antiderivatives of f." Bear in mind that any integrable function has infinitely many antiderivatives, but they all differ by a constant. This is what I mean when I say "family." For example,

$$\int x^2\,dx = \frac{x^3}{3} + C$$

for some constant C. This equation literally means that the antiderivatives of x^2 (with respect to x) are precisely the functions $x^3/3 + C$, where C is any constant. It is an error to omit the "$+C$" at the end, since that would only give one of the antiderivatives and we need them all.

If you know a derivative, you get an antiderivative for free. In particular:

$$\boxed{\text{if} \quad \frac{d}{dx}F(x) = f(x), \qquad \text{then} \qquad \int f(x)\,dx = F(x) + C.}$$

The above example fits this pattern:

$$\frac{d}{dx}\left(\frac{x^3}{3}\right) = x^2, \qquad \text{so} \qquad \int x^2\,dx = \frac{x^3}{3} + C.$$

Similarly, we have

$$\frac{d}{dx}(\sin(x)) = \cos(x), \qquad \text{so} \qquad \int \cos(x)\,dx = \sin(x) + C.$$

One more example for now (there will be many more later!):

$$\frac{d}{dx}(\tan^{-1}(x)) = \frac{1}{1+x^2}, \qquad \text{so} \qquad \int \frac{1}{1+x^2}\,dx = \tan^{-1}(x) + C.$$

Again, the number C is an arbitrary constant. It's just the nature of things that differentiable functions have only one derivative whereas integrable functions have infinitely many antiderivatives.

All the above integrals are examples of *indefinite integrals*. You can tell an indefinite integral from a definite integral by noticing whether or not there are limits of integration. Indefinite integrals don't have limits of integration, while definite integrals do. This might seem like a small difference, but these two objects are very different beasts:

- A definite integral, like $\int_a^b f(x)\,dx$, is a **number**. It represents the signed area of the region bounded by the curve $y = f(x)$, the x-axis, and the lines $x = a$ and $x = b$.
- An indefinite integral, like $\int f(x)\,dx$, is a **family of functions**. This family consists of all functions which are antiderivatives of f (with respect to x). The functions all differ by a constant.

So, for example,

$$\int_1^2 x^2\,dx = \frac{8}{3}, \qquad \text{while} \qquad \int x^2\,dx = \frac{x^3}{3} + C.$$

If it weren't for the Second Fundamental Theorem, it would be crazy to use the same symbol $\int$ for both of these objects. Luckily, the indefinite integral (or antiderivative) is exactly what you need in order to find the definite integral, so it makes a lot of sense to use the symbol in both cases.

Here are two simple facts about indefinite integrals that follow directly from the similar properties for derivatives: if f and g are integrable, and c is a constant, then

$$\boxed{\int (f(x) + g(x))\,dx = \int f(x)\,dx + \int g(x)\,dx}$$

and $\boxed{\int cf(x)\,dx = c \int f(x)\,dx.}$

That is, the integral of the sum is the sum of the integrals, and constant multiples can be pulled through the integral sign. So, in particular,

$$\int (5x^2 + 9\cos(x))\,dx = 5 \int x^2\,dx + 9 \int \cos(x)\,dx = \frac{5x^3}{3} + 9\sin(x) + C.$$

Notice that we only need one constant—even though $5x^3/3$ and $9\sin(x)$ could each get their own constant added to them, you can just combine the two constants into one by adding them up. By the way, what works for sums also works for differences, as well:

$$\int (5x^2 - 9\cos(x))\,dx = 5 \int x^2\,dx - 9 \int \cos(x)\,dx = \frac{5x^3}{3} - 9\sin(x) + C.$$

Again, only one constant is needed.

Before we look at some more examples, I want to make one more comment about the two Fundamental Theorems. The First Fundamental Theorem says that

$$\frac{d}{dx} \int_a^x f(t)\,dt = f(x).$$

In some sense, the derivative of the integral is the original function. You just have to be careful about what you mean by the "integral," bearing in mind that the variable has to be the right-hand limit of integration, not the dummy variable. Now, the Second Fundamental Theorem says that

$$\int_a^b f(x)\,dx = F(x)\Big|_a^b$$

where F is an antiderivative of f. This means that $f(x) = \frac{d}{dx}F(x)$. We can therefore rewrite the above equation as

$$\int_a^b \frac{d}{dx}F(x)\,dx = F(x)\Big|_a^b$$

which can be interpreted as saying that the integral of the derivative is the original function. Again, it's not really the original function: it's the difference between the evaluations of the original function at the endpoints a and b. Even with all this vagueness, it should still be clear that differentiation and integration are essentially opposite operations.

Now, let's see how to use the Fundamental Theorems to solve problems.

17.5 How to Solve Problems: The First Fundamental Theorem

Think about how you'd find the following derivative:

$$\frac{d}{dx} \int_3^x \sin(t^2)\,dt.$$

You could try to find the indefinite integral $\int \sin(t^2)\,dt$, then plug in x and 3 and take the difference; this will give

$$\int_3^x \sin(t^2)\,dt,$$

which you could finally differentiate. Why go to all that work when the derivative and integral effectively cancel each other out? After all, if you wanted to find $(\sqrt{54756})^2$, you wouldn't waste time looking for $\sqrt{54756}$ when you just have to square it again. You'd just write down the answer 54756 and be done with it. Similarly, we can use the First Fundamental Theorem from above to say that

$$\frac{d}{dx} \int_3^x \sin(t^2)\,dt = \sin(x^2).$$

All you have to do is take the integrand $\sin(t^2)$ and change t to x. The number 3 doesn't even come into it (see Section 17.1 above for a discussion of this).

By the way, it would be a mistake to put a "+C" at the end: you are finding a **derivative**, after all, not an antiderivative!

Of course, you have to be versatile—the letters can change around. For example, what is

$$\frac{d}{dz} \int_{-e}^{z} 2^{\cos(w^2 \ln(w+5))} \, dw?$$

Just replace w by z in the integrand and see that

$$\frac{d}{dz} \int_{-e}^{z} 2^{\cos(w^2 \ln(w+5))} \, dw = 2^{\cos(z^2 \ln(z+5))}.$$

Note that $-e$ is a constant, but once again this could have been replaced by any other constant and the answer would be the same. (By the way, the integral only makes sense if $z > -5$.)

That's really all there is to the basic version, where the variable (that you're differentiating with respect to) is just sitting there on the right-hand limit of integration. All you have to do is replace the dummy variable in the integrand with the real variable. There are four variations that can arise, however: let's look at them one at a time.

17.5.1 Variation 1: variable left-hand limit of integration

Consider

$$\frac{d}{dx} \int_{x}^{7} t^3 \cos(t \ln(t)) \, dt.$$

The problem is that the variable x is now the left-hand limit of integration, not the right-hand one we've been used to. No problem—just switch the x and 7 around, introducing a minus sign to compensate for this (see Section 16.3 in the previous chapter to remind yourself why this works). You get

$$\frac{d}{dx} \int_{x}^{7} t^3 \cos(t \ln(t)) \, dt = \frac{d}{dx} \left(- \int_{7}^{x} t^3 \cos(t \ln(t)) \, dt \right).$$

Now pull out the minus sign and use the First Fundamental Theorem to see that this is equal to

$$-x^3 \cos(x \ln(x)),$$

if $x > 0$. In effect, all we are doing is taking the integrand, replacing the dummy variable t by x, and putting a minus sign out front. It's important to justify the minus sign and the use of the First Fundamental Theorem by first switching the limits of integration, as we did in the above example.

17.5.2 Variation 2: one tricky limit of integration

Here's another example:

$$\frac{d}{dx} \int_{0}^{x^2} \tan^{-1}(t^7 + 3t) \, dt.$$

Because the right-hand limit of integration is x^2, not x, we can't just use the First Fundamental Theorem directly. We're going to need the chain rule as well. Start off by letting y be the quantity we want to differentiate:

$$y = \int_0^{x^2} \tan^{-1}(t^7 + 3t)\, dt.$$

We want to find dy/dx. Since y is really a function of x^2, not x directly, we should let $u = x^2$. This means that

$$y = \int_0^u \tan^{-1}(t^7 + 3t)\, dt.$$

The chain rule says that

$$\frac{dy}{dx} = \frac{dy}{du}\frac{du}{dx},$$

while the First Fundamental Theorem says that

$$\frac{dy}{du} = \frac{d}{du}\int_0^u \tan^{-1}(t^7 + 3t)\, dt = \tan^{-1}(u^7 + 3u).$$

Also, since $u = x^2$, we have $du/dx = 2x$. Altogether,

$$\frac{dy}{dx} = \frac{dy}{du}\frac{du}{dx} = \left(\tan^{-1}(u^7 + 3u)\right)(2x).$$

Now all we have to do is replace u by x^2 to see that

$$\frac{dy}{dx} = 2x\tan^{-1}((x^2)^7 + 3(x^2)) = 2x\tan^{-1}(x^{14} + 3x^2).$$

In summary,

$$\frac{d}{dx}\int_0^{x^2} \tan^{-1}(t^7 + 3t)\, dt = 2x\tan^{-1}(x^{14} + 3x^2).$$

Not so bad when you break it down into little pieces.

Let's look at one more example of this sort of problem: what is

$$\frac{d}{dq}\int_4^{\sin(q)} \tan(\cos(a))\, da?$$

Well, let y be the integral in question:

$$y = \int_4^{\sin(q)} \tan(\cos(a))\, da,$$

and remind yourself that you're looking for dy/dq. Now set $u = \sin(q)$, so

$$y = \int_4^u \tan(\cos(a))\, da.$$

By the chain rule, we have

$$\frac{dy}{dq} = \frac{dy}{du}\frac{du}{dq}.$$

By the First Fundamental Theorem,

$$\frac{dy}{du} = \frac{d}{du} \int_4^u \tan(\cos(a))\, da = \tan(\cos(u)).$$

Since $u = \sin(q)$, we have $du/dq = \cos(q)$, so the chain rule equation above becomes

$$\frac{dy}{dq} = \frac{dy}{du}\frac{du}{dq} = \tan(\cos(u))\cos(q).$$

Finally, replace u by $\sin(q)$ to see that

$$\frac{d}{dq} \int_4^{\sin(q)} \tan(\cos(a))\, da = \tan(\cos(\sin(q)))\cos(q).$$

You might also encounter both of the above variations in the same problem. For example, to find

$$\frac{d}{dq} \int_{\sin(q)}^4 \tan(\cos(a))\, da,$$

start by switching the limits of integration, introducing a minus sign as you do so:

$$\frac{d}{dq} \int_{\sin(q)}^4 \tan(\cos(a))\, da = -\frac{d}{dq} \int_4^{\sin(q)} \tan(\cos(a))\, da.$$

Now you can find the right-hand side as we did above; the final answer will be the same, except for that minus sign out front:

$$\frac{d}{dq} \int_{\sin(q)}^4 \tan(\cos(a))\, da = -\frac{d}{dq} \int_4^{\sin(q)} \tan(\cos(a))\, da$$
$$= -\tan(\cos(\sin(q)))\cos(q).$$

17.5.3 Variation 3: two tricky limits of integration

Here's an even harder example:

$$\frac{d}{dx} \int_{x^5}^{x^6} \ln(t^2 - \sin(t) + 7)\, dt.$$

Now there are functions of x in **both** the left-hand and right-hand limits of integration. The way to handle this is to split the integral into two pieces at some number. It actually doesn't matter where you split it, as long as it is at a constant (where the function is defined). So, pick your favorite number—say 0—and split the integral there:

$$\frac{d}{dx} \int_{x^5}^{x^6} \ln(t^2 - \sin(t) + 7)\, dt$$
$$= \frac{d}{dx}\left(\int_{x^5}^0 \ln(t^2 - \sin(t) + 7)\, dt + \int_0^{x^6} \ln(t^2 - \sin(t) + 7)\, dt \right).$$

We've reduced the problem to two easier derivatives. The first one is a combination of the first two variations above. Just switch the limits of integration, introducing the minus sign, to write

$$\frac{d}{dx} \int_{x^5}^{0} \ln(t^2 - \sin(t) + 7)\, dt = -\frac{d}{dx} \int_{0}^{x^5} \ln(t^2 - \sin(t) + 7)\, dt.$$

Now use the chain rule by setting $u = x^5$ and following the method from the previous section. You should check that the above derivative works out to be

$$-5x^4 \ln((x^5)^2 - \sin(x^5) + 7) = -5x^4 \ln(x^{10} - \sin(x^5) + 7).$$

As for the other derivative above,

$$\frac{d}{dx} \int_{0}^{x^6} \ln(t^2 - \sin(t) + 7)\, dt,$$

there's no need to switch the limits of integration—just set $v = x^6$ and apply the chain rule once again. You should check that the above derivative is equal to

$$6x^5 \ln((x^6)^2 - \sin(x^6) + 7) = 6x^5 \ln(x^{12} - \sin(x^6) + 7).$$

Putting it all together, we have shown that

$$\frac{d}{dx} \int_{x^5}^{x^6} \ln(t^2 - \sin(t) + 7)\, dt$$
$$= -5x^4 \ln(x^{10} - \sin(x^5) + 7) + 6x^5 \ln(x^{12} - \sin(x^6) + 7).$$

17.5.4 Variation 4: limit is a derivative in disguise

Here's an example which looks a little different:

$$\lim_{h \to 0} \frac{1}{h} \int_{x}^{x+h} \log_3(\cos^6(t) + 2)\, dt.$$

This isn't a derivative—it's a limit. Actually, it is a derivative in disguise (see Section 6.5 in Chapter 6 for a discussion of these types of limits). The trick is to set

$$F(x) = \int_{a}^{x} \log_3(\cos^6(t) + 2)\, dt$$

for some constant a. You can put in a specific constant if you like, or you can just leave it as a. It doesn't matter, because in any case we have

$$F(x+h) - F(x) = \int_{x}^{x+h} \log_3(\cos^6(t) + 2)\, dt.$$

Check that you believe this, or look back at Section 17.2 above. In any case, in terms of our function F, we have

$$\lim_{h \to 0} \frac{1}{h} \int_{x}^{x+h} \log_3(\cos^6(t) + 2)\, dt = \lim_{h \to 0} \frac{F(x+h) - F(x)}{h} = F'(x).$$

So actually, we have

$$\lim_{h \to 0} \frac{1}{h} \int_{x}^{x+h} \log_3(\cos^6(t) + 2)\, dt = \frac{d}{dx} \int_{a}^{x} \log_3(\cos^6(t) + 2)\, dt$$

for any a you like. See, I told you that the limit was a derivative in disguise! To finish the problem, just apply the First Fundamental Theorem in its basic form to see that the above limit is just $\log_3(\cos^6(x) + 2)$.

17.6 How to Solve Problems: The Second Fundamental Theorem

To find a definite integral using the Second Fundamental Theorem—and this is how you want to find definite integrals, believe me—you need to find the indefinite integral first, then substitute in the endpoints and take the difference. So let's spend a little time discussing how to find indefinite integrals (that is, antiderivatives), then look at some examples of how to find definite integrals. This is only the beginning of the story; in the next two chapters, we'll look at many more ways of finding indefinite integrals.

17.6.1 Finding indefinite integrals

As we saw in Section 17.4 above, whenever you know a derivative, you get an antiderivative for free. We gave some examples there, but here's another: since

$$\frac{d}{dx}(x^4) = 4x^3,$$

we immediately know that

$$\int 4x^3\, dx = x^4 + C.$$

Since constants just pass through the integral sign, we can write this as

$$4 \int x^3\, dx = x^4 + C.$$

Now divide by 4:

$$\int x^3\, dx = \frac{x^4}{4} + \frac{C}{4}.$$

This is fine, but the quantity $C/4$ is a bit silly. It's some arbitrary constant divided by 4, which is another arbitrary constant. So we can just replace the constant $C/4$ by some other constant, which we'll also call C, and get

$$\int x^3\, dx = \frac{x^4}{4} + C.$$

Let's repeat this for any power of x. Start off by noting that

$$\frac{d}{dx}(x^{a+1}) = (a+1)x^a;$$

this means that

$$\int (a+1)x^a \, dx = x^{a+1} + C.$$

If $a \neq -1$, then $a + 1 \neq 0$; so we can divide through by $(a+1)$ and write

$$\boxed{\int x^a \, dx = \frac{x^{a+1}}{a+1} + C.}$$

(Once again, we replaced $C/(a+1)$ by simply C; this is OK since C is just an arbitrary constant.) Now, what happens when $a = -1$? The above method doesn't work on $\int x^{-1} \, dx$, which is just

$$\int \frac{1}{x} \, dx.$$

On the other hand, we do know from Section 9.3 of Chapter 9 that

$$\frac{d}{dx}(\ln(x)) = \frac{1}{x}, \qquad \text{so} \qquad \int \frac{1}{x} \, dx = \ln(x) + C.$$

This is fine, but actually we can do better. You see, $1/x$ is defined everywhere except at $x = 0$, while $\ln(x)$ is only defined when $x > 0$. We can rectify this by writing

$$\int \frac{1}{x} \, dx = \ln|x| + C.$$

Let's check that this works. We need to show that

$$\frac{d}{dx} \ln|x| = \frac{1}{x}$$

for all $x \neq 0$. When $x > 0$, the left-hand side is just $\ln(x)$ and there's no problem. If $x < 0$, then $|x|$ is actually equal to $-x$, so the left-hand side becomes

$$\frac{d}{dx} \ln(-x).$$

It looks a bit weird, but remember that $-x$ is actually positive when $x < 0$. In any case, by the chain rule, the above derivative is

$$\frac{d}{dx} \ln(-x) = -\frac{1}{-x} = \frac{1}{x}.$$

So we have proved the formula

$$\boxed{\int \frac{1}{x} \, dx = \ln|x| + C.}$$

See Section 17.7 below for a technicality involving this formula. In the meantime, we can now summarize most of the basic derivatives and corresponding antiderivatives that we've seen so far in one big table.

Derivatives and integrals to learn:

$$\frac{d}{dx}x^a = ax^{a-1} \qquad\qquad \int x^a \, dx = \frac{x^{a+1}}{a+1} + C \quad \text{(if } a \neq -1)$$

$$\frac{d}{dx}\ln(x) = \frac{1}{x} \qquad\qquad \int \frac{1}{x} \, dx = \ln|x| + C$$

$$\frac{d}{dx}e^x = e^x \qquad\qquad \int e^x \, dx = e^x + C$$

$$\frac{d}{dx}b^x = b^x \ln(b) \qquad\qquad \int b^x \, dx = \frac{b^x}{\ln(b)} + C$$

$$\frac{d}{dx}\sin(x) = \cos(x) \qquad\qquad \int \cos(x) \, dx = \sin(x) + C$$

$$\frac{d}{dx}\cos(x) = -\sin(x) \qquad\qquad \int \sin(x) \, dx = -\cos(x) + C$$

$$\frac{d}{dx}\tan(x) = \sec^2(x) \qquad\qquad \int \sec^2(x) \, dx = \tan(x) + C$$

$$\frac{d}{dx}\sec(x) = \sec(x)\tan(x) \qquad\qquad \int \sec(x)\tan(x) \, dx = \sec(x) + C$$

$$\frac{d}{dx}\cot(x) = -\csc^2(x) \qquad\qquad \int \csc^2(x) \, dx = -\cot(x) + C$$

$$\frac{d}{dx}\csc(x) = -\csc(x)\cot(x) \qquad\qquad \int \csc(x)\cot(x) \, dx = -\csc(x) + C$$

$$\frac{d}{dx}\sin^{-1}(x) = \frac{1}{\sqrt{1-x^2}} \qquad\qquad \int \frac{1}{\sqrt{1-x^2}} \, dx = \sin^{-1}(x) + C$$

$$\frac{d}{dx}\tan^{-1}(x) = \frac{1}{1+x^2} \qquad\qquad \int \frac{1}{1+x^2} \, dx = \tan^{-1}(x) + C$$

$$\frac{d}{dx}\sec^{-1}(x) = \frac{1}{|x|\sqrt{x^2-1}} \qquad\qquad \int \frac{1}{|x|\sqrt{x^2-1}} \, dx = \sec^{-1}(x) + C$$

$$\frac{d}{dx}\sinh(x) = \cosh(x) \qquad\qquad \int \cosh(x) \, dx = \sinh(x) + C$$

$$\frac{d}{dx}\cosh(x) = \sinh(x) \qquad\qquad \int \sinh(x) \, dx = \cosh(x) + C$$

As we've seen, if you replace x by the constant multiple ax in any of the above differentiation formulas, you just have to multiply the corresponding formula by a. For example,

$$\frac{d}{dx}\tan(7x) = 7\sec^2(7x).$$

What if you integrate instead? Now the rule of thumb is that if you replace x by ax, then you have to **divide** by a. For example,

$$\int \sec^2(7x) = \frac{1}{7}\tan(7x) + C.$$

You can see this directly from the previous equation by dividing by 7. Here's another example:

$$\int e^{-x/3} \, dx.$$

You can think of x as having been replaced by $-1/3$ times x; so divide by $-1/3$, like this:

$$\int e^{-x/3} \, dx = \frac{1}{-1/3} e^{-x/3} + C = -3e^{-x/3} + C.$$

How about one more for good measure? Consider

$$\int \frac{1}{1 + 2x^2} \, dx.$$

This can be written as

$$\int \frac{1}{1 + (\sqrt{2}x)^2} \, dx,$$

and now you can consider x as being replaced by $\sqrt{2}x$. So divide by $\sqrt{2}$ to get

$$\int \frac{1}{1 + (\sqrt{2}x)^2} \, dx = \frac{1}{\sqrt{2}} \tan^{-1}(\sqrt{2}x) + C.$$

There are many more complicated techniques for finding antiderivatives which we'll look at in the next two chapters, but it certainly doesn't hurt to remember this simple one, since constant multiples do come up often in integrands.

17.6.2 Finding definite integrals

The Second Fundamental Theorem tells us that to find

$$\int_a^b f(x) \, dx,$$

just find an antiderivative, plug in $x = b$ and $x = a$, and take the difference. We've already looked at some examples of this in Section 17.3 above; let's look at five more. First, consider

$$\int_{-1}^2 x^4 \, dx.$$

By the formula

$$\int x^a \, dx = \frac{x^{a+1}}{a+1} + C,$$

we know that an antiderivative of x^4 is $x^5/5$. No need for the constant— you can choose **any** antiderivative, so just choose the one with $C = 0$ for simplicity. So, we have

$$\int_{-1}^2 x^4 \, dx = \frac{x^5}{5} \Big|_{-1}^2 = \left(\frac{2^5}{5}\right) - \left(\frac{(-1)^5}{5}\right) = \left(\frac{32}{5}\right) - \left(\frac{-1}{5}\right) = \frac{33}{5}.$$

It's important to use parentheses to make sure you don't screw up the minus signs! Now, you might be wondering what happens if you did happen to use a different antiderivative. Well, the constant will just cancel out. For example, if you chose the antiderivative $x^5/5 - 1001$ instead, you'd get

$$\int_{-1}^{2} x^4 \, dx = \left(\frac{x^5}{5} - 1001 \right) \Bigg|_{-1}^{2} = \left(\frac{2^5}{5} - 1001 \right) - \left(\frac{(-1)^5}{5} - 1001 \right)$$

$$= \left(\frac{2^5}{5} \right) - 1001 - \left(\frac{(-1)^5}{5} \right) + 1001.$$

Notice that the -1001 and $+1001$ terms cancel and we're left with exactly what we had before. The moral of the story is to omit the constant C when calculating a definite integral.

Here's our second integral:

$$\int_{-e^2}^{-1} \frac{4}{x} \, dx.$$

The factor 4 can just pass through the integral sign, so we need to use the formula

$$\int \frac{1}{x} \, dx = \ln|x| + C$$

from the above table to see that $4 \ln|x|$ is an antiderivative for $4/x$. So we have

$$\int_{-e^2}^{-1} \frac{4}{x} \, dx = 4 \ln|x| \Bigg|_{-e^2}^{-1} = (4 \ln|-1|) - (4 \ln|-e^2|) = 4 \ln(1) - 4 \ln(e^2) = -8.$$

Here we have used the facts that $\ln(1) = 0$ and $\ln(e^2) = 2 \ln(e) = 2$.

The third example is

$$\int_{0}^{\pi/3} \left(\sec^2(x) - 5 \sin\left(\frac{x}{2} \right) \right) \, dx.$$

You should mentally split up the integrand into two components, $\sec^2(x)$ and $\sin(x/2)$, ignoring the constant 5 outside the second integral. By the above table, an antiderivative of $\sec^2(x)$ is $\tan(x)$. As for $\sin(x/2)$, an antiderivative is $-\cos(x/2)$ divided by $\frac{1}{2}$, since x has been replaced by the constant multiple $\frac{1}{2}x$. This works out to be $-2 \cos(x/2)$ (since dividing by $\frac{1}{2}$ is the same as multiplying by 2). Altogether, we have

$$\int_{0}^{\pi/3} \left(\sec^2(x) - 5 \sin\left(\frac{x}{2} \right) \right) \, dx = \left(\tan(x) - 5 \times \left(-2 \cos\left(\frac{x}{2} \right) \right) \right) \Bigg|_{0}^{\pi/3}.$$

Simplifying and substituting, we get

$$\left(\tan(\pi/3) + 10 \cos\left(\frac{\pi/3}{2} \right) \right) - \left(\tan(0) + 10 \cos\left(\frac{0}{2} \right) \right);$$

you should check that this works out to be $6\sqrt{3} - 10$.

Here's the fourth example:

$$\int_{4}^{9} \frac{1}{x\sqrt{x}} \, dx.$$

The trick here is to write the integrand as $x^{-3/2}$; make sure you believe this! Now we can just use the formula for $\int x^a \, dx$ from our big table in the previous section to get

$$\int_4^9 \frac{1}{x\sqrt{x}} \, dx = \int_4^9 x^{-3/2} \, dx = \left. \frac{1}{-1/2} x^{-1/2} \right|_4^9 = (-2(9)^{-1/2}) - (-2(4)^{-1/2})$$

$$= -\frac{2}{3} + \frac{2}{2} = \frac{1}{3}.$$

Now, our final example for this section is

$$\int_0^{1/6} \frac{dx}{\sqrt{1 - 9x^2}}.$$

Don't let the dx on the top worry you—this is just an alternate way of writing

$$\int_0^{1/6} \frac{1}{\sqrt{1 - 9x^2}} \, dx.$$

Express the $9x^2$ term as $(3x)^2$ to see that

$$\int_0^{1/6} \frac{dx}{\sqrt{1 - 9x^2}} = \int_0^{1/6} \frac{1}{\sqrt{1 - (3x)^2}} \, dx = \left. \frac{1}{3} \sin^{-1}(3x) \right|_0^{1/6}.$$

We have used the integral

$$\int \frac{1}{\sqrt{1 - x^2}} \, dx = \sin^{-1}(x) + C$$

from the above table, except that we have divided by 3, since x was replaced by $3x$. Now let's substitute to see that our integral becomes

$$\left(\frac{1}{3} \sin^{-1}\left(3 \times \frac{1}{6}\right) \right) - \left(\frac{1}{3} \sin^{-1}(3 \times 0) \right) = \left(\frac{1}{3} \times \frac{\pi}{6} \right) - (0) = \frac{\pi}{18}.$$

Here we've used the fact that $\sin^{-1}(\frac{1}{2}) = \pi/6$.

17.6.3 Unsigned areas and absolute values

In Section 16.1.1 of the previous chapter, we saw that

$$\int_{-\pi}^{\pi} \sin(x) \, dx = 0$$

because the area above the axis cancels the area below the axis. Here's a recap of the graph of the situation:

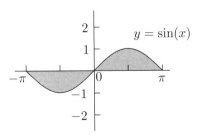

We can check the above integral using antiderivatives:

$$\int_{-\pi}^{\pi} \sin(x)\, dx = -\cos(x)\Big|_{-\pi}^{\pi} = (-\cos(\pi)) - (-\cos(-\pi)) = -(-1) + (-1) = 0.$$

How about finding the unsigned, actual area in the above picture? We looked at a method for doing this in Section 16.4.1 of the previous chapter: the actual area in square units is equal to

$$\int_{-\pi}^{\pi} |\sin(x)|\, dx.$$

Our method calls for splitting the original integral

$$\int_{-\pi}^{\pi} \sin(x)\, dx$$

at the x-intercept 0, then taking the absolute value of each piece. That is,

$$\int_{-\pi}^{\pi} |\sin(x)|\, dx = \left|\int_{-\pi}^{0} \sin(x)\right| + \left|\int_{0}^{\pi} \sin(x)\right|.$$

I leave it to you to use the antiderivative $-\cos(x)$ to show that these two integrals are -2 and 2, respectively. If you just add these numbers, you get the signed area 0 square units; but if you take the absolute values first, you get the actual area, which is $|-2| + |2| = 4$ square units.

Now, let's look at an example of finding the area between two curves. We already saw how to do this in Section 16.4.2 of the previous chapter, but now we have the power of the Second Fundamental Theorem at our disposal, so we can find more exotic areas like this one:

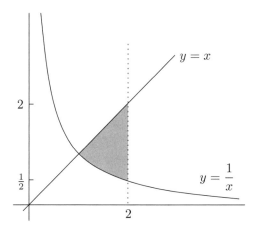

We're looking for the area between the curves $y = x$, $y = 1/x$, and the line $x = 2$. We'll need to find where $y = x$ and $y = 1/x$ intersect: set $x = 1/x$ and we see that $x^2 = 1$. This means that $x = 1$ or $x = -1$. In the above picture, the x-coordinate of the intersection point is positive, so we need $x = 1$. Since

$y = x$ is above $y = 1/x$, we take the top function x minus the bottom function $1/x$ and integrate:

$$\text{shaded area} = \int_1^2 \left(x - \frac{1}{x} \right) \, dx.$$

An antiderivative of x is $x^2/2$, as we can easily see by using the formula $\int x^a = x^{a+1}/(a+1) + C$ with $a = 1$; also, an antiderivative of $1/x$ is $\ln|x|$, as we saw above. So the above integral is equal to

$$\left(\frac{x^2}{2} - \ln|x| \right) \Big|_1^2 = \left(\frac{2^2}{2} - \ln|2| \right) - \left(\frac{1^2}{2} - \ln|1| \right) = 2 - \ln(2) - \frac{1}{2} + \ln(1);$$

This simplifies to $3/2 - \ln(2)$, so the area we want is $3/2 - \ln(2)$ square units. Now, consider what happens if the area we actually want to find is this instead:

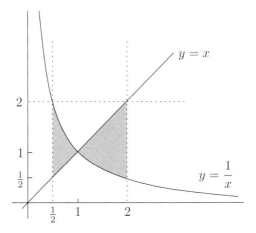

It's tempting to write this area as

$$\text{new shaded area} \stackrel{?}{=} \int_{1/2}^2 \left(x - \frac{1}{x} \right) \, dx,$$

but that would be a load of bull. You see, the curve $y = x$ isn't on top of $y = 1/x$ between $1/2$ and 1. We discussed this point in Section 16.4.2 of the previous chapter, and saw that we actually need to take absolute values:

$$\text{new shaded area} = \int_{1/2}^2 \left| x - \frac{1}{x} \right| \, dx.$$

Since the only intersection point is at $x = 1$, split the integral up into two pieces there and take the absolute value of each piece to get

$$\int_{1/2}^2 \left| x - \frac{1}{x} \right| \, dx = \left| \int_{1/2}^1 \left(x - \frac{1}{x} \right) \, dx \right| + \left| \int_1^2 \left(x - \frac{1}{x} \right) \, dx \right|.$$

We already saw that the second of these integrals is $3/2 - \ln(2)$, which is a positive quantity, since $\ln(2) < \ln(e) = 1$. As for the first integral, we have

$$\int_{1/2}^{1} \left(x - \frac{1}{x} \right) dx = \left(\frac{x^2}{2} - \ln|x| \right) \Big|_{1/2}^{1}$$

$$= \left(\frac{1^2}{2} - \ln|1| \right) - \left(\frac{(1/2)^2}{2} - \ln\left|\frac{1}{2}\right| \right)$$

$$= \frac{1}{2} - \ln(1) - \frac{1}{8} + \ln\left(\frac{1}{2} \right) = \frac{3}{8} - \ln(2).$$

Here we have used the fact that $\ln(1/2) = -\ln(2)$, which you can see either by writing $\ln(1/2) = \ln(1) - \ln(2)$ or $\ln(1/2) = \ln(2^{-1})$, then using one of the log rules from Section 9.1.4 in Chapter 9. Notice that the quantity $3/8 - \ln(2)$ is negative. You can see this by noting that x is actually less than $1/x$ when x is in $[1/2, 1]$, so the integrand $x - 1/x$ is negative there. So when we take the absolute value of $3/8 - \ln(2)$, we actually get $\ln(2) - 3/8$. Altogether, we have

$$\left| \int_{1/2}^{1} \left(x - \frac{1}{x} \right) dx \right| + \left| \int_{1}^{2} \left(x - \frac{1}{x} \right) dx \right| = \left| \frac{3}{8} - \ln(2) \right| + \left| \frac{3}{2} - \ln(2) \right|$$

$$= \left(\ln(2) - \frac{3}{8} \right) + \left(\frac{3}{2} - \ln(2) \right) = \frac{9}{8}.$$

The shaded area we're looking for is $9/8$ square units. Actually, we could have worked this out without using calculus at all. You see, both $y = x$ and $y = 1/x$ are symmetric in the line $y = x$, so if you flip the wedge-shaped region in the line $y = x$, then it fills in a triangle, like this:

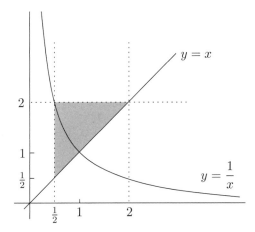

This triangle has base and height equal to $3/2$ units, so its area works out to be $9/8$ square units, agreeing with our above answer!

17.7 A Technical Point

In Section 17.6.1 above, we saw that

$$\int \frac{1}{x}\,dx = \ln|x| + C.$$

Although everyone writes the formula like this, technically it's not correct! You see, we want to find **all** antiderivatives of $1/x$. Sure, $\ln|x| + C$ is an antiderivative for each constant C, but actually there are more. To see why, let's start off with the graph of $y = \ln|x|$:

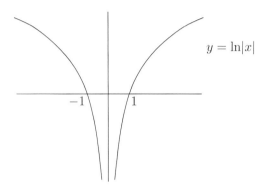

This has two pieces, either of which can be shifted up or down without affecting the derivative. For example, if we shift the left piece up by 1 and the right piece down by $1/2$, the graph looks something like this:

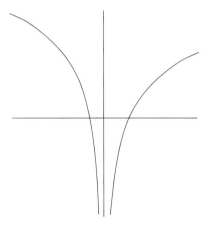

This function isn't of the form $\ln|x| + C$, but its derivative is still $1/x$. So we really need to allow **two** constants, possibly different—one for each of the two pieces of the curve:

$$\int \frac{1}{x}\,dx = \begin{cases} \ln|x| + C_1 & \text{if } x < 0, \\ \ln|x| + C_2 & \text{if } x > 0. \end{cases}$$

The reason we can get away without this level of formality, at least most of the time, is that we only really use one of the constants at a time. Consider the following three integrals:

$$\int_1^e \frac{1}{x}\,dx, \qquad \int_{-e}^{-1} \frac{1}{x}\,dx, \qquad \text{and} \qquad \int_{-1}^{e} \frac{1}{x}\,dx.$$

In the first integral, you are only using the right-hand piece of the curve $y = 1/x$. Similarly, in the second integral, only the left-hand piece is relevant. Try doing both integrals and make sure you get 1 and -1, respectively. As for the third integral, now we are using both pieces of $y = 1/x$, but there's a problem: the vertical asymptote at $x = 0$ lies in our interval $[-1, e]$. We don't know how to handle that. In fact, we will learn how to deal with this sort of thing when we look at improper integrals in Chapter 20. In this case, it turns out that the third integral above doesn't even make sense because of that vertical asymptote. So the only time that definite integrals of the form

$$\int_a^b \frac{1}{x}\,dx$$

make sense is when a and b are both positive or both negative. In either case, only one of the pieces of $\ln|x|$ is involved, and there's no need to mess around with two different constants!

17.8 Proof of the First Fundamental Theorem

In Section 17.2 above, we gave an intuitive proof of the First Fundamental Theorem of Calculus. Let's tighten it up. Recall that

$$F(x) = \int_a^x f(t)\,dt,$$

and we want to find $F'(x)$. We have already seen that

$$F(x + h) - F(x) = \int_x^{x+h} f(t)\,dt.$$

Suppose that $h > 0$. By the Mean Value Theorem for integrals (see Section 16.6.1 of the previous chapter), there is some number c lying in the interval $[x, x + h]$ such that

$$\int_x^{x+h} f(t)\,dt = ((x + h) - x)f(c).$$

That is, we have

$$F(x + h) - F(x) = \int_x^{x+h} f(t)\,dt = hf(c)$$

for some c in $[x, x + h]$. Actually, this is also true if $h < 0$, except that the interval is $[x + h, x]$ instead, since $x + h < x$ in that case. Anyway, divide the above equation by h to get

$$\frac{F(x + h) - F(x)}{h} = f(c).$$

The important thing is that when x is a fixed number (for the moment), the number c depends on h only, and it lies between x and $x + h$. Perhaps we should really rewrite the above equation as

$$\frac{F(x + h) - F(x)}{h} = f(c_h)$$

to emphasize that c depends on h. Now, what happens when $h \to 0$? The quantity c_h is sandwiched between x and $x + h$, so as $h \to 0$, the sandwich principle (see Section 3.6 of Chapter 3) says that $c_h \to x$ as $h \to 0$. On the other hand, since f is continuous, we must also have $f(c_h) \to f(x)$ as $h \to 0$. That is,

$$\lim_{h \to 0} \frac{F(x + h) - F(x)}{h} = \lim_{h \to 0} f(c_h) = f(x).$$

This shows that $F'(x) = f(x)$, wrapping up the proof of the First Fundamental Theorem. As for the Second Fundamental Theorem, we actually already proved it in Section 17.3 above, so we're good to go!

CHAPTER 18 _____

Techniques of Integration, Part One

Let's kick off the process of building up a virtual toolkit of techniques to find antiderivatives. In this chapter, we'll look at the following three techniques:

- the method of substitution (otherwise known as "change of variables");
- integration by parts; and
- using partial fractions to integrate rational functions.

Then, in the next chapter, we'll look at some more techniques involving trig functions.

18.1 Substitution

Using the chain rule, we can easily differentiate e^{x^2} with respect to x and see that

$$\frac{d}{dx}\left(e^{x^2}\right) = 2xe^{x^2}.$$

The factor $2x$ is the derivative of x^2, which appears in the exponent. Now, as we saw in Section 17.4 of the previous chapter, we can flip the equation around to get

$$\int 2xe^{x^2}\,dx = e^{x^2} + C$$

for some constant C. So we can integrate $2xe^{x^2}$ with respect to x. How about just e^{x^2}? You'd think it would be just as easy, if not easier, to find

$$\int e^{x^2}\,dx.$$

It turns out that it's not just hard to find this—it's impossible! Well, not quite impossible, but the fact is, there's no "nice" expression for an antiderivative of e^{x^2}. (You have to resort to infinite series, definite integrals, or some other sort of roundabout device.) Perhaps you think that $e^{x^2}/2x$ works? Nope—use the quotient rule to differentiate this (with respect to x) and you'll see that you get something quite different from e^{x^2}.

What saves us in the case of $\int 2xe^{x^2}\,dx$ is the presence of the $2x$ factor, which is exactly what popped out when we used the chain rule to differentiate e^{x^2}. Now, imagine starting with an indefinite integral like this:

$$\int x^2 \cos(x^3)\,dx.$$

We're taking the cosine of the somewhat nasty quantity x^3, but there's a ray of hope: the derivative of this quantity is $3x^2$. This almost matches the factor x^2 in the integrand—it's only the constant 3 that makes things a little more difficult. Still, constants can move in or out of integrals, so that shouldn't be a problem.

Start off by setting $t = x^3$, so that the $\cos(x^3)$ factor becomes $\cos(t)$. Our aim will be to replace everything that has to do with x in the above integral by stuff in t alone. You might say that the above integral is in x-land and we'd like to migrate it over to t-land. We've already taken care of $\cos(x^3)$, but we still have x^2 and dx to worry about.

In fact, the dx factor is really important. You can't just change it to dt! After all, $t = x^3$, so $dt/dx = 3x^2$. If there's any justice in the world, then we should be able to rewrite this as $dt = 3x^2\,dx$. Let's not worry about what this means; we'll leave that until Section 18.1.3 below. Instead, suppose we divide both sides by 3 to get $\frac{1}{3}\,dt = x^2\,dx$. Then we can get rid of the x^2 and dx pieces from our integral at the same time, replacing both by $\frac{1}{3}\,dt$, like this:

$$\int x^2 \cos(x^3)\,dx = \int \cos(x^3)\,(x^2\,dx) = \int \cos(t)\left(\frac{1}{3}\,dt\right).$$

The middle step isn't really necessary, but it helps to see x^2 and dx next to each other so that you can justify replacing them by $\frac{1}{3}dt$. Anyway, now we can drag the factor of $\frac{1}{3}$ outside the integral, then integrate; altogether, we have

$$\int x^2 \cos(x^3)\,dx = \int \cos(t)\frac{1}{3}\,dt = \frac{1}{3}\int \cos(t)\,dt = \frac{1}{3}\sin(t) + C.$$

It's pretty lazy to leave the answer as $\frac{1}{3}\sin(t) + C$. We started in x-land, then migrated over to t-land; now we have to come back to x-land. This isn't hard to do: just replace t by x^3 once again. We have shown that

$$\int x^2 \cos(x^3)\,dx = \frac{1}{3}\sin(x^3) + C.$$

Check that this is true by differentiating $\frac{1}{3}\sin(x^3)$ with respect to x.

Let's look at some more examples. First, consider

$$\int e^{2x}\sec^2(e^{2x})\,dx.$$

Since we're taking $\sec^2$ of the annoying quantity e^{2x}, let's replace that quantity by t. So substitute $t = e^{2x}$. Differentiate this to see that $dt/dx = 2e^{2x}$. Now throw the dx onto the right-hand side to see that $dt = 2e^{2x}\,dx$. That's almost what we have in the integral—we just need to get rid of the factor of 2. So divide by 2 to get $\frac{1}{2} dt = e^{2x} dx$. Moving the above integral to t-land, we get

$$\int e^{2x} \sec^2(e^{2x}) \, dx = \int \sec^2(e^{2x}) \left(e^{2x} \, dx\right) = \int \sec^2(t) \left(\frac{1}{2} dt\right).$$

Now pull out the factor of $\frac{1}{2}$ and integrate to get $\tan(t) + C$. Finally, move back to x-land by replacing t with e^{2x}. We have proved that

$$\int e^{2x} \sec^2(e^{2x}) \, dx = \frac{1}{2} \tan(e^{2x}) + C.$$

Again, you should check this by differentiating the right-hand side.
    Here's another example:

$$\int \frac{3x^2 + 7}{x^3 + 7x - 9} \, dx.$$

This looks pretty difficult. Fortunately, if you differentiate the denominator $x^3 + 7x - 9$, you get the numerator $3x^2 + 7$. This suggests that we substitute $t = x^3 + 7x - 9$. Since $dt/dx = 3x^2 + 7$, we can write $dt = (3x^2 + 7) \, dx$. In t-land, our integral is

$$\int \frac{3x^2 + 7}{x^3 + 7x - 9} \, dx = \int \frac{1}{x^3 + 7x - 9}((3x^2 + 7) \, dx) = \int \frac{1}{t} \, dt = \ln|t| + C.$$

Now switch back to x-land by replacing t with $x^3 + 7x - 9$; this shows that

$$\int \frac{3x^2 + 7}{x^3 + 7x - 9} \, dx = \ln|x^3 + 7x - 9| + C.$$

Actually, this is a special case of a nice fact: if f is a differentiable function, then

$$\boxed{\int \frac{f'(x)}{f(x)} \, dx = \ln|f(x)| + C.}$$

So if the top is the derivative of the bottom, then the integral is just the log of the bottom (with absolute values and the $+C$). We can prove this in general by making the substitution $t = f(x)$. Then $dt/dx = f'(x)$, so we can write $dt = f'(x) \, dx$. See if you can follow each step in this chain of equations which migrate from x-land to t-land, then back:

$$\int \frac{f'(x)}{f(x)} \, dx = \int \frac{1}{f(x)}(f'(x) \, dx) = \int \frac{1}{t} \, dt = \ln|t| + C = \ln|f(x)| + C.$$

This fact means that in the above example,

$$\int \frac{3x^2 + 7}{x^3 + 7x - 9} \, dx,$$

you can just write down the answer $\ln|x^3 + 7x - 9| + C$, since the top is the derivative of the bottom. Sometimes the top is a multiple of the derivative of the bottom, like this:

$$\int \frac{x}{x^2 + 8} \, dx.$$

The derivative of the bottom is $2x$, but we only have x on the top. No problem—multiply and divide by 2, like this:

$$\int \frac{x}{x^2+8}\,dx = \frac{1}{2}\int \frac{2x}{x^2+8}\,dx.$$

Now you can just write down the answer $\frac{1}{2}\ln|x^2+8|+C$, since the top $(2x)$ is the derivative of the bottom (x^2+8). Finally, consider

$$\int \frac{1}{x\ln(x)}\,dx.$$

The nicest way to do this is to rewrite the integral as

$$\int \frac{1/x}{\ln(x)}\,dx,$$

then notice that the derivative of the bottom $(\ln(x))$ is the top $(1/x)$. By the formula in the box above, the integral is just $\ln|\ln(x)|+C$. That is,

$$\int \frac{1}{x\ln(x)}\,dx = \ln|\ln(x)|+C.$$

18.1.1   Substitution and definite integrals

You can also use the substitution method on definite integrals. There are two legitimate ways to do this. For example, to find

$$\int_0^{\sqrt[3]{\pi/2}} x^2\cos(x^3)\,dx,$$

you could find the indefinite integral $\int x^2\cos(x^3)\,dx$ first, then plug in the limits of integration. We actually found this indefinite integral in the previous section; to recap, we made the substitution $t = x^3$, noting that $dt = 3x^2\,dx$ so $\frac{1}{3}\,dt = x^2\,dx$, then wrote

$$\int x^2\cos(x^3)\,dx = \int \cos(t)\frac{dt}{3} = \frac{1}{3}\int \cos(t)\,dt = \frac{1}{3}\sin(t)+C = \frac{1}{3}\sin(x^3)+C.$$

It's really important to go back to x-land at the last step. Anyway, the important thing is that we have found an antiderivative for $x^2\cos(x^3)$, and we can use the Second Fundamental Theorem from Section 17.3 of the previous chapter to write

$$\int_0^{\sqrt[3]{\pi/2}} x^2\cos(x^3)\,dx = \frac{1}{3}\sin(x^3)\Big|_0^{\sqrt[3]{\pi/2}} = \left(\frac{1}{3}\sin((\sqrt[3]{\pi/2})^3)\right) - \left(\frac{1}{3}\sin(0^3)\right),$$

which works out to be $\frac{1}{3}$. So one way to use the substitution method on a definite integral is to focus on the indefinite integral first, then after you've found it, plug in the limits of integration.

There's a snazzier method, though! You can keep the whole thing as a definite integral the whole way through, provided that you also move the limits of integration over to t-land as well. In our example, we substituted $t = x^3$ and used $\frac{1}{3} dt = x^2 dx$ to help move the integral to t-land. Now, when $x = 0$, we have $t = 0^3 = 0$, so we can leave the left-hand limit of integration as 0. On the other hand, when $x = \sqrt[3]{\pi/2}$, we have $t = \left(\sqrt[3]{\pi/2}\right)^3 = \pi/2$. This means that we must change the right-hand limit of integration to $\pi/2$. Altogether, here's the effect of the substitution:

$$\int_0^{\sqrt[3]{\pi/2}} x^2 \cos(x^3)\, dx = \frac{1}{3} \int_0^{\pi/2} \cos(t)\, dt.$$

We'll finish this soon, but first note that it would be a major error to write

$$\frac{1}{3} \int_0^{\sqrt[3]{\pi/2}} \cos(t)\, dt$$

on the right-hand side instead. Since we're integrating with respect to t, not x, the limits of integration must refer to relevant values of t. In fact, we can make things clearer by writing out the limits of integration in terms of the variable of integration, like this:

$$\int_{x=0}^{x=\sqrt[3]{\pi/2}} x^2 \cos(x^3)\, dx = \frac{1}{3} \int_{t=0}^{t=\pi/2} \cos(t)\, dt.$$

This really highlights what's going on: when $x = 0$, also $t = 0$; but when $x = \sqrt[3]{\pi/2}$, we see that $t = \pi/2$. So, all in all, we've substituted **three** things:

1. the dx bit—that became something to do with dt, burning up some of the other x stuff in order to make the change;
2. all the remaining terms in the integrand involving x, so that they became terms in t;
3. the limits of integration.

Let's finish the problem. The best way to set it out is to make a working column at the left of your page, like this:

$t = x^3$

$dt = 3x^2\, dx$, so $x^2\, dx = \frac{1}{3} dt$

when $x = 0$, $t = 0$

when $x = \sqrt[3]{\pi/2}$, $t = \pi/2$

$$\int_0^{\sqrt[3]{\pi/2}} x^2 \cos(x^3)\, dx = \int_0^{\pi/2} \frac{1}{3} \cos(t)\, dt$$
$$= \frac{1}{3} \sin(t) \Big|_0^{\pi/2}$$
$$= \left(\frac{1}{3} \sin(\pi/2)\right) - \left(\frac{1}{3} \sin(0)\right) = \frac{1}{3}.$$

Note that the entire left-hand column is filled in before we even get to the first equality of the right-hand column, since we have to use all the information there to get to t-land.

Here's a trickier one:

$$\int_{1/\sqrt{2}}^{\sqrt{3}/2} \frac{1}{\sin^{-1}(x)\sqrt{1-x^2}}\, dx.$$

Ask yourself this: do you see any term somewhere whose derivative is also present? Hopefully, you do: the derivative of $\sin^{-1}(x)$ is $1/\sqrt{1-x^2}$. So try the substitution $t = \sin^{-1}(x)$. Yes indeed, $dt/dx = 1/\sqrt{1-x^2}$, so we have

$$dt = \frac{1}{\sqrt{1-x^2}}\,dx.$$

We also have to transfer the limits of integration to t-land by plugging in $x = 1/\sqrt{2}$ and $x = \sqrt{3}/2$ into the equation $t = \sin^{-1}(x)$, one at a time. You should get $t = \pi/4$ and $t = \pi/3$, respectively, provided that you remember your inverse trig basics! (See Chapter 10 to refresh your memory.) Putting everything together, we get:

$$t = \sin^{-1}(x)$$

$$dt = \frac{1}{\sqrt{1-x^2}}\,dx$$

$$\text{when } x = \frac{1}{\sqrt{2}},\ t = \sin^{-1}\!\left(\frac{1}{\sqrt{2}}\right) = \frac{\pi}{4}$$

$$\text{when } x = \frac{\sqrt{3}}{2},\ t = \sin^{-1}\!\left(\frac{\sqrt{3}}{2}\right) = \frac{\pi}{3}$$

$$\int_{1/\sqrt{2}}^{\sqrt{3}/2} \frac{1}{\sin^{-1}(x)\sqrt{1-x^2}}\,dx$$

$$= \int_{\pi/4}^{\pi/3} \frac{1}{t}\,dt = \ln|t|\Big|_{\pi/4}^{\pi/3}$$

$$= \ln\left|\frac{\pi}{3}\right| - \ln\left|\frac{\pi}{4}\right| = \ln\left(\frac{4}{3}\right).$$

To get the final simplified answer, notice that we had to know the log rules (see Section 9.1.4 of Chapter 9). It's a really good idea to have these at your fingertips.

By the way, if you're particularly eagle-eyed, you might notice that the above substitution is actually a special case of the rule from the end of the previous section. This provides an alternative way of finding our integral

$$\int_{1/\sqrt{2}}^{\sqrt{3}/2} \frac{1}{\sin^{-1}(x)\sqrt{1-x^2}}\,dx.$$

Let's start with the indefinite integral, and rewrite it like this:

$$\int \frac{1}{\sin^{-1}(x)\sqrt{1-x^2}}\,dx = \int \frac{1/\sqrt{1-x^2}}{\sin^{-1}(x)}\,dx.$$

Notice that the top is the derivative of the bottom, so we just have to take the log of the absolute value of the bottom to see that

$$\int \frac{1}{\sin^{-1}(x)\sqrt{1-x^2}}\,dx = \ln|\sin^{-1}(x)| + C.$$

Now to find the definite integral, you can substitute the original limits of integration, $\sqrt{3}/2$ and $1/\sqrt{2}$, one at a time into the expression $\ln|\sin^{-1}(x)|$, then take the difference. I leave the details to you.

Here's a different sort of problem involving substitution. At the end of Section 16.1.1 of Chapter 16, we claimed that

$$\text{if } f \text{ is an odd function, then} \qquad \int_{-a}^{a} f(x)\,dx = 0 \qquad \text{for any } a.$$

How would you prove that this is true? Start off by splitting up the integral at $x = 0$:

$$\int_{-a}^{a} f(x)\, dx = \int_{-a}^{0} f(x)\, dx + \int_{0}^{a} f(x)\, dx.$$

In the first integral on the right-hand side, let's substitute $t = -x$. Then $dt = -dx$; also, when $t = -a$, we see that $x = a$, and when $t = 0$, $x = 0$ as well. So we have

$$\int_{-a}^{0} f(x)\, dx = -\int_{a}^{0} f(-t)\, dt = \int_{0}^{a} f(-t)\, dt.$$

In this last step, we used the minus sign to switch the bounds of integration. Now, since f is odd, we know that $f(-t) = -f(t)$. This shows that

$$\int_{0}^{a} f(-t)\, dt = -\int_{0}^{a} f(t)\, dt.$$

Now if we switch the dummy variable back to x, we see we've proved the following nice result:

$$\int_{-a}^{0} f(x)\, dx = -\int_{0}^{a} f(x)\, dx.$$

This is only true when f is an odd function! Anyway, we can finish by going back to our first equation and using our nice result:

$$\int_{-a}^{a} f(x)\, dx = \int_{-a}^{0} f(x)\, dx + \int_{0}^{a} f(x)\, dx = -\int_{0}^{a} f(x)\, dx + \int_{0}^{a} f(x)\, dx = 0.$$

We're all done!

18.1.2 How to decide what to substitute

How do you choose the substitution? Good question. The basic idea is to look for some component of the integrand whose derivative is also present as a factor of the integrand. In the integral

$$\int \frac{1}{\sin^{-1}(x)\sqrt{1 - x^2}}\, dx,$$

the substitution $t = \sin^{-1}(x)$ works because its derivative $1/\sqrt{1 - x^2}$ is right there, waiting for us to use it. The same substitution would work on any of the integrals

$$\int \frac{\sin^{-1}(x)}{\sqrt{1 - x^2}}\, dx, \qquad \int \frac{e^{\sin^{-1}(x)}}{\sqrt{1 - x^2}}\, dx \qquad \text{and} \qquad \int \frac{1}{\sqrt{\sin^{-1}(x)(1 - x^2)}}\, dx.$$

In t-land, these integrals become

$$\int t\, dt, \qquad \int e^t\, dt, \qquad \text{and} \qquad \int \frac{1}{\sqrt{t}}\, dt,$$

respectively. This is pretty easy to see for the first two; for the third, you just have to split up the square root and write

$$\int \frac{1}{\sqrt{\sin^{-1}(x)(1-x^2)}}\,dx = \int \frac{1}{\sqrt{\sin^{-1}(x)}}\frac{1}{\sqrt{1-x^2}}\,dx$$

to see how the substitution works. Now make sure you can complete all three above integrals in t-land and change back to x-land. (For the third integral, it might help to write $1/\sqrt{t}$ as $t^{-1/2}$.) In any case, you should get

$$\frac{(\sin^{-1}(x))^2}{2}+C, \qquad e^{\sin^{-1}(x)}+C, \qquad \text{and} \qquad 2\sqrt{\sin^{-1}(x)}+C,$$

respectively. Try differentiating each one of these to check your answers.

Sometimes the substitution is not obvious at all. For example, how would you find

$$\int \frac{e^x}{e^{2x}+1}\,dx?$$

Reasonable choices for the substitution are $t=e^x$, $t=e^{2x}$, or $t=e^{2x}+1$. The last two don't work very well, because $dt=2e^{2x}\,dx$ in both cases, and there's no e^{2x} term in the numerator of the integrand. So let's try $t=e^x$. We then have $dt=e^x\,dx$, which does care of the numerator of the fraction. As for the denominator, the trick here is to notice that e^{2x} is $(e^x)^2$, which is precisely t^2. So

$$\int \frac{e^x}{e^{2x}+1}\,dx = \int \frac{1}{t^2+1}\,dt,$$

which is just $\tan^{-1}(t)+C$. Moving back to x-land, we find that

$$\int \frac{e^x}{e^{2x}+1}\,dx = \tan^{-1}(e^x)+C$$

for some constant C. Check this by differentiating the right-hand side.

Let's look at one more example:

$$\int x\sqrt[5]{3x+2}\,dx.$$

There is a nice technique for dealing with integrals involving terms such as $\sqrt[n]{ax+b}$. You simply set $t=\sqrt[n]{ax+b}$, but take nth powers before you differentiate to find dt. So:

> to deal with $\sqrt[n]{ax+b}$, set $t=\sqrt[n]{ax+b}$ and differentiate both sides of $t^n=ax+b$.

So in our example, substitute $t=\sqrt[5]{3x+2}$. To find dt, first take 5th powers to get $t^5=3x+2$. Now differentiate both sides with respect to the appropriate variables (this is justified by the chain rule) and get $5t^4\,dt=3\,dx$. Here $5t^4$ is the derivative with respect to t of t^5, and 3 is the derivative with respect to x of $3x+2$. So, we have a nice expression for $3\,dx$ in terms of t, and we can make it a nice expression for dx by dividing by 3. Specifically, we have

$$dx = \frac{5}{3}t^4\,dt.$$

(You could also have seen this by solving for x all the way down to $x = \frac{1}{3}(t^5 - 2)$ and then differentiating with respect to t.) Now let's look back at the integral. There are three pieces: x, $\sqrt[5]{3x + 2}$, and dx. The second piece is just t itself, and we have just worked out the third piece in the above equation. How about the first piece, x? Well, we know $t^5 = 3x + 2$, so we can rearrange this to get $x = \frac{1}{3}(t^5 - 2)$. All in all the integral becomes

$$\int x\sqrt[5]{3x + 2}\,dx = \int \frac{1}{3}(t^5 - 2)(t) \times \frac{5}{3}t^4\,dt.$$

Now we can multiply and integrate to see that this equals

$$\frac{5}{9}\int \left(t^{10} - 2t^5\right)\,dt = \frac{5}{99}t^{11} - \frac{5}{27}t^6 + C.$$

Back to x-land: resubstituting $t = (3x + 2)^{1/5}$ gives

$$\frac{5}{99}(3x + 2)^{11/5} - \frac{5}{27}(3x + 2)^{6/5} + C.$$

You should try working this problem on your own, setting your answer out using a working column on the left, as we've been doing previously. Also, you should check that if you differentiate the answer above, you get the original integrand $x\sqrt[5]{3x + 2}$. By the way, did you notice anything different about this substitution from all the others we've done so far? It's a subtle point, but in all the other examples, we had an equation like $dt = (x\text{-stuff})\,dx$, whereas here, we have $dx = \frac{5}{3}t^4\,dt$. This worked out quite nicely, since we just replaced dx directly. In all the other examples, we had to find a constant multiple of the x-stuff already present in order to have much of a chance. In Section 19.3 of the next chapter, we'll see some other examples of integrals where we can replace dx directly.

In general, there are no hard and fast rules about what to substitute. You just have to go along with your instinct, which will be accurate only if you have done plenty of practice problems. You can always try any substitution you like. If the new integral is worse than the original one, or you can't see how to migrate everything to t-land, then don't panic: just go back to the original integral and try something else.

Now, before we move onto integration by parts, there are two things I want to deal with. One is a justification of the substitution method; I'll do this in the next section. The other is to summarize the method of substitution:

- for **indefinite** integrals, change everything to do with x and dx to stuff involving t and dt, do the new integral, then change back to x stuff;
- for **definite** integrals, change everything to do with x and dx to stuff involving t and dt, **and** change the limits of integration to the corresponding t values as well, then do the new integral (no need to go back to x-land here). Alternatively, treat the integral as an indefinite integral and when you get the final answer, then substitute in the limits of integration.

18.1.3   Theoretical justification of the substitution method

 Suppose you want to make the substitution $t = x^2$ in some integral. You'd note that $dt/dx = 2x$, so you write $dt = 2x\,dx$. In some sense, this is a meaningless statement—after all, what are dt and dx? We know that dt/dx is a derivative, but dt and dx have only been defined as differentials in Chapter 13. So what does $dt = 2x\,dx$ actually mean? A good way to think of it is that a change in x produces a change in t which is $2x$ times as large. We actually looked at this sort of thing all the way back in Section 5.2.7 of Chapter 5. You can run with this observation and see what it does to a Riemann sum, but there's a better way: just use the chain rule.

Here's how to justify everything. Imagine you have done a substitution $t = g(x)$, and you work your magic to end up in t-land with $\int f(t)\,dt$, which works out to be $F(t) + C$ for some constant C. So the t-land part of the calculation looks like this:

$$\int f(t)\,dt = F(t) + C.$$

Since $t = g(x)$, and we have decreed that $dt = g'(x)\,dx$, the above equation means the same thing in x-land as

$$\int f(g(x))g'(x)\,dx = F(g(x)) + C.$$

All I did was replace both t's by $g(x)$ and dt by $g'(x)\,dx$. So, if we want to prove that substitution is a valid method, we need to show that the above equation is true. Let $h(x) = F(g(x))$; by the chain rule (see Version 1 in Section 6.2.5 of Chapter 6), it's true that $h'(x) = F'(g(x))g'(x)$. We can write this in terms of indefinite integrals like this:

$$\int F'(g(x))g'(x)\,dx = h(x) + C.$$

Since $h(x) = F(g(x))$, we have

$$\int F'(g(x))g'(x)\,dx = F(g(x)) + C.$$

Now, since $\int f(t)\,dt = F(t) + C$, we know that $F'(t) = f(t)$. Since $t = g(x)$, we have $F'(g(x)) = f(g(x))$. The above equation becomes

$$\int f(g(x))g'(x)\,dx = F(g(x)) + C,$$

which is exactly the equation we wanted to prove!

By the way, this nice equation allows us to prove the alternative method of substitution, which was discussed after the last example in the previous section above. (We'll also use it over and over when we look at trig substitutions in Section 19.3 of the next chapter.) In the alternative method, instead of setting $t = g(x)$, we set $x = g(t)$ for some other function g, and replaced dx by $g'(t)\,dt$. In that case, our original integral $\int f(x)\,dx$ now supposedly becomes

$$\int f(g(t))g'(t)\,dt.$$

We are now supposed to work this out and try to move back to x-land. Well, by our nice equation, with x replaced by t, we see that the above integral is equal to $F(g(t)) + C$, where F is an antiderivative of f. This is just $F(x) + C$, which is exactly what we want. So this method works as well, and we have justified the method of substitution.

18.2 Integration by Parts

We saw how to reverse the chain rule by using the method of substitution. There is also a way to reverse the product rule—it's called integration by parts. Let's recall the product rule from Section 6.2.3 of Chapter 6: if u and v depend on x, then

$$\frac{d}{dx}(uv) = v\frac{du}{dx} + u\frac{dv}{dx}.$$

Let's rearrange this equation and then integrate both sides with respect to x. We get

$$\int u\frac{dv}{dx}\,dx = \int \frac{d}{dx}(uv)\,dx - \int v\frac{du}{dx}\,dx.$$

The first term on the right-hand side is the antiderivative of the derivative of uv, so it's just equal to $uv + C$. The $+C$ is unnecessary, though, because the second term on the right-hand side is already an indefinite integral: it includes a $+C$ automatically. So we have shown that

$$\int u\frac{dv}{dx}\,dx = uv - \int v\frac{du}{dx}\,dx.$$

This is the formula for integration by parts. It's perfectly usable in this form, but there's an abbreviated form which is even more convenient. If we replace $\frac{dv}{dx}\,dx$ by dv, and replace $\frac{du}{dx}\,dx$ by du, we get the formula

$$\boxed{\int u\,dv = uv - \int v\,du.}$$

 Again, this is just an abbreviation for the real formula, but it is pretty useful. Let's see how it works in practice. Suppose we want to find

$$\int xe^x\,dx.$$

Substitution seems useless (try it and see), so let's try integration by parts. We'd love to get the integral in the form $\int u\,dv$ so we can apply the integration by parts formula. There are a number of ways to do this, but here's one that works: set $u = x$ and $dv = e^x\,dx$. Then we certainly have $\int xe^x\,dx = \int u\,dv$.

Now, to apply the integration by parts formula, we need to be able to find du and v as well. The first one is easy: we know $u = x$, so $du = dx$. How about the second one? We have $dv = e^x\,dx$, so what is v? Just integrate both sides: $\int dv = \int e^x\,dx$. This means that $v = e^x + C$. Actually, we don't need a general v like this—we just need one v that gives $dv = e^x\,dx$. So we can ignore the $+C$ in this situation and just set $v = e^x$.

We are now ready to apply the formula for integration by parts, with $u = x$, $du = dx$, $v = e^x$, and $dv = e^x\,dx$. The easiest way to use the formula is to write a small version of it with generous spacing, then do the substitutions underneath, like this:

$$\int u \quad dv \quad = u\;v\; - \int v\;du$$

$$\int x\;\overbrace{e^x\,dx}\; = x\;e^x\; - \int e^x\,dx.$$

Now we still have one integral left, but it's just $\int e^x\,dx$, which is $e^x + C$. Plugging this in, we see that $\int xe^x\,dx = xe^x - e^x + C$. (Technically it should be $-C$, not $+C$, but minus a constant is just another constant and there's no need to distinguish.)

In order to set out the calculation for du and v, I recommend writing the following:

$$u = x \qquad\qquad v =$$
$$du = \qquad\qquad dv = e^x\,dx,$$

and then filling in the blanks by differentiating u and integrating dv:

$$u = x \qquad\qquad v = e^x$$
$$du = dx \qquad\qquad dv = e^x\,dx.$$

Then you can easily substitute into the integration by parts formula, since you have everything you need at your fingertips.

Now, how on earth did we know to choose $u = x$ and $dv = e^x\,dx$? Why couldn't we have chosen $u = e^x$ and $dv = x\,dx$? Well, we could have. In that case, we would have

$$u = e^x \qquad\qquad v = \tfrac{1}{2}x^2$$
$$du = e^x\,dx \qquad\qquad dv = x\,dx;$$

note that we integrated $dv = x\,dx$ to get $v = \tfrac{1}{2}x^2$ (remember, we don't need $+C$ here). Then by the integration by parts formula, we have

$$\int u \quad dv \quad = u\;v\; - \int v \quad du$$

$$\int xe^x\,dx = \int e^x\;\overbrace{x\,dx}\; = e^x \cdot \tfrac{1}{2}x^2 - \int \tfrac{1}{2}x^2\;\overbrace{e^x\,dx}\,.$$

There's nothing wrong with this, but it's not very useful. You see, the last integral on the right-hand side is nastier than the original integral! So we'd better stick with the first way above. In general, if you see e^x in there, treat it well—it is your friend, since its integral is also e^x. The moral is that if e^x is present, you should normally let $dv = e^x\,dx$ so that v is simply equal to e^x.

18.2.1 Some variations

A few complications can arise. Sometimes you need to integrate by parts twice or more. For example, how would you find

$$\int x^2 \sin(x)\,dx?$$

Well, it's a product, and substitution doesn't seem to work, so let's try integration by parts. There's no e^x, but there is a $\sin(x)$ which is almost as good. Let's try $u = x^2$ and $dv = \sin(x)\,dx$. We get

$$u = x^2 \qquad\qquad v = -\cos(x)$$
$$du = 2x\,dx \qquad\qquad dv = \sin(x)\,dx;$$

here we integrated $dv = \sin(x)\,dx$ to get $v = \int \sin(x)\,dx = -\cos(x)$ (remember, no $+C$ is needed). So we have

$$\overset{\int u}{\int} x^2 \overset{dv}{\overbrace{\sin(x)\,dx}} = \overset{=\; u}{x^2} \overset{v}{\overbrace{(-\cos(x))}} - \overset{\int}{\int} \overset{v}{\overbrace{(-\cos(x))}} \overset{du}{\overbrace{2x\,dx}}$$

$$= -x^2\cos(x) + \int \cos(x) \cdot 2x\,dx.$$

Now we can pull out the 2 from the last integral and we would be finished if only we knew what the integral $\int x\cos(x)\,dx$ was. This is a little simpler than the first integral, since we now have x instead of x^2, and after all, the cosine and sine functions are pretty darn similar. So we integrate by parts again. Let's try $U = x$, and $dV = \cos(x)\,dx$; I'm using capital letters since I already used u and v in this problem. We now have

$$U = x \qquad\qquad V = \sin(x)$$
$$dU = dx \qquad\qquad dV = \cos(x)\,dx,$$

so substituting in, we get

$$\overset{\int U}{\int} x \overset{dV}{\overbrace{\cos(x)\,dx}} = \overset{=\; U}{x} \overset{V}{\sin(x)} - \overset{\int}{\int} \overset{V}{\sin(x)} \overset{dU}{\,dx}.$$

How about that—we know that $\int \sin(x)\,dx = -\cos(x) + C$, so we get

$$\int x\cos(x)\,dx = x\sin(x) + \cos(x) + C.$$

We're almost done. We just have to plug this back in above and get

$$\int x^2\sin(x)\,dx = -x^2\cos(x) + 2x\sin(x) + 2\cos(x) + C.$$

(Once again, I didn't write $+2C$ because it's just a constant.)

Sometimes you can integrate by parts twice but things don't seem to get simpler. In this case, if you're lucky, then you might just get a multiple of the original integral back at the end. Then unless you are actually unlucky, you can throw it over to the other side and solve, which is a neat trick. (If you are unlucky, then the integrals cancel out, which doesn't help at all!) To see what on earth I'm talking about, here's an example:

$$\int \cos(x)e^{2x}\,dx.$$

In the integrand, the cosine bit and the exponential bit are both nice, but the exponential bit is nicer, so I'll set $u = \cos(x)$ and $dv = e^{2x}\,dx$. We get

$$u = \cos(x) \qquad\qquad v = \tfrac{1}{2}e^{2x}$$
$$du = -\sin(x)\,dx \qquad\quad dv = e^{2x}\,dx.$$

(Don't forget to divide by 2 when you integrate e^{2x} to get v.) This gives

$$
\int \underbrace{\cos(x)}_{u}\ \overbrace{e^{2x}\,dx}^{dv} \;=\; \underbrace{\cos(x)}_{u}\ \overbrace{\tfrac{1}{2}e^{2x}}^{v} \;-\; \int \overbrace{\tfrac{1}{2}e^{2x}}^{v}\ \underbrace{(-\sin(x))\,dx}_{du}
$$

$$
= \tfrac{1}{2}\cos(x)e^{2x} + \tfrac{1}{2}\int \sin(x)e^{2x}\,dx.
$$

Now the new integral on the right is about the same level of difficulty as the first one, so it's not clear we've gained anything at all. Nevertheless we persevere and integrate by parts again, this time setting $U = \sin(x)$ and $dV = e^{2x}\,dx$. Let's see what we get:

$$U = \sin(x) \qquad\qquad V = \tfrac{1}{2}e^{2x}$$
$$dU = \cos(x)\,dx \qquad\quad dV = e^{2x}\,dx.$$

Integrating by parts, we find that

$$
\int \underbrace{\sin(x)}_{U}\ \overbrace{e^{2x}\,dx}^{dV} \;=\; \underbrace{\sin(x)}_{U}\ \overbrace{\tfrac{1}{2}e^{2x}}^{V} \;-\; \int \overbrace{\tfrac{1}{2}e^{2x}}^{V}\ \underbrace{\cos(x)\,dx}_{dU}
$$

$$
= \tfrac{1}{2}\sin(x)e^{2x} - \tfrac{1}{2}\int \cos(x)e^{2x}\,dx.
$$

All in all, then, we have

$$
\int \cos(x)e^{2x}\,dx = \frac{1}{2}\cos(x)e^{2x} + \frac{1}{2}\left(\frac{1}{2}\sin(x)e^{2x} - \frac{1}{2}\int \cos(x)e^{2x}\,dx \right)
$$

$$
= \frac{1}{2}\cos(x)e^{2x} + \frac{1}{4}\sin(x)e^{2x} - \frac{1}{4}\int \cos(x)e^{2x}\,dx.
$$

Does this help? Well, yes—if we notice that the same integral appears on both sides, and then put both integrals on the left-hand side. In fact, we can add $\frac{1}{4}$ of the integral to both sides to eliminate it from the right-hand side, and put in a $+C$ to get

$$
\frac{5}{4}\int \cos(x)e^{2x}\,dx = \frac{1}{2}\cos(x)e^{2x} + \frac{1}{4}\sin(x)e^{2x} + C.
$$

Now we just multiply by 4/5 to see that

$$
\int \cos(x)e^{2x}\,dx = \frac{2}{5}\cos(x)e^{2x} + \frac{1}{5}\sin(x)e^{2x} + C.
$$

(Once again, we don't write $+\frac{4}{5}C$; we just relabel the constant and write $+C$.)

There's one other type of integral that needs integration by parts but is in disguise. In particular, the integrand doesn't appear to be a product. Some integrals that fall into this category are

$$\int \ln(x)\,dx, \quad \int (\ln(x))^2\,dx, \quad \int \sin^{-1}(x)\,dx, \quad \text{and} \quad \int \tan^{-1}(x)\,dx.$$

That is, the integrand is any inverse trig function (by itself) or a power of $\ln(x)$. In this case, you should let u be the integrand itself, and let $dv = dx$. For example, to find

$$\int_0^1 \tan^{-1}(x)\,dx,$$

let $u = \tan^{-1}(x)$ and $dv = dx$. We then have

$$u = \tan^{-1}(x) \qquad\qquad v = x$$
$$du = \frac{1}{1+x^2}\,dx \qquad\qquad dv = dx,$$

and so (ignoring the limits of integration for the moment)

$$\int \quad u \qquad dv \quad = \qquad u \qquad v \quad - \quad \int v \qquad du$$

$$\int \tan^{-1}(x)\,dx \;=\; \tan^{-1}(x)\;\; x \;-\; \int x \;\; \frac{1}{1+x^2}\,dx$$

$$= x\tan^{-1}(x) - \int \frac{x}{1+x^2}\,dx.$$

Using the method from the end of Section 18.1 above, the right-hand integral works out to be equal to $\frac{1}{2}\ln(1+x^2) + C$ (make sure you agree with this!), so we have

$$\int_0^1 \tan^{-1}(x)\,dx = \left(x\tan^{-1}(x) - \frac{1}{2}\ln(1+x^2) \right)\Bigg|_0^1 = \frac{\pi}{4} - \frac{1}{2}\ln(2).$$

How do you get the last answer? Know thy logs and inverse trig functions! Make sure that you believe that the above answer is correct. Also, notice that we found the indefinite integral first in order to find the definite integral (as opposed to trying to migrate the limits of integration to u-and-v-land!). This is a good idea in general. That is, when solving a definite integral by integrating by parts, find the indefinite integral first, then substitute the limits of integration at the end.

18.3 Partial Fractions

Let's focus our attention on how to integrate a rational function. So we want to find an integral like

$$\int \frac{p(x)}{q(x)}\,dx,$$

where p and q are polynomials. This covers a whole slew of integrals, for example,

$$\int \frac{x^2+9}{x^4-1}\,dx, \quad \int \frac{x}{x^3+1}\,dx, \quad \text{or} \quad \int \frac{1}{x^3-2x^2+3x-7}\,dx.$$

These seem a little complicated. Here are some simpler ones:

$$\int \frac{1}{x-3}\,dx, \quad \int \frac{1}{(x+5)^2}\,dx, \quad \int \frac{1}{x^2+9}\,dx, \quad \text{and} \quad \int \frac{3x}{x^2+9}\,dx.$$

The last four integrands are all rational functions, but they are a lot simpler. Try to work out all of these integrals using substitution. (Hint: some substitutions which work are $t = x - 3$, $t = x + 5$, $t = x/3$, and $t = x^2 + 9$ for the four integrals, respectively.) The first two of these integrals have denominators which are powers of linear functions, whereas the last two have quadratic denominators which cannot be factored.

So, here's the idea: first we'll see how to take a general rational function and do some algebra to bust it up into a sum of simpler rational functions; then we'll see how to integrate the simpler types of rational functions. The simpler functions I'm talking about are all like the four above: they either look like a constant over a linear power, or they look like a linear function over a quadratic. We'll look at the algebra first, then the calculus. Finally, we'll give a summary and look at a big example.

18.3.1 The algebra of partial fractions

Our goal is to break up a rational function into simpler pieces. The first step in this process is to make sure that the numerator of the function has degree less than the denominator. If not, we'll have to start off with a long division. So in the examples

$$\int \frac{x+2}{x^2-1}\,dx \quad \text{and} \quad \int \frac{5x^2+x-3}{x^2-1}\,dx,$$

the first is fine, since the degree of the top (1) is less than the degree of the bottom (2). The second example isn't so great, because the degrees of the top and bottom are equal (to 2). We'd have the same trouble if there were a cubic or higher-degree polynomial on the top. So, we have to do a long division. To do this, write

$$\text{denominator} \overline{)\,\text{numerator}}$$

In our example of

$$\int \frac{5x^2+x-3}{x^2-1}\,dx,$$

here's what the long division looks like:

$$
\begin{array}{r}
5 \\
x^2-1 \,\overline{)\,5x^2 + x - 3} \\
\underline{5x^2 - 5} \\
x + 2
\end{array}
$$

The division shows that we get a quotient of 5 and a remainder of $x + 2$. So we have

$$\frac{5x^2+x-3}{x^2-1} = 5 + \frac{x+2}{x^2-1}.$$

If we integrate both sides with respect to x, we get

$$\int \frac{5x^2 + x - 3}{x^2 - 1}\, dx = \int \left(5 + \frac{x+2}{x^2 - 1}\right) dx.$$

Now we can break up the integral into two pieces, and actually do the integral in the first piece, to see that our original integral is equal to

$$\int 5\, dx + \int \frac{x+2}{x^2 - 1}\, dx = 5x + \int \frac{x+2}{x^2 - 1}\, dx.$$

The new integral has a degree of 1 on the top and 2 on the bottom, which is the way we like it. We're now ready to proceed.

Next, we'll factor the denominator. If the denominator is a quadratic, check the discriminant: as we saw in Section 1.6 of Chapter 1, if this is negative, you can't factor the quadratic. Otherwise, you can factor it by hand or by using the quadratic formula. If your denominator is more complicated, you may have to guess a root and do a long division.

After factoring the denominator, the next step is to write down something called the "form." This is made by adding together one or more terms for each factor of the denominator, according to the following rules:

1. If you have a linear factor $(x + a)$, then the form has a term like

$$\frac{A}{x+a}.$$

2. If you have the square of a linear factor $(x+a)^2$, then the form has terms like

$$\frac{A}{(x+a)^2} + \frac{B}{x+a}.$$

3. If you have a quadratic factor $(x^2 + ax + b)$, then the form has a term like

$$\frac{Ax + B}{x^2 + ax + b}.$$

Those are the most common ones. Here are some rarer beasts:

4. If you have the cube of a linear factor $(x+a)^3$, then the form has terms like

$$\frac{A}{(x+a)^3} + \frac{B}{(x+a)^2} + \frac{C}{x+a}.$$

5. If you have the fourth power of a linear factor $(x + a)^4$, then the form has terms like

$$\frac{A}{(x+a)^4} + \frac{B}{(x+a)^3} + \frac{C}{(x+a)^2} + \frac{D}{x+a}.$$

Notice that **the form only depends on the denominator**. The numerator is irrelevant! Also, when I use constants like A, B, C, and D above, bear in mind that you can't reuse constants in different terms. So you need to keep advancing along the alphabet. In our example

$$\int \frac{x+2}{x^2 - 1}\, dx$$

from above, the denominator factors as $(x - 1)(x + 1)$; so we have two linear factors, and the form is

$$\frac{A}{x - 1} + \frac{B}{x + 1}.$$

We can't use A twice, so we used B for the second term. By the way, you're playing with fire if you write the constants as C_1, C_2, C_3 instead of A, B, C, and so on. You're less likely to make a careless mistake if you can actually tell the difference between the constants without having to look at tiny little numbers in subscripts.

Here's another example of finding the form. What would the form of

$$\frac{\text{any old junk}}{(x - 1)(x + 4)^3(x^2 + 4x + 7)(3x^2 - x + 1)}$$

be? The answer is

$$\frac{A}{x - 1} + \frac{B}{(x + 4)^3} + \frac{C}{(x + 4)^2} + \frac{D}{x + 4} + \frac{Ex + F}{x^2 + 4x + 7} + \frac{Gx + H}{3x^2 - x + 1}.$$

You may write these terms in a different order, or switch the constants A through H around; that's OK.

Once you've found the form, you should write down that the integrand equals the form, then multiply through by the denominator. For example, we just found that the form for the integrand of

$$\int \frac{x + 2}{x^2 - 1} \, dx$$

is given by

$$\frac{A}{x - 1} + \frac{B}{x + 1};$$

so we write

$$\frac{x + 1}{x^2 - 1} = \frac{A}{x - 1} + \frac{B}{x + 1}.$$

Actually, you're better off writing the denominator on the left-hand side in the factored manner, like this:

$$\frac{x + 2}{(x - 1)(x + 1)} = \frac{A}{x - 1} + \frac{B}{x + 1}.$$

Now multiply through by the denominator $(x - 1)(x + 1)$ to get

$$x + 2 = A(x + 1) + B(x - 1).$$

Notice that the factor $(x - 1)$ cancels in the first term on the right-hand side, and the factor $(x + 1)$ cancels out in the second term. Anyway, now there are two different ways we can proceed. The first way is to substitute clever values of x. If you put $x = 1$, then the $B(x - 1)$ term goes away, and you get

$$1 + 2 = A(1 + 1).$$

That is, $A = \frac{3}{2}$. Now if instead you put $x = -1$ in the original equation, the $A(x+1)$ term goes away:

$$-1 + 2 = B(-1 - 1).$$

So $B = -\frac{1}{2}$. Alternatively, another way of finding A and B is to take our original equation $x + 2 = A(x + 1) + B(x - 1)$ and rewrite it as

$$x + 2 = (A + B)x + (A - B).$$

Now we can equate coefficients of x to see that $1 = A + B$. We can also equate the constant coefficients to get $2 = A - B$. It's easy to solve these simultaneously and find that $A = \frac{3}{2}$ and $B = -\frac{1}{2}$ as before.

You might have noticed that in both of the ways we found A and B, we needed two facts. For the substitution method, we put $x = 1$ and then $x = -1$, whereas for the method of equating coefficients, we equated the coefficients of x and also the constant coefficients. We actually could have used one instance of each method. For example, if you put $x = 1$, you find that $A = \frac{3}{2}$ as above; then if you equate coefficients of x, you find that $1 = A + B$, so $B = -\frac{1}{2}$. In general, however many constants you have to find, that's how many times you have to apply one or both of the methods, mixing and matching as you choose.

All that's left is to rewrite your integrand as equal to the form again, but this time with the constants filled in. So in our example,

$$\frac{x + 2}{x^2 - 1} = \frac{A}{x - 1} + \frac{B}{x + 1} = \frac{3/2}{x - 1} + \frac{-1/2}{x + 1}.$$

Now integrate both sides, pulling out the constant factors as you split up the integral:

$$\int \frac{x + 2}{x^2 - 1}\, dx = \frac{3}{2} \int \frac{1}{x - 1}\, dx - \frac{1}{2} \int \frac{1}{x + 1}\, dx.$$

We have successfully busted up our original integral into two integrals which are much simpler. We'll solve these integrals very soon.

So far, we've seen that we do a long division unless the degree of the top is less than the degree of the bottom; then we factor the denominator; then we write down the form; then we use one of two methods to find the unknown constants. Finally, we write down the integrals of the various pieces. We'll see another example of how to do all this in Section 18.3.3 below. In the meantime, let's do some integration.

18.3.2 Integrating the pieces

We need to see how to integrate the various pieces which remain after you break up the original integral. The simplest type of integral is of the form

$$\int \frac{1}{ax + b}\, dx.$$

To do this, just substitute $t = ax + b$. For example, at the end of the previous section, we saw that

$$\int \frac{x + 2}{x^2 - 1}\, dx = \frac{3}{2} \int \frac{1}{x - 1}\, dx - \frac{1}{2} \int \frac{1}{x + 1}\, dx.$$

You can let $t = x - 1$ to do the first integral, and $t = x + 1$ to do the second. In both cases $dt = dx$, so it's easy to see that

$$\int \frac{x+2}{x^2-1}\,dx = \frac{3}{2}\log|x-1| - \frac{1}{2}\log|x+1| + C.$$

Here's another example: to find

$$\int \frac{1}{4x+5}\,dx,$$

put $t = 4x + 5$ so that $dt = 4\,dx$; then when the integral migrates to t-land, it simply becomes $\frac{1}{4}\int 1/t\,dt$, which is $\frac{1}{4}\ln|t| + C$. Finally, substitute back for t to see that the above integral works out to be $\frac{1}{4}\ln|4x+5| + C$.

The same trick works for a power of a linear factor in the denominator; for example, to find

$$\int \frac{1}{(4x+5)^2}\,dx,$$

substitute $t = 4x + 5$ once again. The integral becomes $\frac{1}{4}\int 1/t^2\,dt$, which is $-\frac{1}{4}(1/t) + C$; going back to x-land, we have shown that

$$\int \frac{1}{(4x+5)^2}\,dx = -\frac{1}{4} \times \frac{1}{4x+5} + C = -\frac{1}{4(4x+5)} + C.$$

The difficult case involves a quadratic in the bottom, like this:

$$\int \frac{Ax+B}{ax^2+bx+c}\,dx.$$

Beware! If the quadratic can be factored, then you need to do this first. This was the case in our previous example,

$$\int \frac{x+2}{x^2-1}\,dx.$$

We factored the denominator as $(x-1)(x+1)$; this eventually led to two integrals whose integrands had linear denominators. So there was no need to integrate anything with a quadratic on the bottom. Even the previous example, with $(4x+5)^2$ on the bottom, posed no problem, since we just had to deal with the square of a linear term.

So, what's left? The only possibility is that the quadratic on the bottom cannot be factored. That is, its discriminant $b^2 - 4ac$ is negative. An example of such an integral is

$$\int \frac{x+8}{x^2+6x+13}\,dx.$$

The denominator is a quadratic with discriminant $6^2 - 4(13)$, which is negative. We actually don't have to do any of the algebra from the previous section in this case, since the denominator can't be factored. There's no need to use any form at all; we just have to do the integral. Here's how: complete the square on the bottom, then make a substitution. (See Section 1.6 in Chapter 1 for a review of completing the square.) Let's complete the square in our example:

$$x^2 + 6x + 13 = x^2 + 6x + 9 + 13 - 9 = (x+3)^2 + 4.$$

So we have

$$\int \frac{x+8}{x^2+6x+13}\, dx = \int \frac{x+8}{(x+3)^2+4}\, dx.$$

Now substitute $t = x + 3$, so that $x = t - 3$ and $dx = dt$:

$$\int \frac{x+8}{x^2+6x+13}\, dx = \int \frac{x+8}{(x+3)^2+4}\, dx = \int \frac{(t-3)+8}{t^2+4}\, dt = \int \frac{t+5}{t^2+4}\, dt.$$

The next step is to break this last integral into two integrals and pull out the factor of 5, so the above integral becomes

$$\int \frac{t}{t^2+4}\, dt + 5\int \frac{1}{t^2+4}\, dt.$$

The first integral is just like the ones at the end of Section 18.1 above. You put a factor of 2 on the top and bottom, then recognize that the derivative of the bottom is just the top, so you get the log of the bottom:

$$\int \frac{t}{t^2+4}\, dt = \frac{1}{2}\int \frac{2t}{t^2+4}\, dt = \frac{1}{2}\ln|t^2+4| + C.$$

Actually, since $t^2 + 4$ is always positive, we can drop the absolute values. Anyway, to do the second integral, which is

$$5\int \frac{1}{t^2+4}\, dt,$$

just remember the useful formula

$$\int \frac{1}{t^2+a^2}\, dt = \frac{1}{a}\tan^{-1}\left(\frac{t}{a}\right) + C.$$

(You should try to prove this by differentiating the right-hand side, or by substituting $t = au$ in the left-hand side.) Anyway, with $a = 2$, this formula becomes

$$5\int \frac{1}{t^2+4}\, dt = 5 \times \frac{1}{2}\tan^{-1}\left(\frac{t}{2}\right) + C.$$

So, we have evaluated our integral as

$$\frac{1}{2}\ln(t^2+4) + \frac{5}{2}\tan^{-1}\left(\frac{t}{2}\right) + C.$$

Now just replace t by $x + 3$ once again to see that

$$\frac{1}{2}\ln((x+3)^2+4) + \frac{5}{2}\tan^{-1}\left(\frac{x+3}{2}\right) + C.$$

The expression $(x+3)^2+4$ immediately simplifies to $x^2+6x+13$, our original denominator. There's actually no need to expand it—just look back to where we completed the square and you'll find the equation you need. So, we have finally shown that

$$\int \frac{x+8}{x^2+6x+13}\, dx = \frac{1}{2}\ln(x^2+6x+13) + \frac{5}{2}\tan^{-1}\left(\frac{x+3}{2}\right) + C.$$

If the original quadratic on the bottom isn't monic, I suggest that you pull out the leading coefficient before completing the square. So, to find

$$\int \frac{x+8}{2x^2 + 12x + 26}\, dx,$$

pull out a factor of 2 in the bottom to write the integral as

$$\frac{1}{2} \int \frac{x+8}{x^2 + 6x + 13}\, dx.$$

This is the same integral as before, except for the factor of $\frac{1}{2}$ out front, so it simplifies to

$$\frac{1}{4} \ln(x^2 + 6x + 13) + \frac{5}{4} \tan^{-1}\left(\frac{x+3}{2}\right) + C.$$

Now, let's summarize the whole partial fraction method, then see a big example of the whole darn thing.

18.3.3 The method and a big example

Here's the complete method for finding the integral of a rational function:

Step 1—check degrees, divide if necessary: check to see if the degree of the numerator is less than the degree of the denominator. If it is, then you're golden—go on to step 2. If not, do a long division, then proceed to step 2.

Step 2—factor the denominator: use the quadratic formula, or guess roots and divide, to factor the denominator of your integrand.

Step 3—the form: write down the "form," with undetermined constants, as described on page 399 above. Write down an equation like

$$\text{integrand} = \text{form}.$$

Step 4—evaluate constants: multiply both sides of this equation by the denominator, then find the constants by (a) substituting clever values of x; (b) equating coefficients; or some combination of (a) and (b). Now you can express your integral as the sum of rational functions which either have constants on the top and powers of linear functions on the bottom, **or** look like a linear function divided by a quadratic function.

Step 5—integrate terms with linear powers on the bottom: solve any integrals whose denominators are powers of linear functions; the answers will involve logs or negative powers of the linear term.

Step 6—integrate terms with quadratics on the bottom: for each integral with a nonfactorable quadratic term in the denominator, complete the square, make a change of variables, then possibly split up into two integrals. The first one will involve logs and the second should involve $\tan^{-1}$. If there's only one integral, it could involve either logs or $\tan^{-1}$. This formula is very useful most of the time:

$$\int \frac{1}{t^2 + a^2}\, dt = \frac{1}{a}\tan^{-1}\left(\frac{t}{a}\right) + C.$$

Remember, you don't always need to use all six steps. Sometimes you can go directly to the last step, such as in our example

$$\int \frac{x + 8}{x^2 + 6x + 13}\, dx$$

from the previous section. Now, here's a nasty example that does involve all the steps:

$$\int \frac{x^5 - 7x^4 + 19x^3 - 10x^2 - 19x + 18}{x^4 - 5x^3 + 9x^2}\, dx.$$

Here's how to apply the above method to solve this example.

Step 1—check degrees, divide if necessary: in the above integral the degree of the top is 5, but the degree of the bottom is only 4. Bummer—we have to do a long division:

$$
\begin{array}{r}
x - 2 \\
x^4 - 5x^3 + 9x^2 \, \overline{\smash{)}\, x^5 - 7x^4 + 19x^3 - 10x^2 - 19x + 18} \\
\underline{x^5 - 5x^4 + 9x^3} \\
-2x^4 + 10x^3 - 10x^2 \\
\underline{-2x^4 + 10x^3 - 18x^2} \\
8x^2 - 19x + 18
\end{array}
$$

Check the details! In any case, we have shown that

$$\frac{x^5 - 7x^4 + 19x^3 - 10x^2 - 19x + 18}{x^4 - 5x^3 + 9x^2} = x - 2 + \frac{8x^2 - 19x + 18}{x^4 - 5x^3 + 9x^2}.$$

Now integrate both sides to see that

$$\int \frac{x^5 - 7x^4 + 19x^3 - 10x^2 - 19x + 18}{x^4 - 5x^3 + 9x^2}\, dx = \int \left(x - 2 + \frac{8x^2 - 19x + 18}{x^4 - 5x^3 + 9x^2}\right) dx.$$

The first two terms in the right-hand integral are easy: they integrate to $\frac{1}{2}x^2 - 2x$ (we'll put in the $+C$ at the very end). So now we have to find

$$\int \frac{8x^2 - 19x + 18}{x^4 - 5x^3 + 9x^2}\, dx.$$

Now the degree of the top is only 2, which is less than the degree of the bottom (which is still 4). We're ready for the next step.

Step 2—factor the denominator: we have a quartic in the denominator, but it has an obvious factor of x^2. So we'll factor the denominator as

$$x^4 - 5x^3 + 9x^2 = x^2(x^2 - 5x + 9).$$

The quadratic $x^2 - 5x + 9$ has discriminant $(-5)^2 - 4(9) = -11$; because this is negative, the quadratic can't be factored. So we're done with step 2.

Step 3—the form: we have two factors, x^2 and $x^2 - 5x + 9$. Don't think of the first factor x^2 as a quadratic; instead, think of it as the square of a linear factor. It might be better to write x^2 as $(x - 0)^2$ to clarify this point. So the x^2 factor contributes

$$\frac{A}{x^2} + \frac{B}{x}$$

to the form. On the other hand, the factor $x^2 - 5x + 9$ contributes

$$\frac{Cx + D}{x^2 - 5x + 9}.$$

Altogether, we have

$$\frac{8x^2 - 19x + 18}{x^2(x^2 - 5x + 9)} = \frac{A}{x^2} + \frac{B}{x} + \frac{Cx + D}{x^2 - 5x + 9}.$$

Step 4—evaluate constants: now we have to find the values of A, B, C, and D. First we multiply both sides of the above equation by the denominator $x^2(x^2 - 5x + 9)$ to get

$$8x^2 - 19x + 18 = A(x^2 - 5x + 9) + Bx(x^2 - 5x + 9) + (Cx + D)x^2.$$

Notice that the bits of the denominator that appear in each term of the right-hand side are precisely the bits that don't appear in the original form. For example, when you multiply the B/x term by $x^2(x^2 - 5x + 9)$, you knock out a factor of x to get $Bx(x^2 - 5x + 9)$.

Let's try substituting a clever value of x in the above equation. The only value of x that will kill off much of this equation is $x = 0$. If we put $x = 0$, the above equation becomes

$$18 = A(9),$$

so we immediately know that $A = 2$. We still need to find three more constants, so we'd better equate coefficients of three different powers of x. Let's start off by expanding the above equation, then grouping together the different powers of x:

$$8x^2 - 19x + 18 = Ax^2 - 5Ax + 9A + Bx^3 - 5Bx^2 + 9Bx + Cx^3 + Dx^2$$
$$= (B + C)x^3 + (A - 5B + D)x^2 + (-5A + 9B)x + 9A.$$

Now we can equate coefficients of x^3, x^2 and x, one at a time:

$$\begin{aligned}
\text{coefficient of } x^3: \quad 0 &= B + C \\
\text{coefficient of } x^2: \quad 8 &= A - 5B + D \\
\text{coefficient of } x^1: \quad -19 &= -5A + 9B.
\end{aligned}$$

Note that the coefficient of x^3 on the left-hand side is 0, since the left-hand side $8x^2 - 19x + 18$ doesn't have an x^3 term. (By the way, if you equate the constant coefficients, you get $18 = 9A$, which is the same equation we got when we substituted $x = 0$ above. Can you see why this happens?)

Anyway, we have some simultaneous equations to solve; starting at the last one and working back using the fact that $A = 2$, it's pretty easy to see that $B = -1$, $D = 1$, and $C = 1$. Substituting into the form that we got at the end of step 3, we have:

$$\frac{8x^2 - 19x + 18}{x^2(x^2 - 5x + 9)} = \frac{2}{x^2} + \frac{-1}{x} + \frac{x+1}{x^2 - 5x + 9}.$$

This means that

$$\int \frac{8x^2 - 19x + 18}{x^2(x^2 - 5x + 9)}\, dx = 2\int \frac{1}{x^2}\, dx - \int \frac{1}{x}\, dx + \int \frac{x+1}{x^2 - 5x + 9}\, dx.$$

Instead of one nasty integral, we have three simpler integrals. Let's work them all out.

Step 5—integrate terms with linear powers on the bottom: The first two of our integrals are pretty easy:

$$2\int \frac{1}{x^2}\, dx - \int \frac{1}{x}\, dx = -\frac{2}{x} - \ln|x| + C.$$

So, there's really not a lot to step 5 in this case. Unfortunately, step 6 is a lot more involved....

Step 6—integrate terms with quadratics on the bottom: We need to find the third integral, which is

$$\int \frac{x+1}{x^2 - 5x + 9}\, dx.$$

Start by completing the square:

$$x^2 - 5x + 9 = \left(x^2 - 5x + \frac{25}{4}\right) + 9 - \frac{25}{4} = \left(x - \frac{5}{2}\right)^2 + \frac{11}{4}. \qquad (\star\star)$$

Now let's rewrite our integral using the fruits of our completing-the-square labors:

$$\int \frac{x+1}{x^2 - 5x + 9}\, dx = \int \frac{x+1}{\left(x - \frac{5}{2}\right)^2 + \frac{11}{4}}\, dx.$$

We can substitute $t = x - \frac{5}{2}$ to make life a lot easier. Indeed, then $x = t + \frac{5}{2}$ and $dt = dx$, so the integral becomes

$$\int \frac{t + \frac{5}{2} + 1}{t^2 + \frac{11}{4}}\, dt = \int \frac{t + \frac{7}{2}}{t^2 + \frac{11}{4}}\, dt$$

in t-land. Now break it up into two new integrals:

$$\int \frac{t}{t^2 + \frac{11}{4}}\, dt \qquad \text{and} \qquad \frac{7}{2}\int \frac{1}{t^2 + \frac{11}{4}}\, dt.$$

To do the first of these integrals, multiply and divide by 2 to get

$$\int \frac{t}{t^2 + \frac{11}{4}}\, dt = \frac{1}{2}\int \frac{2t}{t^2 + \frac{11}{4}}\, dt = \frac{1}{2}\ln\left|t^2 + \frac{11}{4}\right| + C.$$

Once again, the absolute value signs aren't necessary, since $t^2 + \frac{11}{4}$ must be positive. To change back to x-land, we need to replace t by $x - \frac{5}{2}$ again:

$$\frac{1}{2}\ln\left(t^2 + \frac{11}{4}\right) + C = \frac{1}{2}\ln\left(\left(x - \frac{5}{2}\right)^2 + \frac{11}{4}\right) + C.$$

Don't bother multiplying out this last expression—just look at the equation marked ($\star\star$) on the previous page, where we completed the square, to see that everything simplifies to $\frac{1}{2}\ln(x^2 - 5x + 9) + C$. That takes care of the first of our new integrals.

We still have to worry about the second integral, which is

$$\frac{7}{2}\int \frac{1}{t^2 + \frac{11}{4}}\,dt.$$

Let's use the formula

$$\int \frac{1}{t^2 + a^2}\,dt = \frac{1}{a}\tan^{-1}\left(\frac{t}{a}\right) + C$$

with $a = \sqrt{11/4}$, which in fact is equal to $\sqrt{11}/2$:

$$\frac{7}{2}\int \frac{1}{t^2 + \frac{11}{4}}\,dt = \frac{7}{2} \times \frac{2}{\sqrt{11}}\tan^{-1}\left(\frac{t}{\sqrt{11}/2}\right) + C = \frac{7}{\sqrt{11}}\tan^{-1}\left(\frac{2t}{\sqrt{11}}\right) + C.$$

Now put $t = x - \frac{5}{2}$ again to see that this works out to be

$$\frac{7}{\sqrt{11}}\tan^{-1}\left(\frac{2x - 5}{\sqrt{11}}\right) + C.$$

Altogether, then, our two pieces give us the final answer for the step 6 part:

$$\int \frac{x + 1}{x^2 - 5x + 9}\,dx = \frac{1}{2}\ln(x^2 - 5x + 9) + \frac{7}{\sqrt{11}}\tan^{-1}\left(\frac{2x - 5}{\sqrt{11}}\right) + C.$$

Guess what? We're ready to assemble all the pieces for our big-ass integral! The first four steps established that

$$\int \frac{x^5 - 7x^4 + 19x^3 - 10x^2 - 19x + 18}{x^4 - 5x^3 + 9x^2}\,dx$$
$$= \int \left(x - 2 + \frac{2}{x^2} - \frac{1}{x} + \frac{x + 1}{x^2 - 5x + 9}\right) dx.$$

This is the complete partial fraction decomposition. Now, using steps 5 and 6 to do the actual integration, the above integral works out to be

$$\frac{x^2}{2} - 2x - \frac{2}{x} - \ln|x| + \frac{1}{2}\ln(x^2 - 5x + 9) + \frac{7}{\sqrt{11}}\tan^{-1}\left(\frac{2x - 5}{\sqrt{11}}\right) + C.$$

We're finally done with our big example! Admittedly, it was pretty nasty, but if you can do something that difficult, you should have no trouble with easier integrals. As an exercise, see if you can come back to this problem tomorrow and work it out from scratch without looking at these pages.

CHAPTER 19 _____

Techniques of Integration, Part Two

In this chapter, we'll finish gathering our techniques of integration by taking an extensive look at integrals involving trig functions. Sometimes one has to use trig identities to solve these types of problems; on other occasions there are no trig functions present, so you have to introduce some by making a trig substitution. After we finish all this trigonometry, there'll be a quick wrap-up of the techniques from this and the previous chapter so that you can keep it all together. So, this is what we'll look at in this chapter:

- integrals involving trig identities;
- integrals involving powers of trig functions, and reduction formulas;
- integrals involving trig substitutions; and
- a summary of all the techniques of integration we've seen so far.

19.1 Integrals Involving Trig Identities

There are three families of trig identities which are particularly useful in evaluating integrals. The first family arises from the double-angle formula for $\cos(2x)$. In Section 2.4 of Chapter 2, we saw that $\cos(2x) = 2\cos^2(x) - 1$ and also that $\cos(2x) = 1 - 2\sin^2(x)$. (Remember, you get one of these from the other by using $\sin^2(x) + \cos^2(x) = 1$.) For use in integration, it turns out that the best way to use the formulas is to solve the relevant equation for $\cos^2(x)$ or $\sin^2(x)$. So, we have

$$\boxed{\cos^2(x) = \frac{1}{2}\left(1 + \cos(2x)\right)} \qquad \text{and} \qquad \boxed{\sin^2(x) = \frac{1}{2}\left(1 - \cos(2x)\right).}$$

It is well worth remembering these identities! In particular, if you ever have to take a square root of $1 + \cos(\text{anything})$ or $1 - \cos(\text{anything})$, these identities save the day. For example,

$$\int_0^{\pi/2} \sqrt{1 - \cos(2x)}\, dx$$

looks pretty nasty, but in fact

$$\int_0^{\pi/2} \sqrt{1 - \cos(2x)}\, dx = \int_0^{\pi/2} \sqrt{2\sin^2(x)}\, dx$$

by the second boxed identity above. (We had to multiply the identity by 2 before using it.) Anyway, it's very tempting to replace $\sqrt{2\sin^2(x)}$ directly by $\sqrt{2}\sin(x)$, but let's do a quick reality check. The square root of A^2 isn't actually A, it's $|A|$. So the above integral becomes

$$\sqrt{2} \int_0^{\pi/2} |\sin(x)|\, dx.$$

Luckily, when x is between 0 and $\pi/2$, the values of $\sin(x)$ are always greater than or equal to zero, so we can drop the absolute value signs after all! We have reduced things to

$$\sqrt{2} \int_0^{\pi/2} \sin(x)\, dx;$$

 I leave it to you to show that this is just $\sqrt{2}$.

Sometimes you have to be a little more versatile. Consider

$$\int_\pi^{2\pi} \sqrt{1 + \cos(x)}\, dx.$$

It looks like we want to use the first identity in the box above, but that has a factor of $1 + \cos(2x)$ on one side and we need $1 + \cos(x)$. No problem—if you replace x by $x/2$ in the identity, and multiply through by 2, you get

$$2\cos^2\left(\frac{x}{2}\right) = 1 + \cos(x).$$

This is exactly what we need! Check 'dis:

$$\int_\pi^{2\pi} \sqrt{1 + \cos(x)}\, dx = \int_\pi^{2\pi} \sqrt{2\cos^2\left(\frac{x}{2}\right)}\, dx = \sqrt{2} \int_\pi^{2\pi} \left|\cos\left(\frac{x}{2}\right)\right|\, dx.$$

Now we have to be very careful. When x is between π and 2π, we see that $x/2$ is between $\pi/2$ and π, but $\cos(x)$ is less than or equal to zero on the interval $[\pi/2, \pi]$ (draw the graph to check this). So the above integral is actually equal to

$$\sqrt{2} \int_\pi^{2\pi} \left(-\cos\left(\frac{x}{2}\right)\right)\, dx;$$

 I leave it to you to show that this works out to be $2\sqrt{2}$. By the way, if you incorrectly replace $|\cos(x/2)|$ by $\cos(x/2)$ instead of $-\cos(x/2)$, you'll get the answer $-2\sqrt{2}$. This cannot be correct: the original integrand $\sqrt{1 + \cos(x)}$ is always positive, so the integral must be positive too.

Let's move on to the second family of trig identities. These are the Pythagorean identities:

$$\boxed{\sin^2(x) + \cos^2(x) = 1} \qquad \boxed{\tan^2(x) + 1 = \sec^2(x)} \qquad \boxed{1 + \cot^2(x) = \csc^2(x).}$$

These identities are valid for any x, as we saw in Section 2.4 of Chapter 2. Sometimes they are obviously helpful. For example,

$$\int_0^\pi \sqrt{1 - \cos^2(x)} \, dx$$

should just be written as

$$\int_0^\pi \sqrt{\sin^2(x)} \, dx = \int_0^\pi |\sin(x)| \, dx.$$

Since $\sin(x) \geq 0$ when x is between 0 and π, we can drop the absolute values to get

$$\int_0^\pi \sin(x) \, dx,$$

which is just 2. (Check this!) Compare this example, $\int_0^\pi \sqrt{1 - \cos^2(x)} \, dx$, with the example $\int_0^\pi \sqrt{1 - \cos(x)} \, dx$ we just did. They may look similar, but the trig identities we used are different.

Now, sometimes you have to apply a devious trick in order to use the above identities. If you see $1 + \text{trig}(x)$ or $1 - \text{trig}(x)$, where "trig" is some trig function (specifically sine, cosine, secant, or cosecant), in the denominator of an integral, consider multiplying by the conjugate expression. For example, to find

$$\int \frac{1}{\sec(x) - 1} \, dx,$$

multiply top and bottom by the conjugate expression of the denominator, which in this case is $\sec(x) + 1$. That is,

$$\int \frac{1}{\sec(x) - 1} \, dx = \int \frac{1}{\sec(x) - 1} \times \frac{\sec(x) + 1}{\sec(x) + 1} \, dx$$

Now you can use the difference of squares formula $(a - b)(a + b) = a^2 - b^2$ on the denominator to write the integral as

$$\int \frac{\sec(x) + 1}{\sec^2(x) - 1} \, dx.$$

Aha, the bottom is just $\tan^2(x)$, by one of our trig identities in the boxes above. Rewriting the integral using this, then splitting it into two integrals, we find that our integral is

$$\int \frac{\sec(x) + 1}{\tan^2(x)} \, dx = \int \frac{\sec(x)}{\tan^2(x)} \, dx + \int \frac{1}{\tan^2(x)} \, dx.$$

The first of these integrals looks a little nasty, but you can save the day by converting everything to sines and cosines. Specifically,

$$\int \frac{\sec(x)}{\tan^2(x)} \, dx = \int \frac{1/\cos(x)}{\sin^2(x)/\cos^2(x)} \, dx = \int \frac{\cos(x)}{\sin^2(x)} \, dx.$$

The next step is to substitute $t = \sin(x)$, since $dt = \cos(x)\,dx$ is on the top. Try this and see what you get. A fancier way is to rewrite $\cos(x)/\sin^2(x)$ as $\csc(x)\cot(x)$, so

$$\int \frac{\cos(x)}{\sin^2(x)}\,dx = \int \csc(x)\cot(x)\,dx = -\csc(x) + C,$$

since the derivative of $\csc(x)$ is $-\csc(x)\cot(x)$. Now we still have to deal with the second integral,

$$\int \frac{1}{\tan^2(x)}\,dx.$$

No problem—rewrite this as $\int \cot^2(x)\,dx$, then use another of the trig identities from the boxes above to express this as

$$\int \left(\csc^2(x) - 1\right)\,dx = -\cot(x) - x + C.$$

(Did you remember the integral of $\csc^2(x)$? It is a close cousin of the integral of $\sec^2(x)$, which is $\tan(x) + C$. Just put "co-" in front of everything and throw in a minus sign to get the $\csc^2(x)$ version!) In any event, we put these two pieces together to conclude that

$$\int \frac{1}{\sec(x) - 1}\,dx = -\csc(x) - \cot(x) - x + C.$$

Pretty tricky stuff.

Let's look at the third family of identities, the so-called products-to-sums identities:

$$\cos(A)\cos(B) = \frac{1}{2}(\cos(A - B) + \cos(A + B))$$
$$\sin(A)\sin(B) = \frac{1}{2}(\cos(A - B) - \cos(A + B))$$
$$\sin(A)\cos(B) = \frac{1}{2}(\sin(A - B) + \sin(A + B)).$$

It's quite a pain in the butt to remember these. Actually, they all follow from the expressions for $\cos(A \pm B)$ and $\sin(A \pm B)$ (which are also in Section 2.4 of Chapter 2), so if you have those down, you can reverse engineer the above identities from them. These identities are quite indispensable for finding integrals like

$$\int \cos(3x)\sin(19x)\,dx.$$

Indeed, it looks like we need the third formula above with $A = 19x$ and $B = 3x$. (Don't let the order of the cos and sin fool you here! The integral is the same as $\int \sin(19x)\cos(3x)\,dx$.) So we use the identity to get

$$\int \cos(3x)\sin(19x)\,dx = \frac{1}{2}\int \left(\sin(19x - 3x) + \sin(19x + 3x)\right)\,dx$$

$$= \frac{1}{2}\int \left(\sin(16x) + \sin(22x)\right)\,dx$$

$$= \frac{1}{2}\left(-\frac{\cos(16x)}{16} - \frac{\cos(22x)}{22}\right) + C$$

$$= -\frac{\cos(16x)}{32} - \frac{\cos(22x)}{44} + C.$$

19.2 Integrals Involving Powers of Trig Functions

Now we'll see how to find certain integrals which have powers of trig functions in their integrands. For example, how would you find $\int \cos^7(x)\sin^{10}(x)\,dx$ or $\int \sec^6(x)\,dx$? Unfortunately, these types of integrals require different techniques, depending on which trig function or functions you're dealing with. So, let's take them one at a time.

19.2.1 Powers of sin and/or cos

Our example $\int \cos^7(x)\sin^{10}(x)\,dx$ from above fits into this category. Here's the golden rule: if one of the powers of $\sin(x)$ or $\cos(x)$ is odd, then grab it and don't let it get away—it is your friend! (If they are both odd, then take the one with the lowest power as your friend.) If you've grabbed your odd power, then you need to pull out one power to go with the dx; then deal with what's left (which is now an even power) by using one of the identities

$$\boxed{\cos^2(x) = 1 - \sin^2(x)} \quad \text{or} \quad \boxed{\sin^2(x) = 1 - \cos^2(x).}$$

Note that these are just rearrangements of the identity $\sin^2(x) + \cos^2(x) = 1$. Anyway, the best way to see how the technique of pulling out one power from the odd power works is by looking at an example. In particular, to find $\int \cos^7(x)\sin^{10}(x)\,dx$, note that 7 is odd, so we grab $\cos^7(x)$ and pull out just one $\cos(x)$ to go with the dx. We get

$$\int \cos^7(x)\sin^{10}(x)\,dx = \int \cos^6(x)\sin^{10}(x)\cos(x)\,dx.$$

So what? Well, we need to deal with the $\cos^6(x)$ which is left over. Now 6 is even, so we can write $\cos^6(x) = (\cos^2(x))^3 = (1 - \sin^2(x))^3$, and the integral becomes

$$\int (1 - \sin^2(x))^3 \sin^{10}(x)\cos(x)\,dx.$$

Now if we put $t = \sin(x)$, then $dt = \cos(x)\,dx$, so it's easy to get this integral over to t-land—it's just

$$\int (1 - t^2)^3 t^{10}\,dt = \int (1 - 3t^2 + 3t^4 - t^6)t^{10}\,dt = \int (t^{10} - 3t^{12} + 3t^{14} - t^{16})\,dt,$$

which works out to be

$$\frac{t^{11}}{11} - \frac{3t^{13}}{13} + \frac{t^{15}}{5} - \frac{t^{17}}{17} + C.$$

Converting back to x-land, we get our answer:

$$\int \cos^7(x) \sin^{10}(x)\, dx = \frac{\sin^{11}(x)}{11} - \frac{3\sin^{13}(x)}{13} + \frac{\sin^{15}(x)}{5} - \frac{\sin^{17}(x)}{17} + C.$$

You see how stealing one power of $\cos(x)$ allowed us to change the rest of the integrand so that it only involved $\sin(x)$, leaving the $\cos(x)$ to take care of the dx via the substitution $t = \sin(x)$.

Now, what if neither power is odd? Well, if both powers are even—for example, if you had to work out $\int \cos^2(x) \sin^4(x)\, dx$—you should use the double-angle formulas. We just saw them in the previous section, but here they are again for reference:

$$\boxed{\cos^2(x) = \frac{1}{2}(1 + \cos(2x))} \qquad \text{and} \qquad \boxed{\sin^2(x) = \frac{1}{2}(1 - \cos(2x)).}$$

Now you can just replace everything in sight, and you'll get a whole bunch of simpler integrals which are various powers of cosines. You then need to find them using the same techniques as we have just used, depending on whether the power in each integral is even or odd. In our example, we need to think of $\sin^4(x)$ as $\left(\sin^2(x)\right)^2$, so we get

$$\int \cos^2(x) \sin^4(x)\, dx = \int \frac{1}{2}(1 + \cos(2x)) \left(\frac{1}{2}(1 - \cos(2x))\right)^2 dx.$$

Now we expand and multiply to get

$$\frac{1}{8} \int \left(1 - \cos(2x) - \cos^2(2x) + \cos^3(2x)\right) dx.$$

We need to break this up into four integrals. Let's not worry about the $\frac{1}{8}$ out front or the minus signs for the moment; we'll take care of them later. The first two integrals are easy, since $\int 1\, dx = x + C$ and $\int \cos(2x) = \frac{1}{2}\sin(2x) + C$. How do we find $\int \cos^2(2x)\, dx$? It's an even power, so we need to use the double-angle formulas again, but with x replaced by $2x$:

$$\int \cos^2(2x)\, dx = \int \frac{1}{2}(1 + \cos(4x))\, dx = \frac{1}{2}\left(x + \frac{1}{4}\sin(4x)\right) + C.$$

How about $\int \cos^3(2x)\, dx$? Well, now we have an odd power (namely 3), so we grab it! Let's write the integral as $\int \cos^2(2x) \cos(2x)\, dx$ and replace $\cos^2(2x)$ by $(1 - \sin^2(2x))$. Substituting $t = \sin(2x)$, we have $dt = 2\cos(2x)\, dx$, so the integral $\int \cos^3(2x)\, dx$ is

$$\int (1 - \sin^2(2x)) \cos(2x)\, dx = \frac{1}{2} \int (1 - t^2)\, dt = \frac{1}{2}\left(t - \frac{t^3}{3}\right) + C$$

$$= \frac{\sin(2x)}{2} - \frac{\sin^3(2x)}{6} + C.$$

(Pause to catch breath.) Now we put it all together and simplify a little; you should check that we get

$$\int \cos^2(x) \sin^4(x)\, dx$$

$$= \frac{1}{8}\left(x - \frac{\sin(2x)}{2} - \frac{x}{2} - \frac{\sin(4x)}{8} + \frac{\sin(2x)}{2} - \frac{\sin^3(2x)}{6} \right) + C$$

$$= \frac{x}{16} - \frac{\sin(4x)}{64} - \frac{\sin^3(2x)}{48} + C.$$

Make sure you can fill in all the details.

19.2.2 Powers of tan

Consider $\int \tan^n(x)\, dx$, where n is some integer. Let's look at the first couple of cases. For $n = 1$, we need to know how to do $\int \tan(x)\, dx$. This is a pretty standard integral, which you can solve by setting $t = \cos(x)$, noting that $dt = -\sin(x)\, dx$:

$$\int \tan(x)\, dx = \int \frac{\sin(x)}{\cos(x)}\, dx = -\int \frac{dt}{t} = -\ln(t) + C = -\ln|\cos(x)| + C.$$

The answer can also be written as $\ln|\sec(x)| + C$. (Why?)

How about $n = 2$? For this case, and indeed other cases, it's essential to use the Pythagorean identity

$$\boxed{\tan^2(x) = \sec^2(x) - 1}$$

which we looked at in the previous section. So we have

$$\int \tan^2(x)\, dx = \int \left(\sec^2(x) - 1 \right)\, dx = \tan(x) - x + C.$$

To do higher powers ($n \geq 3$), you have to extract $\tan^2(x)$ and change it into $(\sec^2(x) - 1)$. This gives you two integrals. The first can be done by substituting $t = \tan(x)$ and using $dt = \sec^2(x)\, dx$. The second is a lower power of $\tan(x)$ and you can just repeat the method. For example, how would you find $\int \tan^6(x)\, dx$? Let's see:

$$\int \tan^6(x)\, dx = \int \tan^4(x) \tan^2(x)\, dx = \int \tan^4(x) \left(\sec^2(x) - 1 \right)\, dx$$

$$= \int \tan^4(x) \sec^2(x)\, dx - \int \tan^4(x)\, dx.$$

So now we have to work out two integrals. To do the first one, set $t = \tan(x)$; as we just said, $dt = \sec^2(x)\, dx$. This gives

$$\int \tan^4(x) \sec^2(x)\, dx = \int t^4\, dt = \frac{t^5}{5} + C = \frac{\tan^5(x)}{5} + C.$$

Now, the second integral is $\int \tan^4(x)\,dx$, so we have to repeat the whole process. Take out a factor of $\tan^2(x)$ and change it to $(\sec^2(x) - 1)$:

$$\int \tan^4(x)\,dx = \int \tan^2(x)\tan^2(x)\,dx = \int \tan^2(x)\left(\sec^2(x) - 1\right)\,dx$$
$$= \int \tan^2(x)\sec^2(x)\,dx - \int \tan^2(x)\,dx.$$

Once again, we have two integrals. To do the first, let $t = \tan(x)$, so that $dt = \sec^2(x)\,dx$ (sound familiar?). So

$$\int \tan^2(x)\sec^2(x)\,dx = \int t^2\,dt = \frac{t^3}{3} + C = \frac{\tan^3(x)}{3} + C.$$

Meanwhile, we saw above that

$$\int \tan^2(x)\,dx = \int \left(\sec^2(x) - 1\right)\,dx = \tan(x) - x + C.$$

Putting it all together (being careful not to forget the minus signs), we see that

$$\int \tan^6(x)\,dx = \frac{\tan^5(x)}{5} - \frac{\tan^3(x)}{3} + \tan(x) - x + C.$$

What a pain. Still, it could be worse:

19.2.3 Powers of sec

Yup, this one really sucks, except for $\int \sec^2(x)\,dx$, which is easy. Let's start with the first power, $\int \sec(x)\,dx$. There are many ways of finding this integral. The easiest involves a cool trick that is well worth remembering, as it's a real timesaver. Unfortunately it's the sort of trick that is completely counterintuitive, and it boggles the mind that anyone even thought of it in the first place. The idea is to multiply top and bottom by the bizarre quantity $(\sec(x) + \tan(x))$. Watch and be amazed:

$$\int \sec(x)\,dx = \int \sec(x) \times \frac{\sec(x) + \tan(x)}{\sec(x) + \tan(x)}\,dx = \int \frac{\sec^2(x) + \sec(x)\tan(x)}{\sec(x) + \tan(x)}\,dx$$
$$= \ln|\sec(x) + \tan(x)| + C,$$

since the derivative of the denominator $\sec(x) + \tan(x)$ is miraculously equal to the numerator.

How about the second power of $\sec(x)$? Not much to this one:

$$\int \sec^2(x)\,dx = \tan(x) + C.$$

That was easy. Unfortunately, it gets pretty messy for larger powers. The standard idea is to pull out $\sec^2(x)$ (which is similar to what we did with powers of $\tan(x)$) and integrate by parts, using $dv = \sec^2(x)\,dx$ and u as the rest of the powers of $\sec(x)$. This means that $v = \tan(x)$ (remember, we don't need a constant here). When you do the integration by parts, you will of course get a new integral; the integrand should be a lower power of $\sec(x)$ multiplied by $\tan^2(x)$. Once again, we have to use $\tan^2(x) = \sec^2(x) - 1$ and get two integrals. One of them is a multiple of the original integral! You have to put this back on the left-hand side. The other one is a lower power of $\sec(x)$, and you have to repeat the whole process until you get down to $\int \sec(x)\,dx$ or $\int \sec^2(x)\,dx$, both of which we now know how to do.

That was quite a technical explanation. Let's see a formidable example: find $\int \sec^6(x)\,dx$. Start off by breaking out $\sec^2(x)$, like this:

$$\int \sec^6(x)\,dx = \int \sec^4(x)\sec^2(x)\,dx.$$

Now integrate by parts with $u = \sec^4(x)$ and $dv = \sec^2(x)\,dx$. By differentiating u and integrating dv as usual, we find that

$$du = 4\sec^3(x)\sec(x)\tan(x)\,dx = 4\sec^4(x)\tan(x)\,dx \qquad \text{and} \qquad v = \tan(x).$$

So now we can integrate by parts to get

$$\int \underset{u}{\sec^4(x)}\ \overbrace{\underset{dv}{\sec^2(x)\,dx}} = \underset{u}{\sec^4(x)}\ \underset{v}{\tan(x)} - \int \underset{v}{\tan(x)}\ \overbrace{\underset{du}{4\sec^4(x)\tan(x)\,dx}}.$$

Let's look at the integral on the right-hand side. We can write this as

$$4\int \sec^4(x)\tan^2(x)\,dx = 4\int \sec^4(x)\left(\sec^2(x) - 1\right)\,dx$$

$$= 4\left(\int \sec^6(x)\,dx - \int \sec^4(x)\,dx\right).$$

Putting it all together, we have

$$\int \sec^6(x)\,dx = \sec^4(x)\tan(x) - 4\int \sec^6(x)\,dx + 4\int \sec^4(x)\,dx.$$

Now comes the sexy part: transfer the first integral on the right-hand side over to the left-hand side to get

$$5\int \sec^6(x)\,dx = \sec^4(x)\tan(x) + 4\int \sec^4(x)\,dx.$$

We can divide this equation by 5 to get

$$\int \sec^6(x)\,dx = \frac{1}{5}\sec^4(x)\tan(x) + \frac{4}{5}\int \sec^4(x)\,dx.$$

Are we done? No, we still need to know how to do $\int \sec^4(x)\,dx$! We just have to repeat the whole darn process again. Here's where it's your turn to repeat all the above steps. If you don't screw up, you should get

$$\int \sec^4(x)\,dx = \frac{1}{3}\sec^2(x)\tan(x) + \frac{2}{3}\int \sec^2(x)\,dx.$$

Now we need $\int \sec^2(x)\, dx$, but we've finally knocked this down to something we can do—it's just $\tan(x) + C$, as we've already seen. Putting it all together, we have

$$\int \sec^6(x)\, dx = \frac{1}{5}\sec^4(x)\tan(x) + \frac{4}{5}\left(\frac{1}{3}\sec^2(x)\tan(x) + \frac{2}{3}\tan(x)\right) + C$$

$$= \frac{1}{5}\sec^4(x)\tan(x) + \frac{4}{15}\sec^2(x)\tan(x) + \frac{8}{15}\tan(x) + C.$$

Man, I'm exhausted just writing about this. Look, the idea with powers of both $\tan(x)$ and $\sec(x)$ is to knock the power down by 2 and then repeat; keep going until you either get down to the first or second power, which you can just do directly. By the way, how would you do

$$\int \frac{dx}{\cos^6(x)}?$$

That's right, you write it as $\int \sec^6(x)\, dx$, of course (which we just worked out!). How about

$$\int \frac{\sin^2(x)}{\cos^3(x)}\, dx?$$

Write the numerator as $1 - \cos^2(x)$ and break up the integral:

$$\int \frac{\sin^2(x)}{\cos^3(x)}\, dx = \int \frac{1 - \cos^2(x)}{\cos^3(x)}\, dx = \int \sec^3(x)\, dx - \int \sec(x)\, dx.$$

Now use the techniques above to find these two integrals involving powers of $\sec(x)$.

19.2.4 Powers of cot

These work just like powers of $\tan(x)$. You pull out $\cot^2(x)$ and use the Pythagorean identity

$$\boxed{\cot^2(x) = \csc^2(x) - 1.}$$

Just beware that when you set $t = \cot(x)$, you have $dt = -\csc^2(x)\, dx$. That is, don't forget the minus sign! Now try doing a few for practice. For example, try $\int \cot^6(x)\, dx$ and compare your answer with the solution to $\int \tan^6(x)\, dx$ in Section 19.2.2 above. You will see that they are very similar indeed.

19.2.5 Powers of csc

These work just like powers of $\sec(x)$. You pull out $\csc^2(x)$ and integrate by parts, using $dv = \csc^2(x)\, dx$. Beware: you now have $v = -\cot(x)$, and du also involves a minus sign which you have to worry about. Again, try some examples. If you work out $\int \csc^6(x)\, dx$ and compare your solution to the worked example $\int \sec^6(x)\, dx$ from Section 19.2.3 above, you should see more than a passing resemblance.

19.2.6 Reduction formulas

The methods of the last four sections all involve knocking the power of the trig function you're dealing with down by 2, then repeating the process. For example, in Section 19.2.2, we saw that we can integrate a power of $\tan(x)$ by extracting $\tan^2(x)$ and replacing it by $\sec^2(x) - 1$. Let's try to write out the method in general. First, we're dealing with $\int \tan^n(x)\,dx$, so we'll give it a name: I_n (for integral number n). That is,

$$I_n = \int \tan^n(x)\,dx.$$

We already know that

$$I_0 = \int \tan^0(x)\,dx = \int 1\,dx = x + C \qquad \text{and}$$

$$I_1 = \int \tan(x)\,dx = -\ln|\cos(x)| + C.$$

Now, when $n \geq 2$, we can steal $\tan^2(x)$ away from $\tan^n(x)$, leaving behind $\tan^{n-2}(x)$; then we can use our trig identity and split up the integral to get

$$I_n = \int \tan^n(x)\,dx = \int \tan^{n-2}(x) \tan^2(x)\,dx = \int \tan^{n-2}(x)(\sec^2(x) - 1)\,dx$$

$$= \int \tan^{n-2}(x) \sec^2(x)\,dx - \int \tan^{n-2}(x)\,dx.$$

The second integral in this last expression, $\int \tan^{n-2}(x)\,dx$, is just I_{n-2}. As for the first, if you put $t = \tan(x)$ so that $dt = \sec^2(x)\,dx$, you'll see it becomes $\int t^{n-2}\,dt$, which is just $t^{n-1}/(n-1) + C$. Replacing t by $\tan(x)$, we have shown that

$$I_n = \frac{1}{n-1} \tan^{n-1}(x) - I_{n-2}.$$

There's no need for a constant, since both I_n and I_{n-2} are indefinite integrals. The above equation is called a *reduction formula*, since it helps us reduce the number n to a smaller number $n - 2$.

Let's see how to use the formula to find $\int \tan^6(x)\,dx$. This is just I_6. So, put $n = 6$ in the reduction formula to get

$$I_6 = \frac{1}{5} \tan^5(x) - I_4.$$

OK, so we need I_4. Let's write out the reduction formula again, this time with $n = 4$:

$$I_4 = \frac{1}{3} \tan^3(x) - I_2.$$

Once again, but with $n = 2$:

$$I_2 = \frac{1}{1} \tan^1(x) - I_0 = \tan(x) - x + C,$$

where we have used the above formula for I_0. So we now know I_2, and we can work backward to get I_4:

$$I_4 = \frac{1}{3} \tan^3(x) - I_2 = \frac{1}{3} \tan^3(x) - \tan(x) + x + C.$$

Finally, we can find our desired integral, which is none other than I_6:

$$\int \tan^6(x)\,dx = I_6 = \frac{1}{5}\tan^5(x) - I_4 = \frac{1}{5}\tan^5(x) - \frac{1}{3}\tan^3(x) + \tan(x) - x + C.$$

This agrees with our answer from Section 19.2.2. Now try to repeat this for powers of secant, cosecant, and cotangent—the methods are given above, and all you have to do is rewrite them as reduction formulas.

The method also works for definite integrals. For example, how would you find the definite integral $\int_0^{\pi/2}\cos^8(x)\,dx$? You could use the double-angle formulas, as described in Section 19.2.1 above, but that would be a pain in the ass. (Try it if you don't believe me!) Instead, let's set

$$I_n = \int_0^{\pi/2}\cos^n(x)\,dx$$

and make a mental note that we eventually want to find I_8. The trick now is to pull out one factor of $\cos(x)$, like this:

$$I_n = \int_0^{\pi/2}\cos^n(x)\,dx = \int_0^{\pi/2}\cos^{n-1}(x)\cos(x)\,dx.$$

Now integrate by parts with $u = \cos^{n-1}(x)$ and $dv = \cos(x)\,dx$. This means, of course, that $v = \sin(x)$. (See Section 18.2 in the previous chapter for more about integration by parts.) I leave it to you to show that we get

$$I_n = \cos^{n-1}(x)\sin(x)\Big|_0^{\pi/2} + \int_0^{\pi/2}(n-1)\cos^{n-2}(x)\sin^2(x)\,dx.$$

If $n \geq 2$, then the first expression on the right-hand side is 0, since we have $\cos(\pi/2) = 0$ and $\sin(0) = 0$. On the other hand, we can replace $\sin^2(x)$ by $1 - \cos^2(x)$ in the integral to see that

$$I_n = \int_0^{\pi/2}(n-1)\cos^{n-2}(x)(1 - \cos^2(x))\,dx$$

$$= (n-1)\int_0^{\pi/2}\cos^{n-2}(x)\,dx - (n-1)\int_0^{\pi/2}\cos^n(x)\,dx.$$

Now what? Well, notice that the last two integrals are just I_{n-2} and I_n, respectively. So

$$I_n = (n-1)I_{n-2} - (n-1)I_n.$$

Solving for I_n by adding $(n-1)I_n$ to both sides and dividing by n, we arrive at the following reduction formula:

$$I_n = \frac{n-1}{n}I_{n-2}.$$

That should make life a lot easier! In particular, we are looking for I_8, so by using the above formula over and over again, with $n = 8$, then $n = 6$, then $n = 4$, and finally $n = 2$, we get

$$I_8 = \frac{7}{8}I_6 = \frac{7}{8}\cdot\frac{5}{6}I_4 = \frac{7}{8}\cdot\frac{5}{6}\cdot\frac{3}{4}I_2 = \frac{7}{8}\cdot\frac{5}{6}\cdot\frac{3}{4}\cdot\frac{1}{2}I_0.$$

Now we need to find I_0. Since $\cos^0(x)$ is just 1, we have $I_0 = \int_0^{\pi/2} 1\, dx = \pi/2$. Simplifying the above fraction, we have shown that

$$\int_0^{\pi/2} \cos^8(x)\, dx = \frac{7 \cdot 5 \cdot 3 \cdot 1}{8 \cdot 6 \cdot 4 \cdot 2} \times \frac{\pi}{2} = \frac{35\pi}{256}.$$

As a bonus, we can easily find $\int_0^{\pi/2} \cos^n(x)\, dx$ for any other positive integer n. (You'll need to note that $I_1 = \int_0^{\pi/2} \cos(x)\, dx = 1$ in order to get the odd powers.)

By the way, reduction formulas don't have to involve trig functions. For example, if

$$I_n = \int x^n e^x\, dx,$$

then you can integrate by parts with $u = x^n$ and $dv = e^x\, dx$ (so $v = e^x$) to get

$$I_n = x^n e^x - \int n x^{n-1} e^x\, dx.$$

This gives the reduction formula $I_n = x^n e^x - n I_{n-1}$. Incidentally, unlike the situation with all the trig function examples, this time I_n is expressed in terms of I_{n-1}, not I_{n-2}. So you only need to know I_0 at the end of the chain, which isn't hard to find: $I_0 = \int e^x\, dx = e^x + C$.

19.3 Integrals Involving Trig Substitutions

Now let's look at how to do integrals involving an odd power of the square root of a quadratic. Here are some examples of the type of integral we're considering:

$$\int \frac{dx}{x^3 \sqrt{x^2 - 4}} \qquad \text{or} \qquad \int \frac{x^2}{(9 - x^2)^{3/2}}\, dx \qquad \text{or} \qquad \int (x^2 + 15)^{-5/2}\, dx.$$

The basic idea is that there are three types, corresponding to whether you have to worry about $a^2 - x^2$, $x^2 + a^2$, or $x^2 - a^2$. Here a is just some number. For example, the first integral above involves $x^2 - a^2$ with $a = 2$, the second involves $a^2 - x^2$ with $a = 3$, and the third involves $x^2 + a^2$ with $a = \sqrt{15}$. Each of these three types requires a different substitution. Most of the time, after substituting, you end up with an integral involving powers of trig functions, which is where the previous section comes in. Let's look at the three types of integrals one at a time; then we'll summarize the whole situation at the end.

19.3.1 Type 1: $\sqrt{a^2 - x^2}$

If you have an integral involving an odd power of $\sqrt{a^2 - x^2}$, the correct substitution to use is $x = a \sin(\theta)$. (You could use $x = a \cos(\theta)$ if you prefer, but there would be no advantage to it, so stick with sine.) The reason that this substitution is effective is that

$$a^2 - x^2 = a^2 - a^2 \sin^2(\theta) = a^2 \left(1 - \sin^2(\theta)\right) = a^2 \cos^2(\theta),$$

and now you can easily take a square root. Remember that if you are changing variables from x to θ, you have to go from x-land to θ-land. That is, everything about the integral has to be in terms of θ, not x. In particular, we'll need to replace dx by something in θ and $d\theta$. No problem—just differentiate the equation $x = a\sin(\theta)$ to get $dx = a\cos(\theta)\,d\theta$. (This sort of substitution, where the equation is solved for x instead of the substituting variable, was discussed at the ends of Sections 18.1.2 and 18.1.3 of the previous chapter.) Anyway, now we can hopefully do the integral in θ-land, but in the end we have to change the answer back to x-land. To do this, it will be useful to draw the following right-angled triangle with one angle equal to θ:

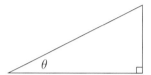

Now we know $\sin(\theta) = x/a$, so we can fill in two of the sides as shown:

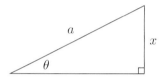

Finally, we can use Pythagoras' Theorem to see that the third side is $\sqrt{a^2 - x^2}$, so we complete the triangle as follows:

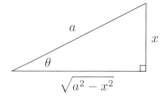

Now we can easily read off from this triangle the values of $\cos(\theta)$, $\tan(\theta)$, or any other trig function of θ, and get back to x-land without too much trouble.

Let's see how it works in practice. We'll use an example from above:

$$\int \frac{x^2}{(9 - x^2)^{3/2}}\,dx.$$

We make the substitution $x = 3\sin(\theta)$, so $dx = 3\cos(\theta)\,d\theta$. Also, we see that $9 - x^2 = 9 - 9\sin^2(\theta) = 9\cos^2(\theta)$. So the integral becomes

$$\int \frac{(3\sin(\theta))^2}{(9\cos^2(\theta))^{3/2}} \cdot 3\cos(\theta)\,d\theta = \frac{3^2 \times 3}{9^{3/2}} \int \frac{\sin^2(\theta)}{\cos^3(\theta)}\cos(\theta)\,d\theta = \int \tan^2(\theta)\,d\theta,$$

since $9^{3/2} = 27$. Now we use the techniques from Section 19.2.2 above to see that

$$\int \tan^2(\theta)\, d\theta = \int \left(\sec^2(\theta) - 1\right) d\theta = \tan(\theta) - \theta + C.$$

We just have to get back to x-land. Since $\sin(\theta) = x/3$, the relevant triangle looks like this:

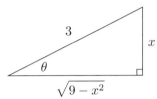

We can read off from the triangle that $\tan(\theta) = x/\sqrt{9 - x^2}$. Also, since $\sin(\theta) = x/3$, we have $\theta = \sin^{-1}(x/3)$. Substituting into the answer above, we see that

$$\int \frac{x^2}{(9 - x^2)^{3/2}}\, dx = \frac{x}{\sqrt{9 - x^2}} - \sin^{-1}\left(\frac{x}{3}\right) + C.$$

If you didn't use the triangle, you might be tempted to write $\tan(\theta)$ as the messy expression

$$\tan\left(\sin^{-1}\left(\frac{x}{3}\right)\right),$$

but I hope you agree that our actual answer above is preferable.

Before we go on to Type 2, do you see that we've been a little careless here? We had to work out $(9\cos^2(\theta))^{3/2}$ and just claimed that it is $27\cos^3(\theta)$. Certainly $9^{3/2} = 27$, but is it always true that $(\cos^2(\theta))^{3/2} = \cos^3(\theta)$? Actually, this is only true if $\cos(\theta) \geq 0$. The problem is that raising a quantity to the power $3/2$ actually involves taking a positive square root. Indeed, for any positive number A, we have $A^{3/2} = (A^{1/2})^3 = (\sqrt{A})^3$. So we should really have written

$$(\cos^2(\theta))^{3/2} = (\sqrt{\cos^2(\theta)})^3 = |\cos^3(\theta)|.$$

Luckily, the absolute value signs turn out to be unnecessary for Type 1 and also for Type 2 below (but not for Type 3), so we were right all along. This point will be discussed in gory detail in Section 19.3.6 below.

19.3.2 Type 2: $\sqrt{x^2 + a^2}$

If an integral involves an odd power of $\sqrt{x^2 + a^2}$, the correct substitution is $x = a\tan(\theta)$. This works because

$$x^2 + a^2 = a^2\tan^2(\theta) + a^2 = a^2(\tan^2(\theta) + 1) = a^2\sec^2(\theta).$$

Also, we'll need to know that $dx = a\sec^2(\theta)\, d\theta$. Since $\tan(\theta) = x/a$, the triangle now looks like this:

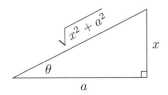

And now we're ready for an example:

$$\int (x^2 + 15)^{-5/2}\, dx.$$

Here the substitution is $x = \sqrt{15}\tan(\theta)$. We have $dx = \sqrt{15}\sec^2(\theta)\, d\theta$, and we also note that $x^2 + 15 = 15\tan^2(\theta) + 15 = 15\sec^2(\theta)$. The integral becomes

$$\int \left(15\sec^2(\theta)\right)^{-5/2} \sqrt{15}\sec^2(\theta)\, d\theta = \frac{15^{1/2}}{15^{5/2}} \int (\sec(\theta))^{-5}\sec^2(\theta)\, d\theta$$

$$= (15)^{-2} \int \cos^3(\theta)\, d\theta.$$

(Once again, we have done something dubious: we replaced $(15\sec^2(\theta))^{-5/2}$ by $15^{-5/2}\sec^{-5}(\theta)$, completely neglecting to use absolute value signs. If you like, check out Section 19.3.6 below to see why this is OK.) We still need to find $15^{-2}\int \cos^3(\theta)\, d\theta$. Let's use the techniques from Section 19.2.1 above. We notice that the integrand is an odd power of $\cos(\theta)$, so we grab it, pull out one power of $\cos(\theta)$, and then substitute for $\sin(\theta)$:

$$(15)^{-2} \int \cos^3(\theta)\, d\theta = (15)^{-2} \int \left(1 - \sin^2(\theta)\right)\cos(\theta)\, d\theta$$

$$= (15)^{-2} \left(\sin(\theta) - \frac{\sin^3(\theta)}{3}\right) + C.$$

(I omitted the details of the substitution here—make sure you can fill them in.) Now, back to x-land. Since $\tan(\theta) = x/\sqrt{15}$, the following triangle applies:

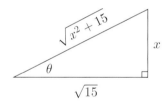

From this triangle, you can simply read off the fact that $\sin(\theta) = x/\sqrt{x^2 + 15}$, which means that

$$\int (x^2 + 15)^{-5/2}\, dx = (15)^{-2} \left(\sin(\theta) - \frac{\sin^3(\theta)}{3}\right) + C$$

$$= \frac{1}{225} \left(\frac{x}{\sqrt{x^2 + 15}} - \frac{x^3}{3(x^2 + 15)^{3/2}}\right) + C.$$

(Can you see why $\sin^3(\theta) = x^3/(x^2 + 15)^{3/2}$? Just rewrite $\sin(\theta)$ in terms of x as $x/(x^2 + 15)^{1/2}$.)

19.3.3 Type 3: $\sqrt{x^2 - a^2}$

Finally, how about integrals involving an odd power of $\sqrt{x^2 - a^2}$? Now the correct substitution is $x = a\sec(\theta)$, since

$$x^2 - a^2 = a^2\sec^2(\theta) - a^2 = a^2(\sec^2(\theta) - 1) = a^2\tan^2(\theta),$$

and you can easily take square roots. To make the substitution, we'll also need the fact that $dx = a\sec(\theta)\tan(\theta)\,d\theta$. Since $\sec(\theta) = x/a$, the triangle looks like this:

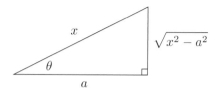

For example, to find

$$\int \frac{dx}{x^3\sqrt{x^2 - 4}},$$

set $x = 2\sec(\theta)$, so $dx = 2\sec(\theta)\tan(\theta)\,d\theta$ and $x^2 - 4 = 4\tan^2(\theta)$. The integral becomes

$$\int \frac{2\sec(\theta)\tan(\theta)}{(2\sec(\theta))^3\sqrt{4\tan^2(\theta)}}\,d\theta = \int \frac{2\sec(\theta)\tan(\theta)}{8\sec^3(\theta) \times 2\tan(\theta)}\,d\theta$$

$$= \frac{1}{8}\int \frac{1}{\sec^2(\theta)}\,d\theta = \frac{1}{8}\int \cos^2(\theta)\,d\theta.$$

Actually, this time it's wrong to replace $\sqrt{4\tan^2(\theta)}$ by $2\tan(\theta)$; this is only correct if $x > 0$ in the original integral, as we'll see in Section 19.3.6 below. So let's make that assumption. Now we need to find $\frac{1}{8}\int \cos^2(\theta)\,d\theta$. The power of cosine is even, so we have to use the double-angle formula from Section 19.2.1 above:

$$\frac{1}{8}\int \cos^2(\theta)\,d\theta = \frac{1}{8}\int \frac{1}{2}(1 + \cos(2\theta))\,d\theta = \frac{\theta}{16} + \frac{\sin(2\theta)}{32} + C.$$

OK, we just have to get back to x-land. This is a little tricky, even using the appropriate triangle:

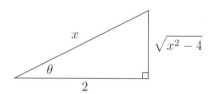

The problem is that we need to know what $\sin(2\theta)$ is. To do this, we use the identity

$$\sin(2\theta) = 2\sin(\theta)\cos(\theta).$$

Then we can use the above triangle to see that $\sin(\theta) = \sqrt{x^2 - 4}/x$ and $\cos(\theta) = 2/x$, substitute everything in, and get

$$\int \frac{dx}{x^3\sqrt{x^2 - 4}} = \frac{1}{16}\sec^{-1}\left(\frac{x}{2}\right) + \frac{1}{32} \cdot 2 \cdot \frac{\sqrt{x^2 - 4}}{x} \cdot \frac{2}{x} + C$$

$$= \frac{1}{16}\sec^{-1}\left(\frac{x}{2}\right) + \frac{\sqrt{x^2 - 4}}{8x^2} + C.$$

Remember, this only applies when $x > 0$. We'll revisit this example in Section 19.3.6 to see how to take care of the case when $x \leq 0$.

19.3.4 Completing the square and trig substitutions

Now, one other important point before we summarize the situation. From time to time, you might want to solve an integral involving an odd power of $\sqrt{\pm x^2 + ax + b}$. That is, you now have a linear term ax to complicate matters. The technique is simple: complete the square first and substitute to get it into one of the three types that we've investigated. For example, to evaluate

$$\int (x^2 - 4x + 19)^{-5/2} \, dx,$$

first complete the square (see Section 1.6 of Chapter 1 for a reminder of how to do this):

$$x^2 - 4x + 19 = (x^2 - 4x + 4) - 4 + 19 = (x - 2)^2 + 15.$$

So the integral we want is actually

$$\int ((x - 2)^2 + 15)^{-5/2} \, dx.$$

Now let $t = x - 2$, so $dt = dx$, and in t-land the integral becomes

$$\int (t^2 + 15)^{-5/2} \, dt,$$

which we have already done earlier in Section 19.3.2! The answer was (replacing the old x by t)

$$\frac{1}{225} \left(\frac{t}{\sqrt{t^2 + 15}} - \frac{t^3}{3(t^2 + 15)^{3/2}} \right) + C,$$

so replacing t now by $x - 2$, we see that

$$\int (x^2 - 4x + 19)^{-5/2} \, dx = \frac{1}{225} \left(\frac{x - 2}{\sqrt{x^2 - 4x + 19}} - \frac{(x - 2)^3}{3(x^2 - 4x + 19)^{3/2}} \right) + C.$$

The moral of the story, both here and when using partial fractions, is that a quadratic with a linear term can be made into a quadratic without one by completing the square and substituting.

19.3.5 Summary of trig substitutions

To summarize the three main types we've looked at, here's a table that shows the appropriate substitutions and triangles for each type:

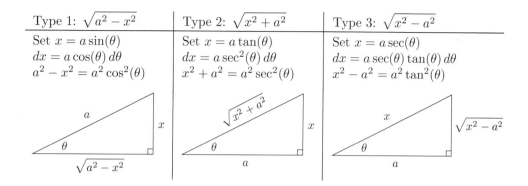

Type 1: $\sqrt{a^2 - x^2}$	Type 2: $\sqrt{x^2 + a^2}$	Type 3: $\sqrt{x^2 - a^2}$
Set $x = a\sin(\theta)$	Set $x = a\tan(\theta)$	Set $x = a\sec(\theta)$
$dx = a\cos(\theta)\,d\theta$	$dx = a\sec^2(\theta)\,d\theta$	$dx = a\sec(\theta)\tan(\theta)\,d\theta$
$a^2 - x^2 = a^2\cos^2(\theta)$	$x^2 + a^2 = a^2\sec^2(\theta)$	$x^2 - a^2 = a^2\tan^2(\theta)$

The next section discusses the technical point about when (and why) you can drop the absolute value signs when you take square roots of quantities like $a^2\cos^2(\theta)$ or $a^2\tan^2(\theta)$. It's the sort of thing that you may want to skim over first, then come back to later if you have time.

19.3.6 Technicalities of square roots and trig substitutions

You have been warned: this section gets a little messy. Still with me? Good. Now, think back to Type 1 above. We simplified $\sqrt{a\cos^2(\theta)}$ down to $a\cos(\theta)$, completely ignoring the need to use absolute values around the $\cos(\theta)$. Actually, when we write $x = a\sin(\theta)$, we really mean that $\theta = \sin^{-1}(x/a)$.

So where is θ? Well, from Section 10.2.1 in Chapter 10, we know that the range of $\sin^{-1}$ is $[-\pi/2, \pi/2]$; this means that θ is in the first or fourth quadrant, so $\cos(\theta)$ is always nonnegative. We don't need any absolute values!

The same goes for Type 2. In that case, we'd really like to simplify $\sqrt{a^2\sec^2(\theta)}$ as $a\sec(\theta)$. Can we do this without using absolute value signs? We have $x = a\tan(\theta)$, so $\theta = \tan^{-1}(x/a)$. The range of $\tan^{-1}$ is $(-\pi/2, \pi/2)$, so θ is once again in the first or fourth quadrant. This means that $\sec(\theta)$ is always positive, so again, we don't need absolute values.

Everything goes wrong in Type 3, unfortunately. Here we need to deal with $\sqrt{a^2\tan^2(\theta)}$, but this isn't always equal to $a\tan(\theta)$. You see, since $x = a\sec(\theta)$, we have $\theta = \sec^{-1}(x/a)$. If you look back at Section 10.2.4 in Chapter 10, you'll see that the range of $\sec^{-1}$ is the interval $[0, \pi]$, except for the point $\pi/2$. So θ is in the first or second quadrant, and $\tan(\theta)$ could be positive or negative. At least it has the same sign as x does, as you can see by looking at the graph of $y = \sec^{-1}(x)$.

So, it's correct to write $\sqrt{a^2\tan^2(\theta)} = a\tan(\theta)$ when $x > 0$. On the other hand, if $x < 0$ then you have to write $-a\tan(\theta)$ instead. In that case, the triangle actually looks like this:

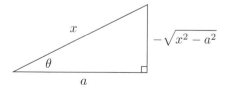

I agree that it's freaky that this triangle has two negative sides (x and $-\sqrt{x^2 - a^2}$), but it works as a neat memory device, since all the signs of the trig functions are correct. In our example

$$\int \frac{dx}{x^3 \sqrt{x^2 - 4}}$$

from Section 19.3.3 above, we saw that the integral works out to be

$$\frac{1}{16} \sec^{-1}\left(\frac{x}{2}\right) + \frac{\sqrt{x^2 - 4}}{8x^2} + C$$

when $x > 0$. (Actually, if $x > 0$, then x has to be greater than 2, or else the $\sqrt{x^2 - 4}$ factor in the denominator really screws up the situation.) Now let's redo the problem for the case when $x < 0$. We still substitute $x = 2\sec(\theta)$, but now we must replace $\sqrt{4\tan^2(\theta)}$ by $-2\tan(\theta)$. The only difference from before is the minus sign:

$$\int \frac{dx}{x^3 \sqrt{x^2 - 4}} = \int \frac{2\sec(\theta)\tan(\theta)}{(2\sec(\theta))^3 \sqrt{4\tan^2(\theta)}} \, d\theta$$

$$= \int \frac{2\sec(\theta)\tan(\theta)}{8\sec^3(\theta) \times (-2\tan(\theta))} \, d\theta$$

$$= -\frac{1}{8}\int \cos^2(\theta) \, d\theta = -\frac{\theta}{16} - \frac{2\sin(\theta)\cos(\theta)}{32} + C.$$

Migrating back to x-land, we have to use a modified triangle:

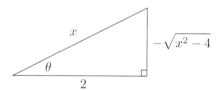

So in fact $\sin(\theta) = -\sqrt{x^2 - 4}/x$ and $\cos(\theta) = 2/x$. Notice that $\sin(\theta)$ is actually greater than 0, since $x < 0$. Anyway, substituting back into the above integral, we see that

$$\int \frac{dx}{x^3 \sqrt{x^2 - 4}} = -\frac{1}{16}\sec^{-1}\left(\frac{x}{2}\right) - \frac{1}{32} \cdot 2 \cdot \frac{-\sqrt{x^2 - 4}}{x} \cdot \frac{2}{x} + C$$

$$= -\frac{1}{16}\sec^{-1}\left(\frac{x}{2}\right) + \frac{\sqrt{x^2 - 4}}{8x^2} + C.$$

So, that's the answer when $x < 0$. It's almost the same as the previous answer, but the inverse secant term needs a minus sign out front. Also, the constant C is potentially different from the other C which arises when $x > 0$. Why? Because we are looking for a function whose derivative is $1/x^3\sqrt{x^2 - 4}$, which itself has domain $(-\infty, -2) \cup (2, \infty)$. So the antiderivative is also in two pieces, either of which can be shifted up or down independently of the other. All in all, the complete answer is

$$\int \frac{dx}{x^3 \sqrt{x^2 - 4}} = \begin{cases} \dfrac{1}{16}\sec^{-1}\left(\dfrac{x}{2}\right) + \dfrac{\sqrt{x^2 - 4}}{8x^2} + C_1 & \text{when } x > 2, \\[4mm] -\dfrac{1}{16}\sec^{-1}\left(\dfrac{x}{2}\right) + \dfrac{\sqrt{x^2 - 4}}{8x^2} + C_2 & \text{when } x < -2. \end{cases}$$

Here C_1 and C_2 are potentially different constants. Actually, we've already encountered an integral where two constants should be involved: $\int 1/x\,dx$. See Section 17.7 in Chapter 17 for more details. In practice, problems involving Type 3 are often phrased (or intended to be phrased) with the condition that $x > 0$. This allows one to avoid all the above mess and take square roots without a care in the world. Just beware: if $x < 0$, then you need to be a lot more careful. . . .

19.4 Overview of Techniques of Integration

We've now built up quite a toolkit of techniques of integration. Now the question is, given an integral, which technique do you use? Sometimes it's not easy, and you may have to try several different methods until you hit upon the right one. Sometimes you even need to combine the methods. Here are some general guidelines to help you out:

- If an "obvious" substitution comes to mind, try it. For example, if one factor of the integrand is the derivative of another piece of the integrand, try substituting t for that other piece.
- If something like $\sqrt[n]{ax+b}$ appears in the integrand, try substituting $t = \sqrt[n]{ax+b}$, as described in Section 18.1.2 of the previous chapter.
- To integrate a rational function (that is, a quotient of polynomials), see if the top is a multiple of the derivative of the bottom. If so, you can just substitute $t = $ denominator. Otherwise, use partial fractions (Section 18.3 of the previous chapter).
- After checking that no obvious substitution looks as if it will work, use the techniques from the beginning of this chapter to find integrals involving:
  - functions containing $\sqrt{1 + \cos(x)}$ or $\sqrt{1 - \cos(x)}$: in this case, use the double-angle formula;
  - functions involving one of $1 - \sin^2(x)$, $1 - \cos^2(x)$, $1 + \tan^2(x)$, $\sec^2(x) - 1$, $\csc^2(x) - 1$, or $1 + \cot^2(x)$: in this case, use one of the Pythagorean identities $\sin^2(x) + \cos^2(x) = 1$, $\tan^2(x) + 1 = \sec^2(x)$, or $1 + \cot^2(x) = \csc^2(x)$;
  - functions with $1 \pm \sin(x)$ (or similar) in the denominator: in this case, multiply and divide by the conjugate expression and try to use the Pythagorean identities;
  - functions containing products like $\cos(mx)\cos(nx)$, $\sin(mx)\sin(nx)$, or $\sin(mx)\cos(nx)$: in this case, use the products-to-sums identities; or
  - powers of trig functions: you'll just have to learn the individual techniques in Sections 19.2.1 through 19.2.5 above.
- If the integrand involves $\sqrt{x^2 - a^2}$ or any odd power of this (for example $(x^2 - a^2)^{3/2}$, $(x^2 - a^2)^{5/2}$, and so on), or $\sqrt{x^2 + a^2}$ or $\sqrt{a^2 - x^2}$ or an odd power of any of these last two, then use a trig substitution (after checking that there's no obvious substitution). If the quadratic includes a linear term, complete the square first. See Section 19.3 above for more details.

- If the integrand is a product and no obvious substitution comes to mind, try integration by parts. (See Section 18.2 of the previous chapter for more details.)
- If no substitution appeals, then a good rule of thumb is that functions involving a power of $\ln(x)$ or an inverse trig function should be integrated by parts. In that case, let u be the power of $\ln(x)$ or the inverse trig function as appropriate. For example, how would you find

$$\int \frac{\ln(1 + x^2)}{x^2}\,dx?$$

First check that no substitution appeals; since nothing springs to mind, think of integration by parts. Wait a second, it's not a product! Wait another second, quotients are products too! Just rewrite the integral as

$$\int \ln(1 + x^2) \times \frac{1}{x^2}\,dx,$$

then integrate by parts with $u = \ln(1 + x^2)$ and $dv = (1/x^2)\,dx$. Try it now—you should get the answer

$$-\frac{\ln(1 + x^2)}{x} + 2\tan^{-1}(x) + C.$$

Even if you memorize all the above techniques, you will be lost in a sea of confusion unless you practice a whole load of problems. Make sure that at some stage you tackle a mixed bag of integrals so that you can be confident of which method to use on which integral. Then you will truly be a bad-ass integrator.

CHAPTER 20 _____

Improper Integrals: Basic Concepts

This is a difficult topic, so I'm devoting two chapters to it. This chapter serves as an introduction to improper integrals. The next chapter gets into the details of how to solve problems involving improper integrals. If you are reading this chapter for the first time, you should probably take care to try to understand all the points in it. On the other hand, if you are reviewing for a test, most likely you'll want to skim over the chapter, noting the boxed formulas and the sections marked as important, and concentrate on the next chapter. Here's what we'll actually look at in this chapter:

- the definition of improper integrals, convergence, and divergence;
- improper integrals over unbounded regions; and
- the theoretical basis for the comparison test, the limit comparison test, the p-test, and the absolute convergence test.

We'll revisit all four of these tests in the next chapter and see many examples of how to apply them.

20.1   Convergence and Divergence

What is an improper integral, anyway? In Chapter 16, we saw that the integral

$$\int_a^b f(x)\, dx$$

certainly makes sense if f is a bounded function on $[a, b]$ which is continuous except at a finite number of places. If f has infinitely many discontinuities, the integral might still make sense, or it might be totally screwed up (see Section 16.7 of Chapter 16 for an example). What if f isn't bounded? This means that the values of $f(x)$ manage to get really large (positively or negatively or both) while x is in the interval $[a, b]$. This sort of thing typically happens when f has a vertical asymptote somewhere in this interval: the function blows up there and can't be bounded. This causes the above integral to be improper.

There's a different type of unboundedness that can occur even if f is bounded. The interval $[a, b]$ can actually be infinite—something like $[0, \infty)$, $[-7, \infty)$, $(-\infty, 3]$ or even $(-\infty, \infty)$. This also makes the above integral improper.

So, the integral $\int_a^b f(x)\,dx$ is *improper* if any of the following conditions apply:

1. f isn't bounded in the **closed** interval $[a, b]$;
2. $b = \infty$; or
3. $a = -\infty$.

For now, let's concentrate on what happens if the first of these conditions fails; we'll return to the other two conditions in Section 20.2 below. As I said, the typical way that a function fails to be unbounded is if it has a vertical asymptote somewhere, although there can be more exotic types of behavior. (An example is $f(x) = \frac{1}{x}\sin(\frac{1}{x})$, which oscillates really wildly as x approaches 0.) If $f(x)$ is unbounded for x near some number c, we'll say that f has a *blow-up point* at $x = c$. Again, in most situations, this is the same thing as a vertical asymptote.

So let's look at the simple case of when our function f has a vertical asymptote at $x = a$. The situation looks something like this:

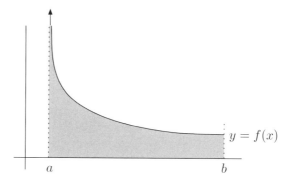

I'd be lying through my teeth if I claimed that $\int_a^b f(x)\,dx$ is the area (in square units) of the shaded region. The problem is that the region actually extends up the page, then past the top of the page, going on and on forever, as the arrow is trying to indicate. The region does get skinnier as it goes up, though, because of the vertical asymptote.

Since the region never stops going up, surely its area should be infinite, right? Not necessarily. A mathematical miracle can occur if the region is skinny enough, and the area can actually be finite. To see how a region can be unbounded yet have a finite area, we'll use limits once again. Here's the idea: let ε be a small positive number; then you can integrate f over the region $[a + \varepsilon, b]$, since f is bounded there. You'll get some nice finite number. Now, replay the situation but with an even smaller ε. You get a new finite number. The situation now looks something like this:

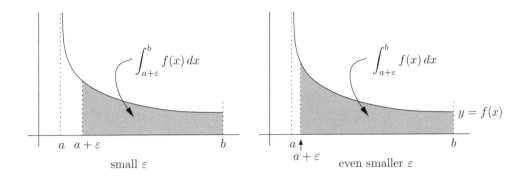

small ε even smaller ε

The smaller ε is, the closer our (bounded) approximating region is to the actual unbounded region. This suggests that we should continue the process with smaller and smaller ε, and see if the numbers we get have a limit L as $\varepsilon \to 0^+$. If so, then we interpret L square units to be the value of the area we're looking for. In that case, we say that the integral $\int_a^b f(x)\,dx$ *converges* to L. If there's no limit, then we can't find a meaningful answer for the area, so we give up and say that the above integral *diverges*. Note that **if the integral isn't improper, it automatically converges**! In practice, this means that if your function is bounded and the region of integration $[a, b]$ is bounded, then there's no issue: the integral converges since it's not even improper. It's just some nice finite number, no sweat.

Now, here's a summary of the situation when you have a blow-up point at $x = a$:

> if $f(x)$ is unbounded for x near a only, then set
> $$\int_a^b f(x)\,dx = \lim_{\varepsilon \to 0^+} \int_{a+\varepsilon}^b f(x)\,dx,$$

provided that the limit exists. If it does, then the integral converges; if not, the integral diverges. Just like any limit, the above one may fail to exist because it might be ∞ or $-\infty$, or things might oscillate around too much as ε tends to 0^+.

This brings us to an important point. When we look at an improper integral, the most important thing we need to find out is whether it converges or diverges. It's much less important to know what the integral converges to (assuming it converges). In practice, you can use computational techniques to estimate the value, but only if you know that the integral converges. If the integral diverges, you can get some whacked-out results if you try to use a computer to approximate your integral. Computers don't really understand infinities or crazy oscillations (yet!).

20.1.1 Some examples of improper integrals

Consider the integrals

$$\int_0^1 \frac{1}{x}\,dx \qquad \text{and} \qquad \int_0^1 \frac{1}{\sqrt{x}}\,dx.$$

These are both improper because their integrands have vertical asymptotes at $x = 0$. So we'll use the formula in the box above. In the first case, we have

$$\int_0^1 \frac{1}{x}\, dx = \lim_{\varepsilon \to 0^+} \int_\varepsilon^1 \frac{1}{x}\, dx = \lim_{\varepsilon \to 0^+} \ln|x| \Big|_\varepsilon^1 = \lim_{\varepsilon \to 0^+} (\ln(1) - \ln(\varepsilon)) = \infty.$$

(We have used the facts that $\ln(1) = 0$ and that $\ln(\varepsilon) \to -\infty$ as $\varepsilon \to 0^+$.) Since we got ∞, the improper integral $\int_0^1 1/x\, dx$ must diverge. How about the other integral? Using the formula again, we have

$$\int_0^1 \frac{1}{\sqrt{x}}\, dx = \lim_{\varepsilon \to 0^+} \int_\varepsilon^1 \frac{1}{x^{1/2}}\, dx = \lim_{\varepsilon \to 0^+} 2x^{1/2} \Big|_\varepsilon^1 = \lim_{\varepsilon \to 0^+} (2\sqrt{1} - 2\sqrt{\varepsilon}) = 2.$$

We got a nice finite number, so the integral $\int_0^1 1/\sqrt{x}\, dx$ converges. As it happens, we've shown that the integral converges to 2, but as I said at the end of the last section, we don't care that much. Our main focus is to decide whether an improper integral converges, without worrying what it actually converges to.

What's really going on here? Why should the improper integral $\int_0^1 1/x\, dx$ diverge but $\int_0^1 1/\sqrt{x}\, dx$ converge? After all, when you think about it, the graphs of $y = 1/x$ and $y = 1/\sqrt{x}$ look roughly the same—something like this:

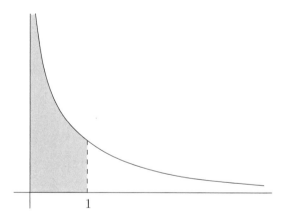

Of course, the integrands are not the same. Indeed, $1/x$ is greater than $1/\sqrt{x}$ when $0 < x < 1$. Geometrically, the graph of $y = 1/\sqrt{x}$ is actually a little closer to the y-axis than $y = 1/x$ is. It turns out that $y = 1/\sqrt{x}$ is close enough to the y-axis to make the corresponding integral converge; while $y = 1/x$ isn't close enough to the y-axis and its integral diverges. Unfortunately, there's no surefire way to classify all the functions with vertical asymptotes at $x = 0$ to decide which ones are close enough to the asymptote and which ones aren't. Most of the time, you just have to look at each improper integral on its own merits.

 Now, here's a really important point. Suppose you have an improper integral $\int_a^b f(x)\, dx$, where f has a vertical asymptote at $x = a$ only, and you want to know if the integral converges or diverges. Then the value of b doesn't matter! You can change it to any finite number bigger than a, so long as you don't pick up any new vertical asymptotes or blow-up points. To see why, first note that, by definition,

$$\int_a^b f(x)\,dx = \lim_{\varepsilon \to 0^+} \int_{a+\varepsilon}^b f(x)\,dx,$$

provided that the limit exists. Now let's change b to some other number c which is bigger than a. If f still only blows up at $x = a$, then we have

$$\int_a^c f(x)\,dx = \lim_{\varepsilon \to 0^+} \int_{a+\varepsilon}^c f(x)\,dx,$$

again provided that the limit exists. We can split this last integral at $x = b$ (the technique is described in Section 16.3 of Chapter 16) to get

$$\int_a^c f(x)\,dx = \lim_{\varepsilon \to 0^+} \left(\int_{a+\varepsilon}^b f(x)\,dx + \int_b^c f(x)\,dx \right).$$

The second integral doesn't depend on ε at all; in fact, since f is bounded between b and c inclusive, that integral converges to some nice number M. So we have shown that

$$\int_a^c f(x)\,dx = \lim_{\varepsilon \to 0^+} \int_{a+\varepsilon}^b f(x)\,dx + M.$$

If the limit on the right-hand side exists, then $\int_a^b f(x)\,dx$ converges. Adding M still keeps everything finite, so $\int_a^c f(x)\,dx$ also converges. If instead the limit doesn't exist, then adding M doesn't change that, so both $\int_a^b f(x)\,dx$ and $\int_a^c f(x)\,dx$ must diverge.

We have shown that the convergence or divergence of an improper integral over a bounded region depends only on what the integrand does very close to its blow-up points. In particular, since we know that $\int_0^1 1/x\,dx$ diverges, we can also conclude that

$$\int_0^2 \frac{1}{x}\,dx, \qquad \int_0^{100} \frac{1}{x}\,dx, \qquad \text{and} \qquad \int_0^{0.0000001} \frac{1}{x}\,dx$$

all diverge. On the other hand, since $\int_0^1 1/\sqrt{x}\,dx$ converges, we get for free that

$$\int_0^2 \frac{1}{\sqrt{x}}\,dx, \qquad \int_0^{100} \frac{1}{\sqrt{x}}\,dx, \qquad \text{and} \qquad \int_0^{0.0000001} \frac{1}{\sqrt{x}}\,dx$$

all converge. All the action goes on really near the asymptote $x = 0$.

20.1.2 Other blow-up points

In the integral $\int_a^b f(x)\,dx$, if f has only one blow-up point at the right-hand limit of integration b (instead of a), then we can play the same game as we did above. The only difference is that this time we have to approach b from the left instead of the right. So

---

if $f(x)$ is unbounded for x near b only, then set

$$\int_a^b f(x)\,dx = \lim_{\varepsilon \to 0^+} \int_a^{b-\varepsilon} f(x)\,dx,$$

---

if the limit exists; if it doesn't exist, then as before, the integral diverges.

Ah, but what if f has a blow-up point at some number c in the interior of the interval? In this case, if f is bounded everywhere on $[a,b]$ except near some point c in the interior (a,b), we have to split the integral into the two pieces

$$\int_a^c f(x)\,dx \qquad \text{and} \qquad \int_c^b f(x)\,dx.$$

We actually know how to define both of these by using limits—using the formulas from the boxes above, we can see that the above integrals are

$$\lim_{\varepsilon \to 0^+} \int_a^{c-\varepsilon} f(x)\,dx \qquad \text{and} \qquad \lim_{\varepsilon \to 0^+} \int_{c+\varepsilon}^b f(x)\,dx,$$

respectively. Here's the essential point: the whole integral $\int_a^b f(x)\,dx$ only converges if both pieces above converge. If either piece diverges, so does the whole thing. After all, how can you add something that doesn't exist to anything else, whether that other quantity exists or not? It can't be done.

This example inspires our first main technique: to investigate an improper integral, split it up into pieces, if necessary. Each piece has to have at most one problem spot, which must be at one of the limits of integration. (For the moment, the term "problem spot" means the same thing as "blow-up point," but in the next section we'll see a different sort of problem spot that isn't a blow-up point.)

For example, to analyze the integral

$$I = \int_0^3 \frac{1}{x(x-1)(x+1)(x-2)}\,dx,$$

we see that the integrand has problem spots at $x = 0$, 1, 2, and -1. The last of these doesn't matter since we're only integrating from 0 to 3. The other three do matter. We need to pick numbers between the problem spots—we may as well pick $1/2$ and $3/2$, since it doesn't matter—and now we need to split the original integral into the following five integrals:

$$I_1 = \int_0^{1/2} \frac{1}{x(x-1)(x+1)(x-2)}\,dx, \quad I_2 = \int_{1/2}^1 \frac{1}{x(x-1)(x+1)(x-2)}\,dx,$$

$$I_3 = \int_1^{3/2} \frac{1}{x(x-1)(x+1)(x-2)}\,dx, \quad I_4 = \int_{3/2}^2 \frac{1}{x(x-1)(x+1)(x-2)}\,dx,$$

$$\text{and} \qquad I_5 = \int_2^3 \frac{1}{x(x-1)(x+1)(x-2)}\,dx.$$

Notice that none of these integrals has more than one problem spot, and that all the problem spots are at one of the limits of integration. The integrals I_1, I_3, and I_5 have their only problem spot at their left-hand limits of integration, while I_2 and I_4 have their problem spot on the right. The only way that original integral I can converge is if all five pieces I_1 through I_5 converge. If they do all converge, then the value of I is the sum of the values of I_1 through I_5. (In fact, none of the five pieces converge! We'll see why in Section 21.5 of the next chapter.)

20.2 Integrals over Unbounded Regions

Now, we still have to look at what happens when one or both of the limits of integration are infinite; this means that the region of integration is *unbounded*. To handle

$$\int_a^\infty f(x)\,dx,$$

where a is any finite number and f has no blow-up points in $[a, \infty)$, let's use another limiting technique. This time, we integrate over the region $[a, N]$, where N is a massively large number. This will give us a nice finite value. Then repeat but with an even larger N to get a new value. Continue onward and see what happens to the values of the integrals. If they have a limit, then the integral converges. Otherwise, it diverges. In symbols, we are defining

$$\boxed{\int_a^\infty f(x)\,dx = \lim_{N \to \infty} \int_a^N f(x)\,dx,}$$

provided that the limit exists; in this case, the integral converges. Otherwise, it diverges. For reasons similar to those described at the end of Section 20.1.1 above, the value of a is irrelevant. So long as you don't pick up any new blow-up points of f, the value of a doesn't affect whether the improper integral converges or diverges. The only thing that really matters is how $f(x)$ behaves when x is very large indeed.

In a similar manner to the above definition, if f has no blow-up points in $(-\infty, b]$, then

$$\boxed{\int_{-\infty}^b f(x)\,dx = \lim_{N \to \infty} \int_{-N}^b f(x)\,dx.}$$

What if f has no blow-up points anywhere and we want to find

$$\int_{-\infty}^\infty f(x)\,dx?$$

Although there are no blow-up points, there are still two problem spots: ∞ and $-\infty$. That's right: we are regarding ∞ and $-\infty$ as problem spots whenever they show up, since we have to treat them separately. So we have to split the above integral into two pieces so that each one has only one problem spot. Pick your favorite number (mine is 0 for the moment), and consider the integrals

$$\int_{-\infty}^0 f(x)\,dx \qquad \text{and} \qquad \int_0^\infty f(x)\,dx.$$

We know what both of these mean, and of course the whole integral converges if and only if both pieces do. It doesn't matter if you have a different favorite number from 0, since the convergence or divergence of the above integrals doesn't depend on the endpoint.

Here are some examples involving an unbounded region of integration. Consider the integrals

$$\int_1^\infty \frac{1}{x}\,dx \qquad \text{and} \qquad \int_1^\infty \frac{1}{x^2}\,dx.$$

The first one is

$$\lim_{N\to\infty} \int_1^N \frac{1}{x}\,dx = \lim_{N\to\infty} \ln|x|\Big|_1^N = \lim_{N\to\infty} (\ln(N) - \ln(1)) = \infty,$$

while the second one is

$$\lim_{N\to\infty} \int_1^N \frac{1}{x^2}\,dx = \lim_{N\to\infty} -\frac{1}{x}\Big|_1^N = \lim_{N\to\infty} \left(-\frac{1}{N} + 1\right) = 1.$$

So the first integral diverges while the second converges.

Here's a question: do the following integrals converge or diverge?

$$\int_0^\infty \frac{1}{x}\,dx \qquad \text{and} \qquad \int_0^\infty \frac{1}{x^2}\,dx?$$

Since both integrals have problem spots at 0 and ∞, we'll have to split them both up. For the first one, we can look at

$$\int_0^1 \frac{1}{x}\,dx \qquad \text{and} \qquad \int_1^\infty \frac{1}{x}\,dx.$$

Note that the choice of 1 as the split point is up to you. It doesn't matter at all what you pick (so long as it's a positive number)! Anyway, we've already seen that both of these integrals diverge, so certainly the integral $\int_0^\infty 1/x\,dx$ diverges.

As for $\int_0^\infty 1/x^2\,dx$, we split it up into the two pieces

$$\int_0^1 \frac{1}{x^2}\,dx \qquad \text{and} \qquad \int_1^\infty \frac{1}{x^2}\,dx.$$

Now we already saw that the second of these pieces converges. To examine the first piece, we could use our formula involving limits, but there's a sneakier way. The idea is that we already saw that $\int_0^1 1/x\,dx$ diverges to infinity. But if you think about it, $1/x^2$ is greater than $1/x$ when x is between 0 and 1. (Right? Hmm ... x^2 is less than x in the region $(0,1)$, so the reciprocals are the other way around.) So if the area under $1/x$ above $[0,1]$ is infinite, the area under $1/x^2$ above $[0,1]$ is even bigger—so still infinite! Without doing any more work, we can already say that $\int_0^1 1/x^2\,dx$ diverges. It follows that the whole integral $\int_0^\infty 1/x^2\,dx$ diverges, and that the real problem is due to the left-hand endpoint 0, not the right-hand endpoint ∞ in this case. Notice how we compared $1/x^2$ with $1/x$; this is a special case of the so-called comparison test, which we'll look at now.

20.3 The Comparison Test (Theory)

Suppose we have two functions which are never negative, at least in some region of interest. If the first function is bigger than the second function, and the integral of the second function (over our region) diverges, then the integral of the first function (over the same region) also diverges. Mathematically, it looks like this. Let's say we want to know something about $\int_a^b f(x)\,dx$, but we only know something about $\int_a^b g(x)\,dx$. If $f(x) \geq g(x) \geq 0$ for x in the interval (a, b), and we know that $\int_a^b g(x)\,dx$ diverges, then so does $\int_a^b f(x)\,dx$. In fact, since $f(x) \geq g(x)$, we can write

$$\int_a^b f(x)\,dx \geq \int_a^b g(x)\,dx = \infty.$$

So the first integral also diverges. In our example above, we'd just write

$$\int_0^1 \frac{1}{x^2}\,dx \geq \int_0^1 \frac{1}{x}\,dx = \infty,$$

and conclude that the left-hand integral diverges. Of course, we had to know that the right-hand integral diverges, but we already saw that earlier.

The situation is even clearer when one looks at a picture:

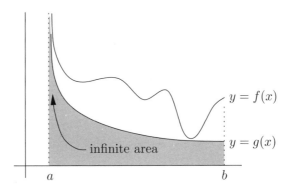

In this picture, the area under $y = g(x)$ between $x = a$ and $x = b$ is supposed to be infinite. The curve $y = f(x)$ sits above $y = g(x)$, so the area under it (between $x = a$ and $x = b$) should be even greater. More than infinite is still infinite, so $\int_a^b f(x)\,dx$ also diverges.

What if $\int_a^b g(x)\,dx$ diverges but $f(x) \leq g(x)$ instead? What can you say about $\int_a^b f(x)\,dx$? The answer is: diddly-squat. Bubkes. Nothing at all. Let's see how the math would go:

$$\int_a^b f(x)\,dx \leq \int_a^b g(x)\,dx = \infty.$$

So the integral we are interested in, $\int_a^b f(x)\,dx$, is less than or equal to infinity. That is, either it is less than infinity, so it converges, or it is equal to infinity, so it diverges. Great—we now know that it either converges or diverges. Whoop-di-doo. Yup, we haven't accomplished anything. So don't do this.

On the other hand, for convergence, it **is** the other way around. Here, if we want to know about $\int_a^b f(x)\,dx$ and we know that $\int_a^b g(x)\,dx$ converges, we'd better hope that $f(x) \leq g(x)$. You might say that we want f to be "controlled" by g. Well, then we'd get convergence (still assuming that both functions are positive). So, if $0 \leq f(x) \leq g(x)$ on (a, b) and $\int_a^b g(x)\,dx$ converges, then so does $\int_a^b f(x)\,dx$. Mathematically,

$$\int_a^b f(x)\,dx \leq \int_a^b g(x)\,dx < \infty,$$

so both integrals converge (noting that the left-hand integral is positive, so it can't diverge down to $-\infty$). The picture looks like this:

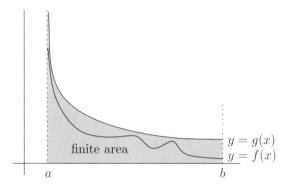

The shaded area under $y = g(x)$ between $x = a$ and $x = b$ is assumed to be finite. You can clearly see from the picture that the area we want, which is under $y = f(x)$ between $x = a$ and $x = b$, is less than the finite shaded area. Since the area we want is positive and less than a finite number, it must also be finite.

Beware: suppose you know that $\int_a^b g(x)\,dx$ converges, but you have the inequality $f(x) \geq g(x)$ instead. Now the curve you want ($y = f(x)$) sits above the other curve ($y = g(x)$). This is no good: you'd only be able to say that

$$\int_a^b f(x)\,dx \geq \int_a^b g(x)\,dx.$$

So the integral we are interested in on the left-hand side is greater than or equal to some finite number. Our integral is therefore finite or infinite. Great—no info whatsoever. Hey, we're in the whoop-di-doo case again! So don't go there.

It's true that I haven't really justified the comparison test, mathematically speaking. Actually there's not that much to it. Some splitting of integrals (and hairs) is required, but we've already seen the basic idea. For example, if f and g both have a vertical asymptote at $x = a$ and have no blow-up points anywhere else, and $0 \leq f(x) \leq g(x)$ for all x in the interval $[a, b]$, then we can

say that

$$0 \le \int_{a+\varepsilon}^{b} f(x)\, dx \le \int_{a+\varepsilon}^{b} g(x)\, dx$$

for any $\varepsilon > 0$. Now take limits. If the improper integral $\int_{a}^{b} g(x)\, dx$ converges, then the right-hand side becomes finite. Everything now depends on the middle integral. Since $f(x)$ is always positive, the middle integral gets bigger as ε tends toward 0 from above. It's getting bigger and bigger, but it can't get past the barrier at $\int_{a}^{b} g(x)\, dx$, which is a nice finite number. The only possibility is that the middle integral converges to some finite number as $\varepsilon \to 0^{+}$.* In other words, $\int_{a}^{b} f(x)\, dx$ converges. That proves the comparison test in its convergence version (the second of the two versions we looked at above), in the special case where f and g only have problems at $x = a$. It is now up to you to prove the divergence version and also to work out how to deal with problems at $x = b$. There's really not much difference. Of course, if the problems are in the middle somewhere, or there are multiple problems, you have to split the integral into pieces before using the comparison test anyway.

We'll look at many examples involving the comparison test in the next chapter. Now it's time to look at another test.

20.4 The Limit Comparison Test (Theory)

The comparison test uses the improper integral of one function to get information about an improper integral of another function. The limit comparison test does the same thing, except that we don't actually need one function to be bigger than the other. Instead, we need the two functions to be just about the same. Here's the basic idea: suppose that two functions f and g are very close to each other at the blow-up point $x = a$ (and have no other blow-up points). Then $\int_{a}^{b} f(x)\, dx$ and $\int_{a}^{b} g(x)\, dx$ either both diverge or both converge. Their behavior is identical. Intuitively, it makes sense; let's get down to details by specifying what we really mean when we say that two functions are "very close" to each other.

20.4.1 Functions asymptotic to each other

Suppose we have two functions f and g such that

$$\lim_{x \to a} \frac{f(x)}{g(x)} = 1.$$

This means that when x is near a, the ratio $f(x)/g(x)$ is close to 1. If the ratio were equal to 1, then $f(x)$ would equal $g(x)$. Since the ratio is only close to 1, then $f(x)$ is "very close" to $g(x)$. This doesn't mean that the difference between $f(x)$ and $g(x)$ is small! For example, $f(x)$ could be a trillion and $g(x)$

---

*Actually, this statement, which seems obvious, is quite profound. The statement is pretty much what distinguishes $\mathbb{R}$ from any of its proper subsets which contain every rational number.

could be a trillion plus a million (for the same value of x); in that case, the ratio $f(x)/g(x)$ would be a little under 1, while the difference between $f(x)$ and $g(x)$ is still a million! On the other hand, the two numbers are relatively close to each other, since a million is a small difference relative to the size of the numbers.

So, we'll say that $f(x) \sim g(x)$ as $x \to a$ if the limit of the ratio is 1. That is,

$$\boxed{f(x) \sim g(x) \text{ as } x \to a \qquad \text{means the same thing as} \qquad \lim_{x \to a} \frac{f(x)}{g(x)} = 1.}$$

This doesn't mean that $f(x)$ is approximately equal to $g(x)$ when x is near a: it means that the ratio of $f(x)$ to $g(x)$ is near 1 when x is near a. We say that f and g are *asymptotic* to each other as $x \to a$. Of course, you could replace $x \to a$ by $x \to \infty$, or even $x \to a^+$; all you have to do is make the same replacement in the limit too.

All this is useless unless we have limits of the form

$$\lim_{x \to a} \frac{f(x)}{g(x)} = 1.$$

Actually, we've seen many of these types of limits! Here are some examples:*

$$\lim_{x \to \infty} \frac{3x^3 - 1000x^2 + 5x - 7}{3x^3} = 1, \qquad \lim_{x \to 0} \frac{\sin(x)}{x} = 1,$$

$$\lim_{x \to 0} \frac{e^x - 1}{x} = 1, \quad \text{and} \quad \lim_{x \to 0} \frac{\ln(1+x)}{x} = 1.$$

The first limit above can be written as $3x^3 - 1000x^2 + 5x - 7 \sim 3x^3$ as $x \to \infty$. That is, $3x^3 - 1000x^2 + 5x - 7$ and $3x^3$ are asymptotic to each other as $x \to \infty$. Similarly, the second limit says that $\sin(x) \sim x$ as $x \to 0$. The third and fourth limits show that $e^x - 1$ and $\ln(1+x)$ are also both asymptotic to x as $x \to 0$; that is, $e^x - 1 \sim x$ and $\ln(1+x) \sim x$ as $x \to 0$.

All we've done is to rewrite each limit in a different form, but it is a very convenient form. Indeed, you can take powers of asymptotic relations and get new ones. For example, knowing that $\sin(x) \sim x$ as $x \to 0$, we can immediately write that $\sin^3(x) \sim x^3$ as $x \to 0$, or even that $1/\sin(x) \sim 1/x$ as $x \to 0$. You can also replace x by any other quantity that goes to 0 as x does, such as a power of x. For example, starting with $\sin(x) \sim x$ as $x \to 0$ once again, we can replace x by $4x^7$ to see that $\sin(4x^7) \sim 4x^7$ as $x \to 0$. You can even multiply or divide two relations by each other, provided that the limit is at the same value of x for both asymptotic relations. For example, we know that $\tan(x) \sim x$ as $x \to 0$ since

$$\lim_{x \to 0} \frac{\tan(x)}{x} = 1.$$

So we can multiply $\tan(x) \sim x$ and $\sin(x) \sim x$ (both as $x \to 0$) together to get the asymptotic relation $\tan(x)\sin(x) \sim x^2$ as $x \to 0$.

---

*The examples can be found in Section 4.3 of Chapter 4, Section 7.1.1 of Chapter 7, and Sections 9.4.2 and Sections 9.4.3 of Chapter 9, respectively.

What you cannot do is add or subtract these relations. For example, if you start with $\tan(x) \sim x$ and $\sin(x) \sim x$ as $x \to 0$, you can't just subtract the second relation from the first to get $\tan(x) - \sin(x) \sim x - x$. Indeed, $x - x$ is just 0, and nothing can be asymptotic to 0. Why not? Well, if $f(x) \sim 0$ as $x \to a$, then we'd need

$$\lim_{x \to a} \frac{f(x)}{0} = 1.$$

That's clearly garbage, since the left-hand side doesn't make any sense. So, by all means, multiply, divide, and take powers of asymptotic relations, but don't add or subtract them.

20.4.2 The statement of the test

OK, so we have this notion of two functions being asymptotic to each other, and we have some examples too (like $\sin(x) \sim x$ as $x \to 0$). So what? Well, suppose you have some function f with a problem spot only at a, and you're trying to see if the improper integral $\int_a^b f(x)\,dx$ converges or diverges. If you can find a function g which behaves like f when the argument x is near a, then you can just replace f by g and see if $\int_a^b g(x)\,dx$ converges or diverges. Whatever you find for g also holds for f.

More formally, if $f(x) \sim g(x)$ as $x \to a$, and neither function has any problem spots anywhere else on the interval $[a, b]$, then the integrals $\int_a^b f(x)\,dx$ and $\int_a^b g(x)\,dx$ both diverge or both converge. (If they both converge, then the values they converge to may be different.) This is one case of the *limit* *comparison test*. Here's a sneak preview of its power; we'll see many more examples in the next chapter. Suppose we want to know whether

$$\int_0^1 \frac{1}{\sin(\sqrt{x})}\,dx$$

converges or diverges. It seems difficult to find an antiderivative of $1/\sin(\sqrt{x})$. Luckily, we don't have to. Since $\sin(x) \sim x$ as $x \to 0$, we can replace the small quantity x by another small quantity $\sqrt{x}$ to see that $\sin(\sqrt{x}) \sim \sqrt{x}$ as $x \to 0^+$. (We need to use $x \to 0^+$ because $\sqrt{x}$ only makes sense when $x \geq 0$.) Taking reciprocals, we have

$$\frac{1}{\sin(\sqrt{x})} \sim \frac{1}{\sqrt{x}} \qquad \text{as } x \to 0^+.$$

Also note that $1/\sin(\sqrt{x})$ and $1/\sqrt{x}$ have no blow-up points in $(0, 1]$. So, the limit comparison test says that the two integrals

$$\int_0^1 \frac{1}{\sin(\sqrt{x})}\,dx \qquad \text{and} \qquad \int_0^1 \frac{1}{\sqrt{x}}\,dx$$

either both converge or both diverge. We have replaced a difficult integral with a much easier one, $\int_0^1 1/\sqrt{x}\,dx$. We already know from Section 20.1.1 above that this easier integral converges, so we can immediately conclude that the integral we want (on the left) also converges.

Of course, there are cases of the test which apply when the blow-up point is at b, or when the region of integration is unbounded. We'll list all the versions in Section 21.2 of the next chapter. In the meantime, let's see why the test works in the above case. Since $f(x) \sim g(x)$ as $x \to a$, we know that

$$\lim_{x \to a} \frac{f(x)}{g(x)} = 1.$$

In particular, provided we get close enough to a, the ratio $f(x)/g(x)$ must be at least $\frac{1}{2}$ and no more than 2. That is, we can pick some c between a and b such that

$$\frac{1}{2} \le \frac{f(x)}{g(x)} \le 2 \qquad \text{for all } x \text{ in } (a, c].$$

This inequality can be rewritten as

$$\frac{1}{2} g(x) \le f(x) \le 2 g(x) \qquad \text{for all } x \text{ in } (a, c].$$

Now we can use the comparison test. For example, if $\int_a^b g(x)\,dx$ diverges, then so does $\int_a^c g(x)\,dx$ (as we've seen above). In fact, so does $\frac{1}{2} \int_a^c g(x)\,dx$, informally since one-half of infinity is still infinity! So, the fact that $f(x)$ is greater than $\frac{1}{2} g(x)$ means that the integral $\int_a^c f(x)\,dx$ diverges, and it follows that

$\int_a^b f(x)\,dx$ diverges too. On the other hand, if $\int_a^b g(x)\,dx$ actually converges, then so does $2 \int_a^c g(x)\,dx$ and we can again use the comparison test (you can fill in the details) to show that $\int_a^b f(x)\,dx$ converges as well.

A quick comment: most textbooks have a different statement of the limit comparison test. In particular, the limit of $f(x)/g(x)$ doesn't actually have to be 1—it could be any positive number and the above argument would still work (after a slight modification). On the other hand, allowing a limit other than 1 doesn't really gain anything, and it loses the ability to use the intuitive $\sim$ notation. As we'll see in the next chapter, we'll get by very nicely with our version of the test.

20.5 The p-test (Theory)

Now that we have the comparison test and limit comparison test, we need to know how to use them. Our basic strategy, which will be greatly elaborated upon in the next chapter, will be to pick a function g which we can compare our function f with. Hopefully g is simple enough that we can at least say whether its integral (over the region under consideration) converges or diverges.

The question is, what are some functions we could choose as g? Well, the most useful are the functions $1/x^p$ for some $p > 0$. For example, we have already looked at some integrals involving $1/x$, $1/\sqrt{x}$, and $1/x^2$, which correspond to $p = 1$, $\frac{1}{2}$, and 2, respectively. Since these functions are so easy to integrate, we can use the limit formulas to get the p-test:

- (**p-test, $\int^{\infty}$ version**) For any finite $a > 0$, the integral

$$\int_a^{\infty} \frac{1}{x^p}\, dx$$

converges if $p > 1$ and diverges if $p \leq 1$.

- (**p-test, $\int_0$ version**) For any finite $a > 0$, the integral

$$\int_0^a \frac{1}{x^p}\, dx$$

converges if $p < 1$ and diverges if $p \geq 1$.

Notice that the two versions of the test are basically opposites: except for when $p = 1$, one of the integrals

$$\int_0^a \frac{1}{x^p}\, dx \qquad \text{and} \qquad \int_a^{\infty} \frac{1}{x^p}\, dx$$

converges and the other one diverges. The case $p = 1$ corresponds to $1/x$, and as we already know, both of the integrals diverge in this case.

Now, this p-test is really useful and comes up often in practice, so it's really important that you don't get the two versions of the test mixed up! One way to remember the correct version of the test is to remember what happens with $1/x^2$ and $1/\sqrt{x}$. I just remember the two little facts:

$$\int_a^{\infty} \frac{1}{x^2}\, dx \text{ converges, and so does } \int_0^a \frac{1}{\sqrt{x}}\, dx.$$

From these two facts, I can remember the whole of the p-test! How does it work? Well, from the first fact, and the knowledge that what goes on near ∞ is opposite from what goes on near 0, I know that

$$\int_0^a \frac{1}{x^2}\, dx$$

diverges. Similarly, from the second fact, I know that

$$\int_a^{\infty} \frac{1}{\sqrt{x}}\, dx$$

also diverges. What about other exponents? Well, any exponent higher than 1 (for example, $\frac{3}{2}$, 2, or 70) behaves in the same manner as $1/x^2$, and any exponent lower than 1 (for example, $\frac{1}{2}$, $\frac{2}{3}$, or 0.999) behaves exactly like $1/\sqrt{x}$ (remember, this is the same as $1/x^{1/2}$).

It might also help to examine the following diagram:

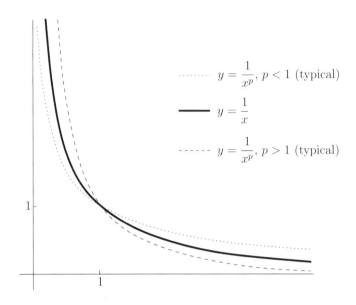

$$y = \frac{1}{x^p},\ p < 1\ \text{(typical)}$$

$$y = \frac{1}{x}$$

$$y = \frac{1}{x^p},\ p > 1\ \text{(typical)}$$

In this picture, the dotted and dashed curves are typical of what the graphs of $y = 1/x^p$ look like for $p < 1$ and $p > 1$, respectively. The solid curve is $y = 1/x$, which isn't quite close enough to the y-axis for $\int_0^1 1/x\,dx$ to converge, nor close enough to the x-axis for $\int_1^\infty 1/x\,dx$ to converge. On the other hand, for any $p < 1$, the dotted curve is close enough to the y-axis for $\int_0^1 1/x^p\,dx$ to converge. The situation is reversed when you look for proximity to the x-axis: there we need to look at the dashed curve, representing $y = 1/x^p$ for $p > 1$, to get close enough for $\int_1^\infty 1/x^p\,dx$ to converge.

Notice that since $1.0000001 > 1$, the integral

$$\int_1^\infty \frac{1}{x^{1.0000001}}\,dx$$

converges even though $\int_1^\infty 1/x\,dx$ diverges! Just nudging the power of x from 1 up to 1.0000001 was enough to make all the difference. This shows how incredibly delicate this whole issue of convergence and divergence really is.

Now, let's prove the p-test. Luckily, this is just a computation using the formulas from Section 20.1 above. First, consider

$$\int_a^\infty \frac{1}{x^p}\,dx$$

for some constant $a > 0$. If $p = 1$, the integrand becomes $1/x$, and we've already seen that the integral diverges in that case. Otherwise, we have

$$\int_a^\infty \frac{1}{x^p}\,dx = \lim_{N\to\infty} \int_a^N x^{-p}\,dx = \lim_{N\to\infty} \frac{1}{1-p} x^{1-p}\Big|_a^N$$

$$= \frac{1}{1-p}\left(\left(\lim_{N\to\infty} N^{1-p}\right) - a^{1-p}\right).$$

Now if the limit

$$\lim_{N\to\infty} N^{1-p}$$

exists, then the whole integral $\int_a^\infty 1/x^p \, dx$ converges. If on the other hand the limit doesn't exist, the integral diverges. So, write the above limit as

$$\lim_{N \to \infty} \frac{1}{N^{p-1}}.$$

If $p > 1$, then $p - 1 > 0$, so N^{p-1} gets very large as N gets large. Its reciprocal becomes very small, and so the limit is therefore 0 and our original integral converges. On the other hand, if $p < 1$, then $1 - p > 0$, so N^{1-p} gets very large and the limit blows up to ∞, proving that the original integral diverges. This proves one half of the p-test. The proof of the other half is almost the same; you just have to use $\varepsilon \to 0^+$ instead of $N \to \infty$. I'll leave the details to you.

20.6 The Absolute Convergence Test

One of the assumptions in the comparison test is that the functions f and g are always nonnegative. What if you want to investigate the behavior of a function which is sometimes negative? Well, if the function is always negative, you could just pull out a minus sign and reduce it to the case of a positive function. We'll see an example of this in the next chapter. On the other hand, if the function keeps oscillating between positive and negative values throughout the region of integration, you can appeal to the **absolute convergence test**. Here's what it says:

$$\text{if} \quad \int_a^b |f(x)| \, dx \quad \text{converges, then so does} \quad \int_a^b f(x) \, dx.$$

This also works on infinite regions of integration (such as $[a, \infty)$ instead of $[a, b]$). Watch out: if the absolute-value version of the original integral diverges, then the original integral could still converge! Such examples are pretty cool but they're beyond the scope of this book. On the other hand, we'll see something similar when we look at alternating series in Section 23.7 of Chapter 23.

Why is the above test useful? Well, for one thing, $|f(x)|$ is always non-negative, so you **can** use the comparison test on improper integrals involving it. For example, consider the improper integral

$$\int_1^\infty \frac{\sin(x)}{x^2} \, dx.$$

The integrand $\sin(x)/x^2$ oscillates between positive and negative values as x gets larger and larger without bound. So we can't use the comparison test* or the limit comparison test yet. Let's try the absolute convergence test first.

---

*Direct comparison won't work, since the integrals $\int_1^N \sin(x)/x^2 \, dx$ are not increasing in N as N gets bigger. The idea of the argument at the end of Section 20.3 above fails since it depends on the integrals getting bigger and bigger without bumping into the ceiling provided by the integral of g.

We need to consider this integral instead:

$$\int_1^\infty \left| \frac{\sin(x)}{x^2} \right| \, dx.$$

This can be rewritten as

$$\int_1^\infty \frac{|\sin(x)|}{x^2} \, dx$$

since x^2 can't be negative. Now we can use the comparison test. You see, $|\sin(x)| \le 1$ for all x, so it follows that

$$\frac{|\sin(x)|}{x^2} \le \frac{1}{x^2}$$

for all x. The comparison test says that

$$\int_1^\infty \frac{|\sin(x)|}{x^2} \, dx \le \int_1^\infty \frac{1}{x^2} \, dx.$$

Since the right-hand integral converges by the p-test, so does the left-hand integral. Finally, we can use the absolute convergence test to say that

$$\int_1^\infty \frac{|\sin(x)|}{x^2} \, dx \quad \text{converges, so} \quad \int_1^\infty \frac{\sin(x)}{x^2} \, dx \quad \text{also converges.}$$

It's a little subtle, but you really do need to use those absolute values. Here's another example:

$$\int_0^\infty \cos(x) \, dx.$$

The integrand $\cos(x)$ oscillates between positive and negative values, so maybe we should look at the absolute value "version" of the integral:

$$\int_0^\infty |\cos(x)| \, dx.$$

Unfortunately, there's not a hope in hell that this new integral converges. To see why, draw the graph of $y = |\cos(x)|$ right now and you'll see that it's a series of identical humps, one after the other. There's no way you can add up the areas of infinitely many identical humps and get a finite value. So the absolute value version diverges. This means that we **cannot** use the absolute convergence test! The only time that this test can be used is when the absolute value version of the integral converges.

We have learned nothing from these shenanigans: we are back to square one. We don't know whether our original integral converges or diverges. So, let's try using the definition of the improper integral with problem spot at ∞:

$$\int_0^\infty \cos(x) \, dx = \lim_{N \to \infty} \int_0^N \cos(x) \, dx = \lim_{N \to \infty} \sin(x) \Big|_0^N$$

$$= \lim_{N \to \infty} (\sin(N) - \sin(0)) = \lim_{N \to \infty} \sin(N).$$

This last limit doesn't exist, since $\sin(N)$ keeps on oscillating between -1 and 1, never making up its mind even as N becomes larger and larger without bound. So our original integral $\int_0^\infty \cos(x)$ diverges, not because it goes to ∞ or $-\infty$, but because it oscillates too much.

Oscillating integrals like this are extremely tricky to deal with. If you're lucky, you can use the formal definition as we did above. Most of the time this doesn't work. Many mathematicians have spent a whole lot of time trying to understand what's going on. For the moment, just bear the above example in mind. We'll have more than enough to deal with in the next chapter, where we return to the tests and see how to solve problems involving improper integrals.

Before we do this, let's take a quick look at why the absolute convergence test works. Suppose we know that

$$\int_a^b |f(x)|\,dx$$

converges. Now comes a nice trick: set $g(x) = |f(x)| + f(x)$ for all x in $[a,b]$ where f is defined. Then g has two important properties: first, $g(x) \geq 0$, and second, $g(x) \leq 2|f(x)|$. (In both cases, x is any number in $[a,b]$ which is in the domain of f.) In fact, if you think about it, you can see that $g(x)$ actually equals $2f(x)$ whenever $f(x) \geq 0$, and that $g(x)$ actually equals 0 whenever $f(x) < 0$. Try to show that the two important properties follow from this. Anyway, we can now use the comparison test on g:

$$0 \leq \int_a^b g(x)\,dx \leq 2 \int_a^b |f(x)|\,dx < \infty.$$

The conclusion is that

$$\int_a^b g(x)\,dx$$

converges as well. So what? Well, notice that $f(x) = g(x) - |f(x)|$, so

$$\int_a^b f(x)\,dx = \int_a^b g(x)\,dx - \int_a^b |f(x)|\,dx.$$

Both integrals on the right converge—the first because we just showed it, and the second because we are assuming it—so the left-hand side converges as well.

CHAPTER 21 _____

Improper Integrals: How to Solve Problems

Let's get practical and look at a lot of examples of improper integrals. As we go along, we'll summarize the main methods. In the previous chapter, we introduced some tests that will turn out to be really useful. To use them effectively, you have to understand how some common functions behave, especially near 0 and near ∞. By "common functions," I mean our usual suspects: polynomials, trig functions, exponentials, and logarithms. So, here's the game plan for this chapter:

- what to do when you first see an improper integral, including how to deal with multiple problem spots and functions which aren't always positive;
- summary of the comparison test, limit comparison test, and p-test;
- the behavior of common functions near ∞ and $-\infty$;
- the behavior of common functions near 0; and
- how to handle problem spots at finite values other than 0.

21.1   How to Get Started

OK, so you have an improper integral, $\int_a^b f(x)\,dx$. (We'll always assume that f is continuous or has finitely many discontinuities.) You know that your integral is improper because the integrand f has at least one problem spot in $[a, b]$. Problem spots occur at blow-up points of f, like vertical asymptotes, and also at ∞ and $-\infty$, if applicable. For example, the integral

$$\int_{-\infty}^{\infty} \frac{1}{x^2 - 1}\,dx$$

has problem spots at ∞ and $-\infty$ (since these are always problem spots if they are involved), and also at $x = 1$ and $x = -1$ (since the integrand is undefined there).

As we said in Section 20.1.2 of the previous chapter, it makes sense to concentrate on one problem spot at a time. Also, we'd like to arrange matters so that the integrand is always positive, at least when x is near the problem spot. So, our first task is to split up the integral as appropriate, and our second task will be to deal with what happens if f is sometimes negative.

21.1.1 Splitting up the integral

Here's the basic plan of attack:

1. **Identify all the problem spots** in the region $[a, b]$.
2. **Split up the integral** into enough pieces so that each new piece has at most one problem spot, which occurs at one of the endpoints of the integral.
3. **Look at each piece individually. If any one piece diverges, so does the whole thing.** The only way the original improper integral can converge is if each piece converges.

How do you split up an integral into the correct pieces? If there's only one problem spot which is either at a or b, there's nothing to do. On the other hand, consider this classic example: does

$$\int_0^\infty \frac{1}{\sqrt{x} + x^2} \, dx$$

converge or diverge? Well, the integrand has a vertical asymptote at $x = 0$, and ∞ is always a problem spot, so we have problem spots at the endpoints 0 and ∞. That's two problem spots, and we can only cope with one per integral. So pick your favorite number between 0 and ∞—mine's 5 at the moment—and split up the integral into two pieces:

$$\int_0^5 \frac{1}{\sqrt{x} + x^2} \, dx \qquad \text{and} \qquad \int_5^\infty \frac{1}{\sqrt{x} + x^2} \, dx.$$

We'll finish off this example in Section 21.4.1 below. For the moment, notice that both of these integrals have one problem spot only, and in both cases the problem spot is at the left or right endpoint of the region of integration. It doesn't matter where you split the integral up—this point was discussed at length at the end of Section 20.1.1 in the previous chapter. You should also check out the example on page 436 involving an improper integral which needs to be split up into five different pieces.

As for the third step above, we'll spend the rest of this chapter looking at how to deal with individual pieces with only one problem spot occurring at one of the endpoints. The key point is that all the pieces have to converge in order for the whole integral to converge. So if you break up an improper integral into five pieces, for example, and you look at one of the pieces and find that it diverges, don't waste your time looking at the other four pieces—you already know that the whole integral diverges.

Here's an important case: what if there are no problem spots? That is, suppose you have an integral $\int_a^b f(x) \, dx$ such that the region $[a, b]$ of integration is bounded (so no ∞ or $-\infty$), and f is bounded on all of the closed interval $[a, b]$. Well, then, as we saw in Section 20.1 of the previous chapter, f has no problem spots, so we know that the integral $\int_a^b f(x) \, dx$ converges. In summary, **if there are no problem spots, the integral automatically converges!** So, for example,

$$\int_0^{100} \frac{\ln(x+1)}{x^4 + x^2 + 1}\, dx$$

converges since the integrand is bounded on the bounded region $[0, 100]$—that is, there are no problem spots. Don't get suckered into using any fancy tests in an example like this.

21.1.2 How to deal with negative function values

If $f(x)$ takes on negative values for some x in $[a, b]$, which often happens when trig functions or logs are present, you need to take special care. Luckily you can often reduce matters to integrals with only positive integrands. Here are three ways to deal with negative function values:

1. If the integrand $f(x)$ is both positive and negative as x ranges over $[a, b]$, you should consider trying the *absolute convergence test*. As we saw in Section 20.6 of the previous chapter, this says that

   $$\text{if} \quad \int_a^b |f(x)|\, dx \quad \text{converges, then so does} \quad \int_a^b f(x)\, dx.$$

   This test is particularly useful for investigating improper integrals involving trig functions when the region of integration is unbounded. The example

   $$\int_1^\infty \frac{\sin(x)}{x^2}\, dx$$

   from Section 20.6 of the previous chapter is of this type. Recall that the way to start is to consider the absolute version of the integral, namely,

   $$\int_1^\infty \frac{|\sin(x)|}{x^2}\, dx;$$

   you don't need absolute values around the denominator since it's always positive. Then show that this new integral converges (see page 448 for the details) and conclude that the original integral converges as well, by the absolute convergence test. In general, don't forget this important point: the absolute convergence test only helps you show that an integral converges. That is, **you cannot use the absolute convergence test to show that an integral diverges!**

2. Suppose that the integrand $f(x)$ is **always** negative (or zero) on $[a, b]$. That is, $f(x) \le 0$ on $[a, b]$. If this is true, you can write

   $$\int_a^b f(x)\, dx = -\int_a^b (-f(x))\, dx.$$

   So what? Well, $-f(x)$ is now always nonnegative, so you can use the comparison test or the p-test to see whether $\int_a^b (-f(x))\, dx$ converges or diverges. Of course, if this integral converges, so does $\int_a^b f(x)\,dx$, and similarly if $\int_a^b (-f(x))\,dx$ diverges, so does $\int_a^b f(x)\,dx$. Here's an example: consider

$$\int_0^{1/2} \frac{1}{x^2 \ln(x)}\,dx.$$

There's certainly a problem spot at $x = 0$. The thing to realize is that $\ln(x)$ is actually negative for x between 0 and 1, so it's a good idea to start out by writing

$$\int_0^{1/2} \frac{1}{x^2 \ln(x)}\,dx = -\int_0^{1/2} \frac{-1}{x^2 \ln(x)}\,dx.$$

Actually, since $\ln(x)$ is the negative part here, you can replace $-\ln(x)$ by $|\ln(x)|$ as follows:

$$\int_0^{1/2} \frac{1}{x^2 \ln(x)}\,dx = -\int_0^{1/2} \frac{1}{x^2 |\ln(x)|}\,dx.$$

Now we can just worry about

$$\int_0^{1/2} \frac{1}{x^2 |\ln(x)|}\,dx.$$

Unfortunately you'll have to wait until page 474 to see that this last integral diverges. The conclusion will then be that the original integral diverges as well. Note that the absolute convergence test doesn't work in this case, since that test can only be used to show an improper integral converges.

3. If neither of the previous two cases seems to apply, you may be able to use the formal definition of the improper integral to see what's going on. An example of this is

$$\int_0^\infty \cos(x)\,dx,$$

which we looked at on page 448.

This is not the end of the story. There are slightly freaky improper integrals which converge, but which are not absolutely convergent.* These sorts of improper integrals seem to come up quite often in actual physics and engineering applications, but they are beyond the scope of this book. So, it's time to go back and review the integral tests.

21.2 Summary of Integral Tests

The most valuable tools you have at your disposal are the comparison test, the limit comparison test, and the p-test. We looked at these tests from a

---

*For example, $\int_1^\infty \sin(x)/x\,dx$ converges but $\int_1^\infty |\sin(x)|/x\,dx$ diverges. Kudos to you if you can work out why either of these assertions are true.

theoretical point of view in the previous chapter; here are the statements once again, for reference. **In all the tests below, the integrand $f(x)$ is assumed to be positive on the region of integration.**

- **Comparison test, divergence version:** if you think that $\int_a^b f(x)\,dx$ diverges, find a smaller function whose integral also diverges. That is, find a nonnegative function g such that $f(x) \geq g(x)$ on (a, b), and such that $\int_a^b g(x)\,dx$ diverges. Then

$$\int_a^b f(x)\,dx \geq \int_a^b g(x)\,dx = \infty,$$

so $\int_a^b f(x)\,dx$ diverges.

- **Comparison test, convergence version:** if you think that $\int_a^b f(x)\,dx$ converges, find a larger function whose integral also converges. That is, find a function g such that $f(x) \leq g(x)$ for all x in (a, b), and such that $\int_a^b g(x)\,dx$ converges. Then

$$\int_a^b f(x)\,dx \leq \int_a^b g(x)\,dx < \infty,$$

so $\int_a^b f(x)\,dx$ also converges.

Beware of the whoop-di-doo case! This was discussed in Section 20.3 of the previous chapter, and arises if you get the above inequalities the wrong way around. The comparison test just doesn't work if you screw up the direction of the inequalities.

As an alternative to the comparison test, there is the limit comparison test. This is useful when you can find a function which behaves just like the integrand near the problem spot. In Section 20.4.1 in the previous chapter, we made the following definition:

$$f(x) \sim g(x) \text{ as } x \to a \qquad \text{means the same thing as} \qquad \lim_{x \to a} \frac{f(x)}{g(x)} = 1.$$

The definition also applies if you replace both instances of $x \to a$ by $x \to \infty$ (or $x \to -\infty$). In any case, if your integrand f is really nasty and you can find a nicer function g such that $f(x) \sim g(x)$ as x approaches the problem spot, you're in business! That's because the limit comparison test says that whatever goes for g also goes for f. More precisely, here are two versions of the test depending on whether the problem spot is infinite or finite:

- **Limit comparison test, ∞ version:** find a simpler nonnegative function g with no problem spots in $[a, \infty)$, such that $f(x) \sim g(x)$ as $x \to \infty$. Then
  - if $\int_a^\infty g(x)\,dx$ converges, so does $\int_a^\infty f(x)\,dx$; whereas
  - if $\int_a^\infty g(x)\,dx$ diverges, so does $\int_a^\infty f(x)\,dx$.

Of course, you can change the region $[a, \infty)$ into $(-\infty, b]$ and everything still works. There's also a version which applies when the problem spot is at some finite value a, which is at the left endpoint of the region of integration:

- **Limit comparison test, finite version:** find a simpler nonnegative function g with no problem spots on $(a, b]$ so that $f(x) \sim g(x)$ as $x \to a$. Then

  - if $\int_a^b g(x)\,dx$ converges, so does $\int_a^b f(x)\,dx$; whereas
  - if $\int_a^b g(x)\,dx$ diverges, so does $\int_a^b f(x)\,dx$.

Needless to say, this is also true if the only problem spot is at the right endpoint $x = b$ instead of $x = a$, provided that $f(x) \sim g(x)$ as $x \to b$ (not a).

So it's up to us to pluck an appropriate function g out of thin air to use as a comparison. It turns out that a lot of problems can be solved simply by taking $g(x)$ to be equal to $1/x^p$ for some appropriately chosen p. The convergence or divergence of the integral of such a function is precisely stated by the p-test:

- **p-test, $\int^\infty$ version:** for any finite $a > 0$, the integral

$$\boxed{\int_a^\infty \frac{1}{x^p}\,dx \text{ converges if } p > 1 \text{ and diverges if } p \leq 1.}$$

- **p-test, $\int_0$ version:** for any finite $a > 0$, the integral

$$\boxed{\int_0^a \frac{1}{x^p}\,dx \text{ converges if } p < 1 \text{ and diverges if } p \geq 1.}$$

Learn all these tests well—they are your friends.

21.3 Behavior of Common Functions near ∞ and $-\infty$

OK, it's now time to answer the most important question of them all: how do you choose the comparison function g? This depends on whether the problem spot is at $\pm\infty$, 0, or some other finite value, so we'll consider these cases separately. In almost all the cases we'll look at, we are just restating limits and inequalities that we've seen earlier, then applying these principles to investigating improper integrals. Now let's start by looking at how common functions behave near ∞ or $-\infty$.

21.3.1 Polynomials and poly-type functions near ∞ and $-\infty$

As far as polynomials are concerned, **the highest power dominates** as $x \to \infty$ or $x \to -\infty$. More precisely, suppose that p is a polynomial; then it's true that

$$\boxed{\text{if the highest-degree term of } p(x) \text{ is } ax^n, \text{ then } p(x) \sim ax^n \text{ as } x \to \infty \text{ or as } x \to -\infty.}$$

For example, we have

$$x^5 + 4x^4 + 1 \sim x^5 \qquad \text{as } x \to \infty.$$

Don't take my word for it: you can check this by showing that the ratio of the quantities $x^5 + 4x^4 + 1$ and x^5 has limit 1 as $x \to \infty$. Here's how it works:

$$\lim_{x \to \infty} \frac{x^5 + 4x^4 + 1}{x^5} = \lim_{x \to \infty} \left(1 + \frac{4}{x} + \frac{1}{x^5} \right) = 1.$$

We also discussed the above principle in Section 4.3 of Chapter 4.

If p is a poly-type function instead of a polynomial, a similar principle applies. (See Section 4.4 in Chapter 4 if you want to learn more about poly-type functions.) For example, to understand the behavior of $3\sqrt{x} - 2\sqrt[3]{x} + 4$ as $x \to \infty$, write it as $3x^{1/2} - 2x^{1/3} + 4$; then since the highest power is $1/2$, we can say that $3\sqrt{x} - 2\sqrt[3]{x} + 4 \sim 3\sqrt{x}$ as $x \to \infty$. (That's not true as $x \to -\infty$, since you can't take the square root of a negative number!)

Sometimes the highest power isn't easily identifiable. Here's an example: $\sqrt{x^4 + 8x^3 - 9} - x^2$ is a poly-type function of x which seems to have highest power 4, but of course you have to take the square root—which knocks the power down to 2. By the time you cancel out the x^2 terms, the highest power is pretty weird. We'll see how to deal with a problem like this at the end of this section.

Since we have many new asymptotic relations, we can use the limit comparison test to analyze a lot of improper integrals. For example, consider

$$\int_1^\infty \frac{1}{2 + 20\sqrt{x}} \, dx \quad \text{and} \quad \int_0^\infty \frac{1}{x^5 + 4x^4 + 1} \, dx.$$

In both cases, ∞ is the only problem spot. Let's look at the first integral. The denominator $2 + 20\sqrt{x}$ may be written as $2 + 20x^{1/2}$; here $1/2$ is the highest power. So it's true that $2 + 20\sqrt{x} \sim 20x^{1/2}$ as $x \to \infty$, and it follows that

$$\frac{1}{2 + 20\sqrt{x}} \sim \frac{1}{20x^{1/2}} \quad \text{as } x \to \infty.$$

Now, the integral

$$\int_1^\infty \frac{1}{20x^{1/2}} \, dx$$

diverges by the p-test, so by the limit comparison test, the integral

$$\int_1^\infty \frac{1}{2 + 20\sqrt{x}} \, dx$$

also diverges. As for the second integral above, since $x^5 + 4x^4 + 1 \sim x^5$ as $x \to \infty$, the same is true for the reciprocals:

$$\frac{1}{x^5 + 4x^4 + 1} \sim \frac{1}{x^5} \quad \text{as } x \to \infty.$$

Now, we have to be careful! We'd like to say that the integral we want behaves exactly like the integral $\int_0^\infty 1/x^5 \, dx$; the difficulty here is that this integral now has an extra problem spot at $x = 0$. In fact, this integral diverges, but only because of the problem spot at 0. This would lead to the wrong answer altogether. In order to avoid these inanities, we should have started by splitting the original integral into the pieces

$$\int_0^1 \frac{1}{x^5 + 4x^4 + 1}\, dx \quad \text{and} \quad \int_1^\infty \frac{1}{x^5 + 4x^4 + 1}\, dx.$$

The first of these integrals converges because there are no problem spots. As for the second, we have

$$\frac{1}{x^5 + 4x^4 + 1} \sim \frac{1}{x^5} \qquad \text{as } x \to \infty;$$

since $\int_1^\infty 1/x^5\, dx$ converges, so does the integral

$$\int_1^\infty \frac{1}{x^5 + 4x^4 + 1}\, dx.$$

Both pieces converge, so our original integral converges too. Beware of this situation—it arises often, so make sure that you split up the integral. Basically, if the "limit comparison function" g has a problem spot that the original function doesn't, you have to split up the original integral to avoid introducing a new problem spot. Normally the new integrand $g(x)$ will be of the form $1/x^p$, so you just need to avoid $x = 0$ when you have a problem spot at ∞, just as in our example.

Here's another example: let's investigate

$$\int_2^\infty \frac{3x^5 + 2x^2 + 9}{x^6 + 22x^4 + \sqrt{4x^{13} + 18x}}\, dx.$$

This is a little more complicated. The only problem spot is at ∞. The numerator of the integrand is easy to handle: $3x^5 + 2x^2 + 9 \sim 3x^5$ as $x \to \infty$. As for the denominator, first note that $\sqrt{4x^{13} + 18x} \sim \sqrt{4x^{13}} = 2x^{13/2}$ as $x \to \infty$. Since $13/2$ is greater than 6, the $\sqrt{4x^{13} + 18x}$ term actually dominates the rest of the denominator, $x^6 + 22x^4$, so the whole denominator is asymptotic to $2x^{13/2}$ as $x \to \infty$. Putting this all together, we get

$$\frac{3x^5 + 2x^2 + 9}{x^6 + 22x^4 + \sqrt{4x^{13} + 18x}} \sim \frac{3x^5}{2x^{13/2}} = \frac{3}{2}\frac{1}{x^{3/2}} \qquad \text{as } x \to \infty.$$

Since the p-test shows that the integral

$$\frac{3}{2}\int_2^\infty \frac{1}{x^{3/2}}\, dx$$

converges, so does our original integral, by the limit comparison test.

Finally, consider

$$\int_9^\infty \frac{1}{\sqrt{x^4 + 8x^3 - 9} - x^2}\, dx.$$

As we discussed above, the highest power in the denominator is difficult to pin down, since $\sqrt{x^4}$ and $-x^2$ cancel out. So, we have to multiply top and bottom by the conjugate expression of the denominator. (We've used this trick many times before; see Section 4.2 in Chapter 4 for some examples.) We get:

$$\int_9^\infty \frac{1}{\sqrt{x^4 + 8x^3 - 9} - x^2} \, dx$$

$$= \int_9^\infty \frac{1}{\sqrt{x^4 + 8x^3 - 9} - x^2} \times \frac{\sqrt{x^4 + 8x^3 - 9} + x^2}{\sqrt{x^4 + 8x^3 - 9} + x^2} \, dx;$$

I leave it to you to simplify this to

$$\int_9^\infty \frac{\sqrt{x^4 + 8x^3 - 9} + x^2}{8x^3 - 9} \, dx.$$

The denominator is easy to handle: $8x^3 - 9 \sim 8x^3$ as $x \to \infty$. How about the numerator? Well, $x^4 + 8x^3 - 9 \sim x^4$, so $\sqrt{x^4 + 8x^3 - 9} \sim x^2$, and finally $\sqrt{x^4 + 8x^3 - 9} + x^2 \sim 2x^2$ (all as $x \to \infty$). The last statement was a little tricky, since you're not allowed to add or subtract asymptotic relations. To justify the statement, we need to show that the ratio of the quantities $\sqrt{x^4 + 8x^3 - 9} + x^2$ and $2x^2$ goes to 1 as $x \to \infty$. Here's how:

$$\lim_{x \to \infty} \frac{\sqrt{x^4 + 8x^3 - 9} + x^2}{2x^2} = \lim_{x \to \infty} \frac{1}{2} \left(\frac{\sqrt{x^4 + 8x^3 - 9}}{x^2} + \frac{x^2}{x^2} \right).$$

Now drag the x^2 on the denominator into the square root (as x^4) and simplify to see that the above limit is

$$\lim_{x \to \infty} \frac{1}{2} \left(\sqrt{\frac{x^4 + 8x^3 - 9}{x^4}} + 1 \right) = \lim_{x \to \infty} \frac{1}{2} \left(\sqrt{1 + \frac{8}{x} - \frac{9}{x^4}} + 1 \right)$$

$$= \frac{1}{2} (\sqrt{1 + 0 - 0} + 1) = 1.$$

This proves that $\sqrt{x^4 + 8x^3 - 9} + x^2 \sim 2x^2$ as $x \to \infty$. Now we can return to our original integrand and write

$$\frac{1}{\sqrt{x^4 + 8x^3 - 9} - x^2} = \frac{\sqrt{x^4 + 8x^3 - 9} + x^2}{8x^3 - 9} \sim \frac{2x^2}{8x^3} = \frac{1}{4x} \qquad \text{as } x \to \infty.$$

Let's use the limit comparison test; since $\int_9^\infty 1/4x \, dx$ diverges, so does the original integral. By the way, would you have guessed that the original integrand is asymptotic to $1/4x$ as $x \to \infty$? It's not so easy to see ... so if you want to use the fact that the highest power dominates, make sure you have one and only one clear highest power!

21.3.2 Trig functions near ∞ and $-\infty$

Perhaps the only really useful thing we can say here is that

$$\boxed{|\sin(A)| \le 1} \qquad \text{and} \qquad \boxed{|\cos(A)| \le 1}$$

for **any** real number A. It's not much, but it's better than nothing. (The other trig functions have too many vertical asymptotes, so they don't satisfy similar inequalities.) There are two main applications of the above inequalities. One is that you can use the comparison test in many cases. For example, does the integral

$$\int_5^\infty \frac{|\sin(x^4)|}{\sqrt{x}+x^2}\,dx$$

converge or diverge? Well, let's start by using $|\sin(x^4)| \le 1$. Note that it doesn't matter that we are taking the sine of x^4 instead of A—the sine (or cosine) of **anything** is no more than 1 in absolute value. So, we have

$$\int_5^\infty \frac{|\sin(x^4)|}{\sqrt{x}+x^2}\,dx \le \int_5^\infty \frac{1}{\sqrt{x}+x^2}\,dx.$$

Great—we got rid of all the trig in the expression. The only problem spot in the right-hand integral is at ∞. Since the highest power dominates for large x, we have $\sqrt{x}+x^2 \sim x^2$ as $x \to \infty$. Now take reciprocals to see that

$$\frac{1}{\sqrt{x}+x^2} \sim \frac{1}{x^2} \qquad \text{as } x \to \infty.$$

By the p-test, we know that $\int_5^\infty 1/x^2\,dx$ converges, so the limit comparison test tells us that

$$\int_5^\infty \frac{1}{\sqrt{x}+x^2}\,dx$$

also converges. Finally, we see that

$$\int_5^\infty \frac{|\sin(x^4)|}{\sqrt{x}+x^2}\,dx \le \int_5^\infty \frac{1}{\sqrt{x}+x^2}\,dx < \infty,$$

so our original integral converges by the comparison test.

The other nice application of the facts that $|\sin(A)| \le 1$ and $|\cos(A)| \le 1$ is that you can treat the sine or cosine of anything as inconsequential compared to any positive power of x, at least as $x \to \infty$ or $x \to -\infty$. For example,

$$2x^3 - 3x^{0.1} + \sin(100x^{200}) \sim 2x^3 \qquad \text{as } x \to \infty.$$

Why? Because the sine term is laughably small compared to $2x^3$ when x is a large number. To be more precise, we have

$$\lim_{x\to\infty} \frac{2x^3 - 3x^{0.1} + \sin(100x^{200})}{2x^3} = \lim_{x\to\infty} \left(1 - \frac{3}{2x^{2.9}} + \frac{\sin(100x^{200})}{2x^3} \right).$$

The term $3/2x^{2.9}$ goes to 0 as $x \to \infty$; the main point is that you can use the sandwich principle to show that

$$\lim_{x\to\infty} \frac{\sin(100x^{200})}{2x^3} = 0.$$

I'll leave the details to you, because we looked at similar examples way back in Section 7.1.3 of Chapter 7. In any case, we have shown that

$$\lim_{x\to\infty} \frac{2x^3 - 3x^{0.1} + \sin(100x^{200})}{2x^3} = 1.$$

This proves that

$$2x^3 - 3x^{0.1} + \sin(100x^{200}) \sim 2x^3 \qquad \text{as } x \to \infty$$

after all. This would be useful if you want to understand whether or not the following integral converges:

$$\int_8^\infty \frac{1}{2x^3 - 3x^{0.1} + \sin(100x^{200})}\, dx.$$

By the limit comparison test and the above asymptotic relation, the integral behaves the same as $\int_8^\infty 1/2x^3\, dx$ does. Since this last integral converges by the p-test, so does our original integral above.

21.3.3 Exponentials near ∞ and $-\infty$

Here's a really useful principle: **exponentials grow faster than polynomials**. We first saw this in Section 9.4.4 of Chapter 9. There we expressed the principle in the form

$$\lim_{x \to \infty} \frac{x^n}{e^x} = 0,$$

where n is any positive number, even a very large one. Now consider the function f defined by $f(x) = x^n/e^x$. We know that $f(0) = 0$; also, the above limit says that $f(x) \to 0$ as $x \to \infty$. So how large could $f(x)$ possibly be when $x \ge 0$? It starts at 0, has no vertical asymptotes, and goes back down to have a horizontal asymptote at $y = 0$. There must be some maximum height that the graph of $y = f(x)$ gets to. Let's call it C; this means that $f(x) = x^n/e^x \le C$ for all $x \ge 0$. (Note that you get a different C for each n, but that doesn't really affect us at all.) Now, writing $1/e^x$ as e^{-x} and dividing both sides by x^n, we get the useful inequality

$$\boxed{e^{-x} \le \frac{C}{x^n} \qquad \text{for all } x > 0.}$$

As we noted in Section 9.4.4 of Chapter 9, the same is true if you replace e^{-x} by $e^{-p(x)}$, where $p(x)$ is any polynomial-type expression that goes to infinity when $x \to \infty$, and also if the base e is replaced by any other number greater than 1. For example, the same inequality is true if e^{-x} is replaced by $2^{-5x^5 + \sqrt{x^3 + 3}}$. The important point is that you get to choose any n you like, and you often have to be careful that you make it large enough. For example, consider

$$\int_1^\infty x^3 e^{-x}\, dx.$$

The good news is that the integrand is positive and there are no problem spots except for ∞. The bad news is that the x^3 factor grows quickly as $x \to \infty$. However, the e^{-x} factor decays (to 0) very fast and actually beats the x^3 factor to a pulp. To see this, we'll notice that

$$e^{-x} \le \frac{C}{x^5}.$$

This is just the above boxed inequality, with n chosen to be 5. Why 5? Because it works:

$$\int_1^\infty x^3 e^{-x}\, dx \leq \int_1^\infty x^3 \frac{C}{x^5}\, dx = C \int_1^\infty \frac{1}{x^2}\, dx < \infty.$$

We have used the p-test to show that $C \int_1^\infty 1/x^2\, dx$ converges. The comparison test now shows that the original integral converges as well. Now, how did I know to use x^5? What would happen if I used, say, $e^{-x} \leq C/x^4$ instead? It doesn't work:

$$\int_1^\infty x^3 e^{-x}\, dx \leq \int_1^\infty x^3 \frac{C}{x^4}\, dx = C \int_1^\infty \frac{1}{x}\, dx = \infty.$$

We are firmly in whoop-di-doo territory here, since we have just shown that the original integral is either finite or infinite, that is, we have shown absolutely nothing. On the other hand, if we'd used $x^{4.0001}$, it would have worked. Why? Convince yourself that the exponent you choose can be any number greater than 4, and the argument still works. In practice, it's good to choose a number 2 more than the power you are trying to kill. Here we wanted to kill x^3, so we used $e^{-x} \leq C/x^5$.

An important point: it is wrong, wrong, wrong to write $x^3 e^{-x} \sim e^{-x}$ as $x \to \infty$. It simply isn't true! If it were, then you could cancel out the positive quantity e^{-x} to conclude that $x^3 \sim 1$ as $x \to \infty$, and this is just crazy talk. So you should use the comparison test, not the limit comparison test, in the previous example.

Now look at this integral:

$$\int_{10}^\infty (x^{1000} + x^2 + \sin(x)) e^{-x^2 + 6}\, dx.$$

Here we need to do a bit of work. The integrand looks as if it might be oscillating between positive and negative values because of the $\sin(x)$ term, but that's not true because $\sin(x)$ isn't big enough to affect the positivity of $x^{1000} + x^2$ when $x \geq 10$. In any case, the first observation is that we have $x^{1000} + x^2 + \sin(x) \sim x^{1000}$ as $x \to \infty$, since the x^2 and $\sin(x)$ terms get their butts kicked by the x^{1000} term. (See the previous section if you want to learn how to provide a slightly more technical explanation!) So we can multiply by $e^{-x^2 + 6}$ to see that

$$(x^{1000} + x^2 + \sin(x)) e^{-x^2 + 6} \sim x^{1000} e^{-x^2 + 6} \qquad \text{as } x \to \infty.$$

Using the limit comparison test, we only need to know whether

$$\int_{10}^\infty x^{1000} e^{-x^2 + 6}\, dx$$

converges or diverges; our original integral will do the same thing. Now we have to be careful, since the exponential term $e^{-x^2 + 6}$ doesn't obey a useful asymptotic rule. We have to use basic comparison here. You see, x^{1000} really grows, but $e^{-x^2 + 6}$ really really really decays. Let's use

$$e^{-x^2 + 6} \leq \frac{C}{x^{1002}}$$

(see, 1002 is 2 more than 1000) to get that

$$x^{1000}e^{-x^2+6} \leq x^{1000} \times \frac{C}{x^{1002}} = \frac{C}{x^2}.$$

So, using the comparison test,

$$\int_{10}^{\infty} x^{1000}e^{-x^2+6}\,dx \leq C\int_{10}^{\infty}\frac{1}{x^2}\,dx < \infty$$

(where the last convergence follows by the p-test). Untangling our logic, we now know that the integral

$$\int_{10}^{\infty} x^{1000}e^{-x^2+6}\,dx$$

converges, so by the limit comparison test,

$$\int_{10}^{\infty} (x^{1000}+x^2+\sin(x))e^{-x^2+6}\,dx$$

also converges.

How about e^x near $-\infty$? Well, this is the same thing as understanding the behavior of e^{-x} near ∞! For example, to investigate

$$\int_{-\infty}^{-4} x^{1000}e^x\,dx,$$

first make the change of variables $t = -x$. Since $dt = -dx$, we have

$$\int_{-\infty}^{-4} x^{1000}e^x\,dx = -\int_{\infty}^{4}(-t)^{1000}e^{-t}\,dt = \int_{4}^{\infty} t^{1000}e^{-t}\,dt.$$

Here we used the minus sign provided by the dt to switch the bounds of integration around. I leave it to you to show that this last integral converges.

Here's a trick question: does

$$\int_{4}^{\infty} x^{1000}e^x\,dx$$

converge or diverge? Well, both factors of the integrand blow up as $x \to \infty$, so of course it diverges! To be really precise, you can easily say that $x^{1000}e^x \geq 1$ whenever $x \geq 4$ (in fact, that is the understatement of the century). So we have

$$\int_{4}^{\infty} x^{1000}e^x\,dx \geq \int_{4}^{\infty} 1\,dx = \infty.$$

Make sure you believe that the right-hand integral diverges. (It should be pretty self-evident, but you can check it using the formal definition or even the p-test with $p = 0$.) In any case, the comparison test now shows that the original integral diverges.

Let's also consider what happens when you add an exponential and a polynomial. As you might expect, if the exponential becomes large, then it dominates the polynomial. For example, to analyze

$$\int_{9}^{\infty} \frac{x^{10}}{e^x - 5x^{20}}\,dx,$$

first take a look at the denominator $e^x - 5x^{20}$. The e^x term should dominate the $5x^{20}$ term, so we should have $e^x - 5x^{20} \sim e^x$ as $x \to \infty$. We can prove this by looking at the limit of the ratio:

$$\lim_{x \to \infty} \frac{e^x - 5x^{20}}{e^x} = \lim_{x \to \infty} \left(1 - \frac{5x^{20}}{e^x}\right) = 1 - 0 = 1.$$

(Here we used the limit from the very beginning of this section.) Anyway, since $e^x - 5x^{20} \sim e^x$ as $x \to \infty$, we also have

$$\frac{x^{10}}{e^x - 5x^{20}} \sim \frac{x^{10}}{e^x} \qquad \text{as } x \to \infty;$$

so let's look at

$$\int_9^\infty \frac{x^{10}}{e^x}\, dx = \int_9^\infty e^{-x} x^{10}\, dx$$

instead. I leave it to you to show that this integral converges by using the comparison test along with the inequality $e^{-x} \le C/x^{12}$; so our original integral converges by the limit comparison test.

Finally, consider the following integral:

$$\int_{18}^\infty \frac{x^2}{7^x - 4^x}\, dx.$$

We'd better work out what happens to the denominator $7^x - 4^x$. Here both terms 7^x and 4^x are exponentials, but the one with the highest base should dominate. That is, $7^x - 4^x \sim 7^x$ as $x \to \infty$. To see why, look at the limit of the ratio:

$$\lim_{x \to \infty} \frac{7^x - 4^x}{7^x} = \lim_{x \to \infty} \left(1 - \frac{4^x}{7^x}\right) = \lim_{x \to \infty} \left(1 - \left(\frac{4}{7}\right)^x\right).$$

In Section 9.4.4 of Chapter 9, we saw that

$$\lim_{x \to \infty} r^x = 0 \qquad \text{if } 0 \le r < 1.$$

This is what we need to show that $(4/7)^x \to 0$ as $x \to \infty$: just replace r by $4/7$. So we have

$$\lim_{x \to \infty} \frac{7^x - 4^x}{7^x} = \lim_{x \to \infty} \left(1 - \left(\frac{4}{7}\right)^x\right) = 1 - 0 = 1.$$

This shows that $7^x - 4^x \sim 7^x$ as $x \to \infty$, as we wanted. So we also get an asymptotic relation for our original integrand:

$$\frac{x^2}{7^x - 4^x} \sim \frac{x^2}{7^x} \qquad \text{as } x \to \infty.$$

I now leave it to you to use the inequality $7^{-x} \le C/x^4$ to show that

$$\int_{18}^\infty \frac{x^2}{7^x}\, dx = \int_{18}^\infty 7^{-x} x^2\, dx$$

converges, so our original integral also converges by the limit comparison test.

21.3.4 Logarithms near ∞

First, notice that we don't consider logarithms near $-\infty$, because you can't take the log of a negative number! So it's futile to ask what happens to $\ln(x)$ as $x \to -\infty$.

On the other hand, **logs grow slowly at** ∞. In fact, they grow more slowly than any positive power of x. In symbols, we can say that if $\alpha > 0$ is some positive number of your choosing, then no matter how small it is, we have

$$\lim_{x \to \infty} \frac{\ln(x)}{x^\alpha} = 0.$$

We looked at this principle in some detail in Section 9.4.5 of Chapter 9. By a similar argument to the one we used at the beginning of Section 21.3.3 above, you can show that there must a constant C such that

$$\boxed{\ln(x) \le Cx^\alpha \qquad \text{for all } x > 1.}$$

The same is true for logs of any base greater than 1, or if $\ln(x)$ is replaced by the log of a polynomial with positive leading coefficient.

For example, what do you make of

$$\int_2^\infty \frac{\ln(x)}{x^{1.001}} \, dx?$$

Without the $\ln(x)$ term, it would converge by the p-test. The idea is that the $\ln(x)$ term barely affects anything since it grows really slowly. That's pretty waffly, although it is definitely the right conceptual idea. To nail this question, you have to use $\ln(x) \le Cx^\alpha$, where α is so small that the x^α term doesn't destroy a nice property that the number 1.001 has: it is bigger than 1. For example, if we try $\ln(x) \le Cx^{0.5}$, we get

$$\int_2^\infty \frac{\ln(x)}{x^{1.001}} \, dx \le \int_2^\infty \frac{Cx^{0.5}}{x^{1.001}} \, dx = C \int_2^\infty \frac{1}{x^{0.501}} \, dx = \infty$$

by the p-test. Yep, it's whoop-di-doo all over again. The integral we want is less than or equal to ∞, which says nothing. Let's be more subtle and use $\ln(x) \le Cx^{0.0005}$. Now 0.0005 is a very small number—so small that when you subtract it from 1.001, you get a number which is still bigger than 1. Let's see how it works:

$$\int_2^\infty \frac{\ln(x)}{x^{1.001}} \, dx \le \int_2^\infty \frac{Cx^{0.0005}}{x^{1.001}} \, dx = C \int_2^\infty \frac{1}{x^{1.0005}} \, dx < \infty.$$

The convergence of the right-hand integral above follows from the p-test, since 1.0005 is greater than 1. Now we know that the left-hand integral converges by the comparison test. You see how subtle it is? The methodology is very similar to how we handled exponentials in Section 21.3.3 above.

Mind you, the principle that logs grow slowly isn't useful in every improper integral involving logs. Here are six improper integrals to consider:

$$\int_2^\infty \frac{\ln(x)}{x^{1.001}} \, dx, \quad \int_2^\infty \frac{1}{x^{1.001} \ln(x)} \, dx, \quad \int_2^\infty \frac{\ln(x)}{x} \, dx,$$

$$\int_2^\infty \frac{1}{x \ln(x)} \, dx, \quad \int_{3/2}^\infty \frac{\ln(x)}{x^{0.999}} \, dx, \quad \text{and} \quad \int_2^\infty \frac{1}{x^{0.999} \ln(x)} \, dx.$$

We just looked at the first one and found that it converges. Now look at the second example:

$$\int_2^\infty \frac{1}{x^{1.001}\ln(x)}\,dx.$$

Here, the integral would still converge without the $\ln(x)$ factor, but this factor actually helps when it's on the bottom! That is, when you throw the $\ln(x)$ into the denominator, you are making the denominator larger than it was before, which makes the whole integrand smaller. This helps the integral to converge. How do you write this down effectively? You need to express the idea that $\ln(x)$ is **bounded from below** when x gets large. In this case, the region of integration is $[2,\infty)$. So how small can $\ln(x)$ possibly be on this region? Since $\ln(x)$ is increasing in x, we find that $\ln(x)$ is smallest on the region $[2,\infty)$ when $x = 2$. So all we need to write is $\ln(x) \ge \ln(2)$ when $x \ge 2$. How does that help? Take reciprocals to find that

$$\frac{1}{\ln(x)} \le \frac{1}{\ln(2)}$$

when $x \ge 2$. Now divide through by $x^{1.001}$ to get our integrand on the left-hand side:

$$\frac{1}{x^{1.001}\ln(x)} \le \frac{1}{x^{1.001}\ln(2)}.$$

The comparison test now saves the day, since

$$\int_2^\infty \frac{1}{x^{1.001}\ln(x)}\,dx \le \int_2^\infty \frac{1}{x^{1.001}\ln(2)}\,dx = \frac{1}{\ln(2)}\int_2^\infty \frac{1}{x^{1.001}}\,dx < \infty.$$

Remember, $\ln(2)$ is a constant, so it can be pulled out of the integral, and the integral converges by the p-test since 1.001 is bigger than 1. So the second of the above six integrals converges. By the way, the precise number $\ln(2)$ is irrelevant—we could have just replaced $\ln(2)$ by some positive constant C without worrying about what C actually is, and the proof would still have been correct.

How about the third of our above integrals? Look at

$$\int_2^\infty \frac{\ln(x)}{x}\,dx.$$

What happens if you take out the $\ln(x)$ factor from the numerator? We know that $\int_2^\infty 1/x\,dx$ diverges. Putting the $\ln(x)$ back in the numerator just makes this worse. So the above integral should diverge. To nail this, let's use the inequality $\ln(x) \ge \ln(2)$ for $x \ge 2$ once more (or if you prefer, you could replace $\ln(2)$ by some constant $C > 0$). We get

$$\int_2^\infty \frac{\ln(x)}{x}\,dx \ge \int_2^\infty \frac{\ln(2)}{x}\,dx = \ln(2)\int_2^\infty \frac{1}{x}\,dx = \infty.$$

By the comparison test, our integral diverges.

As for the fourth integral,

$$\int_2^\infty \frac{1}{x\ln(x)}\,dx,$$

here you have to do something completely different. You see, everything is very finely balanced. Without the $\ln(x)$ factor, the integral would diverge. Since the $\ln(x)$ factor is in the denominator, it helps the integral to have a chance to converge. Does it help it enough? We'd like to use $\ln(x) \le Cx^\alpha$, but no matter how small you make α, you'll never get a comparison that works. (Try it and see!) Instead, let's use a change of variables. Let $t = \ln(x)$, so that $dt = 1/x \, dx$. When $x = 2$, we see that $t = \ln(2)$, and as $x \to \infty$, also $t \to \infty$. So

$$\int_2^\infty \frac{1}{x \ln(x)} \, dx = \int_{\ln(2)}^\infty \frac{dt}{t} = \infty,$$

where the last integral diverges by the p-test. So our original integral diverges. On the other hand, let's change the upper endpoint of the above integral from ∞ to e^{e^8}, like this:

$$\int_2^{e^{e^8}} \frac{1}{x \ln(x)} \, dx.$$

The number e^{e^8} is actually really big. My computer says that it's approximately 4×10^{1294}, which means 4 followed by 1,294 zeroes. This is an unbelievably huge number, which is essentially infinite so far as our poor human brains can comprehend. Since the integral diverges if the upper endpoint is actually ∞, you'd think that the value of the above integral should be enormous. So let's work it out. Using $t = \ln(x)$ once again, we get

$$\int_2^{e^{e^8}} \frac{1}{x \ln(x)} \, dx = \int_{\ln(2)}^{e^8} \frac{1}{t} \, dt = \ln(t) \Big|_{\ln(2)}^{e^8} = \ln(e^8) - \ln(\ln(2)) = 8 - \ln(\ln(2)).$$

Here we have used the fact that when $x = e^{e^8}$, we have $t = \ln(e^{e^8}) = e^8$. In any case, the final answer is a little under 8. That isn't large at all! This might make you think that our improper integral

$$\int_2^\infty \frac{1}{x \ln(x)} \, dx$$

converges, but as we just saw, it actually diverges. It just diverges really really slowly.

Now let's consider

$$\int_2^\infty \frac{1}{x(\ln(x))^{1.1}} \, dx.$$

If you use the substitution $t = \ln(x)$ once again, you get

$$\int_2^\infty \frac{1}{x(\ln(x))^{1.1}} \, dx = \int_{\ln(2)}^\infty \frac{dt}{t^{1.1}} < \infty,$$

where this last integral now converges by the p-test. So the new integral converges. Just throwing in a tiny extra power of $\ln(x)$ on the bottom, namely $(\ln(x))^{0.1}$, is enough to cause convergence. That's pretty whacked out. We still have two more integrals to look at. The first is

$$\int_{3/2}^\infty \frac{\ln(x)}{x^{0.999}} \, dx.$$

This is very similar to the third integral above. Without the $\ln(x)$ factor in the numerator, it would diverge, and $\ln(x)$ just makes things worse. We can't say that $\ln(x) \geq \ln(2)$ for all x in the region of integration, since that's now $[3/2, \infty)$. Who cares—just use $\ln(x) \geq \ln(3/2)$ instead:

$$\int_{3/2}^{\infty} \frac{\ln(x)}{x^{0.999}}\, dx \geq \int_{3/2}^{\infty} \frac{\ln(3/2)}{x^{0.999}}\, dx = \ln(3/2) \int_{3/2}^{\infty} \frac{1}{x^{0.999}}\, dx = \infty$$

where the last divergence follows from the p-test. Now the comparison test shows that the original integral diverges. (Once again, you could write C instead of $\ln(3/2)$ everywhere above, noting that $C > 0$.)

At last we come to the final integral in this section:

$$\int_{2}^{\infty} \frac{1}{x^{0.999} \ln(x)}\, dx.$$

One way to do this is by direct comparison with the fourth improper integral on our above list. Specifically, $x^{0.999} < x$ when $x \geq 2$, so we can take reciprocals, reversing this inequality, to get

$$\int_{2}^{\infty} \frac{1}{x^{0.999} \ln(x)}\, dx > \int_{2}^{\infty} \frac{1}{x \ln(x)}\, dx.$$

Now we already know from above this last integral diverges, so the comparison test shows that our original integral does as well. Alternatively, there's a more direct method. You see, looking at the original integral

$$\int_{2}^{\infty} \frac{1}{x^{0.999} \ln(x)}\, dx,$$

what happens if you take away the $\ln(x)$ factor? It diverges by the p-test. Putting in the $\ln(x)$ factor on the denominator helps the integral try to converge, but not very much. In fact, not enough. So you can use the principle that logs grow slowly: indeed, $\ln(x) \leq C x^{0.0005}$, so taking reciprocals, we have

$$\frac{1}{\ln(x)} \geq \frac{1}{C} \times \frac{1}{x^{0.0005}}.$$

Divide this inequality by $x^{0.999}$ and you get

$$\frac{1}{x^{0.999} \ln(x)} \geq \frac{1}{C} \times \frac{1}{x^{0.999} x^{0.0005}} = \frac{1}{C} \times \frac{1}{x^{0.9995}}.$$

Finally,

$$\int_{2}^{\infty} \frac{1}{x^{0.999} \ln(x)}\, dx \geq \frac{1}{C} \int_{2}^{\infty} \frac{1}{x^{0.9995}}\, dx = \infty,$$

where the last integral diverges by the p-test. So the original integral diverges too. Notice that we again had to pick the power 0.0005 small enough; we could have used any small but positive number so that when you add it to 0.999, you don't get something greater than or equal to 1. Otherwise you will be in whoop-di-doo territory again.

21.4 Behavior of Common Functions near 0

We now know all about how polynomials, trig functions, exponentials, and logarithms behave at infinity. Now let's see what happens to them near zero.

21.4.1 Polynomials and poly-type functions near 0

For polynomials, **the lowest power dominates** as $x \to 0$. This is the opposite of what happens as $x \to \infty$! To be more precise, suppose that p is a polynomial; then it's true that

> if the lowest-degree term of $p(x)$ is bx^m, then $p(x) \sim bx^m$ as $x \to 0$.

For example, $5x^4 - x^3 + 2x^2 \sim 2x^2$ as $x \to 0$. Let's check this by showing that the limit of the ratio is 1:

$$\lim_{x \to 0} \frac{5x^4 - x^3 + 2x^2}{2x^2} = \lim_{x \to 0} \left(\frac{5x^2}{2} - \frac{x}{2} + 1 \right) = 0 - 0 + 1 = 1.$$

For poly-type functions, it's not always easy to find the lowest-degree term, but the general principle still holds water. So, for example, $x^2 + \sqrt{x} \sim \sqrt{x}$ as $x \to 0^+$, since $\sqrt{x} = x^{1/2}$ and $1/2$ is smaller than 2. (By the way, it's as $x \to 0^+$ because you can't take the square root of a negative number.) The principle even works if constants are present—they are really multiples of x^0, which is a very low-degree term! So, for example, $2x^{1/3} + 4 \sim 4$ as $x \to 0$, as $4x^0$ has a lower exponent than $2x^{1/3}$.

Let's look at some examples of improper integrals. First, consider

$$\int_0^5 \frac{1}{x^2 + \sqrt{x}}\, dx.$$

The only problem spot is at $x = 0$. Now we know that

$$\frac{1}{x^2 + \sqrt{x}} \sim \frac{1}{\sqrt{x}} \qquad \text{as } x \to 0^+;$$

since $\int_0^5 1/\sqrt{x}\, dx$ converges (by the p-test), so does

$$\int_0^5 \frac{1}{x^2 + \sqrt{x}}\, dx$$

(by the limit comparison test). So our integral converges, and it's all because of the $\sqrt{x}$ term. Without it, we'd only have $1/x^2$, and the integral of this over $[0, 5]$ diverges. So the $\sqrt{x}$ term saves the day. But wait! At this point I want you to look back at page 460 and see how we saw that the integral

$$\int_5^\infty \frac{1}{x^2 + \sqrt{x}}\, dx$$

also converges. What's important in this last integral is the x^2 term, not the $\sqrt{x}$ term. Without the x^2 term, this last integral would diverge. So the full integral we looked at right at the beginning of Section 21.1.1 above,

$$\int_0^\infty \frac{1}{x^2 + \sqrt{x}}\, dx,$$

converges because both the following pieces converge:

$$\int_0^5 \frac{1}{x^2 + \sqrt{x}}\, dx \quad \text{and} \quad \int_5^\infty \frac{1}{x^2 + \sqrt{x}}\, dx.$$

The problem spot at 0 is OK because of the $\sqrt{x}$ term and the problem spot at ∞ is OK because of the x^2 term. Nice, huh?

How about this one:

$$\int_0^1 \frac{x+3}{x + x^5}\, dx?$$

Well, the problem spot is again at $x = 0$. Now $x + 3 \sim 3$ and $x + x^5 \sim x$ as $x \to 0$, so

$$\frac{x+3}{x + x^5} \sim \frac{3}{x} \quad \text{as } x \to 0.$$

The improper integral $\int_0^1 3/x\, dx$ diverges by the p-test; the limit comparison test now shows that our original integral

$$\int_0^1 \frac{x+3}{x + x^5}\, dx$$

diverges as well.

21.4.2 Trig functions near 0

Here are some very useful facts:

$$\boxed{\sin(x) \sim x, \qquad \tan(x) \sim x, \qquad \text{and} \qquad \cos(x) \sim 1 \qquad \text{as } x \to 0.}$$

These are just restatements of limits we've already looked at in Chapter 7:

$$\lim_{x \to 0} \frac{\sin(x)}{x} = 1, \qquad \lim_{x \to 0} \frac{\tan(x)}{x} = 1, \qquad \text{and} \qquad \lim_{x \to 0} \cos(x) = 1.$$

(If the cosine limit bothers you, write $\cos(x)$ as $\cos(x)/1$ to see that $\cos(x) \sim 1$ as $x \to 0$ after all.) Beware: these asymptotic relations only work with products and quotients, **not** sums and differences. For instance, you cannot write $\sin(x) - x \sim 0$ as $x \to 0$; see the end of Section 20.4.1 in the previous chapter for a more thorough discussion of this.

Let's look at some examples. Consider

$$\int_0^1 \frac{1}{\tan(x)}\, dx \quad \text{and} \quad \int_0^1 \frac{1}{\sqrt{\tan(x)}}\, dx.$$

These look pretty similar, but appearances can be deceptive. We're going to use $\tan(x) \sim x$ (as $x \to 0$) for both integrals. I'll let you fill in the details, but here's the basic idea: for the first integral, use $1/\tan(x) \sim 1/x$ (as $x \to 0$) and the limit comparison test to see that the integral diverges. On the other hand, to do the second integral, use $1/\sqrt{\tan(x)} \sim 1/\sqrt{x}$ (as $x \to 0^+$) and the limit comparison test to see that this integral converges.

Here's another example: how about

$$\int_0^1 \frac{\sin(x)}{x^{3/2}}\, dx?$$

Without the $\sin(x)$ factor, we don't have a hope of convergence, since $3/2$ is greater than 1 and the integral would diverge by the p-test. But the $\sin(x)$ factor saves the day:

$$\frac{\sin(x)}{x^{3/2}} \sim \frac{x}{x^{3/2}} = \frac{1}{x^{1/2}} \qquad \text{as } x \to 0^+.$$

Since $\int_0^1 1/x^{1/2}\, dx$ converges, the limit comparison test shows that our original integral converges. What's interesting about this example is that the integral

$$\int_1^\infty \frac{\sin(x)}{x^{3/2}}\, dx$$

also converges, but for completely different reasons. Here the problem spot is at ∞, and we have to use an absolute integral instead. A direct comparison of the absolute integral gives

$$\int_1^\infty \frac{|\sin(x)|}{x^{3/2}}\, dx \le \int_1^\infty \frac{1}{x^{3/2}}\, dx < \infty,$$

so our integral converges (we have used the p-test, the comparison test, and the absolute convergence test). Note that the power $3/2$ is good at ∞ ($1/2$ would be bad!) and that this time the sine function didn't help (or hurt, for that matter). Incidentally, we have now shown that

$$\int_0^\infty \frac{\sin(x)}{x^{3/2}}\, dx$$

converges—can you see why?

A word of warning: just because we're looking at the behavior as $x \to 0$ doesn't mean that the problem spot has to be at 0. It might even be at ∞, as the following example shows:

$$\int_1^\infty \sin\left(\frac{1}{x}\right) dx.$$

Here the problem spot is at ∞, but $1/x$ becomes very small as $x \to \infty$. So in the relation $\sin(x) \sim x$ as $x \to 0$, replace x by $1/x$ to see that $\sin(1/x) \sim 1/x$ as $1/x \to 0$. Of course, as $x \to \infty$, we know that $1/x \to 0$, so we have shown that

$$\sin\left(\frac{1}{x}\right) \sim \frac{1}{x} \qquad \text{as } x \to \infty.$$

Now you can use the limit comparison test to say that the above integral diverges, since $\int_1^\infty 1/x\, dx$ diverges.

21.4.3 Exponentials near 0

In some sense, **exponentials have no effect at 0**. More precisely,

$$\boxed{e^x \sim 1 \qquad \text{and} \qquad e^{-x} \sim 1 \qquad \text{as } x \to 0.}$$

This is just another way of saying that

$$\lim_{x \to 0} e^x = 1 \qquad \text{and} \qquad \lim_{x \to 0} e^{-x} = 1.$$

For example, the improper integral

$$\int_0^1 \frac{e^x}{x \cos(x)}\, dx$$

diverges, because

$$\frac{e^x}{x \cos(x)} \sim \frac{1}{x \cdot 1} = \frac{1}{x} \qquad \text{as } x \to 0.$$

(You get to fill in the rest of the details.) Beware: this only applies to the exponential of a small quantity (like x or $-x$). An example of a tricky integral where you could trip up is

$$\int_0^1 \frac{e^{-1/x}}{x^5}\, dx.$$

It would be wrong to write $e^{-1/x} \sim 1$, since $1/x \to \infty$ as $x \to 0^+$. We should really use the techniques from Section 21.3.3 above. In particular, there we saw that

$$e^{-\text{large stuff}} \le \frac{C}{(\text{same large stuff})^n}$$

for any n. If the large stuff is $1/x$ (remember, x is small and positive so $1/x$ is large), then this becomes

$$e^{-1/x} \le \frac{C}{(1/x)^n} = C x^n$$

for any n. Now I leave it to you to see that any choice of n which is greater than 4 will work. For example, taking $n = 5$, you get

$$\int_0^1 \frac{e^{-1/x}}{x^5}\, dx \le \int_0^1 \frac{C x^5}{x^5}\, dx = C \int_0^1 1\, dx < \infty,$$

where the last integral obviously converges because there are no problem spots (in fact, the integral is just 1). That was a pretty tough question, by the way.

Here's another possible trap. In the integral

$$\int_0^2 \frac{dx}{\sqrt{e^x - 1}},$$

you might be tempted to use the relation $e^x \sim 1$ as $x \to 0$ to try and write $e^x - 1 \sim 0$ as $x \to 0$. This last relation can't be true, since you're not allowed to divide by 0. We need to be cleverer. In Section 20.4.1 of the previous chapter, we used the classic limit

$$\lim_{x \to 0} \frac{e^x - 1}{x} = 1$$

from Section 9.4.2 of Chapter 9 to conclude that

$$\boxed{e^x - 1 \sim x \qquad \text{as } x \to 0.}$$

It follows that

$$\frac{1}{\sqrt{e^x - 1}} \sim \frac{1}{\sqrt{x}} \qquad \text{as } x \to 0^+.$$

Now the limit comparison test shows that the original integral converges.

21.4.4 Logarithms near 0

Here the principle is that **logs go to $-\infty$ slowly** as $x \to 0^+$. Let's make things go to ∞ instead by taking absolute values, remembering that $\ln(x)$ is negative when $0 < x < 1$. So the idea is that no matter how small $\alpha > 0$ is, there's some constant C such that

$$\boxed{|\ln(x)| \le \frac{C}{x^\alpha} \qquad \text{for all } 0 < x < 1.}$$

This follows from the limit

$$\lim_{x \to 0^+} x^\alpha \ln(x) = 0,$$

which we looked at in Section 9.4.6 of Chapter 9 (except we used a instead of α). The argument is very similar to the one we used at the beginning of Section 21.3.3 above.

So, to understand

$$\int_0^1 \frac{|\ln(x)|}{x^{0.9}} \, dx,$$

we use a new variety of the same trick that we've used several times before. Without the $|\ln(x)|$ term, the integral would converge. We need a power so small so that when you add it to 0.9 you are still below the critical power 1. Let's try $\alpha = 0.05$. The above boxed inequality now says that we have $|\ln(x)| \le C/x^{0.05}$, so

$$\frac{|\ln(x)|}{x^{0.9}} \le \frac{C/x^{0.05}}{x^{0.9}} = \frac{C}{x^{0.9}x^{0.05}} = \frac{C}{x^{0.95}}.$$

Now you can use the comparison test and p-test to finish off the problem and show that the above integral converges. I want you to convince yourself that if we picked α to be anything greater than or equal to 0.1, we'd be in the whoop-di-doo case. By the way, we have now automatically seen that

$$\int_0^1 \frac{\ln(x)}{x^{0.9}} \, dx$$

converges, since it's just the negative of the original integral.

For another example, consider

$$\int_0^{1/2} \frac{1}{x^2 |\ln(x)|}\, dx.$$

If the $|\ln(x)|$ factor weren't there, this would diverge by the p-test. The $|\ln(x)|$ tries to help the integral to converge, but it can't help very much, since it's only a log, and logarithms grow slowly. So we still expect the integral to diverge. To get the math right, note that since $|\ln(x)| \le C/x^\alpha$, we can take reciprocals to see that $1/|\ln(x)| \ge x^\alpha/C$. Once again we have to choose α to be small enough so that we avoid the whoop-di-doo case. We have

$$\frac{1}{x^2 |\ln(x)|} \ge \frac{x^\alpha}{Cx^2},$$

so we will be OK as long as $\alpha \le 1$. (Why?) In fact, with $\alpha = 1$, the right-hand side becomes $1/Cx$, and you can proceed from here to see that the integral diverges. Note that the integral

$$\int_0^{1/2} \frac{1}{x^2 \ln(x)}\, dx$$

also diverges (to $-\infty$) since it is the negative of the original integral.
One final example: how about

$$\int_0^{1/2} \frac{1}{x^{0.9} |\ln(x)|}\, dx?$$

Now the integral converges without the $|\ln(x)|$ factor, but throwing this large quantity into the denominator just helps the integral converge faster. So we just need to find the minimum of $|\ln(x)|$ on $(0, 1/2]$; think about it and convince yourself that the minimum occurs when $x = 1/2$, and so whenever $0 < x \le 1/2$, we have $|\ln(x)| \ge |\ln(1/2)| = \ln(2)$. Finally, take reciprocals and divide by $x^{0.9}$ to get

$$\frac{1}{x^{0.9} |\ln(x)|} \le \frac{1}{x^{0.9} \ln(2)}$$

for all $0 < x \le 1/2$. Now you just need to apply the comparison test and the p-test to see that the original integral converges.

21.4.5 The behavior of more general functions near 0

In Section 24.2.2 of Chapter 24, we'll learn about Maclaurin series. If you haven't seen this yet, don't worry about it! Make a note to come back and read this section after you've learned all about Maclaurin series. Anyway, the basic idea is that if a function has a Maclaurin series which converges to the function near 0, then the function is asymptotic to the lowest-order term in the series as $x \to 0$. That is,

$$\text{if } f(x) = a_n x^n + a_{n+1} x^{n+1} + \cdots, \text{ then } f(x) \sim a_n x^n \text{ as } x \to 0.$$

Consider the following examples:

$$\int_0^1 \frac{dx}{1 - \cos(x)} \qquad \text{and} \qquad \int_0^1 \frac{dx}{(1 - \cos(x))^{1/3}}.$$

We know that $\cos(x) \sim 1$ as $x \to 0$, but that doesn't tell us a thing about $1 - \cos(x)$. One way to deal with this quantity is to use the Maclaurin series for $\cos(x)$:

$$\cos(x) = 1 - \frac{x^2}{2!} + \frac{x^4}{4!} - \cdots ;$$

this can be rearranged to write

$$1 - \cos(x) = \frac{x^2}{2} - \frac{x^4}{24} + \cdots .$$

So, by the above principle, the lowest-degree term on the right-hand side dominates and we can write

$$1 - \cos(x) \sim \frac{x^2}{2} \qquad \text{as } x \to 0.$$

By the way, this agrees with an example in Section 7.1.2 of Chapter 7 where we showed that

$$\lim_{x \to 0} \frac{1 - \cos(x)}{x^2} = \frac{1}{2}.$$

In any case, I leave it to you as an exercise to use the above asymptotic relation to show that the first of our above integrals diverges whereas the second one converges.

21.5 How to Deal with Problem Spots Not at 0 or ∞

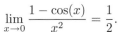

If a problem spot occurs at some finite value other than 0, do a substitution. Specifically:

- If the only problem spot in $\int_a^b f(x) \, dx$ occurs at $x = a$, make the substitution $t = x - a$. Note that $dt = dx$. The new integral has a problem spot at 0 only.
- If the only problem spot in $\int_a^b f(x) \, dx$ occurs at $x = b$, make the substitution $t = b - x$. Note that $dt = -dx$. Use the minus sign to switch the limits of integration. The new integral should have a problem spot at 0 only.

For example, on page 436, we looked at

$$\int_0^3 \frac{1}{x(x - 1)(x + 1)(x - 2)} \, dx.$$

We split this into five integrals, each with only one problem spot, and claimed that they all diverge. One such piece (we called it I_5) is

$$\int_2^3 \frac{1}{x(x - 1)(x + 1)(x - 2)} \, dx.$$

Here the problem spot is at $x = 2$, so let's substitute $t = x - 2$. Since $x = t + 2$, the integral becomes

$$\int_0^1 \frac{1}{(t+2)(t+1)(t+3)t} \, dt.$$

The bounds of integration are now 0 and 1, and the problem spot has been shifted over to 0. Now we can use the fact that the lowest-degree term in any polynomial dominates near 0 to write

$$t + 2 \sim 2, \qquad t + 1 \sim 1, \qquad \text{and} \qquad t + 3 \sim 3 \qquad \text{as } t \to 0.$$

We can combine these facts to see that

$$\frac{1}{(t+2)(t+1)(t+3)t} \sim \frac{1}{2 \times 1 \times 3 \times t} = \frac{1}{6t} \qquad \text{as } t \to 0.$$

The limit comparison test and p-test now show that the above integral diverges.

Another piece of the original integral (we called it I_4) is

$$\int_{3/2}^2 \frac{1}{x(x-1)(x+1)(x-2)} \, dx.$$

Now the problem spot is at $x = 2$, which is the right-hand limit of integration. So substitute $t = 2 - x$. When $x = 3/2$, we see that $t = 1/2$, and when $x = 2$, $t = 0$. Since $dt = -dx$ and $x = 2 - t$, we have

$$\int_{3/2}^2 \frac{1}{x(x-1)(x+1)(x-2)} \, dx = -\int_{1/2}^0 \frac{1}{(2-t)(1-t)(3-t)(-t)} \, dt$$

$$= \int_0^{1/2} \frac{1}{(2-t)(1-t)(3-t)(-t)} \, dt.$$

In this last integral, we have used the minus sign from the equation $dx = -dt$ in order to switch the limits of integration (as described in Section 16.3 of Chapter 16). Anyway, it's not too hard to see that

$$\frac{1}{(2-t)(1-t)(3-t)(-t)} \sim -\frac{1}{6t} \qquad \text{as } t \to 0,$$

so the above integral diverges (again by the limit comparison test and p-test—you get to fill in the details, taking care to handle the negative integrand correctly). In fact, you should now try to show that the other three integrals (I_1, I_2, and I_3 on page 436) diverge.

CHAPTER 22 _____

Sequences and Series: Basic Concepts

Here's the good news: infinite series are pretty similar to improper integrals. So a lot, but not all, of the relevant techniques are shared and we don't need to reinvent the wheel. In order to define what an infinite series is, we'll also need to look at sequences. Just as in the case of improper integrals, I'm devoting two chapters to sequences and series: this first chapter covers general principles, while the next one is more practical and contains methods for solving problems. If you're reading this for the first time, go ahead and check out the details of this chapter. For review, a quick glance over the main points should suffice before moving on to the examples in the next chapter. Here are the topics for this chapter:

- convergence and divergence of sequences;
- two important sequences;
- the connection between limits of sequences and limits of functions;
- convergence and divergence of series, and how to handle geometric series;
- the nth term test for series;
- the connection between series and improper integrals; and
- an introduction to the ratio test, root test, integral test, and alternating series test.

Again, **this chapter is mostly theoretical!** If it's examples you want, most are in the next chapter.

22.1   Convergence and Divergence of Sequences

A sequence is a collection of numbers in order. It might have a finite number of terms, or it might go on forever, in which case it is called an *infinite sequence*. For example,

$$0, 1, -1, 2, -2, 3, -3, \ldots$$

is an infinite sequence which incidentally includes every integer, positive and negative. Sequences are normally written using subscript notation, where a_1 denotes the first element of the series, a_2 the second, a_3 the third, and so on.

(Sometimes a_0 is the first element, a_1 the second, and so on. Also, we don't have to use a; for example, b_n or any other letter is fair game.) So in the above example, $a_1 = 0$, $a_2 = 1$, $a_3 = -1$, $a_4 = 2$, and so on. Often a sequence is given by a formula, such as

$$a_n = \frac{\sin(n)}{n^2}$$

for $n = 1, 2, \ldots$ This defines the sequence

$$\frac{\sin(1)}{1^2}, \frac{\sin(2)}{2^2}, \frac{\sin(3)}{3^2}, \frac{\sin(4)}{4^2}, \ldots$$

Given an infinite sequence, our main focus is going to be on the limiting behavior of the values of the sequence as the index n tends to infinity. That is, what happens to the sequence as you look farther and farther along it? In math notation, does

$$\lim_{n \to \infty} a_n$$

exist, and if so, what is it? By the way, we haven't really defined the above limit, but the definition is not much different from the definition of $\lim_{x \to \infty} f(x)$ for a function f. (See Section A.3.3 of Appendix A for the actual definition.) The basic idea is that the statement

$$\lim_{n \to \infty} a_n = L$$

means that a_n might wander around for a little while, but eventually gets very close—as close as you like—to L and stays at least as close to L for ever after. If there's such a number L, then the sequence $\{a_n\}$ *converges*; otherwise it *diverges*. Just like functions, sequences can diverge to ∞ or $-\infty$, or they can oscillate around (possibly crazily) and not get close to any particular value. For example, the above sequence $0, 1, -1, 2, -2, \ldots$ diverges; it does not diverge to ∞ or $-\infty$, but instead oscillates between positive and negative numbers of bigger and bigger absolute value.

By the way, as we did with functions, we sometimes say that $a_n \to L$ as $n \to \infty$. This means the same thing as saying $\lim_{n \to \infty} a_n = L$.

22.1.1   The connection between sequences and functions

Consider the sequence given by

$$a_n = \frac{\sin(n)}{n^2},$$

which we looked at earlier. This is closely related to the function f defined by

$$f(x) = \frac{\sin(x)}{x^2}.$$

In fact, a_n is equal to $f(n)$ for each positive integer n. So if we can establish that $\lim_{x \to \infty} f(x)$ exists, then we'll know that the sequence $\{a_n\}$ has the same limit. The sequence inherits the limiting properties of the function. There's also a connection to horizontal asymptotes: remember that if $\lim_{x \to \infty} f(x) = L$, then the graph of $y = f(x)$ has a horizontal asymptote at $y = L$.

Inspired by these observations, we can easily extend some other properties of limits of functions to the case of sequences. For example, if you have two convergent sequences $\{a_n\}$ and $\{b_n\}$, such that $a_n \to L$ and $b_n \to M$ as $n \to \infty$, then the sum $a_n + b_n$ gives a new sequence which converges to $L + M$. The same goes for differences, products, quotients (provided that $M \neq 0$, since you can't divide by 0), and constant multiples. This isn't very profound, but it's pretty darn useful.

Another useful fact is that the sandwich principle, otherwise known as the squeeze principle, also works for sequences. (See Section 3.6 in Chapter 3 for a full review of the sandwich principle.) Specifically, suppose you have a sequence $\{a_n\}$ which you suspect converges to some limit L. Try to find a bigger sequence $\{b_n\}$ and a smaller sequence $\{c_n\}$ which both converge to L, and you now have what you want. In math-speak, if $c_n \leq a_n \leq b_n$ and both $b_n \to L$ and $c_n \to L$ as $n \to \infty$, then $a_n \to L$ as $n \to \infty$ as well. For our sequence given by

$$a_n = \frac{\sin(n)}{n^2}$$

from above, you can use the sandwich principle by dividing the classic inequalities $-1 \leq \sin(n) \leq 1$ by n^2 to get

$$\frac{-1}{n^2} \leq \frac{\sin(n)}{n^2} \leq \frac{1}{n^2}$$

for all n. The sequences given by $b_n = 1/n^2$ and $c_n = -1/n^2$ both converge to 0 as $n \to \infty$, so our sequence a_n is squished between them and goes to 0 as well. That is,

$$\lim_{n \to \infty} \frac{\sin(n)}{n^2} = 0.$$

Another property which transfers over from functions is that **continuous functions respect limits**. What the heck does that mean? Well, suppose that $a_n \to L$ as $n \to \infty$. Then if f is a function which is continuous at $x = L$, we can say that $f(a_n) \to f(L)$ as $n \to \infty$. The limit relation is preserved when you hit everything with f. For example, what is

$$\lim_{n \to \infty} \cos\left(\frac{\sin(n)}{n^2}\right)?$$

We have just seen that

$$\frac{\sin(n)}{n^2} \to 0 \qquad \text{as } n \to \infty;$$

since the cosine function is continuous at 0, we can hit both sides with cosine to get

$$\cos\left(\frac{\sin(n)}{n^2}\right) \to \cos(0) = 1 \qquad \text{as } n \to \infty.$$

One more useful tool that we can borrow from the theory of functions is l'Hôpital's Rule. (See Section 14.1 in Chapter 14.) The problem with using the rule on a sequence is that you can't differentiate the quantity a_n with respect to the variable n, since n has to be an integer. Indeed, when you differentiate a function f with respect to a variable x, the idea is that you wobble x around a little and see what happens to $f(x)$. You can't wobble an integer around because it wouldn't be an integer any more. So, if you want to use l'Hôpital's Rule, you have to embed the sequence in a suitable function first. For example, if $a_n = \ln(n)/\sqrt{n}$, you can find $\lim_{n\to\infty} a_n$ by letting

$$f(x) = \frac{\ln(x)}{\sqrt{x}}$$

and then finding $\lim_{x\to\infty} f(x)$ by using l'Hôpital's Rule. Note that this is an ∞/∞ case, so you can use the rule here. Differentiate the top and bottom separately to get

$$\lim_{x\to\infty} \frac{\ln(x)}{\sqrt{x}} \overset{\text{l'H}}{=} \lim_{x\to\infty} \frac{1/x}{1/2\sqrt{x}} = \lim_{x\to\infty} \frac{2}{\sqrt{x}} = 0.$$

Since the function limit is 0, the sequence a_n also converges to 0 as $n \to \infty$. (We could also have used the fact that logs grow slowly at ∞ to find the above limit; just apply the formula at the beginning of Section 21.3.4 in the previous chapter with $\alpha = 1/2$.)

22.1.2 Two important sequences

Pick some constant number r and consider the sequence given by $a_n = r^n$ starting at $n = 0$. This is a *geometric progression*. Notice that each term is a constant multiple of the previous one. Let's look at a few examples of geometric progressions:

- if $r = 0$, the sequence is just $0, 0, 0, \ldots$, which clearly converges to 0;
- if $r = 1$, the sequence is just $1, 1, 1, \ldots$, which clearly converges to 1;
- if $r = 2$, the sequence is $1, 2, 4, 8, \ldots$, which evidently diverges to ∞;
- if $r = -1$, the sequence is $1, -1, 1, -1, 1, \ldots$, which diverges, but not to ∞ or $-\infty$, because it keeps on oscillating back and forth between -1 and 1—in other words, the limit does not exist (DNE);
- if $r = -2$, the sequence is $1, -2, 4, -8, \ldots$, which diverges in the same way (the limit does not exist)—in fact, this time the oscillations are even wilder;
- if $r = 1/2$, the sequence is $1, 1/2, 1/4, 1/8, \ldots$, which converges to 0; and finally,
- if $r = -1/2$, the sequence is $1, -1/2, 1/4, -1/8, \ldots$, which also converges to 0, despite the oscillations, since these oscillations eventually become as small as you like.

These are all special cases of the general rule, which is as follows:

$$\lim_{n\to\infty} r^n \begin{cases} = 0 & \text{if } -1 < r < 1, \\ = 1 & \text{if } r = 1, \\ = \infty & \text{if } r > 1, \\ \text{DNE} & \text{if } r \leq -1. \end{cases}$$

Here's how to justify the above limit. First, when $r \geq 0$, the limit follows from the similar limit involving r^x that we looked at in Section 9.4.4 of Chapter 9 (see the middle box). The tricky case occurs when $r < 0$, since the resulting sequence oscillates. To deal with the oscillations, notice that

$$-|r|^n \leq r^n \leq |r|^n$$

for all n. The nice thing about this is that the sequences $\{-|r|^n\}$ and $\{|r|^n\}$ aren't oscillating. In fact, if $-1 < r < 0$, then $|r| < 1$, so we already know that both of the sequences converge to 0; now we can just use the sandwich principle to see that $r^n \to 0$ as well. Finally, if $r \leq -1$, then r^n cannot possibly converge, since it keeps flipping between positive numbers greater than equal to 1 and negative numbers less than or equal to -1. The resulting limit does not exist (DNE) due to these oscillations. (The situation here is similar to the limit $\lim_{x \to \infty} \sin(x)$ which we looked at in Section 3.4 of Chapter 3; also check out Section A.3.4 of Appendix A.)

Geometric progressions don't have to start at 1. If we set $a_n = ar^n$, where a is some constant, then the first term a_0 is equal to a. You can find $\lim_{n \to \infty} ar^n$ by multiplying the values of $\lim_{n \to \infty} r^n$ in the box above by a. Most important, if $-1 < r < 1$, then $\lim_{n \to \infty} ar^n$ is 0 regardless of the value of a.

Having spent a lot of time on geometric progressions, let's look at the limit of another sequence very quickly. In particular, if k is any constant, then

$$\lim_{n \to \infty} \left(1 + \frac{k}{n}\right)^n = e^k.$$

This follows directly from the limit at the beginning of Section 9.2.3 in Chapter 9. It's really useful to know this limit in the context of sequences, however.

22.2 Convergence and Divergence of Series

A series is just a sum. We'd like to add up all of the terms of a sequence a_n. So, instead of putting commas between the elements, you put plus signs. If the sequence is infinite, things get a little hairy—after all, what does it even mean to add up infinitely many numbers? For example, if the sequence a_n is the geometric progression $1, 1/2, 1/4, 1/8, \ldots$, then the corresponding series is $1 + 1/2 + 1/4 + 1/8 + \cdots$. We need to do something clever to handle the dots at the end, which indicate that the series goes on forever.

In general, we'd like to understand what

$$a_1 + a_2 + a_3 + \cdots$$

means. To deal with this infinite sum, let's chop it off after some large number of terms. We'll call the number of terms N, so the chopped-off series looks like this:

$$a_1 + a_2 + a_3 + \cdots + a_{N-1} + a_N.$$

This is just a sum of finitely many quantities, so it makes sense. Now, here's what we'd like to say:

$$a_1 + a_2 + a_3 + \cdots = \lim_{N \to \infty} (a_1 + a_2 + a_3 + \cdots + a_{N-1} + a_N).$$

The right-hand side looks a little weird, since the number of terms is changing as N gets larger. So let's define a new sequence, which we'll call $\{A_N\}$, by setting

$$A_N = a_1 + a_2 + a_3 + \cdots + a_{N-1} + a_N.$$

This new sequence is called the sequence of *partial sums*. The weird equation now looks like this:

$$a_1 + a_2 + a_3 + \cdots = \lim_{N \to \infty} A_N.$$

Now the right-hand side isn't so weird—it's just the limit of a sequence. If the limit exists and equals L, then we'll say that the series on the left-hand side *converges* to L. If the limit doesn't exist, then the series *diverges*.

Here's a nice analogy to understand all this stuff. I want you to imagine that you're standing at a rest stop on a long, straight highway which extends in both directions—the way you're going and the way you've just come from. The rest stop is at position 0. (We've seen this old highway before, for example in Section 5.2.2 of Chapter 5.) Unfortunately you have lost all your free will, and some guy with a megaphone is commanding you every minute to walk a certain number of feet. You can only move when he says so. If he calls out a negative number, you actually walk backward. Each time you move, we'll call it a step. (Hopefully the guy won't ask you to move 100 feet in a single step!)

The first number that megaphone man calls out is a_1, so you move from position 0 to position a_1 (the units are in feet, but I won't say that every time). The next number is a_2, so you walk forward a_2 feet. Where does that put you? At position $a_1 + a_2$, since you started at a_1. After the third number he calls out, which is of course a_3, you'll be at position $a_1 + a_2 + a_3$. The pattern should be pretty clear: after N steps of sizes a_1, a_2, a_3, and so on up to a_N, you will be at position

$$a_1 + a_2 + a_3 + \cdots + a_{N-1} + a_N.$$

This is exactly the value of the partial sum A_N which we defined above! In other words, A_N is your position after you take N steps. So, when we write

$$a_1 + a_2 + a_3 + \cdots = \lim_{N \to \infty} A_N,$$

we're saying that you can add up all the steps, provided that you eventually start homing in on a particular point on the highway. You have to get really really close to that point, never straying far away from it. You'll be making tiny little steps, tiptoeing around this point. Otherwise, there's no hope of adding up all the steps and the series will diverge.

Now it's time to bust out some sigma notation. (We looked at this in Section 15.1 in Chapter 15.) The formula for A_N becomes

$$A_N = a_1 + a_2 + a_3 + \cdots + a_{N-1} + a_N = \sum_{n=1}^{N} a_n.$$

The infinite series is written as

$$a_1 + a_2 + a_3 + \cdots = \sum_{n=1}^{\infty} a_n.$$

So, here's how to define the value of an infinite series using sigma notation:

$$\sum_{n=1}^{\infty} a_n = \lim_{N \to \infty} \sum_{n=1}^{N} a_n.$$

If the limit on the right-hand side doesn't exist, then the series on the left-hand side diverges. Remember, the right-hand side is really the limit of a sequence, so the above equation isn't as obvious as the notation makes it appear to be.

Let's just review the scenario once more before we move on. You begin with an infinite **sequence**

$$\{a_n\} = a_1, a_2, a_3, \ldots$$

and use it to construct an infinite **series**:

$$\sum_{n=1}^{\infty} a_n = a_1 + a_2 + a_3 + \cdots.$$

To understand the limiting behavior of this series, make a **new sequence** of partial sums:

$$A_N = \sum_{n=1}^{N} a_n = a_1 + a_2 + a_3 + \cdots + a_{N-1} + a_N.$$

By definition, the limit of the series is the same as the limit of the new sequence of partial sums, if the limit exists; otherwise the series diverges. Since there are two sequences and one series floating around here, make sure you understand what's what!

 By the way, we don't need to begin our series at $n = 1$. You can begin at any number, even $n = 0$. All you have to do is change the starting term in the partial sums and everything works out. Now, here's an important point: whether a series converges or diverges has nothing to do with the starting point of the series! For example, we'll see in Section 22.4.3 below that the series

$$\sum_{n=1}^{\infty} \frac{1}{n}$$

diverges. This immediately tells us that all the following series diverge as well:

$$\sum_{n=5}^{\infty} \frac{1}{n}, \qquad \sum_{n=89}^{\infty} \frac{1}{n}, \qquad \text{and even} \qquad \sum_{n=1000000}^{\infty} \frac{1}{n}.$$

To see why the first of these series diverges, just break out the first four terms of the original series, like this:

$$\sum_{n=1}^{\infty} \frac{1}{n} = \frac{1}{1} + \frac{1}{2} + \frac{1}{3} + \frac{1}{4} + \sum_{n=5}^{\infty} \frac{1}{n} = \frac{25}{12} + \sum_{n=5}^{\infty} \frac{1}{n}.$$

So the series starting at $n = 1$ and the series starting at $n = 5$ differ only by the finite constant $25/12$. Since the series starting at $n = 1$ diverges to

∞, subtracting $25/12$ isn't going to affect this at all. The series starting at $n = 5$ must also diverge. Of course, there's nothing special about 5: the same argument works for any starting point. Similarly, we'll see in Section 22.4.3 below that

$$\sum_{n=1}^{\infty} \frac{1}{n^2}$$

actually converges. This means that all of the following series automatically converge as well:

$$\sum_{n=4}^{\infty} \frac{1}{n^2}, \qquad \sum_{n=101}^{\infty} \frac{1}{n^2}, \qquad \text{and even} \qquad \sum_{n=5000000}^{\infty} \frac{1}{n^2}.$$

See if you can prove this by splitting up the original sum.

One more thing before we go on to geometric series: consider the series

$$\sum_{n=0}^{\infty} \frac{1}{n^2}.$$

We've just changed the starting point to $n = 0$, but now something annoying happens: the first term is $1/0^2$, which doesn't exist. So the above series is whacked out. It's not that it diverges; it just doesn't make sense, since the first term isn't defined. We'll always try to avoid this situation by starting at a large enough value of n so that all the terms of the series are actually defined.

22.2.1 Geometric series (theory)

Let's look at an important example of an infinite series. Suppose we start with the geometric progression $1, r, r^2, r^3, \ldots$, which we looked at in Section 22.1.2 above. We can use this sequence as the terms of an infinite series:

$$1 + r + r^2 + r^3 + \cdots = \sum_{n=0}^{\infty} r^n.$$

This is called a *geometric series*. The question is, does it converge, and if so, to what?

To find out, we'd better look at the partial sums. Pick a number N; then the partial sum A_N is given by

$$A_N = 1 + r + r^2 + r^3 + \cdots + r^{N-1} + r^N.$$

In sigma notation, we have

$$A_N = \sum_{n=0}^{N} r^n.$$

Hopefully, in your previous math studies you've seen that the above expression can be simplified as follows:

$$A_N = 1 + r + r^2 + r^3 + \cdots + r^{N-1} + r^N = \frac{1 - r^{N+1}}{1 - r}$$

as long as $r \neq 1$. (In any case, there's a proof of this formula at the bottom of this page.) Now we need to take the limit of A_N as $N \to \infty$.

First, suppose that $-1 < r < 1$. Then we saw in the first box of Section 22.1.2 above that $\lim_{N \to \infty} r^N = 0$, so replace N by $N+1$ to get $\lim_{N \to \infty} r^{N+1} = 0$ as well. So

$$\lim_{N \to \infty} A_N = \lim_{N \to \infty} \frac{1 - r^{N+1}}{1 - r} = \frac{1}{1 - r}.$$

Our geometric series converges to $1/(1 - r)$. Here's how the whole argument looks on one line, using sigma notation:

$$\sum_{n=0}^{\infty} r^n = \lim_{N \to \infty} \sum_{n=0}^{N} r^n = \lim_{N \to \infty} \frac{1 - r^{N+1}}{1 - r} = \frac{1}{1 - r}.$$

How about when r isn't between -1 and 1? It turns out that the geometric series must diverge in this case; we'll see why at the end of the next section. So, in summary:

$$\sum_{n=0}^{\infty} r^n = \frac{1}{1 - r} \quad \text{if } -1 < r < 1;$$

otherwise, if $r \geq 1$ or $r \leq -1$, the series diverges.

In the above geometric series, the first term is always 1, since $r^0 = 1$. If you start at some other number a instead, then the terms are a, ar, ar^2, and so on. So you can multiply everything by a to get a more general form of the above principle:

$$\sum_{n=0}^{\infty} ar^n = \frac{a}{1 - r} \quad \text{if } -1 < r < 1;$$

otherwise, if $r \geq 1$ or $r \leq -1$, the series diverges.

We'll see plenty of examples of how to deal with geometric series in Section 23.1 of the next chapter. Meanwhile, I promised that I'd prove the formula

$$A_N = \sum_{n=0}^{N} r^n = \frac{1 - r^{N+1}}{1 - r}$$

from above. Here's how: first, multiply the sum on the left by $(1 - r)$ to get

$$A_N(1 - r) = (1 - r) \sum_{n=0}^{N} r^n.$$

Now pull the factor of $(1 - r)$ through the sum and simplify to see that

$$A_N(1 - r) = \sum_{n=0}^{N} r^n(1 - r) = \sum_{n=0}^{N} (r^n - r^{n+1}).$$

The right-hand sum is a telescoping series—see Section 15.1.2 in Chapter 15 for a review of this—so the sum works out to be $r^0 - r^{N+1}$, or $1 - r^{N+1}$. So $A_N(1 - r) = 1 - r^{N+1}$; now all you have to do to get our formula is to divide by $(1 - r)$, which is nonzero since we assumed that $r \neq 1$.

22.3 The nth Term Test (Theory)

For a series to converge, the sequence of partial sums has to have a limit. Remember that the partial sum after N steps represents your position after you have taken N steps according to the megaphone dude's orders. (See Section 22.2 above if you don't have any idea what I'm talking about.) Anyway, if your position is going to converge to some special limiting position as you keep on taking more and more steps, then your steps have to become really really small. Otherwise you'll blunder about and not stay consistently close to the special position. It's not good enough to keep moving back and forth, close to the special position: you have to get really close, and stay really close.

So, your step sizes, which are just given by the sequence $\{a_n\}$, eventually have to become very small, at least if you want your series to converge. Mathematically, this means that you need to have $a_n \to 0$ as $n \to \infty$. This leads us to the *nth term test*:

> **nth term test:** if $\lim_{n\to\infty} a_n \neq 0$, or the limit doesn't exist, then the series $\sum_{n=1}^{\infty} a_n$ diverges.

 If $\lim_{n\to\infty} a_n = 0$, then the series may converge or it may diverge, and you have to do more work to resolve the issue. Just beware: **the nth term test cannot be used to show that a series converges!**

So this test is a sort of reality check: if the terms a_n don't tend to 0, stop right there—your series diverges. Otherwise, the problem is still open and you need to do more work. For example, we'll soon see that

$$\sum_{n=1}^{\infty} \frac{1}{n^2} \text{ converges, but } \sum_{n=1}^{\infty} \frac{1}{\sqrt{n}} \text{ diverges.}$$

In both sums, the terms converge to 0:

$$\lim_{n\to\infty} \frac{1}{n^2} = 0 \quad \text{and} \quad \lim_{n\to\infty} \frac{1}{\sqrt{n}} = 0.$$

 The nth term test doesn't apply in either case! It's only when the limits are not zero that you can use the test to say that your series diverges. Here are some examples where the test is good:

$$\sum_{n=0}^{\infty} 2^n, \quad \sum_{n=0}^{\infty} (-3)^n, \quad \text{and} \quad \sum_{n=0}^{\infty} 1.$$

You see, we have

$$\lim_{n\to\infty} 2^n = \infty, \quad \lim_{n\to\infty} (-3)^n \text{ DNE}, \quad \text{and} \quad \lim_{n\to\infty} 1 = 1.$$

All three series above diverge by the nth term test, since in each case the limit of the terms isn't 0. Actually, the series are all geometric series, with ratios 2, -3, and 1, respectively. In general, if you have a geometric series $\sum_{n=0}^{\infty} r^n$ with $r \geq 1$ or $r \leq -1$, then the terms r^n don't go to 0 as $n \to \infty$. (We saw this in Section 22.1.2 above—check out the formula in the big box.) So the nth term test tells us that any geometric series with ratio not strictly between -1 and 1 diverges.

In a convergent series, although the terms a_n must go to 0, that doesn't mean that the limit of the series is 0. For example, the geometric progression $1, 1/2, 1/4, 1/8, \ldots$ with ratio $r = 1/2$ converges to 0, and we can actually work out the value of the associated series using the formula from the previous section:

$$\sum_{n=0}^{\infty} \left(\frac{1}{2}\right)^n = \frac{1}{1-r} = \frac{1}{1-\frac{1}{2}} = 2.$$

So the underlying sequence converges to 0, but the series converges to 2. It couldn't be the other way around—if a sequence converges to 2, then by the nth term test the associated series would diverge automatically.

We'll see some other examples of the nth term test in Section 23.2 in the next chapter. Meanwhile, it's time to look at some more tests.

22.4 Properties of Both Infinite Series and Improper Integrals

It turns out that there are some connections between infinite series and improper integrals, particularly improper integrals with a problem spot at ∞. One of these connections is expressed in the integral test, which we'll look at in Section 22.5.3 below. In this section, I want to show you that all four of the tests we have for improper integrals also work for infinite series. Let's look at them one at a time.

22.4.1 The comparison test (theory)

Suppose that you have a series $\sum_{n=1}^{\infty} a_n$, where all the terms a_n are nonnegative. If you suspect that the series diverges, find a smaller series $\sum_{n=1}^{\infty} b_n$ which also diverges and your suspicion is confirmed. That is, if $0 \le b_n \le a_n$ for all n, and $\sum_{n=1}^{\infty} b_n$ diverges, so does $\sum_{n=1}^{\infty} a_n$. If instead you suspect that your original series converges, find a bigger series $\sum_{n=1}^{\infty} b_n$ which also converges, and your suspicion is confirmed. That is, if $b_n \ge a_n \ge 0$ for all n, and $\sum_{n=1}^{\infty} b_n$ converges, then so does $\sum_{n=1}^{\infty} a_n$.

This is basically the same as the comparison test for improper integrals. The justification of the series version of the test is virtually identical to that of the integral version, so I'll leave it to you to fill in the details if you feel sufficiently motivated.

 By the way, the first term in the series doesn't have to be $n = 1$: it could be anything at all. For example, consider

$$\sum_{n=3}^{\infty} \left(\frac{1}{2}\right)^n |\sin(n)|.$$

This is quite easy to deal with, using the comparison test. You see, $|\sin(n)| \le 1$

for any n, so we can write

$$\sum_{n=3}^{\infty} \left(\frac{1}{2}\right)^n |\sin(n)| \le \sum_{n=3}^{\infty} \left(\frac{1}{2}\right)^n < \infty.$$

The last sum converges as it is a geometric progression with ratio $1/2$, which is between -1 and 1. So we can use the comparison test and claim that our original series converges. We'll see some more examples of the comparison test in the next chapter.

22.4.2 The limit comparison test (theory)

In Section 20.4.1 of Chapter 20, we made the following definition:

$$f(x) \sim g(x) \text{ as } x \to \infty \qquad \text{means the same thing as} \qquad \lim_{x \to \infty} \frac{f(x)}{g(x)} = 1.$$

There's a version of this for sequences that looks almost the same:

$$a_n \sim b_n \text{ as } n \to \infty \qquad \text{means the same thing as} \qquad \lim_{n \to \infty} \frac{a_n}{b_n} = 1.$$

The limit comparison test then says that if $a_n \sim b_n$ as $n \to \infty$, and all terms a_n and b_n are finite, then $\sum_{n=1}^{\infty} a_n$ and $\sum_{n=1}^{\infty} b_n$ both converge or both diverge. You can't have one without the other. Of course, you don't have to start at $n = 1$; you could start at $n = 0$, $n = 19$, or any other finite value of n that you like. Once again, the justification of this test is almost identical to the justification of the limit comparison test for improper integrals, so I'll omit it. You can fill in the details if you like. By the way, if $a_n \sim b_n$ as $n \to \infty$, we say that the sequences are *asymptotic* to each other.

 All the properties of functions we looked at in Chapter 21 are still good for sequences. For example, consider

$$\sum_{n=0}^{\infty} \sin\left(\frac{1}{2^n}\right).$$

When n is large, $1/2^n$ becomes very small (that is, close to 0). We know that $\sin(x) \sim x$ as $x \to 0$ (see Section 21.4.2 in the previous chapter); replacing x by $1/2^n$, we see that

$$\sin\left(\frac{1}{2^n}\right) \sim \frac{1}{2^n} \qquad \text{as} \qquad \frac{1}{2^n} \to 0.$$

Now, we can rewrite $1/2^n$ as $(1/2)^n$, and also note that $1/2^n \to 0$ is equivalent to $n \to \infty$. So the above relation can be written as

$$\sin\left(\frac{1}{2^n}\right) \sim \left(\frac{1}{2}\right)^n \qquad \text{as } n \to \infty.$$

The limit comparison test then says that the two series

$$\sum_{n=0}^{\infty} \sin\left(\frac{1}{2^n}\right) \qquad \text{and} \qquad \sum_{n=0}^{\infty} \left(\frac{1}{2}\right)^n$$

both converge or both diverge. Now we know that the right-hand series converges, as it is a geometric series with ratio 1/2 (which is less than 1 in absolute value). So the left-hand series converges as well. By the way, the right-hand series converges to 2 (as we saw in Section 22.3 above); this does not mean that the left-hand series also converges to 2. We don't know what it converges to, only that it converges.

22.4.3 The p-test (theory)

There's also a p-test for series. It's basically the same as the p-test for improper integrals with problem spot at ∞. In particular, it says that

$$\sum_{n=a}^{\infty} \frac{1}{n^p} \quad \begin{cases} \text{converges} & \text{if } p > 1, \\ \text{diverges} & \text{if } p \le 1. \end{cases}$$

 The easiest proof of this uses the integral test, so I'll postpone it to Section 22.5.3 below. Some simple examples of the p-test are that

$$\sum_{n=1}^{\infty} \frac{1}{n^2} \text{ converges,} \quad \text{but } \sum_{n=1}^{\infty} \frac{1}{\sqrt{n}} \text{ diverges.}$$

The power 2 in the first series is greater than 1, so the series converges. On the other hand, since $\sqrt{n} = n^{1/2}$, we have a power of 1/2 in the second series; since 1/2 is less than or equal to 1, the series diverges.

Before we move on to the absolute convergence test, just consider the so-called *harmonic series*

$$\sum_{n=1}^{\infty} \frac{1}{n}$$

for a few minutes. This series diverges by the p-test, but we can actually show that it diverges directly. The idea is to write out a whole bunch of terms of the series and then group them in a clever way. Specifically, the above series can be written out like this:

$$1 + \frac{1}{2} + \left(\frac{1}{3} + \frac{1}{4}\right) + \left(\frac{1}{5} + \frac{1}{6} + \frac{1}{7} + \frac{1}{8}\right)$$
$$+ \left(\frac{1}{9} + \frac{1}{10} + \frac{1}{11} + \frac{1}{12} + \frac{1}{13} + \frac{1}{14} + \frac{1}{15} + \frac{1}{16}\right) + \cdots.$$

Except for the 1 and 1/2 at the beginning, each grouping has twice as many terms as the previous grouping. Now here's the main deal: the last term in each grouping is the smallest. So the above sum is bigger than

$$1 + \frac{1}{2} + \left(\frac{1}{4} + \frac{1}{4}\right) + \left(\frac{1}{8} + \frac{1}{8} + \frac{1}{8} + \frac{1}{8}\right)$$
$$+ \left(\frac{1}{16} + \frac{1}{16} + \frac{1}{16} + \frac{1}{16} + \frac{1}{16} + \frac{1}{16} + \frac{1}{16} + \frac{1}{16}\right) + \cdots.$$

In this new series, there is one term of size 1, one term of size 1/2, two terms of size 1/4, four terms of size 1/18, eight terms of size 1/16, and so on. That is, apart from the first term, each grouping adds up to exactly 1/2. So the above series is really equal to

$$1 + \frac{1}{2} + \frac{1}{2} + \frac{1}{2} + \frac{1}{2} + \cdots,$$

which diverges! Finally, the comparison test shows that the harmonic series diverges, since it is bigger than the above divergent series. Now we get for free that $\sum_{n=1}^{\infty} 1/n^p$ diverges when $p \le 1$, since $1/n^p \ge 1/n$ and you can use the comparison test again. (Try filling in the details.)

22.4.4 The absolute convergence test

Suppose you have a series $\sum_{n=1}^{\infty} a_n$ with terms a_n which are sometimes positive and sometimes negative. This kind of sucks; it makes life more difficult (or more interesting, depending on your point of view). If eventually all the terms a_n become positive, then there's no problem—you can just ignore all the terms at the beginning and start the series at the point where all the terms are positive. Remember, the beginning terms of a series have no impact on whether the series converges or diverges. Similarly, if the terms eventually become negative, you can ignore the beginning terms and end up with a series with only negative terms. Then consider the series $\sum_{n=m}^{\infty}(-a_n)$, which has all positive terms: if it converges, so does the original series, and if it diverges, so does the original series. This is because this new series is just the negative of the original series.

So, what if the series keeps switching between positive and negative terms? Some examples of this are

$$\sum_{n=3}^{\infty} \sin(n) \left(\frac{1}{2}\right)^n, \qquad \sum_{n=1}^{\infty} \frac{(-1)^n}{n^2}, \qquad \text{and} \qquad \sum_{n=1}^{\infty} \frac{(-1)^n}{n}.$$

The second and third of these series are actually *alternating series*. This means that the terms alternate between positive and negative numbers. For example, the third series can be expanded as

$$-1 + \frac{1}{2} - \frac{1}{3} + \frac{1}{4} - \frac{1}{5} + \cdots,$$

and you can clearly see that every other term is negative. On the other hand, the first series above is not alternating. Sometimes $\sin(n)$ is positive and sometimes it's negative, but it doesn't alternate. For example, $\sin(1)$, $\sin(2)$, and $\sin(3)$ are all positive (since 1, 2, and 3 are all between 0 and π), whereas $\sin(4)$, $\sin(5)$ and $\sin(6)$ are all negative.

Anyway, there is a special test to deal with alternating series, which we'll look at in Section 22.5.4 below. We still have the absolute convergence test, however, which says that if $\sum_{n=1}^{\infty} |a_n|$ converges, so does $\sum_{n=1}^{\infty} a_n$. Again, the series can start at any value of n, not necessarily $n = 1$. Let's see how this works for our above examples. For the first one,

$$\sum_{n=3}^{\infty} \sin(n) \left(\frac{1}{2}\right)^n,$$

consider the absolute version of the series:

$$\sum_{n=3}^{\infty} |\sin(n)| \left(\frac{1}{2}\right)^n.$$

Note that we only needed absolute value signs around $\sin(n)$, since the factor $(1/2)^n$ is always positive. Anyway, we already used the comparison test in Section 22.4.1 above to show that the above series converges. The absolute convergence test then says that the original series above (without the absolute values) converges too. In fact, we say that the original series *converges absolutely*. More on this in Section 22.5.4 below.

For the second series,

$$\sum_{n=1}^{\infty} \frac{(-1)^n}{n^2},$$

the absolute version is

$$\sum_{n=1}^{\infty} \frac{1}{n^2}.$$

This converges by the p-test (since $2 > 1$), so the original series converges absolutely, by the absolute convergence test.

For the third series,

$$\sum_{n=1}^{\infty} \frac{(-1)^n}{n},$$

the absolute version is

$$\sum_{n=1}^{\infty} \frac{1}{n}.$$

This diverges by the p-test, so you **cannot** apply the absolute convergence test. That is, you cannot conclude that the original series

$$\sum_{n=1}^{\infty} \frac{(-1)^n}{n}$$

diverges. All you can say is that this series does not converge absolutely. In fact, in Section 22.5.4 below, we'll see that the series does in fact converge, even though its absolute version diverges! Before we do that, however, we have a few other tests to look at.

22.5 New Tests for Series

Let's look at four tests for convergence of series which have no corresponding improper integral version: the ratio test, the root test, the integral test, and the alternating series test. We'll examine them one at a time before seeing how to apply them in the next chapter.

22.5.1   The ratio test (theory)

Here's a really really useful test which only works for series, not improper integrals. It's called the *ratio test* because it involves the ratio of successive terms of a sequence. Let's set the scene: suppose we have a series $\sum_{n=1}^{\infty} a_n$. We'd like the terms to go to 0 fast enough for this series to converge. Here's one way this can happen: suppose we consider a new sequence, which we'll call b_n, of the absolute value of ratios of successive terms of the series. That is, we let

$$b_n = \left| \frac{a_{n+1}}{a_n} \right|$$

for each n. This is a sequence, so maybe it converges to something. Now here's the result: if the sequence $\{b_n\}$ converges to a number less than 1, then we can immediately conclude that the series $\sum_{n=1}^{\infty} a_n$ converges. In fact, it converges absolutely: that is, $\sum_{n=1}^{\infty} |a_n|$ also converges. On the other hand, if the sequence $\{b_n\}$ converges to a number greater than 1, then the series $\sum_{n=1}^{\infty} a_n$ diverges. If the sequence $\{b_n\}$ converges to 1, or if it doesn't converge, then we can't say anything about the original series.

We'll look at a lot of examples of the ratio test in the next chapter, so let's just see if we can justify the test. This is a tricky argument, so don't worry if you get lost—just skip to the next section. Let's give it a try, though. We might as well assume that $a_n \geq 0$ for all n, so we can drop the absolute values. Suppose that b_n converges to a number L which is less than 1. That is, suppose that

$$\frac{a_{n+1}}{a_n} \to L < 1 \qquad \text{as } n \to \infty.$$

Well, this means that when n is large, the ratio a_{n+1}/a_n is approximately equal to L. If this ratio were exactly equal to L, then the series would be a geometric series with ratio L, which converges since $L < 1$. Since it's only equal to L in the limit, we have to be a bit more clever.

The idea is to let r be equal to the average of L and 1. Since $L < 1$, the average r lies between L and 1, so r is also less than 1. That is, $L < r < 1$. So what? Well, since the ratio a_{n+1}/a_n converges to L, eventually it must always be less than r. That is, the ratios can wander around doing whatever they like for a while, but then they get serious and start getting close to L. You can't get close to L without being less than r, since r is bigger than L. So, the point is that if you throw away enough of the series at the beginning, you can always say that a_{n+1}/a_n is less than r.

Let's see where we're at: we started off with $\sum_{n=1}^{\infty} a_n$, but we've thrown away a whole bunch of the terms at the beginning to get $\sum_{n=m}^{\infty} a_n$ for some number m. This throwing-away routine doesn't affect whether the series converges. On the other hand, it helps because we are sure that $a_{n+1}/a_n < r$ for all $n \geq m$. Another way of writing this is $a_{n+1} < ra_n$ for all $n \geq m$.

Now comes the real meat: the sequence $\{a_n\}$ is dominated by a geometric progression with ratio r. After all, to advance from one term a_n to the next term a_{n+1}, you multiply by some number less than r (since $a_{n+1} < ra_n$). On the other hand, to advance from one term of a geometric progression with ratio r to the next term, you actually do multiply by r. So if the geometric progression starts at a_m, then it pulls ahead of our sequence $\{a_n\}$ and stays ahead. (All this can be justified by using induction. Assume that $a_n < Ar^n$. Then multiply both sides by r to get $ra_n < Ar^{n+1}$. Since $a_{n+1} < ra_n$, we have $a_{n+1} < Ar^{n+1}$. Now you just have to choose A so that $a_m < Ar^m$ as well; any number greater than a_m/r^m will do.)

OK, we've shown that $a_n < Ar^n$ for some number A. This means that

$$\sum_{n=m}^{\infty} a_n \leq \sum_{n=m}^{\infty} Ar^n.$$

Since $0 \leq r < 1$, the right-hand side converges, so, by the comparison test, the left-hand side converges too. Finally, by the absolute convergence test, $\sum_{n=m}^{\infty} a_n$ also converges even if some of the terms a_n are negative after all.

Not so easy. Luckily the divergence version isn't so bad. Suppose that the ratios $|a_{n+1}/a_n|$ converge to a number L which is bigger than 1. Now if we throw away enough of the series, we can just look at $\sum_{n=m}^{\infty} |a_n|$, where m is large enough to force $|a_{n+1}/a_n| > 1$ for all $n \geq m$. This means that $|a_{n+1}| > |a_n|$ for all $n \geq m$. The terms $|a_n|$ are actually getting bigger as n gets larger, so we can't possibly have $\lim_{n \to \infty} a_n = 0$. Now we can just use the nth term test to say that $\sum_{n=m}^{\infty} a_n$ diverges, so $\sum_{n=1}^{\infty} a_n$ also diverges.

Now all that's left is to convince ourselves that everything breaks down if $L = 1$. Here's a good example of what can go wrong: consider the series $\sum_{n=1}^{\infty} 1/n^p$. Let's work out the ratio of successive terms:

$$\left| \frac{a_{n+1}}{a_n} \right| = \frac{\dfrac{1}{(n+1)^p}}{\dfrac{1}{n^p}} = \frac{n^p}{(n+1)^p} = \left(\frac{n}{n+1} \right)^p.$$

We were able to drop the absolute value signs since everything is positive. In any case, as $n \to \infty$, it's easy to see that $n/(n+1) \to 1$, so the pth power also goes to 1. That is,

$$\lim_{n \to \infty} \left| \frac{a_{n+1}}{a_n} \right| = \lim_{n \to \infty} \left(\frac{n}{n+1} \right)^p = 1^p = 1.$$

So the limit L of the ratios is 1, regardless of what p is. Now, we know that $\sum_{n=1}^{\infty} 1/n^p$ converges if $p > 1$ and diverges if $p \leq 1$. The limiting ratio $L = 1$ cannot distinguish between these two possibilities. This one example is enough to show that if $L = 1$, then the original series could converge or it could diverge: you just can't tell.

22.5.2 The root test (theory)

The root test (also called the nth root test) is a close cousin of the ratio test. Instead of considering ratios of successive terms, just consider the nth root of the absolute value of the nth term. That is, starting with a series $\sum_{n=1}^{\infty} a_n$, let's make a new sequence given by

$$b_n = |a_n|^{1/n}.$$

(Remember, raising a quantity to the power $1/n$ is the same as taking the nth root.) Now you see whether the sequence $\{b_n\}$ converges and try to find the limit. If the limit is less than 1, then the series $\sum_{n=1}^{\infty} a_n$ converges (in fact, converges absolutely). If the limit is greater than 1, the series diverges. If the limit equals 1, then you can't tell what the heck is going on and have to try something else.

Again, we'll look at an example in the next chapter. Let's try to justify what's going on here. If this seems a little nasty, just skip to the next section. Anyway, the main idea is that the test is again inspired by looking at a geometric series. Suppose that $a_n = r^n$. Then the nth root of $|a_n|$ is exactly $|r|$. So the series converges if $|r| < 1$ and diverges otherwise. Now, we don't exactly have a geometric series but it's pretty close. Let's start off with the assumption that

$$\lim_{n\to\infty} |a_n|^{1/n} = L < 1 \qquad \text{as } n \to \infty.$$

By the same logic we used in the justification of the ratio test, we let r be the average of L and 1 and realize that eventually $|a_n|^{1/n} < r$. That is, after a certain point $n = m$ in the series, $|a_n| < r^n$. So we have

$$\sum_{n=m}^{\infty} |a_n| \leq \sum_{n=m}^{\infty} r^n.$$

Since $r < 1$, the right-hand series converges and we can use the comparison test to show that the left-hand series converges as well; so $\sum_{n=1}^{\infty} a_n$ converges absolutely.

On the other hand, suppose that the limit L is greater than 1, that is,

$$\lim_{n\to\infty} |a_n|^{1/n} = L > 1 \qquad \text{as } n \to \infty.$$

Eventually for large enough n, it's always true that $|a_n|^{1/n} > 1$, which means that $|a_n| > 1$. So $\sum_{n=1}^{\infty} a_n$ diverges by the nth term test, since the terms can't go to 0.

If the limit L is exactly 1, the test is still utterly useless. Again the example $\sum_{n=1}^{\infty} 1/n^p$ illustrates this pretty clearly. I leave it to you to show that

$$\lim_{n\to\infty} \left| \frac{1}{n^p} \right|^{1/n} = \lim_{n\to\infty} n^{-p/n} = 1.$$

(Treat it as a l'Hôpital Type **C** problem; see Section 14.1.5 of Chapter 14 to learn about this type of problem.) We know the series $\sum_{n=1}^{\infty} 1/n^p$ diverges for some values of p and converges for other values of p. It follows that the root test can't possibly give any useful information, since the above limit is 1, no matter what p is.

22.5.3   The integral test (theory)

We already saw in Section 22.4 above that there's a connection between improper integrals and infinite series. The integral test really nails down this connection. In particular, suppose you have a series $\sum_{n=1}^{\infty} a_n$ whose terms a_n are **positive** and **decreasing**. By "decreasing," I mean that $a_{n+1} \leq a_n$

for all n. (Technically, I should say "nonincreasing" since the inequality isn't strict.) An example of such a series is $\sum_{n=1}^{\infty} 1/n^p$ for any $p > 0$: the terms are certainly positive, and it's easy to see that they are also decreasing. Let's draw a picture of the general situation:

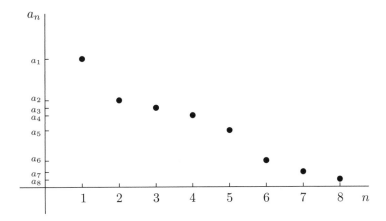

The axes are actually labeled n and a_n instead of x and y. The idea is that the height of the dot above the number n is the value of a_n. Notice that all the dots are above the x-axis (actually, the n-axis!) since all the terms a_n are positive; also, the heights are getting smaller, so the terms a_n are decreasing.

Now, imagine you can find some continuous function f that is decreasing and connects the dots:

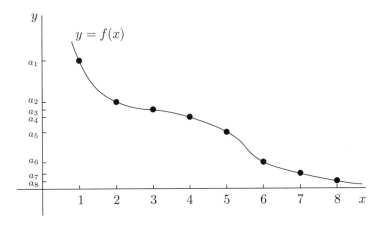

Since the curve $y = f(x)$ passes through every dot, we have $f(n) = a_n$ for all positive integers n. Now consider the integral

$$\int_1^{\infty} f(x)\, dx.$$

If that integral converges, so does the series $\sum_{n=1}^{\infty} a_n$. Why is it so? Well, let's draw some sneaky lines in the picture:

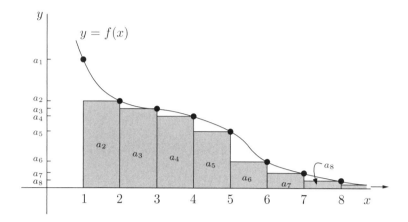

We have drawn a bunch of rectangles here which lie below the curve. Each rectangle has a base of length 1 unit, and the heights of the rectangles are a_2 units, a_3 units, a_4 units, and so on. (Poor old a_1 doesn't get a rectangle here.) The total area of all the rectangles (in square units) is $\sum_{n=2}^{\infty} a_n$. This must be some finite number by the comparison test, since

$$0 \leq \sum_{n=2}^{\infty} a_n \leq \int_1^{\infty} f(x)\,dx < \infty.$$

So $\sum_{n=2}^{\infty} a_n$ converges, and of course so does $\sum_{n=1}^{\infty} a_n$. (Remember, the beginning terms of a series do not affect the convergence!)

On the other hand, suppose that $\int_1^{\infty} f(x)$ diverges. Well, this time we need to draw different rectangles:

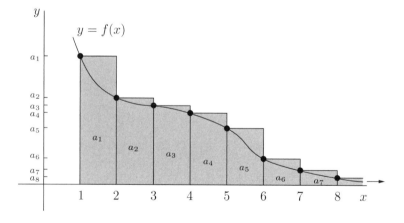

The rectangles lie above the curve in this case. The base of each rectangle still has length 1 unit, and the heights are a_1 units, a_2 units, a_3 units, and so on. (This time a_1 gets a rectangle of its own!) Since the rectangles lie above the curve, we have

$$\sum_{n=1}^{\infty} a_n \geq \int_1^{\infty} f(x)\,dx = \infty,$$

and the comparison test now shows that $\sum_{n=1}^{\infty} a_n$ diverges.

In summary, we have the **integral test**: if f is a decreasing positive function such that $f(n) = a_n$ for all positive integers n, then

$$\int_1^{\infty} f(x)\,dx \qquad \text{and} \qquad \sum_{n=1}^{\infty} a_n$$

either both converge or both diverge. Again, the series can start at any number, not just $n = 1$; just change the lower bound of the integral to match. We'll see some examples of how to use the integral test in the next chapter, but for the moment we can at least use it to prove the p-test for series, which we first saw in Section 22.4.3 above.

So, to investigate the convergence of $\sum_{n=1}^{\infty} 1/n^p$, first suppose that $p > 0$ and consider the function f defined by $f(x) = 1/x^p$ for $x > 0$. This function clearly agrees with $1/n^p$ when $x = n$, and it's also decreasing. (One way of showing this is by considering the derivative. In this case, $f'(x) = -px^{p-1}$ which is a negative quantity for $x > 0$, so f is decreasing.) Anyway, we can now use the integral test to say that

$$\int_1^{\infty} \frac{1}{x^p}\,dx \qquad \text{and} \qquad \sum_{n=1}^{\infty} \frac{1}{n^p}$$

either both converge or both diverge. Which is it? Well, when $p > 1$, the integral converges by the integral p-test, so the series does as well. When $0 < p \le 1$, the integral diverges by the integral p-test, so the series diverges as well.

How about when $p < 0$? Then you can't use the integral test, since the function f given by $f(x) = 1/x^p$ is actually increasing. You see, if $p < 0$, then we can write $p = -q$ for some $q > 0$. Then

$$\sum_{n=1}^{\infty} \frac{1}{n^p} = \sum_{n=1}^{\infty} \frac{1}{n^{-q}} = \sum_{n=1}^{\infty} n^q.$$

This last series diverges by the nth term test, since $n^q \to \infty$ (not 0) as $n \to \infty$. Finally, if $p = 0$, the series $\sum_{n=1}^{\infty} 1/n^p$ is just $\sum_{n=1}^{\infty} 1 = 1 + 1 + 1 + \cdots$, which clearly diverges. Putting everything together, we see that $\sum_{n=1}^{\infty} 1/n^p$ converges when $p > 1$ and diverges when $p \le 1$, which is exactly the p-test for series!

22.5.4   The alternating series test (theory)

Suppose you have a series $\sum_{n=1}^{\infty} a_n$ where the terms can't make up their minds whether to be positive and negative but instead keep switching sign. We already saw some examples of this in Section 22.4.4 above. Sometimes the absolute convergence test saves the day here, since if the absolute version $\sum_{n=1}^{\infty} |a_n|$ converges, so does our original series. But what if the absolute version diverges? What on earth do you do then?

This is quite a question. There's no easy answer, in general. This is a tricky little topic which has inspired much thought and discussion over the years. Let's be happy with a simple test that comes up surprisingly often in applications. Suppose that your series is alternating. Remember, this means that every second term is positive and every other term is negative. If you take any series with positive terms and multiply each term by $(-1)^n$, then you get an alternating series. (You could use $(-1)^{n+1}$ instead.) Two of the series we looked at above,

$$\sum_{n=1}^{\infty} \frac{(-1)^n}{n^2} \quad \text{and} \quad \sum_{n=1}^{\infty} \frac{(-1)^n}{n},$$

are alternating. We have already seen (in Section 22.4.4) that the first of these series converges absolutely, so it converges. The second one is more interesting. It **doesn't** converge absolutely, since its absolute version $\sum_{n=1}^{\infty} 1/n$ diverges. Amazingly, it turns out that the original series $\sum_{n=1}^{\infty} (-1)^n/n$ converges! When a series converges, but its absolute version diverges, we say that the series *converges conditionally*. So $\sum_{n=1}^{\infty} (-1)^n/n$ converges conditionally. Let's see why.

The *alternating series test* says that if a series $\sum_{n=1}^{\infty} a_n$ is alternating, and the absolute values of its terms are decreasing to 0, then the series converges. That is, we need a_n to be alternately positive and negative, and $|a_n|$ to be decreasing, and $\lim_{n \to \infty} |a_n| = 0$. In that case, the series converges. So, for example, the above series $\sum_{n=1}^{\infty} (-1)^n/n$ converges since it is alternating, and the absolute values of the terms are $\{1/n\}$, which is a decreasing sequence with limit 0. We'll summarize the test and see some more examples of the alternating series test in Section 23.7 of the next chapter.

Why does the test work? Well, first let's just do a reality check. One of the conditions is that the limit of the terms of the series has to be 0. If that isn't true, then the series diverges by the nth term test! So that condition's a no-brainer. Now, here's how the rest of it works. Consider the partial sums $\{A_N\}$, where $A_N = \sum_{n=1}^{N} a_n$. Because a_n keeps alternating between positive and negative values, the partial sums A_N wobble back and forth. Think back to the idea of the megaphone guy telling you to step back and forth: every second call he makes is a forward step, and every other call is a backward step. You might take all the forward steps with your right foot and the backward steps with your left foot. On the other hand, the step sizes (which are $|a_n|$) are getting smaller and in fact are tending toward 0. So you find yourself taking shorter and shorter steps back and forth. This means that your left and right feet are coming together. Every time you step with your left foot, your left foot is farther forward than it was before. Every time you step with your right foot, it's farther back than last time. In the limit, your feet come together at the same point, so the series converges!

We can write this mathematically by supposing that $a_1, a_3, a_5, \ldots$ are all positive and $a_2, a_4, a_6, \ldots$ are all negative. Now consider the odd partial sums A_1, A_3, A_5, and so on. That's the position of your right foot as you keep on stepping. I claim that this is a decreasing sequence. Indeed, $A_1 = a_1$, whereas $A_3 = a_1 + a_2 + a_3$, which we can also write as $A_1 + a_2 + a_3$. Now a_2 is negative, a_3 is positive, and $|a_2| \geq |a_3|$ by our assumption that the step sizes are decreasing. This means that a_2 is more negative than a_3 is positive, so $a_2 + a_3 \leq 0$. That is, $A_3 = A_1 + a_2 + a_3 \leq A_1$. Let's just repeat this argument for A_5 so we can see what's going on. You see, A_5 is the sum of the first five values of a_n. But A_3 is the sum of the first three values, so we can write $A_5 = A_3 + a_4 + a_5$. (If you know where you are after three steps—namely A_3—then you can just take the next two steps of signed length a_4 and a_5 to see where you are after five steps, which is A_5.) Anyway, $a_4 + a_5 \leq 0$, since a_4 is negative, a_5 is positive, and $|a_4| \geq |a_5|$. This means that $A_5 \leq A_3$. If you continue this process, then you find that

$$A_1 \geq A_3 \geq A_5 \geq A_7 \geq \cdots ,$$

so your right foot is indeed moving farther back as time goes by.

You can repeat the same argument (but in the opposite direction) with the even terms A_2, A_4, A_6, and so on. Try it and see if you can show that

$$A_2 \leq A_4 \leq A_6 \leq A_8 \leq \cdots ,$$

so your left foot is moving forward as time goes by. Now, here's the main point: the odd sequence $A_1, A_3, A_5, \ldots$ is decreasing, so either it drops off to $-\infty$, or it converges to some finite value. It can't drop off to $-\infty$, though, because all these terms are bigger than A_2. (Why is that true?) Similarly, the even sequence $A_2, A_4, A_6, \ldots$ is increasing, so either it blows up to ∞ or it converges. It can't blow up to ∞ since all these terms are less than A_1. (Again, why?) So both the odd and the even series converge. Since the differences $|a_n|$ between odd and even terms are getting smaller, the limits of both series must be the same! That is, the odd series decreases to the same limit that the even series increases to: your feet are moving closer and closer together until they are arbitrarily close together. That's all that you need to show that the full sequence $\{A_N\}$ of partial sums converges, which means that the original series $\sum_{n=1}^{\infty} a_n$ converges too.

So the alternating series test works. It's important that you only use it after checking that your given series is not absolutely convergent. We'll see how this works in the next chapter when we look at lots of examples.

CHAPTER 23 _____

How to Solve Series Problems

The scenario: you are given a series $\sum_{n=1}^{\infty} a_n$, and you want to know whether or not it converges. If it does converge, then perhaps you'd like to know its value (that is, what it converges to). The series has to be pretty special in order to find a nice expression for its value. Of course, the series may not start at $n = 1$ as in the above series—it could be $n = 0$ or some other value of n.

This chapter is all about giving you a blueprint of how to proceed. Here's a possible flowchart for how to approach a series:

1. **Is the series geometric?** If your series only involves exponentials like 2^n or e^{3n}, it might be a geometric series, or it might be the sum of one or more geometric series. See Section 23.1 below to see how to deal with this case.

2. **Do the terms go to 0?** If the series isn't geometric, try the **nth term test**. Check that the terms converge to 0; otherwise the series diverges by the nth term test. See Section 23.2 below for more details.

3. **Are there negative terms in the series?** If so, you may have to use the **absolute convergence test** or the **alternating series test**. See Section 23.7 at the end of this chapter for more information.

4. **Are factorials involved?** If so, use the **ratio test**. The test is also useful when there are exponentials involved but the series isn't geometric. See Section 23.3 below.

5. **Are there tricky exponentials with n in the base and the exponent?** If so, try the **root test**. In general, if it is easy to take the nth root of the term a_n, the root test is probably a winner; check out Section 23.4 below for more details.

6. **Do the terms have a factor of exactly $1/n$ as well as logarithms?** In that case, the **integral test** is probably what you want. We'll look at this test in Section 23.5 below.

7. **Do none of the above tests seem to work?** You may have to use the **comparison test** or the **limit comparison test** in conjunction with the **p-test**, as well as all the understanding of the behavior of functions

which we looked at in Chapter 21. We'll see how to apply these tests in Section 23.6 below.

The above blueprint will help guide your way through a lot of different series. It's not perfect! There are always tricks and traps that could arise. Hopefully these will be pretty rare. My advice is to master all this material, then worry about the once-in-a-blue-moon cases as you come across them in your studies. Anyway, let's get on with the details.

23.1   How to Evaluate Geometric Series

If your series only involves exponentials like 2^n or e^{3n}, it might be the sum of one or more geometric series. As we saw in the previous chapter, geometric series are simple enough that you can actually find their values (if they converge). The general form of a geometric series is $\sum_{n=m}^{\infty} ar^n$, where r is the common ratio. On page 485, we saw how to find the value of the series. Rather than learn the formula in mathematical language, I recommend learning it in words:

$$\boxed{\text{sum of infinite geometric series} = \frac{\text{first term}}{1 - \text{common ratio}}, \quad \text{if } -1 < \text{ratio} < 1.}$$

If the common ratio isn't between -1 and 1, then the series diverges. Let's see how it works. Suppose you want to find

$$\sum_{n=5}^{\infty} \frac{4}{3^n}.$$

This is a geometric series, since you can write

$$\frac{4}{3^n} = 4 \left(\frac{1}{3} \right)^n.$$

From this, we can see that the common ratio is $1/3$. This ratio is between -1 and 1, so the series converges. To what, you ask? Well, the first term occurs when $n = 5$, so it is $4/3^5$. So

$$\sum_{n=5}^{\infty} \frac{4}{3^n} = \sum_{n=5}^{\infty} 4 \left(\frac{1}{3} \right)^n = \frac{4/3^5}{1 - 1/3},$$

which works out to be $2/81$.

Here's a trickier example:

$$\sum_{n=2}^{\infty} \frac{2^{2n} - (-7)^n}{11^n}.$$

This is not a geometric series, but it can be split up into the difference of two geometric series:

$$\sum_{n=2}^{\infty} \frac{2^{2n} - (-7)^n}{11^n} = \sum_{n=2}^{\infty} \frac{2^{2n}}{11^n} - \sum_{n=2}^{\infty} \frac{(-7)^n}{11^n}.$$

Why are both these pieces geometric series? In the first series, you can replace 2^{2n} by 4^n, then express $4^n/11^n$ as $(4/11)^n$. This last trick also works in the second series, so we have

$$\sum_{n=2}^{\infty} \frac{2^{2n} - (-7)^n}{11^n} = \sum_{n=2}^{\infty} \left(\frac{4}{11}\right)^n - \sum_{n=2}^{\infty} \left(\frac{-7}{11}\right)^n.$$

Both these series converge, since their common ratios are $4/11$ and $-7/11$ (respectively) and both these numbers are between -1 and 1. So we can use the above formula. The first terms occur when $n = 2$, so they are $(4/11)^2$ and $(-7/11)^2$, respectively. All in all, the series works out to be

$$\frac{(4/11)^2}{1 - (4/11)} - \frac{(-7/11)^2}{1 - (-7/11)},$$

which simplifies to $-5/126$.

How about if we change the problem slightly? Consider

$$\sum_{n=2}^{\infty} \frac{2^{2n} - (-13)^n}{11^n}.$$

Again, we can split up the sum and group terms to rewrite this as

$$\sum_{n=2}^{\infty} \left(\frac{4}{11}\right)^n - \sum_{n=2}^{\infty} \left(\frac{-13}{11}\right)^n.$$

Don't even bother working out the first series—just notice that it converges, but the second one diverges since the ratio $-13/11$ isn't between -1 and 1. The sum of a divergent series and a convergent series must diverge!

As we've seen, geometric series are fairly easy to deal with. If your series isn't geometric, keep working your way down this list, beginning with the nth term test.

23.2   How to Use the nth Term Test

Always try the nth term test first! The test says:

> if $\lim\limits_{n\to\infty} a_n \neq 0$, or the limit doesn't exist, then the series $\sum\limits_{n=1}^{\infty} a_n$ diverges.

If the terms of your series don't tend to 0, the series must diverge. If the terms do tend to 0, the series might converge or it might diverge: you have to do more work. **This test cannot be used to show that a series converges.** Anyway, it goes without saying that you should quickly check to see that the terms of your series do tend to 0 before wasting time with other methods. For

example, to investigate the series

$$\sum_{n=1}^{\infty} \frac{n^2 - 3n + 7}{4n^2 + 2n + 1},$$

don't screw around with any of the tests below—simply note that

$$\lim_{n \to \infty} \frac{n^2 - 3n + 7}{4n^2 + 2n + 1} = \frac{1}{4},$$

so the terms of the series don't go to 0 and the original series diverges by the nth term test.

If the terms of your series do tend to 0, you'll have to try one of the other tests to see what's going on. Before going any further, take a quick look at whether your series has any negative terms. This can happen if the terms involve regular old minus signs, or factors like $(-1)^n$, or trig functions (especially $\sin(n)$ or $\cos(n)$). If there are negative terms, check out Section 23.7 first to see how to handle them. Otherwise, if everything's positive, proceed through the tests below.

23.3   How to Use the Ratio Test

Use the ratio test whenever factorials are involved. Remember, factorials involve exclamation points, such as in $n!$ or $(2n+5)!$. The ratio test is also often useful when there are exponentials around, such as 2^n or $(-5)^{3n}$. Here's the statement of the test, summarized from what we found in Section 22.5.1 of the previous chapter:

> if $L = \lim_{n \to \infty} \left| \dfrac{a_{n+1}}{a_n} \right|$, then $\displaystyle\sum_{n=1}^{\infty} a_n$ converges absolutely if $L < 1$,
>
> and diverges if $L > 1$; but if $L = 1$ or the limit doesn't exist, then the ratio test tells you nothing.

To use the ratio test, always start with the following framework:

$$\lim_{n \to \infty} \left| \frac{a_{n+1}}{a_n} \right| = \lim_{n \to \infty} \left| \frac{n\text{th term with } n \text{ replaced by } (n+1)}{n\text{th term}} \right|.$$

Make sure you use a bigass fraction bar, since you may have to write a fraction over a fraction. The nth term of the series is just a_n, whereas if you replace n by $(n+1)$ wherever you see it, you get a_{n+1} instead. Anyway, now you have to find the above limit; let's say you've done that and got an answer L. There are three possibilities:

1. If $L < 1$, then the original series $\sum_{n=1}^{\infty} a_n$ converges; in fact, it converges absolutely.
2. If $L > 1$, then the original series diverges.
3. If $L = 1$, or the limit doesn't exist, then the ratio test is useless. Try something else.

Now let's look at some examples. First, consider

$$\sum_{n=1}^{\infty} \frac{n^{1000}}{2^n}.$$

It's not a geometric series because the numerator is a polynomial. Since exponentials grow faster than polynomials (see Section 21.3.3 in Chapter 21), the limit of the nth term is zero:

$$\lim_{n \to \infty} \frac{n^{1000}}{2^n} = 0.$$

So we can't use the nth term test. Since the series involves exponentials, let's try the ratio test. Following the standard framework, we start with

$$\lim_{n \to \infty} \left| \frac{a_{n+1}}{a_n} \right| = \lim_{n \to \infty} \left| \frac{\dfrac{(n+1)^{1000}}{2^{n+1}}}{\dfrac{n^{1000}}{2^n}} \right|.$$

Notice that the denominator is just the nth term, copied directly from the original series. The numerator is the same as the denominator, except that we have replaced every occurrence of n by $(n+1)$. Now, it's good technique to simplify the above expression by inverting and multiplying, grouping similar terms together as you do so. The above expression works out to be

$$\lim_{n \to \infty} \left| \frac{(n+1)^{1000}}{n^{1000}} \frac{2^n}{2^{n+1}} \right| = \lim_{n \to \infty} \left(\frac{n+1}{n} \right)^{1000} \frac{1}{2} = 1^{1000} \times \frac{1}{2} = \frac{1}{2}.$$

Note that we dropped the absolute values (everything's positive), and we also grouped the 1000th powers together and used the fact that $\lim_{n \to \infty} (n+1)/n = 1$. Anyway, the above limit is $1/2$, which is less than 1, so the original series converges by the ratio test. End of story.

Now consider

$$\sum_{n=2}^{\infty} \frac{3^n}{n \ln(n)}.$$

You should be able to show that the terms go to ∞ as $n \to \infty$, so the series diverges by the nth term test. Suppose that you just try the ratio test right off the bat. This still works:

$$\lim_{n \to \infty} \left| \frac{a_{n+1}}{a_n} \right| = \lim_{n \to \infty} \left| \frac{\dfrac{3^{n+1}}{(n+1) \ln(n+1)}}{\dfrac{3^n}{n \ln(n)}} \right| = \lim_{n \to \infty} \frac{3^{n+1}}{3^n} \frac{n}{n+1} \frac{\ln(n)}{\ln(n+1)} = 3.$$

We used $\lim_{n \to \infty} n/(n+1) = 1$, which is easy, and also $\lim_{n \to \infty} \ln(n)/\ln(n+1) = 1$, which is not. You should try using l'Hôpital's Rule to convince yourself that this last limit is true. Anyway, the limiting ratio above is 3, and since $3 > 1$, the original series diverges. So even though we didn't use the nth term test, the ratio test sufficed anyway.

The ratio test is particularly useful when dealing with factorials. Remember that $n!$ is the product of the numbers from 1 to n inclusive:

$$n! = 1 \times 2 \times 3 \times \cdots \times (n-1) \times n.$$

When using the ratio test with factorials, you will often have to consider ratios such as

$$\frac{n!}{(n+1)!}.$$

The only reliable way to simplify this is to expand the factorials and cancel:

$$\frac{n!}{(n+1)!} = \frac{1 \times 2 \times \cdots \times (n-1) \times n}{1 \times 2 \times \cdots \times (n-1) \times n \times (n+1)} = \frac{1}{n+1}.$$

That's not so bad, but it is possible to run into trouble when looking at something like $(2n)!$. This is **not** the same as $2 \times n!$—that's a common mistake. Consider the ratio

$$\frac{(2(n+1))!}{(2n)!} = \frac{(2n+2)!}{(2n)!}.$$

The numerator is the product of the first $2n+2$ numbers, whereas the denominator is the product of only the first $2n$ numbers. So, the ratio is

$$\frac{1 \times 2 \times \cdots \times (2n-1) \times (2n) \times (2n+1) \times (2n+2)}{1 \times 2 \times \cdots \times (2n-1) \times (2n)} = (2n+1)(2n+2).$$

This sort of calculation comes up pretty often. For example, consider the following series:

$$\sum_{n=1}^{\infty} \frac{(2n)!}{(n!)^2}.$$

Does this converge or diverge? It's not even clear if the terms go to 0. There are factorials involved, so let's jump straight in and try the ratio test:

$$\lim_{n\to\infty} \left| \frac{a_{n+1}}{a_n} \right| = \lim_{n\to\infty} \left| \frac{\dfrac{(2(n+1))!}{((n+1)!)^2}}{\dfrac{(2n)!}{(n!)^2}} \right| = \lim_{n\to\infty} \frac{(2n+2)!}{(2n)!} \left(\frac{n!}{(n+1)!} \right)^2.$$

Note that we have rearranged the fractions and powers to our best advantage. By what we have just seen above, this simplifies to

$$\lim_{n\to\infty} (2n+2)(2n+1) \left(\frac{1}{n+1} \right)^2 = \lim_{n\to\infty} \frac{4n^2 + 6n + 2}{n^2 + 2n + 1} = 4.$$

So the limit is greater than 1 and the series diverges. To give you an idea of how sensitive this stuff is, let's just modify the example slightly by putting an extra factor of 5^n on the bottom, like this:

$$\sum_{n=1}^{\infty} \frac{(2n)!}{5^n (n!)^2}.$$

Now you should try to calculate the ratio; you should get an extra factor of $1/5$ coming out, so you'll find that the limit is $4/5$, which is less than 1, and so the modified series converges.

Now consider the series

$$\sum_{n=1}^{\infty} \frac{n!}{(n+3)^n}.$$

This involves a factorial, so let's try the ratio test. We get

$$\lim_{n\to\infty} \left| \frac{a_{n+1}}{a_n} \right| = \lim_{n\to\infty} \left| \frac{\dfrac{(n+1)!}{((n+1)+3)^{n+1}}}{\dfrac{n!}{(n+3)^n}} \right| = \lim_{n\to\infty} \frac{(n+1)!}{n!} \frac{(n+3)^n}{(n+4)^{n+1}}.$$

Now we know from above that $(n+1)!/n!$ simplifies down to $(n+1)$, so the above quantity is

$$\lim_{n\to\infty} (n+1)\frac{(n+3)^n}{(n+4)^{n+1}}.$$

Now what the heck do you do? This is pretty tricky. How about writing the denominator as $(n+4) \times (n+4)^n$ so that we match the power of n in the numerator? Then we can group the terms like this:

$$\lim_{n\to\infty} (n+1)\frac{(n+3)^n}{(n+4)^{n+1}} = \lim_{n\to\infty} \frac{n+1}{n+4}\frac{(n+3)^n}{(n+4)^n} = \lim_{n\to\infty} \frac{n+1}{n+4}\left(\frac{n+3}{n+4}\right)^n.$$

Now the plot thickens. The first factor, $(n+1)/(n+4)$, clearly tends to 1 as $n \to \infty$, but the second factor is trickier. One way to handle it is to replace n by x and consider the limit

$$\lim_{x\to\infty} \left(\frac{x+3}{x+4}\right)^x.$$

Following the l'Hôpital's Rule Type **C** method (see Section 14.1.5 in Chapter 14), we find the limit of the logarithm (after a bit of clever algebra):

$$\lim_{x\to\infty} \ln\left(\frac{x+3}{x+4}\right)^x = \lim_{x\to\infty} x\ln\left(\frac{x+3}{x+4}\right) = \lim_{x\to\infty} \frac{\ln\left(\frac{x+3}{x+4}\right)}{1/x}$$
$$= \lim_{x\to\infty} \frac{\ln(x+3) - \ln(x+4)}{1/x}.$$

The numerator goes to 0 as $x \to \infty$, since $(x+3)/(x+4) \to 1$ and $\ln(1) = 0$. The denominator also goes to 0, so I leave it to you to use l'Hôpital's Rule to show that

$$\lim_{x\to\infty} \ln\left(\frac{x+3}{x+4}\right)^x = -1.$$

Exponentiating and changing x back into n, we have shown that

$$\lim_{n\to\infty} \left(\frac{n+3}{n+4}\right)^n = e^{-1}.$$

So, we now have all the pieces of the puzzle at our disposal. The limiting ratio above works out to be

$$\lim_{n\to\infty} \left|\frac{a_{n+1}}{a_n}\right| = \lim_{n\to\infty} \frac{n+1}{n+4}\left(\frac{n+3}{n+4}\right)^n = 1 \times e^{-1} = \frac{1}{e}.$$

Since this limit is less than 1, the original series converges.

How about

$$\sum_{n=2}^{\infty} \frac{1}{n\ln(n)}?$$

The terms certainly go to 0 as $n \to \infty$. Let's try the ratio test:

$$\lim_{n\to\infty} \left|\frac{a_{n+1}}{a_n}\right| = \lim_{n\to\infty} \left|\frac{\dfrac{1}{(n+1)\ln(n+1)}}{\dfrac{1}{n\ln(n)}}\right| = \lim_{n\to\infty} \frac{n}{n+1}\frac{\ln(n)}{\ln(n+1)} = 1.$$

(Once again, you should use l'Hôpital's Rule to get the limit of the ratio of the logs.) We've just shown that the limiting ratio is equal to 1. What does this mean? It means that the ratio test fails to give us any useful information. We don't know anything more about the series than when we started (except that the ratio test doesn't work!). So we have to try something else. It turns out that the integral test is the best one to use in this example—we'll check it out a little later, in Section 23.5.

23.4 How to Use the Root Test

Use the root test when there are a lot of tricky exponentials around involving functions of n. It's especially useful when the terms of your series look like A^B, where both A and B are functions of n. Here's the statement of the test, fresh from Section 22.5.2 of the previous chapter:

> if $L = \lim_{n\to\infty} |a_n|^{1/n}$, then $\sum_{n=1}^{\infty} a_n$ converges absolutely if $L < 1$,
>
> and diverges if $L > 1$; but if $L = 1$ or the limit doesn't exist, then the ratio test tells you nothing.

To use the root test, always start off with the following expression:

$$\lim_{n\to\infty} |a_n|^{1/n},$$

and then replace a_n by the general term of the series. Find the limit (if it exists) and call it L. Then you have three possibilities, which are identical to the possibilities which arise in the ratio test. The conclusions are luckily the same as well:

1. If $L < 1$, then the original series $\sum_{n=1}^{\infty} a_n$ converges; in fact, it converges absolutely.
2. If $L > 1$, then the original series diverges.
3. If $L = 1$, or the limit doesn't exist, then the root test is useless. Try something else.

For example, consider the series

$$\sum_{n=1}^{\infty} \left(1 - \frac{2}{n}\right)^{n^2}.$$

Since the terms involved have exponents involving powers of n, this series is just begging you to use the root test on it. Let's try it:

$$\lim_{n\to\infty} |a_n|^{1/n} = \lim_{n\to\infty} \left|\left(1 - \frac{2}{n}\right)^{n^2}\right|^{1/n} = \lim_{n\to\infty} \left(1 - \frac{2}{n}\right)^{n^2 \times \frac{1}{n}} = \lim_{n\to\infty} \left(1 - \frac{2}{n}\right)^{n}$$

$$= e^{-2} < 1.$$

Note that we removed the absolute values since everything's positive, and used the important limit at the end of Section 22.1.2 of the previous chapter (with k replaced by -2). So the limiting ratio is e^{-2}, which is certainly less than 1; by the root test, the original series above converges.

23.5 How to Use the Integral Test

Use the integral test when the series involves both $1/n$ and $\ln(n)$.
In Section 22.5.3 of the previous chapter, we saw that if N is any positive integer, then we can say:

---

if $a_n = f(n)$ for some continuous decreasing function f, then

$$\sum_{n=N}^{\infty} a_n \text{ and } \int_N^{\infty} f(x)\,dx \text{ either both converge or both diverge.}$$

In practice, here are the steps involved in using the integral test.

- Replace n by x, change $\sum_{n=1}^{\infty}$ into $\int_1^{\infty}$, and put a dx at the end. Of course, if the series begins at $n = 2$, then you use $\int_2^{\infty}$ instead, for example.

- Check that the integrand is decreasing; you can do that by showing that the derivative is negative, or just by inspecting the integrand directly.

- Now deal with the improper integral from the first step. The main advantage of integrals over series is that you can use a substitution (or change of variables, if you prefer) in an integral. The most common substitution in this context is $t = \ln(x)$.

- If the improper integral converges, so does the series. If the integral diverges, the series diverges too.

For example, consider

$$\sum_{n=2}^{\infty} \frac{1}{n \ln(n)}.$$

We have already looked at this series—in fact, we tried to use the ratio test at the end of Section 23.3 above, with no success. Let's try the integral test instead, which is suggested by the presence of the factor $1/n$ and the presence of $\ln(n)$. Change the variable n to x, and the sum to an integral, to get

$$\int_2^{\infty} \frac{1}{x \ln(x)}\,dx$$

instead. The integrand $1/(x \ln(x))$ is indeed decreasing in x; you can show this by differentiating and seeing that the derivative is negative, or more directly by observing that x and $\ln(x)$ are both increasing in x, so their product $x \ln(x)$ is as well, so the reciprocal $1/x \ln(x)$ is decreasing in x. Anyway, we have already looked at the above improper integral in Chapter 21, but here's the solution outline once again: substitute $t = \ln(x)$, so $dt = 1/x\,dx$, and the integral becomes

$$\int_{\ln(2)}^{\infty} \frac{1}{t}\,dt,$$

which diverges by the p-test for integrals. Since the integral diverges, so does the original series (by the integral test).

On the other hand, let's modify the series slightly: consider

$$\sum_{n=2}^{\infty} \frac{1}{n(\ln(n))^2}.$$

Again, we have a factor $1/n$ and logarithms are involved, so try the integral test. Replace n by x and turn the series into an integral to get

$$\int_2^{\infty} \frac{1}{x(\ln(x))^2} \, dx.$$

Try to convince yourself that the integrand is decreasing in x. Substitute $t = \ln(x)$, and this time the integral becomes

$$\int_{\ln(2)}^{\infty} \frac{1}{t^2} \, dt,$$

which converges by the p-test. So this time the series converges (by the integral test). Looking at this example and the previous one together, we can really see just how subtle this whole business of convergence of series is. We know $\ln(n)$ is pretty small compared to any positive power of n as n gets large, but the above examples together demonstrate that a log can make a difference. One extra measly power of $\ln(n)$ thrown into the denominator of $\sum_{n=2}^{\infty} 1/n\ln(n)$ turns it from a divergent series into a convergent series. (We looked at a similar example in Section 21.3.4 of Chapter 21.)

23.6 How to Use the Comparison Test, the Limit Comparison Test, and the p-test

Use these tests for series with positive terms when none of the other tests seem to apply. You definitely want to try the nth term test first, then use the ratio test if factorials are involved, the root test if the terms have exponentials where the base and exponent are both functions of n, or the integral test if you have a factor of $1/n$ and logarithms are involved. What does that leave? Basically the same tools as you have for integrals: the comparison test, the limit comparison test, the p-test, and an understanding of how common functions behave near ∞ and near 0. You really need to review Chapter 21 before looking at this section, since the techniques are almost identical. In any case, here are the tests once more. (For the comparison and limit comparison tests, we assume all the terms a_n are nonnegative.)

1. **Comparison test, divergence version:** if you think $\sum_{n=1}^{\infty} a_n$ diverges, find a smaller series which also diverges. That is, find a positive sequence $\{b_n\}$ such that $a_n \ge b_n$ for all n, and such that $\sum_{n=1}^{\infty} b_n$ diverges. Then

$$\sum_{n=1}^{\infty} a_n \ge \sum_{n=1}^{\infty} b_n = \infty,$$

so $\sum_{n=1}^{\infty} a_n$ diverges.

2. **The comparison test, convergence version:** if you think $\sum_{n=1}^{\infty} a_n$ converges, find a larger series which also converges. That is, find $\{b_n\}$ such that $a_n \le b_n$ for all n, and such that $\sum_{n=1}^{\infty} b_n$ converges. Then

$$\sum_{n=1}^{\infty} a_n \le \sum_{n=1}^{\infty} b_n < \infty,$$

so $\sum_{n=1}^{\infty} a_n$ converges.

3. **Limit comparison test:** find a simpler series $\sum_{n=1}^{\infty} b_n$ so that $a_n \sim b_n$ as $n \to \infty$. Then if $\sum_{n=1}^{\infty} b_n$ converges, so does $\sum_{n=1}^{\infty} a_n$. On the other hand, if $\sum_{n=1}^{\infty} b_n$ diverges, then so does $\sum_{n=1}^{\infty} a_n$. (Remember that "$a_n \sim b_n$ as $n \to \infty$" means the same thing as "$\lim_{n \to \infty} a_n/b_n = 1$.")

4. **p-test:** if $a \ge 1$, the series

$$\sum_{n=a}^{\infty} \frac{1}{n^p} \quad \begin{cases} \text{converges} & \text{if } p > 1, \\ \text{diverges} & \text{if } p \le 1. \end{cases}$$

This is the same as the $\int^{\infty}$ version of the p-test for integrals.

Now let's look at some examples. In each example below, you could replace the sum by an integral and get an improper integral (with problem spot at ∞) instead of a series. The solutions to the improper integral problems are identical to the corresponding solutions for the series below. In each case, you should try to write down the equivalent problem and solution for the improper integral version. It's also a good idea to look back at Chapter 21 and try to convert every improper integral with problem spot at ∞ to a series. Almost all of them can be solved using the above tests. (The exceptions are the problems whose solutions involve the change of variables $t = \ln(x)$; for those problems, you'd need to use the integral test in order to solve the corresponding series problems.) Anyway, consider the series

$$\sum_{n=1}^{\infty} \frac{2n^2 + 3n + 7}{n^4 + 2n^3 + 1}.$$

To examine this, note that the highest term in each polynomial dominates, since n is getting larger and larger. (See Section 21.3.1 of Chapter 21 for more details.) So we have

$$\frac{2n^2 + 3n + 7}{n^4 + 2n^3 + 1} \sim \frac{2n^2}{n^4} = \frac{2}{n^2} \quad \text{as } n \to \infty.$$

By the p-test, $\sum_{n=1}^{\infty} 2/n^2$ converges (the constant 2 is irrelevant); so by the limit comparison test, the original series above converges as well.

A slight technicality arises in the almost identical example

$$\sum_{n=0}^{\infty} \frac{2n^2 + 3n + 7}{n^4 + 2n^3 + 1}.$$

The only difference between this and the previous example is that the sum now begins at $n = 0$. If you use the same argument as in the previous example, you may find that you are comparing the above series with the series $\sum_{n=0}^{\infty} 2/n^2$. This last series isn't well-defined, since its first term looks like $2/0$, which really sucks. You can avoid this issue by one of two methods: you could just say that you're changing the first term $n = 0$ into something else, like $n = 1$, and this doesn't affect the convergence. Alternatively, you could break off the term $n = 0$ from the sum. Indeed, when $n = 0$, the quantity $(2n^2 + 3n + 7)/(n^4 + 2n^3 + 1)$ is just 7, so

$$\sum_{n=0}^{\infty} \frac{2n^2 + 3n + 7}{n^4 + 2n^3 + 1} = 7 + \sum_{n=1}^{\infty} \frac{2n^2 + 3n + 7}{n^4 + 2n^3 + 1}.$$

The series on the right-hand side converges, so the series on the left-hand side does as well. Adding the finite number 7 makes no difference. In general, if your sum begins at $n = 0$ and you want to use the limit comparison test, break off the first term so that you can consider the series starting at $n = 1$ instead.

Now let's look at

$$\sum_{n=1}^{\infty} \frac{\sqrt[3]{27n^6 + 9n^2 + 4}}{n^3 + 9n^2 + 4}.$$

By our standard ideas about higher powers dominating, we have

$$\frac{\sqrt[3]{27n^6 + 9n^2 + 4}}{n^3 + 9n^2 + 4} \sim \frac{\sqrt[3]{27n^6}}{n^3} = \frac{3n^2}{n^3} = \frac{3}{n} \qquad \text{as } n \to \infty.$$

By the p-test, $\sum_{n=1}^{\infty} 3/n$ diverges, so by the limit comparison test, the original series diverges as well.

How about

$$\sum_{n=1}^{\infty} 2^{-n} n^{1000}?$$

In Section 23.3 above, we used the ratio test to solve this problem (we actually wrote $n^{1000}/2^n$ instead of $2^{-n} n^{1000}$, but of course they are the same thing!). Let's now solve the problem using the comparison test. To do it this way, we should use the idea that exponentials grow quickly. Using the same methods as those described in Section 21.3.3 of Chapter 21, we write

$$2^{-n} \le \frac{C}{n^{1002}};$$

we have picked the exponent 1002 since it's 2 bigger than the exponent 1000 in the question. We now have

$$\sum_{n=1}^{\infty} 2^{-n} n^{1000} \le \sum_{n=1}^{\infty} \frac{C}{n^{1002}} n^{1000} = C \sum_{n=1}^{\infty} \frac{1}{n^2} < \infty,$$

where the last series converges by the p-test. So the original series converges by the comparison test.

Now consider

$$\sum_{n=2}^{\infty} \frac{\ln(n)}{n^{1.001}}.$$

This is just the series version of an example on page 465. In fact, you could use the integral test to convert this series problem into the improper integral problem there, since the integrand is decreasing, but what would be the point? We might as well just solve it directly. As we did in the improper integral example, we use $\ln(n) \leq Cn^{0.0005}$, where we have cunningly chosen the exponent 0.0005 so small that you can take it away from the exponent 1.001 (which arises in our series) and still be greater than 1. So we have

$$\sum_{n=1}^{\infty} \frac{\ln(n)}{n^{1.001}} \leq \sum_{n=1}^{\infty} \frac{Cn^{0.0005}}{n^{1.001}} = C \sum_{n=1}^{\infty} \frac{1}{n^{1.0005}} < \infty,$$

where the last series converges by the p-test. So our original series converges by the comparison test.

The series

$$\sum_{n=1}^{\infty} \frac{|\sin(n)|}{n^2}$$

is pretty easy to deal with. Remembering that $|\sin(n)| \leq 1$, we see that

$$\sum_{n=1}^{\infty} \frac{|\sin(n)|}{n^2} \leq \sum_{n=1}^{\infty} \frac{1}{n^2} < \infty.$$

So the series converges by the comparison test.

Now consider the series

$$\sum_{n=1}^{\infty} \sin\left(\frac{1}{n}\right).$$

It may look like some of the terms of the series might be negative, but that's a load of bull. Indeed, when n starts at 1 and works its way up the positive integers, the numbers $1/n$ start at 1 and work their way down toward 0. That is, $1/n$ is always between 0 and 1. Since $\sin(x)$ is positive when x is between 0 and 1, the series has all positive terms! So what? We still haven't done the problem. How do we proceed? In Section 21.4.2 of Chapter 21, we saw that $\sin(x) \sim x$ as $x \to 0$. Replacing x by $1/n$, we see that $\sin(1/n) \sim 1/n$ as $1/n \to 0$. Wait a second—when $1/n \to 0$, we must have $n \to \infty$. That is, we have shown that $\sin(1/n) \sim 1/n$ as $n \to \infty$. This is exactly what we need! Since the series $\sum_{n=1}^{\infty} 1/n$ diverges, the limit comparison test shows that our original series above diverges too. (Compare this example with the last example in Section 21.4.2 of Chapter 21.)

On the other hand, the series

$$\sum_{n=1}^{\infty} \sin^2\left(\frac{1}{n}\right)$$

converges, since $\sin^2(1/n) \sim 1/n^2$ as $n \to \infty$; you get to fill in the details.

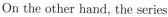

Finally, a really nasty series:

$$\sum_{n=2}^{\infty} \cos^2(n) \tan\left(\frac{(n^2 + 4n - 3)\ln(n)}{\sqrt{n^7 + 2n^4 + 3n}}\right).$$

How do you approach this? Consider the pieces of this series. As $n \to \infty$, the $(n^2 + 4n - 3)$ factor is asymptotic to n^2, and the $\sqrt{n^7 + 2n^4 + 3n}$ factor is asymptotic to $\sqrt{n^7}$, which is just $n^{7/2}$. So we can say that

$$\frac{(n^2 + 4n - 3)\ln(n)}{\sqrt{n^7 + 2n^4 + 3n}} \sim \frac{n^2 \ln(n)}{n^{7/2}} = \frac{\ln(n)}{n^{3/2}} \qquad \text{as } n \to \infty.$$

On the other hand, both sides of this above relation go to 0 as n gets large (remember, logs grow slowly!). So we can use the relation $\tan(x) \sim x$ as $x \to 0$, with x replaced by the horrible quantity $(n^2 + 4n - 3)\ln(n)/\sqrt{n^7 + 2n^4 + 3n}$, to get

$$\tan\left(\frac{(n^2 + 4n - 3)\ln(n)}{\sqrt{n^7 + 2n^4 + 3n}}\right) \sim \frac{(n^2 + 4n - 3)\ln(n)}{\sqrt{n^7 + 2n^4 + 3n}} \sim \frac{\ln(n)}{n^{3/2}} \qquad \text{as } n \to \infty.$$

Now let's concentrate for a moment on the series

$$\sum_{n=2}^{\infty} \frac{\ln(n)}{n^{3/2}}.$$

We need to use the fact that logs grow slowly to make the $\ln(n)$ insignificant compared to the $n^{3/2}$ term (see Section 21.3.4 of Chapter 21 for more details). Specifically, we like the power $3/2$ in the denominator and don't want this to be 1 or smaller. So let's use $\ln(n) < Cn^{1/4}$ (the power here just needs to be less than $1/2$) and see that

$$\frac{\ln(n)}{n^{3/2}} \le \frac{Cn^{1/4}}{n^{3/2}} = \frac{C}{n^{5/4}}.$$

So, summing everything up, we see that

$$\sum_{n=2}^{\infty} \frac{\ln(n)}{n^{3/2}} \le C \sum_{n=2}^{\infty} \frac{1}{n^{5/4}} < \infty$$

by the p-test. So

$$\sum_{n=2}^{\infty} \frac{\ln(n)}{n^{3/2}}$$

converges by the comparison test. Now look back at the asymptotic relation way back above; the limit comparison test now implies that

$$\sum_{n=2}^{\infty} \tan\left(\frac{(n^2 + 4n - 3)\ln(n)}{\sqrt{n^7 + 2n^4 + 3n}}\right)$$

also converges. Great—we're nearly done. How about that $\cos^2(n)$ factor? That's not too helpful, since it keeps oscillating. We do know that it's less than or equal to 1, and that it's positive (since it's a square). So we'll just use $\cos^2(n) \le 1$ and see what we get. In fact,

$$\sum_{n=2}^{\infty} \cos^2(n) \tan\left(\frac{(n^2 + 4n - 3)\ln(n)}{\sqrt{n^7 + 2n^4 + 3n}}\right) \le \sum_{n=2}^{\infty} \tan\left(\frac{(n^2 + 4n - 3)\ln(n)}{\sqrt{n^7 + 2n^4 + 3n}}\right) < \infty,$$

as we have just shown that the right-hand series converges. So, our original series converges by the comparison test. How about that—we used the comparison test twice, the limit comparison test once, and the p-test twice. Tricky stuff; but if you can do that sort of problem on your own, then you should be able to do just about any problem involving these three tests.

23.7 How to Deal with Series with Negative Terms

Suppose some of the numbers a_n which appear as terms in your series are negative. Here are some ways to handle this situation:

1. **If all the terms a_n are negative, then modify the series by putting a minus sign in front of all the terms.** The modified series is $\sum_{n=1}^{\infty}(-a_n)$, which has all positive terms. Then you can use the techniques above to work out whether the modified series converges or diverges. Then if the modified series diverges, so does the original one we are interested in, whereas if the modified series converges, then the original one also converges. In fact, if the modified series converges to L, then the original one converges to $-L$, since the modified series is just the negative of the original series. For example, does the series

$$\sum_{n=3}^{\infty} \ln\left(\frac{1}{n}\right)\frac{1}{\sqrt{n}}$$

converge or diverge? Well, $1/n$ is near 0 when n is large, so taking log of it will give a negative number. (Remember, $\ln(x) < 0$ if $0 < x < 1$.) It's therefore easier to consider the modified series

$$\sum_{n=3}^{\infty} -\ln\left(\frac{1}{n}\right)\frac{1}{\sqrt{n}},$$

which is actually the same thing as

$$\sum_{n=3}^{\infty} \ln(n)\frac{1}{\sqrt{n}},$$

since $-\ln(1/n) = -(\ln(1) - \ln(n)) = \ln(n)$. Now, what's our intuition about this? If this series were just

$$\sum_{n=3}^{\infty} \frac{1}{\sqrt{n}},$$

it would diverge by the p-test. Normally logs don't make much difference, but this isn't always true—remember the examples from the integral test above. In any case, this particular log actually helps the series to diverge, since it blows up as $n \to \infty$. The basic logic is that as n ranges from 3 upward, the least that $\ln(n)$ can be is $\ln(3)$, so we have

$$\ln(n) \geq \ln(3)$$

for any $n \geq 3$. In our series, it follows that

$$\sum_{n=3}^{\infty} \ln(n)\frac{1}{\sqrt{n}} \geq \sum_{n=3}^{\infty} \frac{\ln(3)}{\sqrt{n}} = \ln(3)\sum_{n=3}^{\infty} \frac{1}{\sqrt{n}} = \infty$$

by the p-test (with $p = 1/2$). So the modified series diverges to ∞; we conclude that the original series diverges to $-\infty$.

2. If some terms are positive and some terms are negative, try the nth term test first. That is, check that the terms tend to 0 as $n \to \infty$; otherwise you know right away that the series diverges. For example,

$$\sum_{n=1}^{\infty}(-1)^n n^2$$

diverges because the limit of the terms $(-1)^n n^2$ isn't zero. (In fact, the limit doesn't exist, because the sequence oscillates wildly between bigger and bigger positive and negative numbers.) There's no need to use any other tests here.

3. If some terms are positive and some terms are negative, and the terms converge to 0 as $n \to \infty$, next try the absolute convergence test:

$$\text{if} \quad \sum_{n=1}^{\infty}|a_n| \quad \text{converges, then so does} \quad \sum_{n=1}^{\infty}a_n.$$

In this case, we say that the sequence is *absolutely convergent* or that it *converges absolutely*. For example, the series

$$\sum_{n=1}^{\infty}\frac{\sin(n)}{n^2}$$

converges absolutely, since we already saw on page 513 above that

$$\sum_{n=1}^{\infty}\frac{|\sin(n)|}{n^2}$$

converges. So, given a series with positive and negative terms that doesn't obviously fail the nth term test, you should always check to see if the series is absolutely convergent. If it's absolutely convergent, then the series converges. On the other hand, if it doesn't converge absolutely, don't give up on it—go on to the next step.

4. If the series doesn't converge absolutely, try the alternating series test. As we saw in Section 22.5.4 of the previous chapter,

> if the absolute values of the terms of an alternating series decrease to 0 monotonically as $n \to \infty$, the series converges.

So there are actually three things to check if you want to use the test on a series $\sum_{n=1}^{\infty} a_n$:

- the terms a_n alternate between positive and negative (that is, the signs of the terms are, in order, $+, -, +, -, \ldots$, or perhaps $-, +, -, +, \ldots$);
- the quantities $|a_n|$ tend to 0 as n gets large; that is,

$$\lim_{n \to \infty} |a_n| = 0;$$

- the absolute values of the terms $|a_n|$ are decreasing in n (so the underlying sequence is getting smaller and smaller, in terms of absolute value).

If all three of these properties are true, then the series converges. **Note: you should always try the absolute convergence test first. If the series converges absolutely, do not use the alternating series test!** Also, notice that the second property is just the nth term test in disguise; this follows from the fact that $\lim_{n\to\infty} |a_n| = 0$ if and only if $\lim_{n\to\infty} a_n = 0$. So even if you forget to try the nth term test first, you have to do it anyway as part of the alternating series test.

Here's a classic example:
$$\sum_{n=1}^{\infty} \frac{(-1)^n}{n}.$$

The absolute version is $\sum_{n=1}^{\infty} 1/n$, which diverges by the p-test; so our series does not converge absolutely. Let's dive straight in to the alternating series test. We need to check the three properties. First, is the series alternating? Yes. A series is automatically alternating if it has terms which look like $(-1)^n$ or $(-1)^{n+1}$ multiplied by a positive number. In this case, the nth term is $(-1)^n$ multiplied by the positive number $1/n$. How about the second property? We need to show that

$$\lim_{n\to\infty} \left| \frac{(-1)^n}{n} \right| = 0.$$

This is obviously true, since $|(-1)^n/n| = 1/n$. As for the third property, we need to show that $\{|(-1)^n/n|\}$ is a decreasing sequence. This is pretty straightforward, again since $|(-1)^n/n| = 1/n$ and we know that $1/n$ is decreasing in n. So the alternating series test applies and shows that the original series

$$\sum_{n=1}^{\infty} \frac{(-1)^n}{n}$$

converges. Since we've already checked that it doesn't converge absolutely, we know that it converges conditionally.

On the other hand, consider the series

$$\sum_{n=1}^{\infty} \frac{(-1)^n}{n^2}.$$

The absolute version is $\sum_{n=1}^{\infty} 1/n^2$, which converges by the p-test. So the above series converges absolutely, and there's no need to waste time with the alternating series test.

Let's see a couple more examples. First, let's look at

$$\sum_{n=1}^{\infty} (-1)^n \sin\left(\frac{1}{n}\right).$$

This is very similar to an example we looked at on page 513 above, except that we now have an extra factor of $(-1)^n$. The first thing to do with our series is to check whether or not it is absolutely convergent. The absolute version is

$$\sum_{n=1}^{\infty} \left| (-1)^n \sin\left(\frac{1}{n}\right) \right| = \sum_{n=1}^{\infty} \sin\left(\frac{1}{n}\right).$$

We saw in that previous example that the quantity $\sin(1/n)$ is nonnegative when $n \geq 1$, so this justifies dropping the absolute value signs. We also saw that this last series diverges, so our original series above doesn't converge absolutely. On the other hand, the terms of this series are clearly alternating, and we've already seen that they go to 0 as $n \to \infty$ since $\sin(1/n)$ does. Now consider $|a_n|$, which is just $\sin(1/n)$. Does this decrease in n? You could differentiate $\sin(1/x)$ with respect to x to get $-\cos(1/x)/x^2$, and show that this is negative whenever $x \geq 1$; or you could just argue that $1/n$ decreases in n, and $\sin(x)$ is increasing in x when x is near 0, so $\sin(1/n)$ decreases in n. Either way, we have now verified all three properties, so the alternating series test shows that our series converges. Since it doesn't converge absolutely, it must converge conditionally.

One final example. Consider the series

$$\sum_{n=1}^{\infty} (-1)^n \left(1 + \frac{1}{n}\right)^n.$$

The series is certainly alternating, but what is the limit of the nth term? We need it to be 0 for the series to have any hope at all of converging. That seems problematic in this case—by the boxed limit at the end of Section 22.1.2 of the previous chapter with k replaced by 1, we have

$$\lim_{n \to \infty} \left(1 + \frac{1}{n}\right)^n = e^1 = e,$$

so the alternating version of this sequence (with the $(-1)^n$ term in there) oscillates between numbers close to e and $-e$. This means that

$$\lim_{n \to \infty} (-1)^n \left(1 + \frac{1}{n}\right)^n \qquad \text{does not exist (DNE).}$$

Since this limit isn't 0 (it doesn't even exist!), the nth term test shows that the original series

$$\sum_{n=1}^{\infty} (-1)^n \left(1 + \frac{1}{n}\right)^n$$

must diverge. Don't fall into the trap of using the alternating series test here to conclude that the series converges!

So as you have seen, this series business isn't too easy. What's more, we're going to have to use these techniques again when we look at power series and Taylor series in the next chapter, so you really need to understand the stuff in this chapter or there will be a large chunk of material coming up that will elude you. Of course, doing as many problems as you can will really help.

CHAPTER 24 _____

Introduction to Taylor Polynomials, Taylor Series, and Power Series

We now come to the important topics of power series and Taylor polynomials and series. In this chapter, we'll see a general overview of these topics. The following two chapters will deal with problem-solving techniques in the context of the material in this chapter. Here's what we'll look at first:

- approximations, Taylor polynomials, and a Taylor approximation theorem;
- how good our approximations are, and the full Taylor Theorem;
- the definition of power series;
- the definition of Taylor series and Maclaurin series; and
- convergence issues involving Taylor series.

24.1 Approximations and Taylor Polynomials

Here's a nice fact: for any real number x, we have

$$e^x \cong 1 + x + \frac{x^2}{2} + \frac{x^3}{6}.$$

Also, the closer x is to 0, the better the approximation.

Let's play around with this for a little bit. Start off with $x = 0$. Actually, both sides are then equal to 1, so the approximation is perfect! What about when x isn't 0? Let's try $x = -1/10$. The above equation says that

$$e^{-1/10} \cong 1 - \frac{1}{10} + \frac{1/100}{2} - \frac{1/1000}{6},$$

which simplifies to

$$e^{-1/10} \cong \frac{5429}{6000}.$$

My calculator says that $e^{-1/10}$ is equal to 0.9048374180 (to ten decimal places), while 5429/6000 is equal to 0.9048333333 (also to ten decimal places).

These numbers are pretty close to each other! In fact, the difference is only about 0.0000040847.

How on earth did I come up with the polynomial $1 + x + x^2/2 + x^3/6$? It's clearly not just any old polynomial; it's specially related to e^x. Rather than concentrate on e^x itself, let's get a little more general and consider other functions. Also, there's nothing special about the degree 3 of our polynomial: we could have used any degree. So let's start with degree 1 and see what happens.

24.1.1 Linearization revisited

Let's say we have some function f which is very smooth, so that it can be repeatedly differentiated as many times as you like without causing any problems. Here's a question we asked back in Section 13.2 of Chapter 13: what is the equation of the line which best approximates the curve $y = f(x)$ near the point $(a, f(a))$? The answer to this question is that the line we're looking for is the tangent line to the curve at the point $(a, f(a))$, and its equation is

$$y = f(a) + f'(a)(x - a).$$

This is precisely the linearization of f at $x = a$. The right-hand side is a polynomial of degree 1. In the picture below, the tangent line to the curve $y = f(x)$ at $x = a$ is drawn in, and looks like a pretty lousy approximation to the whole curve:

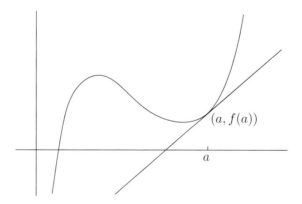

Nevertheless, it is a pretty good approximation to the curve near $(a, f(a))$. In fact, let's zoom in near $(a, f(a))$:

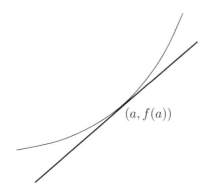

Now you can see that there's not that much difference between the tangent line and the curve $y = f(x)$. The more we zoom in near $x = a$, the smaller this difference becomes.

24.1.2 Quadratic approximations

Why stick to lines, though? Let's ask the same question we did at the beginning of the previous section, but with parabolas instead. Here is our question: what is the equation of the quadratic which best approximates the curve $y = f(x)$ near $(a, f(a))$? Using the same function as in the picture above, here's a guess as to what the quadratic should look like:

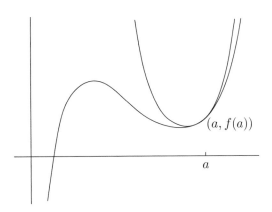

It turns out that the formula for the quadratic which best approximates the curve $y = f(x)$ for x near a (that is, near the point $(a, f(a))$ on the curve) is given by

$$y = f(a) + f'(a)(x - a) + \frac{f''(a)}{2}(x - a)^2.$$

This is actually a quadratic in x, because if you expand $(x - a)^2$, the highest power of x floating around is x^2. Still, it's better to leave it in the above form and say that it's a "quadratic in $(x - a)$." Let's call the quadratic P_2; that is, we set

$$P_2(x) = f(a) + f'(a)(x - a) + \frac{f''(a)}{2}(x - a)^2.$$

Now, let's gather some nice facts about P_2:

1. Plug $x = a$ into the above equation for $P_2(x)$ and you easily see that $P_2(a) = f(a)$. So the values of P_2 and f match when $x = a$. In fact, since the zeroth derivative of a function is just the function itself, you might say that the values of the zeroth derivatives of P_2 and f match when $x = a$.
2. Now differentiate P_2 to see that $P_2'(x) = f'(a) + f''(a)(x - a)$. Again, if you plug in $x = a$, you see that $P_2'(a) = f'(a)$. The values of the first derivatives P_2' and f' also match when $x = a$.
3. Differentiate once more to get $P_2''(x) = f''(a)$. When $x = a$, this becomes $P_2''(a) = f''(a)$, so even the values of the second derivatives match when $x = a$.

4. On the other hand, since $f''(a)$ is constant, $P_2'''(x) = 0$ for all x. The same is true for all higher derivatives. (After all, P_2 is a quadratic, and the third and higher derivatives of **any** quadratic must be zero everywhere!)

So P_2 shares the zeroth, first, and second derivatives with f at $x = a$; but the third and higher derivatives of P_2 are always 0. You might say that P_2 is the distillation of all the information about f at $x = a$ up to and including the second derivative.

Here's another nice fact about P_2: if you ignore the last term on the right-hand side of the above equation for $P_2(x)$, you just get $f(a) + f'(a)(x - a)$. This is exactly the linearization from the previous section. So you can think of the last term $\frac{1}{2}f''(a)(x-a)^2$ as a so-called *second-order correction term*. This means that we should actually be able to do a better job of approximation than just by using the tangent line. The second-order correction term helps us get even closer to the curve, at least for x near a. (An exception to this occurs when $f''(a) = 0$, in which case P_2 is actually the linearization and we haven't gotten any closer.)

24.1.3 Higher-degree approximations

Let's continue the same pattern, except that we'll use some arbitrary degree N instead of just 1 or 2. So, here's our question: which polynomial of degree N or less gives the best approximation to $f(x)$ for values of x near a? The answer is provided by the following theorem.

> **A Taylor approximation theorem:** if f is smooth at $x = a$, then of all the polynomials of degree N or less, the one which best approximates $f(x)$ for x near a is given by
> $$P_N(x) = f(a) + f'(a)(x - a) + \frac{f''(a)}{2!}(x - a)^2$$
> $$+ \frac{f^{(3)}(a)}{3!}(x - a)^3 + \cdots + \frac{f^{(N)}(a)}{N!}(x - a)^N.$$

In sigma notation, the formula looks like this:

$$P_N(x) = \sum_{n=0}^{N} \frac{f^{(n)}(a)}{n!}(x - a)^n.$$

In this formula, remember that $0! = 1$, that $f^{(0)}(a)$ means the same thing as $f(a)$ (zero derivatives), and that $f^{(1)}(a)$ means the same thing as $f'(a)$ (one derivative).

We call the polynomial P_N the Nth-order Taylor polynomial of $f(x)$ at $x = a$. Note that the degree of P_N might be less than N; for example, if $f^{(N)}(a) = 0$, then the last term in the above sum vanishes and the degree of P_N could be at most $N - 1$. This is why we call it an Nth-order Taylor polynomial, not an Nth-degree Taylor polynomial. (By the way, the polynomial $P_N(x)$ is sometimes written as $P_N(x; a)$ to emphasize that you get a different polynomial for each choice of N and a. I'll just write $P_N(x)$, since we're only dealing with one choice of a at a time anyway.)

Once again, the important property of P_N is that

$$P_N^{(n)}(a) = f^{(n)}(a)$$

for all $n = 0, 1, \ldots, N$. That is, the values of all the derivatives of P_N and f match when $x = a$, up to and including the Nth derivative; but all higher derivatives of P_N must be zero everywhere. The function P_N is the distillation of all the information about f which comes from its derivatives up to order N at $x = a$.

Of course, when $N = 1$, we just get $P_1(x) = f(a) + f'(a)(x - a)$, which is the linearization of f at $x = a$, and when $N = 2$, we recover the formula for $P_2(x)$ from the previous section. Let's see how this works for $f(x) = e^x$, setting $a = 0$. By the above formula with $N = 3$ and $a = 0$, we have

$$P_3(x) = f(0) + f'(0)x + \frac{f''(0)}{2!}x^2 + \frac{f^{(3)}(0)}{3!}x^3.$$

Luckily, all the derivatives of e^x with respect to x are just e^x, so we can see that $f(0)$, $f'(0)$, $f''(0)$, and $f^{(3)}(0)$ are all e^0, which is equal to 1. Since $2! = 2$ and $3! = 6$, the above formula becomes

$$P_3(x) = 1 + x + \frac{1}{2}x^2 + \frac{1}{6}x^3.$$

This is exactly the cubic polynomial from the beginning of Section 24.1 above! Of all degree three or lower polynomials, this one gives the best approximation for e^x when x is near 0. Why 0? Because that's the value we chose for a. If we chose a different value of a, we'd get a different polynomial which would approximate e^x really well for x near a. By getting rid of the cubic term $x^3/6$, we can see that $P_2(x) = 1 + x + x^2/2$; then by throwing away the quadratic term $x^2/2$, we get the linearization $P_1(x) = 1 + x$. Another way to look at this is that $P_2(x)$ improves upon $P_1(x)$ by adding on the second-order correction term $x^2/2$, while $P_3(x)$ improves upon $P_2(x)$ by adding on a third-order correction term $x^3/6$. Each time you increase N by 1, you are making the approximation better by adding on another correction term.

The Taylor approximation theorem actually depends on Taylor's Theorem, which we'll look at in the next section. There's something ambiguous about the statement of the approximation theorem, as well: what on earth does it mean to be the "best approximation," anyway? We'll explore this more in the next section, but the real answer is contained in Section A.7 of Appendix A, along with the proof of the theorem.

24.1.4 Taylor's Theorem

In Section 24.1 above, we saw that

$$e^x \cong 1 + x + \frac{x^2}{2} + \frac{x^3}{6}.$$

In particular, we noticed that when $x = -1/10$, the above approximation becomes

$$e^{-1/10} \cong 1 - \frac{1}{10} + \frac{1/100}{2} - \frac{1/1000}{6} = \frac{5429}{6000}.$$

How good is this approximation? One way to measure this is to consider the difference between the true quantity $e^{-1/10}$ and the approximation $5429/6000$. We'll call this quantity the *error* in the approximation, since it shows how wrong we are to use our approximation instead of the true value. Here's what the error is in our case:

$$\text{error} = \text{true value} - \text{approximate value} = e^{-1/10} - \frac{5429}{6000}.$$

If the error is small, then the approximation is good. In Section 24.1, we saw that the difference was 0.0000040847 to 10 decimal places—but we needed to use a calculator, which defeats the entire purpose of doing our own approximation. Remember that the number the calculator gives you is also an approximation! Besides, how do you think the calculator works? It probably finds its approximation of $e^{-1/10}$ using a Taylor polynomial.

What we'd really like is another formula for the error. That's where Taylor's theorem comes in. Rather than just specializing to the case of e^x, let's get more general once again. We're dealing with a smooth function f and its Nth-order Taylor polynomial about $x = a$; as we saw in the previous section, this polynomial is given by

$$P_N(x) = \sum_{n=0}^{N} \frac{f^{(n)}(a)}{n!}(x-a)^n.$$

We want to use the value of $P_N(x)$ to approximate the true value $f(x)$, so we consider the error term, which is the difference between the true value and the approximate value:

$$R_N(x) = f(x) - P_N(x).$$

Actually, $R_N(x)$ is called the *Nth-order error term*; it's also referred to as the *Nth-order remainder term*, since it's all that remains when you take $P_N(x)$ away from $f(x)$. As promised above, Taylor's Theorem gives an alternative formula for $R_N(x)$:

Taylor's Theorem: the Nth-order remainder term $R_N(x)$ about $x = a$ is

$$R_N(x) = \frac{f^{(N+1)}(c)}{(N+1)!}(x-a)^{N+1}$$

where c is some number which lies between x and a.

Note that the number c depends on what x and N are, and cannot be determined in general! Since $f(x) = P_N(x) + R_N(x)$, we can write the whole kit and caboodle as

---

$$f(x) = \sum_{n=0}^{N} \frac{f^{(n)}(a)}{n!}(x-a)^n + \frac{f^{(N+1)}(c)}{(N+1)!}(x-a)^{N+1}.$$

This seems pretty nasty. And what on earth is with this number c, anyway? Actually, we've seen something like this before. Take a look back at our discussion of Mean Value Theorem (MVT) in Section 11.3 of Chapter 11. The MVT says that if f is smooth enough on an interval $[a, b]$, then there is a number c in $[a, b]$ (which cannot be determined in general) such that

$$f'(c) = \frac{f(b) - f(a)}{b - a}.$$

If you replace b by x and solve for $f(x)$, you get

$$f(x) = f(a) + f'(c)(x - a),$$

where c is between a and x. Now let's look back to the last equation in Taylor's Theorem and put $N = 0$. What is $P_0(x)$? It's just $f(a)$. How about $R_0(x)$? According to Taylor's Theorem,

$$R_0(x) = \frac{f^{(1)}(c)}{1!}(x - a)^1 = f'(c)(x - a),$$

where c is between x and a. Hey, so Taylor's Theorem (with $N = 0$) says that

$$f(x) = P_0(x) + R_0(x) = f(a) + f'(c)(x - a),$$

which is exactly what the MVT says! So, Taylor's Theorem is basically the Mean Value Theorem on steroids. By the way, the reason we say that c is between x and a instead of writing $a \le c \le x$ is that x might actually be less than a, so then we would have $x \le c \le a$.

Now let's put $N = 1$ instead of $N = 0$. The main formula in the box above becomes

$$f(x) = f(a) + f'(a)(x - a) + \frac{f''(c)}{2!}(x - a)^2 = L(x) + R_1(x);$$

here $L(x) = f(a) + f'(a)(x - a)$ is the linearization of f about $x = a$, and $R_1(x) = \frac{1}{2}f''(c)(x - a)^2$ is the first-order error term. This agrees with the formula for the error term $r(x)$ which we gave in Section 13.2.4 of Chapter 13.

We still have to go back to our approximation for e^x. When we wrote

$$e^x \cong 1 + x + \frac{x^2}{2} + \frac{x^3}{6},$$

we now understand that this is just saying that $e^x \cong P_3(x)$, where P_3 is the third-order Taylor polynomial of $f(x) = e^x$ about $x = 0$. Taylor's Theorem above says that R_3 is then given by

$$R_3(x) = \frac{f^{(4)}(c)}{4!}x^4,$$

where c is between 0 and x. (I just plugged $N = 3$ and $a = 0$ into the formula for $R_N(x)$ in the box above.) Since any derivative of e^x (with respect to x) is e^x, we know that $f^{(4)}(c) = e^c$; also $4! = 24$, so we actually have

$$R_3(x) = \frac{e^c}{24}x^4.$$

In other words,

$$e^x = 1 + x + \frac{x^2}{2} + \frac{x^3}{6} + \frac{e^c}{24}x^4.$$

We have changed our approximation into an exact equation, but we don't know what c is! Still, we do get something very useful from this, because we know that c lies between 0 and x. For example, if you put $x = -1/10$ once again, you get

$$e^{-1/10} \cong 1 - \frac{1}{10} + \frac{1/100}{2} - \frac{1/1000}{6} + \frac{e^c}{24}(1/10000),$$

which reduces to

$$e^{-1/10} = \frac{5429}{6000} + \frac{e^c}{240000}.$$

This time, we know c lies between 0 and $x = -1/10$, so we actually have $-1/10 < c < 0$. Since e^c is increasing in c, it's clearly biggest when c is as big as possible; this means that c would have to be 0, and so e^c can't be bigger than $e^0 = 1$. So the error term is at most $1/240000$. In other words, when we write $e^{-1/10} \cong 5249/6000$, we know that the approximation is accurate to better than $1/240000$, which is about 0.0000041667. (Compare this with the actual value of the difference in Section 24.1 above.)

We'll look at some examples of using Taylor's Theorem in Section 25.3 in the next chapter. Now it's time to check out power and Taylor series.

24.2 Power Series and Taylor Series

Here's another fact:

$$e^x = 1 + x + \frac{x^2}{2!} + \frac{x^3}{3!} + \frac{x^4}{4!} + \frac{x^5}{5!} + \cdots$$

for all real numbers x. You might notice that it looks similar to the approximation at the beginning of Section 24.1 above, but there are two big differences. First, we're no longer dealing with an approximation, and second, there's an infinite series on the right-hand side! Whenever you have an infinite series, you've got to be careful.

So, let's see if we can understand what the above equation actually means. Suppose we start with the right-hand side,

$$1 + x + \frac{x^2}{2!} + \frac{x^3}{3!} + \frac{x^4}{4!} + \frac{x^5}{5!} + \cdots.$$

This looks like a polynomial, but it isn't, since there's no highest-degree term. It just keeps on going forever. In fact, it's an example of a power series. If you replace x by any particular value, you get a regular old series. For example, if $x = -1/10$, you get the series

$$1 - \frac{1}{10} + \frac{1/100}{2!} - \frac{1/1000}{3!} + \frac{1/10000}{4!} - \frac{1/100000}{5!} + \cdots,$$

which you could rewrite as

$$1 - \frac{1}{10} + \frac{1}{100 \times 2!} - \frac{1}{1000 \times 3!} + \frac{1}{10000 \times 4!} - \frac{1}{100000 \times 5!} + \cdots.$$

This series might converge, or it might diverge. So which is it? The answer is that it converges, and what's more, we even know that it converges to $e^{-1/10}$. That's the power of knowing that our above equation is valid for **any** real x:

$$e^x = 1 + x + \frac{x^2}{2!} + \frac{x^3}{3!} + \frac{x^4}{4!} + \frac{x^5}{5!} + \cdots .$$

It just means that if you plug any particular x into the right-hand side, you get a series which converges to the number e^x. We'll prove that this is actually true in Section 24.2.3 below; in the meantime, here are some more examples of what happens when you plug in a few values of x, one at a time:

$$x = 2: \qquad 1 + 2 + \frac{2^2}{2!} + \frac{2^3}{3!} + \frac{2^4}{4!} + \frac{2^5}{5!} + \cdots \qquad \text{converges to } e^2,$$

$$x = -5: \qquad 1 - 5 + \frac{5^2}{2!} - \frac{5^3}{3!} + \frac{5^4}{4!} - \frac{5^5}{5!} + \cdots \qquad \text{converges to } e^{-5},$$

$$x = 0: \qquad 1 + 0 + 0 + 0 + 0 + \cdots \qquad \text{converges to } 1.$$

I could give you a million more examples—actually, infinitely more. This single power series gives us information about infinitely many regular series, one for each value of x. By the way, it's pretty obvious that the last series above converges to 1. There's something special about setting $x = 0$: it makes all the terms vanish except for the constant term. We'll address this point soon; first, let's look at general power series.

24.2.1 Power series in general

A *power series about $x = 0$* is an expression of the form

$$a_0 + a_1 x + a_2 x^2 + a_3 x^3 + a_4 x^4 + \cdots ,$$

where the numbers a_n are fixed constants. Even though a power series isn't a polynomial, we'll still refer to a_n as the *coefficient* of x^n in the power series. The above series can also be written using sigma notation as

$$\sum_{n=0}^{\infty} a_n x^n .$$

In our example from the previous section, the series is

$$1 + x + \frac{x^2}{2!} + \frac{x^3}{3!} + \frac{x^4}{4!} + \frac{x^5}{5!} + \cdots$$

which can be written in sigma notation as

$$\sum_{n=0}^{\infty} \frac{1}{n!} x^n .$$

So this is a power series with coefficients given by $a_n = 1/n!$ for each non-negative integer n. Notice that x is the only true variable here; n is just a dummy variable which goes away when you actually expand the sum. An even simpler power series than the above one, written in expanded form and also using sigma notation, is

$$1 + x + x^2 + x^3 + x^4 + \cdots = \sum_{n=0}^{\infty} x^n.$$

In this series, the coefficients a_n are all equal to 1. Hopefully you can recognize this as a geometric series with first term 1 and ratio x.

We often want to write an equation like

$$f(x) = a_0 + a_1 x + a_2 x^2 + a_3 x^3 + a_4 x^4 + \cdots$$

for some given range of x. This means that when you plug in one of the allowed values of x, the power series becomes a regular old series which converges to the value $f(x)$. For example, we have said (but not yet proved) that

$$e^x = 1 + x + \frac{x^2}{2!} + \frac{x^3}{3!} + \frac{x^4}{4!} + \frac{x^5}{5!} + \cdots$$

for all x. On the other hand, when we looked at how to find the sum of a geometric progression in Section 22.2 of Chapter 22, we saw that

$$1 + r + r^2 + r^3 + r^4 + \cdots = \sum_{n=0}^{\infty} r^n = \frac{1}{1-r} \qquad \text{provided that } -1 < r < 1.$$

Let's replace r by x:

$$1 + x + x^2 + x^3 + x^4 + \cdots = \sum_{n=0}^{\infty} x^n = \frac{1}{1-x} \qquad \text{provided that } -1 < x < 1.$$

That is, we are claiming that

$$\frac{1}{1-x} = 1 + x + x^2 + x^3 + x^4 + \cdots$$

when $-1 < x < 1$. If you replace x by any such number, you get a regular series on the right and the value it converges to on the left. On the other hand, what if $x > 1$ or $x \le -1$? The left-hand side makes sense, but the right-hand side doesn't since the series diverges for these values of x. (Both sides are undefined if x is actually equal to 1.)

Something nice happens to the power series

$$a_0 + a_1 x + a_2 x^2 + a_3 x^3 + a_4 x^4 + \cdots$$

when you set $x = 0$: all the terms vanish except for the a_0 at the beginning, so the series automatically converges (to a_0, of course!). This doesn't tell us anything about whether the series converges for any other value of x. For example, the geometric series only converges when $-1 < x < 1$, while we'll show in Section 26.1.2 in Chapter 26 that the following power series only converges when $x = 0$:

$$\sum_{n=0}^{\infty} n! x^n$$

Now 0 is a pretty funky number, admittedly, but it doesn't need to be more special than the rest of the real numbers. Let's transfer this special property over to some other number a. All we have to do is replace x by $(x - a)$. So here is the general expression for a *power series about $x = a$*:

$$a_0 + a_1(x - a) + a_2(x - a)^2 + a_3(x - a)^3 + a_4(x - a)^4 + \cdots .$$

In sigma notation, this looks like

$$\sum_{n=0}^{\infty} a_n(x - a)^n.$$

This series converges for sure when $x = a$, since all the terms except a_0 vanish. The number a is called the *center* of the power series. When would you want to consider a power series with a center other than 0? One example might be if you wanted to find a power series which converges to $\ln(x)$. This quantity isn't defined at $x = 0$, so it would be silly to try to find a power series about $x = 0$ which converges to $\ln(x)$. On the other hand, we can find a power series with center 1 which converges to $\ln(x)$, at least for some values of x. Indeed, at the end of Section 26.2.1 of Chapter 26, we'll see that the equation

$$\sum_{n=1}^{\infty} \frac{(-1)^{n-1}}{n}(x - 1)^n = \ln(x)$$

is valid for $-1 < (x - 1) < 1$, that is, for $0 < x < 2$. (It's actually even true for $x = 2$:

$$\sum_{n=1}^{\infty} \frac{(-1)^{n-1}}{n} = 1 - \frac{1}{2} + \frac{1}{3} - \frac{1}{4} + \frac{1}{5} - \cdots = \ln(2).$$

This isn't so easy to prove, however!)

24.2.2 Taylor series and Maclaurin series

In the previous section, we saw that a general power series about $x = a$ is given (using sigma notation and also in expanded form) by

$$\sum_{n=0}^{\infty} a_n(x - a)^n = a_0 + a_1(x - a) + a_2(x - a)^2 + a_3(x - a)^3 + a_4(x - a)^4 + \cdots .$$

This converges for $x = a$, and might converge for other values of x. In Section 26.1.2 of Chapter 26, we'll look at some methods for finding which values of x make the series converge. We could then plug in all these values of x one at a time, find what the series converges to in each case, and call that $f(x)$. So, starting with a power series, we have defined a function.

Suppose that we instead start off with some smooth function f. We're going to define a special power series about $x = a$ by using all the derivatives of f:

$$\sum_{n=0}^{\infty} \frac{f^{(n)}(a)}{n!}(x - a)^n.$$

When you expand the sigma notation, this becomes

$$f(a) + f'(a)(x-a) + \frac{f''(a)}{2!}(x-a)^2 + \frac{f^{(3)}(a)}{3!}(x-a)^3 + \frac{f^{(4)}(a)}{4!}(x-a)^4 + \cdots.$$

The coefficients of this power series are given by $a_n = f^{(n)}(a)/n!$. The series is called the *Taylor series of f about $x = a$*. So, starting with a function, we have defined a power series.

Take a closer look at the definition of the Taylor series above. It should look familiar. In fact, the formula is very similar to the definition of the Taylor polynomial $P_N(x)$ from Section 24.1.3 above. The only difference is that the sum doesn't stop at $n = N$: it keeps on going to ∞. In other words, the Taylor polynomial $P_N(x)$ is the Nth partial sum of the Taylor series.

We'll explore the connection between Taylor polynomials and Taylor series in the next section. First, we have just one more definition: the *Maclaurin series of f* is just another name for the Taylor series of f about $x = 0$. So it's given by

$$\sum_{n=0}^{\infty} \frac{f^{(n)}(0)}{n!} x^n,$$

or in expanded form by

$$f(0) + f'(0)x + \frac{f''(0)}{2!}x^2 + \frac{f^{(3)}(0)}{3!}x^3 + \frac{f^{(4)}(0)}{4!}x^4 + \cdots.$$

Whenever you see the words "Maclaurin series," mentally replace them by "Taylor series with $a = 0$" and you'll do just fine.

24.2.3 Convergence of Taylor series

OK, let's review the situation. We started out with a function f and a number a, and we constructed the Taylor series of f about $x = a$:

$$\sum_{n=0}^{\infty} \frac{f^{(n)}(a)}{n!}(x-a)^n.$$

This is a power series with center a, but it's not just any old power series: it encapsulates the values of all the derivatives of f at $x = a$. It would be really cool if we could write

$$f(x) = \sum_{n=0}^{\infty} \frac{f^{(n)}(a)}{n!}(x-a)^n,$$

since then we'd know that the Taylor series converges for any x and also that it converges to the original function value $f(x)$. The problem is, the above equation isn't always valid. The series could diverge for some values of x, or even every value of x (except $x = a$: as we've seen, a power series always converges at its center). Even worse, the series could converge to something other than $f(x)$! Luckily, in our examples, we'll avoid this weird possibility.*

---

*I will just mention a classic example of a whacked-out Taylor series: if $f(x) = e^{-1/x^2}$ when $x \neq 0$, and we also define $f(0) = 0$, then all the derivatives of f at 0 are equal to 0, so the Taylor series of f with center 0 is just 0. This is not the same thing as $f(x)$ at all, except when $x = 0$.

So, how do you know if and when a Taylor series actually converges to its underlying function? Start by writing

$$f(x) = P_N(x) + R_N(x),$$

as we did in Section 24.1.4 above. Remember,

$$P_N(x) = \sum_{n=0}^{N} \frac{f^{(n)}(a)}{n!}(x-a)^n \qquad \text{and} \qquad R_N(x) = \frac{f^{(N+1)}(c)}{(N+1)!}(x-a)^{N+1}.$$

This expresses $f(x)$ as its approximate value $P_N(x)$ plus the error, or remainder, $R_N(x)$. Now here's the clever part: we let N get larger and larger. This should hopefully make the approximation $P_N(x)$ get closer and closer to the true value $f(x)$. This is the same thing as saying that hopefully the error $R_N(x)$ gets smaller and smaller.

Let's try to write down some equations to describe all this. Suppose that for some x, we know that

$$\lim_{N \to \infty} R_N(x) = 0.$$

In the equation $f(x) = P_N(x) + R_N(x)$, take limits as $N \to \infty$:

$$\lim_{N \to \infty} f(x) = \lim_{N \to \infty} P_N(x) + \lim_{N \to \infty} R_N(x) = \lim_{N \to \infty} P_N(x).$$

Since $f(x)$ doesn't depend on N, the left-hand side is just $f(x)$, so we know that

$$f(x) = \lim_{N \to \infty} P_N(x) = \lim_{N \to \infty} \sum_{n=0}^{N} \frac{f^{(n)}}{n!}(x-a)^n = \sum_{n=0}^{\infty} \frac{f^{(n)}}{n!}(x-a)^n.$$

So $f(x)$ equals its Taylor series! In other words, **if you want to prove that a function equals its Taylor series at some number x, try to show that $R_N(x) \to 0$ as $N \to \infty$.**

Let's do exactly this for $f(x) = e^x$ with $a = 0$. By adapting some stuff we looked at in Section 24.1.4 above, you should be able to see that

$$P_N(x) = \sum_{n=0}^{N} \frac{x^n}{n!} = 1 + x + \frac{x^2}{2!} + \frac{x^3}{3!} + \cdots + \frac{x^N}{N!},$$

and that

$$R_N(x) = \frac{e^c}{(N+1)!}x^{N+1}$$

for some c between x and 0. Now, we need to find the limit of $R_N(x)$ as $N \to \infty$ and show that it's zero:

$$\lim_{N \to \infty} R_N(x) = \lim_{N \to \infty} e^c \frac{x^{N+1}}{(N+1)!}.$$

In Section 24.3 below, I'll prove that

$$\lim_{N \to \infty} \frac{x^{N+1}}{(N+1)!} = 0$$

for any x. We have to be a little careful about the e^c factor, since c depends on N. The question is, how big can e^c be? Remember that c is between 0 and x. If x is negative, the biggest e^c could be is if $c = 0$, which means that $e^c \leq 1$. If x is positive, the biggest e^c could be is if $c = x$, which means that $e^c \leq e^x$. In either case, since x is fixed (that is, treated as constant), we can write that $0 \leq e^c \leq C$, where C is another constant. This is true no matter what N is, even though c is wobbling around all over the place as N is changing. Anyway, hopefully you believe this, in which case you might believe that

$$0 \leq e^c \frac{|x|^{N+1}}{(N+1)!} \leq C \frac{|x|^{N+1}}{(N+1)!}.$$

Now the left-hand and right-hand sides go to 0 as N tends to ∞, so we can apply the sandwich principle to see that the middle quantity does too. So, we've proved that

$$\lim_{N \to \infty} R_N(x) = 0$$

for any real x. This means that we have finally proved that

$$e^x = 1 + x + \frac{x^2}{2!} + \frac{x^3}{3!} + \frac{x^4}{4!} + \frac{x^5}{5!} + \cdots$$

for all real x.

 Let's try to see everything in one self-contained example by finding the Maclaurin series of $f(x) = \cos(x)$ and showing that it converges to $f(x)$ for all x. First, we need to differentiate f over and over again, then plug in 0 for each derivative and see what happens. Well, when you differentiate $\cos(x)$ with respect to x, over and over again, you get $-\sin(x)$, then $-\cos(x)$, then $\sin(x)$, then $\cos(x)$, then $-\sin(x)$, then $-\cos(x)$, and so on. Clearly this cycle will keep on going. When you plug in $x = 0$, the $\sin(x)$ terms go away, and the $\pm \cos(x)$ terms become ± 1. So the sequence of numbers $f^{(n)}(0)$ looks like this:

$$1, 0, -1, 0, 1, 0, -1, 0, 1, 0, -1, 0, \ldots.$$

If you plug these numbers into the Maclaurin series formula

$$f(0) + f'(0)x + \frac{f''(0)}{2!}x^2 + \frac{f^{(3)}(0)}{3!}x^3 + \frac{f^{(4)}(0)}{4!}x^4 + \frac{f^{(5)}(0)}{5!}x^5 + \frac{f^{(6)}(0)}{6!}x^6 + \cdots,$$

all the odd-degree terms go away and you get

$$1 - \frac{1}{2!}x^2 + \frac{1}{4!}x^4 - \frac{1}{6!}x^6 + \cdots,$$

which you can rewrite more compactly as

$$1 - \frac{x^2}{2!} + \frac{x^4}{4!} - \frac{x^6}{6!} + \cdots.$$

This is the Maclaurin series for $\cos(x)$, or if you prefer, the Taylor series for $\cos(x)$ about $x = 0$. To get the corresponding Taylor polynomials, all you have to do is chop off the series at the right place. For example,

$$P_4(x) = 1 - \frac{1}{2!}x^2 + \frac{1}{4!}x^4.$$

By the way, the formula for $P_5(x)$ is the same as for $P_4(x)$, since there's no fifth-degree term in the above Maclaurin series. This is a good example of why we need the word "order": P_5 is of order 5, but degree only 4.

Now, all that's left is to prove that $\cos(x)$ actually equals its Maclaurin series for all real x:

$$\cos(x) = 1 - \frac{x^2}{2!} + \frac{x^4}{4!} - \frac{x^6}{6!} + \cdots.$$

To do this, we need to show that

$$\lim_{N \to \infty} R_N(x) = 0.$$

We know that

$$R_N(x) = \frac{f^{(N+1)}(c)}{(N+1)!} x^{N+1},$$

where c is between 0 and x. Let's take absolute values:

$$|R_N(x)| = \frac{|f^{(N+1)}(c)|}{(N+1)!} |x|^{N+1}.$$

All the derivatives of f are equal to either $\pm\cos(x)$ or $\pm\sin(x)$, so the quantity $|f^{(N+1)}(c)|$ is either $|\cos(c)|$ or $|\sin(c)|$. In either case, this quantity is less than or equal to 1, so we have

$$0 \le |R_N(x)| \le \frac{1}{(N+1)!} |x|^{N+1}.$$

Once again, we'll show in the next section that

$$\lim_{N \to \infty} \frac{x^{N+1}}{(N+1)!} = 0.$$

Now you can use the sandwich principle to show that

$$\lim_{N \to \infty} |R_N(x)| = 0,$$

which means that also

$$\lim_{N \to \infty} R_N(x) = 0.$$

We have proved that

$$\cos(x) = 1 - \frac{x^2}{2!} + \frac{x^4}{4!} - \frac{x^6}{6!} + \cdots$$

for all real x. Let's celebrate by expressing the above series in sigma notation. (What, isn't that how you celebrate solving a tough problem?) Anyway, how do you get only even powers of x? The answer is to use $2n$ instead of n (see the end of Section 15.1 in Chapter 15 for a discussion of this sort of thing). Since the factorial on the bottom matches the degree, we might guess that the Maclaurin series can be written as

$$\sum_{n=0}^{\infty} \frac{x^{2n}}{(2n)!}.$$

The problem is, this series doesn't alternate. So insert a factor of $(-1)^n$:

$$\sum_{n=0}^{\infty} \frac{(-1)^n x^{2n}}{(2n)!}.$$

If you expand this, you'll find that it works. Here's a summary of what we found:

$$\cos(x) = \sum_{n=0}^{\infty} \frac{(-1)^n x^{2n}}{(2n)!} = 1 - \frac{x^2}{2!} + \frac{x^4}{4!} - \frac{x^6}{6!} + \cdots$$

for all real x.

24.3 A Useful Limit

This section isn't about power series at all—it just contains a proof of a limit we needed twice in the previous section:

$$\lim_{N \to \infty} \frac{x^{N+1}}{(N+1)!} = 0$$

for any real number x. By letting $n = N + 1$ (think of it like a substitution in an integral), this is exactly the same as showing that

$$\lim_{n \to \infty} \frac{x^n}{n!} = 0$$

for any real number x. There are several ways to prove this last statement, but here's a sneaky way. Let me explain the logic I'm going to use, then actually do it. I'm going to prove that the series

$$\sum_{n=0}^{\infty} \frac{x^n}{n!}$$

converges, regardless of what x is. (Yes, we "know" that it actually converges to e^x, but not until after we finish showing that the above limit is zero after all!) Anyway, it doesn't matter what the series converges to; simply knowing that it converges is enough. Why? Because then the nth term $x^n/n!$ **must** go to 0 as n goes to ∞, or else the nth term test would fail. That is, if the terms didn't go to 0 as n goes to ∞, then the series would diverge. So let's use the ratio test to show that the series converges for all x. Let's fix x for once and for all and, with $a_n = x^n/n!$, simply look at the limiting ratio:

$$L = \lim_{n \to \infty} \left| \frac{a_{n+1}}{a_n} \right| = \lim_{n \to \infty} \left| \frac{x^{n+1}/(n+1)!}{x^n/n!} \right| = \lim_{n \to \infty} \left| \frac{x^{n+1}}{x^n} \frac{n!}{(n+1)!} \right|.$$

Now we know that $n!/(n+1)!$ boils down to $1/(n+1)$, so this last limit is

$$\lim_{n \to \infty} |x| \frac{1}{n+1} = 0,$$

since $|x|$ is fixed and $1/(n+1)$ goes to 0. The limit L is 0, which is less than 1, so the series converges and we have, as a by-product, shown that the useful limit is correctly stated above. By the way, the technique of fixing x and then applying the ratio test to see whether the series converges for that particular x will be used many times in Section 26.1.2 of Chapter 26.

CHAPTER 25 _____

How to Solve Estimation Problems

In the previous chapter, we showed how Taylor polynomials can be used to estimate (or approximate, if you prefer) certain quantities. We also saw that the remainder term could be used to get an idea of how good the approximation actually is. In this chapter, we'll develop these techniques and look an number of examples. So, here's the plan for the chapter:

- a review of the most important facts about Taylor polynomials and series;
- how to find Taylor polynomials and series;
- estimation problems; and
- a different method for analyzing the error.

25.1 Summary of Taylor Polynomials and Series

Here are the most important facts about Taylor polynomials and series, all of which were developed in the previous chapter:

1. Of all the polynomials of degree N or less, the one which best approximates the smooth function f for x near a is called the Nth-order Taylor polynomial about $x = a$, and is given by

$$P_N(x) = f(a) + f'(a)(x - a) + \frac{f''(a)}{2!}(x - a)^2$$
$$+ \frac{f^{(3)}(a)}{3!}(x - a)^3 + \cdots + \frac{f^{(N)}(a)}{N!}(x - a)^N.$$

Using sigma notation, this can be written as

$$P_N(x) = \sum_{n=0}^{N} \frac{f^{(n)}(a)}{n!}(x - a)^n.$$

2. The polynomial P_N has the same derivatives as f at $x = a$, up to and

including order N. That is,

$$P_N(a) = f(a), \quad P_N'(a) = f'(a), \quad P_N''(a) = f''(a), \quad P_N^{(3)}(a) = f^{(3)}(a),$$

and so on up to $P_N^{(N)}(a) = f^{(N)}(a)$. The above equations aren't true in general if a is replaced by any other number, or for derivatives of order higher than N. (In fact, the derivatives of P_N of order higher than N are identically 0, since P_N is a polynomial of degree N.)

3. The Nth-order remainder term $R_N(x)$, otherwise known as the Nth-order error term, is simply the difference $f(x) - P_N(x)$. It follows that

$$\boxed{f(x) = P_N(x) + R_N(x)}$$

for any N. The remainder term is given by

$$\boxed{R_N(x) = \frac{f^{(N+1)}(c)}{(N+1)!}(x-a)^{N+1}}$$

where c is some number between x and a which cannot be computed in general.

4. So, the complete expression for $f(x)$ is given by

$$\boxed{f(x) = \sum_{n=0}^{N} \frac{f^{(n)}(a)}{n!}(x-a)^n + \frac{f^{(N+1)}(c)}{(N+1)!}(x-a)^{N+1}.}$$

5. The infinite series

$$\boxed{\sum_{n=0}^{\infty} \frac{f^{(n)}(a)}{n!}(x-a)^n}$$

is called the *Taylor series* of $f(x)$ about $x = a$. For any particular x, this series may or may not converge. If for any particular x the remainder term $R_N(x)$ converges to 0 as $N \to \infty$, then we can write

$$f(x) = \sum_{n=0}^{\infty} \frac{f^{(n)}(a)}{n!}(x-a)^n$$

for that x. That is, $f(x)$ is equal to its Taylor series representation (about $x = a$) at the point x.

6. In the special case where $a = 0$, the Taylor series is

$$\sum_{n=0}^{\infty} \frac{f^{(n)}(0)}{n!}x^n.$$

This is called the *Maclaurin series* of $f(x)$. So, whenever you see the words "Maclaurin series," you can mentally replace them by "Taylor series about $x = 0$."

25.2 Finding Taylor Polynomials and Series

Suppose you want to find a certain Taylor polynomial or series. If you're lucky, you can take a Taylor polynomial or series you already know, manipulate it, and get the polynomial or series you want. We'll see some techniques of how to do this in Section 26.2 of the next chapter. Unfortunately, this doesn't always work: sometimes, you need to bust out the formula for the Taylor series of f about $x = a$ from the above summary:

$$\sum_{n=0}^{\infty} \frac{f^{(n)}(a)}{n!}(x-a)^n.$$

Knowing the number a and the function f, you have to find the values of all the derivatives of f, evaluated at $x = a$, and then plug them into the above formula. This can be a real pain in the butt, however! Differentiating once or twice is bad enough, but differentiating hundreds and thousands of times is ridiculous. Things aren't so bad if you only want to find a Taylor polynomial of low degree, since then you only have to calculate a few derivatives. We'll also see some nice tricks in Section 26.2 that can help you avoid the above formula altogether, if you're lucky.

On the other hand, some functions are really easy to differentiate. One such example is the function f defined by $f(x) = e^x$; we looked at the Maclaurin series of this function in the previous chapter. What if you don't want the Maclaurin series of f, but instead you want the Taylor series about $x = -2$? Well, put $a = -2$ instead of 0 in the above formula to see that we are looking for

$$\sum_{n=0}^{\infty} \frac{f^{(n)}(-2)}{n!}(x+2)^n.$$

We need the values of $f^{(n)}(-2)$ for many values of n, so it's really helpful to set up a table of derivatives. In general, the template should look like this:

n	$f^{(n)}(x)$	$f^{(n)}(a)$
0		
1		
2		
3		

The middle column of this table should be filled in first. Start off with the function itself in the top row, then just keep differentiating. Each time you differentiate, put the result in the next row of the table (still in the middle column). When the middle column is all filled in, substitute $x = a$ into each entry in the middle column and enter the value in the same row in the third column. Note that you may have to use more rows—it depends how big n is, or how soon you can work out the pattern. In our example, $a = -2$ and all of the derivatives of $f(x)$ are e^x, so the filled-in table looks like this:

n	$f^{(n)}(x)$	$f^{(n)}(-2)$
0	e^x	e^{-2}
1	e^x	e^{-2}
2	e^x	e^{-2}
3	e^x	e^{-2}

The pattern is pretty clear: $f^{(n)}(-2) = e^{-2}$ for all n. If you plug that into our formula

$$\sum_{n=0}^{\infty} \frac{f^{(n)}(-2)}{n!}(x+2)^n$$

from above, you get the Taylor series for e^x about $x = -2$:

$$\sum_{n=0}^{\infty} \frac{e^{-2}}{n!}(x+2)^n.$$

It's a great idea to make sure you can expand this without using sigma notation:

$$e^{-2} + e^{-2}(x+2) + \frac{e^{-2}}{2!}(x+2)^2 + \frac{e^{-2}}{3!}(x+2)^3 + \cdots.$$

Here's another example: find the Taylor series of $\sin(x)$ about $x = \pi/6$, showing terms up to fourth order. We start off with a table of derivatives:

n	$f^{(n)}(x)$	$f^{(n)}(\pi/6)$
0	$\sin(x)$	$1/2$
1	$\cos(x)$	$\sqrt{3}/2$
2	$-\sin(x)$	$-1/2$
3	$-\cos(x)$	$-\sqrt{3}/2$
4	$\sin(x)$	$1/2$

It looks similar to the table we used to find the Maclaurin series, but now we are evaluating the derivatives at $\pi/6$ instead of at 0. So whip out the standard formula for a Taylor series:

$$\sum_{n=0}^{\infty} \frac{f^{(n)}(a)}{n!}(x-a)^n.$$

Let's expand this:

$$f(a) + f'(a)(x-a) + \frac{f''(a)}{2!}(x-a)^2 + \frac{f^{(3)}(a)}{3!}(x-a)^3 + \frac{f^{(4)}(a)}{4!}(x-a)^4 + \cdots.$$

Now put $a = \pi/6$ and plug in the values from the table above to see that the Taylor series of $\sin(x)$ about $x = \pi/6$ is

$$\frac{1}{2} + \frac{\sqrt{3}}{2}\left(x - \frac{\pi}{6}\right) + \frac{-1/2}{2!}\left(x - \frac{\pi}{6}\right)^2 + \frac{-\sqrt{3}/2}{3!}\left(x - \frac{\pi}{6}\right)^3 + \frac{1/2}{4!}\left(x - \frac{\pi}{6}\right)^4 + \cdots.$$

This is a lot harder to write out in sigma notation, so we'll just tidy it up a bit and leave it like this:

$$\frac{1}{2} + \frac{\sqrt{3}}{2}\left(x - \frac{\pi}{6}\right) - \frac{1}{2 \times 2!}\left(x - \frac{\pi}{6}\right)^2 - \frac{\sqrt{3}}{2 \times 3!}\left(x - \frac{\pi}{6}\right)^3 + \frac{1}{2 \times 4!}\left(x - \frac{\pi}{6}\right)^4 + \cdots.$$

Of course, to find the fourth-order Taylor polynomial $P_4(x)$ (still about the center $x = \pi/6$), just drop the "$+ \cdots$" at the end. If you only want $P_3(x)$, then also drop the last term, so that the final power is 3:

$$P_3(x) = \frac{1}{2} + \frac{\sqrt{3}}{2}\left(x - \frac{\pi}{6}\right) - \frac{1}{4}\left(x - \frac{\pi}{6}\right)^2 - \frac{\sqrt{3}}{12}\left(x - \frac{\pi}{6}\right)^3.$$

(Now we replaced 2! by 2 and 3! by 6.) On the other hand, if you actually wanted $P_5(x)$, you'd have to add another row at the end of the above table corresponding to $n = 5$, so that you get the extra term in $(x - \pi/6)^5$ that you need.

One more example: what is the Maclaurin series of $(1 + x)^{1/2}$? Since we want a Maclaurin series, we need to set $a = 0$. Let's draw up a table of derivatives up to fourth order:

n	$f^{(n)}(x)$	$f^{(n)}(0)$
0	$(1+x)^{1/2}$	1
1	$\frac{1}{2}(1+x)^{-1/2}$	$1/2$
2	$-\frac{1}{4}(1+x)^{-3/2}$	$-1/4$
3	$\frac{3}{8}(1+x)^{-5/2}$	$3/8$
4	$-\frac{15}{16}(1+x)^{-7/2}$	$-15/16$

Now, let's write down the general formula for the Maclaurin series,

$$f(0) + f'(0)x + \frac{f''(0)}{2!}x^2 + \frac{f^{(3)}(0)}{3!}x^3 + \frac{f^{(4)}(0)}{4!}x^4 + \cdots,$$

then plug in the numbers for the derivatives from the above table to get

$$1 + \frac{1}{2}x + \frac{-1/4}{2!}x^2 + \frac{3/8}{3!}x^3 + \frac{-15/16}{4!}x^4 + \cdots.$$

Let's simplify this as

$$1 + \frac{x}{2} - \frac{x^2}{8} + \frac{x^3}{16} - \frac{5x^4}{128} + \cdots.$$

In fact, it turns out that the remainder term goes to 0 when x is between -1 and 1 (this is tricky to prove!), so we actually have

$$(1+x)^{1/2} = 1 + \frac{x}{2} - \frac{x^2}{8} + \frac{x^3}{16} - \frac{5x^4}{128} + \cdots$$

when $-1 < x < 1$. This is a special case of the *binomial theorem*, which says that

$$(1+x)^a = 1 + ax + \frac{a(a-1)}{2!}x^2 + \frac{a(a-1)(a-2)}{3!}x^3$$
$$+ \frac{a(a-1)(a-2)(a-3)}{4!}x^4 + \cdots$$

for $-1 < x < 1$. The series on the right-hand side diverges when $x > 1$ or $x < -1$ unless a happens to be a nonnegative integer. (In that case, the right-hand side is actually a polynomial. Can you see why?)

25.3 Estimation Problems Using the Error Term

In Section 24.1.4 of the previous chapter, we used a third-order Taylor polynomial P_3 to estimate $e^{-1/10}$; then we used the remainder term R_3 to get an idea of how good our approximation was. Let's revisit these methods and generalize them.

To set the scene, consider the following two similar examples:

1. Estimate $e^{1/3}$ using a Taylor polynomial of order 2, and also estimate the error.

2. Estimate $e^{1/3}$ with an error no more than $1/10000$.

The second problem is more difficult than the first one. You see, in the first problem, we know that we're dealing with a Taylor polynomial of order 2, so we can set $N = 2$ in our formulas. In the second problem, we actually have to find N, which is one more thing to worry about.

 With these two types of problems in mind, check out the general method for solving estimation (or approximation) problems:

1. Look at what you want to estimate, and pick a relevant function f. In our examples above, we want to estimate $e^{\text{something}}$, so set $f(x) = e^x$. Later on, we will set $x = 1/3$, since $f(1/3) = e^{1/3}$, the quantity we want to estimate.

2. Pick a number a which is pretty close to this value of x, and so that $f(a)$ is really nice. This means that you should be able to write down $f(a)$ exactly, as well as $f'(a)$, $f''(a)$, and so on. In our example, we'll put $a = 0$, since that's pretty close to $1/3$ and also e^0 is easy to compute.

3. Make a table of derivatives of f, just like we did in the previous section. It should have three columns which show the values of n, $f^{(n)}(x)$, and $f^{(n)}(a)$. If you know the order of the Taylor polynomial to use, that's the value of N you'll need; make sure to go up to the $(N+1)$th derivative in the table. Otherwise, just write down as many rows as you can be bothered to; you can always fill in more later if you need to.

4. If you don't care about the error in your estimate, skip to step 8. Otherwise, write down the formula for $R_N(x)$:

$$R_N(x) = \frac{f^{(N+1)}(c)}{(N+1)!}(x-a)^{N+1}$$

making sure to write "c is between a and x." As you're writing, replace a by its true value on the fly, including in your comment about c.

5. If you know the order of the Taylor polynomial to use, replace N by this number in the above formula. If not, make an educated guess based on how small you need the error to be. The smaller, the higher N should be. For many problems, $N = 2$ or 3 will do nicely. If you're wrong, you'll know soon enough; you'll just have to repeat this step and the next two steps with a higher value of N.

6. Now, replace x by the value you want in the formula for $R_N(x)$. No unknown variables should be left except for c, and you should write

down the possible range of c as an inequality. In our case, with $a = 0$ and $x = 1/3$, we know that c lies in between, so we'd write $0 < c < 1/3$.

7. Find the maximum value of $|R_N(x)|$, where c lies in the appropriate interval. This is how big the error can possibly be. If you know the value of N, you're all done with the error estimate. If not, compare the actual error with the one you want. If your actual error is smaller, that's great—you have found a good value of N. Otherwise, you're a little bit screwed—you have to go back to step 5 and try again. (We'll look at some techniques for maximizing $|R_N(x)|$ in Section 25.3.6 below.)

8. Finally, it's time to find the actual estimate! Write down the formula for $P_N(x)$:

$$P_N(x) = f(a) + f'(a)(x - a) + \frac{f''(a)}{2!}(x - a)^2$$
$$+ \frac{f^{(3)}(a)}{3!}(x - a)^3 + \cdots + \frac{f^{(N)}(a)}{N!}(x - a)^N.$$

Now replace a and N by the values from above to get a formula in terms of x alone. Finally, write down the approximation

$$\boxed{f(x) \cong P_N(x)}$$

and plug in the actual value of x that you need. The left-hand side will be the quantity you want, and the right-hand side will be the approximation.

9. One other piece of information is available if you want it: if $R_N(x)$ is positive, your estimate is an underestimate; if $R_N(x)$ if negative, the estimate is an overestimate. These facts follow from the equation

$$\boxed{f(x) = P_N(x) + R_N(x).}$$

Now, let's look at five examples of these types of problems.

25.3.1 First example

We'd better start with the two questions from the previous section. In the first problem, we want to estimate $e^{1/3}$ by using a second-order Taylor polynomial. This is actually quite similar to our example involving $e^{-1/10}$ from Section 24.1.4 of the previous chapter. Anyway, let's follow the above method. We start by picking f; since we're exponentiating, let's set $f(x) = e^x$ and note that our quantity $e^{1/3}$ is just $f(1/3)$. Eventually, we'll put $x = 1/3$, but not yet. We also need to pick a close to $1/3$ so that e^a is nice; as I mentioned, 0 is a natural choice.

Now, it's time for a table of derivatives:

n	$f^{(n)}(x)$	$f^{(n)}(0)$
0	e^x	1
1	e^x	1
2	e^x	1
3	e^x	1

I went up to 3, since that's one more than 2, and we need a second-order Taylor polynomial (that is, $N = 2$). OK, so moving on, the error term is

$$R_N(x) = \frac{f^{(N+1)}(c)}{(N+1)!}x^{N+1},$$

where c is between 0 and x. Notice that I replaced a by 0 in the standard formula for $R_N(x)$. Now, we know that $N = 2$, so we actually need

$$R_2(x) = \frac{f^{(3)}(c)}{3!}x^3 = \frac{e^c}{6}x^3.$$

I read $f^{(3)}(c) = e^c$ from the last row of the middle column in the table above, replacing x by c. Now, let's replace x by $1/3$ to see that

$$R_2(1/3) = \frac{e^c}{6}(1/3)^3 = \frac{e^c}{162};$$

here c is between 0 and $x = 1/3$, so $0 < c < 1/3$. Let's take absolute values:

$$|R_2(1/3)| = \left|\frac{e^c}{162}\right| = \frac{e^c}{162},$$

since e^c must be positive. Next, we need to maximize $|R_2(1/3)|$. Since e^c is increasing in c, the largest value occurs when $c = 1/3$. This shows that

$$|R_2(1/3)| = \frac{e^c}{162} < \frac{e^{1/3}}{162}.$$

We seem to have a problem, since we don't know what $e^{1/3}$ is. That's actually the whole point of the question in the first place! Never mind, let's make a gross overestimate for $e^{1/3}$. You see, $e < 8$, so $e^{1/3} < 8^{1/3}$ which is just 2. Why did I choose 8? Because I can take the cube root of it without thinking too much! Anyway, using the inequality $e^{1/3} < 2$, the above inequality for $|R_2(1/3)|$ becomes

$$|R_2(1/3)| = \frac{e^c}{162} < \frac{e^{1/3}}{162} < \frac{2}{162} = \frac{1}{81}.$$

So the error is no more than $1/81$. We still need to find the estimate. Let's write down the formula for $P_2(x)$, using the fact that $a = 0$:

$$P_2(x) = f(0) + f'(0)x + \frac{f''(0)}{2!}x^2.$$

From the above table, we can replace all of $f(0)$, $f'(0)$, and $f''(0)$ by 1:

$$P_2(x) = 1 + x + \frac{1}{2}x^2.$$

Finally, put $x = 1/3$ to get

$$P_2(1/3) = 1 + \frac{1}{3} + \frac{1}{2}\left(\frac{1}{3}\right)^2 = \frac{25}{18}.$$

Since $f(x) \cong P_2(x)$, we have

$$f(1/3) \cong P_2(1/3).$$

Using the fact that $f(x) = e^x$, we see that

$$e^{1/3} = f(1/3) \cong P_2(1/3) = \frac{25}{18}.$$

We have already shown that $|R_2(1/3)| < 1/81$, so our estimate is accurate to at least $1/81$. In fact, since $R_2(1/3)$ is positive, our estimate $25/18$ is an underestimate to the true value $e^{1/3}$.

25.3.2 Second example

Now we'll do the second example from Section 25.3 above: estimate $e^{1/3}$ with an error less than $1/10000$. Just as in the previous example, we'll set $f(x) = e^x$, $a = 0$, and eventually we'll put $x = 1/3$. Once again, we have

$$R_N(x) = \frac{f^{(N+1)}(c)}{(N+1)!} x^{N+1},$$

where c is between 0 and x. We already know from the previous example that $N = 2$ won't work, since we got a maximum error of $1/81$ and we need the error to be less than $1/10000$. So, let's see if $N = 3$ will work. The error term is now

$$R_3(x) = \frac{f^{(4)}(c)}{4!} x^4 = \frac{e^c}{24} x^4,$$

where c is between 0 and x. Put $x = 1/3$ to get

$$R_3(1/3) = \frac{e^c}{24} \left(\frac{1}{3}\right)^4 = \frac{e^c}{24 \times 81},$$

where $0 < c < 1/3$. Again, we can use the fact from the previous section that $e^c < 2$ if c is between 0 and $1/3$:

$$|R_3(1/3)| = \frac{|e^c|}{24 \times 81} < \frac{2}{24 \times 81} = \frac{1}{972}.$$

This is not less than $1/10000$, so $N = 3$ is not big enough. Let's try $N = 4$. Repeating the above steps, we have

$$R_4(x) = \frac{f^{(5)}(c)}{5!} x^5 = \frac{e^c}{120} x^5,$$

so plugging in $x = 1/3$, we see that

$$R_4(1/3) = \frac{e^c}{120} \left(\frac{1}{3}\right)^5 = \frac{e^c}{120 \times 243}.$$

Again c is between 0 and $1/3$, and again $e^c < 2$ there, so

$$|R_4(1/3)| < \frac{2}{120 \times 243} = \frac{1}{14580}.$$

(If you think you need a calculator to work out the last fraction, think again—you can reduce $2/120$ down to $1/60$, then work out 6×243, multiply it by 10, and stick it in the denominator.) In any case, we know that $|R_4(1/3)|$ is plenty less than $1/10000$, so we're golden: we can take $N = 4$. So what is the estimate? We need to find $P_4(1/3)$. In general, when $a = 0$, the fourth-order Taylor polynomial P_4 is given by

$$P_4(x) = 1 + x + \frac{x^2}{2!} + \frac{x^3}{3!} + \frac{x^4}{4!},$$

so

$$P_4\left(\frac{1}{3}\right) = 1 + \frac{1}{3} + \frac{(1/3)^2}{2} + \frac{(1/3)^3}{6} + \frac{(1/3)^4}{24} = 1 + \frac{1}{3} + \frac{1}{18} + \frac{1}{162} + \frac{1}{1944} = \frac{2713}{1944}.$$

That is,

$$e^{1/3} = f(1/3) \cong P_4(1/3) = \frac{2713}{1944}.$$

So, we can replace our estimate $25/18$ from the previous example by a much better estimate, namely $2713/1944$. This new estimate is guaranteed to be within $1/10000$ of the true value $e^{1/3}$. As a test, I did use a calculator to see that $2713/1944$ is 1.39558 to five decimal places, whereas $e^{1/3}$ is 1.39561 to five decimal places. These quantities are therefore at most 0.00004 apart, which is well within the allowed tolerance of $1/10000 = 0.0001$.

25.3.3 Third example

Here's a question: estimate $\sqrt{27}$ with an error of no more than $1/250$. According to the above method, we have to select an appropriate function f and values of a and x. A good choice of the function would be given by $f(x) = \sqrt{x}$, or if you prefer, $f(x) = x^{1/2}$. Then we want to estimate $f(27) = \sqrt{27}$, so eventually we'll set $x = 27$. Now we need a number close to 27 that we can easily take the square root of. It seems as if 25 is pretty good, so let's take $a = 25$. That takes care of the first step. Moving on to step 2, let's draw up a table of derivatives:

n	$f^{(n)}(x)$	$f^{(n)}(25)$
0	$x^{1/2}$	5
1	$\frac{1}{2}x^{-1/2}$	$1/10$
2	$-\frac{1}{4}x^{-3/2}$	$-1/500$
3	$\frac{3}{8}x^{-5/2}$	$3/8 \times 1/5^5$

Remember, to fill in this table, we put the entry $x^{1/2}$ in the top row of the middle column, and then differentiated a few times, putting the results in each successive row in the middle column. Finally, the entries in the right-hand column come from substituting the value $a = 25$. The difficulty is that we don't know how much of this table we need. Perhaps we'll even need more rows.

So let's look at the error term, which is given by

$$R_N(x) = \frac{f^{(N+1)}(c)}{(N+1)!}(x - 25)^{N+1},$$

where c is between x and 25. Since we care about $x = 27$, let's substitute that in:

$$R_N(27) = \frac{f^{(N+1)}(c)}{(N+1)!}(27 - 25)^{N+1} = \frac{f^{(N+1)}(c)}{(N+1)!}2^{N+1},$$

where $25 \le c \le 27$. Now, how lucky do you feel? Maybe $N = 0$ will be good enough! Let's try it and see:

$$|R_0(27)| = \left|\frac{f'(c)}{1!}(27 - 25)^1\right| = \frac{1}{2}c^{-1/2} \times 2 = c^{-1/2},$$

where we have used the above table to find $f'(c)$ and dropped absolute values since everything's positive. Now the big question is, how big could $c^{-1/2}$ be, given that $25 \le c \le 27$? Notice that $c^{-1/2}$ is decreasing in c, so the maximum occurs when $c = 25$. Then $c^{-1/2}$ equals $25^{-1/2} = 1/5$. So, we have

$$|R_0(27)| = c^{-1/2} \le 1/5.$$

So the error could be as much as $1/5$. This is too high: we need the error to be no more than $1/250$. So, the choice $N = 0$ was obviously wildly optimistic! We need to do better than that. Let's try $N = 1$. Then

$$|R_1(27)| = \left|\frac{f''(c)}{2!}(27 - 25)^2\right| = \left|-\frac{1}{4}c^{-3/2} \times \frac{1}{2!} \times 2^2\right| = \frac{c^{-3/2}}{2}.$$

Again, we used our table of derivatives from above to find $f''(c)$. This time we did need the absolute values, since $R_1(27)$ is actually negative (yup, we're headed for an overestimate here). Once again, $c^{-3/2}$ is biggest when c is smallest, namely $c = 25$, in which case the expression equals $25^{-3/2} = 1/125$. So,

$$|R_1(27)| = \frac{c^{-3/2}}{2} \le \frac{1}{125} \times \frac{1}{2} = \frac{1}{250}.$$

Hey, that means that our error is no more than $1/250$, which is what we want. So, we can take $N = 1$, and now all we need to do is find $P_1(27)$. (By the way, since $N = 1$, we're actually just using the linearization here.) Anyway, we know that

$$P_1(x) = f(25) + f'(25)(x - 25) = 5 + \frac{1}{10}(x - 25),$$

where we used the above table to get the values of $f(25)$ and $f'(25)$; putting $x = 27$, we have

$$P_1(27) = 5 + \frac{1}{10}(27 - 25) = \frac{26}{5}.$$

We conclude that $\sqrt{27}$ is approximately equal to $26/5$; in fact, these two numbers are within $1/250$ of each other, and $26/5$ is an overestimate for $\sqrt{27}$ (since the error term $R_1(27)$ is negative). Indeed, my calculator says that $\sqrt{27}$ is about 5.19615, which is within $1/250$ of $26/5 = 5.2$.) It wouldn't have been wrong to try $N = 2$ or any higher value—the estimate would then have been even better, but the numbers would have been a little messier.

25.3.4 Fourth example

To see just what we're up against here, let's suppose that we change the previous question slightly. Instead of $\sqrt{27}$, let's say we want to estimate $\sqrt{23}$ within a tolerance of $1/250$. This can't be too different from the previous example, right? Well, not quite. Let's see what happens. We're still going to use the Taylor series for $f(x) = x^{1/2}$ with $a = 25$, but now we have to put $x = 23$ instead of $x = 27$. Let's see what happens with the remainder term R_1 which worked so well in the previous example:

$$|R_1(23)| = \left| \frac{f''(c)}{2!}(23 - 25)^2 \right| = \left| -\frac{1}{4}c^{-3/2} \times \frac{1}{2!} \times (-2)^2 \right| = \frac{c^{-3/2}}{2}.$$

This is exactly what the error term was in the previous example! There is an important difference: now c is between 23 and 25. So, how big can $\frac{1}{2}c^{-3/2}$ be? Well, again this quantity is decreasing in c, so it's biggest when c is as small as possible, namely when $c = 23$. This leads to the following estimate:

$$|R_1(23)| = \frac{c^{-3/2}}{2} \le \frac{23^{-3/2}}{2}.$$

Unfortunately, $23^{-3/2}$ isn't as easy to compute as $25^{-3/2}$. The one thing we can be sure of is that this isn't good enough. You see, $\frac{1}{2} \cdot 25^{-3/2} = 1/250$, but $\frac{1}{2} \cdot 23^{-3/2}$ is bigger than this, so it's too big. So $N = 1$ isn't going to fly; we have to try $N = 2$.

OK, so taking $N = 2$ and using the table on page 544 above, we have

$$|R_2(23)| = \left| \frac{f^{(3)}(c)}{3!}(23 - 25)^3 \right| = \left| -\frac{3}{8}c^{-5/2} \times \frac{1}{3!} \times (-2)^3 \right| = \frac{c^{-5/2}}{2},$$

where $23 \le c \le 25$. Once again, $c^{-5/2}$ is biggest when $c = 23$, so we have

$$|R_2(23)| = \frac{c^{-5/2}}{2} \le \frac{23^{-5/2}}{2}.$$

Is this good enough? Not having a calculator available, we have to come up with some way of estimating $23^{-5/2}$. Man, how are we going to do that? The best way I can think of is to come up with a number that is less than 23 that we can easily raise to the power $-5/2$. That would be 16. Now $16^{-5/2} = 1/4^5 = 1/1024$, so

$$|R_2(23)| \le \frac{23^{-5/2}}{2} \le \frac{16^{-5/2}}{2} = \frac{1}{1024} \times \frac{1}{2} = \frac{1}{2048}.$$

This is certainly smaller than $1/250$, so taking $N = 2$ works and we can use $P_2(23)$. Now

$$P_2(x) = f(25) + f'(25)(x - 25) + \frac{f''(25)}{2!}(x - 25)!$$

$$= 5 + \frac{1}{10}(x - 25) - \frac{1}{500 \times 2}(x - 25)^2$$

(using the table once more), so replacing x by 23, we have

$$P_2(23) = 5 + \frac{1}{10}(23 - 25) - \frac{1}{1000}(23 - 25)^2 = 5 - \frac{2}{10} - \frac{4}{1000} = \frac{1199}{250}.$$

So our estimate for $\sqrt{23}$ is $1199/250$. Now, my calculator says that this last fraction is equal to 4.796 exactly, whereas it says that $\sqrt{23}$ is about 4.79583. These two numbers are indeed within $1/250$ of each other.

25.3.5 Fifth example

Let's look at one more example: estimate $\cos(\pi/3 - 0.01)$ using a third-order Taylor series, and determine how good the estimate is. Well, we need to choose a function; the obvious one is given by $f(x) = \cos(x)$, so we'll need to put $x = \pi/3 - 0.01$ in the end. What's a number close to this value of x that we can easily take the cosine of? It seems like $a = \pi/3$ is a natural candidate. So we set up a table as follows:

n	$f^{(n)}(x)$	$f^{(n)}(\pi/3)$
0	$\cos(x)$	$1/2$
1	$-\sin(x)$	$-\sqrt{3}/2$
2	$-\cos(x)$	$-1/2$
3	$\sin(x)$	$\sqrt{3}/2$
4	$\cos(x)$	not needed

The error term $R_3(x)$ is given by

$$R_3(x) = \frac{f^{(4)}(c)}{4!}\left(x - \frac{\pi}{3}\right)^4 = \frac{\cos(c)}{24}\left(x - \frac{\pi}{3}\right)^4,$$

where c is between x and $\pi/3$. Notice that we need $f^{(4)}(c)$, not $f^{(4)}(\pi/3)$; that explains the use of "not needed" in the above table. Now, when $x = \pi/3 - 0.01$, we have

$$R_3\left(\frac{\pi}{3} - 0.01\right) = \frac{\cos(c)}{24}\left(\frac{\pi}{3} - 0.01 - \frac{\pi}{3}\right)^4 = \frac{\cos(c)}{24}(-0.01)^4 = \frac{\cos(c)}{24 \times 10^8}.$$

(Here we have used $(-0.01)^4 = (0.01)^4 = (10^{-2})^4 = 10^{-8}$.) Now we just need to estimate the absolute value of this error term; since $|\cos(c)| \le 1$, we see that

$$\left|R_3\left(\frac{\pi}{3} - 0.01\right)\right| = \frac{|\cos(c)|}{24 \times 10^8} \le \frac{1}{24 \times 10^8} = \frac{1}{2400000000}.$$

Great—we know that using $P_3(\pi/3 - 0.01)$ to estimate $\cos(\pi/3 - 0.01)$ will be accurate to within the tiny number $1/2400000000$. So what is $P_3(\pi/3 - 0.01)$? In general,

$$P_3(x) = f\left(\frac{\pi}{3}\right) + f'\left(\frac{\pi}{3}\right)\left(x - \frac{\pi}{3}\right) + \frac{1}{2!}f''\left(\frac{\pi}{3}\right)\left(x - \frac{\pi}{3}\right)^2 + \frac{1}{3!}f^{(3)}\left(\frac{\pi}{3}\right)\left(x - \frac{\pi}{3}\right)^3.$$

Using the above table of derivatives, this becomes

$$P_3(x) = \frac{1}{2} - \frac{\sqrt{3}}{2}\left(x - \frac{\pi}{3}\right) - \frac{1}{2} \times \frac{1}{2}\left(x - \frac{\pi}{3}\right)^2 + \frac{1}{6} \times \frac{\sqrt{3}}{2}\left(x - \frac{\pi}{3}\right)^3.$$

Put $x = \pi/3 - 0.01$ and simplify; the result is

$$P_3\left(\frac{\pi}{3} - 0.01\right) = \frac{1}{2} - \frac{\sqrt{3}}{2}(-0.01) - \frac{1}{4}(-0.01)^2 + \frac{\sqrt{3}}{12}(-0.01)^3$$

$$= \frac{1}{2} + \frac{\sqrt{3}}{200} - \frac{1}{40000} - \frac{\sqrt{3}}{12000000}.$$

This might seem like a nasty expression, but it's really not too bad. The only tricky quantity is $\sqrt{3}$, but that's pretty easy to estimate by itself. At least there are no trig functions to deal with. Anyway, since $f(\pi/3 - 0.01)$ is approximately equal to $P_3(\pi/3 - 0.01)$, we have

$$\cos\left(\frac{\pi}{3} - 0.01\right) = f\left(\frac{\pi}{3} - 0.01\right) \cong \frac{1}{2} + \frac{\sqrt{3}}{200} - \frac{1}{40000} - \frac{\sqrt{3}}{12000000},$$

accurate to within $1/2400000000$.

25.3.6 General techniques for estimating the error term

In all the above examples, we had to estimate the quantity $|f^{(N+1)}(c)|$ for c in some given range. Here are some general tips for doing this:

1. Regardless of the value of c, you can always use the standard inequalities $|\sin(c)| \le 1$ and $|\cos(c)| \le 1$.

2. If the function $f^{(N+1)}$ is increasing, then its value is biggest at the right-hand endpoint. In the first two examples above, we needed to find the largest value of e^c, where $0 < c < 1/3$. Since e^c is increasing in c, we can say that $e^c < e^{1/3}$. On the other hand, in the example from Section 24.1.4 of the previous chapter, we also needed to maximize e^c, but this time $-1/10 < c < 0$. Again, since e^c is increasing in c, this maximum value is just $e^0 = 1$. That is, $e^c < e^0 = 1$.

3. If the function $f^{(N+1)}$ is decreasing, then the greatest value of $f^{(N+1)}(c)$ occurs at the left-hand endpoint of the interval. For example, if you know that c is between 1 and 5, then the greatest value of $1/(3+c)^4$ occurs at the left-hand endpoint of the interval $[1, 5]$, since $1/(3+c)^4$ is decreasing in c. So the above expression is biggest when $c = 1$, and its value then is $1/4^4 = 1/256$.

4. In general, you might have to find the critical points of the function $f^{(N+1)}$ in order to maximize it. (See Section 11.1.1 of Chapter 11 for a reminder on how to do this.)

25.4 Another Technique for Estimating the Error

Cast your mind back to the alternating series test (see Section 22.5.4 in Chapter 22). This test says that if a series is alternating, and has terms whose absolute values are decreasing to 0, then the series converges. The reason this is true is that the partial sums form a sort of yoyo about the actual limit: one is bigger, the next one is smaller, the next is bigger, and so on. Each time, the partial sums do get closer to the actual limit, so the yoyo is losing steam. The idea is that at each point in the series, adding the next term overshoots the actual value, so the entire error is less than the next term in absolute value.

Let's see what this looks like in symbols. Suppose you start off with some function f, and find its Taylor series about $x = a$. If you also happen to know that the series converges to $f(x)$ for some particular value of x (as it often does for the sorts of functions we look at), then you can write

$$f(x) = \sum_{n=0}^{\infty} \frac{f^{(n)}(a)}{n!}(x - a)^n.$$

For the particular value of x that you're interested in, if the above series is alternating with terms whose absolute values decrease to 0, then the error is less than the next term. That is,

$$|R_N(x)| \leq \left| \frac{f^{(N+1)}(a)}{(N + 1)!}(x - a)^{N+1} \right|.$$

There's no nasty c to worry about, which is more than enough reason to use this nice fact. Remember, it only works if the series satisfies the three conditions for the alternating series test!

 Here's an example of where this method really shines. Suppose we'd like to use a Maclaurin series to find an estimate for the definite integral

$$\int_0^1 \frac{1 - \cos(t)}{t^2}\, dt$$

with an error no greater than $1/3000$. By the way, this looks like an improper integral, with problem spot at $t = 0$, but actually $t = 0$ isn't a problem spot at all. You see, by l'Hôpital's Rule,

$$\lim_{t \to 0} \frac{1 - \cos(t)}{t^2} \overset{\text{l'H}}{=} \lim_{t \to 0} \frac{\sin(t)}{2t} = \frac{1}{2}.$$

That is, the integrand doesn't blow up at $t = 0$ after all, so the integral isn't improper. Anyway, that's just an observation. Now we have to solve the problem.

The first useful idea is that we can form a function that looks something like the integral by setting

$$f(x) = \int_0^x \frac{1 - \cos(t)}{t^2}\, dt.$$

The integral we want to estimate is then $f(1)$. We need to find the Maclaurin series for f. To do this, replace $\cos(t)$ by its Maclaurin series, which we found in Section 24.2.3 of the previous chapter. That is,

$$f(x) = \int_0^x \frac{1 - \left(1 - \dfrac{t^2}{2!} + \dfrac{t^4}{4!} - \dfrac{t^6}{6!} + \dfrac{t^8}{8!} - \cdots \right)}{t^2}\, dt.$$

If you simplify things a little, you should be able to reduce this to

$$f(x) = \int_0^x \left(\frac{1}{2!} - \frac{t^2}{4!} + \frac{t^4}{6!} - \frac{t^6}{8!} + \cdots \right) dt.$$

Now do the integration and evaluate at the endpoints:

$$f(x) = \left(\frac{t}{2!} - \frac{t^3}{3 \times 4!} + \frac{t^5}{5 \times 6!} - \frac{t^7}{7 \times 8!} + \cdots \right) \Big|_0^x$$

$$= \frac{x}{2!} - \frac{x^3}{3 \times 4!} + \frac{x^5}{5 \times 6!} - \frac{x^7}{7 \times 8!} + \cdots.$$

By the way, it's a good exercise to try writing this series in sigma notation. Anyway, we can now put in $x = 1$ to see that

$$f(1) = \int_0^1 \frac{1 - \cos(t)}{t^2}\, dt = \frac{1}{2!} - \frac{1}{3 \times 4!} + \frac{1}{5 \times 6!} - \frac{1}{7 \times 8!} + \cdots.$$

Truth be told, I pulled a couple of fast ones on you here. First, I replaced $\cos(t)$ by its Maclaurin series. Well, that's ok—we've seen in Section 24.2.3 of the previous chapter that we can do this for all t. Second, I integrated an infinite series term by term and claimed that the new series converges to f for all x. We'll see in Section 26.2.3 of the next chapter that this sort of thing is valid (although we won't prove it). Anyway, the above equation is correct; we now have an exact expression for our integral in terms of an infinite series.

Now the only question is, how many terms do we have to take to get an approximation that is within $1/3000$ of the true value? Well, notice that the series is alternating and that its terms are decreasing and go to 0. So we can use the idea that the absolute value of the next term is bigger than the error. For example, if you approximate the integral by the first term $1/2!$, the error is no bigger than $1/3 \times 4!$, which equals $1/72$. That is much too big. How about if you approximate the integral using the first two terms? That is, what if you use

$$\int_0^1 \frac{1 - \cos(t)}{t^2}\, dt \cong \frac{1}{2!} - \frac{1}{3 \times 4!} = \frac{35}{72}?$$

Then the error is less than the absolute value of the next term:

$$|\text{error}| \le \frac{1}{5 \times 6!} = \frac{1}{5 \times 720} = \frac{1}{3600}.$$

This is less than our tolerance of $1/3000$, so it's all good. We can safely say that the integral is approximately equal to $35/72$, with an error less than $1/3000$. (We can even tell that $35/72$ is an underestimate. Why?) By the way, I tried the integral on a computer program that can handle such things and it told me that the value of the integral is approximately 0.486385, whereas my calculator says that $35/72$ equals 0.486111 (to six decimal places); these two numbers are indeed within $1/3000$ of each other.

Now as an exercise, you should try approximating

$$\int_0^{1/2} \frac{\sin(t)}{t}\, dt$$

within a tolerance of $1/1000$, using the same method as above. (You'll need the Maclaurin series for $\sin(t)$, which you can find in Section 26.2 of the next chapter.)

CHAPTER 26 _____

Taylor and Power Series: How to Solve Problems

In this chapter, we'll look at how to solve four different classes of problems involving Taylor series, Taylor polynomials and power series:

- how to find where power series converge or diverge;
- how to manipulate Taylor series to get other Taylor series or Taylor polynomials;
- using Taylor series or Taylor polynomials to find derivatives; and
- using Maclaurin series to find limits.

26.1 Convergence of Power Series

Let's say we have a power series about $x = a$:

$$\sum_{n=0}^{\infty} a_n (x - a)^n.$$

As we saw in the case of geometric series, a power series might converge for some x and diverge for other x. The question that we want to ask is this: given our power series, for which x does it converge, and for which x does it diverge? Furthermore, if the series converges for a specific x, it would be nice to know whether the convergence is absolute or merely conditional. So, let's see what could possibly happen, and then we'll take advantage of these observations.

26.1.1 Radius of convergence

We want to find out for which x the power series $\sum_{n=0}^{\infty} a_n (x - a)^n$ converges. On the face of it, it seems like we have to answer infinitely many questions here, since there are infinitely many values of x to substitute in and test to see whether the series converges or not. Let's draw a number line representing different values of x. For each x such that our power series converges, we'll put a check mark above it, whereas if the power series diverges for a particular x, we'll put a cross instead. (Of course, we won't do this for every single x, since the diagram would get crowded! We'll just do enough to get the idea.) For example, the geometric series $\sum_{n=0}^{\infty} x^n$ converges when $-1 < x < 1$ and diverges otherwise, so its picture looks like this:

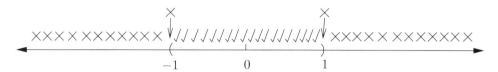

Note that I took special care to indicate the divergence at the endpoints 1 and -1.

On the other hand, we've seen that the series

$$\sum_{n=0}^{\infty} \frac{x^n}{n!}$$

converges for all x (to e^x, of course), so its picture looks like this:

It seems like this could be pretty unpredictable. One thing that we can say for sure is that the power series always converges at $x = a$. In fact, if you substitute $x = a$ into

$$\sum_{n=0}^{\infty} a_n (x-a)^n = a_0 + a_1(x-a) + a_2(x-a)^2 + \cdots,$$

you can see that all the terms vanish except a_0. So, the series evidently converges (to a_0). Unfortunately, the value $x = a$ is the only value for which we can predict the convergence for certain. How about the other values? Maybe it would be possible to get a hodgepodge of checks and crosses, like this:

It turns out that the above picture can't happen for power series. Specifically, there are only three possibilities that can occur:

1. There is some number $R > 0$, called the *radius of convergence* of the power series, such that the picture looks like this:

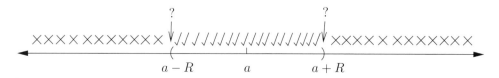

The explanation of this diagram is as follows:

- The power series converges absolutely in the region $|x - a| < R$ (you can write this condition as $a - R < x < a + R$ if you prefer), so there are check marks there.
- The power series diverges in the region $|x - a| > R$ (you can write this as $x < a - R$ or $x > a + R$), so there are crosses there.
- At the two specific points where $|x - a| = R$, (that is, at $x = a + R$ and $x = a - R$), the power series might converge absolutely, converge conditionally, or diverge. You have to check both these points separately to see what happens there, so there are question marks at these two points in the above diagram. I'll refer to these points as the "endpoints."

An example of this is the geometric series $\sum_{n=0}^{\infty} x^n$. This is a power series with $a = 0$ which converges absolutely when $|x| < 1$ and diverges otherwise. The radius of convergence is therefore equal to 1, and the series diverges at the endpoints 1 and -1.

2. The power series might converge absolutely for all x, in which case the diagram looks like this:

In this case, we say that the radius of convergence is ∞. As we saw above, an example of this is the power series for e^x,

$$\sum_{n=0}^{\infty} \frac{x^n}{n!}.$$

Other examples include the Maclaurin series for $\sin(x)$ and $\cos(x)$.

3. The power series might converge absolutely only for $x = a$ and diverge for all other x. In this case, the radius of convergence is 0. We'll soon see that this is the case for the series

$$\sum_{n=0}^{\infty} n! x^n,$$

for example. The picture for this case looks like this:

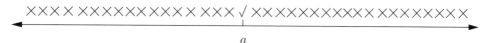

Of course, I haven't said why these are the only possibilities. This should become clear very soon!

26.1.2 How to find the radius and region of convergence

Given a power series, how do we find the radius of convergence? The answer is to use the **ratio test**. Sometimes the root test will be more effective, but for most problems the ratio test is better. (See Sections 23.3 and 23.4 in Chapter 23 for more about the ratio and root tests, respectively.) Here's the general approach:

1. Write down the limiting absolute ratio; this should always look like

$$\lim_{n \to \infty} \left| \frac{a_{n+1}(x-a)^{n+1}}{a_n(x-a)^n} \right| = \lim_{n \to \infty} \left| \frac{a_{n+1}}{a_n} \right| |x-a|.$$

   If instead you use the root test, you should get

$$\lim_{n \to \infty} |a_n(x-a)^n|^{1/n} = \lim_{n \to \infty} |a_n|^{1/n} |x-a|.$$

2. Work out the limit. It's important to note that the limit is as $n \to \infty$, not $x \to \infty$. There's a big difference! Regardless of whether you use the ratio test or the root test, the answer should be of the form $L|x-a|$, where L might be a finite number, 0, or even ∞. The important point is that there is a factor of $|x-a|$ present.

3. In either the ratio test or the root test, the important thing is whether the limit $L|x-a|$ is less than 1, greater than 1, or equal to 1. So, if L is positive, then divide by L to understand everything: if $|x-a| < 1/L$, the power series converges absolutely; if $|x-a| > 1/L$, then the power series diverges; whereas if $|x-a| = 1/L$, then we can't tell and need to check the two endpoints. We are in the first situation from the previous section, and the radius of convergence is $1/L$.

4. If $L = 0$, then the limiting ratio is always 0 regardless of the value of x. Since $0 < 1$, this means that the power series converges absolutely for all x, so we are in the second case from the previous section and the radius of convergence is ∞.

5. If $L = \infty$, then it looks like the power series never converges. In fact, the series must converge when $x = a$, but it will diverge for every other x and so we are in the third case from the previous section: the radius of convergence is 0.

This more or less shows why we must get one of the three cases of the previous section. It's still pretty abstract, though—we need to illustrate this with a whole bunch of examples.

First, consider the power series

$$\sum_{n=2}^{\infty} \frac{x^n}{n \ln(n)}.$$

Let's use the ratio test. We start off by taking the standard term $x^n/n \ln(n)$ and putting it in the denominator of a big fraction; then to get the numerator of our fraction, start with the standard term $x^n/n \ln(n)$ again, but this time replace each occurrence of n by $(n+1)$. Finally, take absolute values, then limits as $n \to \infty$. So we are looking for

$$\lim_{n\to\infty} \left| \frac{\dfrac{x^{n+1}}{(n+1)\ln(n+1)}}{\dfrac{x^n}{n\ln(n)}} \right|.$$

This can be dealt with in the same way as ratio test problems for plain vanilla series: just group together like terms. You get

$$\lim_{n\to\infty} \left| \frac{\dfrac{x^{n+1}}{(n+1)\ln(n+1)}}{\dfrac{x^n}{n\ln(n)}} \right| = \lim_{n\to\infty} \left| \frac{x^{n+1}}{x^n} \frac{n}{n+1} \frac{\ln(n)}{\ln(n+1)} \right|$$

$$= \lim_{n\to\infty} |x| \frac{n}{n+1} \frac{\ln(n)}{\ln(n+1)} = |x|.$$

Again, the limit is as $n \to \infty$, which is why we replaced $n/(n+1)$ and $\ln(n)/\ln(n+1)$ by 1. (Use l'Hôpital's Rule to deal with the logarithms; I'll leave the details to you.) Anyway, the limiting ratio is $|x|$, so by the ratio test, our power series converges absolutely when $|x| < 1$ and diverges when $|x| > 1$. That is, the radius of convergence is 1. We still have to check what happens when $x = 1$ and $x = -1$. Let's do $x = 1$ first. Substituting $x = 1$, the original power series becomes

$$\sum_{n=2}^{\infty} \frac{1^n}{n\ln(n)} = \sum_{n=2}^{\infty} \frac{1}{n\ln(n)}.$$

Does this converge? I leave it to you to use the integral test to see that it diverges (or see Section 23.5 in Chapter 23). Now let's put $x = -1$ in the original power series above to get

$$\sum_{n=2}^{\infty} \frac{(-1)^n}{n\ln(n)}.$$

This doesn't converge absolutely—in fact, the series obtained by replacing the terms by their absolute values is exactly the series when $x = 1$, which we just saw diverges. On the other hand, the above series for $x = -1$ converges by the alternating series test (you supply the details—use the methods of Section 23.7 of Chapter 23). So, we have conditional convergence at the point $x = -1$. Summarizing, the power series converges absolutely when $-1 < x < 1$, converges conditionally when $x = -1$, and diverges for all other x. The picture looks like this:

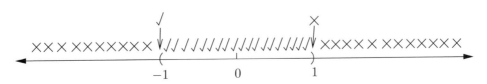

Now consider

$$\sum_{n=2}^{\infty} \frac{x^n}{n(\ln(n))^2}.$$

This is almost the same question as the previous one, but let's see what happens. We have

$$\lim_{n\to\infty} \left| \frac{\dfrac{x^{n+1}}{(n+1)(\ln(n+1))^2}}{\dfrac{x^n}{n(\ln(n))^2}} \right| = \lim_{n\to\infty} \left| \frac{x^{n+1}}{x^n} \frac{n}{n+1} \frac{(\ln(n))^2}{(\ln(n+1))^2} \right|$$

$$= \lim_{n\to\infty} |x| \frac{n}{n+1} \left(\frac{\ln(n)}{\ln(n+1)} \right)^2,$$

which again simplifies to $|x|$. So once again the power series converges absolutely when $|x| < 1$ and diverges when $|x| > 1$. The radius of convergence is therefore 1. As for the endpoints, let's put in $x = 1$:

$$\sum_{n=2}^{\infty} \frac{1^n}{n(\ln(n))^2} = \sum_{n=2}^{\infty} \frac{1}{n(\ln(n))^2}.$$

As we've seen in Section 23.5 of Chapter 23, you can use the integral test to see that this converges; since all the terms are positive, the convergence is absolute. Now, when $x = -1$, we get

$$\sum_{n=2}^{\infty} \frac{(-1)^n}{n(\ln(n))^2}.$$

The series of absolute values of these terms is

$$\sum_{n=2}^{\infty} \frac{1}{n(\ln(n))^2},$$

which is the same as the series when $x = 1$, so it converges absolutely. We conclude that the power series converges absolutely when $-1 \le x \le 1$ and diverges for all other x, giving the following picture:

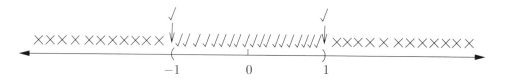

So, it's the same as the previous example, except for different behavior at the endpoints 1 and -1.

How about

$$\sum_{n=1}^{\infty} n! x^n?$$

We have

$$\lim_{n\to\infty}\left|\frac{(n+1)!x^{n+1}}{n!x^n}\right|=\lim_{n\to\infty}\left|\frac{(n+1)!}{n!}\frac{x^{n+1}}{x^n}\right|=\lim_{n\to\infty}(n+1)|x|.$$

What is this last limit? Well, if $x = 0$, then this is just the limit as $n \to \infty$ of $0(n + 1) = 0$, which is of course 0. (You may notice that the quantity x^{n+1}/x^n isn't well-defined in this case, though!) However, for any other value of x, we're screwed—the limit is ∞, which is certainly bigger than 1. We conclude that the series only converges when $x = 0$ (remember, it has to converge at $x = a$, which is 0 in this case). So, the radius of convergence is 0 and the picture looks like this:

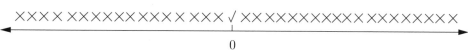

$$0$$

 Now consider

$$\sum_{n=1}^{\infty}\frac{(-2)^n}{\sqrt{n}}(x-7)^n.$$

 This is a power series with $a = 7$, so that point must be at the center of the region of convergence. In any case, check that we have

$$\lim_{n\to\infty}\left|\frac{\dfrac{(-2)^{n+1}(x-7)^{n+1}}{\sqrt{n+1}}}{\dfrac{(-2)^n(x-7)^n}{\sqrt{n}}}\right|=\lim_{n\to\infty}\left|\frac{(-2)^{n+1}}{(-2)^n}\frac{(x-7)^{n+1}}{(x-7)^n}\sqrt{\frac{n}{n+1}}\right|$$

$$=2|x-7|.$$

So the power series converges absolutely when $2|x - 7| < 1$ and diverges when $2|x - 7| > 1$. Dividing through by 2, we see that it converges when $|x - 7| < \frac{1}{2}$ and diverges when $|x - 7| > \frac{1}{2}$. The radius of convergence is therefore $\frac{1}{2}$, so our picture looks like this so far:

$$6\tfrac{1}{2} \qquad 7 \qquad 7\tfrac{1}{2}$$

We still have to check the endpoints. Let's try $x = 7\frac{1}{2}$. Then the series is

$$\sum_{n=1}^{\infty}\frac{(-2)^n}{\sqrt{n}}(7\tfrac{1}{2}-7)^n=\sum_{n=1}^{\infty}\frac{(-2)^n}{\sqrt{n}}\frac{1}{2^n}=\sum_{n=1}^{\infty}\frac{(-1)^n}{\sqrt{n}}.$$

 Make sure you realize why $(-2)^n/2^n$ can be simplified to $(-1)^n$. Anyway, I leave it to you to show that this last series doesn't converge absolutely (use the p-test) but that it does converge conditionally (use the alternating series test). Now, when $x = 6\frac{1}{2}$, we get

$$\sum_{n=1}^{\infty} \frac{(-2)^n}{\sqrt{n}} (6\tfrac{1}{2} - 7)^n = \sum_{n=1}^{\infty} \frac{(-2)^n}{\sqrt{n}} \left(-\frac{1}{2}\right)^n = \sum_{n=1}^{\infty} \frac{(-2)^n}{\sqrt{n}} \frac{1}{(-2)^n} = \sum_{n=1}^{\infty} \frac{1}{\sqrt{n}},$$

which diverges. So, we conclude that the power series converges absolutely when $6\frac{1}{2} < x < 7\frac{1}{2}$ and conditionally when $x = 7\frac{1}{2}$, and diverges otherwise. The full picture is as follows:

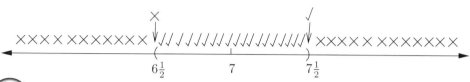

 Consider the series

$$\sum_{n=1}^{\infty} \frac{3^n}{2^{n^2}} (x + 2)^n.$$

Now, this is a good candidate for the root test because of the complicated factor 2^{n^2}. You can work it out with the ratio test, but the root test is better. Consider the limit of the nth root of the absolute value of the nth term:

$$\lim_{n \to \infty} \left| \frac{3^n}{2^{n^2}} (x + 2)^n \right|^{1/n} = \lim_{n \to \infty} \frac{(3^n)^{1/n}}{(2^{n^2})^{1/n}} (|x + 2|^n)^{1/n} = \lim_{n \to \infty} \frac{3}{2^n} |x + 2|.$$

Now, regardless of the value of x, this limit is equal to 0, which is less than 1; by the root test, the power series converges absolutely for all x. That is, the radius of convergence is ∞ and the picture looks like this:

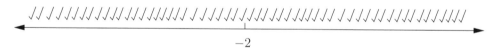

Just one more comment, before we move on to the next section: note that when the radius of convergence is positive, you might get convergence at both endpoints, at neither endpoint, at the left endpoint only, or at the right endpoint only. We've seen examples of all four possibilities above.

26.2 Getting New Taylor Series from Old Ones

Let's look at some techniques for finding Taylor series. One way to find the Taylor series about $x = a$ of a given function f is to use the formula directly, as we did in Section 25.2 of the previous chapter. To use the formula, you have to find all the derivatives of f, at least at $x = a$. For most functions, this is a pain. Often a better idea is to use some common Taylor series to synthesize new ones. Of course, you have to know some Taylor series first! It is really useful to have the following five Maclaurin series (Taylor series about $x = 0$) at your fingertips:

1. For $f(x) = e^x$:

$$e^x = \sum_{n=0}^{\infty} \frac{x^n}{n!} = 1 + x + \frac{x^2}{2!} + \frac{x^3}{3!} + \cdots$$

which is true for all real x.

2. For $f(x) = \sin(x)$:

$$\sin(x) = \sum_{n=0}^{\infty} \frac{(-1)^n x^{2n+1}}{(2n+1)!} = x - \frac{x^3}{3!} + \frac{x^5}{5!} - \frac{x^7}{7!} + \cdots$$

which is true for all real x.

3. For $f(x) = \cos(x)$:

$$\cos(x) = \sum_{n=0}^{\infty} \frac{(-1)^n x^{2n}}{(2n)!} = 1 - \frac{x^2}{2!} + \frac{x^4}{4!} - \frac{x^6}{6!} + \cdots$$

which is true for all real x.

4. For $f(x) = 1/(1-x)$:

$$\frac{1}{1-x} = \sum_{n=0}^{\infty} x^n = 1 + x + x^2 + x^3 + \cdots$$

which is true only for $-1 < x < 1$.

5. For $f(x) = \ln(1+x)$ or $f(x) = \ln(1-x)$:

$$\ln(1+x) = \sum_{n=1}^{\infty} -\frac{(-1)^n x^n}{n} = x - \frac{x^2}{2} + \frac{x^3}{3} - \frac{x^4}{4} + \cdots$$

$$\ln(1-x) = \sum_{n=1}^{\infty} -\frac{x^n}{n} = -x - \frac{x^2}{2} - \frac{x^3}{3} - \frac{x^4}{4} - \cdots$$

which are true for $-1 < x < 1$. (Actually, the first formula is also true for $x = 1$ as well, and the second formula is true for $x = -1$, but this gets a little complicated!)

So far, we've proved formulas #1 and #3 (in Section 24.2.3 of Chapter 24) as well as #4 (in Section 22.2 of Chapter 22). We'll deal with #2 and #5 in Sections 26.2.2 and 26.2.3 below, respectively.

Anyway, suppose that you've learned all five series. Here's how to manipulate them to get new power series.*

---

*The proofs that the following techniques work are a little beyond the scope of this book.

26.2.1 Substitution and Taylor series

The most useful technique is substitution. In a Maclaurin series, you can replace x by a multiple of x^n, where n is an integer, to get a new Maclaurin series. For example, we know that

$$e^x = 1 + x + \frac{x^2}{2!} + \frac{x^3}{3!} + \frac{x^4}{4!} + \cdots$$

for any x; so if you want to find the Maclaurin series for $f(x) = e^{x^2}$, simply replace x by x^2 in the above series to get

$$e^{x^2} = 1 + x^2 + \frac{(x^2)^2}{2!} + \frac{(x^2)^3}{3!} + \frac{(x^2)^4}{4!} + \cdots,$$

which you can simplify down to

$$e^{x^2} = 1 + x^2 + \frac{x^4}{2!} + \frac{x^6}{3!} + \frac{x^8}{4!} + \cdots.$$

Since the original series holds for any x, so does this one.

Let's look at another common example: what is the Maclaurin series for $f(x) = 1/(1+x^2)$? To do this, start with the geometric series

$$\frac{1}{1-x} = \sum_{n=0}^{\infty} x^n = 1 + x + x^2 + x^3 + \cdots,$$

which is valid for $-1 < x < 1$; then replace x by $-x^2$ to get

$$\frac{1}{1+x^2} = \sum_{n=0}^{\infty} (-x^2)^n = \sum_{n=0}^{\infty} (-1)^n x^{2n} = 1 - x^2 + x^4 - x^6 + \cdots,$$

which is valid for $-1 < -x^2 < 1$. Notice that we also replaced x by $-x^2$ in this "valid for" inequality! This isn't important here, since the inequalities reduce to $-1 < x < 1$ anyway; but suppose instead we wanted to work out the Maclaurin series for $1/(1+2x^2)$. Then we would have replaced x by $-2x^2$ instead. This gives

$$\frac{1}{1+2x^2} = \sum_{n=0}^{\infty} (-2x^2)^n = \sum_{n=0}^{\infty} (-1)^n 2^n x^{2n} = 1 - 2x^2 + 4x^4 - 8x^6 + \cdots,$$

but this is valid only for $-1 < -2x^2 < 1$. Convince yourself that this inequality reduces to $-1/\sqrt{2} < x < 1/\sqrt{2}$. (By the way, all the series in these examples are geometric series.)

Now, suppose you start with the following equation, which is true for all real x:

$$\sin(x) = x - \frac{x^3}{3!} + \frac{x^5}{5!} - \frac{x^7}{7!} + \cdots.$$

The right-hand side is the Maclaurin series, or Taylor series about $x = 0$, of $\sin(x)$. If you replace x by $(x - 18)$, you get a Taylor series about $x = 18$ instead:

$$\sin(x - 18) = (x - 18) - \frac{(x-18)^3}{3!} + \frac{(x-18)^5}{5!} - \frac{(x-18)^7}{7!} + \cdots.$$

The right-hand side is not the Taylor series about $x = 18$ of $\sin(x)$, because the left-hand side is no longer $\sin(x)$—it's $\sin(x-18)$. So our substitution has translated the original function as well. We have actually found the Taylor series about $x = 18$ of $\sin(x-18)$. To find the Taylor series of $\sin(x)$ about $x = 18$, you have to use the formula in Taylor's Theorem. (We looked at a similar problem at the end of Section 25.2 in the previous chapter.)

The moral of this last example is that if you replace x by $(x-a)$, then you get a Taylor series about $x = a$ instead of a Maclaurin series, but the function is different. This can still be useful. For example, to find the Taylor series of $\ln(x)$ about $x = 1$, start with one of the formulas from the previous section:

$$\ln(1+x) = \sum_{n=1}^{\infty} -\frac{(-1)^n x^n}{n} = x - \frac{x^2}{2} + \frac{x^3}{3} - \frac{x^4}{4} + \cdots \qquad \text{for } -1 < x < 1.$$

Now, let's replace x by $(x-1)$. The quantity $\ln(1+x)$ becomes $\ln(1+(x-1))$, or just $\ln(x)$; so we get

$$\ln(x) = \sum_{n=1}^{\infty} -\frac{(-1)^n (x-1)^n}{n} = (x-1) - \frac{(x-1)^2}{2} + \frac{(x-1)^3}{3} - \frac{(x-1)^4}{4} + \cdots$$

$$\text{for } -1 < (x-1) < 1.$$

Notice that I also replaced x by $(x-1)$ in the original inequality $-1 < x < 1$, arriving at $-1 < (x-1) < 1$. This looks a bit silly, so add 1 everywhere to get $0 < x < 2$. We end up with

$$\ln(x) = \sum_{n=1}^{\infty} -\frac{(-1)^n (x-1)^n}{n} = (x-1) - \frac{(x-1)^2}{2} + \frac{(x-1)^3}{3} - \frac{(x-1)^4}{4} + \cdots$$

$$\text{for } 0 < x < 2.$$

We have used the Maclaurin series of $\ln(1+x)$ to get the Taylor series about $x = 1$ of $\ln(x)$.

By the way, the substitution technique can also be used to find Taylor polynomials, but you have to be careful to get the order right. For example, if you take $f(x) = e^x$ and $a = 0$, the Taylor polynomial of order 3 is

$$P_3(x) = 1 + x + \frac{x^2}{2!} + \frac{x^3}{3!}.$$

Now if $g(x) = e^{x^2}$, it's a mistake to replace x by x^2 in the above polynomial and claim the third-order Taylor polynomial of g is

$$P_3(x) = 1 + x^2 + \frac{x^4}{2!} + \frac{x^6}{3!}.$$

This is actually the **sixth-order** Taylor polynomial of g about 0, so the left-hand side should say $P_6(x)$ instead of $P_3(x)$. To get the correct formula for $P_3(x)$, just drop all the terms of degree greater than 3. This means that $P_3(x) = 1 + x^2$. Of course, this is also $P_2(x)$ as well! Be careful with your degrees. That's an order. (At least, if you want to pass calculus and get your degree ... ouch. OK, no more puns, I promise.)

26.2.2 Differentiating Taylor series

If a power series converges to a differentiable function f on an open interval (a, b), then it turns out that you can differentiate the series term-by-term to get a new series which converges to $f'(x)$ on the same interval. The situation at the endpoints a and b is a little trickier: the differentiated series might diverge even if the original series converges.* So check the endpoints separately.

Our first example is to find the Maclaurin series for $\sin(x)$, assuming that we know the Maclaurin series for $\cos(x)$ is given by

$$\cos(x) = 1 - \frac{x^2}{2!} + \frac{x^4}{4!} - \frac{x^6}{6!} + \frac{x^8}{8!} - \cdots ;$$

the formula is valid for all x. (We proved this in Section 24.2.3 of Chapter 24.) If you differentiate both sides, term-by-term on the right, you get

$$-\sin(x) = -\frac{2x}{2!} + \frac{4x^3}{4!} - \frac{6x^5}{6!} + \frac{8x^7}{8!} - \cdots .$$

We need to multiply both sides by -1 to get rid of the minus sign on the left-hand side, but there's another simplification to be made. We have to deal with quantities like $2/2!$, $4/4!$, $6/6!$ and $8/8!$. Consider $4/4!$ for a second. Since $4!$ is actually $3! \times 4$, you can reduce $4/4!$ to $1/3!$ by canceling out a factor of 4. Similarly, $6! = 5! \times 6$, so we have $6/6! = 1/5!$, and also $8! = 7! \times 8$, so $8/8! = 1/7!$. Altogether, the above equation becomes

$$\sin(x) = x - \frac{x^3}{3!} + \frac{x^5}{5!} - \frac{x^7}{7!} + \cdots .$$

Since the series for $\cos(x)$ is valid for all x, so is the differentiated series above. That is, the Maclaurin series for $\sin(x)$ is given by the above equation, which is valid for all x. This proves formula #2 in Section 26.2 above.

Here's another example of differentiating a power series. Suppose you want to find the Maclaurin series for $f(x) = 1/(1 + x)^2$. The best way would be to start with the series for $1/(1 + x)$, which is obtained from the standard geometric series (#4 above) by replacing x by $-x$:

$$\frac{1}{1 + x} = 1 - x + x^2 - x^3 + x^4 - \cdots ;$$

this is valid for $-1 < x < 1$. Then differentiate both sides, term-by-term on the right-hand side, to get

$$-\frac{1}{(1 + x)^2} = 0 - 1 + 2x - 3x^2 + 4x^3 - \cdots .$$

All that's left is to take negatives of both sides to get

$$\frac{1}{(1 + x)^2} = 1 - 2x + 3x^2 - 4x^3 + \cdots = \sum_{n=0}^{\infty} (-1)^n (n + 1) x^n.$$

---

*By the way, if the differentiated series converges at one (or both) of the endpoints, then the original series converges there too.

This is valid for $-1 < x < 1$. (You should check that the expression in sigma notation is correct, and that the series doesn't converge at the endpoints $x = \pm 1$.)

Once again, you can apply these ideas to Taylor polynomials; you just have to be careful with orders, once again. Since differentiating a polynomial knocks the degree down by one, the differentiated Taylor polynomial is order one less than the original polynomial. For example, the third-order Taylor polynomial about 0 of $1/(1+x)$ is $1 - x + x^2 - x^3$, as you can see from the previous example; if you differentiate and multiply by -1, you see that the **second-order** Taylor polynomial about 0 of $1/(1+x)^2$ is $1 - 2x + 3x^2$.

26.2.3 Integrating Taylor series

You can also integrate a power series term-by-term. The new series converges in the same interval as the old one (except perhaps at the endpoints of the interval of convergence). If you use an indefinite integral, don't forget the constant! Let's see a few examples. First, let's try to prove the following formula for $\ln(1-x)$, which we first stated as part of formula #5 in Section 26.2 above but never proved:

$$\ln(1 - x) = \sum_{n=1}^{\infty} -\frac{x^n}{n} = -x - \frac{x^2}{2} - \frac{x^3}{3} - \frac{x^4}{4} - \cdots$$

for $-1 < x < 1$. To do it, we'll use the geometric series formula, which is #4 in Section 26.2:

$$\frac{1}{1 - x} = \sum_{n=0}^{\infty} x^n = 1 + x + x^2 + x^3 + \cdots,$$

valid for $-1 < x < 1$. Then integrate everything with respect to x:

$$\int \frac{1}{1 - x}\, dx = \int \sum_{n=0}^{\infty} x^n \, dx = \int (1 + x + x^2 + x^3 + \cdots)\, dx.$$

(Note that I have used both sigma notation and expanded notation here, but you would normally only use one of the two.) Now integrate term-by-term:

$$-\ln(1 - x) = C + \sum_{n=0}^{\infty} \frac{x^{n+1}}{n + 1} = C + x + \frac{x^2}{2} + \frac{x^3}{3} + \frac{x^4}{4} + \cdots.$$

It's a good idea to put the constant first instead of as $+C$ at the end, since it's really the zeroth-dgree term in the power series. Now we have to find out what C actually is. The best way is to substitute $x = 0$. In this case, we get

$$-\ln(1 - 0) = C + 0 + \frac{0^2}{2} + \frac{0^3}{3} + \frac{0^4}{4} + \cdots,$$

which reduces to $C = 0$. Substituting in and taking negatives of both sides, we get our series for $\ln(1 - x)$ as before:

$$\ln(1 - x) = \sum_{n=1}^{\infty} -\frac{x^n}{n} = -x - \frac{x^2}{2} - \frac{x^3}{3} - \frac{x^4}{4} - \cdots.$$

Since the original series (for $1/(1-x)$) converges for $-1 < x < 1$, so does the integrated series (for $-\ln(1-x)$, hence also $\ln(1-x)$). Actually, the series for $\ln(1-x)$ does also converge when $x = -1$; however, as I said, integrating power series term by term doesn't give any information about the endpoints of the interval of convergence. By the way, now you can replace x by $-x$ to get the expansion of $\ln(1+x)$ from formula #5 of Section 26.2 above.

Another example: how would you find the Maclaurin series for $\tan^{-1}(x)$? This would be a real pain to differentiate over and over (just try it and see!), but we can be really sneaky and integrate a series we already know. Let's see, $\tan^{-1}(x)$ is an antiderivative of $1/(1+x^2)$, and we saw in Section 26.2.1 above that we have

$$\frac{1}{1+x^2} = 1 - x^2 + x^4 - x^6 + \cdots$$

when $-1 < x < 1$. We can now integrate both sides:

$$\int \frac{1}{1+x^2}\,dx = \int \left(1 - x^2 + x^4 - x^6 + \cdots\right) dx.$$

Integrating term-by-term on the right-hand side gives

$$\tan^{-1}(x) = C + x - \frac{x^3}{3} + \frac{x^5}{5} - \frac{x^7}{7} + \cdots.$$

Now we substitute $x = 0$ to find out what C is:

$$\tan^{-1}(0) = C + 0 - \frac{0^3}{3} + \frac{0^5}{5} - \frac{0^7}{7} + \cdots,$$

which simplifies to $C = \tan^{-1}(0) = 0$. So, we have

$$\tan^{-1}(x) = x - \frac{x^3}{3} + \frac{x^5}{5} - \frac{x^7}{7} + \cdots = \sum_{n=0}^{\infty} \frac{(-1)^n x^{2n+1}}{2n+1}.$$

(Check that you believe the sigma-notation version on the right-hand side.) Since the original series for $1/(1-x^2)$ converges when $-1 < x < 1$, so does the series for $\tan^{-1}(x)$.*

Let's look at an example of a definite integral. Suppose that a function f is defined by

$$f(x) = \int_0^x \sin(t^3)\,dt.$$

What is its Maclaurin series? We should start by finding the series for $\sin(t^3)$. To do this, substitute $x = t^3$ in the Maclaurin series for $\sin(x)$ to get

$$
\begin{aligned}
\sin(t^3) &= t^3 - \frac{(t^3)^3}{3!} + \frac{(t^3)^5}{5!} - \frac{(t^3)^7}{7!} + \cdots \\
&= t^3 - \frac{t^9}{3!} + \frac{t^{15}}{5!} - \frac{t^{21}}{7!} + \cdots.
\end{aligned}
$$

---

*In fact, the series for $\tan^{-1}(x)$ also converges when $x = 1$ (or $x = -1$) by the alternating series test, eventually leading to the cute formula

$$1 - \frac{1}{3} + \frac{1}{5} - \frac{1}{7} + \cdots = \tan^{-1}(1) = \frac{\pi}{4}.$$

Since the series for $\sin(x)$ is valid for all real x, the series for $\sin(t^3)$ is valid for all real t. Now, we can integrate both sides from 0 to x to get

$$f(x) = \int_0^x \sin(t^3)\, dt = \int_0^x \left(t^3 - \frac{t^9}{3!} + \frac{t^{15}}{5!} - \frac{t^{21}}{7!} + \cdots \right) dt;$$

integrating the right-hand side term-by-term, we get

$$f(x) = \left(\frac{t^4}{4} - \frac{t^{10}}{10 \cdot 3!} + \frac{t^{16}}{16 \cdot 5!} - \frac{t^{22}}{22 \cdot 7!} + \cdots \right) \Big|_0^x$$

$$= \frac{x^4}{4} - \frac{x^{10}}{10 \cdot 3!} + \frac{x^{16}}{16 \cdot 5!} - \frac{x^{22}}{22 \cdot 7!} + \cdots ;$$

this is valid for all real x. (You should try to convert this series to sigma notation. The answer is given in Section 26.3 below.)

You can also apply the above integration techniques to Taylor polynomials; this time the order of the Taylor polynomial increases by 1.

26.2.4 Adding and subtracting Taylor series

If you know the Taylor series about $x = a$ for two functions f and g, then the Taylor series for the sum $f(x) + g(x)$ is of course the sum of the two respective Taylor series, at least in the overlap of the regions where the Taylor series converge. The same goes for the difference $f(x) - g(x)$. The only thing you need to do in practice is to group terms of the same degree together, and worry about where the resulting series converges. For example, the Maclaurin series for $\sin(x) - e^x$ is given by

$$\left(x - \frac{x^3}{3!} + \frac{x^5}{5!} - \frac{x^7}{7!} + \cdots \right) - \left(1 + x + \frac{x^2}{2!} + \frac{x^3}{3!} + \frac{x^4}{4!} + \frac{x^5}{5!} + \frac{x^6}{6!} + \frac{x^7}{7!} + \cdots \right),$$

which should be simplified; after canceling, the series looks like

$$-1 - \frac{x^2}{2!} - \frac{2x^3}{3!} - \frac{x^4}{4!} - \frac{x^6}{6!} - \frac{2x^7}{7!} - \cdots ,$$

at least up to terms of order 7. Since the series for $\sin(x)$ and e^x are valid for all x, so is the series for $\sin(x) - e^x$.

If you want to deal with Taylor polynomials, you have to be careful to take the order to be the lesser of the two orders. For example, we know that the third-order Taylor polynomial about 0 of $1/(1-x)$ is

$$1 + x + x^2 + x^3,$$

while the fourth-order polynomial of e^x about 0 is

$$1 + x + \frac{x^2}{2!} + \frac{x^3}{3!} + \frac{x^4}{4!}.$$

If you set $f(x) = 1/(1-x) + e^x$ and look for its Taylor polynomial about 0, it's no good taking the sum of the above two polynomials. The problem is that you have a fourth-order term in the polynomial for e^x, but no fourth-order term for $1/(1-x)$. It's like comparing apples and oranges. You pretty much have to ignore the $x^4/4!$ term in the polynomial for e^x to get the third-order Taylor polynomial

$$1 + x + \frac{x^2}{2!} + \frac{x^3}{3!}.$$

Now you can add $1 + x + x^2 + x^3$ to the above polynomial to see that the third-order Taylor polynomial about $x = 0$ for $1/(1-x) + e^x$ is

$$(1 + x + x^2 + x^3) + \left(1 + x + \frac{x^2}{2!} + \frac{x^3}{3!}\right),$$

which simplifies to

$$2 + 2x + \frac{3x^2}{2} + \frac{7x^3}{6}.$$

26.2.5 Multiplying Taylor series

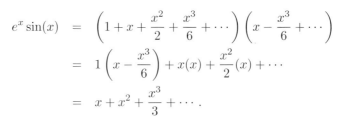

You can also multiply two Taylor series to get a new one which converges to the product of the two relevant functions, at least in the intersection of the regions where the Taylor series converge. Writing this in sigma notation can get pretty messy and usually involves double sums. Normally one is interested in the first few terms of a series. For example, let's find the terms of the Maclaurin series up to and including third order for $f(x) = e^x \sin(x)$. To do this, write out the series for e^x and $\sin(x)$ up to third order, multiply out, and ignore any terms greater than third order:

$$\begin{aligned} e^x \sin(x) &= \left(1 + x + \frac{x^2}{2} + \frac{x^3}{6} + \cdots\right)\left(x - \frac{x^3}{6} + \cdots\right) \\ &= 1\left(x - \frac{x^3}{6}\right) + x(x) + \frac{x^2}{2}(x) + \cdots \\ &= x + x^2 + \frac{x^3}{3} + \cdots. \end{aligned}$$

There's a skill in ignoring terms you don't need. For example, I left out the product of the terms x and $-x^3/6$ from the first and second sums, respectively; this is because I realized that this would give a term in x^4, which I don't care about since we only need terms up to third order. If I wanted terms of up to fourth order, then of course I would have had to worry about more terms.

Actually, it's important not to pay attention to terms of order higher than the ones you've actually written down for the original functions. For example, take the second-order Taylor polynomial of e^x about 0, which is

$$1 + x + \frac{x^2}{2};$$

now multiply it by the second-order Taylor polynomial of e^{-x} about 0, which is

$$1 - x + \frac{x^2}{2}.$$

You get

$$\left(1 + x + \frac{x^2}{2}\right)\left(1 - x + \frac{x^2}{2}\right),$$

which simplifies to

$$1 + \frac{x^4}{4}.$$

If you look at this and claim that this is the fourth-order Taylor polynomial about 0 for the product $(e^x)(e^{-x})$, you'd be wrong! After all, the product is just 1, so **all** of its Taylor polynomials are just 1. The correct thing to do is to ignore all terms in the product of degree greater than 2. After all, we only started with second-order polynomials, so why should we expect anything higher when we multiply these polynomials together? In the above polynomial $1 + x^4/4$, the term $x^4/4$ is of degree higher than 2, so it's not accurate and should be ignored. The second-order polynomial for the product is 1, and that's all you can tell from the product of the two second-order Taylor polynomials we started with. Don't bite off more degrees than you can chew!

26.2.6 Dividing Taylor series

You can do exactly the same thing with quotients by using long division. The trick is to ignore all but the terms of order up to the one you are interested in. For example, to find the Maclaurin series for $f(x) = \sec(x)$ up to fourth order, first write $\sec(x)$ as $1/\cos(x)$, then set up a long division just as you do with polynomials. The main difference here is that you should write the terms so that the degrees are increasing, instead of in the normal decreasing manner. Since we're interested in terms up to fourth order, we'll use

$$\cos(x) = 1 - \frac{x^2}{2} + \frac{x^4}{24} - \cdots$$

in the long division for $1/\cos(x)$:

$$
\begin{array}{r}
1 \qquad + \frac{1}{2}x^2 \qquad\quad + \frac{5}{24}x^4 + \cdots \\
\hline
1 + 0x - \frac{1}{2}x^2 + 0x^3 + \frac{1}{24}x^4 - \cdots \,\big)\; 1 + 0x + 0x^2 + 0x^3 + 0x^4 + \cdots \\
1 + 0x - \frac{1}{2}x^2 + 0x^3 + \frac{1}{24}x^4 + \cdots \\
\hline
\frac{1}{2}x^2 + 0x^3 - \frac{1}{24}x^4 + \cdots \\
\frac{1}{2}x^2 + 0x^3 - \frac{1}{4}x^4 + \cdots \\
\hline
\frac{5}{24}x^4 + \cdots
\end{array}
$$

So the Maclaurin series for $\sec(x)$ is $1 + x^2/2 + 5x^4/24 + \cdots$, up to terms of fourth order.

If instead we would like to find the Maclaurin series for $\tan(x)$ up to fourth order, we could proceed similarly, since $\tan(x) = \sin(x)/\cos(x)$. Using $\sin(x) = x - x^3/6 + \cdots$ and $\cos(x) = 1 - x^2/2 + x^4/24 - \cdots$, the division would begin like this:

$$1 + 0x - \frac{1}{2}x^2 + 0x^3 + \frac{1}{24}x^4 - \cdots \,\big)\; 0 + x + 0x^2 - \frac{1}{6}x^3 + 0x^4 + \cdots$$

I leave it to you to do the calculation and see that $\tan(x) = x + x^3/3 + \cdots$ up to terms of fourth order (note that the fourth order term here is actually 0).

So the moral of the story is that you may not have to differentiate over and over again and use the formula for Taylor series. If you're lucky, you can instead use some of the five basic series, plus one or more of the techniques of substitution, differentiation, integration, addition, subtraction, multiplication, and division.

26.3 Using Power and Taylor Series to Find Derivatives

Recall the formula for the nth coefficient of the Taylor series of $f(x)$ about $x = a$:

$$a_n = \frac{f^{(n)}(a)}{n!}.$$

Let's multiply through by $n!$ to arrive at the following formula:

$$\boxed{f^{(n)}(a) = n! \times a_n}$$

In words, this means that

$$\boxed{f^{(n)}(a) = n! \times \begin{pmatrix} \text{the coefficient of } (x-a)^n \text{ in the} \\ \text{Taylor series of } f(x) \text{ about } x = a \end{pmatrix}.}$$

So if you know the Taylor series of a function about some point a, you can easily find the derivatives of that function at a. This is all you get! There's no information about the value of the derivatives at any other value of x; it's only $x = a$. (Actually, to find the nth derivative, you only need a Taylor polynomial at $x = a$ of order n or more, not the whole Taylor series.)

To use the above equation, you need to start by finding an appropriate Taylor series for your function. The techniques from the previous few sections can be really useful for this. For example, suppose that $f(x) = e^{x^2}$, and we want to find $f^{(100)}(0)$ and $f^{(101)}(0)$. We kick off by finding the Maclaurin series for e^{x^2}:

$$e^{x^2} = \sum_{n=0}^{\infty} \frac{(x^2)^n}{n!} = \sum_{n=0}^{\infty} \frac{x^{2n}}{n!} = 1 + x^2 + \frac{x^4}{2!} + \frac{x^6}{3!} + \cdots.$$

By the boxed formula above,

$$f^{(100)}(0) = 100! \times (\text{coefficient of } x^{100} \text{ in the above Maclaurin series}).$$

So what is the coefficient of x^{100} in the Maclaurin series, anyway? You can look at it and just see that it's $1/(50!)$, or if you want to be more formal about it, you can work out which value of n will give you x^{100}. In particular, we want to locate the term $x^{2n}/n!$ that is a multiple of x^{100}. This means that $2n = 100$, so $n = 50$, and the term is $x^{100}/(50!)$. So the coefficient is $1/(50!)$. This means that

$$f^{(100)}(0) = 100! \times \frac{1}{50!} = \frac{100!}{50!}.$$

(Don't make the mistake of trying to simplify this last expression down to 2!; factorials don't work that way.) Now, how about finding $f^{(101)}(0)$? This is equal to 101! times the coefficient of x^{101} in the above series. What is that coefficient? Hang on, there are no odd powers at all in the series! Put another way, what value of n would give you x^{101}? It would have to solve $2n = 101$, but n has to be an integer, so the power x^{101} isn't present. That means that the coefficient of x^{101} is 0, so

$$f^{(101)}(0) = 101! \times 0 = 0.$$

All right, let's see a more difficult example. In Section 26.2.3 above, we found that the Maclaurin series of the function f, defined by

$$f(x) = \int_0^x \sin(t^3)\, dt,$$

is given by

$$\frac{x^4}{4} - \frac{x^{10}}{10 \cdot 3!} + \frac{x^{16}}{16 \cdot 5!} - \frac{x^{22}}{22 \cdot 7!} + \cdots ;$$

this series converges to $f(x)$ for all real x. I now ask you this: what is $f^{(50)}(0)$? How about $f^{(52)}(0)$? To do this, we are going to need the coefficients of x^{50} and x^{52} in the above series for $f(x)$. Remember, $f^{(50)}(0)$ is 50! times the coefficient of x^{50} in the Maclaurin series of $f(x)$, and of course the same is true for $f^{(52)}(0)$ except with 52 instead of 50 everywhere.

Now, to find the coefficients of x^{50} and x^{52} in the above series, you could keep on writing it out until you got far enough. A better way is to change the series to sigma notation. I challenged you to do this as an exercise earlier; here's how you do it, in any case. Note that the powers of x go 4, 10, 16, 22, and so on. This means that they go up by 6 every time, starting at 4. So, the exponents are given by $6n + 4$, where n runs through the numbers 0, 1, 2, 3, and so on. Now, let's look at the denominator. It's the product of the quantity $6n + 4$ and a factorial of an odd number. The odd numbers go 1, 3, 5, 7, ..., so they are given by $2n+1$. So, the denominator is $(6n+4)(2n+1)!$. Finally, the terms alternate, beginning with positive sign, so there should be a $(-1)^n$ in there as well. We have now seen that

$$f(x) = \sum_{n=0}^{\infty} \frac{(-1)^n x^{6n+4}}{(6n+4)(2n+1)!}.$$

Now we can finally find the coefficients of x^{50} and x^{52}. For the first one, try to solve $6n + 4 = 50$. This would give $n = 23/3$, which is not an integer, so the coefficient of x^{50} is 0. This means that

$$f^{(50)}(0) = 50! \times \left(\text{coefficient of } x^{50}\right) = 50! \times 0 = 0.$$

On the other hand, for x^{52}, try to solve $6n + 4 = 52$. This gives $n = 8$, so we can get the coefficient of x^{52} by looking at what happens when we put $n = 8$. The term in the sum given by $n = 8$ is

$$\frac{(-1)^8 x^{6\times 8+4}}{(6 \times 8 + 4)(2 \times 8 + 1)!} = \frac{x^{52}}{52 \times 17!},$$

so the coefficient is $1/(52 \times 17!)$. Finally,

$$f^{(52)}(0) = 52! \times \left(\text{coefficient of } x^{52}\right) = 52! \times \frac{1}{52 \times 17!} = \frac{51!}{17!}.$$

Notice that I did a little canceling here: $52!/52 = 51!$. Convince yourself that this is true before proceeding!

Sometimes a function is already defined by a power series about $x = a$, and you may need to find certain derivatives of the function at a. This is even easier than the above examples, since you don't have to find the Taylor series first. For example, suppose $f(x)$ is defined by

$$f(x) = \sum_{n=0}^{\infty} \frac{(-1)^{n+1} n^3 (x-6)^{3n}}{n!},$$

which converges for all x (why?!?). Say that you want to evaluate $f^{(300)}(6)$. Well, the power series is about $x = 6$, so we can use the formula

$$f^{(300)}(6) = 300! \times \left(\text{coefficient of } (x-6)^{300} \text{ in the above series}\right).$$

To see what the coefficient is, we should find out which value of n gives the correct term. Looking at the above series, the general exponent of $(x-6)$ is $3n$, so we need the term where $3n = 300$. Thus $n = 100$, and substituting, we see that the correct term is

$$\frac{(-1)^{100+1} 100^3 (x-6)^{300}}{100!} = \frac{-1000000}{100!} (x-6)^{300}.$$

So the coefficient is $-1000000/100!$. If you want to get really fancy, you can write $100!$ as $100 \times 99!$ and cancel out a factor of 100 to see that the coefficient is $-10000/99!$. Anyway, this shows that

$$f^{(300)}(6) = 300! \times \frac{-10000}{99!} = -\frac{300! \times 10000}{99!}.$$

What if you wanted to find $f^{(301)}(6)$? I leave it to you to show that there is no term $(x-6)^{301}$ appearing in the power series, so the answer is 0.

26.4 Using Maclaurin Series to Find Limits

You can also use some Taylor series to find certain limits. In particular, if you have a limit like

$$\lim_{x \to 0} \frac{f(x)}{g(x)},$$

where both the numerator and the denominator are 0 when $x = 0$, then you could use l'Hôpital's Rule; however, if you wanted to evaluate

$$\lim_{x \to 0} \frac{e^{-x^2} + x^2 \cos(x) - 1}{1 - \cos(2x^3)},$$

you'd have to be stark raving mad to do it that way. The numerator and denominator are no fun to differentiate once, let alone the six or so times you'd actually have to do it (as it turns out). So, the correct method is to replace everything in sight by enough terms of the appropriate Maclaurin series. What do I mean by "enough terms"? Well, we expect that some terms might cancel, and we don't want to be left with 0 in the numerator or the denominator. Let's try going up to eighth order first. Let's write down Maclaurin series for everything involved. First, since

$$e^x = 1 + x + \frac{x^2}{2} + \frac{x^3}{6} + \frac{x^4}{24} + \cdots,$$

replacing x by $-x^2$, we get

$$e^{-x^2} = 1 - x^2 + \frac{x^4}{2} - \frac{x^6}{6} + \frac{x^8}{24} - \cdots.$$

Now, since

$$\cos(x) = 1 - \frac{x^2}{2} + \frac{x^4}{24} - \frac{x^6}{6!} + \cdots,$$

we can get a series for $x^2 \cos(x)$ by multiplying through by x^2:

$$x^2 \cos(x) = x^2 - \frac{x^4}{2} + \frac{x^6}{24} - \frac{x^8}{6!} + \cdots.$$

If instead we go back to the series for $\cos(x)$ and replace x by $2x^3$, we get

$$\cos(2x^3) = 1 - \frac{(2x^3)^2}{2} + \frac{(2x^3)^4}{24} - \cdots = 1 - 2x^6 + \frac{2}{3}x^{12} - \cdots,$$

where we don't even need this last term, let alone any higher ones, since we have decided to go up to order 8. Still, it doesn't hurt to put it in, so we'll leave it. Anyway, if we put all this together, the numerator is

$e^{-x^2} + x^2 \cos(x) - 1$

$$= \left(1 - x^2 + \frac{x^4}{2} - \frac{x^6}{6} + \frac{x^8}{24} - \cdots\right) + \left(x^2 - \frac{x^4}{2} + \frac{x^6}{24} - \frac{x^8}{6!} + \cdots\right) - 1$$

$$= -\frac{1}{8}x^6 + \left(\frac{1}{24} - \frac{1}{720}\right)x^8 + \cdots,$$

whereas the denominator becomes

$$1 - \cos(2x^3) = 1 - \left(1 - 2x^6 + \frac{2}{3}x^{12} - \cdots\right) = 2x^6 - \frac{2}{3}x^{12} + \cdots.$$

Now, substituting into the limit, we have

$$\lim_{x \to 0} \frac{e^{-x^2} + x^2 \cos(x) - 1}{1 - \cos(2x^3)} = \lim_{x \to 0} \frac{-\frac{1}{8}x^6 + \left(\frac{1}{24} - \frac{1}{720}\right)x^8 + \cdots}{2x^6 - \frac{2}{3}x^{12} + \cdots}.$$

Divide top and bottom by the lowest power, x^6, and plug in $x = 0$ to see that this limit is equal to

$$\lim_{x \to 0} \frac{-\frac{1}{8} + \left(\frac{1}{24} - \frac{1}{720}\right)x^2 + \cdots}{2 - \frac{2}{3}x^6 + \cdots} = \frac{-1/8}{2} = -\frac{1}{16}.$$

So, as you can see, the terms involving order higher than 6 didn't come into it at all (which is incidentally why I never bothered simplifying the expression $1/24 - 1/720$). Basically, if everything cancels out, you haven't used enough terms, whereas if something is still left, you've gone far enough and can proceed. If you'd only gone up to terms of order 5 (or less), then you would have gotten $0/0$ again, so you wouldn't have gone far enough.

Let's look at one more example: find

$$\lim_{x \to 0} \left(\frac{1}{\sin(x)} - \frac{1}{e^x - 1} \right).$$

This doesn't look like a fraction, so the first step is to do some algebra. Take a common denominator, just as we did in the case of l'Hôpital Type **B1** limits in Section 14.1.3 of Chapter 14, to write the limit as

$$\lim_{x \to 0} \frac{e^x - 1 - \sin(x)}{\sin(x)(e^x - 1)}.$$

Now we have

$$e^x - 1 = x + \frac{x^2}{2} + \frac{x^3}{6} + \cdots,$$

and

$$\sin(x) = x - \frac{x^3}{6} + \cdots.$$

Putting all this in, the limit becomes

$$\lim_{x \to 0} \frac{\left(x + \dfrac{x^2}{2} + \dfrac{x^3}{6} + \cdots \right) - \left(x - \dfrac{x^3}{6} + \cdots \right)}{\left(x - \dfrac{x^3}{6} + \cdots \right) \left(x + \dfrac{x^2}{2} + \dfrac{x^3}{6} + \cdots \right)}$$

$$= \lim_{x \to 0} \frac{\dfrac{x^2}{2} + \dfrac{x^3}{3} + \cdots}{\left(x - \dfrac{x^3}{6} + \cdots \right) \left(x + \dfrac{x^2}{2} + \dfrac{x^3}{6} + \cdots \right)}.$$

Now, once again, the lowest power dominates as $x \to 0$; to see this, divide top and bottom by x^2. Let's be sneaky about it, though: on the bottom, we want to divide both factors by x, which is the same as dividing the whole thing by x^2. The limit becomes

$$\lim_{x \to 0} \frac{\dfrac{1}{2} + \dfrac{x}{3} + \cdots}{\left(1 - \dfrac{x^2}{6} + \cdots \right) \left(1 + \dfrac{x}{2} + \dfrac{x^2}{6} + \cdots \right)} = \frac{1/2}{(1)(1)} = \frac{1}{2}.$$

Once again, it doesn't hurt if you write extra terms—I only used up to third order here, but higher orders would be fine. Actually, the third-order terms didn't even come into it at all, and in the denominator we only needed the first-order terms. Unless you are psychic or have a really good intuition about such things, it's pretty hard to guess how many terms you need. So, it's better to use more terms rather than fewer; you can always ignore them later, whereas if you use too few terms, you can't even solve the problem.

Here's the real reason all the above limits work: if f has a Maclaurin series with lowest-degree term $a_N x^N$, then

$$f(x) \sim a_N x^N \qquad \text{as } x \to 0.$$

We mentioned this fact way back in Section 21.4.5 of Chapter 21; it's useful in conjunction with the limit comparison test. In fact, the above equation is true even if the Maclaurin series for f doesn't converge for $x \neq 0$. So there's no need to work with the complete Maclaurin series: the lowest-order nonzero Taylor polynomial for f about $x = 0$ is good enough. There's just one technical condition, which is that the $(N+1)$th derivative of f has to be bounded near 0. Here's how the whole thing works: by Taylor's Theorem, we have

$$f(x) = a_N x^N + R_N(x) = a_N x^N + \frac{f^{(N+1)}(c)}{(N+1)!} x^{N+1},$$

where c is between 0 and x. Now divide both sides by $a_N x^N$ to get

$$\frac{f(x)}{a_N x^N} = 1 + \frac{f^{(N+1)}(c)}{a_N (N+1)!} x.$$

The quantity $f^{(N+1)}(c)/(a_N(N+1)!)$ on the right-hand side is bounded in absolute value as $x \to 0$, since the denominator is constant and we've assumed the numerator is bounded. Now you can use the sandwich principle to show that the last term on the right-hand side of the above equation goes to 0 as $x \to 0$. That is,

$$\lim_{x \to 0} \frac{f(x)}{a_N x^N} = 1.$$

This is the same as saying that

$$f(x) \sim a_N x^N \qquad \text{as } x \to 0,$$

and we have proved our claim. So what? Well, not only do we get a handy tool to use with the limit comparison test, but we've actually proved that all the above limits work. For example, to really nail the above limit

$$\lim_{x \to 0} \frac{e^x - 1 - \sin(x)}{\sin(x)(e^x - 1)},$$

we should note that $e^x - 1 - \sin(x)$ has a Maclaurin series beginning with $x^2/2$, so $e^x - 1 - \sin(x) \sim x^2/2$ as $x \to 0$; similarly, $\sin(x) \sim x$ as $x \to 0$, and $e^x - 1 \sim x$ as $x \to 0$. Since you can multiply and divide these asymptotic relations (but not add or subtract them!), we can say that

$$\frac{e^x - 1 - \sin(x)}{\sin(x)(e^x - 1)} \sim \frac{x^2/2}{(x)(x)} \qquad \text{as } x \to 0.$$

The right-hand side is just $1/2$, so we have proved that

$$\lim_{x \to 0} \frac{e^x - 1 - \sin(x)}{\sin(x)(e^x - 1)} = \frac{1}{2}.$$

In reality, the above method (using the full series with the $+\cdots$ notation) is generally accepted, even though technically it dances around the true issue. What's really going on is shown in the above argument involving the remainder term R_N.

CHAPTER 27 _____

Parametric Equations and Polar Coordinates

So far, we've sketched the graphs of many equations of the form $y = f(x)$ with respect to Cartesian coordinates. Now we're going to look at things in a different way: first, we'll look at what happens when the coordinates x and y are not directly related, but are instead related by a common parameter; and then we'll see what happens when we replace the whole darn coordinate system with something entirely different. Of course, we have to do some calculus too. So here's the program for this chapter:

- parametric equations, graphs and finding tangents;
- converting from polar coordinates to Cartesian coordinates, and vice versa;
- finding tangents to polar curves; and
- finding areas enclosed by polar curves.

27.1 Parametric Equations

When you write an equation like $y = x^2 \sin(x)$, you are expressing y as a function of x. So if you have a particular value of x in mind, then you can easily find the corresponding value of y by plugging that value of x into the above equation. On the other hand, consider the relation $x^2 + y^2 = 9$. Now if you have a particular value of x in mind, you have to work a little harder to find the corresponding value of y. In fact, there may be multiple values of y which correspond to your value of x, or there may be none at all. Of course, you can write $y = \pm\sqrt{9 - x^2}$; this means that there are actually two values of y corresponding to x if $-3 < x < 3$, but only one value of y if $x = \pm 3$ and no values of y otherwise.

Now let's try a different approach: suppose that both x and y are functions of another variable t. For example, we could set

$$x = 3\cos(t) \quad \text{and} \quad y = 3\sin(t).$$

So I'm asking you think of x as a function of t; if you like, you could even write $x(t) = 3\cos(t)$ to emphasize this. The same goes for y. If you pick a particular value of t, then you can get corresponding values for **both** x and y by plugging your value of t into the above equations. The variable t is called a *parameter*, and the above equations are called *parametric equations*.

What does the graph of the above pair of parametric equations look like? Let's try plotting points. Instead of the normal technique of picking some values of x and finding the corresponding values of y, we instead pick values of t and find the corresponding values of both x and y. To plot the points, only use the values of x and y—there is no t-axis involved! Anyway, since there are trig functions around, we should make sure that all of our test values involve π. Indeed, suppose we try the following values of t:

t	0	$\pi/6$	$\pi/4$	$\pi/3$	$\pi/2$
x					
y					

If we work out the corresponding values of x and y using the above equations $x = 3\cos(t)$ and $y = 3\sin(t)$, we can fill in the table like this:

t	0	$\pi/6$	$\pi/4$	$\pi/3$	$\pi/2$
x	3	$3\sqrt{3}/2$	$3/\sqrt{2}$	$3/2$	0
y	0	$3/2$	$3/\sqrt{2}$	$3\sqrt{3}/2$	3

So $t = 0$ corresponds to the point $(3, 0)$, and $t = \pi/6$ corresponds to the point $(3\sqrt{3}/2, 3/2)$, for example. Here's a graph showing all five points:

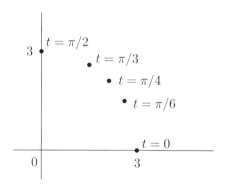

It seems as if we are dealing with a quarter-arc of a circle of radius 3 units centered at the origin. This should come as no surprise, knowing what we know about trigonometry! (Of course, for any value of t, it is true that $x^2 + y^2 = (3\cos(t))^2 + (3\sin(t))^2 = 9(\cos^2(t) + \sin^2(t)) = 9$.) Now if you continue the above table up to $t = \pi$, you describe a semicircle, whereas if you go all the way to $t = 2\pi$, you get the full circle. What happens if you keep going? Well, you just start to retrace the circle. The same thing happens if instead you start at $t = 0$ and make t go more negative, except that now you move around the circle clockwise instead of counterclockwise. Notice that if you pick a point (x, y) on the circle, there isn't just one value of t which corresponds to that point! There are infinitely many, all separated by multiples of 2π. For example, if n is any integer, then $t = 2\pi n$ corresponds to $x = 3$ and $y = 0$, that is, the point $(3, 0)$.

So, the above pair of parametric equations describes the circle $x^2 + y^2 = 9$, at least if you let t range over a large enough interval—for example, $[0, 2\pi)$. You can say that

$$x = 3\cos(t) \quad \text{and} \quad y = 3\sin(t), \quad \text{where } 0 \le t < 2\pi$$

is a *parametrization* of $x^2 + y^2 = 9$. Now, I ask you this: is the graph of $x^2 + y^2 = 9$ the same as the graph of the above parametrization? Yes and no. Certainly the two graphs look like the same circle, but the parametric version tells you a little more: it tells you how the circle is drawn. If you start at $t = 0$ and move continuously up to $t = 2\pi$, then you trace out the circle by starting at $(3, 0)$, then drawing counterclockwise at a constant speed until you're back at the starting point.

The whole thing is sort of like looking at a slime trail left by a snail, compared with actually watching the snail move and leave the trail. Just looking at the trail isn't enough to tell you in which direction the snail moved—it might even have backtracked! You also can't tell how fast it was moving at different times along the trail. (No, "at a snail's pace" is not a scientific description of how fast it was moving.) Having a parametrization is like knowing where the snail is at each time; it allows you to find the extra information of direction and speed.

So is the above parametrization the only possible one for $x^2 + y^2 = 9$? No way. There are many other ways to draw the same circle. For example, you could put $x = 3\cos(2t)$ and $y = 3\sin(2t)$. Now you only need t to range from 0 to π to cover the whole circle, and in fact you go around twice as fast as you did before. Alternatively, you could try $x = 3\sin(t)$ and $y = 3\cos(t)$ for $0 \le t < 2\pi$. Now you're back to normal speed, but this time you start at $(0, 3)$ and go clockwise around the circle instead of counterclockwise. Convince yourself that these facts are true by plotting a few points.

How would you find a parametrization for $x^2 + 4y^2 = 9$? Sketching this curve gives an ellipse through $(\pm 3, 0)$ and $(0, \pm 3/2)$. If you set $Y = 2y$, then $x^2 + Y^2 = 9$. This is a circle in the new coordinates (x, Y), so we can use our above parametrization: $x = 3\cos(\theta)$ and $Y = 3\sin(\theta)$ for $0 \le \theta < 2\pi$. Now we just have to write $y = Y/2$ to get the parametrization

$$x = 3\cos(t) \quad \text{and} \quad y = \frac{3}{2}\sin(t), \quad \text{where} \quad 0 \le t < 2\pi$$

for the ellipse. This is not the only possible parametrization, of course!

How about $x^6 + y^6 = 64$? I leave it to you to sketch this curve and see that it looks like a bloated circle of "radius" $64^{1/6} = 2$ units. This should inspire us to adapt the above parametrization of the circle. First, we need to change the radius to 2 units: indeed, $x = 2\cos(t)$ and $y = 2\sin(t)$ would do the circle $x^2 + y^2 = 4$ but it fails for the bloated circle, since it's not true in general that $\cos^6(t) + \sin^6(t) = 1$. How do we fix this? Well, let's replace $\cos(t)$ by some power of itself so that when we take the 6th power, we get $\cos^2(t)$. That would have to be $\cos^{1/3}(t)$. So if we try $x = 2\cos^{1/3}(t)$ and $y = 2\sin^{1/3}(t)$, then this should work. Let's test it:

$$x^6 + y^6 = (2\cos^{1/3}(t))^6 + (2\sin^{1/3}(t))^6 = 64\cos^2(t) + 64\sin^2(t) = 64,$$

which is what we wanted. To get the whole curve, we let t range from 0 to 2π as before.

27.1.1 Derivatives of parametric equations

This is a calculus book, so we'd better do some calculus with this parametric stuff. To find the equation of a tangent line to the curve, we'll need a derivative, of course. Since x and y are both functions of t, we have to use the chain rule. This says that

$$\frac{dy}{dt} = \frac{dy}{dx}\frac{dx}{dt};$$

now divide through by dx/dt and rearrange to get

$$\boxed{\frac{dy}{dx} = \frac{dy/dt}{dx/dt}.}$$

If you are thinking of x as $x(t)$ and similarly for y, then you can rewrite this equation as

$$\frac{dy}{dx} = \frac{y'(t)}{x'(t)}.$$

Let's look at three examples of how to use this.

First, suppose that we want the slope and equation of the tangent line at the point corresponding to $t = 1/2$ on the parametric curve defined by

$$x = e^{-2t}, \qquad y = \sin^{-1}(t), \qquad -1 < t < 1.$$

Differentiating, we find that

$$\frac{dx}{dt} = -2e^{-2t} \qquad \text{and} \qquad \frac{dy}{dt} = \frac{1}{\sqrt{1-t^2}}.$$

Since we only care about the point $t = 1/2$, we might as well evaluate the above derivatives at $t = 1/2$ right now to get

$$\frac{dx}{dt} = -2e^{-1} = -\frac{2}{e} \qquad \text{and} \qquad \frac{dy}{dt} = \frac{1}{\sqrt{1-1/4}} = \frac{2}{\sqrt{3}}.$$

So at $t = 1/2$, we have

$$\frac{dy}{dx} = \frac{dy/dt}{dx/dt} = \frac{2/\sqrt{3}}{-2/e} = -\frac{e}{\sqrt{3}}.$$

Great—we've found the slope. How about the tangent line? Well, this line passes through (x, y) and has slope dy/dx. We know what the slope is, but what about x and y? Well, we have to put $t = 1/2$ in the original equations for x and y above to see that $x = e^{-2\cdot(1/2)} = 1/e$ and $y = \sin^{-1}(1/2) = \pi/6$. So the equation of the line is

$$y - \frac{\pi}{6} = -\frac{e}{\sqrt{3}}\left(x - \frac{1}{e}\right),$$

which simplifies slightly to

$$y = -\frac{e}{\sqrt{3}}x + \frac{1}{\sqrt{3}} + \frac{\pi}{6}.$$

Now for a trickier example. Suppose we want to find the equation of the tangent to the curve $x^6 + y^6 = 64$ at the point $(-2^{5/6}, 2^{5/6})$. (You should check by substituting that this point is actually on the curve.) This can be done by implicit differentiation, but let's try using our parametrization $x = 2\cos^{1/3}(t)$ and $y = 2\sin^{1/3}(t)$ from the end of the previous section; here $0 \le t < 2\pi$. We get

$$\frac{dx}{dt} = -\frac{2}{3}\cos^{-2/3}(t)\sin(t) \qquad \text{and} \qquad \frac{dy}{dt} = \frac{2}{3}\sin^{-2/3}(t)\cos(t).$$

So by the chain rule,

$$\frac{dy}{dx} = \frac{dy/dt}{dx/dt} = \frac{\frac{2}{3}\sin^{-2/3}(t)\cos(t)}{-\frac{2}{3}\cos^{-2/3}(t)\sin(t)} = -\frac{\cos^{5/3}(t)}{\sin^{5/3}(t)}.$$

We want to know what happens at $(-2^{5/6}, 2^{5/6})$. Let's set $x = -2^{5/6}$; since $x = 2\cos^{1/3}(t)$, we see that $2\cos^{1/3}(t) = -2^{5/6}$, so $\cos(t) = -1/\sqrt{2}$. If you play the same game with y, you'll find that $\sin(t) = 1/\sqrt{2}$. You could now find t if you like—if you think about it, you should be able to see that $t = 3\pi/4$ is the only solution between 0 and 2π. But in any case, you don't even have to find t, believe it or not! Knowing just the values of $\sin(t)$ and $\cos(t)$ is enough to substitute into the above expression for dy/dx to get

$$\frac{dy}{dx} = -\frac{\cos^{5/3}(t)}{\sin^{5/3}(t)} = -\frac{(-1/\sqrt{2})^{5/3}}{(1/\sqrt{2})^{5/3}} = 1.$$

So we have found that the slope of the tangent line is 1. To find the equation of the line, we know it passes through $(x, y) = (-2^{5/6}, 2^{5/6})$ and has slope 1, so its equation is

$$y - 2^{5/6} = 1(x - (-2^{5/6}));$$

make sure you see why this can be simplified to

$$y = x + 2^{11/6}.$$

Now for our trickiest example (conceptually speaking, at least). Suppose that we are given the following parametric equations:

$$x = 4t^2 - 4 \qquad \text{and} \qquad y = 2t - 2t^3 \qquad \text{for all real } t.$$

These equations describe a curve in the x,y-plane; let's find the equation of any tangent line to this curve at the origin. Notice that I said "any" instead of "the." There's a reason for this! Let's try to work out which value of t corresponds to the origin. At the origin, both x and y are 0, so we'll need $x = 4(t^2 - 1) = 0$ and $y = 2(t - t^3) = 0$. The first of these equations holds only when $t^2 = 1$, so t must be ± 1. Both of these values satisfy the second equation as well. The conclusion is that the curve passes through the origin at $t = 1$ and also at $t = -1$. Now we know that

$$\frac{dy}{dx} = \frac{dy/dt}{dx/dt} = \frac{2 - 6t^2}{8t} = \frac{1}{4t} - \frac{3t}{4}.$$

When $t = 1$, we have $dy/dx = -1/2$; so the tangent line in this case is a line through the origin with slope $-1/2$. Its equation must therefore be $y = -x/2$. On the other hand, when $t = -1$, we have $dy/dx = 1/2$, so now the tangent line is $y = x/2$. Let's see why this is plausible by sketching the curve. To do this, let's take some values of t and work out the corresponding values of x and y:

t	-2	$-\dfrac{3}{2}$	-1	$-\dfrac{1}{2}$	0	$\dfrac{1}{2}$	1	$\dfrac{3}{2}$	2
x	12	5	0	-3	-4	-3	0	5	12
y	12	$\dfrac{15}{4}$	0	$-\dfrac{3}{4}$	0	$\dfrac{3}{4}$	0	$-\dfrac{15}{4}$	-12

Plotting these points and making an educated guess, the curve should look something like this:

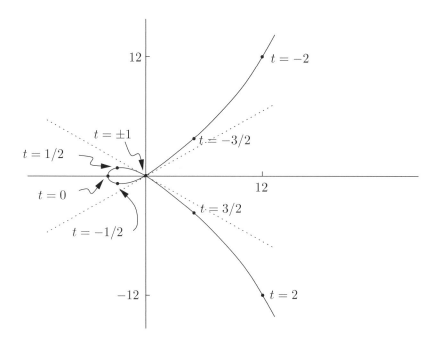

So we see that there are indeed two tangent lines at the origin, and their slopes of $1/2$ and $-1/2$ look reasonable.

Suppose that now we want to find the **second** derivative of the above parametric equations at $t = 1$. The secret to finding d^2y/dx^2 is to consider it as dy'/dx. That is, think of the second derivative as the derivative of y', which itself is the derivative of y with respect to x. Then the problem becomes easy. We already saw above that

$$y' = \frac{dy}{dx} = \frac{1}{4t} - \frac{3t}{4},$$

so without substituting $t = 1$ yet, we now use the chain rule (and the fact that $x = 4t^2 - 4$) to write

$$\frac{d^2y}{dx^2} = \frac{dy'}{dx} = \frac{dy'/dt}{dx/dt} = \frac{\frac{d}{dt}\left(\frac{1}{4t} - \frac{3t}{4}\right)}{\frac{d}{dt}(4t^2 - 4)} = \frac{\frac{-1}{4t^2} - \frac{3}{4}}{8t} = -\frac{1}{32t^3} - \frac{3}{32t}.$$

Now we can finally substitute $t = 1$ to see that

$$\frac{d^2y}{dx^2} = -\frac{1}{32} - \frac{3}{32} = -\frac{1}{8}.$$

As a reality check, look at the above graph. The relevant portion of the curve when $t = 1$ is actually the top half of the loop to the left of the y-axis, moving down through the origin into the fourth quadrant. If you just focus on this part of the curve near the origin, you can see that it is indeed concave down, so at least we have convinced ourselves that the second derivative should be negative, as we found above.

27.2 Polar Coordinates

Suppose your friend is standing in a big flat field at a point that you both agree will be the origin. You'd like to tell him or her how to get to another spot in the field. If you use Cartesian coordinates, then you might tell your friend to go to the point (x, y), where this means that your friend should walk x units to the east and then y units north. (You'll have to agree on what units you're using in advance.) Of course, if x or y is negative, this means that your friend has to walk backward for the appropriate amount. Also, your friend could walk y units north and then x units east—that still gets him or her to the same place.

Instead, you could tell your friend to face due east, then call out an angle for him or her to turn in the counterclockwise direction (while staying at the origin). If the angle is negative, that means your friend should turn clockwise instead. After that, you call out a distance for your friend to march in the direction he or she is facing. If the distance is negative, it's a backward march. So instead of coordinates in Cartesian form (x, y), your friend will get (r, θ); here θ is the amount to turn and r units is the distance to march.

If the point you want to describe is actually the origin, then you could tell your friend $(0, \theta)$ for any angle θ. It doesn't matter how much he or she turns—there will be no marching, so your friend just stays at the origin. Also observe that you could add 2π onto the angle θ and it wouldn't make a difference. Your friend would simply spin around a full revolution in addition to θ. The same thing goes for 4π, 6π, or any other integer multiple of 2π, even negative multiples—it just depends how sadistic you want to be, making your friend spin around many times without purpose just to make him or her dizzy! Anyway, now it's time to look at some formulas.

27.2.1   Converting to and from polar coordinates

Consider the point (r, θ) in polar coordinates, which could look something like this:

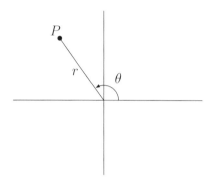

Remember, your friend started at the origin facing toward the positive direction on the x-axis, then turned counterclockwise an angle θ, then marched forward r units to get to the point P. What are the Cartesian coordinates (x, y) of P? Well, we know that $\cos(\theta) = x/r$ and $\sin(\theta) = y/r$, so that gives us

$$\boxed{x = r\cos(\theta) \qquad \text{and} \qquad y = r\sin(\theta).}$$

(Compare this with the example $x = 3\cos(t)$, $y = 3\sin(t)$ from Section 27.1 above.) Anyway, these equations show us how to convert from polar to Cartesian coordinates. For example, what are the Cartesian coordinates of the point given in polar coordinates by $(2, 11\pi/6)$? First, it's not a bad idea to draw a picture:

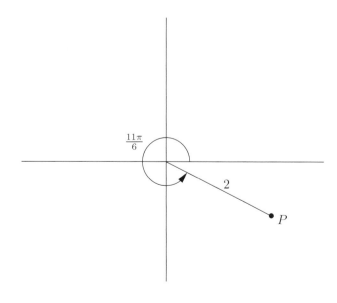

The picture shows that the reference angle is $2\pi - 11\pi/6$, which equals $\pi/6$. We are in the fourth quadrant, so cosines are positive and sines are negative; we therefore have $x = 2\cos(11\pi/6) = 2 \cdot (\sqrt{3}/2) = \sqrt{3}$, and we also have $y = 2\sin(11\pi/6) = 2 \cdot (-1/2) = -1$. That is, the Cartesian coordinates are $(\sqrt{3}, -1)$.

It's always easier translating from a foreign language into your native language than the other way around; the same thing happens with polar coordinates. It's a little harder getting from Cartesian coordinates to polar coordinates. The easy part is r, since by Pythagoras' Theorem, $r^2 = x^2 + y^2$. (You can also see this by squaring both equations in the box above, then adding them together and using $\cos^2(x) + \sin^2(x) = 1$.) How about θ? We know $\tan(\theta) = y/x$ provided that $x \neq 0$, but that doesn't tell us exactly what θ is. You could always add any integer multiple of π to θ without changing the value of $\tan(\theta)$. So you should draw a picture to see what's going on. Here's a summary of the situation:

$$r^2 = x^2 + y^2 \quad \text{and} \quad \tan(\theta) = \frac{y}{x} \quad \text{if } x \neq 0, \quad \text{but check the quadrant!}$$

Let's look at an example: suppose we want to write $(-1, -1)$ in polar coordinates. If you put $x = -1$ and $y = -1$ in the above formulas, you get $r^2 = (-1)^2 + (-1)^2 = 2$ and $\tan(\theta) = (-1)/(-1) = 1$. So it looks like $r = \sqrt{2}$ and $\theta = \tan^{-1}(1) = \pi/4$. This can't be right, though! Check out the following picture:

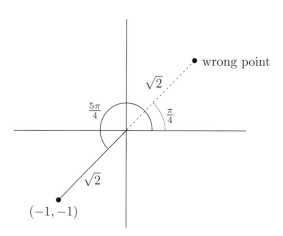

The point with polar coordinates $(\sqrt{2}, \pi/4)$ is the wrong point, since it's in the first quadrant. The correct point is in the third quadrant, and as you can see from the picture, its polar coordinates should be $(\sqrt{2}, 5\pi/4)$.

So, where did we go wrong? Well, we said that $\tan(\theta) = 1$, so $\theta = \pi/4$. We forgot about the solution $\theta = 5\pi/4$. Actually, we also said that $r^2 = 2$, so $r = \sqrt{2}$, neglecting the solution $r = -\sqrt{2}$. If you look at the above picture again, you can see that the point $(-1, -1)$ could also be written in polar coordinates as $(-\sqrt{2}, \pi/4)$. If your friend is standing at the origin, facing toward the wrong point, but then walks backward for $\sqrt{2}$ units, he or she will be at the correct point after all.

We now have two ways of writing $(-1, -1)$ in polar coordinates: $(\sqrt{2}, 5\pi/4)$ and $(-\sqrt{2}, \pi/4)$. That's not all, though—we could also add any integer multiple of 2π to θ without changing the situation. So, the complete list of points in polar coordinates we could use is as follows:

$$\left(\sqrt{2}, \frac{5\pi}{4} + 2\pi n\right), \qquad \left(-\sqrt{2}, \frac{\pi}{4} + 2\pi n\right) \qquad \text{where } n \text{ is an integer.}$$

There are infinitely many pairs (r, θ) in the above list, lying in two families—and all of them describe the same point $(-1, -1)$ in the plane! Luckily, in almost every case, we just want one of pairs (r, θ), and the convention is usually to choose the one where $r \geq 0$ and θ lies between 0 and 2π. So, it would be fine to say that $(-1, -1)$ has polar coordinates $(\sqrt{2}, 5\pi/4)$, provided that you understand that this is not the only way of writing it.

A few more examples: what are polar coordinates for the points with Cartesian coordinates $(0, 1)$, $(-2, 0)$, and $(0, -3)$? Let's plot these points on the same set of axes:

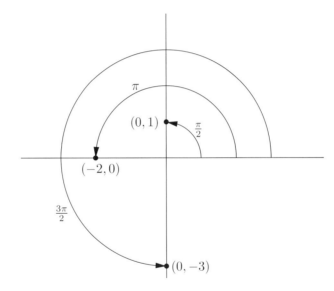

You can get in some trouble using the formula $\tan(\theta) = y/x$ from above. For example, at the point $(0, 1)$, you'd get $\tan(\theta) = 1/0$, which is undefined. Forget it! Just look at the picture and see that the angle we want is $\pi/2$, so $(0, 1)$ has polar coordinates $(1, \pi/2)$. Similarly, $(-2, 0)$ has polar coordinates $(2, \pi)$, and $(0, -3)$ has polar coordinates $(3, 3\pi/2)$. Of course, there are infinitely many different answers. For example, the point $(0, -3)$ would often be written in polar coordinates as $(3, -\pi/2)$ instead of $(3, 3\pi/2)$. Anyway, the thing to do is practice converting lots of points to and from polar coordinates until you get the hang of it.

Let's just pause for thought before we move on. You know those radar screens that you always see in movies involving submarines—the glowing green ones that make a "bip … bip … bip" noise? The screens look something like this:

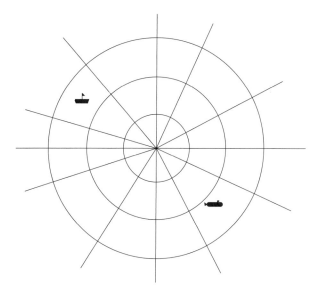

This is just a "grid" in polar coordinates. You see, a regular grid consists of some lines where x is constant (the vertical lines) and some lines where y is constant (the horizontal lines). If instead you work in polar coordinates, then you should draw some of the curves where r is constant, and some of the curves where θ is constant. The points where r is equal to some constant C map out a circle centered at the origin of radius C units; while the points where θ is constant trace out a ray that starts at the origin. Some of these circles and rays appear in the above picture. So, you've probably already seen polar coordinates and never realized it!

27.2.2 Sketching curves in polar coordinates

Suppose you know that $r = f(\theta)$ for some function f, and you want to sketch the graph of all points (r, θ) in polar coordinates where $r = f(\theta)$ for θ in some given range. This isn't so easy to do. Probably the best way to proceed is to draw up a table of values and plot points. It can also be helpful to sketch $r = f(\theta)$ in Cartesian coordinates first. For example, to sketch $r = 3\sin(\theta)$ in polar coordinates, where $0 \le \theta \le \pi$, let's first sketch $r = 3\sin(\theta)$ with respect to Cartesian axes labeled θ and r:

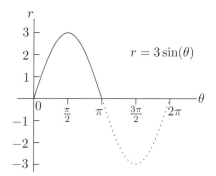

This shows that as the angle turns from 0 to π, the distance r increases from 0 up to 3, then heads back down to 0 by the time we get back to π. So the curve we want looks something like this:

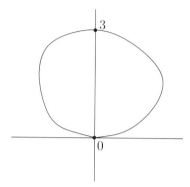

This looks a little pathetic. To tighten it up, we can write down the following table of values:

θ	0	$\pi/6$	$\pi/4$	$\pi/3$	$\pi/2$	$2\pi/3$	$3\pi/4$	$5\pi/6$	π
r	0	$3/2$	$3/\sqrt{2}$	$3\sqrt{3}/2$	3	$3\sqrt{3}/2$	$3/\sqrt{2}$	$3/2$	0

Plotting these points leads to the following picture:

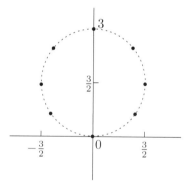

So it actually looks like a real circle, not a pathetic one like our first attempt. In fact, we can check that it is a circle by converting to Cartesian coordinates. Indeed, since $y = r\sin(\theta)$, and we have $r = 3\sin(\theta)$, we can eliminate θ and get $r^2 = 3y$. On the other hand, $r^2 = x^2 + y^2$, so we get $x^2 + y^2 = 3y$. Putting the $3y$ on the left-hand side and completing the square in y, we get $x^2 + (y - 3/2)^2 = (3/2)^2$, which is the equation of the circle with center $(0, 3/2)$ and radius $3/2$ units. This agrees with the above picture. Now, you should try to convince yourself that if θ goes from π to 2π, you just end up retracing the same circle again.

Let's look at another example. Suppose that we want to sketch the curve $r = 1 + 2\cos(\theta)$, where $0 \le \theta \le 2\pi$. First, note that the Cartesian graph looks like this:

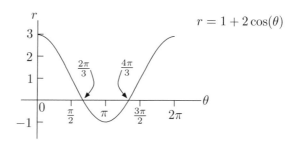

It's important to work out where the above graph intersects the θ-axis (that's the horizontal axis normally known as the x-axis!). You see, when that happens, we have $r = 0$, so the graph in polar coordinates will return to the origin then. In our case, we have $1 + 2\cos(\theta) = 0$, which means $\cos(\theta) = -1/2$. Since $\cos(\theta)$ is negative, θ must be in the second or third quadrant. Also, the reference angle is $\cos^{-1}(1/2)$, which is $\pi/3$. We conclude that $r = 0$ when $\theta = 2\pi/3$ or $4\pi/3$, as shown on the above graph.

Now, let's start drawing the polar graph of $r = 1 + 2\cos(\theta)$. As θ goes from 0 to $2\pi/3$, the distance r decreases from 3 to 0, passing through 1 when $\theta = \pi/2$. Here's what we've got so far:

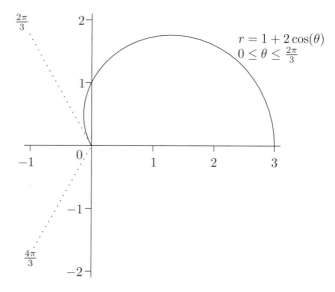

Now, as θ continues from $2\pi/3$ to π, the distance r goes down to -1. This means that, instead of staying in the second quadrant, we have to move backward into the fourth quadrant. The following picture tells the story:

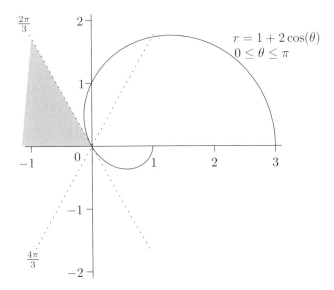

As θ goes from $2\pi/3$ to π, the graph should be contained within the shaded region, but because r is negative then, the graph busts into the fourth quadrant instead. Anyway, we could continue up to $\theta = 2\pi$ in this way, or simply note that the Cartesian graph of $r = 1 + 2\cos(\theta)$ is symmetric about the line $\theta = \pi$. This means that the completed graph we're looking for is just the mirror image (in the horizontal axis) of what we have so far:

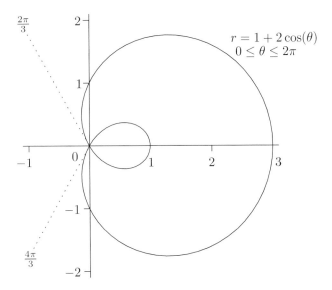

Let's finish off this section by looking at a selection of polar curves. You might want to cover up the graphs and try sketching them first, or alternatively see if you can convince yourself that the graph is correct in each case. In any event, you should try sketching a lot of polar curves until you feel you're going round in circles.

$$r = 1 + \cos(\theta)$$
$$0 \le \theta \le 2\pi$$

$$r = 1 + \tfrac{3}{4}\cos(\theta)$$
$$0 \le \theta \le 2\pi$$

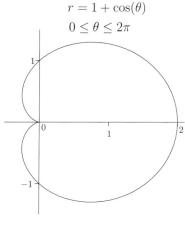
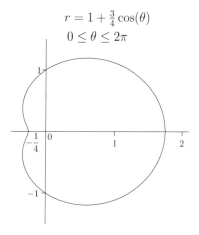

$$r = \sin(2\theta)$$
$$0 \le \theta \le 2\pi$$

$$r = \sin(3\theta)$$
$$0 \le \theta \le \pi$$

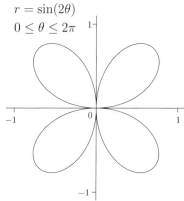
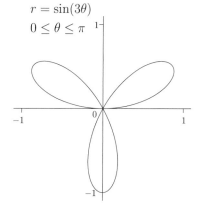

$$r = \frac{1}{\pi}\theta$$
$$0 \le \theta \le 4\pi$$

$$r = \frac{2}{1 + \sin(\theta)}$$
$$-\frac{\pi}{4} \le \theta \le \frac{5\pi}{4}$$

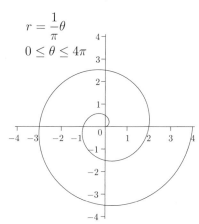
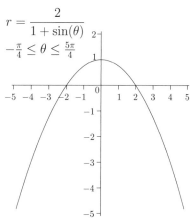

Some facts about the above curves:

1. The curve given by $r = 1 + \cos(\theta)$ is called a *cardioid*. The curve $r = 1 + \frac{3}{4}\cos(\theta)$ is an example of a *limaçon*, of which the cardioid is a special case.

2. In the above graph of $r = \sin(3\theta)$, the angle θ only goes from 0 to π. As θ goes from π to 2π, the graph is retraced, just as in the case of the circle $r = \sin(\theta)$.

3. The curve given by $r = \theta/\pi$ is an example of a *spiral of Archimedes*. This is not periodic: as θ increases, the spiral gets bigger and bigger.

4. The curve given by $r = 2/(1 + \sin(\theta))$ looks like a parabola. In fact, you should try to show that the above equation becomes $x^2 = 4 - 4y$ in Cartesian coordinates.

27.2.3 Finding tangents to polar curves

Luckily, finding tangents to polar curves is just a special case of finding tangents to curves given by parametric equations. We've seen how to do this in general in Section 27.1.1 above. Let's see how it works in the case of polar coordinates.

We have $r = f(\theta)$, and we'd like to find the tangent to the curve at some point on the curve. Using $x = r\cos(\theta)$ and $y = r\sin(\theta)$, we can write

$$x = f(\theta)\cos(\theta) \qquad \text{and} \qquad y = f(\theta)\sin(\theta);$$

this means that x and y are parametrized by θ. By the formula from Section 27.1.1 above, we have

$$\frac{dy}{dx} = \frac{dy/d\theta}{dx/d\theta}.$$

This gives the slope of the tangent in general. Finally, we just have to plug in the value of θ we care about. That's all there is to it, but let's see what happens when we look at some examples.

Consider the curve given in polar coordinates by $r = 1 + 2\cos(\theta)$. We sketched this in the previous section. Suppose we want the equation of the tangent through the point with polar coordinates $(2, \pi/3)$. First, let's do a reality check: does this point even lie on the curve? Well, when $\theta = \pi/3$, we have $1 + 2\cos(\theta) = 1 + 2\cos(\pi/3) = 2$, which is the given value of r. So the point does lie on the curve after all. Next, we have to find the slope of the tangent, dy/dx. We have $x = r\cos(\theta) = (1 + 2\cos(\theta))\cos(\theta)$, and $y = r\sin(\theta) = (1 + 2\cos(\theta))\sin(\theta)$. We need to find $dy/d\theta$ and $dx/d\theta$. Unfortunately, this involves the product rule, but it's not too bad. I leave it to you to check that

$$\frac{dy}{d\theta} = -2\sin^2(\theta) + (1 + 2\cos(\theta))\cos(\theta) \qquad \text{and} \qquad \frac{dx}{d\theta} = -\sin(\theta)(1 + 4\cos(\theta)),$$

so we have

$$\frac{dy}{dx} = \frac{dy/d\theta}{dx/d\theta} = \frac{-2\sin^2(\theta) + (1 + 2\cos(\theta))\cos(\theta)}{-\sin(\theta)(1 + 4\cos(\theta))}.$$

We want to know what happens when $\theta = \pi/3$, so plug that in. You should get

$$\frac{dy}{dx} = \frac{-2(3/4) + (1 + 2(1/2))(1/2)}{-(\sqrt{3}/2)(1 + 4(1/2))} = \frac{1}{3\sqrt{3}}.$$

So we know the slope of the line we're looking for. Now we just need a point the line goes through. That point is obviously $(2, \pi/3)$ in polar coordinates, but we need it in Cartesian coordinates. So, just use $x = r\cos(\theta)$ and $y = r\sin(\theta)$ to get $x = 2\cos(\pi/3) = 1$ and $y = 2\sin(\pi/3) = \sqrt{3}$. Great—we need the line through $(1, \sqrt{3})$ with slope $1/3\sqrt{3}$. That line is given by

$$y - \sqrt{3} = \frac{1}{3\sqrt{3}}(x - 1),$$

which simplifies a little to the answer we're looking for,

$$y = \frac{1}{3\sqrt{3}}(x + 8).$$

How about the tangent line to the same curve at the origin? Looking at the graph of $r = 1 + 2\cos(\theta)$ on page 588, you can see that there should in fact be two tangent lines! We can still find their equations, however. Indeed, we know that the curve hits the origin when $r = 0$, and we saw in the previous section that this happens when $\theta = 2\pi/3$ or $\theta = 4\pi/3$. Check that substituting these values of θ one at a time into the above equation for dy/dx gives $-\sqrt{3}$ and $\sqrt{3}$, respectively. Since both tangent lines pass through the origin, they must have equations $y = -\sqrt{3}x$ and $y = \sqrt{3}x$. In fact, these lines complete the rays corresponding to $\theta = 2\pi/3$ and $\theta = 4\pi/3$, shown as dotted lines in the graph (again, it's on page 588).

27.2.4 Finding areas enclosed by polar curves

If we want to find the area enclosed by the polar curve $r = f(\theta)$, where f is assumed to be continuous, then we're going to have to integrate something. But what? We just have to set up the correct Riemann sum. (See Section 16.2 in Chapter 16 for a review of Riemann sums.) Suppose we take a small chunk of angle between θ and $\theta + d\theta$. Then as we move counterclockwise along this chunk of angle, r meanders from $f(\theta)$ to $f(\theta + d\theta)$. If $d\theta$ is very small, then r doesn't have a chance to move far away from $f(\theta)$, so we can approximate the wedge we're looking for by a thin slice of pie of radius $r = f(\theta)$ units and angle $d\theta$, centered at the origin, as shown in the following diagram:

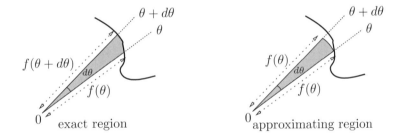

The area of a sector is one half of the radius squared, multiplied by the angle of the sector (in radians, of course!). So, we can approximate the area of the wedge (in square units) by $\frac{1}{2}(f(\theta))^2\, d\theta$, which is just $\frac{1}{2}r^2 d\theta$. The total area, as θ varies from θ_0 to θ_1 is found by adding up the areas of all the wedges and letting $d\theta$ go down to 0, leading* to the following integral:

$$\left(\text{area inside } r = f(\theta) \text{ between } \theta = \theta_0 \text{ and } \theta = \theta_1\right) = \int_{\theta_0}^{\theta_1} \tfrac{1}{2}r^2\, d\theta.$$

As usual, the area is given in square units.

Let's try out this formula on the curve $r = 3\sin(\theta)$, where $0 \le \theta \le \pi$. We saw in Section 27.2.2 above that this is a circle of radius $3/2$ units, so its area

*To prove the formula, one needs to set up upper and lower sums for the area by considering the maximum and minimum values of $f(\theta)$, where θ varies over a subinterval in a partition of $[\theta_0, \theta_1]$, then show that the upper and lower sums converge to the same value as the mesh of the partition goes to 0.

should be $\pi(3/2)^2$, or $9\pi/4$ square units. Let's verify this. We have

$$\text{area} = \int_0^\pi \frac{1}{2} r^2 \, d\theta = \frac{1}{2} \int_0^\pi (3\sin(\theta))^2 \, d\theta = \frac{9}{2} \int_0^\pi \sin^2(\theta) \, d\theta.$$

This integral can be done using the double-angle formulas, as described at the beginning of Section 19.1 in Chapter 19. Check that you agree that the answer is $9\pi/4$.

Here's a harder example. Let's try to find the area of the croissant-shaped region enclosed by our curve $r = 1 + 2\cos(\theta)$, as shown in the following diagram:

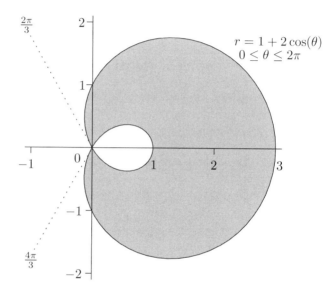

It seems as if we should just be able to use the formula to say that the area we want is given by

$$\int_0^{2\pi} \frac{1}{2} r^2 \, d\theta = \frac{1}{2} \int_0^{2\pi} (1 + 2\cos(\theta))^2 \, d\theta.$$

Again, to do this integral we need the double-angle formulas. I leave it to you to show that

$$\frac{1}{2} \int (1 + 2\cos(\theta))^2 \, d\theta = \frac{3}{2}\theta + 2\sin(\theta) + \frac{1}{2}\sin(2\theta) + C,$$

so the above definite integral can be evaluated by plugging in $\theta = 2\pi$ and $\theta = 0$ and subtracting, giving 3π. Unfortunately, this isn't the correct answer. The problem is that r becomes negative when θ is between $2\pi/3$ and $4\pi/3$. Since the formula for the area involves r^2, there's no way to distinguish between positive and negative area. (This is very different from the situation in Cartesian coordinates, where area below the y-axis is indeed negative.) So what we have actually found is the area inside the curve $r = |1 + \cos(2\theta)|$, which looks like this:

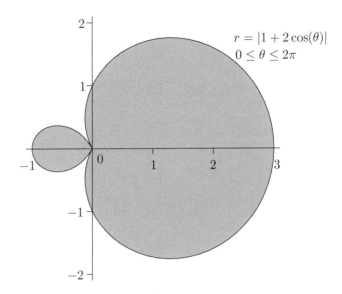

$$r = |1 + 2\cos(\theta)|$$
$$0 \le \theta \le 2\pi$$

To fix up this crappy situation, we need to find the area inside the little loop to the left of the vertical axis, and then take it away twice from our original area. Why twice? Because taking it away once just gives the rest of the shaded area in the previous picture, but we actually want to cut out a little loop from the region to get the area we need. So, how do we find the area inside the little loop? Just repeat the above integral, except from $2\pi/3$ to $4\pi/3$:

$$\text{area of little loop} = \frac{1}{2} \int_{2\pi/3}^{4\pi/3} (1 + 2\cos(\theta))^2 \, d\theta.$$

Now you should use the above antiderivative to show that this integral works out to be $(\pi - 3\sqrt{3}/2)$ square units. So, we can finally express the area we want as 3π square units minus twice the area of the loop, then work out the value of the area:

$$\text{area we want} = 3\pi - 2\left(\pi - \frac{3\sqrt{3}}{2}\right) = (\pi + 3\sqrt{3}) \text{ square units.}$$

As this example shows, you have to be very careful when using the above formula for area in polar coordinates if r can ever be negative.

CHAPTER 28 _____

Complex Numbers

Why should some quadratics have all the fun? The quadratic $x^2 - 1$ gets the privilege of having two roots (1 and -1), but poor old $x^2 + 1$ doesn't have any, since its discriminant is negative. To even things up a little, let's introduce the concept of complex numbers. Using complex numbers, any quadratic has two roots.* (You have to count the double root a of $(x - a)^2$ as two roots.) Anyway, here's what we're going to be doing with complex numbers:

- basic manipulations (adding, subtracting, multiplying, dividing) and solving quadratic equations;
- the complex plane, and Cartesian and polar forms for complex numbers;
- taking large powers of complex numbers;
- solving equations of the form $z^n = w$;
- solving equations of the form $e^z = w$; and
- using some tricks from power series and complex numbers to solve some series questions.

28.1 The Basics

It kind of sucks that you can't take the square root of -1. So, we'll just do it anyway. Let's just create a square root of -1 and call it i. OK, so then we must have $i^2 = -1$. Is i the only square root of -1? No, $-i$ should also be a square root, since if there were any justice in the world, then

$$(-i)^2 = (-1)^2(i)^2 = 1(-1) = -1.$$

(There is in fact justice in the world: this last series of equations is correct.) Since $i^2 + 1 = 0$ and $(-i)^2 + 1 = 0$, we now have two roots for the quadratic

---

*The surprising thing is that this also works for higher-degree polynomials: every polynomial of degree n has n complex roots (counting multiplicities). This is due to the so-called Fundamental Theorem of Algebra, but that's way beyond the scope of this book. You might have to look at a book on complex analysis to learn more about this.

$x^2 + 1$ after all—but they are not real: they are imaginary. How about $2i$? That's also imaginary. In fact, $(2i)^2 = 2^2 i^2 = 4(-1) = -4$, so $(2i)^2$ is a negative number. So, when we say that a number is *imaginary*, we mean that its square is a negative number. The only imaginary numbers are of the form yi where y is a real number not equal to 0. You can also write iy instead of yi.

Now, you can add or subtract real and imaginary numbers, for example $2 - 3i$, but you can't simplify the result. In this way, we get all the *complex numbers*, which are all the numbers of the form $x + iy$, where x and y are real. The set of all complex numbers is normally denoted by the symbol $\mathbb{C}$. Notice that all imaginary numbers are complex numbers; for example, $2i = 0 + 2i$. All real numbers are also complex numbers; for example, $-13 = -13 + 0i$. Every complex number has a real and an imaginary part. If $z = x + iy$, then the real part is x and the imaginary part is y. These are written as $\text{Re}(z)$ and $\text{Im}(z)$, respectively. For example, $\text{Re}(2 - 3i) = 2$ and $\text{Im}(2 - 3i) = -3$. Note that $\text{Im}(2 - 3i)$ is not $-3i$, it's just -3. What is $\text{Re}(2i)$? Well, write $2i$ as $0 + 2i$ to see that the real part is 0. On the other hand, the imaginary part, $\text{Im}(2i)$, is of course 2.

Adding and subtracting complex numbers is pretty easy. Just add (or subtract) the real parts, and then do the imaginary parts. For example,

$$(2 - 3i) + (-6 - 7i) = 2 - 6 - 3i - 7i = -4 - 10i;$$

an example of subtraction is

$$(2 - 3i) - (-6 - 7i) = 2 + 6 - 3i + 7i = 8 + 4i.$$

Multiplication isn't much harder—you just expand, but remember to change i^2 into -1 whenever you see it. For example,

$$
\begin{aligned}
(2 - 3i)(-6 - 7i) &= 2(-6) + 2(-7i) - (3i)(-6) - (3i)(-7i) \\
&= -12 - 14i + 18i + 21i^2 = -12 + 4i - 21 = -33 + 4i.
\end{aligned}
$$

By the way, what is i^3? How about i^4? i^5? Let's start off with i^3. We have $i^3 = i^2 \times i = (-1) \times i = -i$. So i^3 is just $-i$. On the other hand, $i^4 = i^3 \times i = (-i) \times i = 1$. That is, $i^4 = 1$. For i^5, we play the same game: $i^5 = i^4 \times i = 1 \times i = i$. In fact, because $i^4 = 1$, we can see that the powers of i keep on cycling through $1, i, -1, -i$. For example, $i^{101} = i$ since $i^{100} = 1$ (remembering that 100 is divisible by 4).

How about division? That's a little trickier, but not much. The technique is very similar to rationalizing the denominator. It's inspired by the following observation: if you have a complex number $x + iy$ and multiply it by the complex number $x - iy$, you get a real number. When we do the math, we recognize and apply the formula for the difference of two squares:

$$(x + iy)(x - iy) = x^2 - (iy)^2 = x^2 - i^2 y^2 = x^2 + y^2.$$

Now x and y are real, so obviously x^2 and y^2 are as well, and so is their sum. If $z = x + iy$, the related number $x - iy$ is so important that it has a name:

it is called the *complex conjugate* of $x + iy$ and denoted $\bar{z}$. For example, if $z = 2 - 3i$, then $\bar{z} = 2 + 3i$, whereas if $z = 7i$, then $\bar{z} = -7i$. Note that the complex conjugate of a real number is the same number. This is because you just flip the sign of the imaginary part to take the complex conjugate, and real numbers have imaginary part zero. Now as the above formula shows, a number multiplied by its complex conjugate is real; it is the sum of the squares of its real and imaginary parts. Inspired by Pythagoras' Theorem and the above formula, given a complex number $z = x + iy$, let's define the *modulus* of z to be $\sqrt{x^2 + y^2}$. We write the modulus of z as $|z|$. So

$$|x + iy| = \sqrt{x^2 + y^2}.$$

Here are some examples: $|2 - 3i| = \sqrt{2^2 + (-3)^2} = \sqrt{4 + 9} = \sqrt{13}$. Similarly, $|7i| = \sqrt{0^2 + 7^2} = 7$. How about $|-13|$? This equals $\sqrt{(-13)^2 + 0^2} = 13$, which is exactly the same as the absolute value of -13. Our notation for modulus is completely consistent with the previous notation for absolute value. In fact, think of the modulus as a beefed-up version of absolute value. Anyway, the difference of two squares formula above shows that a complex number multiplied by its complex conjugate is the square of its modulus. That is,

$$z\bar{z} = |z|^2.$$

After all these preliminaries, we are ready to see how to divide complex numbers. All you do is multiply top and bottom by the complex conjugate of the bottom, then expand. The new denominator becomes the square of the modulus of the old one. For example,

$$\frac{2 - 3i}{-6 - 7i} = \frac{(2 - 3i)(-6 + 7i)}{(-6 - 7i)(-6 + 7i)}.$$

Now the top needs to be fully multiplied out, but the bottom is just $|-6 - 7i|^2$, so

$$\frac{2 - 3i}{-6 - 7i} = \frac{-12 + 18i + 14i - 21i^2}{(-6)^2 + (-7)^2} = \frac{9 + 32i}{85} = \frac{9}{85} + \frac{32}{85}i.$$

We can conclude that

$$\mathrm{Re}\left(\frac{2 - 3i}{-6 - 7i}\right) = \frac{9}{85} \quad \text{and} \quad \mathrm{Im}\left(\frac{2 - 3i}{-6 - 7i}\right) = \frac{32}{85}.$$

Another example: how would you find

$$\mathrm{Re}\left(\frac{3 + 4i}{i - 1}\right)?$$

This example contains a slight trick to throw you off guard. The denominator should really be written as $-1 + i$. Once you do this, you can see that the complex conjugate of the denominator is $-1 - i$, so

$$\frac{3 + 4i}{i - 1} = \frac{(3 + 4i)(-1 - i)}{(-1 + i)(-1 - i)} = \frac{-3 - 3i - 4i - 4i^2}{(-1)^2 + (1)^2} = \frac{1 - 7i}{2} = \frac{1}{2} - \frac{7}{2}i.$$

So the real part of $(3 + 4i)/(i - 1)$ is just $\frac{1}{2}$, and as a bonus, its imaginary part is $-\frac{7}{2}$.

Now let's see how to solve quadratic equations. For example, let's say that you want to solve $x^2 + 3x + 14 = 0$. Just use the quadratic formula and the fact that $\sqrt{-1} = \pm i$ to write

$$x = \frac{-3 \pm \sqrt{3^2 - 4 \times 1 \times 14}}{2} = \frac{-3 \pm \sqrt{-47}}{2} = -\frac{3}{2} \pm \frac{\sqrt{47}}{2} i.$$

Notice that we have simplified $\pm\sqrt{-47}$ as $\pm\sqrt{47} \cdot i$. Now, how about if you have a quadratic whose coefficients are complex numbers? The quadratic formula still works, but you may well have to take the square root of a complex number, not just a negative number, as in the example we just did. We'll look at an example of this in Section 28.4.1 below.

28.1.1 Complex exponentials

We've discussed how to add and multiply complex numbers. How about exponentiating them? Let's see how we can make sense of something like e^z when z is complex. From Section 24.2.3 in Chapter 24, we know that

$$e^x = \sum_{n=0}^{\infty} \frac{x^n}{n!}$$

for all real x. What happens if we replace x by z on the right-hand side, where z is some complex number? We'll get a series whose terms are complex numbers. Believe it or not, you can still use the ratio test to show that the series converges, no matter what complex number z happens to be. (We only proved the ratio test for real series, but it turns out that once you define what convergence means for complex sequences, the same proof works.) Inspired by all this, we'll define e^z, for any complex number z, by the following equation:

$$e^z = \sum_{n=0}^{\infty} \frac{z^n}{n!}.$$

Certainly this works nicely enough when z is real, since the definition agrees with the above equation for e^x. On the other hand, it would be good to know that our new toy, e^z, does all the nice things that we expect of exponentials. Actually, the critical thing it needs to do is to satisfy the exponential rule $e^z e^w = e^{z+w}$. Once we know that, all the other exponential rules follow more or less immediately.

So, how do we show that $e^z e^w = e^{z+w}$? Here's a sneaky way. We know that $e^x e^y = e^{x+y}$ for any real x and y, so this means that

$$\sum_{n=0}^{\infty} \frac{x^n}{n!} \sum_{m=0}^{\infty} \frac{y^m}{m!} = \sum_{k=0}^{\infty} \frac{(x+y)^k}{k!}.$$

We have just replaced each exponential by its Maclaurin series, using a different dummy variable in each sum. If you multiply out the two series on the left-hand side, you get some double power series in powers of x and y, and the same goes for the right-hand side. The coefficients of $x^n y^m$ on the left- and right-hand sides of the equations must therefore be the same. This will also be true if x and y are replaced by complex numbers like z and w (respectively), so we have proved that $e^z e^w = e^{z+w}$ for any two complex numbers z and w after all!

28.2 The Complex Plane

Real numbers are usually represented as points on a number line, which is one-dimensional. Complex numbers literally have an extra dimension. Indeed, if $z = x + iy$, we can't squish all the information into just one real number. Instead of a real number line, we'll use a complex number plane. The complex number $z = x + iy$ will be represented as the point (x, y) in Cartesian coordinates. It's pretty easy to plot complex numbers like $2 - 3i$, $2i$, and -1:

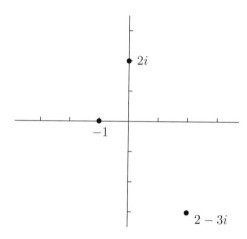

You should think of each point as representing one complex number, rather than as a pair of real numbers.

In the previous chapter, we saw that you can also express every point in the plane in polar coordinates instead. (You should review Section 27.2.1 now if you haven't looked at it for a while.) So suppose you have a point in the complex plane which has polar coordinates (r, θ). What is the complex number represented by that point? Well, we can convert to Cartesian coordinates using $x = r\cos(\theta)$ and $y = r\sin(\theta)$. So the point (r, θ) in polar coordinates represents the complex number $z = x + iy = r\cos(\theta) + ir\sin(\theta)$. In particular, if $r = 1$, then z is just $\cos(\theta) + i\sin(\theta)$.

Now, there's a pretty bizarre and funky identity, due to Euler, which is really important:

$$e^{i\theta} = \cos(\theta) + i\sin(\theta).$$

This is true* for all real θ. This means that the complex number $e^{i\theta}$, as defined in the previous section, has polar coordinates $(1, \theta)$ when you plot it on the complex number plane. So $e^{i\theta}$ lives on the unit circle and has angle θ from the positive x-axis. The following picture shows a few positions of $e^{i\theta}$ for different values of θ:

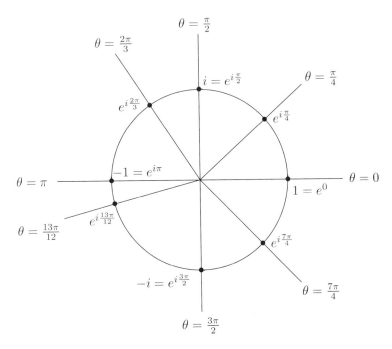

For points not on the unit circle, you just have to multiply by r. Specifically, we saw that if z is represented by the point (r, θ) in polar coordinates, then $z = r\cos(\theta) + ir\sin(\theta)$. By Euler's identity, this means that $z = re^{i\theta}$. So we have shown that

$$\text{if } (x, y) \text{ and } (r, \theta) \text{ are the same point, then } x + iy = re^{i\theta}.$$

Let's say that a complex number like $re^{i\theta}$ is in *polar form*, (as opposed to $x + iy$, which is in *Cartesian form*). For example, in the above diagram, it says that $-1 = e^{i\pi}$; this is because the point $(-1, 0)$ in Cartesian coordinates has polar coordinates $(1, \pi)$, so $-1 + 0i = 1e^{i\pi}$. That is, the polar form of -1 is $e^{i\pi}$. Similarly, the point $(0, 1)$ in Cartesian coordinates can be written in polar coordinates as $(1, \pi/2)$, so we have $0 + 1i = 1e^{\pi/2}$, or $i = e^{i\pi/2}$. This formula looks a bit strange, but it's true enough—the left-hand side is in Cartesian form, while the right-hand side is in polar form. The same thing goes for $-i = e^{i(3\pi/2)}$ (can you see why this true?).

In Section 27.2.1 of the previous chapter, we saw that there are infinitely many ways to write a given point in polar coordinates. Let's agree that when we're dealing with complex numbers, we'll never let r be negative. Still, if you

---

*See the end of this chapter for the proof of this identity.

have found the polar coordinates (r, θ) for your point, then you could add any integer multiple of 2π to θ and it wouldn't make a difference. For example, the point $(0, -1)$ has polar coordinates $(1, 3\pi/2)$, or you can subtract 2π to see that it also has polar coordinates $(1, -\pi/2)$. In terms of complex numbers, this means that $e^{i(3\pi/2)} = e^{-i\pi/2}$. So $e^{i\theta}$ is **periodic** in θ with period 2π. This is an important fact which will come in handy a little later.

We've just seen above that $e^{i\pi} = -1$. Let's just reflect on this for a moment. It's really quite awesome, when you think about it. What have been the fundamental new numbers in your math education so far? Introducing the number -1 opens the door to negative numbers. The number π arises from the geometry of circles. The number e is the natural base for logarithms and is fundamental in the study of calculus. And the number i leads the way to complex numbers and being able to solve quadratic (and higher-degree polynomial) equations. The fact that they are combined into such a simple formula is pretty remarkable, if you ask me. Anyway, enough of this philosophical rambling: let's look at some examples of how to convert complex numbers from polar to Cartesian form and vice versa.

28.2.1 Converting to and from polar form

To convert a complex number from polar to Cartesian form, just use Euler's identity directly (that's $e^{i\theta} = \cos(\theta) + i\sin(\theta)$ in case you have already forgotten!). For example, what is $2e^{i(5\pi/6)}$ in Cartesian form? Well, Euler's identity says that it is $2(\cos(5\pi/6) + i\sin(5\pi/6))$. See why you need to know your trig? Hopefully you can work out that $\cos(5\pi/6) = -\sqrt{3}/2$ and that $\sin(5\pi/6) = 1/2$, so we have

$$2e^{i(5\pi/6)} = 2\left(\cos\left(\frac{5\pi}{6}\right) + i\sin\left(\frac{5\pi}{6}\right)\right) = 2\left(-\frac{\sqrt{3}}{2} + i\frac{1}{2}\right) = -\sqrt{3} + i.$$

On the other hand, converting from Cartesian to polar form is more difficult, as we observed in Section 27.2.1. There we saw that

$$r = \sqrt{x^2 + y^2} \qquad \text{and} \qquad \tan(\theta) = \frac{y}{x},$$

where we have now dropped the possible solution $r = -\sqrt{x^2 + y^2}$ since we want $r \geq 0$ for complex numbers. By the way, we defined the modulus of z to be $|z| = \sqrt{x^2 + y^2}$. So r is the same as $|z|$. The modulus $|z|$ is therefore the distance from the point z to the origin (in the complex number plane). The angle θ is called the *argument* of z and is written $\arg(z)$. (Normally one requires that $0 \leq \arg(z) < 2\pi$ so that there's no ambiguity.*)

So, to convert z from Cartesian to polar coordinates, we just have to find the modulus and argument of z, using the above formulas. (In fact, sometimes the polar form of z is referred to as *mod-arg form*.) For example, how would you convert $z = 1 - i$ into polar form? Well, think of z as being written as

---

*Often this condition is replaced by $-\pi < \arg(z) \leq \pi$ instead.

$1 + (-1)i$; so we need to set $x = 1$ and $y = -1$ in the above formulas. Indeed, if $z = re^{i\theta}$, then $r = \sqrt{1^2 + (-1)^2} = \sqrt{2}$, and $\tan(\theta) = (-1)/1 = -1$. Now, you have to check the quadrant in order to get the correct value of θ. The best way is to draw a diagram:

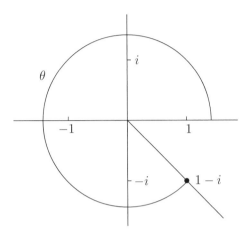

You can easily see that the point $(1, -1)$ is in the fourth quadrant, so θ must be equal to $7\pi/4$. (Alternatively, you could write $\theta = -\pi/4$.) So we just have to package $r = \sqrt{2}$ and $\theta = 7\pi/4$ together as $re^{i\theta}$ to see that $1 - i = \sqrt{2}e^{i(7\pi/4)}$. (If you used $\theta = -\pi/4$ instead, you'll get $1 - i = \sqrt{2}e^{-i(\pi/4)}$. Remember, you could add any integer multiple of 2π to θ and still be correct.)

 Let's revisit a couple of examples that might seem confusing. First, how would you write $2i$ in polar form? Consider $2i$ as being $0 + 2i$, so it is represented by the point $(0, 2)$ in the complex plane. So, if $2i = re^{i\theta}$, then we have $r = \sqrt{0^2 + 2^2} = 2$, whereas $\tan(\theta) = 2/0$. Wait, that's no good—you can't divide by 0. Let's just draw a picture and see what θ should be:

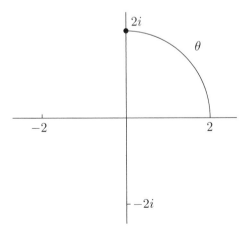

OK, so $\theta = \pi/2$ from the picture. This is actually consistent with our strange value for $\tan(\theta)$ above, since $\tan(\pi/2)$ is undefined. So, we have $2i = 2e^{i\pi/2}$. Of course, this is just twice our formula $i = e^{i\pi/2}$ from the previous section.

How about changing -6 into polar form? Well, now we write -6 as $-6+0i$, and see that $r = \sqrt{(-6)^2 + 0^2} = 6$ and $\tan(\theta) = 0/(-6) = 0$. This means that θ is an integer multiple of π, but to nail it down, let's draw another picture:

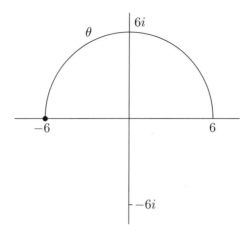

Now we see that $\theta = \pi$ (or if you prefer, $-\pi$, or even 3π, or any odd multiple of π). So, we have $-6 = 6e^{i\pi}$. Incidentally, if we divide by 6, we get the amazing formula $e^{i\pi} = -1$, which we discussed in the previous section.

28.3 Taking Large Powers of Complex Numbers

Why on earth would you want to use the polar form? One reason is that it's really easy to multiply and take powers in polar form. Imagine you wanted to multiply $3e^{i\pi/4}$ by $2e^{-i(3\pi/8)}$. This is pretty simple—you just use the normal exponential rules (see Section 9.1.1 in Chapter 9) to write

$$(3e^{i\pi/4})(2e^{-i(3\pi/8)}) = 6e^{i(\pi/4 - 3\pi/8)} = 6e^{-i\pi/8}.$$

Even better, imagine you want to raise $3e^{i\pi/4}$ to the 200th power. This is just

$$(3e^{i\pi/4})^{200} = 3^{200}e^{(i\pi/4)\times 200} = 3^{200}e^{i(50\pi)}.$$

In fact, by Euler's identity, $e^{i(50\pi)} = \cos(50\pi) + i\sin(50\pi)$. Since 50π is an integer multiple of 2π, we have $\cos(50\pi) = 1$ and $\sin(50\pi) = 0$, so we have proved that $(3e^{i\pi/4})^{200} = 3^{200}$.

A lot of the time, you might want the final answer in Cartesian form. For example, suppose we'd like to compute $(1-i)^{99}$ and give the answer in Cartesian form. Expanding the expression by multiplying out would be crazy, so we won't go there. The correct way to proceed is to translate $1-i$ into polar form, take the 99th power, then translate back into Cartesian form. OK, we saw in the previous section that $1 - i = \sqrt{2}e^{i(7\pi/4)}$ in polar form, so we have

$$(1-i)^{99} = (\sqrt{2}e^{i(7\pi/4)})^{99} = (2^{1/2})^{99}(e^{i(7\pi/4)})^{99} = 2^{99/2}e^{i(693\pi/4)}.$$

Now, we have to go back to Cartesian form. Before we do this, let's look at $e^{i(693\pi/4)}$. This fraction $693\pi/4$ is a bit of a pest. Remember that $e^{i\theta}$ is

2π-periodic in θ, however, so we can knock off any multiple of 2π from the fraction $693\pi/4$ and not affect the answer. So, write $693/4 = 173\frac{1}{4}$. The biggest even number less than this is 172, and the difference between these two numbers is $173\frac{1}{4} - 172 = 5/4$. So, we can think of $693\pi/4$ as $172\pi + 5\pi/4$. Since 172π is a multiple of 2π (this is why we wanted an even number, 172 in this case), we know that $e^{i(693\pi/4)} = e^{i(5\pi/4)}$. That's much nicer. Now we can convert the whole thing to Cartesian form:

$$(1 - i)^{99} = 2^{99/2}e^{i(693\pi/4)} = 2^{99/2}e^{i(5\pi/4)} = 2^{99/2}\left(\cos\left(\frac{5\pi}{4}\right) + i\sin\left(\frac{5\pi}{4}\right)\right)$$

$$= 2^{99/2}\left(-\frac{1}{\sqrt{2}} - i\frac{1}{\sqrt{2}}\right).$$

In fact, this can be further simplified by writing $1/\sqrt{2}$ as $2^{-1/2}$; the final answer should be $-2^{49}(1+i)$. Now, as an exercise, you should check that you can arrive at the same answer by starting off using an alternate polar form, $1 - i = \sqrt{2}e^{-i\pi/4}$.

In summary, to take a large power of a complex number, first convert it to polar form, then take the power. Find the largest even multiple of π less than the angle θ, and take that away from θ and replace θ by that new number. Finally, convert back to Cartesian form.

28.4 Solving $z^n = w$

Let's move onto a trickier subject: how to solve equations of the form $z^n = w$, where n is a given integer and w is a given complex number. This amounts to taking nth roots of w, but we don't just want to say $z = \sqrt[n]{w}$ since that doesn't tell us very much. Instead, we'll try to find a solution directly. Since powers work so well in polar form, that's what we'll use.

For example, to solve $z^5 = -\sqrt{3} + i$, we should use polar coordinates for both z and $w = -\sqrt{3} + i$. Since we don't know what z is, let's put $z = re^{i\theta}$. Now to find z, we just have to find what r and θ are. As for w, let's write $-\sqrt{3} + i = Re^{i\varphi}$ and then find R and φ. (We have to use R and φ instead of r and θ since the last two variables are already taken for z.) Now, let's draw a picture of the situation:

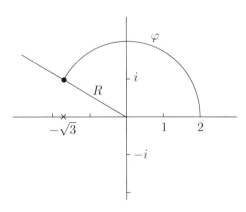

So we have $R = \sqrt{(-\sqrt{3})^2 + (1)^2} = 2$, and $\tan(\varphi) = -1/\sqrt{3}$. Since the point is in the second quadrant, φ must be $5\pi/6$. Great, so we know that $-\sqrt{3} + i = 2e^{i(5\pi/6)}$ in polar form.

Now let's turn our attention to the equation $z^5 = -\sqrt{3} + i$ and convert the whole thing to polar form. On the left, we replace z by $re^{i\theta}$ to get $z^5 = (re^{i\theta})^5 = r^5 e^{i(5\theta)}$, whereas we've just seen that the right-hand side is $2e^{i(5\pi/6)}$. So our equation becomes

$$r^5 e^{i(5\theta)} = 2e^{i(5\pi/6)}.$$

If you take the modulus of both sides, you get $r^5 = 2$ (because the modulus of e^{iA} is always 1 if A is real). Then we can cancel out r^5 and 2, since they are equal, to get $e^{i(5\theta)} = e^{i(5\pi/6)}$. We have dissected the above equation into two separate equations:

$$r^5 = 2 \qquad \text{and} \qquad e^{i(5\theta)} = e^{i(5\pi/6)}.$$

The first is easy to solve: just take the 5th root to get $r = 2^{1/5}$, which is legit since r is a nonnegative real number. As for the second equation, you may be tempted to say $5\theta = 5\pi/6$, but it's not that simple. Remember, $e^{i\theta}$ is 2π-periodic in the variable θ! You can express this fact via the following important principle, which I want you to remember better than you've ever remembered anything before:

> If $e^{iA} = e^{iB}$ for real numbers A and B, then
> $$A = B + 2\pi k, \text{where } k \text{ is an integer.}$$

This principle saves the day. Since $e^{i(5\theta)} = e^{i(5\pi/6)}$, we use the principle to see that

$$5\theta = \frac{5\pi}{6} + 2\pi k,$$

where k is an integer. Dividing by 5, we have

$$\theta = \frac{\pi}{6} + \frac{2\pi k}{5}.$$

So it looks as if there are infinitely many values of θ, and therefore infinitely many values of z that solve our equation. Appearances can be deceptive, however! You see, since $n = 5$, you only need to use the first five values for k, namely, $k = 0, 1, 2, 3, 4$. We'll see why in just a moment; for now, we can calculate that as k goes from 0 through 4 inclusive, the values of θ are

$$\frac{\pi}{6}, \qquad \left(\frac{\pi}{6} + \frac{2\pi}{5}\right) = \frac{17\pi}{30}, \qquad \left(\frac{\pi}{6} + \frac{4\pi}{5}\right) = \frac{29\pi}{30},$$

$$\left(\frac{\pi}{6} + \frac{6\pi}{5}\right) = \frac{41\pi}{30}, \qquad \left(\frac{\pi}{6} + \frac{8\pi}{5}\right) = \frac{53\pi}{30},$$

respectively. Putting these values of θ, along with $r = 2^{1/5}$, into the equation $z = re^{i\theta}$, we get

$$z = 2^{1/5}e^{i\pi/6}, \quad 2^{1/5}e^{i(17\pi/30)}, \quad 2^{1/5}e^{i(29\pi/30)}, \quad 2^{1/5}e^{i(41\pi/30)},$$

$$\text{or} \quad 2^{1/5}e^{i(53\pi/30)}.$$

Of course, it would be nice to change these to Cartesian form. The first solution is pretty easy:

$$2^{1/5}e^{i\pi/6} = 2^{1/5}\left(\cos\left(\frac{\pi}{6}\right) + i\sin\left(\frac{\pi}{6}\right)\right) = 2^{1/5}\left(\frac{\sqrt{3}}{2} + i\frac{1}{2}\right) = 2^{-4/5}(\sqrt{3} + i).$$

As for the others, they don't look too nice. For example, the second solution from the above list works out to be

$$2^{1/5}e^{i(17\pi/30)} = 2^{1/5}\left(\cos\left(\frac{17\pi}{30}\right) + i\sin\left(\frac{17\pi}{30}\right)\right),$$

which can't easily be simplified. (Do you know what $\cos(17\pi/30)$ is? I don't either, and it's not worth working out.) I leave it to you to write out the other three solutions in (unsimplified) Cartesian form.

Now, let's see why you only need to let k go from 0 through 4, discarding all the other possible values of k. Let's see what happens when $k = 5$. Using the equation

$$\theta = \frac{\pi}{6} + \frac{2\pi k}{5}$$

from above, we see that when $k = 5$, we have

$$\theta = \frac{\pi}{6} + \frac{2\pi \times 5}{5} = \frac{\pi}{6} + 2\pi.$$

This is certainly a different value of θ from any of the ones we already listed above, but it doesn't lead to a different value of z. Why? Because

$$2^{1/5}e^{i(\pi/6+2\pi)} = 2^{1/5}e^{i(\pi/6)}.$$

That is, we get the same solution as the case $k = 0$. Similarly, if you try to put $k = 6$, you should get the same value of z as when $k = 1$. In general, any time you increase k by 5, you will simply get the same value of z again. So, the values $k = 0, 5, 10, \ldots$, as well as $k = -5, -10, -15, \ldots$, all lead to the same solution, $z = 2^{1/5}e^{i(\pi/6)}$. Similarly, the values $k = 1, 6, 11, \ldots$ and $k = -4, -9, -14, \ldots$ give the same solution. The same goes for the other three solutions. While you need to appreciate this fact, in practice it is simple to apply: unless $w = 0$, the equation $z^n = w$ has n different solutions, which occur when $k = 0, 1, \ldots, n - 1$. Those are the only values of k you need to use. In our case $n = 5$, so we only needed $k = 0, 1, 2, 3, 4$.

It's interesting to plot the solutions in the complex plane. They all have modulus $2^{1/5}$, which means that they lie on the circle centered at the origin of radius $2^{1/5}$ units. Also, the difference between the arguments (that is, values of θ) of consecutive solutions is $2\pi/5$, which is one-fifth of a complete revolution. This means that the solutions are evenly spaced around the circle; that is, they form a regular pentagon (the solutions are labeled z_0 through z_4):

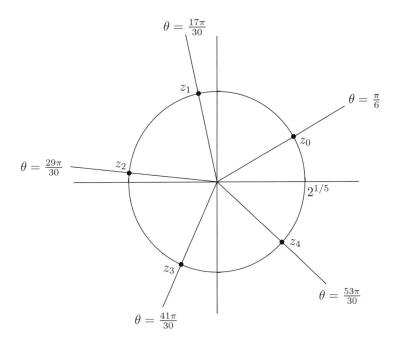

In general, there are n solutions to the equation $z^n = w$, which when plotted form the vertices of a regular n-sided polygon. (The exception is if $w = 0$, in which case $z = 0$ is the only solution, but it is of multiplicity n.)

So, let's outline the main steps in solving $z^n = w$:

1. Write $z = re^{i\theta}$ in polar coordinates. Then $z^n = r^n e^{in\theta}$.

2. Convert w to polar coordinates. Let's say that $w = Re^{i\varphi}$.

3. Since $z^n = w$, we can write the original equation as $r^n e^{in\theta} = Re^{i\varphi}$. Here, the values of n, R, and φ should be filled in with your values, but r and θ are always what we need to find (so they appear as variables).

4. Decompose into two equations: $r^n = R$ and $e^{in\theta} = e^{i\varphi}$.

5. The first is simple to solve: take nth roots to get $r = R^{1/n}$.

6. For the second, use the above triple-boxed principle to get $n\theta = \varphi + 2\pi k$, where k is an integer.

7. Divide this by n, then write out all the different values for θ when $k = 0, 1, 2, \ldots, n - 1$.

8. Substitute the value of r and the different values of θ into $z = re^{i\theta}$ to get n different values for z, which are the solutions.

9. If necessary, change each and every one of those solutions into Cartesian coordinates.

Let's look at one more example: what are the cube roots of i? This question is asking us to solve the equation $z^3 = i$. We start off by writing $z = re^{i\theta}$, so $z^3 = r^3 e^{i(3\theta)}$ (step 1). Now, we have to convert i into polar coordinates (step 2), but we have already seen above that $i = e^{i\pi/2}$. So, since $z^3 = i$, we have $r^3 e^{i(3\theta)} = 1e^{i\pi/2}$ (step 3). This leads to the equations $r^3 = 1$ and $e^{i(3\theta)} = e^{i\pi/2}$ (step 4). Taking cube roots in the first equation gives $r = 1$ (step 5), and our important principle and the second equation show that $3\theta = \pi/2 + 2\pi k$, where k is an integer (step 6). This is the same as $\theta = \pi/6 + 2\pi k/3$; since $n = 3$ in this question, we only need $k = 0, 1, 2$. Writing these out, we have

$$\theta = \frac{\pi}{6}, \qquad \left(\frac{\pi}{6} + \frac{2\pi}{3}\right) = \frac{5\pi}{6}, \qquad \text{or} \qquad \left(\frac{\pi}{6} + \frac{4\pi}{3}\right) = \frac{3\pi}{2}$$

(step 7). This leads to three possibilities for z, which are

$$z = e^{i\pi/6}, \quad e^{i(5\pi/6)} \quad \text{or} \quad e^{i(3\pi/2)}$$

(step 8). Finally, we should convert these into Cartesian form (step 9). The first solution is

$$z = e^{i\pi/6} = \cos\left(\frac{\pi}{6}\right) + i\sin\left(\frac{\pi}{6}\right) = \frac{\sqrt{3}}{2} + i\frac{1}{2}.$$

The second solution is

$$z = e^{i(5\pi/6)} = \cos\left(\frac{5\pi}{6}\right) + i\sin\left(\frac{5\pi}{6}\right) = -\frac{\sqrt{3}}{2} + i\frac{1}{2}.$$

Finally, the third solution is

$$z = e^{i(3\pi/2)} = \cos\left(\frac{3\pi}{2}\right) + i\sin\left(\frac{3\pi}{2}\right) = 0 - i(1) = -i.$$

Let's plot these three solutions and check that they do indeed form an equilateral triangle:

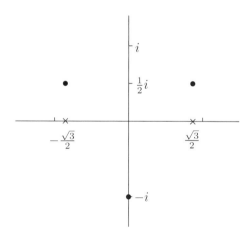

28.4.1 Some variations

 Suppose you want to solve the equation $(z - 2)^3 = i$. No problem—just let $Z = z - 2$, so that the equation is $Z^3 = i$. Solve this exactly as we just did at the end of the previous section to find that

$$Z = z - 2 = \frac{\sqrt{3}}{2} + i\frac{1}{2}, \qquad -\frac{\sqrt{3}}{2} + i\frac{1}{2}, \qquad \text{or} \qquad -i.$$

Finally, add 2 to both sides to get

$$z = 2 + \frac{\sqrt{3}}{2} + i\frac{1}{2}, \qquad 2 - \frac{\sqrt{3}}{2} + i\frac{1}{2}, \qquad \text{or} \qquad 2 - i.$$

There's not much to that, then. Here's a tougher one. Let's try solving the quadratic equation

$$z^2 + \frac{1}{\sqrt{2}}z - \frac{\sqrt{3}i}{8} = 0.$$

We can use the quadratic formula to get

$$z = \frac{\frac{-1}{\sqrt{2}} \pm \sqrt{\frac{1}{2} + i\frac{\sqrt{3}}{2}}}{2}.$$

While this is correct, it's not in Cartesian form (nor polar form) so we should try to simplify it. We need to find the square roots of the complex number $\frac{1}{2} + i\frac{\sqrt{3}}{2}$. How do we do this? By solving the equation $Z^2 = \frac{1}{2} + i\frac{\sqrt{3}}{2}$. Following the above steps, we write $Z = re^{i\theta}$, and I'll leave it to you to see that $\frac{1}{2} + i\frac{\sqrt{3}}{2} = e^{i\pi/3}$ in polar form. So our equation becomes $r^2 e^{i2\theta} = e^{i\pi/3}$. This means that $r^2 = 1$ and $2\theta = \pi/3 + 2\pi k$, where $k = 0$ or 1. (Remember the important principle!) So we have $r = 1$ and $\theta = \pi/6$ or $7\pi/6$, which means that $Z = e^{i\pi/6}$ or $e^{i7\pi/6}$. Again, you should check that these correspond to $Z = \frac{\sqrt{3}}{2} + \frac{1}{2}i$ or $-\frac{\sqrt{3}}{2} - \frac{1}{2}i$ in Cartesian form (respectively). Finally, we can replace $\pm\sqrt{\frac{1}{2} + i\frac{\sqrt{3}}{2}}$ by $\pm(\frac{\sqrt{3}}{2} + \frac{1}{2}i)$ in the equation for z above to get

$$z = \frac{-\frac{1}{\sqrt{2}} \pm \left(\frac{\sqrt{3}}{2} + \frac{1}{2}i\right)}{2}.$$

This simplifies to

$$z = -\frac{1}{2\sqrt{2}} + \frac{\sqrt{3}}{4} + \frac{i}{4} \qquad \text{or} \qquad -\frac{1}{2\sqrt{2}} - \frac{\sqrt{3}}{4} - \frac{i}{4}.$$

One more example. How would you factor $(z^4 - z^2 + 1)$ over the complex numbers? How about over the real numbers? In the first case, we just need to find all four complex solutions of the equation $z^4 - z^2 + 1 = 0$. To do this, first we need to realize that this equation is actually a quadratic equation in z^2. Let's set $Z = z^2$, so that the equation becomes $Z^2 - Z + 1 = 0$. This can be solved using the quadratic formula to get

$$Z = z^2 = \frac{1 \pm \sqrt{-3}}{2} = \frac{1}{2} \pm i\frac{\sqrt{3}}{2}.$$

So we need to find the square roots of $\frac{1}{2} + i\frac{\sqrt{3}}{2}$ and $\frac{1}{2} - i\frac{\sqrt{3}}{2}$. We just did the first one in the previous example, and you can repeat the steps easily enough to handle the second one. Both of these numbers have two square roots each, which work out to be

$$\frac{\sqrt{3} + i}{2}, \qquad \frac{-\sqrt{3} - i}{2}, \qquad \frac{-\sqrt{3} + i}{2}, \qquad \text{and} \qquad \frac{\sqrt{3} - i}{2}.$$

These are the four solutions to $z^4 - z^2 + 1 = 0$. It follows that we can factor $z^4 - z^2 + 1$ as follows:

$$z^4 - z^2 + 1 = \left(z - \frac{\sqrt{3} + i}{2}\right)\left(z - \frac{\sqrt{3} - i}{2}\right)\left(z - \frac{-\sqrt{3} + i}{2}\right)\left(z - \frac{-\sqrt{3} - i}{2}\right).$$

This is the complex factorization. To get the real factorization, we need to use a nice fact: if w is any complex number, then $(z - w)(z - \bar{w})$ has real coefficients when you multiply it out. Indeed, you get $z^2 - (w + \bar{w})z + w\bar{w}$, but it's easy enough to see that $w + \bar{w} = 2\text{Re}(w)$ (which is real), and we've already seen that $w\bar{w} = |w|^2$, which is also real. Anyway, notice that I have cunningly grouped the above four factors so that if we multiply out the first two, we get

$$\left(z - \frac{\sqrt{3} + i}{2}\right)\left(z - \frac{\sqrt{3} - i}{2}\right)$$

$$= z^2 - \left(\frac{\sqrt{3} + i}{2} + \frac{\sqrt{3} - i}{2}\right)z + \left(\frac{\sqrt{3} + i}{2}\right)\left(\frac{\sqrt{3} - i}{2}\right)$$

$$= z^2 - \sqrt{3}z + 1.$$

Similarly, you should check that multiplying out the last two factors gives $z^2 + \sqrt{3}z + 1$. The conclusion is that

$$z^4 - z^2 + 1 = (z^2 - \sqrt{3}z + 1)(z^2 + \sqrt{3}z + 1).$$

Notice that there are no complex numbers here, yet working this out without them would have been pretty darn tricky.

28.5 Solving $e^z = w$

Now it's time to see how to solve equations of the form $e^z = w$ for given w. It'd be nice if we could just write $z = \ln(w)$, but this isn't very helpful. For example, what exactly is $\ln(-\sqrt{3} + i)$? Let's try to answer this question.

Fortunately, solving $e^z = w$ isn't much harder than solving $z^n = w$; in fact, if anything, it's simpler. Before we see how to do this, we need to understand e^z a little better. Let's see what happens if we write $z = x + iy$. We get

$$e^z = e^{x+iy} = e^x e^{iy}.$$

So what? Well, the main point is that this is already in polar form. The modulus is e^x and the argument is y. If you prefer, $r = e^x$ (remember, e^x is real and positive) and $\theta = y$. This means that if z is in Cartesian form $x + iy$, then e^z is **automatically** in polar form: $e^z = e^x e^{iy}$. So, the main difference between solving $e^z = w$ and $z^n = w$ is that you don't need to put z in polar form in the first case, whereas you do in the second case. A sort of by-product of this is that there are infinitely many solutions to the equation $e^z = w$ (unless $w = 0$, in which case there are no solutions).

Let's solve $e^z = -\sqrt{3} + i$. We have already converted the right-hand side to polar coordinates as $2e^{i(5\pi/6)}$ (see page 604). To handle the left-hand side, write $z = x + iy$ in **Cartesian** coordinates, so $e^z = e^x e^{iy}$. So, changing the original equation to polar form, we get

$$e^x e^{iy} = 2e^{i(5\pi/6)}.$$

Now, this separates into two equations:

$$e^x = 2 \qquad \text{and} \qquad e^{iy} = e^{i(5\pi/6)}.$$

To solve the first equation, we have to take logarithms to see that $x = \ln(2)$. The second is handled by our important principle to get $y = 5\pi/6 + 2\pi k$, where k is an integer. Finally, putting these values into $z = x + iy$, we get

$$z = \ln(2) + i\left(\frac{5\pi}{6} + 2\pi k\right),$$

where k is any integer. In this case, we **do** get a different value of z for each value of k, so we need to use them all. Let's plot some of the possible values of z corresponding to $k = -2, -1, 0, 1$, and 2 (I'll use a different scale on the axes for clarity):

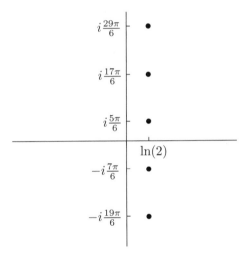

So the solutions are equally spaced on the vertical line $x = \ln(2)$. Incidentally, this means that they form an arithmetic progression of complex numbers. Although only five solutions are shown in the above picture, you should bear in mind that there are actually infinitely many solutions to the original equation $e^z = -\sqrt{3} + i$.

Let's look at one more example. Suppose you want to solve $e^{2iz+3} = i$. The exponent $2iz + 3$ makes this a little more complicated than the previous example, but it's not too bad. We've already seen that the right-hand side in polar coordinates is $e^{i\pi/2}$, but how about the left-hand side? Once again, we write $z = x + iy$, but now we need $2iz + 3 = 2i(x + iy) + 3 = (-2y + 3) + i(2x)$. So, the polar form of the left-hand side is given by

$$e^{2iz+3} = e^{-2y+3} e^{i(2x)}.$$

Notice how the factor of i switched the real and imaginary parts (and also the sign of y). Anyway, translating our equation $e^{2iz+3} = i$ into polar form, we have

$$e^{-2y+3}e^{i(2x)} = 1e^{i\pi/2}.$$

This leads to the equations

$$e^{-2y+3} = 1 \quad \text{and} \quad e^{i(2x)} = e^{i\pi/2}.$$

To solve the first equation, take logs to get $-2y + 3 = \ln(1) = 0$, so $y = \frac{3}{2}$. To solve the second equation, use the boxed principle to get $2x = \pi/2 + 2\pi k$, where k is an integer. This means that $x = \pi/4 + \pi k$, so since $z = x + iy$, we have

$$z = \frac{\pi}{4} + \pi k + \frac{3}{2}i,$$

where k is an integer. Let's plot what these solutions look like for $k = -2$, $-1, 0, 1$, and 2, bearing in mind that these are only five of the infinitely many solutions:

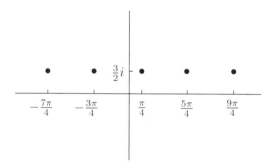

Once again, the solutions are in arithmetic progression, but this time they lie on the horizontal line $y = \frac{3}{2}$.

28.6 Some Trigonometric Series

A *trigonometric series* is a series of the form

$$\sum_{n=0}^{\infty} (a_n \cos(n\theta) + b_n \sin(n\theta))$$

for some coefficients $\{a_n\}$ and $\{b_n\}$. In this section, we'll see that there are a few such series which can be simplified.

For example, consider the trigonometric series

$$\sum_{n=0}^{\infty} \frac{\sin(n\theta)}{n!},$$

where θ is real. Note that this is not a power series in θ, since $\sin(n\theta)$ is not a power of θ. On the other hand, we can make the whole thing into a power series by using the complementary series

$$\sum_{n=0}^{\infty} \frac{\cos(n\theta)}{n!},$$

in a clever way. In fact, we can find both series at once. The key is Euler's identity. Watch carefully, because this is a sneaky trick. Let's find both series at once by combining them like this:

$$\sum_{n=0}^{\infty} \frac{\cos(n\theta)}{n!} + i\sum_{n=0}^{\infty} \frac{\sin(n\theta)}{n!}.$$

OK, so this is one series plus i times the other. So what? Well, by massaging the sums* and then using Euler's identity, this simplifies to

$$\sum_{n=0}^{\infty} \frac{\cos(n\theta) + i\sin(n\theta)}{n!} = \sum_{n=0}^{\infty} \frac{e^{in\theta}}{n!}.$$

Finally, use the exponential rules to write $e^{in\theta}$ as $(e^{i\theta})^n$; the sum becomes

$$\sum_{n=0}^{\infty} \frac{(e^{i\theta})^n}{n!}.$$

Now, the last sum looks familiar. In fact, we saw in Section 28.1.1 above that

$$\sum_{n=0}^{\infty} \frac{z^n}{n!} = e^z$$

for all complex numbers z. Now we just have to substitute $z = e^{i\theta}$ to get

$$\sum_{n=0}^{\infty} \frac{(e^{i\theta})^n}{n!} = e^{e^{i\theta}}.$$

If you've been following this chain of reasoning, you should see that we've proved that

$$\sum_{n=0}^{\infty} \frac{\cos(n\theta)}{n!} + i\sum_{n=0}^{\infty} \frac{\sin(n\theta)}{n!} = e^{e^{i\theta}}.$$

Now what? Well, we need to convert the right-hand side into Cartesian form. To do this, write $e^{i\theta} = \cos(\theta) + i\sin(\theta)$, so

$$e^{e^{i\theta}} = e^{\cos(\theta) + i\sin(\theta)} = e^{\cos(\theta)}e^{i\sin(\theta)}.$$

This is a good start—this is the polar form of $e^{e^{i\theta}}$. To get the Cartesian form, we need to convert $e^{i\sin(\theta)}$ into $\cos(\sin(\theta)) + i\sin(\sin(\theta))$. Putting it all together, we get

$$\sum_{n=0}^{\infty} \frac{\cos(n\theta)}{n!} + i\sum_{n=0}^{\infty} \frac{\sin(n\theta)}{n!} = e^{\cos(\theta)}\cos(\sin(\theta)) + ie^{\cos(\theta)}\sin(\sin(\theta)).$$

---

*This needs some justification. It turns out that everything's OK because both our series converge absolutely.

Now, if two complex numbers are equal, then their real parts must be equal, and also their imaginary parts must be equal. This leads to the following two equations, which are valid for all real θ:

$$\sum_{n=0}^{\infty} \frac{\cos(n\theta)}{n!} = e^{\cos(\theta)} \cos(\sin(\theta)) \qquad \text{and} \qquad \sum_{n=0}^{\infty} \frac{\sin(n\theta)}{n!} = e^{\cos(\theta)} \sin(\sin(\theta)).$$

 Not easy, but this is basically what you have to do. I'll do one more example, without all the explanations. Your task is to follow this and explain each step. The example is to find

$$\sum_{n=0}^{\infty} \frac{\cos(n\theta)}{3^n} \qquad \text{and} \qquad \sum_{n=0}^{\infty} \frac{\sin(n\theta)}{3^n}.$$

Following the pattern of the above example, we have

$$\sum_{n=0}^{\infty} \frac{\cos(n\theta)}{3^n} + i \sum_{n=0}^{\infty} \frac{\sin(n\theta)}{3^n} = \sum_{n=0}^{\infty} \frac{\cos(n\theta) + i\sin(n\theta)}{3^n}$$

$$= \sum_{n=0}^{\infty} \frac{e^{in\theta}}{3^n} = \sum_{n=0}^{\infty} \frac{(e^{i\theta})^n}{3^n} = \sum_{n=0}^{\infty} \left(\frac{e^{i\theta}}{3}\right)^n.$$

Now this is a geometric series with ratio $e^{i\theta}/3$. This last number is in polar form with modulus $1/3$, which is less than 1; so the geometric series should converge. By the formula for the sum of a geometric series (see Section 23.1 of Chapter 23), we have

$$\sum_{n=0}^{\infty} \left(\frac{e^{i\theta}}{3}\right)^n = \frac{1}{1 - \frac{1}{3}e^{i\theta}}.$$

 We now have the wretched task of converting this into Cartesian coordinates. First, try it and see if you can do it. If not, at least try to understand the following steps:

$$\frac{1}{1 - \frac{1}{3}e^{i\theta}} = \frac{1}{1 - \frac{1}{3}\cos(\theta) - i\frac{1}{3}\sin(\theta)}$$

$$= \frac{1}{1 - \frac{1}{3}\cos(\theta) - i\frac{1}{3}\sin(\theta)} \cdot \frac{1 - \frac{1}{3}\cos(\theta) + i\frac{1}{3}\sin(\theta)}{1 - \frac{1}{3}\cos(\theta) + i\frac{1}{3}\sin(\theta)}$$

$$= \frac{1 - \frac{1}{3}\cos(\theta) + i\frac{1}{3}\sin(\theta)}{\left(1 - \frac{1}{3}\cos(\theta)\right)^2 + \left(\frac{1}{3}\sin(\theta)\right)^2}$$

$$= \frac{1 - \frac{1}{3}\cos(\theta) + i\frac{1}{3}\sin(\theta)}{1 - \frac{2}{3}\cos(\theta) + \frac{1}{9}\cos^2(\theta) + \frac{1}{9}\sin^2(\theta)}$$

$$= \frac{1 - \frac{1}{3}\cos(\theta) + i\frac{1}{3}\sin(\theta)}{1 - \frac{2}{3}\cos(\theta) + \frac{1}{9}}$$

$$= \frac{9 - 3\cos(\theta) + i3\sin(\theta)}{10 - 6\cos(\theta)}$$

$$= \frac{9 - 3\cos(\theta)}{10 - 6\cos(\theta)} + i\frac{3\sin(\theta)}{10 - 6\cos(\theta)}.$$

After all this, we're ready to write

$$\sum_{n=0}^{\infty} \frac{\cos(n\theta)}{3^n} + i \sum_{n=0}^{\infty} \frac{\sin(n\theta)}{3^n} = \frac{9 - 3\cos(\theta)}{10 - 6\cos(\theta)} + i\frac{3\sin(\theta)}{10 - 6\cos(\theta)};$$

since the real and imaginary parts must be equal, we conclude that

$$\sum_{n=0}^{\infty} \frac{\cos(n\theta)}{3^n} = \frac{9 - 3\cos(\theta)}{10 - 6\cos(\theta)} \quad \text{and} \quad \sum_{n=0}^{\infty} \frac{\sin(n\theta)}{3^n} = \frac{3\sin(\theta)}{10 - 6\cos(\theta)}$$

for any real number θ. As you see, these problems are quite hard!

28.7 Euler's Identity and Power Series

Let's finish the chapter with a justification of Euler's identity

$$e^{i\theta} = \cos(\theta) + i\sin(\theta)$$

using power series. By the definition of e^z from Section 28.1.1 above, with z replaced by $i\theta$, we see that

$$
\begin{aligned}
e^{i\theta} &= 1 + (i\theta) + \frac{(i\theta)^2}{2!} + \frac{(i\theta)^3}{3!} + \frac{(i\theta)^4}{4!} + \frac{(i\theta)^5}{5!} + \frac{(i\theta)^6}{6!} + \frac{(i\theta)^7}{7!} + \cdots \\
&= 1 + i\theta - \frac{\theta^2}{2!} - i\frac{\theta^3}{3!} + \frac{\theta^4}{4!} + i\frac{\theta^5}{5!} - \frac{\theta^6}{6!} - i\frac{\theta^7}{7!} + \cdots .
\end{aligned}
$$

Since the powers of i keep cycling through the values $1, i, -1, -i$, we conclude that the even powers in the above series all have real coefficients, whereas the odd powers all have imaginary coefficients. Furthermore, every second even-power term is negative and the others are positive; the same is true for the odd powers. So, the real part of $e^{i\theta}$ is

$$1 - \frac{\theta^2}{2!} + \frac{\theta^4}{4!} - \frac{\theta^6}{6!} + \cdots = \cos(\theta),$$

and the imaginary part is

$$\theta - \frac{\theta^3}{3!} + \frac{\theta^5}{5!} - \frac{\theta^7}{7!} + \cdots = \sin(\theta).$$

(See Section 26.2 in Chapter 26 to refresh your memory about these Maclaurin series.) From this last equation, it follows that $e^{i\theta} = \cos(\theta) + i\sin(\theta)$.

CHAPTER 29 _____

Volumes, Arc Lengths, and Surface Areas

We have used definite integrals to find areas. Now we're going to use them to find volumes, lengths of curves, and surface areas. For volumes and surface areas, we'll pay special attention to solids which are formed by revolving a region in the plane about some axis which lies in the plane; such solids are called *solids of revolution*. In the case of volumes, we'll also look at some more general solids. Here, then, is the game plan for this chapter:

- finding volumes of solids of revolution using the disc and shell methods;
- finding volumes of more general solids;
- finding arc lengths of smooth curves and speeds of parametric particles; and
- finding surface areas of solids of revolution.

29.1 Volumes of Solids of Revolution

We'll start with finding volumes of solids of revolution. The idea is that there is some region in the plane, and some axis also in the plane, and a solid is formed by revolving the region about the axis. For our purposes, the axis will always be parallel to the x-axis or the y-axis. (It is possible to deal with diagonal axes, but it's a real pain unless you use techniques from linear algebra.)

Before we put on our 3D glasses, however, let's remind ourselves how definite integrals work. We originally looked at this in Chapter 16, but here's a quick review of some of the main ideas. Let's work in the context of finding the area of the region below the curve

$$y = \sqrt{1 - (x - 3)^2}$$

and above the x-axis. What does this look like? Well, if we square the equation and rearrange, we get $(x - 3)^2 + y^2 = 1$; the graph of this relation is the circle of radius 1 unit centered at $(3, 0)$, so our function is the top half of the circle:

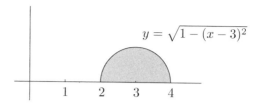

By the definition of the definite integral, we know that the shaded area (in square units) is

$$\int_2^4 \sqrt{1-(x-3)^2}\,dx$$

which can also be written as $\int_2^4 y\,dx$.

On the other hand, to find the area of this semicircle using a Riemann sum, we have to chop the base on the x-axis into little segments, then build the segments up into strips. The strips don't have to have the same width, and the only thing you need to make sure of is that the top of each strip cuts the curve somewhere (or touches the curve at one of its corners). The total area of the strips can easily be worked out, since it is just the sum of areas of rectangles. This area is an approximation for the actual area of the semicircle; the thinner the strips, the better the approximation, as you can see:

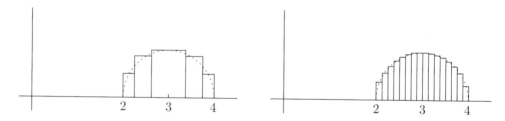

Let's just check out one generic strip. To make things a little easier, we'll assume that the top left-hand corner lies on the curve. As we've seen in Section 16.4 of Chapter 16, it doesn't matter which strips you choose, as long as the tops of all the strips pass through the curve. Anyway, here's what one strip looks like:

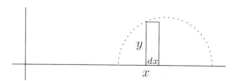

Since this rectangular strip has base length dx units and height y units, its area is $y\,dx$ square units. Now all we have to do is add up the areas of all the little strips, while simultaneously letting the maximum strip width tend to zero. The beauty of the notation is that you can accomplish both simply by putting an integral sign in front of the strip area and using the correct bounds. In our example, x lies in the interval $[2,4]$, and the area of one little strip is $y\,dx$ square units; so the area of **all** the strips, in the limit as the maximum strip width goes to zero, is $\int_2^4 y\,dx$ square units.

So here's the pattern: we make a little strip of width dx units and height y units at position x on the x-axis, work out its area, then put a definite integral sign in front to get the total area we're looking for. This technique doesn't just work for areas—it also works for volumes. In particular, let's see how it works using two different methods for finding volumes of revolution: the disc method and the shell method.

29.1.1 The disc method

Suppose that we revolve the semicircle from the previous section about the x-axis. This will give us a sphere. (Can you see why?) Let's try to work out its volume. We'll start with one strip, just like in the picture at the end of the previous section, and revolve that strip about the x-axis. Here's what we get:

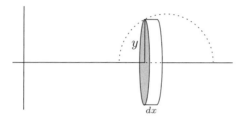

This is a thin disc of width dx units and radius y units. Think of it as a cylinder on its side; the radius is y units and the height is dx units. Since the volume of a cylinder of radius r units and height h units is $\pi r^2 h$ cubic units, the volume of our thin disc is $\pi y^2\,dx$ cubic units. So, now we take a number of strips so that their bases form a partition of our interval $[2,4]$, and revolve them all about the x-axis. For example, if you use five strips, you might get something like this:

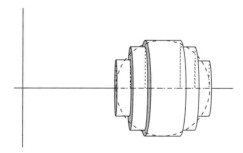

As perfect spheres go, the above object is pretty crappy, but its volume is a decent approximation to the sphere's. And the thinner the discs you use, the better the approximation. In the limit, as the maximum disc thickness goes down to zero, the approximation becomes perfect: the total volume of the discs tends toward the volume of the sphere. Again, the idea of "adding up all the volumes while letting the maximum disc thickness go down to zero" is realized simply by taking the volume of an arbitrary disc ($\pi y^2\,dx$ cubic units) and integrating over the interval we want. In our case, $y = \sqrt{1 - (x-3)^2}$

and x goes from 2 to 4, so we have

$$V = \int_2^4 \pi y^2 \, dx = \pi \int_2^4 \left(1 - (x-3)^2\right) dx.$$

The volume works out to be $\frac{4}{3}\pi$ cubic units (try it!) which is what we'd expect, since we're dealing with a sphere of radius 1 unit here. The method we just used is called the *disc method*; it is also known as the method of *slicing*.

29.1.2 The shell method

Now, let's suppose that we take our favorite semicircular region from before (see page 618 above) but this time we revolve it about the y-axis. Try to imagine what you'd get—it's actually the top half of a bagel (without the poppy seeds). Let's approximate the semicircle by thin strips again, but this time we'll revolve each strip about the y-axis, instead of the x-axis. As we saw before, a typical strip looks like this:

When you revolve it about the y-axis, you don't get a disc—you get a cylindrical shell:

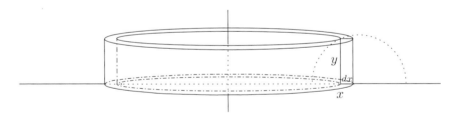

We're going to approximate our bagel half by using a number of shells, then letting the maximum shell thickness decrease to zero. For example, if you use five strips to approximate the region (as we did in the previous section), you might get something like this:

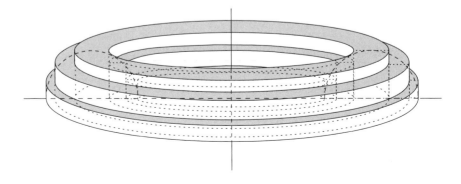

This weird solid is a pretty lumpy bagel half, but its volume is fairly close to what we're looking for. The thinner the maximum shell thickness, the better the approximation. As before, integrating takes care of both the addition of all the shell volumes and also taking the limit as the maximum shell thickness goes to zero.

First we need to find the volume of one generic shell. The easiest way to do this is to think of the shell as a really thin metal can without a top or bottom. As you can see from the picture of the shell on the previous page, the height of the can is y units, the radius is x units, and the thickness is dx units. Imagine cutting the can down the side with some sharp scissors, then unfolding it and flattening it out into a thin rectangle-like piece of metal. It's not actually a rectangle, of course. You see, a rectangle is a 2-dimensional object, whereas the unrolled can is 3-dimensional—although the can is pretty thin, it still has some thickness. (Even a piece of paper has some thickness, or else a ream of paper would be really really thin.) Now it's actually not even a rectangular prism, since the inner radius of the can isn't exactly the same as the outer radius. But the point is, it's **almost** a rectangular prism. The thinner the can gets, the closer it is to a rectangular prism, and when we take limits in the end (using the integral), everything will work out.* So, the idealized version of the unfolded can looks like this:

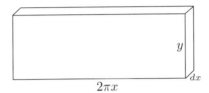

The thickness is dx units, and the side we cut along is still the height of the cylindrical shell, that is, y units. How about the long side? Well, that is equal to the circumference of the shell (think about it!) which is $2\pi x$ units, since the radius of the shell is basically x units. So, the volume of the shell is very close to $2\pi x y\, dx$ cubic units. Now all we have to do is integrate from $x = 2$ to $x = 4$ to see that the volume of the bagel half (in cubic units) is

$$\int_2^4 2\pi x y\, dx = 2\pi \int_2^4 x\sqrt{1 - (x-3)^2}\, dx.$$

Great—we've now reduced the problem to evaluating a definite integral, but it's a bit of a messy one. Start off by substituting $t = x - 3$, so $dt = dx$; also, when $x = 2$, we have $t = -1$, and when $x = 4$, we see that $t = 1$. So in t-land, the integral becomes

$$2\pi \int_{-1}^1 (t+3)\sqrt{1 - t^2}\, dt = 2\pi \left(\int_{-1}^1 t\sqrt{1 - t^2}\, dt + 3 \int_{-1}^1 \sqrt{1 - t^2}\, dt \right).$$

---

*More formally, we can view the volume of the shell as the difference in volumes of the outer shell (of radius $x + dx$ units) and the inner shell (of radius x units). Both shells have height y units, so the volume of the shell is $\pi y((x + dx)^2 - x^2)$, which simplifies to $2\pi x y\, dx + \pi y (dx)^2$ cubic units. When this is integrated, the second term vanishes due to the negligible quantity $(dx)^2$.

The first integral could be done by substituting $u = 1 - t^2$, and the second could be done by a trig substitution. A better way to do them is to note that the first integral is actually equal to 0, since the integrand is an odd function of t and the region of integration $[-1, 1]$ is symmetric about $t = 0$. (We proved this shortcut at the end of Section 18.1.1 of Chapter 18.) Furthermore, the easiest way to do the second integral (ignoring the factor of 3 out front for the moment) is to realize that it's equal to the area in square units of a semicircle of radius 1 unit, which is $\pi/2$. So without too much work, we see that the total answer is $3\pi^2$, therefore the volume of the bagel half is $3\pi^2$ cubic units. The method we just used is, unsurprisingly, called the *shell method* (also known as the method of *cylindrical shells*).

29.1.3 Summary . . . and variations

So far we have seen how to use the disc and shell methods in the special case of our semicircle. The same method works for general regions which are contained between a curve, the x-axis, and two vertical lines:

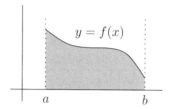

$$y = f(x)$$

$$a \qquad b$$

By the same reasoning that we used above in the special case of the semicircle, we can arrive at the following principles:

- If you revolve the area under the curve $y = f(x)$ between $x = a$ and $x = b$ (as shown above) about the x-axis, then the disc method applies and the volume is equal to

$$\int_a^b \pi y^2 \, dx \quad \text{cubic units.}$$

- If you revolve the area under the curve $y = f(x)$ between $x = a$ and $x = b$ (as shown above) about the y-axis, then the shell method applies and the volume is equal to

$$\int_a^b 2\pi xy \, dx \quad \text{cubic units.}$$

It's not a bad idea to know these formulas by heart, but it's an even better idea to be able to derive them by knowing how to find the volume of a typical disc or shell. This will be especially useful if you encounter one (or more) of the following variations:

1. The region to be revolved might lie between a curve and the y-axis (instead of the x-axis).

2. The region to be revolved might lie between two curves, instead of just being a region under a curve down to an axis.

3. The axis of revolution may be parallel to the x-axis or y-axis, not the axis itself.

Any combination of these cases can be handled by taking a typical strip and revolving it appropriately, then integrating; before we see how, it's important to know how to decide whether to use the disc method or the shell method. Notice that when you use the disc method, the strips are revolved about an axis parallel to their short sides; whereas when you use the shell method, the strips are revolved about an axis perpendicular to their short sides. That is, after you carve up the region into little strips, then:

- if the really thin bit of each strip is **parallel** to the axis of revolution, the **disc method** applies;
- whereas if the really thin bit of each strip is **perpendicular** to the axis of revolution, the **shell method** applies.

Armed with this knowledge, we can now look at our three variations one by one.

29.1.4 Variation 1: regions between a curve and the y-axis

If the region is between the curve and the y-axis, you probably want to take strips lying on their sides, with the thin part of the strip along the y-axis:

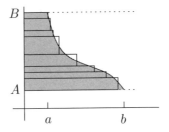

We actually did the same thing when we saw how to find the area of a region bounded by some curve and the y-axis, way back in Section 16.4.3 of Chapter 16. In any case, suppose that you want to find the volume of the solid formed by revolving this region about the y-axis. You should use the disc method, since the thin side of the strip is parallel to that axis. A typical strip at position y has width dy and length x units, so the resulting disc has volume $\pi x^2\, dy$ cubic units. When you integrate this to find the total volume, be very careful that the limits of integration are relevant points on the y-axis, not the x-axis, since the integral is taken with respect to y (because of the dy). In particular, we need the integral to go from A to B, not a to b (see the above diagram), so the volume we want is $\int_A^B \pi x^2\, dy$.

There's another way to look at this. Look at the above picture and rest your head on your right shoulder. The y-axis becomes horizontal, but everything's back to front, so try to visualize what would happen if the page were transparent and you looked at the diagram in reverse (still with your head tipped over). Now the y-axis and x-axis have switched places! This suggests that you can just switch the variables x and y wherever you see them, provided that you also make the bounds of integration refer to points on the

y-axis. Indeed, if we do this to our formula $V = \int_a^b \pi y^2 \, dx$ from Section 29.1.3 above, we see that the volume of a region down to the y-axis revolved about the y-axis is $\int_A^B \pi x^2 \, dy$, which agrees with what we've seen above.

How about if the above region is revolved about the x-axis, instead of the y-axis? Simply adapt the shell formula $V = \int_a^b 2\pi xy \, dx$ from Section 29.1.3 above to see that the volume we want is $\int_A^B 2\pi yx \, dy$. This makes sense, since revolving a typical strip about the x-axis gives a shell with thickness dy, height x, and radius y units. You should draw what happens when you unfold such a strip into a thin shape which is approximately a rectangular prism, calculate its volume, and see that you do indeed get $2\pi yx \, dy$. In summary, then, the rule of thumb is this:

> If the region lies between a curve and the y-axis, switch x and y.

As always, drawing a typical strip, revolving it, calculating the resulting volume, and integrating is the most reliable way; the above rule of thumb is just a guide.

Here's an example of Variation 1. Let R denote the region between the curve $y = \sqrt{x}$, $y = 2$, and the y-axis:

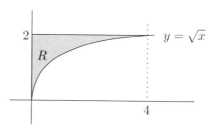

Let's work out the volume of the revolution of R about the y-axis and also about the x-axis. In the first case, we use the disc method, since the region lies between the curve and the y-axis, and we're revolving about the same axis. The volume is then

$$\int_0^2 \pi x^2 \, dy.$$

Since $y = \sqrt{x}$, we have $x = y^2$, so $x^2 = y^4$. This means that the volume is

$$\int_0^2 \pi x^2 \, dy = \pi \int_0^2 y^4 \, dy = \frac{\pi y^5}{5} \Big|_0^2 = \frac{32\pi}{5}$$

cubic units. On the other hand, the volume of revolution of R about the x-axis is done by shells, and we see that it is

$$\int_0^2 2\pi yx \, dy = 2\pi \int_0^2 y^3 \, dy,$$

since $yx = y \times y^2 = y^3$; check that this works out to be 8π cubic units. Please make sure you can draw a typical strip in each case and justify the above formulas. Also note that the integrals must go between 0 and 2, **not** 0 and 4: after all, the integration is with respect to y (not x!) and the relevant y-range is $[0, 2]$, as can be seen on the above graph.

29.1.5 Variation 2: regions between two curves

Suppose the region to be revolved lies between two curves. We'll handle this situation in the same way as finding the area of a region between two curves in Section 16.4.2 of Chapter 16. The general idea is to take the top curve and revolve the region under it all the way to the axis, to get a bigger solid than you want. Now take the bottom curve and revolve the region under it all the way to the axis, to get a solid which you actually need to cut out of the big solid and throw away to get the desired solid. Finally, subtract the small volume from the big one. Indeed, consider the following three regions:

The region we want to revolve is shown in the left-hand picture; it is the set difference of the region under the top curve down to the x-axis (in the middle picture above) and the region under the bottom curve down to the x-axis (in the right-hand picture). Now, regardless of whether you revolve about the x-axis or the y-axis, the volume of revolution of the region we want is equal to the difference between the volume of revolution of the big region and the volume of revolution of the small region. For example, if you revolve the region about the x-axis, then you get a cone-like structure with chopped-off ends and a weird-shaped hole going through the middle of it from left to right. The solid is the set difference of the filled-in version (with no hole) and the hole itself:

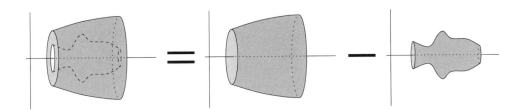

So, here's what we conclude:

> If the region lies between two curves, find the difference between the two corresponding volumes of revolution.

 Let's look at a concrete example. Consider the finite region between the curves $y = 2x^3$ and $y = x^4$, as shown on the next page. What is the volume of the solid formed by revolving the region about the x-axis?

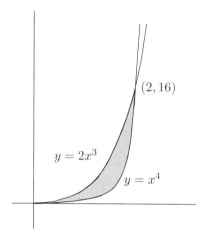

To find the intersection point, we have to set $2x^3 = x^4$, which gives $x = 0$ or $x = 2$. So the intersection points are the origin and $(2, 16)$, as shown in the above diagram, therefore the range of x we are concerned with is $[0, 2]$. Now the curve $y = 2x^3$ lies above the curve $y = x^4$ for this range of x, so we'll find the volume of revolution for $y_1 = 2x^3$ and then subtract the volume of revolution for $y_2 = x^4$. Note that it's really useful to use y_1 and y_2 instead of calling them both y and getting confused. Now use the disc method on each of the two curves to see that the volume we want is

$$\int_0^2 \pi y_1^2 \, dx - \int_0^2 \pi y_2^2 \, dx = \pi \int_0^2 (2x^3)^2 \, dx - \pi \int_0^2 (x^4)^2 \, dx.$$

 You should work this out and check that the answer is $1024\pi/63$ cubic units.

How about revolving the same region about the y-axis? Since we're just taking the area between two curves, we don't have a particular bias toward one axis or the other, so we should actually be able to do this either by the disc method or by shells. Let's see both ways in action. First, the disc method. Suppose we chop up the region so that the thin sides of the strips are parallel to the y-axis:

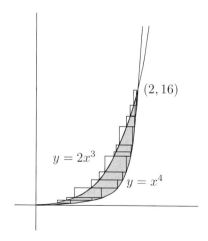

The volume we want is the difference between the volumes of revolution of $y = x^4$ and $y = 2x^3$. The first of these volumes is bigger than the second, since x^4 is to the right of $2x^3$; so let's solve for x and put $x_1 = y^{1/4}$ and $x_2 = (y/2)^{1/3}$. Using the disc method, with x and y switched (as in Variation 1 above) and integrating between $y = 0$ and $y = 16$ (not from 0 to 2!), we see that the volume we want is

$$\int_0^{16} \pi x_1^2 \, dy - \int_0^{16} \pi x_2^2 \, dy = \pi \int_0^{16} (y^{1/4})^2 \, dy - \pi \int_0^{16} \left((y/2)^{1/3}\right)^2 \, dy$$

$$= \pi \int_0^{16} y^{1/2} \, dy - 2^{-2/3}\pi \int_0^{16} y^{2/3} \, dy.$$

This works out to be $64\pi/15$ cubic units after a bit of fiddling, which you should definitely try for practice.

   Let's try to find the same volume by using shells. This time, we slice the region vertically:

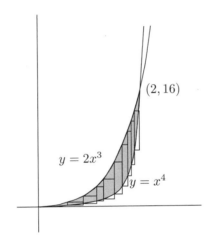

$(2, 16)$

$y = 2x^3$

$y = x^4$

Since $y_1 = 2x^3$ is above $y_2 = x^4$, we take the difference of volumes as follows:

$$\int_0^2 2\pi x y_1 \, dx - \int_0^2 2\pi x y_2 \, dx = 2\pi \int_0^2 2x^4 \, dx - 2\pi \int_0^2 x^5 \, dx,$$

which is $64\pi/15$ cubic units—the same answer as the one we just found using the disc method, of course! Note that when we use the disc method, we are thinking of the solid we want as being formed by one bowl with another bowl hollowed out of it, whereas the shell method is more like a basin with another slightly smaller basin removed. You should try to sketch some pictures to see what's going on here.

   This variation also applies when the area doesn't go all the way down to the axis. For example, suppose we want to find the volume of revolution when the region between the curve $y = 1 + \sqrt{25 - x^2}$ and the line $y = 1$ is revolved about the x-axis. Note that the curve is the top half of the circle $x^2 + (y - 1)^2 = 25$ of radius 5 units centered at $(0, 1)$, so the region looks like this:

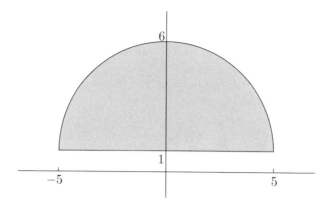

When we revolve this about the x-axis, we get a bead-like shape—it's a roundish solid with a hole down the center. What's its volume? Well, you could use the shell method with respect to the y-variable—you should try this as an exercise.* A more sensible way is to use the disc method. We should consider our region as lying between the curves $y_1 = 1 + \sqrt{25 - x^2}$ and $y_2 = 1$, so the volume is given by

$$\int_{-5}^{5} \pi \left(1 + \sqrt{25 - x^2}\right)^2 dx - \int_{-5}^{5} \pi (1)^2 dx.$$

The second integral is just 10π, which is not coincidentally the volume in cubic units of a cylinder of height 10 units and radius 1 unit—precisely the shape of the empty core of the bead. I leave the first integral to you, reminding you that $\int_{-5}^{5} \sqrt{25 - x^2}\, dx$ is much easier than you think—no calculus required, since it's just the area (in square units) of a semicircle of radius 5 units. In any case, you should check that the answer is $25\pi^2 + 500\pi/3$ cubic units.

29.1.6 Variation 3: revolving about axes parallel to the coordinate axes

Finally, let's see how to handle revolution about the axis $x = h$ or $y = h$, where h is some number not necessarily equal to 0. We'll start with $y = h$, which is parallel to the x-axis but is at height h. Suppose we want to revolve the region between the curve $y = f(x)$ and the lines $y = h$, $x = a$, and $x = b$ about the line $y = h$:

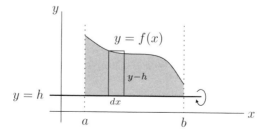

---

*If you do try it, you have to be careful because the area isn't just "under" one curve with respect to the y-axis. Best is just to work out the volume if the right-hand half of the semicircle is revolved about the x-axis, and then double your answer.

A typical strip is shown in the above picture. The width is dx, but the height isn't y: it's $y - h$. In the picture, h is shown as a positive number, so $y - h$ is of course less than y, as it should be. If h happens to be negative, then the height of the strip is more than y ... but of course then $y - h$ is actually greater than y, since h is negative! Regardless of the sign of h, we see that the strip has height $y - h$, so the volume of the corresponding disc is $\pi(y - h)^2 \, dx$, and the volume of the whole solid of revolution is $\int_a^b \pi(y - h)^2 \, dx$.

In fact, the only difference between this formula and the regular disc method is that y has been replaced by the quantity $(y - h)$. As we saw in Section 1.3 of Chapter 1, this has the effect of translating the standard picture, where the region goes down to the x-axis, upward by h units (which is actually downward if h is negative). The only problem with this is that it's possible that the line $y = h$ is actually above the curve, like this:

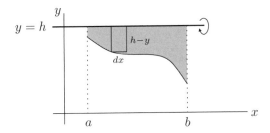

In this case, the height of the strip is $h - y$, not $y - h$. This doesn't really matter in the case of the disc method, since you square the height anyway, but it's good to be careful about these things. Besides, the shell method is a different story.

Indeed, suppose we now want to find the volume of the solid formed by revolving the region below about the axis $x = h$:

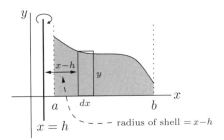

Here we have to use the shell method, since the thin side of the strip is perpendicular to the axis of revolution. A typical shell has height y and thickness dx units, but the radius is now $x - h$ units instead of x units. You should check that you agree that the volume of the shell is $2\pi(x - h)y \, dx$, so the total volume is $\int_a^b 2\pi(x - h)y \, dx$ cubic units. Again, notice that this comes from the standard formula for shells given in Section 29.1.3 above once you replace x by $(x - h)$. This has the effect of translating the standard picture to the right by h units, including the axis of revolution—all we've done is slide the picture over.

How about if the axis is to the right of the region? Consider the following picture:

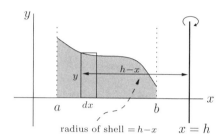

Now the radius of the shell is $h - x$ units, not $x - h$ units, since h is bigger than x for all x in the region of integration $[a, b]$. So this time the volume of revolution is $\int_a^b 2\pi(h - x)y\, dx$ cubic units (check the details).

So, here's the general idea for Variation 3:

> If the axis of revolution is $x = h$, replace x by $(x - h)$ (or $(h - x)$ if $x < h$).
> If the axis of revolution is $y = h$, replace y by $(y - h)$ (or $(h - y)$ if $y < h$).

Let's look at some examples of Variation 3. In all the examples, we'll be dealing with the region between the curve $y = x^3$, the line $x = 2$ and the line $y = 1$:

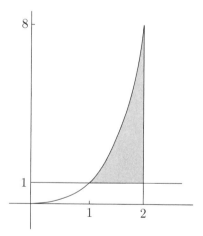

(Note that the scales on the x- and y-axes are different in the diagram so that the picture doesn't look ridiculously thin.) Let's start with finding the volume when the region is revolved about the line $y = 1$. To do this, replace y by $y - 1$ wherever you see it, translating the picture downward by 1 unit. The volume is therefore given by

$$\int_1^2 \pi(y - 1)^2\, dx = \pi \int_1^2 (x^3 - 1)^2\, dx,$$

which easily works out to be $163\pi/14$ cubic units. See if you can justify this by finding the volume of a typical disc (the strips are vertical).

How about revolving the same region about the line $x = 2$? This is actually a combination of Variation 1 and Variation 3, since the revolution is about an axis parallel to the y-axis, so we'll swap x and y, and also replace x by $(2 - x)$ to handle the translation. Note that it's $(2 - x)$ instead of $(x - 2)$, since the region is to the left of the line $x = 2$. Also, the integral will have to be from 1 to 8 since it's with respect to y, not x. The volume is therefore

$$\int_1^8 \pi(2 - x)^2 \, dy = \pi \int_1^8 (2 - y^{1/3})^2 \, dy,$$

which simplifies to $8\pi/5$ cubic units. It's a good idea to make sure that you can also work this out by finding the volume of a typical disc, noting that this

time we have sliced the region into horizontal strips, as in Variation 1.

Now, what about if we revolve the same region about $x = -3$? This is starting to get a little messy. If we use vertical strips, then we'll need the shell method because the thin side of each strip is perpendicular to the axis of revolution. We'll use a combination of Variation 2 and Variation 3. You see, thinking vertically, the region lies between the two curves $y_1 = x^3$ (on the top) and $y_2 = 1$ (on the bottom). Also, the quantity x needs to be replaced by $(x + 3)$ in the standard formula for shells. This means that the volume is given by

$$\int_1^2 2\pi(x+3)y_1 \, dx - \int_1^2 2\pi(x+3)y_2 \, dx = 2\pi \int_1^2 (x+3)x^3 \, dx - 2\pi \int_1^2 (x+3) \, dx,$$

which works out to be $259\pi/10$ cubic units.

Let's repeat the same example, this time taking horizontal strips. Now we have to use the disc method, since the thin part of each strip is parallel to the axis of revolution. We need to switch x and y, since the axis is vertical (Variation 1); we also have to think of the region as lying horizontally between the curves $x_1 = 2$ on the right and $x_2 = y^{1/3}$ on the left (Variation 2); finally, we need to replace x by $x+3$ (Variation 3), which will actually mean replacing x_1 by $x_1 + 3$ and also x_2 by $x_2 + 3$. So this example uses all three of our variations! The standard disc volume is $\pi y^2 \, dx$; change x and y to get $\pi x^2 \, dy$; replace x by $x+3$ to get $\pi(x+3)^2 \, dy$; then integrate this from 1 to 8, separately for x_1 and x_2, and take the difference. This shows that the volume is

$$\int_1^8 \pi(x_1 + 3)^2 \, dy - \int_1^8 \pi(x_2 + 3)^2 \, dy = \pi \int_1^8 (2 + 3)^2 \, dy - \pi \int_1^8 (y^{1/3} + 3)^2 \, dy,$$

which again works out to be $259\pi/10$ cubic units. At least we got the same answer as before! Again, it's a good idea to convince yourself that you can find the volume of a typical disc.

Anyway, that's more than enough theory about volumes of revolution; you have to do a lot of practice problems if you want to master all the variations. For now, it's time to look at finding the volume of more general solids.

29.2 Volumes of General Solids

Most solids can't be formed by revolving some planar area about an axis in that plane. For example, a pyramid has no curvy surfaces, so it isn't a solid of revolution, no matter which way you look at it. One technique for finding the volume of such a solid is the method of *slicing*, which actually generalizes the disc method from Section 29.1.1 above.

Imagine your solid is a vegetable, like a cucumber or a squash. You put it on a cutting board and chop it up into thin, parallel slices. The slices won't all be the same size. Even the two exposed areas of an individual slice won't always be the same. For example, in the case of a cucumber, the slices near the end will be a little skewed. On the other hand, if a slice is very thin, then its two exposed areas will be pretty close. So we're going to approximate the volume of the slice by taking one of the exposed areas—it doesn't matter which one—and multiplying by the thickness of the slice. Then we're going to add up all the volumes and take the limit as the slice thicknesses all go down to zero.

Now, in practice, this procedure is a little complicated. The fact is, there are many ways to cut the solid. For example, if you cut up a cucumber lying on its side, you get thin disc-like slices. If you stand the cucumber on its end, it's more difficult to slice, but you could still do it. You'd end up with slices which look like ovals of different sizes. Or you could tilt the cucumber on an angle and get smaller ovals.

Basically, here is your choice: you need to pick an *axis*, which doesn't necessarily have to go through the solid. All your slices will be perpendicular to this axis. Once you've picked the axis, your way forward is clear: you need to find the cross-sectional area of every slice perpendicular to that axis. Different slices will have different areas. So, on your axis, you need to specify an origin and a positive direction, then work out the cross-sectional area of a slice through x, where x is an arbitrary point on the axis. The last step is to approximate the volume of the slice by the area multiplied by the tiny thickness dx, then integrate; this adds up the volume of all the slices, while simultaneously taking the limit as the maximum slice thickness goes down to zero. In summary, then, here's the plan:

1. Choose an axis.
2. Find a typical cross-sectional area at a point x on the axis; call this area $A(x)$ square units.
3. Then if V is the volume of the solid (in cubic units), we have

$$V = \int_a^b A(x)\, dx,$$

where $[a, b]$ is the range of x which completely covers the solid.

Believe me, you really want to choose this axis so that the cross-sections are as simple as possible. It helps if you can ensure that the cross-sections are in fact similar to each other, that is, different-sized copies of each other. This isn't always possible, though.

Let's use the above technique to find the volume of a "generalized" cone. What this means is that we have some shape in a plane of area A square units, and an apex point P which hovers some distance above the plane:

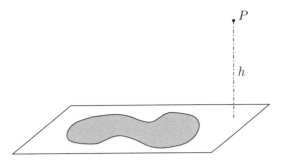

Now we draw a line segment from each point on the edge of the shape up to P. This gives us a surface whose base is the shape we started with. The solid we're interested in is the filled-in version of the surface, or if you prefer, the interior of the surface. Here's sort of what the surface looks like, in skeleton form:

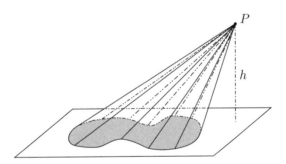

For example, if the base is a circle and the point P sits directly above the center of the circle, then we'll get an ordinary cone. If the base is a square and P is directly above the center of the square (that is, where the diagonals of the square intersect), then we'll get a square pyramid. You should think about what choice of base and point P gives you (a) a regular pyramid, or (b) a skew-cone (which looks like a weird hat—sort of like a witch's hat but it doesn't go straight up). It turns out that the only quantities which are relevant to finding the volume of the solid are the area of the base, A square units, and the perpendicular distance from P to the plane. We'll call this last quantity h units (it's labeled in the above figures).

So, how do we find the volume? We first have to choose an axis. P seems to be a special point, so the line we choose should probably pass through P. Where else should it go? You could try all sorts of things, but the only thing that works is to make the line perpendicular to the plane which contains the base. Let's also set the origin of the axis to lie at P, and the positive direction will be downward. (This might seem a little strange, but there's no reason not to make downward positive. After all, the generalized cone might have been presented to us as balanced on its point, in which case upward would be positive.) This will make our calculations much easier. Let's see what happens if we pick a point x on the axis and take a perpendicular slice through x:

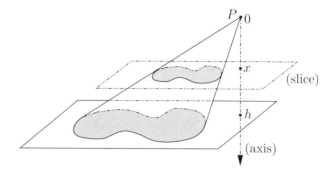

The cross-section is a smaller copy of the original base. In math-talk, the cross-section is similar to the base. Now we have to work out the area of the cross-section. To do this, let's pick any point on the edge of the base and draw the line up to P. This line is on the boundary of the generalized cone, and also passes through the corresponding point on our cross-section. We may as well pick our point so that the line is the right edge of the diagram, but we could have picked any point on the boundary of the base. We also want to draw in some perpendicular line segments, as shown:

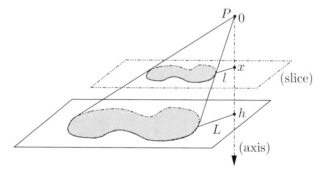

I labeled the lengths of the perpendiculars in the above diagram. Let's just look at the triangles which arise:

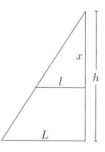

Using similar triangles, we can see that

$$\frac{x}{l} = \frac{h}{L},$$

which means that $l = xL/h$. Let's just do a quick reality check on this equation. If $x = 0$, then the slice is just through the top of the cone (P) and l should be 0, which it is. On the other hand, if $x = h$, then the slice is just the base plane, and the cross-section isn't a smaller copy of the base—it **is** the base. So of course, in that case l should equal L, which it does.

Now let's look at our base and our cross-section, with the line segments of lengths L and l units drawn in:

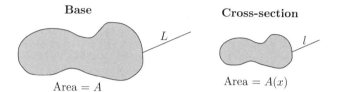

These two figures, including the line segments, are similar—one is an exact magnification of the other. Now here's an important principle of similarity. Say we have two similar figures, and we know the lengths of corresponding line segments, one on each figure. The line segments have to match exactly if we magnify one figure to be the same size as the other one. Then the ratio of areas of the figures is the **square** of the ratio of the two corresponding lengths. For example, if we take two square tiles, one with side length three times that of the other, then the area of the big tile is nine times that of the small one. So, going back to the picture above, the area of the base is A square units and the area of the cross-section is $A(x)$ square units. So the ratio of the areas is the square of the ratio of the corresponding lengths, which are L and l units in our case:

$$\frac{A}{A(x)} = \left(\frac{L}{l}\right)^2.$$

Simplifying and using our above expression for l, we get

$$A(x) = \frac{Al^2}{L^2} = \frac{A}{L^2} \cdot \left(\frac{xL}{h}\right)^2 = \frac{Ax^2}{h^2}.$$

Once again, a reality check: if $x = 0$, the cross-section is just the point P, which has no area. This checks out, since $A(0) = A \times 0^2/h^2 = 0$. How about when $x = h$? Then we're dealing with the base, so our cross-sectional area should be A square units. No problem: $A(h) = A \times h^2/h^2 = A$.

Finally, we're ready to integrate! The only question is, what's the range of x? Well, as we've seen, $x = 0$ is the top and $x = h$ is the bottom, so that's the correct range of x. So,

$$V = \int_0^h A(x)\, dx = \int_0^h \frac{Ax^2}{h^2}\, dx = \frac{A}{h^2}\int_0^h x^2\, dx = \frac{A}{h^2} \cdot \frac{h^3}{3} = \frac{1}{3}Ah$$

cubic units.

Hey, so we just got the formula for the volume of any sort of pyramid or cone-like object. For example, for the regular old cone, the volume is $\frac{1}{3}\pi r^2 h$ cubic units, which is exactly what we found above since $A = \pi r^2$. Same thing for a square pyramid—the volume is $\frac{1}{3}l^2h$ cubic units (where the side length of the base is l units), which works as well because the base area is given by $A = l^2$.

Let's look at one more example. Take the curve $y = e^x$ between $x = 0$ and $x = \frac{1}{2}$ and consider the region between the curve and the x-axis. It looks something like this:

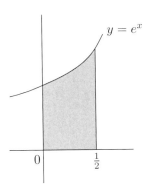

Suppose you have a somewhat bizarre solid sitting on top of the above plane, sticking out of the page, whose base is exactly the shaded region. The solid is shaped in such a way that if you cut it straight down along any line parallel to the y-axis, then the cross-section is a rectangle whose long side lies in the base of the figure, and whose short side is half the length of the long side. Tipping the graph over a little in order to see the perspective, here's what a few of the cross-sections look like:

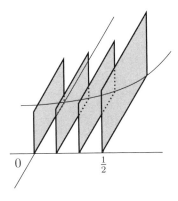

What is the volume of the solid? Let's start by picking an axis. How about the x-axis? That sounds reasonable since we know what the cross-sections perpendicular to this axis look like. We already have an origin and a positive direction, so let's stick with them. At the point x on the axis, the vertical line segment has length e^x units. This is the length of the long side of the rectangle, so the short side has length $\frac{1}{2}e^x$ units (remember, the short side is half the length of the long side). The area of the rectangle is therefore

$$A(x) = e^x \times \frac{1}{2}e^x = \frac{1}{2}e^{2x}$$

square units. So the volume is

$$V = \int_0^{1/2} A(x)\, dx = \frac{1}{2} \int_0^{1/2} e^{2x}\, dx = \frac{1}{2} \frac{e^{2x}}{2} \Big|_0^{1/2} = \frac{1}{4}\,(e-1) \quad \text{cubic units.}$$

29.3 Arc Lengths

Say we have a graph of $y = f(x)$ for some function f, where x ranges from a to b. Take a piece of string and lay it on top of the curve, marking both ends, and then take it off the page, straighten it out, and measure the length between the marks. How do you calculate what the length would be? This length is called the *arc length* of the curve, and we're going to find a formula for it. The strategy will be to get a sort of prototype expression, then to adapt this to get several useful versions of the formula.

So, let's look at a little piece of curve between x and $x + dx$:

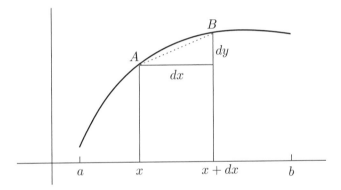

Let's approximate the length of the curve between A and B by the length of the dotted line segment AB. The closer A and B are to each other, the better the approximation. By Pythagoras' Theorem, the length of AB is $\sqrt{(dx)^2 + (dy)^2}$ units. Now we just need to repeat this process with lots of little line segments which flesh out an approximation to the curve, then add up the lengths and take some sort of limit. As usual, the integration takes care of the adding up and limiting parts, but you have to be careful. If you just put an integral sign in front of the little length $\sqrt{(dx)^2 + (dy)^2}$, you'll get

$$\text{arc length} \;=\; \int_?^? \sqrt{(dx)^2 + (dy)^2}.$$

The problem is, this integral doesn't really mean anything! We need to integrate with respect to one variable. Luckily you can adapt the above formula in a variety of cases to produce a meaningful result. For example, you could pull a factor of $(dx)^2$ out of the square root to express the little bit of arc length as $\sqrt{1 + (dy/dx)^2}\, dx$ units, which seems much more promising. (That maneuver actually needs a proof, but the details are a little beyond the scope of this book.) Anyway, in each of the cases below, we'll see how to adapt the above prototypical formula to get a legitimate formula for arc length:

1. If $y = f(x)$ and x ranges from a to b, then take out a factor of $(dx)^2$ in the above integrand (as we just did above) and pull it out of the square root to get

$$\boxed{\text{arc length} \; = \; \int_a^b \sqrt{1 + \left(\frac{dy}{dx}\right)^2}\, dx} \qquad \text{(standard form)}.$$

In terms of f, you can rewrite this as

$$\text{arc length} \; = \; \int_a^b \sqrt{1 + (f'(x))^2}\, dx.$$

2. Suppose that x is given in terms of y. If $x = g(y)$ and y ranges from A to B, then you take out a factor of $(dy)^2$ instead (or if you prefer, swap each occurrence of x and y in the boxed formula above) to get

$$\boxed{\text{arc length} \; = \; \int_A^B \sqrt{1 + \left(\frac{dx}{dy}\right)^2}\, dy} \qquad \text{(in terms of } y\text{)},$$

which can also be written as

$$\text{arc length} \; = \; \int_A^B \sqrt{1 + (g'(y))^2}\, dy.$$

3. How about the parametric form? This means that x and y are functions of a parameter t which ranges from t_0 to t_1. (See Section 27.1 in Chapter 27 for a review of parametric equations.) We can think of the quantity $(dx)^2$ as $(dx/dt)^2(dt)^2$ and similarly for y. We can then pull the $(dt)^2$ out and take its square root to get the useful formula:

$$\boxed{\text{arc length} \; = \; \int_{t_0}^{t_1} \sqrt{\left(\frac{dx}{dt}\right)^2 + \left(\frac{dy}{dt}\right)^2}\, dt} \qquad \text{(parametric version)}.$$

4. A special case of this last formula occurs in the case of polar coordinates. In particular, in Section 27.2.4 of Chapter 27, we saw how to find the area inside the curve $r = f(\theta)$, where θ ranges from θ_0 to θ_1; now let's find the arc length of the same curve. We know that $x = r\cos(\theta)$ and $y = r\sin(\theta)$, so replacing r by $f(\theta)$, we have $x = f(\theta)\cos(\theta)$ and $y = f(\theta)\sin(\theta)$. Here θ acts as a parameter, so we can use the above formula for arc length in parameters (with t replaced by θ). We'll need to know what $dx/d\theta$ and $dy/d\theta$ are. By the product rule,

$$\frac{dx}{d\theta} = f'(\theta)\cos(\theta) - f(\theta)\sin(\theta)$$

and

$$\frac{dy}{d\theta} = f'(\theta)\sin(\theta) + f(\theta)\cos(\theta).$$

Now you have to square both of these things and add them. Go on, try it! You'll find that some terms cancel. Also you have two lots of

$\sin^2(\theta) + \cos^2(\theta)$ terms which can be replaced by 1. Altogether, you should get the formula

$$\text{arc length} = \int_{\theta_0}^{\theta_1} \sqrt{(f(\theta))^2 + (f'(\theta))^2}\, d\theta \qquad \text{(polar, } r = f(\theta)\text{).}$$

By the way, you should express all these arc lengths in units.

Let's look at some examples. Suppose you want to find the arc length of the curve $y = \ln(x)$ where x ranges from $\sqrt{3}$ to $\sqrt{15}$. We use the first formula above to see that

$$\text{arc length} = \int_{\sqrt{3}}^{\sqrt{15}} \sqrt{1 + \left(\frac{dy}{dx}\right)^2}\, dx = \int_{\sqrt{3}}^{\sqrt{15}} \sqrt{1 + \left(\frac{1}{x}\right)^2}\, dx$$

$$= \int_{\sqrt{3}}^{\sqrt{15}} \frac{\sqrt{x^2 + 1}}{x}\, dx.$$

This is actually quite a difficult integral. You should definitely try it as an exercise. If you get stuck, here is the plan of attack: start out with an appropriate trig substitution. If you do it right, the indefinite version of the integral becomes $\int \sec^3(\theta)/\tan(\theta)\, d\theta$. To find this, express the numerator as $\sec(\theta)(1 + \tan^2(\theta))$ and break everything up into two integrals, which can be done using the techniques in Chapter 19. Check that you get an arc length of $2 + \ln(3) - \frac{1}{2}\ln(5)$ units.

How about if you are looking for the arc length of the curve described in parameters as $x = 3t^2 - 12t + 4$ and $y = 8\sqrt{2}t^{3/2}$, where t ranges from 3 to 5? We have to use the parametric version of the formula. Indeed, $dx/dt = 6t - 12$ and $dy/dt = 12\sqrt{2}t^{1/2}$, so

$$\text{arc length} = \int_3^5 \sqrt{\left(\frac{dx}{dt}\right)^2 + \left(\frac{dy}{dt}\right)^2}\, dt = \int_3^5 \sqrt{(6t - 12)^2 + (12\sqrt{2}t^{1/2})^2}\, dt.$$

Now let's look at the innermost part of the integrand. There's a factor of 6^2 which can be pulled out to get

$$(6t - 12)^2 + (12\sqrt{2}t^{1/2})^2 = 6^2((t - 2)^2 + (2\sqrt{2}t^{1/2})^2)$$
$$= 36(t^2 - 4t + 4 + 8t) = 36(t + 2)^2.$$

It is now a simple matter to substitute this into the integrand and do the integration to see that the arc length is 72 units. I'll leave the details to you!

29.3.1 Parametrization and speed

Before we move on to finding surface areas, there's one little fact related to the arc length formula in parametric coordinates that I'd like to look at. Suppose an ant (not a snail, this time!) is crawling around on a flat piece of ground, and we define the ant's position at time t seconds to be $(x(t), y(t))$. What is the speed of the ant at time t? Well, we know that velocity is the derivative of position with respect to time. So the ant's velocity in the x direction is dx/dt and its velocity in the y direction is dy/dt. Its real speed has to involve both of these velocities. In fact, by Pythagoras' Theorem, we should have*:

$$\text{speed} = \sqrt{\left(\frac{dx}{dt}\right)^2 + \left(\frac{dy}{dt}\right)^2}$$

Hey, this is the quantity that we've been integrating to find arc length in the parametric case! Indeed, to find the total distance the ant has traveled, you have to integrate its speed. So we now have a meaning for the integrand in the formula for arc length, at least in the parametric case: it is the instantaneous speed of a particle moving along the curve, as described by the parameters.

Consider the example at the end of the previous section where we have $x = 3t^2 - 12t + 4$ and $y = 8\sqrt{2}t^{3/2}$. From what we observed above,

$$\text{speed} = \sqrt{\left(\frac{dx}{dt}\right)^2 + \left(\frac{dy}{dt}\right)^2} = \sqrt{36(t+2)^2} = 6(t+2),$$

where the answer is in units per second (assuming t is measured in seconds). This means that at time $t = 3$, the speed of a particle, whose position at time t is $(x(t), y(t))$, is $6(3+2) = 30$ units per second; whereas at time $t = 5$, the speed's a little higher at $6(5+2) = 42$ units per second.

In Section 27.1 of Chapter 27, we observed that the parametric equations $x = 3\cos(t)$ and $y = 3\sin(t)$, where $0 \le t < 2\pi$, describe the circle of radius 3 units centered at the origin. The speed of a particle moving as described by these equations is

$$\sqrt{\left(\frac{dx}{dt}\right)^2 + \left(\frac{dy}{dt}\right)^2} = \sqrt{(-3\sin(t))^2 + (3\cos(t))^2} = \sqrt{9} = 3,$$

since $\sin^2(t) + \cos^2(t) = 1$. This means that the particle moves at a constant speed of 3 units per second around the circle (counterclockwise, of course). On the other hand, we also observed that $x = 3\cos(2t)$ and $y = 3\sin(2t)$, this time where $0 \le t < \pi$, also describes the same circle. Now the speed is

$$\sqrt{\left(\frac{dx}{dt}\right)^2 + \left(\frac{dy}{dt}\right)^2} = \sqrt{(-6\sin(2t))^2 + (6\cos(2t))^2} = \sqrt{36} = 6,$$

so a particle observing this new parametrization does indeed move around the same circle twice as fast as the original particle.

29.4 Surface Areas of Solids of Revolution

The last thing we'll consider in this chapter is how to find the surface area of a surface formed by revolving a curve about an axis. The method is a sort of combination of how we found arc lengths and volumes. We start by chopping the curve into small bits of arc, then concentrating on what happens to one of these small bits when we revolve it about the axis. Let's suppose we are

---

*We're getting into vectors here; this really belongs in a book on multivariable calculus.

revolving about the x-axis. What happens to one of these little bits of arc when we revolve it? We get a sort of loop, but the side of it is pretty curvy. If the width of the loop is small enough, we should be able to approximate it by a straighter version. Let's start off by approximating the arc by its secant line segment, just as we did in Section 29.3 above. As we saw, the length of that secant is $\sqrt{(dx)^2 + (dy)^2}$ units. When we revolve that secant instead of the arc length, we get a loop whose outside is straighter, like this:

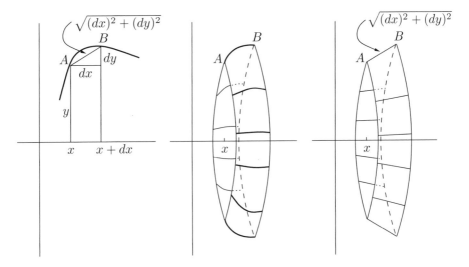

The left-hand picture above shows a piece of the curve and the approximating secant; the middle picture shows the actual curvy ring whose surface area we want to find; and the right-hand picture shows the approximating loop which we're going to use instead. Actually, we are even lazier than that: the side of the loop is not parallel to the x-axis, so our loop is actually part of the surface of a cone. The area of such an object can be computed, but it's really messy. Instead, we are going to do a further approximation and pretend that we are just dealing with a loop with the same side length, but now the loop is cylindrical:

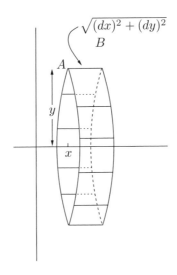

The end result is that we have a cylindrical loop of radius y units, and width $\sqrt{(dx)^2 + (dy)^2}$ units, so it has surface area $2\pi y \sqrt{(dx)^2 + (dy)^2}$ square units. (That's the circumference, $2\pi y$ units, times the width.) It turns out* that the approximation works in the limit as the loop width goes down to zero and we add up the surface areas of the loops, so we are led to the prototypical formula for revolution about the x-axis:

$$\text{surface area} = \int_?^? 2\pi y \sqrt{(dx)^2 + (dy)^2}. \qquad \text{(revolution about x-axis)}$$

Alternatively, if the revolution is about the y-axis, then the loop we use has the same width, but the radius is now x instead of y units, so the prototypical formula for revolution about the y-axis is

$$\text{surface area} = \int_?^? 2\pi x \sqrt{(dx)^2 + (dy)^2}. \qquad \text{(revolution about y-axis)}$$

You can also see this along the lines of Variation 1 from volumes (see Section 29.1.4 above) by switching x and y in the first prototypical formula above.

Anyway, just as in the case of arc lengths, these prototypical formulas can't actually be applied to find any surface areas! Let's see how we can modify the formulas so we can actually use them:

1. Suppose we want to revolve the curve $y = f(x)$ about the x-axis, where x ranges from a to b. We take out a factor of $(dx)^2$ in the integrand of the first prototypical formula and pull it out of the square root, just as we did in the case of arc length, to get

$$\boxed{\text{surface area} = \int_a^b 2\pi y \sqrt{1 + \left(\frac{dy}{dx}\right)^2}\, dx} \qquad \text{(about the x-axis)}.$$

In terms of f, it looks like this:

$$\text{surface area} = \int_a^b 2\pi f(x) \sqrt{1 + (f'(x))^2}\, dx.$$

2. If instead we want to revolve the same curve about the y-axis, the same manipulations applied to the other prototypical formula give

$$\boxed{\text{surface area} = \int_a^b 2\pi x \sqrt{1 + \left(\frac{dy}{dx}\right)^2}\, dx} \qquad \text{(about the y-axis)},$$

or in terms of f,

$$\text{surface area} = \int_a^b 2\pi x \sqrt{1 + (f'(x))^2}\, dx.$$

---

*The computations involved are a little gross—if you want to try it, use the fact that the surface area of a frustum of a cone of radii r and R units and height h units is given by $\pi(R + r)\sqrt{(R - r)^2 + h^2}$ square units.

3. Of course, there's also a parametric form. If x and y are functions of a parameter t which ranges from t_0 to t_1, then dividing and multiplying by dt leads to the following formulas:

$$\text{surface area} = \int_{t_0}^{t_1} 2\pi y \sqrt{\left(\frac{dx}{dt}\right)^2 + \left(\frac{dy}{dt}\right)^2}\, dt \qquad \left(\begin{array}{l}\text{parametric version,}\\ \text{about the } x\text{-axis}\end{array}\right)$$

and

$$\text{surface area} = \int_{t_0}^{t_1} 2\pi x \sqrt{\left(\frac{dx}{dt}\right)^2 + \left(\frac{dy}{dt}\right)^2}\, dt \qquad \left(\begin{array}{l}\text{parametric version,}\\ \text{about the } y\text{-axis.}\end{array}\right)$$

Again, all of these surface areas are in square units.

Here's an example: if the curve $y = \cos(x)$ from $x = 0$ to $x = \pi/2$ is revolved about the x-axis, we need the formula from case 1 above to see that the surface area would be

$$\int_0^{\pi/2} 2\pi y \sqrt{1 + \left(\frac{dy}{dx}\right)^2}\, dx = 2\pi \int_0^{\pi/2} \cos(x)\sqrt{1 + \sin^2(x)}\, dx.$$

To evaluate this integral, first let $t = \sin(x)$, then use a trig substitution to handle the new integral. Try it—the surface area should work out to be $\pi(\sqrt{2} + \ln(1 + \sqrt{2}))$ square units.

On the other hand, the surface area resulting when the parabola $y = x^2/2$ between $x = 0$ and $x = 2\sqrt{2}$ is revolved about the y-axis (not the x-axis) can be found using the formula from case 2 above; since $dy/dx = x$, the surface area is given by

$$\int_0^{2\sqrt{2}} 2\pi x \sqrt{1 + \left(\frac{dy}{dx}\right)^2}\, dx = 2\pi \int_0^{2\sqrt{2}} x\sqrt{1 + x^2}\, dx;$$

it works out to be $52\pi/3$ square units after substituting $t = 1 + x^2$.

Now consider the semicircle which is the upper half of the circle of radius r units centered at the origin. This is parametrized by $x = r\cos(\theta)$ and $y = r\sin(\theta)$, where θ ranges from 0 to π (we stop at π so we only get the upper half). If we revolve this semicircle around the x-axis, we get a sphere, whose surface area is given by the first formula in case 3 above (with t replaced by θ):

$$\int_0^{\pi} 2\pi y \sqrt{\left(\frac{dx}{d\theta}\right)^2 + \left(\frac{dy}{d\theta}\right)^2}\, d\theta = 2\pi \int_0^{\pi} r\sin(\theta)\sqrt{(-r\sin(\theta))^2 + (r\cos(\theta))^2}\, d\theta.$$

You can now use the fact that $\sin^2(\theta) + \cos^2(\theta) = 1$ to see that the surface area works out to be $4\pi r^2$ square units, justifying the classic formula.

Finally, let's consider the surface area analogue of Variation 3 from volumes of revolution (see Section 29.1.6 above). If the axis of revolution is not the x-axis, but is instead the line $y = h$ (which is parallel to the x-axis), then the radius of the cylindrical loop is $y - h$ units, not y units, so the formula from case 1 above needs to be modified appropriately:

$$\text{surface area} = \int_a^b 2\pi(y - h)\sqrt{1 + \left(\frac{dy}{dx}\right)^2} \, dx \qquad \text{(about } y = h\text{)}.$$

(Actually, if the curve is under the line $y = h$, you'd better use $h - y$ instead of $y - h$ or you'll get a negative answer for your surface area!) Again, you shouldn't learn the above formula separately; instead, understand how to derive it from the one you already know. In fact, you should now be able to modify all the other formulas above to allow for revolution about $y = h$ or $x = h$ as appropriate.

CHAPTER 30 _____

Differential Equations

A differential equation is an equation involving derivatives. These things are really useful for describing how quantities change in the real world. For example, if you want to understand how fast a population grows, or even how quickly you can pay off a student loan, a differential equation can help model the situation and give you a decent answer. In this final chapter, we'll see how to solve certain types of differential equations. In particular, here's what we'll look at:

- an introduction to differential equations;
- separable first-order differential equations;
- first-order linear differential equations;
- first- and second-order constant-coefficient differential equations; and
- modeling using differential equations.

30.1 Introduction to Differential Equations

We've already seen an example of a differential equation when we looked at exponential growth and decay, way back in Section 9.6 of Chapter 9. We considered the equation

$$\frac{dy}{dx} = ky,$$

where k is some fixed constant, and claimed that the only solutions to it are of the form $y = Ae^{kx}$ for some constant A. We'll prove this claim in Section 30.2 below. By the way, we shouldn't be surprised to see a constant like A popping up. After all, the original equation involves a derivative; the only way to unravel a derivative is to integrate it, and integration introduces an unknown constant (think $+C$).

The equation $dy/dx = ky$ is an example of a *first-order* differential equation. This is because there's only a first derivative floating around. In general, the *order* of a differential equation is the order of the highest derivative involved. For example, the nasty equation

$$x^2 \frac{d^4y}{dx^4} + \sin(x)\frac{d^2y}{dx^2}\left(\frac{dy}{dx}\right)^7 + e^x y = \tan(x)$$

is a fourth-order differential equation, since there is a fourth derivative involved but no fifth or higher derivative.

Now consider a specific example of the first-order differential equation at the beginning of this section, but with an extra condition:

$$\frac{dy}{dx} = -2y, \qquad y(0) = 5.$$

This means that not only do you need the differential equation to be satisfied by your solution, you also need to ensure that when you set $x = 0$, you get $y = 5$. We know that $y = Ae^{kx}$ is the general solution to the differential equation $dy/dx = ky$; by setting $k = -2$, we see that the general solution of the above differential equation is $y = Ae^{-2x}$ for some constant A. Now put $x = 0$ and $y = 5$ to see that $5 = Ae^{-2(0)}$, or simply $A = 5$. The extra piece of information $y(0) = 5$ has allowed us to pin down the value of A, so the actual solution is $y = 5e^{-2x}$.

What we have just been looking at is an example of an *initial value problem*, or *IVP*. The idea is that you know a starting condition (in this case, $y(0) = 5$) as well as a differential equation that tells you how the situation evolves from there (in this case, $dy/dx = -2y$), and you can use these two facts to find out the exact solution with no undetermined constants. For a second-order differential equation, you effectively need to integrate twice, so you'll get **two** undetermined constants; it follows that you need two pieces of information. Normally these would be the value of $y(0)$ as well as the value of $y'(0)$ (the derivative when $x = 0$). We'll see some examples of this in Section 30.4.2 below.

Now, the study of differential equations is pretty bloody huge. These things are hard to solve. In fact, they are basically impossible, at least in general. Luckily, there are some simple types which can be solved without too much trouble. We're going to look at three such types: first-order separable equations, first-order linear equations, and linear constant-coefficient equations.

30.2   Separable First-order Differential Equations

A first-order differential equation is called *separable* if you can put all the y-stuff on one side (including the dy), and all the x-stuff on the other side (including the dx). For example, the equation $dy/dx = ky$ can be rearranged to read

$$\frac{1}{ky}\,dy = dx,$$

so it is separable. As another example, the equation

$$\frac{dy}{dx} + \cos^2(y)\cos(x) = 0$$

can be rearranged (check out the algebra yourself!) into

$$\sec^2(y)\,dy = -\cos(x)\,dx.$$

Now, the way to continue is simply to whack integral signs on both sides and integrate, then rearrange* to solve for y. In the first example, we get

$$\int \frac{1}{ky}\,dy = \int dx,$$

which becomes

$$\frac{1}{k}\ln|y| = x + C,$$

where C is a constant. To solve for y, multiply by k and take exponentials. We get

$$|y| = e^{kx+kC} = e^{kC}e^{kx}.$$

This means that $y = \pm e^{kC}e^{kx}$. Now, $\pm e^{kC}$ is just some other nonzero constant, so let's call it A, giving the solution $y = Ae^{kx}$ as we expected. (In fact, A can even be 0: indeed, if $y = 0$ for all x, the equation $dy/dx = ky$ is obviously satisfied since both sides are 0. The reason this didn't come up in our solution above is that we divided by y; this assumed that y is never 0.)

As for the second example above, integrating both sides gives

$$\int \sec^2(y)\,dy = \int -\cos(x)\,dx,$$

which leads to

$$\tan(y) = \sin(x) + C,$$

where C is a constant. This is perfectly good as a solution, but maybe you are tempted to write

$$y = \tan^{-1}(\sin(x) + C).$$

The problem with this is, the inverse tangent function has range $(-\pi/2, \pi/2)$ only. We should be able to add any integer multiple of π to the above expression and still get a valid solution. Indeed, $\sec^2(y)$ has period π, so the complete solution should be

$$y = \tan^{-1}(\sin(x) + C) + n\pi,$$

where C is a constant and n is an integer. Maybe we should just avoid these issues by leaving it as $\tan(y) = \sin(x) + C$. (Again, we divided by $\cos^2(y)$ right at the beginning of the solution; this caused us to miss the constant solutions $y = n\pi/2$, where n is an odd number, since that's when $\cos^2(y) = 0$. These solutions arise as $C \to \pm\infty$ in the above solution.)

How about the same example, but as an IVP (initial value problem)? For example, consider the IVP

$$\frac{dy}{dx} + \cos^2(y)\cos(x) = 0, \qquad y(0) = \frac{\pi}{4}.$$

---

*As you might expect, these maneuvers can be fully justified by using the chain rule.

If you solve the differential equation using the above technique, you end up with

$$\tan(y) = \sin(x) + C$$

as before. Now put $x = 0$ and $y = \pi/4$ to get

$$\tan(\pi/4) = \sin(0) + C,$$

which means that $C = 1$. So we have

$$\tan(y) = \sin(x) + 1.$$

Now if we write

$$y = \tan^{-1}(\sin(x) + 1) + n\pi,$$

where n is an integer, we can again put $x = 0$ and $y = \pi/4$ to see that $\pi/4 = \tan^{-1}(1) + n\pi$, which means that $n = 0$. So it's fair to write the solution as

$$y = \tan^{-1}(\sin(x) + 1).$$

To see this a bit more clearly, imagine that the initial condition is $y(0) = 5\pi/4$ instead of $y(0) = \pi/4$. Plugging this into the equation $\tan(y) = \sin(x) + C$ once again leads to $C = 1$, since $\tan(5\pi/4) = 1$. So once more, we find that we have $\tan(y) = \sin(x) + 1$, but it's a mistake to write this as $y = \tan^{-1}(\sin(x) + 1)$. Why? When $x = 0$, we have

$$y = \tan^{-1}(\sin(0) + 1) = \tan^{-1}(0) = \frac{\pi}{4},$$

which isn't what we want. So we need to add π to make it work:

$$y = \tan^{-1}(\sin(x) + 1) + \pi.$$

Now the differential equation is satisfied, and $y(0) = 5\pi/4$ as we wanted. The same precaution would be required if the initial condition were $y(0) = \pi/4 + n\pi$ for any nonzero integer n. These things require a delicate touch!

30.3 First-order Linear Equations

Here's a different type of first-order differential equation:

$$\frac{dy}{dx} + p(x)y = q(x),$$

where p and q are given functions of x. Such an equation is called a *first-order linear* differential equation. It may not be separable, and it may not even look particularly linear! For example,

$$\frac{dy}{dx} + 6x^2 y = e^{-2x^3} \sin(x)$$

doesn't look very linear, yet this equation is indeed first-order linear. The reason is that the powers of y and dy/dx are both one. So something like

$$\frac{dy}{dx} + 6x^2 y^3 = e^{-2x^3} \sin(x)$$

is no good, since y^3 is not first-degree in y. Similarly,

$$\left(\frac{dy}{dx}\right)^2 + 6x^2 y = e^{-2x^3} \sin(x)$$

is also not linear because the quantity dy/dx is squared.

Let's go back to the linear equation from above,

$$\frac{dy}{dx} + 6x^2 y = e^{-2x^3} \sin(x).$$

This equation isn't separable. Try it! You won't be able to get all the y-stuff on one side and all the x-stuff on the other side. Luckily, there's a neat trick that will save the day. Imagine that we multiply both sides by the quantity e^{2x^3}. This certainly makes the right-hand side nicer, as it happens, but there's actually a more interesting effect. Let's see what happens:

$$e^{2x^3} \frac{dy}{dx} + 6x^2 e^{2x^3} y = \sin(x).$$

Watch carefully, now: there's nothing up my sleeve as I rewrite this as

$$\frac{d}{dx}\left(e^{2x^3} y\right) = \sin(x).$$

How is this possible? Well, all I had to do was mentally reverse the product rule while differentiating implicitly! (Piece of cake ...) To see that this is correct, all you have to do is differentiate it out. Indeed, by the product rule, one term is e^{2x^3} times the derivative of y, that is, $e^{2x^3}(dy/dx)$; the other term is y times the derivative of e^{2x^3}, that is, $y \times 6x^2 e^{2x^3}$ (using the chain rule). But that's exactly what the original left-hand side was! So we do indeed have

$$\frac{d}{dx}\left(e^{2x^3} y\right) = \sin(x).$$

Now all we have to do is integrate both sides with respect to x. This cancels out the derivative on the left-hand side, leaving

$$e^{2x^3} y = \int \sin(x)\, dx = -\cos(x) + C.$$

Dividing by e^{2x^3}, we get the solution

$$y = (C - \cos(x))e^{-2x^3},$$

where C is an arbitrary constant. Now try differentiating this and check that it satisfies the original differential equation!

The key to the previous solution was multiplying by e^{2x^3}. When we did this, we were able to wrap the left-hand side into $\frac{d}{dx}(\text{stuff})$, which could be integrated easily. For this reason, the quantity e^{2x^3} is called an *integrating factor*. It turns out that for the general first-order linear differential equation

$$\frac{dy}{dx} + p(x)y = q(x),$$

a good integrating factor is given by the equation

$$\text{integrating factor} = e^{\int p(x)\,dx},$$

where you don't need a $+C$ in the integral. After you multiply the original differential equation by this integrating factor, the left-hand side can be "factored" as

$$\frac{d}{dx}\left(\text{integrating factor} \times y\right).$$

We'll see why in Section 30.3.1 below. In the meantime, let's rework our above example

$$\frac{dy}{dx} + 6x^2 y = e^{-2x^3}\sin(x)$$

using this more general framework. First, find the integrating factor by taking the coefficient of y (which is $6x^2$), integrating it and exponentiating the result:

$$\text{integrating factor} = e^{\int 6x^2\,dx} = e^{2x^3}.$$

Now we can proceed as we did in the original solution above: multiply the differential equation by e^{2x^3} and rewrite the left-hand side as $\frac{d}{dx}(e^{2x^3}y)$, which is the derivative of the product of the integrating factor and y.

By far the best way to learn this is to do a lot of practice problems until you get the hang of it. Here are two more examples. First, how would you solve

$$\frac{dy}{dx} = e^x y + e^{2x}, \qquad y(0) = 2(e-1)?$$

This is an initial value problem (IVP), but we'll worry about that aspect after we solve the differential equation. The first thing to do is put it into *standard form,* meaning that you need all the y-stuff to be on the left, all the pure x-stuff on the right, and the coefficient of dy/dx to be 1. In this case, we just have to subtract $e^x y$ from both sides, to get

$$\frac{dy}{dx} - e^x y = e^{2x}, \qquad y(0) = 2(e-1).$$

The coefficient of y is $-e^x$, so the integrating factor is the exponential of the integral of that quantity:

$$\text{integrating factor} = e^{\int(-e^x)\,dx} = e^{-e^x}.$$

(Remember, you don't need a $+C$ here.) Let's multiply the above differential equation by this integrating factor:

$$e^{-e^x}\frac{dy}{dx} - e^x e^{-e^x}y = e^{-e^x}e^{2x}.$$

As always, the left-hand side is the derivative of y times the integrating factor, so we have

$$\frac{d}{dx}(e^{-e^x}y) = e^{-e^x}e^{2x}.$$

It's not a bad idea to check that this simplification is valid by differentiating the left-hand side. In any case, integrate both sides of the above equation to get

$$e^{-e^x} y = \int e^{-e^x} e^{2x} \, dx.$$

To do this integral, set $t = e^x$, so that $dt = e^x \, dx$. Note that you have to write e^{2x} as $e^x e^x$ to make it work. I leave it to you to do the integral (using integration by parts) and check that the resulting equation is

$$e^{-e^x} y = -e^x e^{-e^x} - e^{-e^x} + C.$$

Finally, divide through by the integrating factor e^{-e^x} to get

$$y = -e^x - 1 + Ce^{e^x}$$

for some constant C. Now all that's left is to solve the IVP. When $x = 0$, we know that $y = 2(e - 1)$, so inserting this into the above equation, we have

$$2(e - 1) = -e^0 - 1 + Ce^{e^0}.$$

You can easily solve this to see that $C = 2$, so the final solution is

$$y = 2e^{e^x} - e^x - 1.$$

Check by differentiating that this satisfies the original differential equation.
Let's quickly go through one more example of a first-order linear differential equation:

$$\tan(x) \frac{dy}{dx} = e^{\sin(x)} - y.$$

First, put the y-stuff on the left and divide by $\tan(x)$ to make the coefficient of dy/dx equal to 1:

$$\frac{dy}{dx} + \cot(x)y = \cot(x)e^{\sin(x)}.$$

The coefficient of y is $\cot(x)$, so

$$\text{integrating factor} = e^{\int \cot(x) \, dx} = e^{\ln(\sin(x))} = \sin(x).$$

(Technically we should have written $|\sin(x)|$, but this complicates things unnecessarily.) Anyway, multiply the differential equation through by $\sin(x)$ to get

$$\sin(x) \frac{dy}{dx} + \cos(x)y = \cos(x)e^{\sin(x)},$$

since $\sin(x) \cot(x) = \cos(x)$. Now the left-hand side factors into the derivative of y times the integrating factor (check it):

$$\frac{d}{dx}(y \sin(x)) = \cos(x)e^{\sin(x)}.$$

Iqntegrate both sides (use a substitution to simplify the right-hand side):

$$y \sin(x) = \int \cos(x)e^{\sin(x)} \, dx = e^{\sin(x)} + C.$$

Finally, divide through by $\sin(x)$ to get

$$y = \csc(x)e^{\sin(x)} + C\csc(x),$$

and we have found the solution to our differential equation.

In summary, here's the method for dealing with first-order linear differential equations:

- Put the stuff involving y on the left-hand side and the stuff involving x on the right-hand side, then divide through by the coefficient of dy/dx to get the equation into the standard form

$$\frac{dy}{dx} + p(x)y = q(x).$$

- Multiply through by the integrating factor, which we'll call $f(x)$, given by

$$\boxed{\text{integrating factor } f(x) = e^{\int p(x)\,dx}}$$

where no $+C$ is needed in the integral in the exponent.

- The left-hand side becomes $\frac{d}{dx}(f(x)y)$, where $f(x)$ is the integrating factor. Rewrite the equation with this new left-hand side.
- Integrate both sides; this time you must put a $+C$ on the right-hand side.
- Divide by the integrating factor to solve for y.

Practice this and you won't regret it!

30.3.1 Why the integrating factor works

Why is the weird expression $e^{\int p(x)\,dx}$ a good integrating factor? Well, suppose we take our general equation

$$\frac{dy}{dx} + p(x)y = q(x)$$

and multiply it by the integrating factor $e^{\int p(x)\,dx}$. We get

$$e^{\int p(x)\,dx}\frac{dy}{dx} + e^{\int p(x)\,dx}p(x)y = \text{stuff in } x.$$

I'm really focusing on the left-hand side for the moment, so I just wrote "stuff in x" on the right. Now, we have claimed that we can rewrite the left-hand side so that the above equation becomes

$$\frac{d}{dx}\left(e^{\int p(x)\,dx}y\right) = \text{stuff in } x;$$

this is much easier to deal with. To prove our claim, use the product rule on the left-hand side to write it as

$$e^{\int p(x)\,dx}\frac{dy}{dx} + \frac{d}{dx}\left(e^{\int p(x)\,dx}\right)y.$$

That's almost what we need; we just have to use the chain rule to write

$$\frac{d}{dx}\left(e^{\int p(x)\,dx}\right) = \frac{d}{dx}\left(\int p(x)\,dx\right) \times e^{\int p(x)\,dx} = p(x)e^{\int p(x)\,dx}.$$

Note that $\frac{d}{dx}\int p(x)\,dx = p(x)$, since $\int p(x)\,dx$ (without the $+C$) is an antiderivative of p. Now if you assemble all the pieces from above, you can see that

$$e^{\int p(x)\,dx}\frac{dy}{dx} + e^{\int p(x)\,dx}p(x)y = \frac{d}{dx}\left(e^{\int p(x)\,dx}y\right)$$

after all. Our method works!

30.4 Constant-coefficient Differential Equations

Now it's time to look at linear differential equations with constant coefficients. These equations look something like this:

$$a_n\frac{d^n y}{dx^n} + \cdots + a_2\frac{d^2 y}{dx^2} + a_1\frac{dy}{dx} + a_0 y = f(x).$$

Here f is some function of x only, and $a_n, \ldots, a_1, a_0$ are just plain old constant real numbers. Notice that the left-hand side of the above equation looks a bit like a polynomial in y, except that instead of taking powers of y, we are taking derivatives.

Let's look at a first-order example. Consider the differential equation

$$3\frac{dy}{dx} - \sin(5x) = 12x - 6y.$$

This can be rearranged to put all the purely x-stuff on the right and the y-stuff (including the derivative) on the left. Finally, divide by 3 to get

$$\frac{dy}{dx} + 2y = 4x + \frac{1}{3}\sin(5x).$$

This is a first-order constant-coefficient linear equation. In fact, you can solve it by means of the techniques described in the previous section on first-order linear equations. If you do it that way, you'll need to use an integrating factor, which is actually a bit of a pain in this case (try it and see!). We'll soon look at another method to deal with such equations; in fact, we'll solve the above example in Section 30.4.6 below.

We'll also examine the second-order case in some detail. In this case we are dealing with equations like

$$a\frac{d^2 y}{dx^2} + b\frac{dy}{dx} + cy = f(x),$$

for example,

$$\frac{d^2 y}{dx^2} - 5\frac{dy}{dx} + 6y = 2x^2 e^x.$$

We'll see how to solve this in Section 30.4.6 below. First, we need to look at some general ideas for solving both first- and second-order constant-coefficient linear equations.*

Let's start by considering a simple case: assume there's no stuff in x on the right-hand side. Two such examples are

$$\frac{dy}{dx} - 3y = 0 \qquad \text{and} \qquad \frac{d^2y}{dx^2} - \frac{dy}{dx} + 20y = 0.$$

Such equations are called *homogeneous*. Let's look at how to solve first-order (like the left-hand example above) and second-order (like the right-hand one) homogeneous equations.

30.4.1 Solving first-order homogeneous equations

This is pretty easy. The solution to

$$\frac{dy}{dx} + ay = 0$$

is just $y = Ae^{-ax}$. (In fact, this equation is simply $dy/dx = ky$ with $k = -a$; see Sections 30.1 and 30.2 above.) For example, given the differential equation

$$\frac{dy}{dx} - 3y = 0,$$

you can simply write down the solution $y = Ae^{3x}$, where A is some constant.

30.4.2 Solving second-order homogeneous equations

This case is a little more involved. We need to solve

$$a\frac{d^2y}{dx^2} + b\frac{dy}{dx} + cy = 0.$$

Although it might seem a little strange, the easiest way to do this is to pluck a quadratic equation seemingly out of thin air. The quadratic equation, called the *characteristic quadratic equation*, is $at^2 + bt + c = 0$. For example, consider the following three differential equations:

(a) $y'' - y' - 20y = 0$ (b) $y'' + 6y' + 9y = 0$ (c) $y'' - 2y' + 5y = 0.$

Notice that we have written y' instead of dy/dx and y'' instead of d^2y/dx^2. In any case, the characteristic quadratic equations of these three examples are $t^2 - t - 20 = 0$, $t^2 + 6t + 9 = 0$, and $t^2 - 2t + 5 = 0$, respectively.

The next thing is to find the roots of the characteristic quadratic. There are three possibilities, depending on whether there are two real roots, one (double) real root or two complex roots. Let's summarize the whole method, then solve the above three examples.

---

*These ideas also work for higher-order equations, but we will concentrate on first- and second-order equations in this book.

How to solve the homogeneous equation $ay'' + by' + cy = 0$:

1. Write down the characteristic quadratic equation $at^2 + bt + c = 0$ and solve it for t.

2. If there are two different real roots α and β, the solution is

$$y = Ae^{\alpha x} + Be^{\beta x}.$$

3. If there is only one (double) real root α, the solution is

$$y = Ae^{\alpha x} + Bxe^{\alpha x}.$$

4. If there are two complex roots, they will be conjugate to each other. That is, they must be of the form $\alpha \pm i\beta$. The solution is

$$y = e^{\alpha x}(A\cos(\beta x) + B\sin(\beta x)).$$

In all three cases (2, 3 and 4), A and B are undetermined constants.

So, for example (a) above, we saw that the characteristic quadratic equation is $t^2 - t - 20 = 0$. If you factor the quadratic as $(t + 4)(t - 5)$, it's clear that the solutions to the equation are $t = -4$ and $t = 5$. By step 2 above, we see that the solution to our equation $y'' - y' - 20y = 0$ is given by

$$y = Ae^{-4x} + Be^{5x},$$

for some constants A and B.

The characteristic quadratic equation $t^2 + 6t + 9 = 0$ in example (b) reduces to $(t + 3)^2 = 0$, so the only solution is $t = -3$. By step 3 above, the solution to the homogeneous equation $y'' + 6y' + 9 = 0$ is

$$y = Ae^{-3x} + Bxe^{-3x}.$$

Finally, if we use the quadratic formula to solve the characteristic quadratic equation $t^2 - 2t + 5 = 0$ of example (c), we get $t = 1 \pm 2i$. (Try it and see!) So, with $\alpha = 1$ and $\beta = 2$, step 4 above says that the solution to $y'' - 2y' + 5y = 0$ is

$$y = e^x(A\cos(2x) + B\sin(2x)).$$

Once again, A and B are undetermined constants.

30.4.3 Why the characteristic quadratic method works

Now let's see why the above method works. (If you don't care why, you'd better move on to the next section!) Otherwise, consider what happens when you put $y = e^{\alpha x}$ in the equation $ay'' + by' + cy = 0$. We have $y' = \alpha e^{\alpha x}$ and $y'' = \alpha^2 e^{\alpha x}$, so

$$ay'' + by' + cy = a\alpha^2 e^{\alpha x} + b\alpha e^{\alpha x} + ce^{\alpha x} = (a\alpha^2 + b\alpha + c)e^{\alpha x}.$$

So, if α is a root of the characteristic quadratic $at^2 + bt + c$, then we have $a\alpha^2 + b\alpha + c = 0$. The above equation now implies that $ay'' + by' + cy = 0$— that is, $y = e^{\alpha x}$ solves our differential equation! Also, any constant multiple of this solves the equation, and if you have another root β, then you can add the two solutions $y = Ae^{\alpha x}$ and $y = Be^{\beta x}$ to get more solutions (try it and see). That takes care of step 2 above.

Let's look at step 4 next. If the two solutions to the quadratic are complex conjugates of the form $\alpha + i\beta$, then by the same argument as for step 2, the solution must be

$$y = Ae^{(\alpha+i\beta)x} + Be^{(\alpha-i\beta)x} = e^{\alpha x}(Ae^{i\beta x} + Be^{-i\beta x}),$$

where A and B can even be complex numbers. Now you can use Euler's identity (see Section 28.2 in Chapter 28) to see that

$$y = e^{\alpha x}(A(\cos(\beta x) + i\sin(\beta x)) + B(\cos(\beta x) - i\sin(\beta x)))$$
$$= e^{\alpha x}((A + B)\cos(\beta x) + (A - B)i\sin(\beta x)).$$

Relabel the constant $(A + B)$ as A and the constant $(A - B)i$ as B to get the correct formula.

Finally, for step 3, suppose the characteristic quadratic has just one root, α. If you substitute $y = xe^{\alpha x}$ into the differential equation $ay'' + by' + cy = 0$, you can use $y' = \alpha xe^{\alpha x} + e^{\alpha x}$ and $y'' = \alpha^2 xe^{\alpha x} + 2\alpha e^{\alpha x}$ to see that

$$ay'' + by' + cy = (a\alpha^2 + b\alpha + c)xe^{\alpha x} + (2a\alpha + b)e^{\alpha x}.$$

If α is a double root of $at^2 + bt + c$, then not only does $a\alpha^2 + b\alpha + c = 0$, but also $2a\alpha + b = 0$.* This leads to the correct solution from step 3 above.

30.4.4 Nonhomogeneous equations and particular solutions

Now let's see what happens if we do have some stuff in x alone, which we put on the right-hand side. For example, consider the differential equation

$$y'' - y' - 20y = e^x.$$

This isn't homogeneous because of the e^x term on the right-hand side. Suppose we try to guess a solution. We know that the derivatives of e^x are all e^x, so let's try $y = e^x$. Then $y' = e^x$ and $y'' = e^x$, so the left-hand side $y'' - y' - 20y$ becomes $e^x - e^x - 20e^x = -20e^x$. That's not equal to the right-hand side, but it's pretty close. We just have to divide by -20. So, let's try again: set $y = -\frac{1}{20}e^x$. Then y' and y'' are also $-\frac{1}{20}e^x$, so we have

$$y'' - y' - 20y = -\frac{1}{20}e^x - \left(-\frac{1}{20}e^x\right) - 20\left(-\frac{1}{20}e^x\right) = e^x.$$

So we have shown that $y = -\frac{1}{20}e^x$ is a solution to our original equation

---

*Here's why $2a\alpha + b = 0$ if the quadratic $at^2 + bt + c = 0$ has a double root at $t = \alpha$: the discriminant is 0, so $b^2 = 4ac$. Then

$$(2a\alpha + b)^2 = 4a^2\alpha^2 + 4ab\alpha + b^2 = 4a^2\alpha^2 + 4ab\alpha + 4ac = 4a(a\alpha^2 + b\alpha + c) = 0.$$

Since $(2a\alpha + b)^2 = 0$, we also have $2a\alpha + b = 0$.

$y'' - y' - 20y = e^x$. It's not the only solution, though. To see why, consider the related homogeneous equation

$$y'' - y' - 20y = 0.$$

This was actually example (a) from Section 30.4.2 above. There we saw that the complete solution is

$$y = Ae^{-4x} + Be^{5x}.$$

So let's play a little game. We'll write this as y_H instead of just y, where the H stands for homogeneous. We have shown that

$$\text{if} \quad y_H = Ae^{-4x} + Be^{5x}, \qquad \text{then} \quad y_H'' - y_H' - 20y_H = 0.$$

On the other hand, we showed above that

$$\text{if} \quad y_P = -\frac{1}{20}e^x, \qquad \text{then} \quad y_P'' - y_P' - 20y_P = e^x.$$

Here I wrote the solution $-\frac{1}{20}e^x$ from above as y_P; this is called a *particular solution*, which explains the subscript P. Now, suppose we add up the equations $y_H'' - y_H' - 20y_H = 0$ and $y_P'' - y_P' - 20y_P = e^x$. Grouping derivatives together, we get

$$y_H'' + y_P'' - y_H' - y_P' - 20y_H - 20y_P = 0 + e^x.$$

In fact, since the sum of the derivatives is the derivative of the sum, and the same for the second derivative, we get

$$(y_H + y_P)'' - (y_H + y_P)' - 20(y_H + y_P) = e^x.$$

So if $y = y_H + y_P$, then y is also a solution to our original differential equation $y'' - y' - 20y = e^x$. In other words, we could take our particular solution

$$y_P = -\frac{1}{20}e^x,$$

which actually solves the differential equation; then add any solution to the homogeneous version of the differential equation; the result is still a solution to the original differential equation. Furthermore, **all** the solutions to the nonhomogeneous equation are in this form.

The same methodology works for both the first-order and the second-order cases. The only issue is how to guess the particular solution. In the next section, we'll see how to make a guess of what the form of the solution should be (this is similar to the partial fraction technique from Section 18.3 of Chapter 18). Then if you're lucky, you can plug in that form and find the unknown constants in order to nail down the particular solution.

Here's a summary of our methods so far:

1. Rearrange the equation into the correct form. That is, put all the x-junk on the right-hand side. You should be able to reduce the equation to

$$\frac{dy}{dx} + ay = f(x)$$

for the first-order case, or

$$a\frac{d^2y}{dx^2} + b\frac{dy}{dx} + cy = f(x)$$

for the second-order case.

2. Using the techniques from Sections 30.4.1 and 30.4.2 above, solve the associated homogeneous equation

$$\frac{dy}{dx} + ay = 0 \qquad \text{or} \qquad a\frac{d^2y}{dx^2} + b\frac{dy}{dx} + cy = 0.$$

The solution, which we'll write as y_H, will have one or two undetermined constants in it (depending on whether the equation is first- or second-order). We call y_H the *homogeneous solution* of the equation.

3. If the original function f is actually 0, then we're already done; the complete solution is $y = y_H$.

4. On the other hand, if the function f is anything other than 0, then write down the form for the particular solution y_P (see Section 30.4.5 below). The form will have some constants which must be determined. Substitute y_P into the original equation and equate coefficients to find the constants.

5. Finally, the solution is $y = y_H + y_P$.

We'll look at what happens if you are dealing with an initial value problem (IVP) in Section 30.4.8 below. Meanwhile, let's see how to find a particular solution.

30.4.5 Finding a particular solution

So far, we have blissfully ignored the stuff involving x which could appear on the right-hand side (it was called $f(x)$ earlier). Now it's time to deal with it. The tactic is to write down the form of the particular solution, then to find the actual solution by plugging the form into the equation. The table on the next page shows how to come up with the correct form. For example, in the differential equation

$$y' - 3y = 5e^{2x},$$

the right-hand side is a multiple of e^{2x}; so the table indicates that the form should be $y_p = Ce^{2x}$, where C is a constant that we have to find by substituting y_P into the original equation. It's easy to see that $y'_P = 2Ce^{2x}$, so we have

$$2Ce^{2x} - 3(Ce^{2x}) = 5e^{2x}.$$

This reduces to $-Ce^{2x} = 5e^{2x}$, so $C = -5$. The particular solution is therefore $y_P = -5e^{2x}$. In fact, since we saw in Section 30.4.1 above that the solution to the homogeneous version $y' - 3y = 0$ is $y_H = Ae^{3x}$, we now know that the full solution to $y' - 3y = 5e^{2x}$ is

$$y = y_H + y_P = Ae^{3x} - 5e^{2x},$$

where A is an unknown constant. Note that the homogeneous solution involves an unknown constant, while the particular solution must have no unknown constants.

Now, here's the table:

If f is a ...	then the form is ...
polynomial of degree n	$y_P = $ **general polynomial of degree n**
e.g., $f(x) = 7$	$y_P = a$
$\quad f(x) = 3x - 2$	$y_P = ax + b$
$\quad f(x) = 10x^2$	$y_P = ax^2 + bx + c$
$\quad f(x) = -x^3 - x^2 + x + 22$	$y_P = ax^3 + bx^2 + cx + d$
multiple of an exponential e^{kx}	$y_P = Ce^{kx}$
e.g., $f(x) = 10e^{-4x}$	$y_P = Ce^{-4x}$
$\quad f(x) = e^x$	$y_P = Ce^x$
multiple of $\cos(kx)$	$y_P = C\cos(kx) + D\sin(kx)$
$\quad$ **+ multiple of $\sin(kx)$**	
e.g., $f(x) = 2\sin(3x) - 5\cos(3x)$	$y_P = C\cos(3x) + D\sin(3x)$
$\quad f(x) = \cos(x)$	$y_P = C\cos(x) + D\sin(x)$
$\quad f(x) = 2\sin(11x)$	$y_P = C\cos(11x) + D\sin(11x)$
a sum or product of one of the above	**the sum or product of forms (if a product, omit a constant)**
e.g., $f(x) = 2x^2 + e^{-6x}$	$y_P = ax^2 + bx + c + Ce^{-6x}$
$\quad f(x) = 2x^2 e^{-6x}$	$y_P = (ax^2 + bx + c)e^{-6x}$
$\quad f(x) = 7e^{2x}\sin(3x)$	$y_P = (C\cos(3x) + D\sin(3x))e^{2x}$
$\quad f(x) = \cos(2x) + 6\sin(x)$	$y_P = C\cos(2x) + D\sin(2x)$
	$\qquad + E\cos(x) + F\sin(x)$
$\quad f(x) = 4x\cos(3x)$	$y_P = (x + b)(C\cos(3x) + D\sin(3x))$
If y_P conflicts with y_H, multiply the form by x or x^2 as appropriate.	

This table should be fairly self-explanatory, except for the last line, which will be explained in Section 30.4.7 below, and also the instruction "if a product, omit a constant." To see what this instruction means, first note that there is a redundant constant if you just multiply two forms together. For example, $2x^2 e^{-6x}$ looks as if it should lead to the form $(ax^2 + bx + c)Ce^{-6x}$, but the constant C is unnecessary and can be omitted, since it can be absorbed into the other constants a, b, and c. The same sort of thing applies to the examples $7e^{2x}\cos(3x)$ and $4x\cos(3x)$ in the above table.

(By the way, the table only shows you what to do if f happens to be a polynomial, an exponential, a sine, a cosine, or some product or sum of one or more of these types of function. Otherwise the method just doesn't work. There is a fancier method called "variation of parameters" which is much more general, but it's outside the scope of this book.)

30.4.6 Examples of finding particular solutions

Once you've written down the form for y_P, you still have to substitute y_P into the original differential equation in order to find the constants. To make the calculation easier, you should first calculate y'_P and y''_P (for the first order case, you actually only need y'_P). Let's look at one example of this; then we'll finally go back and complete the two unresolved examples from Section 30.4 above.

First consider the differential equation

$$y'' - 4y' + 4y = 25e^{3x}\sin(2x).$$

Let's quickly dispense with the homogeneous part; in fact, the characteristic quadratic equation for $y'' - 4y' + 4y = 0$ is $t^2 - 4t + 4 = 0$, which has one solution, namely $t = 2$. So we have $y_H = Ae^{2x} + Bxe^{2x}$, where A and B are constants. Now let's look for a particular solution. Break up the right-hand side of our differential equation, $25e^{3x}\sin(2x)$, into two components: $25e^{3x}$ and $\sin(2x)$. According to the above table, the form for a constant multiple of e^{3x} is Ce^{3x}; and the form for $\sin(2x)$ is $C\cos(2x) + D\sin(2x)$. We need to multiply these together, but we can consolidate the constants as we do so and write

$$y_P = e^{3x}(C\cos(2x) + D\sin(2x))$$

as our form. Now, let's do some fiddly calculations using the product rule many times:

$$\begin{aligned}
y_P &= e^{3x}(C\cos(2x) + D\sin(2x)), \\
y'_P &= e^{3x}(-2C\sin(2x) + 2D\cos(2x)) + 3e^{3x}(C\cos(2x) + D\sin(2x)) \\
&= e^{3x}((3C + 2D)\cos(2x) + (3D - 2C)\sin(2x)), \\
y''_P &= e^{3x}(-2(3C + 2D)\sin(2x) + 2(3D - 2C)\cos(2x)) \\
&\quad + 3e^{3x}((3C + 2D)\cos(2x) + (3D - 2C)\sin(2x)) \\
&= e^{3x}((5C + 12D)\cos(2x) + (5D - 12C)\sin(2x)).
\end{aligned}$$

Now it's time to substitute this mess into the original differential equation $y'' - 4y' + 4y = 25e^{3x}\sin(2x)$. We get the gross-looking equation

$$\begin{aligned}
e^{3x}((5C + 12D)\cos(2x) + (5D - 12C)\sin(2x)) & \\
- 4e^{3x}((3C + 2D)\cos(2x) + (3D - 2C)\sin(2x)) & \\
+ 4e^{3x}(C\cos(2x) + D\sin(2x)) &= 25e^{3x}\sin(2x),
\end{aligned}$$

which mercifully simplifies to

$$e^{3x}(4D - 3C)\cos(2x) + e^{3x}(-4C - 3D)\sin(2x) = 25e^{3x}\sin(2x).$$

To make these expressions equal for all x, the $e^{3x}\cos(2x)$ stuff has to disappear and the coefficient of $e^{3x}\sin(2x)$ must be 25. This means that $4D - 3C = 0$ and $-4C - 3D = 25$. Solving these equations simultaneously, you should get $C = -4$ and $D = -3$. We now know that $y_P = e^{3x}(-4\cos(2x) - 3\sin(2x))$, so the complete solution is

$$y = y_H + y_P = Ae^{2x} + Bxe^{2x} - e^{3x}(4\cos(2x) + 3\sin(2x)),$$

where A and B are constants.

Now it's time to finish off two of the examples from Section 30.4 above, as promised:

$$y' + 2y = 4x + \frac{1}{3}\sin(5x) \qquad \text{and} \qquad y'' - 5y' + 6y = 2x^2 e^x.$$

At this point, you should try to solve them both. Once you've done that, read on.

The left-hand example is a first-order equation. The homogeneous version is $y' + 2y = 0$, which has the solution $y = Ae^{-2x}$, where A is a constant. Upon consulting the above table, we see that the form for a particular solution is $y_P = ax + b + C\cos(5x) + D\sin(5x)$. We'll need the derivative, namely $y_P' = a - 5C\sin(5x) + 5D\cos(5x)$. Substituting y_P' and y_P into the original equation, we get

$$(a - 5C\sin(5x) + 5D\cos(5x)) + 2(ax + b + C\cos(5x) + D\sin(5x)) = 4x + \frac{1}{3}\sin(5x),$$

which reduces to

$$2ax + 2b + a + (5D + 2C)\cos(5x) + (2D - 5C)\sin(5x) = 4x + \frac{1}{3}\sin(5x).$$

Now we have to equate coefficients of various components of this expression. The coefficient of x is $2a$ on the left-hand side and 4 on the right-hand side, so $a = 2$. The constant coefficient on the left is $2b + a$, whereas there's no constant on the right, so $2b + a = 0$. This means that $b = -1$. Meanwhile, there's no term in $\cos(5x)$ on the right, so $5D + 2C = 0$. On the other hand, the $\sin(5x)$ terms must match, so we have $2D - 5C = 1/3$. Solving these last two equations simultaneously (try it!) gives $C = -5/87$ and $D = 2/87$. So, we have

$$y_P = 2x - 1 - \frac{5}{87}\cos(5x) + \frac{2}{87}\sin(5x);$$

putting it all together, we get the solution

$$y = y_H + y_P = Ae^{-2x} + 2x - 1 - \frac{5}{87}\cos(5x) + \frac{2}{87}\sin(5x),$$

where A is a constant.

How about the other example above? That's a second-order equation, with homogeneous version given by $y'' - 5y' + 6y = 0$. The characteristic quadratic equation is $t^2 - 5t + 6 = 0$, which has solutions $t = 2$ and $t = 3$. So, $y_H = Ae^{2x} + Be^{3x}$, where A and B are constants. Now it's time to deal with the particular solution. Since the right-hand side of the original differential equation is $2x^2 e^x$, the form should be $y_P = (ax^2 + bx + c)e^x$; remember that you don't need a constant outside of the e^x, since that constant could be absorbed into a, b and c. Let's differentiate y_P a couple of times:

$$
\begin{aligned}
y_P &= (ax^2 + bx + c)e^x, \\
y_P' &= (ax^2 + bx + c)e^x + (2ax + b)e^x \\
&= (ax^2 + (2a + b)x + (b + c))e^x, \\
y_P'' &= (ax^2 + (2a + b)x + (b + c))e^x + (2ax + (2a + b))e^x \\
&= (ax^2 + (4a + b)x + (2a + 2b + c))e^x.
\end{aligned}
$$

Now substitute into the original equation $y' - 5y' + 6y = 2x^2 e^x$ to get

$$(ax^2 + (4a+b)x + (2a+2b+c))e^x - 5(ax^2 + (2a+b)x + (b+c))e^x$$
$$+ 6(ax^2 + bx + c)e^x = 2x^2 e^x.$$

This simplifies to

$$(2ax^2 + (-6a+2b)x + (2a-3b+2c))e^x = 2x^2 e^x.$$

Now equate coefficients to see that $2a = 2$, $-6a + 2b = 0$ and $2a - 3b + 2c = 0$. This means that $a = 1$, $b = 3$ and $c = \frac{7}{2}$, so $y_P = (x^2 + 3x + \frac{7}{2})e^x$. The solution to the whole equation is therefore

$$y = y_H + y_P = Ae^{2x} + Be^{3x} + \left(x^2 + 3x + \frac{7}{2}\right)e^x,$$

where A and B are constants.

30.4.7 Resolving conflicts between y_P and y_H

The last line of the table in Section 30.4.5 above indicates that there might be conflicts between y_P and y_H. How can this happen? Well, consider the differential equation

$$y'' - 3y' + 2y = 7e^{2x}.$$

The homogeneous version is $y'' - 3y' + 2y = 0$, with characteristic quadratic equation given by $t^2 - 3t + 2 = (t-1)(t-2) = 0$, so the homogeneous solution is

$$y_H = Ae^x + Be^{2x}.$$

Here A and B are unknown constants. Now, since the right-hand side of the differential equation is $7e^{2x}$, our table says that the form for the particular solution is $y_P = Ce^{2x}$. The sad fact, alas, is that this choice will crash and burn. Indeed, this y_P is included in y_H by setting $A = 0$ and $B = C$. This means that if you plug $y_P = Ce^{2x}$ into the differential equation, you will get 0 on the left-hand side (try it!), so it doesn't work. Instead, as the final line of the table indicates, you need to introduce an extra power of x to make it work. So, we'll use $y_P = Cxe^{2x}$ instead. Let's see what happens now. First, note that $y_P' = 2Cxe^{2x} + Ce^{2x}$ and $y_P'' = 4Cxe^{2x} + 4Ce^{2x}$, so when you substitute into the differential equation above, you get

$$(4Cxe^{2x} + 4Ce^{2x}) - 3(2Cxe^{2x} + Ce^{2x}) + 2Cxe^{2x} = 7e^{2x}.$$

The terms in xe^{2x} cancel completely, and you're left with $Ce^{2x} = 7e^{2x}$. So $C = 7$, meaning that $y_P = 7xe^{2x}$. Finally, the complete solution is given by $y = y_H + y_P = Ae^x + Be^{2x} + 7xe^{2x}$.

One more example. If you want to solve

$$y'' + 6y' + 9y = e^{-3x},$$

you'll have to go even further than before. Now the homogeneous equation $y'' + 6y' + 9y = 0$ has characteristic quadratic $t^2 + 6t + 9 = (t+3)^2$, so the homogeneous solution is $y_H = Ae^{-3x} + Bxe^{-3x}$. Since the right-hand side of the differential equation is e^{-3x}, we'd want to take $y_P = Ce^{-3x}$. That won't work, since it's included in y_H (with $A = C$ and $B = 0$). Even $y_P = Cxe^{-3x}$ won't work, since that's also included in y_H (with $A = 0$ and $B = C$). So we have to go all the way up to x^2 and set $y_P = Cx^2e^{-3x}$. Now you can differentiate twice to see that $y'_P = 2Cx^{-3x} - 3Cx^2e^{-3x}$ and $y''_P = 2Ce^{-3x} - 12Cxe^{-3x} + 9Cx^2e^{-3x}$ (check this!). I leave it to you to plug these quantities into the original equation and show that it all simplifies to $2Ce^{-3x} = e^{-3x}$. This means that $C = \frac{1}{2}$, so the solution to the differential equation is $y = y_H + y_P = Ae^{-3x} + Bxe^{-3x} + \frac{1}{2}x^2e^{-3x}$ for some constants A and B.

30.4.8 Initial value problems (constant-coefficient linear)

Let's see how to deal with initial-value problems (IVPs) involving constant-coefficient linear differential equations. As usual, to solve an IVP, first solve the differential equation, then use the initial conditions to find the remaining unknown constants.

Let's modify the last two examples from Section 30.4.6 above to make them into IVPs, then solve them. For the first example, suppose you are given that $y' + 2y = 4x + \frac{1}{3}\sin(5x)$, and that $y(0) = -1$. Well, ignoring the condition $y(0) = -1$ for the moment, we already saw that the general solution is

$$y = Ae^{-2x} + 2x - 1 - \frac{5}{87}\cos(5x) + \frac{2}{87}\sin(5x).$$

Now we also know that $y(0) = -1$, which means that when $x = 0$, $y = -1$. Substituting this in, we get

$$-1 = Ae^0 + 2(0) - 1 - \frac{5}{87}\cos(0) + \frac{2}{87}\sin(0) = A - 1 - \frac{5}{87}.$$

This reduces to $A = 5/87$, so the solution to the IVP is

$$y = \frac{5}{87}e^{-2x} + 2x - 1 - \frac{5}{87}\cos(5x) + \frac{2}{87}\sin(5x).$$

There are no unknown constants.

To modify the second example, let's suppose that $y'' - 5y' + 6y = 2x^2e^x$ and that $y(0) = y'(0) = 0$. As we saw in Section 30.4.6, the general solution (ignoring the initial conditions $y(0) = 0$ and $y'(0) = 0$) is given by

$$y = Ae^{2x} + Be^{3x} + \left(x^2 + 3x + \frac{7}{2}\right)e^x.$$

We'll need to differentiate this once to take advantage of the fact that we know what $y'(0)$ is; check that

$$y' = 2Ae^{2x} + 3Be^{3x} + \left(x^2 + 5x + \frac{13}{2}\right)e^x.$$

So, when $x = 0$, we know that both y and y' are equal to 0; substituting into the equation for y gives

$$0 = Ae^0 + Be^0 + \left(0^2 + 3(0) + \frac{7}{2}\right)e^0 = A + B + \frac{7}{2},$$

whereas substituting into the equation for y' gives

$$0 = 2Ae^0 + 3Be^0 + \left(0^2 + 5(0) + \frac{13}{2}\right)e^0 = 2A + 3B + \frac{13}{2}.$$

Solving these equations simultaneously, we get $A = -4$ and $B = \frac{1}{2}$. This means that the solution to the IVP is

$$y = -4e^{2x} + \frac{1}{2}e^{3x} + \left(x^2 + 3x + \frac{7}{2}\right)e^x.$$

Notice that in both examples there are no constants left: the initial conditions have allowed us to home in on the unique solution. Without initial conditions, there will always be one or two unknown constants.

Let's look at one last IVP example. Suppose that

$$y'' + 6y' + 13y = 26x^3 - 3x^2 - 24x, \qquad y(0) = 1, \quad y'(0) = 2.$$

The homogeneous equation is $y'' + 6y' + 13y = 0$, with characteristic quadratic equation $t^2 + 6t + 13 = 0$. Using the quadratic formula, the solutions to this last equation are $t = (-6 \pm \sqrt{36 - 4 \cdot 13})/2 = -3 \pm 2i$. This means that $y_H = e^{-3x}(A\cos(2x) + B\sin(2x))$. Turning now to the particular solution: since the right-hand side (the x-stuff) of the original equation is a cubic, we should write down the form $y_P = ax^3 + bx^2 + cx + d$. Now we have to find the constants a through d by substituting y_P into the differential equation. Note that $y'_P = 3ax^2 + 2bx + c$ and $y''_P = 6ax + 2b$. Substituting, we get

$$(6ax + 2b) + 6(3ax^2 + 2bx + c) + 13(ax^3 + bx^2 + cx + d) = 26x^3 - 3x^2 - 24x.$$

Equating coefficients (just as we did for partial fractions) for x^3, x^2, x, and 1, we get $13a = 26$, $18a + 13b = -3$, $6a + 12b + 13c = -24$, and $2b + 6c + 13d = 0$, respectively. I leave it to you to solve these equations and see that $a = 2$, $b = -3$, $c = 0$, and $d = 6/13$. So $y_P = 2x^3 - 3x^2 + 6/13$, and therefore

$$y = y_H + y_P = e^{-3x}(A\cos(2x) + B\sin(2x)) + 2x^3 - 3x^2 + \frac{6}{13}$$

for some constants A and B. Now, to find these constants, let's use the initial conditions. Since $y(0) = 1$, we know that $y = 1$ when $x = 0$; substituting, we have

$$1 = e^{-3(0)}(A\cos(0) + B\sin(0)) + 2(0)^3 - 3(0)^2 + \frac{6}{13} = A + \frac{6}{13},$$

so $A = 7/13$. Meanwhile, differentiating the expression for y gives

$$y' = e^{-3x}(-2A\sin(2x) + 2B\cos(2x)) - 3e^{-3x}(A\cos(2x) + B\sin(2x)) + 6x^2 - 6x.$$

Now, since $y'(0) = 2$, we know that $y' = 2$ when $x = 0$; substituting this into the above expression for y', we get

$$\begin{aligned} 2 &= e^0(-2A\sin(0) + 2B\cos(0)) - 3e^0(A\cos(0) + B\sin(0)) + 6(0)^2 - 6(0) \\ &= 2B - 3A. \end{aligned}$$

Since $A = 7/13$, we can solve this last equation to find that $B = 47/26$. Now we plug these values in to find the final answer:

$$y = e^{-3x} \left(\frac{7}{13} \cos(2x) + \frac{47}{26} \sin(2x) \right) + 2x^3 - 3x^2 + \frac{6}{13}.$$

Once again, note that there are no constants involved here: the initial conditions (that is, the values of $y(0)$ and $y'(0)$) pinpoint the explicit solution.

30.5 Modeling Using Differential Equations

Many quantities in the real world can be modeled (that is, theoretically approximated) by differential equations. Examples include heat flow, wave height, inflation, current in electrical circuits, and population growth, to name a few. Here's a simple example of a somewhat realistic situation involving population growth.

A certain culture of bacteria grows exponentially in such a way that its instantaneous hourly rate of increase is equal to twice the number of bacteria in the culture. Suppose that an antibiotic is continuously introduced into the culture at the constant rate of 8 ounces per hour. Each ounce of antibiotic present kills 25,000 of the bacteria per hour. What is the minimum initial population of bacteria that need to be present in order to ensure that the culture is never completely wiped out?

The idea here is that the number of bacteria is increasing as they breed, but the amount of killer antibiotic is increasing too as it gets pumped into the petri dish. Which one wins, the bacteria or the antibiotic? To find out, we need to write down a differential equation that models the situation. In effect, we have to translate the word problem into a differential equation. If there were no antibiotic, you'd have the standard population growth differential equation with $k = 2$:

$$\frac{dP}{dt} = 2P,$$

where P is the population at time t hours. (We looked at this sort of thing in Section 9.6.1 of Chapter 9.) Now we have to modify this to take the antibiotic into account. At time t hours, we know there are $8t$ ounces of antibiotic present, so the death rate due to this amount present is $8t \times 25000 = 200000t$. The correct differential equation is therefore

$$\frac{dP}{dt} = 2P - 200000t.$$

This can be rearranged into standard form as

$$\frac{dP}{dt} - 2P = -200000t.$$

The integrating factor (see Section 30.3 above) for this first-order linear equation is $e^{\int -2\,dt}$, which simplifies to e^{-2t}. Multiplying the equation by the integrating factor, we get

$$e^{-2t} \frac{dP}{dt} - 2e^{-2t} P = -200000e^{-2t}t.$$

As usual, the left-hand side simplifies to the derivative of P times the integrating factor:

$$\frac{d}{dt}(e^{-2t}P) = -200000e^{-2t}t,$$

or just

$$e^{-2t}P = -200000\int e^{-2t}t\,dt.$$

The right-hand side needs to be integrated by parts (see Section 18.2 in Chapter 18); I leave it to you to show that

$$e^{-2t}P = 100000te^{-2t} + 50000e^{-2t} + 200000C.$$

Now we can replace the arbitrary constant $200000C$ by the equally arbitrary constant C and multiply through by e^{2t} to get

$$P = 100000t + 50000 + Ce^{2t}.$$

This is the equation for the population of bacteria at time t. If the initial population is P_0, then we can set $t = 0$ in the equation to get

$$P_0 = 100000(0) + 50000 + Ce^{2(0)} = 50000 + C.$$

This means that $C = P_0 - 50000$, so we can insert that into the equation and get

$$P = 100000t + 50000 + (P_0 - 50000)e^{2t}.$$

Great! So we know a lot about the situation. We still have to answer the question. For what values of P_0 will we ever get down to a population of 0? It seems that 50000 is a pretty critical number. Indeed, if $P_0 = 50000$, the above equation is just $P = 100000t + 50000$; in that case, the bacteria start at 50000 and grow at a constant rate of 100000 per hour, so the population never dies. If $P_0 > 50000$, then you add a positive multiple of e^{2t} to this and so the population grows even faster. How about if $P_0 < 50000$? Then the quantity $P_0 - 50000$ is negative, so we have

$$P = 100000t + 50000 + (\text{negative constant})e^{2t}.$$

Since exponentials eventually dominate, it's a sure thing that if t is large enough, P will eventually get down to 0. For example, even if the initial population is 49999, then we have

$$P = 100000t + 50000 - e^{2t}.$$

Here's the graph of P versus t in this case:

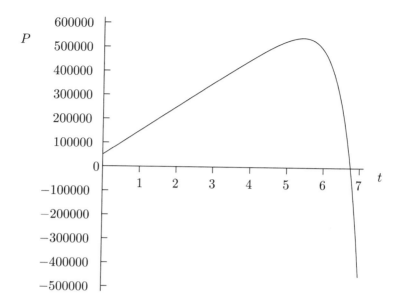

You can see from the graph that the population grows almost linearly for the first 5 hours, then there is a rapid turnaround, and finally the population hits 0 some time between 6.5 and 7 hours. (Of course, once it hits 0, that's the end of the story—the population never goes below 0, since you can't have a negative population! So the above graph doesn't accurately reflect the situation when $P < 0$.) In general, we conclude that if the initial population is under 50000, the bacteria will die out, whereas if it is 50000 or more, the culture will survive; in fact, it will always grow in that case.

APPENDIX A _____

Limits and Proofs

Throughout this book we have used limits extensively, in their own right and also as an essential part of the definitions of the derivative and the integral. Since limits are so important, it's about time that we define them properly. Once we know how they work, we can prove a number of facts that we've been taking for granted. So, here's what's in this appendix:

- the formal definition of a limit (including left-hand and right-hand limits, infinite limits, limits at $\pm\infty$, and limits of sequences);
- combining limits, and a proof of the sandwich principle;
- the relationship between continuity and limits, including a proof of the Intermediate Value Theorem;
- differentiation and limits, including proofs of the product, quotient, and chain rules;
- a proof of a result concerning piecewise-defined functions and derivatives;
- a proof of the existence of e;
- proofs of the Extreme Value Theorem, Rolle's Theorem, the Mean Value Theorem (for derivatives), the formula for the error term in linearization, and l'Hôpital's Rule; and
- a proof of the Taylor approximation theorem.

A.1 Formal Definition of a Limit

We start with a function f and a real number a. In Section 3.1 of Chapter 3 we introduced the notation

$$\lim_{x \to a} f(x) = L,$$

which is used throughout this book. Intuitively, the above equation means that when x is close to a, the values of $f(x)$ get very very close to L. How close? As close as you want them to be. To see what this means, let's play a little game, you and I.

A.1.1 A little game

Here's how the game works. Your move consists of picking an interval on the y-axis with L in the middle. You get to draw lines parallel to the x-axis through the endpoints of your interval. Here's an example of what your move might be:

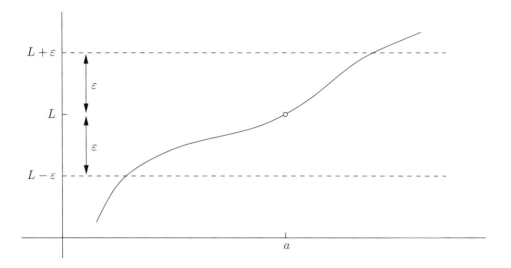

Notice that I labeled the endpoints of the interval $L - \varepsilon$ and $L + \varepsilon$. So both endpoints are a distance ε away from L.

Anyway, the point is, you can't tolerate any bit of the function being outside those two horizontal lines. My move, then, is to throw away some of the function by restricting the domain. I just have to make sure that the new domain is an interval with a at the center, and that every bit of the function remaining lies between your lines, except possibly at $x = a$ itself. Here's one way I could make my move, based on the move you just made:

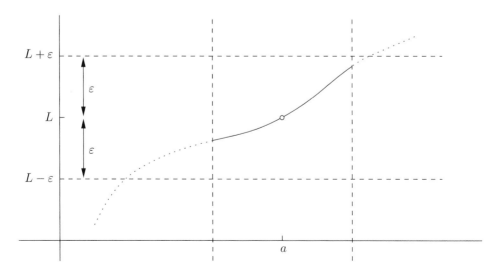

I could have taken away more and it would still have been fine—as long as what's left is between your lines.

Now it's your move again. You have realized that my task is harder when your lines are closer together, so this time you pick a smaller value of ε. Here's the situation after your second move:

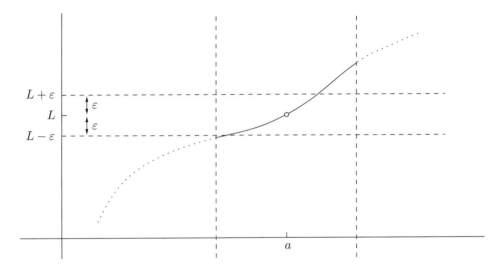

Parts of the curve are outside the horizontal lines again, but I haven't had my second move yet. I'm going to throw away more of the function away from $x = a$, like this:

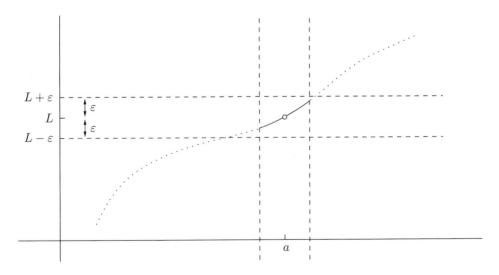

So once again I was able to make a move to counter your move.

When does the game stop? Hopefully, the answer is never! If I can always move, no matter how close together you make the lines, then it will indeed be true that $\lim_{x \to a} f(x) = L$. We will have zoomed in and in, you pushing your lines closer together, I responding by focusing only on the part of the function close enough to $x = a$. On the other hand, if I ever get stuck for a move, then it's not true that $\lim_{x \to a} f(x) = L$. The limit might be something else, or it may not exist, but it's definitely not L.

A.1.2 The actual definition

We need to turn the game into some more symbols. First, notice the interval you choose is $(L - \varepsilon, L + \varepsilon)$. In fact, you can also think of that interval as the set of points y satisfying $|y - L| < \varepsilon$. Why? Because $|y - L|$ is the distance between y and L on a number line (such as the y-axis). So your interval consists of all the points which are less than ε away from L. As you might guess, it will be incredibly useful for you to be able to convert an inequality like $|y - L| < \varepsilon$ into its equivalent form $L - \varepsilon < y < L + \varepsilon$ and back again.

Now it's my move. I always need the function to lie in your interval. This means that after I've thrown away a lot of the domain, all the remaining values of $f(x)$ must be less than ε away from L. So after my move, you'll have to conclude that

$$|f(x) - L| < \varepsilon \quad \text{for all } x \neq a \text{ which are close enough to } a.$$

To be more precise about my move, notice that I am throwing away everything except in an interval centered at a. My interval looks like $(a - \delta, a + \delta)$ for some other number δ, so I can also think of it as all the numbers x such that $|x - a| < \delta$. In fact, since I don't want x to be equal to a, I can write $0 < |x - a| < \delta$.

In summary, then, your move consists of picking $\varepsilon > 0$. (It had better be positive or else there's no tolerance window at all!) My move consists of picking a number $\delta > 0$ such that

$$|f(x) - L| < \varepsilon \quad \text{for all } x \text{ satisfying} \quad 0 < |x - a| < \delta.$$

Remember, this means that whenever x is no more than a distance δ away from a (except for a itself), the value of $f(x)$ is no more than a distance of ε away from L. This quantifies the idea that $f(x)$ is close to L when x is close to a. Now all that's left is to allow you to make your choice of ε as small as you like, and I still have to pick the δ accordingly. So, here's the formal definition we're looking for:

> "$\lim_{x \to a} f(x) = L$" means that for any choice of $\varepsilon > 0$ you make, I can pick $\delta > 0$ such that:
> $|f(x) - L| < \varepsilon$ for all x satisfying $0 < |x - a| < \delta$.

It's very important that I get to move after you do! My choice of δ depends on your choice of ε. Normally I can't make a universal choice of δ that works for every $\varepsilon > 0$. I just have to adapt to your choice.

A.1.3 Examples of using the definition

As a simple example, let's show, without using continuity, that

$$\lim_{x \to 3} x^2 = 9.$$

Tempting as it is to write $3^2 = 9$ and declare victory, that doesn't work because the limit only depends on what happens when x is near, but not equal to, 3. So, we have to play our little game. You choose your $\varepsilon > 0$, which makes a little window $(9 - \varepsilon, 9 + \varepsilon)$ that I have to stay within. Now I get to pick my δ. Suppose that your ε is 8, which is humongous in this context. Then your window is $(1, 17)$. Well, I can easily stay in there by choosing my $\delta = 1$, which means that my window is $(2, 4)$. (Remember, my window is centered at 3, while yours is centered at 9.) Indeed, if you square any number between 2 and 4, you get a number between 4 and 16, so my move is fine. If your ε is even bigger than 8, well, that just widens your interval, but I'll stick with my $\delta = 1$ and be just fine.

Now, if you choose your tolerance ε less than 8, I have to change my tactic. My choice in this case will be ... drumroll ... $\delta = \varepsilon/8$. That is, I'm making my window eight times smaller than yours, no matter how wide you choose it. To see that this works, we have to be clever. Basically, we have to take any number in my interval, square it, and show that it lies in your interval. My interval is $(3 - \varepsilon/8, 3 + \varepsilon/8)$ and yours is $(9 - \varepsilon, 9 + \varepsilon)$.

So let's pick x in my interval. How big could it be? It's got to be less than $3 + \varepsilon/8$. That is, $x < 3 + \varepsilon/8$, which you can also write as $x - 3 < \varepsilon/8$. By the way, since your ε is less than 8, my x is less than 4. So, using both inequalities $x - 3 < \varepsilon/8$ and $x < 4$, we get

$$(x - 3)(x + 3) < \left(\frac{\varepsilon}{8}\right)(4 + 3) = \frac{7\varepsilon}{8}.$$

Since $(x - 3)(x + 3)$ is just $x^2 - 9$, we can add 9 to both sides of the equation and see that

$$x^2 < 9 + \frac{7\varepsilon}{8}.$$

So we're OK on the upper tolerance level (the upper of your two lines). We needed $x^2 < 9 + \varepsilon$, and we have done that. How about the lower one? Well, how small could my x be, given that it lies in my interval $(3 - \varepsilon/8, 3 + \varepsilon/8)$? It's got to be bigger than $3 - \varepsilon/8$, so we have $x > 3 - \varepsilon/8$. This means that $x - 3 > -\varepsilon/8$. Since your ε is less than 8, we also have $x - 3 > -8/8 = -1$, which means that $x > 2$. Again, using both inequalities $x - 3 > -\varepsilon/8$ and $x > 2$, we get

$$(x - 3)(x + 3) > \left(-\frac{\varepsilon}{8}\right)(2 + 3) = -\frac{5\varepsilon}{8}.$$

Once again, $(x - 3)(x + 3) = x^2 - 9$, so we add 9 to both sides and get

$$x^2 > 9 - \frac{5\varepsilon}{8}.$$

This takes care of the lower tolerance level! We have shown that if x lies in the interval $(3 - \varepsilon/8, 3 + \varepsilon/8)$, then x^2 is in the interval $(9 - 5\varepsilon/8, 9 + 7\varepsilon/8)$. Since both 5/8 and 7/8 are less than 1, we can also confidently say that x^2 is in the interval $(9 - \varepsilon, 9 + \varepsilon)$; after all, this interval contains the other one.

Tying it all together, let's set $f(x) = x^2$, and we'll justify the equation

$$\lim_{x \to 3} f(x) = 9.$$

You choose ε, and I respond by picking $\delta = \varepsilon/8$ unless your ε is 8 or more, in which case I just pick $\delta = 1$. We have shown that in either case, if x is in the interval $(3 - \delta, 3 + \delta)$, then $f(x)$ is in the interval $(9 - \varepsilon, 9 + \varepsilon)$. In other words, whenever $|x - 3| < \delta$, then $|f(x) - 9| < \varepsilon$. We can also exclude $x = 3$ if we like and say that if $0 < |x - 3| < \delta$, then $|f(x) - 9| < \varepsilon$. This is exactly what we need—we have justified our equation. Believe it or not, that's pretty much what you have to do if you want to prove that the above limit is true by using the definition!

A.2   Making New Limits from Old Ones

That last example was pretty annoying. Just to show that $x^2 \to 9$ as $x \to 3$, we had to do a lot of work. Luckily it turns out that once you know a couple of limits, you can put them together and get a whole bunch of new ones. For example, you can add, subtract, multiply, and divide limits within reason, and there's also the sandwich principle. Let's see why all this is true.

A.2.1   Sums and differences of limits—proofs

Suppose that we have two functions f and g, and we know that as $x \to a$, we have $f(x) \to L$ and $g(x) \to M$. What should happen to $f(x) + g(x)$ as $x \to a$? Intuitively, it should tend to $L + M$. Let's prove this using the definition. So, we know that

$$\lim_{x \to a} f(x) = L \quad \text{and} \quad \lim_{x \to a} g(x) = M.$$

This means that if you pick $\varepsilon > 0$, I can ensure that $|f(x) - L| < \varepsilon$ by restricting x close enough to a. I can also ensure that $|g(x) - M| < \varepsilon$ if x is close enough to a. The degrees of closeness that I need might be different for f and g, but it doesn't matter—I can just go close enough so that **both** inequalities work.

Now, if $f(x) + g(x)$ is close to $L + M$, this means that the difference between these things should be small. So we'll need to worry about the quantity $|(f(x) + g(x)) - (L + M)|$. We'll write this as $|(f(x) - L) + (g(x) - M)|$. We can then use the so-called *triangle inequality*, which says* that $|a + b| \le |a| + |b|$ for any numbers a and b, as follows:

$$|(f(x) - L) + (g(x) - M)| \le |f(x) - L| + |g(x) - M| < \varepsilon + \varepsilon = 2\varepsilon,$$

provided that x is close enough to a. This is almost good enough, except that you wanted a tolerance of ε, not 2ε! So I have to make my move again

---

*Since we're proving stuff, here's a proof of the triangle inequality. We start off with the observation that $x \le |x|$ for any number x. Indeed, if x is positive or 0, then $x = |x|$; otherwise, the left-hand side is negative and the right-hand side is positive. Replace x by ab to get $ab \le |ab| = |a| \cdot |b|$. Now multiply this by 2 and add $a^2 + b^2$ to both sides. We get $a^2 + b^2 + 2ab \le a^2 + b^2 + 2|a| \cdot |b|$. The left-hand side is just $(a + b)^2$. Since $x^2 = |x|^2$ for any x, we can replace the left-hand side by $|a + b|^2$. Similarly, on the right we have $|a|^2 + |b|^2 + 2|a| \cdot |b|$, or just $(|a| + |b|)^2$. Altogether then, our inequality is $|a + b|^2 \le (|a| + |b|)^2$. Now we just have to take square roots and we're done, since $|a + b|$ and $|a| + |b|$ are both nonnegative.

(sorry about that); this time I'll narrow my focus so that both $|f(x) - L|$ and $|g(x) - M|$ are less than $\varepsilon/2$ instead of ε. This is no problem, since I can deal with any positive number that you pick. Anyway, if you redo the above equation, you'll get ε on the right instead of $\varepsilon/2$, so we have proven that I can find a little window about $x = a$ such that

$$|(f(x) + g(x)) - (L + M)| < \varepsilon$$

whenever x is in my window. (You can use δ if you like to describe the window better, but that doesn't really get us anything extra.) So this proves the following:

if $\quad \lim_{x \to a} f(x) = L \quad$ and $\quad \lim_{x \to a} g(x) = M, \quad$ then $\quad \lim_{x \to a} (f(x) + g(x)) = L + M.$

That is, the limit of the sum is the sum of the limits. Another way of writing this is

$$\lim_{x \to a} (f(x) + g(x)) = \lim_{x \to a} f(x) + \lim_{x \to a} g(x),$$

but here you have to be careful to check that both limits on the right exist and are finite. If either limit is $\pm\infty$ or doesn't exist, the deal's off. Both limits have to be finite to guarantee that you can add them up. You might get lucky if they're not, but there's no guarantee.

How about $f(x) - g(x)$? That should go to $L - M$, and it does:

if $\quad \lim_{x \to a} f(x) = L \quad$ and $\quad \lim_{x \to a} g(x) = M, \quad$ then $\quad \lim_{x \to a} (f(x) - g(x)) = L - M.$

The proof is almost identical to the one we just looked at, except that you need a slightly different form of the triangle inequality: $|a - b| \le |a| + |b|$. Actually, this is just the triangle inequality applied to a and $-b$; indeed, $|a + (-b)| \le |a| + |-b|$, but of course $|-b|$ is equal to $|b|$. I leave it to you to rewrite the above argument but change the plus signs between $f(x)$ and $g(x)$, and between L and M, into minus signs.

A.2.2 Products of limits—proof

Now we once again assume that we have two functions f and g such that

$$\lim_{x \to a} f(x) = L \qquad \text{and} \qquad \lim_{x \to a} g(x) = M.$$

We want to show that

$$\lim_{x \to a} f(x)g(x) = LM.$$

That is, the limit of the product is the product of the limits. Another way of writing this is

$$\lim_{x \to a} f(x)g(x) = \lim_{x \to a} f(x) \times \lim_{x \to a} g(x),$$

again with the understanding that both limits on the right-hand side are already known to exist and be finite. To prove this, we need to show that the difference between $f(x)g(x)$ and the (hopeful) limit LM is small. Let's consider that difference $f(x)g(x) - LM$. The trick is to subtract $Lg(x)$ and add it back on again! That is,

$$f(x)g(x) - LM = f(x)g(x) - Lg(x) + Lg(x) - LM.$$

What does that get us? Let's take absolute values, then use the triangle inequality:

$$|f(x)g(x) - LM| = |(f(x) - L)g(x) + L(g(x) - M)|$$
$$\leq |(f(x) - L)g(x)| + |L(g(x) - M)|.$$

We can tidy this up a little and write

$$|f(x)g(x) - LM| \leq |f(x) - L| \cdot |g(x)| + |L| \cdot |g(x) - M|.$$

Now it's time to play the game. You pick your positive number ε and then I get to work. I concentrate on an interval around $x = a$ so small that $|f(x) - L| < \varepsilon$ and $|g(x) - M| < \varepsilon$. In fact, if you pick $\varepsilon \geq 1$ (a pretty feeble move, if you ask me—you want ε to be small!) then I'm even going to insist that $|g(x) - M| < 1$ in that case. So we know in either case that $|g(x) - M| < 1$, which means that $M - 1 < g(x) < M + 1$ on my interval. In particular, we can see that $|g(x)| < |M| + 1$. The whole point is that we have some nice inequalities on my interval:

$$|f(x) - L| < \varepsilon, \qquad |g(x)| < |M| + 1, \qquad \text{and} \qquad |g(x) - M| < \varepsilon.$$

We can insert these into the inequality for $|f(x)g(x) - LM|$ above:

$$|f(x)g(x) - LM| \leq |f(x) - L| \cdot |g(x)| + |L| \cdot |g(x) - M|$$
$$< \varepsilon \cdot (|M| + 1) + |L| \cdot \varepsilon = \varepsilon(|M| + |L| + 1)$$

for x close enough to a. That's almost what I want! I was supposed to get ε on the right-hand side, but I got an extra factor of $(|M| + |L| + 1)$. This is no problem—you just have to allow me to make my move again, but this time I'll make sure that $|f(x) - L|$ is no more than $\varepsilon/(|M| + |L| + 1)$ and similarly for $|g(x) - M|$. Then when I replay all the steps, ε will be replaced by $\varepsilon/(|M| + |L| + 1)$, and at the very last step, the factor $(|M| + |L| + 1)$ will cancel out and we'll just get our ε! So we have proved the result.

By the way, it's worth noting a special case of the above. If c is constant, then

$$\lim_{x \to a} cf(x) = c \lim_{x \to a} f(x).$$

This is easy to see by setting $g(x) = c$ in our main formula above; I leave the details to you.

A.2.3   Quotients of limits—proof

Now we repeat our exercise. We want to show that if

$$\lim_{x \to a} f(x) = L \qquad \text{and} \qquad \lim_{x \to a} g(x) = M,$$

then we have

$$\lim_{x \to a} \frac{f(x)}{g(x)} = \frac{L}{M}.$$

So the limit of the quotient is the quotient of the limits. For this to work, we'd better have $M \neq 0$ or else we'll be dividing by 0. Another way of writing the above equation is

$$\lim_{x \to a} \frac{f(x)}{g(x)} = \frac{\lim_{x \to a} f(x)}{\lim_{x \to a} g(x)},$$

provided that both limits exist and are finite, and that the g-limit is nonzero.

Here's how the proof goes. We want $f(x)/g(x)$ to be close to L/M, so we consider the difference. Then we'll need to take a common denominator, leaving us with

$$\frac{f(x)}{g(x)} - \frac{L}{M} = \frac{Mf(x) - Lg(x)}{Mg(x)}.$$

Now we do a trick similar to the one we used in for products of limits: we'll subtract and add LM to the numerator, then factor. This gives us

$$\begin{aligned}\frac{f(x)}{g(x)} - \frac{L}{M} &= \frac{Mf(x) - LM + LM - Lg(x)}{Mg(x)} \\ &= \frac{M(f(x) - L)}{Mg(x)} + \frac{L(M - g(x))}{Mg(x)} \\ &= \frac{f(x) - L}{g(x)} - \frac{L(g(x) - M)}{Mg(x)}.\end{aligned}$$

If we take absolute values and then use the triangle inequality in the form $|a - b| \le |a| + |b|$, we get

$$\left|\frac{f(x)}{g(x)} - \frac{L}{M}\right| = \left|\frac{f(x) - L}{g(x)} - \frac{L(g(x) - M)}{Mg(x)}\right| \le \left|\frac{f(x) - L}{g(x)}\right| + \left|\frac{L(g(x) - M)}{Mg(x)}\right|.$$

So you make your move by picking $\varepsilon > 0$, and then I narrow the window of interest around $x = a$ so that $|f(x) - L| < \varepsilon$ and $|g(x) - M| < \varepsilon$ in the little window. Now I need to be even trickier, though. You see, I know that $M - \varepsilon < g(x) < M + \varepsilon$, which means that $|g(x)| > |M| - \varepsilon$. All's well if this right-hand quantity $|M| - \varepsilon$ is positive, but if it's negative, it tells us nothing since we already knew that $|g(x)|$ can't be negative. So if your ε is small enough, then I don't worry, but if it's a little bigger, I need to narrow my window more so that $|g(x)| > |M|/2$ on the window. So altogether we have three inequalities which are true on the little interval:

$$|f(x) - L| < \varepsilon, \qquad |g(x)| > \frac{|M|}{2}, \qquad \text{and} \qquad |g(x) - M| < \varepsilon.$$

This middle inequality can be inverted to read

$$\frac{1}{|g(x)|} < \frac{2}{|M|}.$$

Putting everything together, we have

$$\left|\frac{f(x)}{g(x)} - \frac{L}{M}\right| \le \frac{|f(x) - L|}{|g(x)|} + \frac{|L| \cdot |g(x) - M|}{|M||g(x)|} < \varepsilon \cdot \frac{2}{|M|} + \varepsilon \cdot \frac{|L|}{|M|} \cdot \frac{2}{|M|}.$$

Not quite what we wanted—we have an extra factor of $(2/|M| + 2|L|/|M|^2)$, but we know how to handle this—I just make my move again, but instead of your ε, I use ε divided by this extra factor.

A.2.4 The sandwich principle—proof

In Section 3.6, we looked at the sandwich principle. Now it's time to prove it. We start with functions f, g, and h, such that $g(x) \le f(x) \le h(x)$ for all x close enough to a. We also know that

$$\lim_{x \to a} g(x) = L \qquad \text{and} \qquad \lim_{x \to a} h(x) = L.$$

Intuitively, f is squished between g and h more and more, so that in the limit as $x \to a$, we should have $f(x) \to L$ as well. That is, we need to prove that

$$\lim_{x \to a} f(x) = L.$$

Well, you start off by picking your positive number ε, and then I can focus on an interval centered at a small enough so that $|g(x) - L| < \varepsilon$ and $|h(x) - L| < \varepsilon$ on the interval. I'm also going to need the inequality $g(x) \le f(x) \le h(x)$ to be true on the interval; since that inequality might only be true when x is very near to a, I may have to shrink my original interval.

Anyway, we know that $|h(x) - L| < \varepsilon$ when x is close enough to a; the inequality can be rewritten as

$$L - \varepsilon < h(x) < L + \varepsilon.$$

Actually, we only need the right-hand inequality, $h(x) < L + \varepsilon$; you see, on my little interval, we know that $f(x) \le h(x)$, so we also have

$$f(x) \le h(x) < L + \varepsilon.$$

Similarly, we know that

$$L - \varepsilon < g(x) < L + \varepsilon$$

when x is close enough to a; this time we throw away the right-hand inequality and use $g(x) \le f(x)$ to get

$$L - \varepsilon < g(x) \le f(x).$$

Putting all this together, we have shown that when x is close to a, we have

$$L - \varepsilon < f(x) < L + \varepsilon,$$

or simply $|f(x) - L| < \varepsilon$. That's what we need to show our limit—we've proved the sandwich principle!

A.3 Other Varieties of Limits

Now let's quickly look at the definitions of some other types of limits: infinite limits, left-hand and right-hand limits, and limits at $\pm\infty$.

A.3.1 Infinite limits

Our game isn't going to work if we want to use it to define a limit like this:

$$\lim_{x \to a} f(x) = \infty.$$

When you try to draw your two lines close to the limit, you'll be completely stuck, since the limit is supposed to be ∞ instead of some finite value L. So we have to modify the rules a little bit. My move won't change much, but yours will. Instead of picking a little number ε and then drawing two horizontal lines (at height $L - \varepsilon$ and $L + \varepsilon$), this time you'll pick a large number M and only draw in the line at height M. I still make my move by throwing away most of the function, except for a small bit around $x = a$; this time, though, I have to make sure that what's left is always **above** your line. For example, the following pictures show a move you might make and then a possible response for me:

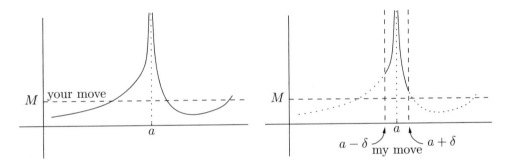

Now here's what happens if you make another move but with a larger value of M:

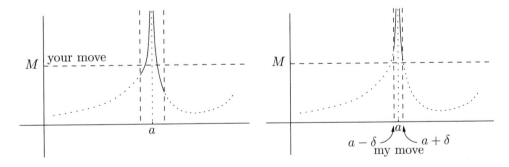

So the idea is that this time you raise your bar higher and higher; if I can always make a move in response, then the limit is indeed ∞. In symbols, I need to be able to ensure that $f(x) > M$ whenever x is close enough to a, no matter how big M is. The definition looks like this:

> "$\lim_{x \to a} f(x) = \infty$" means that for any choice of $M > 0$ you make, I can pick $\delta > 0$ such that:
>
> $f(x) > M$ for all x satisfying $0 < |x - a| < \delta$.

It's very similar to the situation when the limit is some finite number L, except that the inequality $|f(x) - L| < \varepsilon$ is replaced by $f(x) > M$.

For example, suppose that we want to show that

$$\lim_{x \to 0} \frac{1}{x^2} = \infty.$$

You start off by picking your number M; then I have to make sure that $f(x) > M$ when x is close enough to 0. Well, suppose that I throw everything away except for x satisfying $|x| < 1/\sqrt{M}$. For such an x, we have $x^2 < 1/M$, so $1/x^2 > M$ (note that we have assumed that $x \neq 0$). That means that $f(x) > M$ in my interval, which means my move is valid. So for any M you pick, I can make a valid move, and we have proved that the limit is indeed ∞.

How about $-\infty$? Everything is just reversed. You still pick a large positive number M, but this time I need to make my move so that the function is always below the horizontal line of height $-M$. So here's what the definition looks like:

> "$\lim_{x \to a} f(x) = -\infty$" means that for any choice of $M > 0$ you make, I can pick $\delta > 0$ such that:
> $$f(x) < -M \text{ for all } x \text{ satisfying } 0 < |x - a| < \delta.$$

A.3.2 Left-hand and right-hand limits

To define a right-hand limit, we play the same game, except this time before we start, we already throw away everything to the left of $x = a$. The effect is that instead of choosing an interval like $(a - \delta, a + \delta)$ when I make my move, now I just have to worry about $(a, a + \delta)$. Nothing to the left of a is relevant.

Similarly, for a left-hand limit, only the values of x to the left of a matter. This means that my intervals look like $(a - \delta, a)$; I have thrown away everything to the right of $x = a$.

This all means that you can take any of the above definitions in boxes and change the inequality $0 < |x - a| < \delta$ to $0 < x - a < \delta$ to get the right-hand limit. To get the left-hand limit, you change the inequality to $0 < a - x < \delta$ instead. I'll spare you the gory details of writing out all six versions (that's each of the limits with values L, ∞, and $-\infty$ in both left-hand and right-hand versions) but it's not a bad exercise for you to try to do it without looking at these pages.

A.3.3 Limits at ∞ and $-\infty$

Our final variety of limit occurs when the limit is taken at ∞ or $-\infty$ instead of at some finite value a. So we want to define what the following equation means:

$$\lim_{x \to \infty} f(x) = L.$$

The game has to change a little, of course, but we already know how. In fact we just have to adapt the methods from Section A.3.1 above. You'll start by picking your little number $\varepsilon > 0$, establishing your tolerance interval

$(L - \varepsilon, L + \varepsilon)$; then my move will be to throw away the function to the left of some vertical line $x = N$, so that all the function values to the right of the line lie in your tolerance interval. Then you pick a smaller ε, and I move the line rightward if I have to in order to lie within your new, smaller interval. Here's what the first couple of moves for both of us might look like:

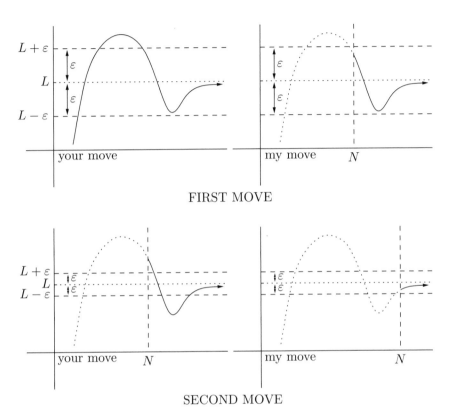

FIRST MOVE

SECOND MOVE

After your first move, my move ensures that all the function values to the right of the line $x = N$ lie in your tolerance interval. You respond by closing in the interval, but then I just move the line to the right until I can meet your new, more restrictive tolerance interval. Again, if I can always make a move in response to you, then the above limit is true.

More formally, my move consists of picking N such that $f(x)$ is in the interval $(L - \varepsilon, L + \varepsilon)$ whenever $x > N$ (so x is to the right of the vertical line $x = N$). Using absolute values, we can write this as follows:

> "$\lim_{x \to \infty} f(x) = L$" means that for any choice of $\varepsilon > 0$ you make, I can pick N such that:
> $$|f(x) - L| < \varepsilon \text{ for all } x \text{ satisfying } x > N.$$

It's worth noting that any limit as $x \to \infty$ is necessarily a left-hand limit—there's nothing to the right of ∞! Anyway, there are still a couple of variations to look at. First, what does $\lim_{x \to \infty} f(x) = \infty$ mean? You just have to adapt the previous definitions. In particular, you can take the above definition and change your move to picking $M > 0$, and now instead of requiring that $|f(x) - L| < \varepsilon$, this changes to $f(x) > M$. If instead you would like to show that $\lim_{x \to \infty} f(x) = -\infty$, you would change the inequality to $f(x) < -M$. Pretty straightforward.

It's also pretty easy to define what

$$\lim_{x \to -\infty} f(x) = L, \qquad \lim_{x \to -\infty} f(x) = \infty, \qquad \text{and} \qquad \lim_{x \to -\infty} f(x) = -\infty$$

mean. The only thing that changes from the respective case where $x \to \infty$ is that my vertical line will be at $x = -N$, and now the function values have to lie in your tolerance region to the left of the line instead of to the right. That is, you just change the inequality $x > N$ to $x < -N$ in all the definitions.

We can actually use the same idea to define the limit of an infinite sequence. In Section 22.1 of Chapter 22, we gave an informal definition, but now we can do better. Start off with an infinite sequence $a_1, a_2, a_3, \ldots$; then

---

"$\lim_{n \to \infty} a_n = L$" means that for any choice of $\varepsilon > 0$ you make, I can pick N such that:

$$|a_n - L| < \varepsilon \text{ for all } n \text{ satisfying } n > N.$$

If you compare this definition with that of

$$\lim_{x \to \infty} f(x) = L$$

above, you'll see that they are almost the same. The only difference is that the continuous variable x has been replaced by the integer-valued variable n. In the case that L is replaced by ∞ (or $-\infty$), then you choose $M > 0$ instead of $\varepsilon > 0$, and the inequality $|a_n - L| < \varepsilon$ changes to $a_n > M$ (or $a_n < -M$, respectively).

Now if you really want a challenge, try writing out the definition of every possible type of limit (there are 18 that we've looked at!), and for an encore, see if you can prove analogues of all the results in Section A.2 above for the other cases.

A.3.4 Two examples involving trig

In Section 3.4 of Chapter 3, we claimed the following limit does not exist (DNE):

$$\lim_{x \to \infty} \sin(x).$$

The intuition is that $\sin(x)$ keeps oscillating between -1 and 1, so it doesn't tend to any one number. Let's use the definition from Section A.3.3 above to prove that the intuition is correct. Suppose that the limit does exist and that it has the value L. You pick your number $\varepsilon > 0$, and then I need to pick a large number N such that $|\sin(x) - L| < \varepsilon$ whenever $x > N$. So let's suppose you pick your ε to be $\frac{1}{2}$. This means that I need to ensure that $|\sin(x) - L| < \frac{1}{2}$ whenever $x > N$. Another way of looking at this is that $\sin(x)$ has to lie in the interval $(L - \frac{1}{2}, L + \frac{1}{2})$ for all $x > N$. Unfortunately, this can't happen, no matter what L and N are! To see why, just pick the first multiple of π bigger than N; let's say that this number is $n\pi$ for some integer n. Then

$\sin(n\pi + \pi/2) = 1$ while $\sin(n\pi + 3\pi/2) = -1$. These two values of $\sin(x)$ are distance 2 apart, so they can't both lie in the interval $(L - \frac{1}{2}, L + \frac{1}{2})$ since that interval is only 1 unit long. So the limit can't be L for any finite number L.

Here's a picture of what's going on for three potential candidates for our hopeful limit L:

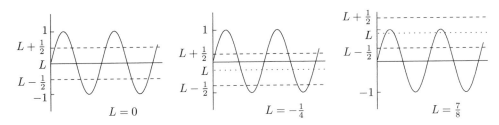

The width of the interval around L is $\frac{1}{2}$ in each case, but in each of the three cases, I can't cram $\sin(x)$ into the interval even if I throw a lot of it away. There's no vertical line I can draw and state that to the right of that line, I am always in your interval, since $\sin(x)$ keeps going out of the interval. The same is true no matter what horizontal stripe of height 1 we look at.

To be completely diligent, we should also make sure that the limit can't be ∞ or $-\infty$. In fact, if the limit were ∞, then you'd pick $M > 0$ and I'd have to make sure that $\sin(x) > M$ whenever $x > N$ for some N. All you have to do to thwart me, though, is to pick $M = 2$. Then I'm screwed, since $\sin(x) > 2$ is never true for any x! The same move works for $-\infty$ (try it and see). So we have indeed shown that the above limit doesn't exist.

 We also claimed, this time in Section 3.3, that

$$\lim_{x \to 0+} \sin\left(\frac{1}{x}\right)$$

does not exist. To show that this is true, you can pick a potential limit L and argue as we did in the previous example. If your move is to pick $\varepsilon = \frac{1}{2}$, then I need to try to pick $\delta > 0$ so that $|\sin(1/x) - L| < \frac{1}{2}$ whenever $0 < x < \delta$. (Here we are using the definition from Section A.3.2 above.) You can now be clever and try to find two tiny values of x that cause this to screw up. Indeed, if you try $x = 1/(n\pi + \pi/2)$ and then $x = 1/(n\pi + 3\pi/2)$ for large enough n, you will be within $0 < x < \delta$, but $\sin(1/x)$ will turn out to be 1 and -1, respectively; this is a problem, since both of them can't lie in the tolerance interval $(L - \frac{1}{2}, L + \frac{1}{2})$ regardless of what L is.

 You should try writing out these details; but there is a simpler way. You see, since we already know that $\lim_{x \to \infty} \sin(x)$ doesn't exist, we can just do a simple substitution of the limiting variable. Indeed, if you let $u = 1/x$, then $x = 1/u$, and we immediately know that

$$\lim_{1/u \to \infty} \sin\left(\frac{1}{u}\right)$$

does not exist. Now, when is it true that $1/u \to \infty$? The only way this can happen is if $u \to 0^+$. It's not hard to justify this switcheroo in general (see

Section A.4.1 below), so we see that

$$\lim_{u \to 0^+} \sin\left(\frac{1}{u}\right) \text{ DNE.}$$

Now just change the dummy variable u to x and we get what we want without any mess!

A.4 Continuity and Limits

As we saw in Section 5.1.1, to say that a function f is continuous at $x = a$ means that

$$\lim_{x \to a} f(x) = f(a).$$

That is, when $x \to a$, we have $f(x) \to f(a)$. So the function f **preserves** limits; this is the essence of continuity. Anyway, we can now use our knowledge of limits to justify that when you add, subtract, multiply, or divide two functions which are both continuous at $x = a$, then the new function is also continuous there. (In the case of division, the denominator can't be 0 at $x = a$.) Indeed, suppose that f and g are both continuous at $x = a$. Then we know that

$$\lim_{x \to a} f(x) = f(a) \qquad \text{and} \qquad \lim_{x \to a} g(x) = g(a).$$

So to show that the function $f + g$ is continuous at $x = a$, all we have to do is split up the limit, which was justified in Section A.2.1 above:

$$\lim_{x \to a} (f(x) + g(x)) = \lim_{x \to a} f(x) + \lim_{x \to a} g(x) = f(a) + g(a).$$

That's all there is to it. Now you can replace the $+$ signs by $-$, $\times$, or $/$ signs to get the similar results for subtraction, multiplication, and division.

A.4.1 Composition of continuous functions

Let's look at something a little trickier. Suppose that f and g are both continuous everywhere; we want to show that the composition $f \circ g$ is continuous too. We need to focus on one particular value of x to make this work. So let's suppose that g is continuous at $x = a$. Where do we need f to be continuous? We want to show that

$$\lim_{x \to a} f(g(x)) = f(g(a)).$$

So it's pointless to worry about whether f is continuous at $x = a$; we need it to be continuous at $g(a)$ instead, since we are evaluating f near and at the point $g(a)$.

So, here's the situation: we know that g is continuous at $x = a$, and that f is continuous at $x = g(a)$, and we want to show that $f \circ g$ is continuous at $x = a$. To do this, we need to add a third player to our game. I will actually play against this new player, who is called Smiddy, and Smiddy will play against you.

Here's how it works. Since f is continuous at $g(a)$, we know that

$$\lim_{y \to g(a)} f(y) = f(g(a)).$$

Note that I used y as a dummy variable instead of x, but that's fine—you could change the y to any letter you please and it means the same thing. Anyway, let's set $L = f(g(a))$. Then you pick your $\varepsilon > 0$, establishing your tolerance interval $(L - \varepsilon, L + \varepsilon)$, and you challenge Smiddy to throw away everything outside a little interval centered at $y = g(a)$ in such a way that all the remaining function values lie in your interval. That is, Smiddy should pick $\lambda > 0$ so that $|f(y) - L| < \varepsilon$ whenever $|y - g(a)| < \lambda$. Because the above limit is true, Smiddy can do this. Why λ instead of δ? Because Smiddy's cool like that.

Now it's my turn to play against Smiddy. This time, we use the fact that g is continuous at $x = a$ to write

$$\lim_{x \to a} g(x) = g(a).$$

Here's the key: instead of ε, which you already used, Smiddy just uses the number λ! So Smiddy's tolerance interval is $(g(a) - \lambda, g(a) + \lambda)$. Now I have to throw away everything outside a little interval centered at $x = a$ so that the remaining function values lie in Smiddy's interval. Because the above limit is true, I can choose $\delta > 0$ such that whenever $|x - a| < \delta$, we have $|g(x) - g(a)| < \lambda$.

All we have to do is put everything together. Because of my game with Smiddy, we know that whenever $|x - a| < \delta$, we also have $|g(x) - g(a)| < \lambda$. Now your game with Smiddy shows that if $|y - g(a)| < \lambda$, then $|f(y) - L| < \varepsilon$. Pushing Smiddy to one side and replacing L by $f(g(a))$ and y by $g(x)$, we see that whenever $|x - a| < \delta$, we have $|f(g(x)) - f(g(a))| < \varepsilon$. This means that if I play against you directly, I can always make a legitimate move, no matter what ε is (as long as it's positive). So we have indeed shown that

$$\lim_{x \to a} f(g(x)) = f(g(a)),$$

provided that g is continuous at a and f is continuous at $g(a)$. Of course, if f and g are continuous everywhere, then so is the composition function $f \circ g$.

The argument can be modified to include the cases where $x \to \infty$ or $x \to -\infty$ instead of a. We have to make a slight change to the statement, since the right-hand side can't be $g(\infty)$. So the best we can do is as follows:

$$\lim_{x \to \infty} f(g(x)) = f\left(\lim_{x \to \infty} g(x)\right),$$

and similarly for the case where $x \to -\infty$. I leave it to you to write out the details of the proofs, but here's the basic idea. Your game with Smiddy will be the same, but mine changes slightly: I pick N instead of δ, and the inequality $|x - a| < \delta$ has to be replaced by $x > N$ or $x < -N$ depending on whether you are in the case of $x \to \infty$ or $x \to -\infty$.

We can now establish the following limit, which appeared in Section 3.4 of Chapter 3:

$$\lim_{x \to \infty} \sin\left(\frac{1}{x}\right) = 0.$$

Indeed, if you set $f(x) = \sin(x)$ and $g(x) = 1/x$, then both f and g are continuous everywhere, except that g isn't continuous at $x = 0$. Since

$$\lim_{x \to \infty} g(x) = \lim_{x \to \infty} \frac{1}{x} = 0,$$

we can use our formula from above to conclude that

$$\lim_{x \to \infty} \sin\left(\frac{1}{x}\right) = \lim_{x \to \infty} f(g(x)) = f\left(\lim_{x \to \infty} g(x)\right) = f(0) = \sin(0) = 0.$$

A more intuitive way of expressing this is that $1/x \to 0$ as $x \to \infty$, so $\sin(1/x) \to \sin(0) = 0$ as $x \to \infty$.

A.4.2 Proof of the Intermediate Value Theorem

In Section 5.1.4, we looked at the Intermediate Value Theorem, which says that if f is continuous on $[a, b]$, and also $f(a) < 0$ and $f(b) > 0$, then there is some number c such that $f(c) = 0$. Now we're going to look at the idea of the proof of this theorem.

Consider the set of values x in the interval $[a, b]$ such that $f(x) < 0$. We know that a is in this set, since $f(a) < 0$, and that b isn't in the set. We'd like to find the largest number c which is in the set, but that might not be possible. For example, what's the largest number less than 0 itself? There isn't one—for any negative number, you can always find a negative number closer to zero, for example, by dividing your number by 2. On the other hand, we can find a number c that is a sort of right-hand bookend of the set. In particular, we can insist that no member of the set is to the right of c, and also that any open interval with right-hand endpoint c includes at least one member of the set. (This is due to a nice property of the real line called *completeness*.) So here's what we know, written in symbols:

1. for any $x > c$, we have $f(x) \geq 0$; and
2. for any interval $(c - \delta, c)$ where $\delta > 0$, there is at least one point x in the interval such that $f(x) < 0$.

Now let's get busy. Here's the big question: what is $f(c)$? Suppose that it's negative. In that case, $c \neq b$ since $f(b) > 0$. Because f is continuous, the values of $f(x)$ should be near $f(c)$ when x is near c; this will be a problem when x is a little to the right of c, because $f(x)$ is supposed to be positive but $f(c)$ is negative. More formally, you can choose $\varepsilon = -f(c)/2$ (which is positive); then your tolerance interval is $(3f(c)/2, f(c)/2)$, which consists only of negative numbers. I can't pick any interval of the form $(c - \delta, c + \delta)$ lying inside $[a, b]$ that works, since any such interval includes an x which is bigger than c. By condition #1 above, we know that $f(x)$ would have to be positive, which means that it doesn't lie in your tolerance region. So it can't be true that $f(c) < 0$. Intuitively, if it is, then your bookend still has books to the right of it!

Perhaps $f(c) > 0$. In this case, we can't have $c = a$ since $f(a) < 0$. Now, the values of $f(x)$ should be near $f(c)$ when x is near c; so in particular they should be positive. This is a problem because of condition #2 above. Specifically, this time you can choose $\varepsilon = f(c)/2$, so that your tolerance interval is $(f(c)/2, 3f(c)/2)$. I need to try to find an interval $(c-\delta, c+\delta)$ within $[a, b]$ such that for any x in my interval, $f(x)$ always lies in your tolerance interval. In particular, $f(x) > 0$. This means that $f(x) > 0$ for all x in the interval $(c - \delta, c)$, which violates condition #2. So $f(c) > 0$ isn't true either; if it were true, then the bookend could be pushed to the left some more, so it wouldn't be at c.

What's left? The only possibility is that $f(c) = 0$, so we have proved our theorem. By the way, it's easy to change the situation to the case when $f(a) > 0$ and $f(b) < 0$ instead; you can either rewrite the proof slightly differently, or you can just set $g(x) = -f(x)$ and apply the theorem to g instead of f.

A.4.3   Proof of the Max-Min Theorem

Now let's prove the Max-Min Theorem, which we looked at in Section 5.1.6. The idea is that we once again have a function f which is continuous on the closed interval $[a, b]$; the claim is that there is some number c in the interval which is a maximum for f. As we saw, this means that $f(c)$ is greater than or equal to every other value of $f(x)$ where x wanders over the whole interval $[a, b]$.

Here's how it's done. The first thing we want to show is that you can plonk down some horizontal line at $y = N$, say, such that the function values $f(x)$ all lie below that line. If you couldn't do that, then the function would somehow grow bigger and bigger somewhere inside $[a, b]$, and it wouldn't have a maximum. So, let's suppose you **can't** draw such a line. Then for every positive integer N, there's some point x_N in $[a, b]$ such that $f(x_N)$ is above the line $y = N$. That is, we have found some points x_N such that $f(x_N) > N$ for every N. Let's mark them on the x-axis with an X.

Now, where are these marked points? There are infinitely many. So if we chop the interval $[a, b]$ in half to get two new intervals, one of them must still have infinitely many marked points. Perhaps they both do, but they can't both have finitely many marked points or else the total would be finite. Let's focus on the half of the original interval that has infinitely many marked points; if they both do, choose your favorite one (it doesn't matter). Now repeat the exercise with the new, smaller interval: chop it in half. One of the halves must have infinitely many marked points. Continue doing this for as long as you like, and you will get a collection of intervals which get smaller and smaller, all nested inside each other, and each of which has infinitely many marked points. Stacking the intervals on top of each other, this is what the situation might look like:

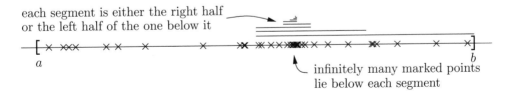

each segment is either the right half or the left half of the one below it

a b

infinitely many marked points lie below each segment

Intuitively, there has to be some real number which is inside every single one of these intervals.* Let's call the number q. What is $f(q)$? We can use the

---

*Again, one needs to use the completeness property of the real line to show this. Actually, there has to be exactly one such number—can you see why?

continuity of f to get some idea of what it should be. Indeed, we know that

$$\lim_{x \to q} f(x) = f(q).$$

So if you pick your ε to be 1, for example, then I should be able to find an interval $(q - \delta, q + \delta)$ so that $|f(x) - f(q)| < 1$ for all x in the interval. The problem is that the interval $(q - \delta, q + \delta)$ contains infinitely many marked points! This is because eventually one of the little nested intervals that we chose will lie within $(q - \delta, q + \delta)$, no matter how small δ is. This is a real problem: we are supposed to have all these marked points inside our interval $(q - \delta, q + \delta)$, but when you take f of any of them, you get a number between $f(q) - 1$ and $f(q) + 1$. So, no matter what $f(q)$ is, we're going to get in trouble: some of the marked points are going to have function values which are much bigger than $f(q) + 1$. The whole thing is out of control. So we were wrong about not being able to draw in a line like $y = N$ which had the whole function beneath it!

We're still not done. We have this line $y = N$ which lies above the graph of $y = f(x)$ on $[a, b]$, but now we need to move it down until it hits the graph in order to find the maximum. So, let's pick N as small as possible so that $f(x) \leq N$ for all x in $[a, b]$. (We have used completeness once again.) Now we need to show that $N = f(c)$ for some c. To do this, we're going to repeat the same trick as we did above with marked points, except this time they'll be circled. Pick a positive integer n; we must be able to find some number c_n in $[a, b]$ such that $f(c_n) > N - 1/n$. If not, then we should have drawn our line at $y = N - 1/n$ (or even lower) instead of $y = N$. So there is such a c_n, and there's one for every positive integer n. Circle all of these points. There are infinitely many of them, and when you apply f to them, the resulting values get closer and closer—arbitrarily close, in fact—to N. (None of the values can be bigger than N because $f(x) \leq N$ for all x!) Now all we have to do is keep bisecting the interval $[a, b]$ over and over again, such that each little interval has infinitely many circled points in it. As before, there is a number c in all the intervals. This number is really surrounded by a fog of circled points.

What is $f(c)$? It can't be more than N, but maybe it can be less than N. Let's suppose that $f(c) = M$, where $M < N$, and let's set $\varepsilon = (N - M)/2$. Since f is continuous, we really need

$$\lim_{x \to c} f(x) = f(c) = M.$$

You have your ε, and so I need to find an interval $(c - \delta, c + \delta)$ so that $f(x)$ lies in $(M - \varepsilon, M + \varepsilon)$ for x in my interval. The problem is that $M + \varepsilon = N - \varepsilon$, and also that there are infinitely many circled points lying in $(c - \delta, c + \delta)$, no matter how I choose $\delta > 0$. Some of them might have function values lying in $(M - \varepsilon, M + \varepsilon)$, but since the function values get closer to N, most of them won't. So I can't make my move. The only way out is that $f(c) = N$ after all. This means that c is a maximum, and we're done!

To get the minimum version of the theorem, just reapply the theorem to $g(x) = -f(x)$. After all, if c is a maximum for g, then it is a minimum for f.

A.5 Exponentials and Logarithms Revisited

In Section 9.2 of Chapter 9, we developed the theory of exponentials and logarithms, culminating in the discovery that

$$\frac{d}{dx}e^x = e^x \qquad \text{and} \qquad \frac{d}{dx}\ln(x) = \frac{1}{x}.$$

There is one loose end: we claimed that

$$\lim_{h \to 0^+} (1+h)^{1/h}$$

exists, and called it e, but we never proved it. It's possible to show directly that the above limit exists, but it's not particularly informative. Instead, I'm going to assume that you've learned about integration and the Fundamental Theorems of Calculus (see Chapters 16 and 17) and take a different approach to the subject at hand. In fact, it all begins with logarithms.

Let's start by defining a function F by the rule

$$F(x) = \int_1^x \frac{1}{t}\,dt$$

for all $x > 0$. This is a function based on the integral of another function; see Section 17.1 of Chapter 17 to remind yourself about this sort of function. Now, I know that you can just write

$$F(x) = \int_1^x \frac{1}{t}\,dt = \ln|t|\Big|_1^x = \ln|x| - \ln|1| = \ln(x),$$

since $x > 0$ and $\ln(1) = 0$. The problem is, we are jumping the gun! If we are really going to do this properly, we're not allowed to use the fact that $\int 1/t\,dt = \ln|t| + C$. Actually, that's one of the things we're trying to show. So for the moment, we can't assume that $F(x) = \ln(x)$; let's start by **proving** that.

So let's write down some interesting properties of this function F. The derivative of F is given by

$$F'(x) = \frac{d}{dx}\int_1^x \frac{1}{t}\,dt = \frac{1}{x},$$

by the First Fundamental Theorem of Calculus. So F is differentiable, which means that it's continuous (see Section 5.2.11 of Chapter 5). Next, set $x = 1$ to see that

$$F(1) = \int_1^1 \frac{1}{t}\,dt = 0,$$

using the property that the integral of any function is 0 if both limits of integration are equal and the function is actually defined there (see Section 16.3 of Chapter 16). How about

$$\lim_{x \to \infty} F(x)?$$

Actually, by the definition of the improper integral (as given in Section 20.2 of Chapter 20), we have

$$\lim_{x \to \infty} F(x) = \lim_{x \to \infty} \int_1^x \frac{1}{t}\, dt = \int_1^\infty \frac{1}{t}\, dt = \infty.$$

We have to be really careful about saying that the improper integral $\int_1^\infty 1/t\, dt$ diverges. When we originally proved this divergence, we used the formula $\int 1/t\, dt = \ln|t| + C$, but we're not allowed to do this! Instead, the way to do it is to use the integral test to say that $\int_1^\infty 1/t\, dt$ and $\sum_{n=1}^\infty 1/n$ either both converge or diverge; then use the argument from Section 22.4.3 of Chapter 22 to show that the series diverges; so the integral diverges too. So we have

$$F(1) = 0 \qquad \text{and} \qquad \lim_{x \to \infty} F(x) = \infty.$$

Since F is continuous, the Intermediate Value Theorem (see Section 5.1.4 of Chapter 5) says that there must be a number e such that $F(e) = 1$. After all, 1 is between 0 and ∞! Also, since $F'(x) = 1/x > 0$ for all $x > 0$, we know that F is always increasing. So there can't be any other number c such that $F(c) = 1$. We have arrived at our official definition of e:

$$e \text{ is the unique number such that } \int_1^e \frac{1}{t}\, dt = 1.$$

Now let's pick a rational number α, and define

$$G(x) = F(x^\alpha) = \int_1^{x^\alpha} \frac{1}{t}\, dt.$$

We can use the Variation 2 technique described in Section 17.5.2 of Chapter 17 to see that

$$G'(x) = \frac{d}{dx} \int_1^{x^\alpha} \frac{1}{t}\, dt = \alpha x^{\alpha-1} \frac{1}{x^\alpha} = \alpha \cdot \frac{1}{x}.$$

(This assumes that we know that $\frac{d}{dx}(x^\alpha) = \alpha x^{\alpha-1}$ without using logarithmic differentiation; see if you can prove this fact for all rational numbers, knowing only that it's true for positive integers, as we saw in Section 6.1 of Chapter 6.) On the other hand, we know that $F'(x) = 1/x$, so the above equation implies that $G'(x) = \alpha F'(x)$. Since α is constant, we see that $G(x) = \alpha F(x) + C$, where C is constant. In particular, if we set $x = 1$, this equation becomes $G(1) = \alpha F(1) + C$. Now $G(1) = F(1^\alpha) = F(1) = 0$, so $C = 0$. Since $G(x) = F(x^\alpha)$, we've shown that $F(x^\alpha) = \alpha F(x)$ for any rational number α and $x > 0$. In fact, since F is continuous, the same thing must be true for any real α at all! Now set $x = e$ to see that $F(e^\alpha) = \alpha F(e) = \alpha$, since $F(e) = 1$. Changing α to x, we have shown that $F(e^x) = x$. So F is the inverse function of e^x, which means that $F(x) = \ln(x)$. Since we know that $F'(x) = 1/x$, we have shown that $\frac{d}{dx} \ln(x) = 1/x$. Now if $y = e^x$, then $x = \ln(y)$, so

$$\frac{dx}{dy} = \frac{1}{y} = \frac{1}{e^x};$$

by the chain rule, $dy/dx = e^x$. So we've differentiated both $\ln(x)$ and e^x from scratch, **and** shown that e exists!

Now all we need to do is show that

$$\lim_{h \to 0^+} (1 + h)^{1/h} = e.$$

This has become pretty easy: let $y = (1 + h)^{1/h}$, so that $\ln(y) = \ln(1 + h)/h$. Then

$$\lim_{h \to 0^+} \ln(y) = \lim_{h \to 0^+} \frac{\ln(1 + h)}{h} = 1$$

by the same argument we used in Section 9.4.3 of Chapter 9 (or just l'Hôpital's Rule). Of course, if $\ln(y) \to 1$ as $h \to 0^+$, then $y \to e^1 = e$ as $h \to 0^+$. This proves the above limit. The key point is that once you know that the derivative with respect to x of $\ln(x)$ is $1/x$, then you're golden: everything else is easy.

A.6 Differentiation and Limits

In this section, we'll prove some results involving derivatives and limits. More specifically, we'll deal with differentiating constant multiples of functions, sums, and differences of functions, and the product, quotient, and chain rules; then we'll prove the Extreme Value Theorem, Rolle's Theorem, the Mean Value Theorem, and the formula for the error term in linearization. We'll finish off by looking at derivatives of piecewise-defined functions and a proof of l'Hôpital's Rule.

A.6.1 Constant multiples of functions

Suppose y is a differentiable function of x and c is some constant. We want to show that

$$\frac{d}{dx}(cy) = c\frac{dy}{dx}.$$

It's pretty easy. Define f by $y = f(x)$; then the left-hand side of the above equation is

$$\lim_{\Delta x \to 0} \frac{cf(x + \Delta x) - cf(x)}{\Delta x}.$$

All you have to do is take out a factor of c from the numerator and drag it out of the limit. This was justified at the end of Section A.2.2 above:

$$\lim_{\Delta x \to 0} \frac{cf(x + \Delta x) - cf(x)}{\Delta x} = \lim_{\Delta x \to 0} \frac{c(f(x + \Delta x) - f(x))}{\Delta x}$$
$$= c \lim_{\Delta x \to 0} \frac{f(x + \Delta x) - f(x)}{\Delta x}.$$

The right-hand side is just $cf'(x)$, which is the same thing as $c(dy/dx)$, and we're all done.

A.6.2 Sums and differences of functions

If u and v are differentiable functions of x, we'd like to show that

$$\frac{d}{dx}(u + v) = \frac{du}{dx} + \frac{dv}{dx},$$

and similarly with both plus signs replaced by minus signs. There's almost nothing to this. If $u = f(x)$ and $v = g(x)$, then the left-hand side of the above equation is

$$\lim_{\Delta x \to 0} \frac{f(x + \Delta x) + g(x + \Delta x) - (f(x) + g(x))}{\Delta x}.$$

All you have to do is rearrange the sum and split up the limit, which was justified in Section A.2.1 above, to see that the above limit is equal to

$$\lim_{\Delta x \to 0} \frac{f(x + \Delta x) - f(x)}{\Delta x} + \lim_{\Delta x \to 0} \frac{g(x + \Delta x) - g(x)}{\Delta x}.$$

But this is just $f'(x) + g'(x)$, which equals the right-hand side of the equation we're trying to prove. The situation with minus signs instead of plus signs is just as easy!

A.6.3 Proof of the product rule

For the proofs of the product and quotient rules, we'll stick with the dy/dx rather than $f'(x)$ notation, as it's easier to understand the concepts using the former version. As we saw in Section 5.2.7, we have

$$\frac{dy}{dx} = \lim_{\Delta x \to 0} \frac{\Delta y}{\Delta x},$$

with the understanding that Δy is the amount y changes when you move x to $x + \Delta x$.

So we want to prove the product rule, which says that

$$\frac{d}{dx}(uv) = v\frac{du}{dx} + u\frac{dv}{dx}.$$

Suppose we change x to $x + \Delta x$. Then u changes to $u + \Delta u$, and v changes to $v + \Delta v$. This means that uv changes to $(u + \Delta u)(v + \Delta v)$. How much of a change is this? Take the difference between the old and new quantities to see that

$$\Delta(uv) = (u + \Delta u)(v + \Delta v) - uv.$$

Expanding and canceling, we end up with

$$\Delta(uv) = v\Delta u + u\Delta v + \Delta u\Delta v.$$

Now divide this equation by Δx. In the case of the last term, we'll even divide by an extra Δx, but then multiply by it once more to make things balance. We end up with

$$\frac{\Delta(uv)}{\Delta x} = v\frac{\Delta u}{\Delta x} + u\frac{\Delta v}{\Delta x} + \frac{\Delta u}{\Delta x}\frac{\Delta v}{\Delta x}\Delta x.$$

If you take limits as $\Delta x \to 0$, then all the ratios go to the corresponding derivatives, but the final factor of Δx goes to 0:

$$\frac{d}{dx}(uv) = v\frac{du}{dx} + u\frac{dv}{dx} + \frac{du}{dx}\frac{dv}{dx} \times 0.$$

Since the last term is 0, we have proved the product rule. Now you should try writing out a proof using the $f(x)$ notation (version 1) instead.

A.6.4 Proof of the quotient rule

Now we want to show that

$$\frac{d}{dx}\left(\frac{u}{v}\right) = \frac{v\dfrac{du}{dx} - u\dfrac{dv}{dx}}{v^2}.$$

Again, when x changes to $x + \Delta x$, we know that u and v change to $u + \Delta u$ and $v + \Delta v$, respectively. This means that u/v changes to $(u + \Delta u)/(v + \Delta v)$. The amount of change is

$$\Delta\left(\frac{u}{v}\right) = \frac{u + \Delta u}{v + \Delta v} - \frac{u}{v}.$$

Taking a common denominator and canceling $uv - uv$ leads to

$$\Delta\left(\frac{u}{v}\right) = \frac{v\Delta u - u\Delta v}{v^2 + v\Delta v}.$$

Dividing this by Δx, and then multiplying and dividing the Δv term in the denominator by Δx, gives

$$\frac{\Delta\left(\dfrac{u}{v}\right)}{\Delta x} = \frac{v\dfrac{\Delta u}{\Delta x} - u\dfrac{\Delta v}{\Delta x}}{v^2 + v\dfrac{\Delta v}{\Delta x}\Delta x}.$$

Now let $\Delta x \to 0$. All fractions become derivatives, and the final factor on the bottom goes to 0, so we end up with

$$\frac{d}{dx}\left(\frac{u}{v}\right) = \frac{v\dfrac{du}{dx} - u\dfrac{dv}{dx}}{v^2 + v\dfrac{dv}{dx} \times 0}.$$

Since the final term in the denominator is just 0, we have proved the quotient rule.

A.6.5 Proof of the chain rule

Suppose that y is a differentiable function of u, which is itself a differentiable function of x. We want to prove that

$$\frac{dy}{dx} = \frac{dy}{du}\frac{du}{dx}.$$

At first glance there's nothing to this using the Δ notation—you just write

$$\frac{\Delta y}{\Delta x} = \frac{\Delta y}{\Delta u}\frac{\Delta u}{\Delta x}$$

and take limits. Unfortunately, Δu might sometimes be 0, which would invalidate the whole equation. So let's use the function notation. Let f and g be differentiable, and set $h(x) = f(g(x))$. We want to show that

$$h'(x) = f'(g(x))g'(x).$$

If g is constant near x, then so is h, so both sides of this equation are 0. Otherwise, we know that

$$h'(x) = \lim_{\Delta x \to 0} \frac{h(x + \Delta x) - h(x)}{\Delta x} = \lim_{\Delta x \to 0} \frac{f(g(x + \Delta x)) - f(g(x))}{\Delta x}.$$

Multiply and divide the fraction by $g(x + \Delta x) - g(x)$, which must be nonzero for infinitely many values of Δx near 0, then split up the limit to get

$$h'(x) = \lim_{\Delta x \to 0} \frac{f(g(x + \Delta x)) - f(g(x))}{g(x + \Delta x) - g(x)} \times \lim_{\Delta x \to 0} \frac{g(x + \Delta x) - g(x)}{\Delta x}.$$

The right-hand limit is just $g'(x)$, but how about the left-hand one? The trick is to set $\varepsilon = g(x + \Delta x) - g(x)$. Then the quantity $g(x + \Delta x)$ in the numerator of the left-hand limit can be written as $g(x) + \varepsilon$ (can you see why?), whereas the denominator is just ε itself. So we have

$$h'(x) = \lim_{\Delta x \to 0} \frac{f(g(x) + \varepsilon) - f(g(x))}{\varepsilon} \times g'(x).$$

Now what happens to ε when $\Delta x \to 0$? Since g is differentiable, we know from Section 5.2.11 that g is continuous. In particular,

$$\lim_{\Delta x \to 0} g(x + \Delta x) = g(x).$$

If you subtract $g(x)$ from both sides, you see that $\varepsilon \to 0$ when $\Delta x \to 0$. This means that in our expression for $h'(x)$, we can replace the $\Delta x \to 0$ by $\varepsilon \to 0$ and get

$$h'(x) = \lim_{\varepsilon \to 0} \frac{f(g(x) + \varepsilon) - f(g(x))}{\varepsilon} \times g'(x).$$

Now the first term is exactly $f'(g(x))$, so $h'(x) = f'(g(x))g(x)$ and we have proved the chain rule.

A.6.6 Proof of the Extreme Value Theorem

In Section 11.1.2 of Chapter 11, we stated the Extreme Value Theorem. This says that if $x = c$ is a local maximum or minimum for a function f, then $x = c$ is a critical point for f. This means that either $f'(c)$ doesn't exist, or $f'(c) = 0$.

To prove this, let's first suppose that $x = c$ is a local minimum for f. If $f'(c)$ doesn't exist, then it's a critical point, which is exactly what we were hoping for. On the other hand, if $f'(c)$ exists, then

$$f'(c) = \lim_{h \to 0} \frac{f(c + h) - f(c)}{h}.$$

Since c is a local minimum, we know that $f(c + h) \geq f(c)$ when $c + h$ is very close to c. Of course, $c + h$ is close to c exactly when h is close to 0. For such h, the numerator $f(c + h) - f(c)$ in the above fraction must be nonnegative. When $h > 0$, the quantity

$$\frac{f(c + h) - f(c)}{h}$$

is positive (or 0), but when $h < 0$, the quantity is negative (or 0). So the right-hand limit

$$\lim_{x \to c^+} \frac{f(c+h) - f(c)}{h}$$

must be greater than or equal to 0, while the same left-hand limit is less than or equal to 0. Since the two-sided limit exists, the left-hand and right-hand limits are equal; the only possibility is that they are both 0. This shows that $f'(c) = 0$, so $x = c$ is once again a critical point for f.

How about if $x = c$ is a local maximum? I leave it to you to repeat the argument. The only difference is that the quantity $f(c+h) - f(c)$ is now negative (or 0) when h is close to 0.

A.6.7 Proof of Rolle's Theorem

Suppose f is continuous on $[a, b]$, differentiable on (a, b), and satisfies the condition $f(a) = f(b)$. Then we want to show that there is a number c in (a, b) such that $f'(c) = 0$. To do this, we use the Max-Min Theorem to say that f has a global maximum and a global minimum in $[a, b]$. If either the maximum or the minimum occurs at some number c in (a, b), then the Extreme Value Theorem says that $f'(c) = 0$. (We know that $f'(c)$ exists since f is differentiable in (a, b).) The only other possibility is that the global maximum and the global minimum both occur at the endpoints a and b. In that case, since $f(a) = f(b)$, the function must be constant, so every number c in (a, b) satisfies $f'(c) = 0$. That's all there is to the proof!

A.6.8 Proof of the Mean Value Theorem

Now we have f which is continuous on $[a, b]$ and differentiable on (a, b), but we don't assume that $f(a) = f(b)$. The Mean Value Theorem says that there is some c in (a, b) with

$$f'(c) = \frac{f(b) - f(a)}{b - a}.$$

To prove this, define a new function g by the equation

$$g(x) = f(x) - \frac{f(b) - f(a)}{b - a}(x - a).$$

It looks a little complicated, but actually all we are doing is subtracting a constant multiple of the linear function $(x - a)$ from $f(x)$ and calling it g. So the function g is also continuous on $[a, b]$ and differentiable on (a, b), and what's more, we have

$$g(a) = f(a) - \frac{f(b) - f(a)}{b - a}(a - a) = f(a) \qquad \text{and}$$

$$g(b) = f(b) - \frac{f(b) - f(a)}{b - a}(b - a) = f(a).$$

So we have shown that $g(a) = g(b)$, which means we can apply Rolle's Theorem! We end up with a number c such that $g'(c) = 0$. Now we just have to differentiate g and see what that means for f. Since the quantities $f(b) - f(a)$ and $b - a$ are constant, we get

$$g'(x) = f'(x) - \frac{f(b) - f(a)}{b - a}.$$

Now plug in $x = c$. Since $g'(c) = 0$, we have

$$0 = f'(c) - \frac{f(b) - f(a)}{b - a}.$$

This means that

$$f'(c) = \frac{f(b) - f(a)}{b - a},$$

which is exactly what we wanted to show!

A.6.9 The error in linearization

Let's tie up another loose end. In Section 13.2 of Chapter 13, we looked at the linearization L of a function f about $x = a$, where a is some number in the domain of f:

$$L(x) = f(a) + f'(a)(x - a).$$

If x is near a, we can use $L(x)$ to estimate the value of $f(x)$. How wrong could we possibly be? According to the formula in Section 13.2.4 of Chapter 13, if f'' exists between x and a, then

$$|\text{error}| = \frac{1}{2}|f''(c)||x - a|^2;$$

here c is some number between x and a. Let's prove this formula. Start off by calling the error term $r(x)$; since $r(x)$ is the difference between the true value $f(x)$ and our guess, which is the linearization $L(x) = f(a) + f'(a)(x - a)$, we have

$$r(x) = f(x) - L(x) = f(x) - f(a) - f'(a)(x - a).$$

Now, the clever idea is to fix x as a constant and let a be the variable. Inspired by this, let

$$g(t) = f(x) - f(t) - f'(t)(x - t).$$

So the error $r(x)$ arises exactly when $t = a$. That is, the error is $g(a)$. Note that

$$g(x) = f(x) - f(x) - f'(x)(x - x) = 0.$$

Let's differentiate g with respect to t. The term $f(x)$ is constant, so its derivative is 0. Also, we need the product rule to deal with $f'(t)(x - t)$. All in all, we get

$$g'(t) = 0 - f'(t) - (f'(t) \times (-1) + f''(t)(x - t)) = -f''(t)(x - t).$$

In particular, we have

$$g'(x) = -f''(x)(x - x) = 0.$$

Everything we've done so far makes a lot of sense. Now we have to do something that seems to be a little whacked out. Remember that we want to show that the error is $\frac{1}{2}f''(c)(x-a)^2$, where c is between x and a. Since the error is $g(a)$, this suggests that $g(t)$ is something like $K(x-t)^2$, where K is some number which does not depend on t, but only on x and a. Even this isn't exactly true, but it might explain why we're going to let

$$h(t) = g(t) - K(x-t)^2.$$

You see, when you differentiate this with respect to t, holding x constant, you get

$$h'(t) = g'(t) + 2K(x-t).$$

So what? Well, we can use the Mean Value Theorem (see Section 11.3 of Chapter 11) to get

$$h'(c) = \frac{h(x) - h(a)}{x - a}$$

for some c between x and a. We can substitute for $h'(c)$, $h(x)$, and $h(a)$ using the above equations:

$$g'(c) + 2K(x-c) = \frac{(g(x) - K(x-x)^2) - (g(a) - K(x-a)^2)}{x - a}$$
$$= \frac{-g(a) + K(x-a)^2}{x - a},$$

since $g(x) = 0$. Since $g'(c) = -f''(c)(x-c)$, this last equation can be rearranged to

$$g(a) - K(x-a)^2 = (x-a)(x-c)(f''(c) - 2K).$$

We're close, but there's still a problem. We can't handle the factor $(x-c)$, since that's nowhere to be found in our error term! The only way we can get rid of it is if the left-hand side is actually 0. That is, we should have chosen K such that $g(a) - K(x-a)^2 = 0$. Indeed, if $K = g(a)/(x-a)^2$, then the above equation becomes

$$0 = (x-a)(x-c)\left(f''(c) - \frac{2g(a)}{(x-a)^2}\right).$$

Since $x \neq a$ and $x \neq c$, we must have

$$f''(c) - \frac{2g(a)}{(x-a)^2} = 0,$$

which means that $g(a) = \frac{1}{2}f''(c)(x-a)^2$. Since $g(a) = r(x)$ is the error we're looking for, we're finished.

A.6.10 Derivatives of piecewise-defined functions

Imagine that f is defined in piecewise fashion as

$$f(x) = \begin{cases} f_1(x) & \text{if } x > a, \\ f_2(x) & \text{if } x \leq a. \end{cases}$$

(You could change $x > a$ to $x \geq a$, and $x \leq a$ to $x < a$; it doesn't make a difference.) Anyway, in Section 6.6 of Chapter 6, we considered the question of whether f is differentiable at a. We have assumed that if the functions f_1 and f_2 match at $x = a$, and also the derivatives f_1' and f_2' match at $x = a$, then f is differentiable at a. How can we justify this? Well, first note that the matching of f_1 and f_2 at $x = a$ means that

$$\lim_{x \to a^+} f_1(x) = \lim_{x \to a^-} f_2(x) = f(a).$$

This ensures that f is at least continuous. Now we are also assuming that the derivatives match: this means that f_1 is differentiable to the immediate right of a, f_2 is differentiable to the immediate left of a, and

$$\lim_{x \to a^+} f_1'(x) = \lim_{x \to a^-} f_2'(x) = L,$$

where L is some nice finite number. So consider the quantity

$$\frac{f(a+h) - f(a)}{h}$$

for some small number $h \neq 0$. If $h > 0$, then we can apply the Mean Value Theorem (see Section 11.3 of Chapter 11) to say that

$$\frac{f(a+h) - f(a)}{h} = f_1'(c),$$

where c is some number between a and $a + h$. (Here we needed the continuity of f on $[a, a+h]$.) By the sandwich principle, as $h \to 0^+$, the number c is sandwiched between a and $a + h$, so $c \to a^+$ as $h \to 0^+$. We now see that

$$\lim_{h \to 0^+} \frac{f(a+h) - f(a)}{h} = \lim_{h \to 0^+} f_1'(c) = \lim_{c \to a^+} f_1'(c) = L.$$

The left-hand limit works the same way, except that we use f_2' instead of f_1' to see that

$$\lim_{h \to 0^-} \frac{f(a+h) - f(a)}{h} = \lim_{h \to 0^-} f_2'(c) = \lim_{c \to a^-} f_2'(c) = L.$$

The left-hand and right-hand limits are both equal to L, so we have shown that $f'(a)$ exists and is also equal to L.

A.6.11 Proof of l'Hôpital's Rule

Let's prove l'Hôpital's Rule (see Chapter 14). Specifically, suppose we have two functions f and g which are both differentiable on some interval containing a point a (but maybe not at a itself); and $f(a) = g(a) = 0$; and also $g'(x) \neq 0$ except maybe at a itself. Then we need to show that

$$\lim_{x \to a} \frac{f(x)}{g(x)} = \lim_{x \to a} \frac{f'(x)}{g'(x)},$$

provided that the limit on the right-hand side exists. We'll need a slightly different version of the Mean Value Theorem, called Cauchy's Mean Value

Theorem: if f and g are continuous on $[A, B]$ and differentiable on (A, B), and also $g'(x) \neq 0$ on (A, B), then there is some C in (A, B) such that

$$\frac{f'(C)}{g'(C)} = \frac{f(B) - f(A)}{g(B) - g(A)}.$$

Let's prove this first, then use it to prove l'Hôpital's Rule. Incidentally, note that if $g(x) = x$ for all x, then $g'(x) = 1$ and the above equation becomes

$$f'(C) = \frac{f(B) - f(A)}{B - A}.$$

This is just the regular Mean Value Theorem! That doesn't really help us, though. Let's go back to the original equation above and look at the denominator on the right-hand side, which is $g(B) - g(A)$. That can't be equal to 0; if it were, then $g(A) = g(B)$, meaning that $g'(C) = 0$ for some C in (A, B) by Rolle's Theorem (see Section 11 in Chapter 11.2). So the right-hand side makes sense. Now, define a new function h by

$$h(x) = f(x) - \left(\frac{f(B) - f(A)}{g(B) - g(A)} \right) g(x)$$

for all x in (A, B). (Compare this with the function we called g in the proof of the ordinary Mean Value Theorem in Section A.6.8 above.) Anyway, let's write down some nice facts about this function. First, let's calculate $h(A)$ and $h(B)$. We have

$$\begin{aligned} h(A) &= f(A) - \left(\frac{f(B) - f(A)}{g(B) - g(A)} \right) g(A) \\ &= \frac{f(A)g(B) - f(A)g(A) - f(B)g(A) + f(A)g(A)}{g(B) - g(A)} \\ &= \frac{f(A)g(B) - f(B)g(A)}{g(B) - g(A)}, \end{aligned}$$

whereas

$$\begin{aligned} h(B) &= f(B) - \left(\frac{f(B) - f(A)}{g(B) - g(A)} \right) g(B) \\ &= \frac{f(B)g(B) - f(B)g(A) - f(B)g(B) + f(A)g(B)}{g(B) - g(A)} \\ &= \frac{f(A)g(B) - f(B)g(A)}{g(B) - g(A)}. \end{aligned}$$

So $h(A) = h(B)$. Also, note that h is differentiable, and since A and B are constant, we have

$$h'(x) = f'(x) - \left(\frac{f(B) - f(A)}{g(B) - g(A)} \right) g'(x).$$

We can use Rolle's Theorem, since $h(A) = h(B)$, to conclude that there's a number C in (A, B) such that $h'(C) = 0$. This means that

$$h'(C) = f'(C) - \left(\frac{f(B) - f(A)}{g(B) - g(A)} \right) g'(C) = 0.$$

If you rearrange this equation, you get the one we want:

$$\frac{f'(C)}{g'(C)} = \frac{f(B) - f(A)}{g(B) - g(A)}.$$

Now we're ready to prove l'Hôpital's Rule. Since $f(a) = g(a) = 0$, we have

$$\lim_{x \to a} \frac{f(x)}{g(x)} = \lim_{x \to a} \frac{f(x) - f(a)}{g(x) - g(a)}.$$

If $x > a$, then we can use Cauchy's Mean Value Theorem (which we just proved) on the interval $[a, x]$ to say that

$$\lim_{x \to a} \frac{f(x)}{g(x)} = \lim_{x \to a} \frac{f(x) - f(a)}{g(x) - g(a)} = \lim_{x \to a} \frac{f'(c)}{g'(c)}$$

for some c in (a, x). Otherwise, if $x < a$, then the same thing is true but c is in (x, a). (Note that we've used the fact that g' isn't 0, except possibly at a; that's one of the conditions of Cauchy's Mean Value Theorem.) Of course, the number c depends on what x is; but we can see that as $x \to a$, also $c \to a$. So we have

$$\lim_{x \to a} \frac{f(x)}{g(x)} = \lim_{x \to a} \frac{f'(c)}{g'(c)} = \lim_{c \to a} \frac{f'(c)}{g'(c)}.$$

All that's left is to treat c as the dummy variable and change it to x, and l'Hôpital's Rule is proved!

Well, sort of. We still haven't proved the ∞/∞ case, nor the case when $x \to \infty$ (or $-\infty$). It's a great exercise to try to adapt the above proof to these cases, if you dare.

A.7 Proof of the Taylor Approximation Theorem

Now let's look at how to prove the Taylor approximation theorem from Section 24.1.3 in Chapter 24. Here's what the theorem says: if f is smooth at $x = a$, then of all the polynomials of degree N or less, the one which best approximates $f(x)$ for x near a is the Nth-order Taylor polynomial P_N, which is given by

$$P_N(x) = f(a) + f'(a)(x - a) + \frac{f''(a)}{2!}(x - a)^2$$
$$+ \frac{f^{(3)}(a)}{3!}(x - a)^3 + \cdots + \frac{f^{(N)}(a)}{N!}(x - a)^N.$$

The plan is to show how this theorem follows from the full Taylor theorem, which we looked at in Section 24.1.4 of Chapter 24. I'm omitting the proof of the full Taylor theorem because you can find it in most textbooks or even by typing "proof of Taylor's Theorem" into a search engine. What you won't find as easily is the proof of the approximation theorem, so let's look at it now.

Let's first simplify matters by setting $a = 0$. Since we're assuming the full Taylor Theorem has been proved, we know that $f(x) = P_N(x) + R_N(x)$, where

$$P_N(x) = \sum_{n=0}^{N} \frac{f^{(n)}(0)}{n!} x^n$$

is a polynomial of degree N, and

$$R_N(x) = \frac{f^{(N+1)}(c)}{(N+1)!} x^{N+1}$$

for some c between 0 and x. (Remember, we have set $a = 0$, so factors like $(x-a)^n$ just become x^n and quantities like $f^{(n)}(a)$ become $f^{(n)}(0)$.) What we want to show is this:

> of all polynomials of degree N or less, P_N gives the best approximation to f near 0.

How on earth do you go about showing something like that? What does "best" even mean in this context, anyway? The trick is to pick some other polynomial of degree no more than N; let's call it Q. Since Q is different from P_N, we know that Q has at least one coefficient which differs from the corresponding coefficient in P_N. We want to show that $P_N(x)$ is closer to $f(x)$ than $Q(x)$ is, at least when x is close to 0. To see how close two quantities are, you look at the difference between the quantities. So what we really want to show is the following inequality:

$$|f(x) - P_N(x)| < |f(x) - Q(x)|$$

when x is close to 0. If this is true, then you can conclude that $P_N(x)$ is indeed closer to the ideal value $f(x)$ than $Q(x)$ is.

To get at our desired inequality above, let's look at both sides individually. The left-hand side is the absolute value of $f(x) - P_N(x)$, which is actually the remainder term $R_N(x)$. We have an expression for $R_N(x)$ above; it has three factors, which are $f^{(N+1)}(c)$, x^{N+1}, and $1/(N+1)!$. We know that c is trapped between 0 and x; as $x \to 0$, by the sandwich principle we must also have $c \to 0$. Since we are assuming f is very smooth, the function $f^{(N+1)}$ is continuous. So, as $x \to 0$, we have $c \to 0$, so it follows that $f^{(N+1)}(c) \sim f^{(N+1)}(0)$. Putting the three factors together and taking absolute values, we have

$$|f(x) - P_N(x)| = |R_N(x)| = \left| \frac{f^{(N+1)}(c)}{(N+1)!} x^{N+1} \right| \sim \frac{|f^{(N+1)}(0)|}{(N+1)!} |x|^{N+1}$$

as $x \to 0$. Actually, we can let $C = f^{(N+1)}(0)/(N+1)!$ and notice that C is just some constant which doesn't depend on x. So we have

$$|f(x) - P_N(x)| \sim |C| |x|^{N+1} \qquad \text{as } x \to 0.$$

Great. Now let's look at the right-hand side of the inequality we're trying to prove. This is the quantity $|f(x) - Q(x)|$. Let's write $f(x) = P_N(x) + R_N(x)$, so that

$$|f(x) - Q(x)| = |P_N(x) + R_N(x) - Q(x)| = |S(x) + R_N(x)|,$$

where we have lumped together P_N with Q by setting $S(x) = P_N(x) - Q(x)$. Let's take a closer look at S. It is the difference between two polynomials of degree no more than N which are not the same polynomial. So S is a polynomial of degree less than or equal to N, but it's not the zero polynomial. Let's suppose that if you write out $S(x)$ in powers of x, it looks something like this:

$$S(x) = a_m x^m + \cdots,$$

where $a_m x^m$ is the lowest-degree term. The number m has to be between 0 and N, since S has degree less than or equal to N. We know that S behaves like its lowest-degree term (see Section 21.4.1 of Chapter 21 for a discussion of this). That is, $S(x) \sim a_m x^m$ as $x \to 0$. On the other hand, we need to look at $S(x) + R_N(x)$ since that is the right-hand side of our desired inequality. We have already seen that $R_N(x) \sim C x^{N+1}$ as $x \to 0$, so the lowest-degree term in $S(x) + R_N(x)$ still looks like $a_m x^m$ (remember, $m \leq N$ so x^m is a lower-degree term than x^{N+1}). So, all up, we have

$$|f(x) - Q(x)| = |S(x) + R_N(x)| \sim |a_m||x^m| \qquad \text{as } x \to 0.$$

Great—we want to prove that the inequality

$$|f(x) - P_N(x)| < |f(x) - Q(x)|$$

is true when x is near 0. We know that $|f(x) - P_N(x)| \sim |C||x|^{N+1}$ and $|f(x) - Q(x)| \sim |a_m||x|^m$ as $x \to 0$. Since $m < N + 1$ (and $|C|$ and $|a_m|$ are constant), it is easy to see that the quantity $|C||x|^{N+1}$ is much smaller than $|a_m||x|^m$ when x is small. Indeed, the ratio of the two quantities is

$$\frac{|C||x|^{N+1}}{|a_m||x|^m} = C_1 |x|^{N+1-m}$$

where $C_1 = |C|/|a_m|$ is just another constant. The right-hand quantity goes to 0 as $x \to 0$. So, the above inequality is indeed true when x is close to 0 and we have finally proved our Taylor approximation theorem!

Actually, there is one little point we didn't cover: we assumed that $a = 0$. To get from this situation to the general situation, all you have to do is replace the quantity x by the translated quantity $(x - a)$ everywhere you see it in the above argument. The only thing you have to note is that $(x - a) \to 0$ is the same thing as $x \to a$. I leave it to you to fill in the details. Well done if you made it through the above proof.

APPENDIX B

Estimating Integrals

Most of the time when we've looked at definite integrals, we've been used to giving an exact answer by using antiderivatives and the Second Fundamental Theorem. In real life, alas, finding antiderivatives in a useful form can be difficult or impossible. Sometimes the best you can do is find an approximation to the value of your integral. So we'll look at three techniques for estimating definite integrals: strips, the trapezoidal rule, and Simpson's rule. In summary, here's the plan for this final appendix:

- estimating definite integrals using strips, the trapezoidal rule, and Simpson's rule; and
- estimating the error in the above approximations.

B.1   Estimating Integrals Using Strips

Here's a perfectly reasonable definite integral:

$$\int_0^2 e^{-x^2}\, dx.$$

It corresponds to the area of the region bounded by the x-axis, the curve $y = e^{-x^2}$, and the lines $x = 0$ and $x = 2$, like this:

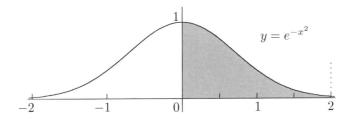

Finding an area like this might seem a little technical, but it's actually incredibly useful. The above curve is commonly known as a bell-shaped curve,[*]

---

[*]Technically **the** bell-shaped curve, or *normal distribution*, is actually given by the equation $y = e^{-x^2/2}/\sqrt{2\pi}$.

and it is fundamental in the study of probability theory. So it's especially annoying that there's no nice, simple way to write down the antiderivative

$$\int e^{-x^2}\, dx.$$

Actually, you can use Maclaurin series to express this integral as an infinite series, but that's not so nice or simple. The cold, hard reality of the situation is that there's no way to write down the exact value of the definite integral at the beginning of this section in a simple, closed form. (We already discussed this point in Section 16.5.1 of Chapter 16.)

On the other hand, we can find an approximate value for the integral—an estimate, if you prefer—by using the definition of the Riemann integral. Indeed, in Section 16.2 of Chapter 16, we looked at partitions, meshes, and Riemann sums. Since the integral is the limit of Riemann sums, we can get an approximation simply by not taking the limit. So, to estimate the integral

$$\int_a^b f(x)\, dx,$$

you can chop up the interval $[a, b]$ into a partition of the form

$$a = x_0 < x_1 < \cdots < x_{n-1} < x_n = b,$$

then choose a point c_1 in $[x_0, x_1]$, a point c_2 in $[x_1, x_2]$, and so on until you choose c_n in $[x_{n-1}, x_n]$. At that point, you're ready to write

$$\int_a^b f(x)\, dx \cong \sum_{j=1}^n f(c_j)(x_j - x_{j-1}).$$

This just says that the integral is approximately equal to one of its Riemann sums.

It all seems pretty abstract. Let's see how it works in the case of our above example. We're integrating from 0 to 2, so we need a partition of the integral $[0, 2]$. The simplest partition of that interval is just the interval $[0, 2]$, which corresponds to the choices $n = 1$, $x_0 = 0$, and $x_1 = 2$. We just need to pick c_1 inside $[0, 2]$. The approximation we'll end up with depends a lot on this choice! For example, if you choose $c_1 = 0$, $c_1 = 1$, or $c_1 = 2$, then your approximations will end up being the areas of the following regions, respectively:

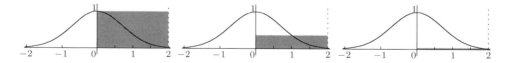

The first one is clearly a huge overestimate, while the third one is an under-estimate. The middle isn't so bad, but still not perfect. In order to work out the values of these three estimates, we'll use the formula:

$$\int_0^2 e^{-x^2}\, dx \cong \sum_{j=1}^n f(c_j)(x_j - x_{j-1}).$$

Replacing n by 1, $f(c_1)$ by $e^{-c_1^2}$, x_0 by 0, and x_1 by 2, we get

$$\int_0^2 e^{-x^2}\,dx \cong e^{-c_1^2}(2-0) = 2e^{-c_1^2}.$$

When c_1 is 0, 1, and 2, these values are 2, $2/e \cong 0.736$, and $2/e^4 \cong 0.037$, respectively. As you can see, there's a lot of difference between these three estimates!

Now let's see if we can do better by using more strips. Suppose we take a five-strip partition of $[0,2]$ that looks like this:

$$0 < \tfrac{1}{2} < 1 < \tfrac{5}{4} < \tfrac{3}{2} < 2.$$

So $n = 5$, and $x_0 = 0$, $x_1 = \tfrac{1}{2}$, $x_2 = 1$, $x_3 = \tfrac{5}{4}$, $x_4 = \tfrac{3}{2}$, $x_5 = 2$. Suppose that we choose our numbers c_j to be at the left-hand endpoint of each little interval. This means that $c_1 = 0$, $c_2 = \tfrac{1}{2}$, $c_3 = 1$, $c_4 = \tfrac{5}{4}$, and $c_5 = \tfrac{3}{2}$. Plugging everything into the above approximation formula, we have

$$\int_0^2 e^{-x^2}\,dx \cong \sum_{j=1}^{n} f(c_j)(x_j - x_{j-1})$$

$$= \sum_{j=1}^{5} e^{-c_j^2}(x_j - x_{j-1})$$

$$= e^{-0^2}(\tfrac{1}{2}-0) + e^{-(1/2)^2}(1-\tfrac{1}{2}) + e^{-1^2}(\tfrac{5}{4}-1)$$

$$+ e^{-(5/4)^2}(\tfrac{3}{2}-\tfrac{5}{4}) + e^{-(3/2)^2}(2-\tfrac{3}{2}).$$

This can be simplified a little if you like, or you can just use a calculator or computer to see that it is equal to 1.0865 to four decimal places. Now, I leave it to you to write down the value of the estimate that we would have gotten had we used the right-hand endpoint of each little interval instead of the left-hand one.

B.1.1 Evenly spaced partitions

It's often convenient to take your partition to be evenly spaced. This means that each of the little intervals has the same width, and it's not too hard to work out what that width is. If the interval of integration is $[a,b]$, then its length is $b - a$ units; so if you chop up the interval into n equal pieces, then each piece has length $(b-a)/n$ units. We'll call this quantity h; so $h = (b-a)/n$. Furthermore, the expression $(x_j - x_{j-1})$, which appears in the definition of the Riemann sum, is just the width of the jth strip, so this is exactly h as well. Our expression

$$\sum_{j=1}^{n} f(c_j)(x_j - x_{j-1})$$

can now be simplified to

$$h \times \sum_{j=1}^{n} f(c_j).$$

You still need to choose the numbers c_j, but things are a lot simpler. For example, let's estimate our integral

$$\int_0^2 e^{-x^2}\, dx$$

using 10 strips of equal width. The width of each strip is $h = (2-0)/10$, or $1/5$, and $n = 10$, so we have

$$\int_0^2 e^{-x^2}\, dx \cong h \times \sum_{j=1}^n f(c_j) = \frac{1}{5}\sum_{j=1}^{10} e^{-c_j^2}.$$

The intervals all have width $\frac{1}{5}$, so starting from 0, we see that they are partitioned as follows:

$$0 < \tfrac{1}{5} < \tfrac{2}{5} < \tfrac{3}{5} < \tfrac{4}{5} < 1 < \tfrac{6}{5} < \tfrac{7}{5} < \tfrac{8}{5} < \tfrac{9}{5} < 2.$$

If we let c_j be at the right-hand endpoint in each case, then we'll have $c_1 = \frac{1}{5}$, $c_2 = \frac{2}{5}$, and so on up to $c_{10} = 2$. Plugging these numbers into the above formula, we have

$$\int_0^2 e^{-x^2}\, dx \cong \frac{1}{5}\left(e^{-(1/5)^2} + e^{-(2/5)^2} + \cdots + e^{-(9/5)^2} + e^{-2^2}\right).$$

There are ten terms in the sum. Since our function f is decreasing between 0 and 2, and we've used the right-hand endpoint for each strip, the above estimate is an underestimate. (Can you see why?) In any case, you can use a calculator or computer to find that the above sum is approximately 0.783670 (to six decimal places).

Now, what if you wanted to use the midpoint of each interval, rather than the left-hand or right-hand boundary? Well, the midpoint of $[0, \frac{1}{5}]$ is $\frac{1}{10}$, the midpoint of $[\frac{1}{5}, \frac{2}{5}]$ is $\frac{3}{10}$, and so on. So another possible approximation is given by

$$\int_0^2 e^{-x^2}\, dx \cong \frac{1}{5}\left(e^{-(1/10)^2} + e^{-(3/10)^2} + \cdots + e^{-(17/10)^2} + e^{-(19/10)^2}\right).$$

This is approximately 0.882202.

B.2 The Trapezoidal Rule

There's quite a bit of a burden involved in picking the numbers c_j. Most of the time, people choose either the left-hand endpoint or the right-hand endpoint, but the midpoint is also a common (and reasonable) choice. Here's another method for estimating integrals that removes the element of choice (once you decide to use the method, of course!) while giving even better estimates. It's called the *trapezoidal rule*.

The idea is very simple: allow the tops of the strips to be nonparallel to the base. The top of each strip will be the line segment joining the two corresponding points on the curve $y = f(x)$. Here's a picture illustrating the difference in the two approaches:

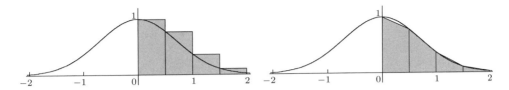

Let's take a closer look at one of the new strips:

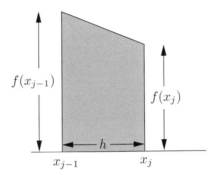

Since there are two parallel sides, the strip is a trapezoid. The base length is $(x_j - x_{j-1})$ units, whereas the heights of the parallel sides are $f(x_{j-1})$ and $f(x_j)$ units. By the formula for the area of a trapezoid, we see that the area of this trapezoidal strip is $\frac{1}{2}(f(x_{j-1}) + f(x_j))(x_j - x_{j-1})$ square units. If we also make sure that we always take our partition to be evenly spaced, then as in the previous section, we see that $x_j - x_{j-1}$ is just $(b-a)/n$. This is exactly the strip width (in units), which we called h, so the area of one strip becomes

$$\frac{h}{2}(f(x_{j-1}) + f(x_j))$$

square units. All that's left is to add up the areas of all the trapezoidal strips. We could just whack a sigma sign outside the above quantity, pulling out the constant factor $h/2$, like this:

$$\int_a^b f(x)\, dx \cong \frac{h}{2} \sum_{j=1}^n (f(x_{j-1}) + f(x_j)).$$

Actually, we can simplify this expression quite a bit. You see, except for the leftmost and rightmost strips, every other pair of adjacent strips shares an edge, like this:

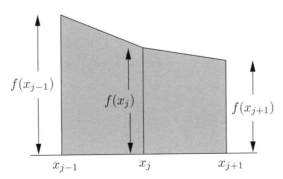

This means that we can collect a lot of the terms together. In particular, except for x_0 and x_n, every term of the form $f(x_j)$ needs to be counted twice. For example, if $n = 4$, we have

$$\int_a^b f(x)\,dx \cong \frac{h}{2}\left((f(x_0) + f(x_1)) + (f(x_1) + f(x_2))\right.$$
$$\left. + f(x_2) + f(x_3)) + (f(x_3) + f(x_4))\right).$$

So we can group all but the first and last terms in the sum into pairs to get

$$\int_a^b f(x)\,dx \cong \frac{h}{2}\left(f(x_0) + 2f(x_1) + 2f(x_2) + 2f(x_3) + f(x_4)\right).$$

The same trick works in general; we end up with:

Trapezoidal rule: if $x_0 < x_1 < \cdots < x_n$ is an evenly spaced partition of $[a, b]$, and $h = (b - a)/n$ is the strip width, then

$$\int_a^b f(x)\,dx \cong \frac{h}{2}\left(f(x_0) + 2f(x_1) + 2f(x_2)\right.$$
$$\left. + \cdots + 2f(x_{n-2}) + 2f(x_{n-1}) + f(x_n)\right).$$

 Let's use this to find an approximate value of our old integral,

$$\int_0^2 e^{-x^2}\,dx.$$

We'll take $n = 5$. Since $[0, 2]$ has length 2 units, the width of each strip is therefore $h = \frac{2}{5}$ units, and the partition is

$$0 < \tfrac{2}{5} < \tfrac{4}{5} < \tfrac{6}{5} < \tfrac{8}{5} < 2.$$

According to the trapezoidal rule, we have

$$\int_0^2 e^{-x^2}\,dx \cong \frac{2/5}{2}\left(e^{-0^2} + 2e^{-(2/5)^2}\right.$$
$$\left. + 2e^{-(4/5)^2} + 2e^{-(6/5)^2} + 2e^{-(8/5)^2} + e^{-2^2}\right).$$

If you want, you can simplify the right-hand side to

$$\frac{1}{5}\left(1 + 2e^{-4/25} + 2e^{-16/25} + 2e^{-36/25} + 2e^{-64/25} + e^{-4}\right).$$

You could also use a calculator or computer to see that the above number is 0.881131 to six decimal places. This is somewhat less than the estimate 1.0865 which we found at the end of Section B.1 above, but quite close to the estimate 0.882202 from the end of Section B.1.1.

B.3 Simpson's Rule

Why stop at trapezoids? They still have a clunky linear top. We can do better by using a curve at the top of the strip, rather than a line segment. Here's how it's done. Start by looking at a pair of adjacent strips, but instead of connecting the tops with two line segments, use a quadratic curve, like this:

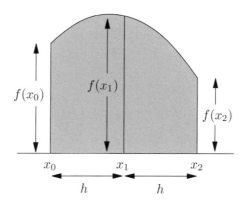

As we'll see in Section B.3.1 below, the shaded area is given by

$$\frac{h}{3}(f(x_0) + 4f(x_1) + f(x_2))$$

square units, where we have once again set $h = (b-a)/n$. Now if we repeat this for each pair of strips and add up all the areas, we'll get our approximation. As in the case of the trapezoidal rule, adjacent pairs of strips share an edge, so there's some doubling up. For example, if there are four strips, then the total would be

$$\frac{h}{3}\left((f(x_0) + 4f(x_1) + f(x_2)) + (f(x_2) + 4f(x_3) + f(x_4))\right);$$

the two terms of the form $f(x_2)$ combine to give $2f(x_2)$, so the total is

$$\frac{h}{3}\left(f(x_0) + 4f(x_1) + 2f(x_2) + 4f(x_3) + f(x_4)\right).$$

The same pattern persists with more strips, so that the coefficient of $f(x_j)$ is equal to 2 if j is even and 4 if j is odd—except for $f(x_0)$ and $f(x_n)$, which both have coefficient 1. All in all, we have:

> **Simpson's rule:** if n is even, $x_0 < x_1 < \cdots < x_n$ is an evenly spaced partition of $[a, b]$, and $h = (b - a)/n$, then
>
> $$\int_a^b f(x)\, dx \cong \frac{h}{3}\left(f(x_0) + 4f(x_1) + 2f(x_2) + 4f(x_3)\right.$$
>
> $$\left. + \cdots + 2f(x_{n-2}) + 4f(x_{n-1}) + f(x_n)\right).$$

Compare this to the trapezoidal rule from the previous section. Instead of coefficients which look like $1, 2, 2, \ldots, 2, 2, 1$, this time the coefficients follow the pattern $1, 4, 2, 4, 2, \ldots, 2, 4, 2, 4, 1$. Also note that the denominator of the constant out front is 3 instead of 2.

Simpson's rule is pretty easy to apply. Let's go back to our old example of

$$\int_0^2 e^{-x^2}\, dx,$$

and use Simpson's rule with $n = 8$. (We can't do $n = 5$, since n has to be even in order to use Simpson's rule.) The length of each strip is $h = (2-0)/8$ units, which is $\frac{1}{4}$, so the partition is

$$0 < \tfrac{1}{4} < \tfrac{1}{2} < \tfrac{3}{4} < 1 < \tfrac{5}{4} < \tfrac{3}{2} < \tfrac{7}{4} < 2.$$

By the above formula, we have

$$\int_0^2 e^{-x^2}\, dx \cong \frac{1/4}{3}\left(e^{-0^2} + 4e^{-(1/4)^2} + 2e^{-(1/2)^2} + 4e^{-(3/4)^2} + 2e^{-1^2} \right.$$
$$\left. + 4e^{-(5/4)^2} + 2e^{-(3/2)^2} + 4e^{-(7/4)^2} + e^{-2^2} \right).$$

This is approximately 0.882066, according to my calculator. This is quite close to our estimate from the previous section; specifically, when we used the trapezoidal rule with $n = 5$, we got the estimate 0.881131. For the record, my computer program says that the correct value of the integral is 0.882081 to six decimal places, so Simpson's rule (with $n = 8$) is better than the trapezoidal rule (with $n = 5$). Of course, a fairer comparison would be to use $n = 8$ in both cases; I leave it to you to repeat the trapezoidal rule computation in this case and compare the result with the corresponding Simpson's rule estimate which we just found.

B.3.1 Proof of Simpson's rule

Let's translate the picture over so that the middle line lies along the y-axis, like this:

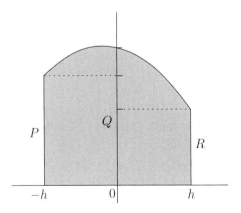

As you can see, this shifts the x-coordinates of the partition endpoints over to $-h$, 0, and h. Instead of writing $f(x_0)$, $f(x_1)$, and $f(x_2)$, let's just write

P, Q, and R, respectively. The top points are connected by some quadratic, but we don't have a clue what it is. Well, let's call it g and suppose that $g(x) = Ax^2 + Bx + C$. We know that $P = g(-h)$, $Q = g(0)$ and $R = g(h)$; this means that

$$P = A(-h)^2 + B(-h) + C,$$
$$Q = A(0)^2 + B(0) + C,$$
$$R = Ah^2 + Bh + C.$$

The middle equation says that $C = Q$; then you can rearrange the other two equations to see that $A = (P + R - 2Q)/(2h^2)$. (We don't need to know what B is!) Now, the shaded area we want is given by

$$\int_{-h}^{h} (Ax^2 + Bx + C)\, dx = \left(\frac{A}{3}x^3 + \frac{B}{2}x^2 + Cx \right)\Big|_{-h}^{h} = \frac{2Ah^3}{3} + 2Ch$$

square units, after simplifying. Substituting the values of A and C from above, the above expression reduces to

$$\frac{2h^3}{3} \times \frac{P + R - 2Q}{2h^2} + 2Qh = \frac{h}{3}(P + 4Q + R).$$

Now all we have to do is translate to a more general position (which doesn't affect the area) and replace P, Q, and R by the function values $f(x_0)$, $f(x_1)$, and $f(x_2)$, respectively, to get the prototype formula from the beginning of the previous section.

B.4 The Error in Our Approximations

The whole point of making an approximation (or estimate, if you prefer that word), is that you end up with something close to the exact quantity you're looking for. If you could actually pin your finger on the exact answer, you'd do that, but sometimes it's just too difficult. So an approximation at least gives you a number close to the exact one. As we've seen a number of times, most notably when we looked at linearization and also Taylor series (see Section 13.2 of Chapter 13 and Section 25.3 of Chapter 25), there's one other important consideration: how good is the approximation? Is your approximate answer at least close to the actual one, or does it suck?

To quantify this, we look once again at the error in the approximation, which is the difference between the actual quantity and the approximation itself. So suppose that we use one of the above techniques—evenly spaced strips, the trapezoidal rule or Simpson's rule—to approximate the integral $\int_a^b f(x)\, dx$. We'd get something like

$$\int_a^b f(x)\, dx = A,$$

where A is our approximate value. The absolute value of the error is then given by

$$|\text{error}| = \left| \int_a^b f(x)\, dx - A \right|.$$

It turns out that we can get some idea of how big the error could possibly be by using the derivatives of f, if they exist. In that case, we can let M_1 be the maximum value of $|f'(x)|$ on $[a, b]$. Similarly, let M_2 be the maximum value of $|f''(x)|$ on $[a, b]$, and finally let M_4 be the maximum value of $|f^{(4)}(x)|$ on $[a, b]$. Then one can show the following bounds on the error term, depending on the method used:

$$\text{for evenly spaced strips,} \qquad |\text{error}| \leq \frac{1}{2}M_1(b - a)h,$$

$$\text{for the trapezoidal rule,} \qquad |\text{error}| \leq \frac{1}{12}M_2(b - a)h^2,$$

$$\text{for Simpson's rule,} \qquad |\text{error}| \leq \frac{1}{180}M_4(b - a)h^4.$$

Here h is the strip width $(b - a)/n$, as usual. Although the above formulas are pretty similar, there are some differences. First, the coefficients out front are different. Second, different derivatives are involved: for strips, the first derivative comes up (in the form of M_1); for the trapezoidal rule, it's the second derivative; while for Simpson's rule, it's the fourth derivative. The most significant difference, however, is the power of h which appears. This shows how much the error decreases as the strip width becomes smaller, which of course happens when you take more strips. As h becomes small, h^4 gets smaller much more quickly than h^2 or h, so Simpson's rule should kick some serious butt over the other methods when you use lots of strips.

B.4.1 Examples of estimating the error

Let's see how these errors turn out for the example

$$\int_0^2 e^{-x^2}\, dx,$$

which we've looked at earlier in this appendix. First, we'll set $f(x) = e^{-x^2}$, and then calculate that

$$f'(x) = -2xe^{-x^2}, \quad f''(x) = (4x^2 - 2)e^{-x^2}, \quad f^{(3)}(x) = -4x(2x^2 - 3)e^{-x^2},$$
$$\text{and} \quad f^{(4)}(x) = 4(4x^4 - 12x^2 + 3)e^{-x^2}.$$

Let's find M_1 first. This means that we have to find the maximum value of $|f'(x)|$, which is actually $-f'(x)$, on $[0, 2]$. Since the second derivative $f''(x)$ is 0 at $x = 1/\sqrt{2}$, and changes sign from negative to positive there, we have a local minimum for $f'(x)$ at $1/\sqrt{2}$. This means that the minimum value of $f'(x)$ on $[0, 2]$ is $-\sqrt{2}e^{-1/2}$, so the maximum value of $|f'(x)|$ is $\sqrt{2}e^{-1/2}$. That is, $M_1 = \sqrt{2}e^{-1/2}$.

Now we can go back to our estimates for the integral from Section B.1.1 above. There we used 10 evenly spaced strips to estimate our integral. Since $a = 0$, $b = 2$, and $h = (2 - 0)/10 = \frac{1}{5}$, we have

$$|\text{error using 10 strips}| \leq \frac{1}{2}M_1(b - a)h = \frac{1}{2} \times \sqrt{2}e^{-1/2}(2 - 0)\frac{1}{5} = \frac{\sqrt{2}}{5}e^{-1/2}.$$

This is approximately 0.171553. Note that it doesn't matter if you use the left-hand endpoint, the right-hand endpoint, or somewhere in between as your choice of c_n. (In Section B.1.1, we used the right-hand endpoints and then the midpoints to get two different estimates, but they are both accurate to about ± 0.171553.)

Let's move on to the trapezoidal rule estimate. In Section B.2 above, we used 5 trapezoids of width $h = 2/5$ to estimate our integral (so $n = 5$). To see how big the error could be, we'll need to find M_2 by maximizing the value of $|f''(x)|$ on $[0, 2]$. To do this, look back at the above formulas for $f^{(2)}(x)$ and $f^{(3)}(x)$. The zeroes of $f^{(3)}(x)$ which lie in $[0, 2]$ are at $x = 0$ and $x = \sqrt{3/2}$, so these are the critical points of $f^{(2)}(x)$. (Remember, the third derivative is the derivative of the second derivative!) So we can test the values of $f''(0)$ and $f''(\sqrt{3/2})$, as well as the value $f(2)$ at the other endpoint 2. We find $f''(0) = -2$, $f''(\sqrt{3/2}) = 4e^{-3/2}$ and $f''(2) = 14e^{-4}$. The largest of these, in absolute value, is $f''(0)$. This means that $M_2 = 2$. Now we can estimate the error (remembering that $h = 2/5$):

$$|\text{error using 5 trapezoids}| \leq \frac{1}{12} M_2 (b - a) h^2 = \frac{1}{12} \times 2(2 - 0) \left(\frac{2}{5} \right)^2 = \frac{4}{75},$$

which is $0.053333\ldots$. This is a lot less than the error using 10 strips, even though we only used 5 trapezoids! Since our previous estimate was approximately 0.881131, we have shown that the approximation

$$\int_0^2 e^{-x^2} \, dx \cong 0.881131$$

is accurate to about ± 0.053333. (This is certainly consistent with my observation at the end of Section B.3 above that the correct value is actually 0.882081, to six decimal places.)

Finally, we'll estimate the error using Simpson's rule. In Section B.3 above, we used Simpson's rule with $n = 8$ to show that

$$\int_0^2 e^{-x^2} \, dx \cong 0.882066.$$

We'll need M_4, which is the maximum value of $|f^{(4)}(x)|$ on $[0, 2]$. This could be very messy, since $f^{(4)}(x) = 4(4x^4 - 12x^2 + 3)e^{-x^2}$. Let's cheat by finding the maximum value of each of the three factors. No problem with 4, and e^{-x^2} is positive and is maximized at $x = 0$ (with a value of 1); so we only have to find the maximum of $|4x^4 - 12x^2 + 3|$ on $[0, 2]$. We have

$$\frac{d}{dx}(4x^4 - 12x^2 + 3) = 16x^3 - 24x = 8x(2x^2 - 3),$$

so the maximum we're looking for could only occur at one of the critical points $x = 0$ and $x = \sqrt{3/2}$, or at the other endpoint $x = 2$. Plugging these numbers in, we find that the greatest value 19 occurs at $x = 2$, which means that

$$|4x^4 - 12x^2 + 3| \leq 19$$

on $[0, 2]$. Putting everything together, we can say that

$$M_4 \leq 4 \times 19 \times 1 = 76.$$

(Actually, $M_4 = 12$, but you need to look at the fifth derivative of f to see this, and enough is enough!) Now we can finally use our formula (with $h = (2 - 0)/8 = 1/4$):

$$|\text{error using Simpson's rule with } n = 8| \leq \frac{1}{180} M_4 (b - a) h^4$$

$$\leq \frac{1}{180} \times 76(2 - 0) \left(\frac{1}{4}\right)^4 = \frac{19}{5760}.$$

This is about 0.003299, which is much lower than the previous two errors we calculated.

B.4.2   Proof of an error term inequality

The proofs of the last two of the three error inequalities in Section B.4 above are a little beyond the scope of this book, but it's not too hard to show that the first one is true:

$$|\text{error using } n \text{ evenly spaced strips of width } h| \leq \frac{1}{2} M_1 (b - a) h,$$

where M_1 is the maximum value of $f'(x)$ in $[a, b]$. Suppose that we use the left-hand endpoints for our estimate. Let's look at just one of the strips. If its base is the interval $[q, q + h]$ (for some q), then it looks something like this:

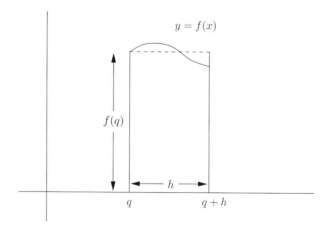

The approximating rectangle has height $f(q)$ and width h units, so the approximate area is $hf(q)$ square units. How bad could this be, in general? It all depends on how much the graph of f deviates from the constant line $y = f(q)$. Here are the two worst-case scenarios:

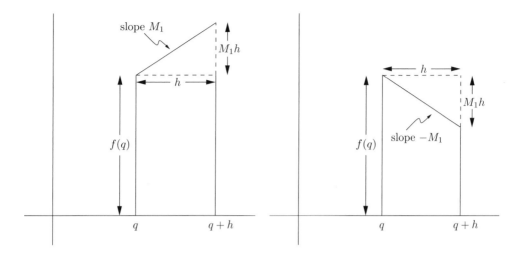

The first shows a line segment of slope M_1 beginning at $(q, f(q))$, while the second shows a line segment of slope $-M_1$ beginning at the same point. In fact, the function must be trapped between these two extremes. Indeed, the first line has equation $y = f(q) + M_1(x - q)$. If $f(x)$ ever gets above this line (for x in the interval $[q, q + h]$), then we have $f(x) > f(q) + M_1(x - q)$, or

$$\frac{f(x) - f(q)}{x - q} > M_1.$$

By the Mean Value Theorem (see Section 11.3 in Chapter 11), the left-hand side is equal to $f'(c)$ for some c in $[q, x]$, so $f'(c) > M_1$. This is impossible, since M_1 is the maximum value of $|f'(x)|$ on $[a, b]$. A similar argument shows that $y = f(x)$ always lies above the down-sloping line.

Now we can look at the error. In the first worst-case scenario, the actual region includes the strip plus a triangle of side lengths h by M_1h units; in the second worst-case scenario, the same triangle is actually removed from the strip. In either case, our area could be off by the area of the triangle, which is $\frac{1}{2}M_1h^2$ square units. All that's left is to multiply this error by the number of strips, which is n, to see that our approximation can't be worse than $\frac{1}{2}M_1h^2n$. In fact, we can steal away one of the factors of h and use the equation $nh = (b - a)$ to rewrite the above expression as $\frac{1}{2}M_1(b - a)h$. This is exactly what we want! Now I leave it to you to repeat the above argument in the case where we don't necessarily choose the left-hand endpoint. (In fact, if you use the midpoint, you can show that the error is actually only $\frac{1}{4}M_1(b - a)h$.)

LIST OF SYMBOLS

Symbol	Meaning	Page(s)
$\mathbb{R}$	set of real numbers	1
$[a, b]$	closed interval from a to b	3
(a, b)	open interval from a to b	3
$(a, b]$	half-open interval from a to b	3
$A \backslash B$	all numbers in A not including those in B	5
$f(x)$	function f evaluated at x	2
f^{-1}	inverse function of f	8
$f \circ g$	composition of f with g	12
Δ	discriminant of quadratic	20
$|x|$	absolute value of x	23
sin, cos, tan	basic trig functions (sine, cosine, tangent)	26
sec, csc, cot	reciprocal trig functions (secant, cosecant, cotangent)	27
$\sin^{-1}, \cos^{-1}, \tan^{-1}$	inverse trig functions (arcsine, arccosine, arctangent)	208–215
$\sec^{-1}, \csc^{-1}, \cot^{-1}$	inverse reciprocal trig functions (arcsecant, arccosecant, arccotangent)	216–218
sinh, cosh, tanh	basic hyperbolic functions (hyperbolic sine, cosine, tangent)	198–200
sech, csch, coth	reciprocal hyperbolic functions (hyperbolic secant, cosecant, cotangent)	198–200
$\sinh^{-1}, \cosh^{-1}, \tanh^{-1}$	inverse trig functions (hyperbolic arcsine, arccosine, arctangent)	220–223
$\text{sech}^{-1}, \text{csch}^{-1}, \coth^{-1}$	inverse reciprocal trig functions (hyperbolic arcsecant, arccosecant, arccotangent)	222–223
$\ln(x), \log_e(x)$	natural logarithm of x	176
$\lim\limits_{x \to a}$	two-sided limit as x approaches a	42, 672
$\lim\limits_{x \to a^+}$	right-hand limit as x approaches a (from above)	44, 680
$\lim\limits_{x \to a^-}$	left-hand limit as x approaches a (from below)	44, 680
DNE	limit does not exist	44
$0/0, \infty/\infty, 0 \times \infty$	indeterminate forms	58, 293–303
$0^0, 1^\infty, \infty^0$	indeterminate forms	293–303
$\overset{\text{l'H}}{=}$	equals, using l'Hôpital's Rule	295
$\sim$	asymptotic functions or sequences	442, 488
$\cong$	approximately equal to	33

Symbol	Meaning	Page(s)
Δx	change in x	91
$f'(x)$	derivative of f with respect to x	90
$f''(x), f^{(2)}(x)$	second derivative of f with respect to x	94
$f^{(n)}(x)$	nth derivative of f with respect to x	94
$\dfrac{dy}{dx}, \dfrac{d}{dx}(y), dy/dx$	derivative of y with respect to x	93
$\dfrac{d^2 y}{dx^2}, \dfrac{d^2}{dx^2}(y)$	second derivative of y with respect to x	94
x, v, a	displacement, velocity, acceleration	114
g	acceleration due to gravity	115
$|AB|$	length of line segment AB	139
ΔABC	triangle with vertices A, B, C	139
e	base of natural logarithm	174
$t_{1/2}$	half-life of radioactive material	197
$\star$	discontinuity (used in table of signs)	246
$L(x)$	linearization	280
df	differential of f	282
$\displaystyle\sum_{j=a}^{b}$	sum from $j = a$ to b of …	307
$F(x)\Big|_a^b$	$F(b) - F(a)$	363
$\displaystyle\int_a^b f(x)\,dx$	definite integral of f with respect to x	326
$\displaystyle\int f(x)\,dx$	indefinite integral (antiderivative) of f with respect to x	364
f_{av}	average value of f	350
I_n	integral number n (reduction formulas)	419
$\{a_n\}$	sequence $a_1, a_2, a_3, \ldots$	478, 483
$\displaystyle\sum_{n=1}^{\infty} a_n$	infinite series $a_1 + a_2 + a_3 + \cdots$	483
$n!$	n factorial ($1 \times 2 \times 3 \times \cdots \times (n-1) \times n$)	505
$P_N(x)$	Nth-order Taylor polynomial	522
$R_N(x)$	Nth-order remainder term	524
(r, θ)	polar coordinates	582
i	$\sqrt{-1}$	595
$z = x + iy$	complex number in Cartesian form	596, 599
$z = re^{i\theta}$	complex number in polar form	600
e^z	complex exponential of z	598
$\mathrm{Re}(z)$	real part of z	596
$\mathrm{Im}(z)$	imaginary part of z	596
$\bar{z}$	complex conjugate of z	597
$|z|$	modulus of z	597
$\arg(z)$	argument of z	601
y_H	homogeneous solution (differential equations)	657
y_P	particular solution (differential equations)	657

INDEX